ANNUAL REVIEW OF PLANT PHYSIOLOGY AND PLANT MOLECULAR BIOLOGY

EDITORIAL COMMITTEE (1995)

ANNUAL REVIEW OF PLANT PHYSIOLOGY AND PLANT MOLECULAR BIOLOGY

VOLUME 46, 1995

RUSSELL L. JONES, *Editor*
University of California, Berkeley

CHRISTOPHER R. SOMERVILLE, *Associate Editor*
Carnegie Institution of Washington, Stanford, California

VIRGINIA WALBOT, *Associate Editor*
Stanford University

ANNUAL REVIEWS INC.   4139 EL CAMINO WAY   P.O. BOX 10139   PALO ALTO, CALIFORNIA   94303-0139

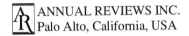

ANNUAL REVIEWS INC.
Palo Alto, California, USA

International Standard Serial Number: 1040-2519
International Standard Book Number: 0-8243-0646-5
Library of Congress Catalog Card Number: A-51-1660

Annual Review and publication titles are registered trademarks of Annual Reviews Inc.

The paper used in this publication meets the minimum requirements of American National Standards for Information Sciences—Permanence of Paper for Printed Library Materials, ANZI Z39.48-1984

Annual Reviews Inc. and the Editors of its publications assume no responsibility for the statements expressed by the contributors to this *Review*.

Typesetting by Ruth McCue-Saavedra and the Annual Reviews Inc. Editorial Staff

PRINTED AND BOUND IN THE UNITED STATES OF AMERICA

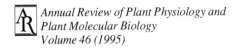
Annual Review of Plant Physiology and Plant Molecular Biology
Volume 46 (1995)

CONTENTS

SOME RELATED ARTICLES IN OTHER *ANNUAL REVIEWS*

From the *Annual Review of Biochemistry,* Volume 64 (1995)

The Nuclear Pore Complex, LI Davis
Plasma Lipid Transfer Proteins, A Tall
6-Phosphofructo-2-Kinase/Fructose-2,6-Bisphosphatase: A Metabolic Signaling Enzyme, SJ Pilkis, TH Claus, IJ Kurland, and AJ Lange
Superoxide Radical and Superoxide Dismutases, I Fridovich
Eukaryotic Phospholipid Biosynthesis, C Kent
Structure and Function of Voltage-Gated Ion Channels, WA Catterall
The Multiplicity of Domains in Proteins, RF Doolittle
Modified Nucleosides, Nucleotides, and Oligonucleotides, BE Eaton and WA Pieken
Protein-RNA Recognition, DE Draper
The Small Nucleolar RNAs, ES Maxwell, and MJ Fournier

From the *Annual Review of Biophysics and Biomolecular Structure,* Volume 23 (1994)

Membrane Proteins: From Sequence to Structure, G von Heijne
Global Statistics of Protein Sequences: Implications for the Origins, Evolution, and Prediction of Structure, SH White
DNA Branched Junctions, NC Seeman and NR Kallenbach
Perspectives on the Physiology and Structure of Inward-Rectifying K$^+$ Channels in Higher Plants: Biophysical Implications for K$^+$ Uptake, JI Schroeder, JM Ward, and W Gassmann
Evolution of the EF-Hand Family of Proteins, S Nakayama and RH Kretsinger
Voltage-Dependent Gating of Ionic Channels, F Bezanilla and E Stefani
Cyclic Nucleotide-Gated Ion Channels and Sensory Transduction in Olfactory Receptor Neurons, F Zufall, S Firestein, and GM Shepherd
Polypeptide Interactions with Molecular Chaperones and Their Relationship to In Vivo Protein Folding, SJ Landry and LM Gierasch
Biomolecular Imaging with the Atomic Force Microscope, HG Hansma and JH Hoh
High Pressure NMR Spectroscopy of Proteins and Membranes, J Jonas and A Jonas
Mass Spectrometry of Macromolecules: Has Its Time Now Come?, MW Senko and FW McLafferty

From the *Annual Review of Cell Biology,* Volume 10 (1994)

Protein Serine/Threonine Phosphatases: New Avenues for Cell Regulation, S Shenolikar
Signal Sequence Recognition and Protein Targeting to the Endoplasmic Reticulum Membrane, P Walter and AE Johnson
Microtubules in Plant Morphogenesis: Role of the Cortical Array, RJ Cyr
Structure of Actin-Binding Proteins: Insights about Function at Atomic Resolution, TD Pollard, S Almo, S Quirk, V Vinson and EE Lattman
Receptor Protein-Tyrosine Kinases and Their Signal Transduction Pathways, P van der Geer, T Hunter, and RA Lindberg

Robert H Burris

Annu. Rev. Plant Physiol. Plant Mol. Biol. 1995. 46:1–19

BREAKING THE N ≡ N BOND

Robert H. Burris

Department of Biochemistry, University of Wisconsin, Madison, Wisconsin 53706

KEY WORDS: nitrogen fixation, isotopes, metabolism

CONTENTS

The Early Years

I was fortunate to grow up in Brookings, South Dakota, far from the diversions of city life, but with the cultural advantages of a college town of 4500 people. My father, Edward T. Burris, had been born in a sod house on the South Dakota prairie on December 31, 1880. The site for Rolvag's *Giants in the Earth* was laid about 60 miles south of my father's birthplace about 13 years earlier, so Rolvag's description of rugged pioneer life on the treeless prairie had real pertinence to the problems the Burris family faced. Grandmother Margaret Bruce Burris, the 20-year-old daughter of a Presbyterian minister from civilized Wisconsin, must have wondered about the sanity of the family's move to South Dakota after a few winters on the prairie. After finishing high school, my father taught in a one-room country school, and then became a printer.

My mother, Mabel Clara Harza, was born on a homestead about 20 miles northeast of Brookings. Grandfather Harza (transliterated from the German name *Herse*) had come from Germany at age 9, and Grandmother Jolley had lived in southeast Minnesota. The family moved into Brookings when my mother was a small child. She went to school there and gave piano lessons in the area. The first 22 years of my life were spent in a house half a block from

0066-4294/95/0601-0001$05.00

the South Dakota State College (now University) campus. The spot of my birth now is commemorated with a parking lot.

Brookings supported two weekly newspapers, but the Burris Printery was a job shop not associated with the newspapers. The shop provided a modest living, and I worked there all through high school and much of college. My father always had one or two college students working in the shop, and they were a great bunch. The shop was small, and all of us there were skilled in all operations. We all had to "stick" type (there was no Linotype) and run the platen presses. Anyone who had to stick type soon was able to read upside down and backwards as well as in the normal orientation.

My high school years (from 1928 to 1932) were anything but prosperous, and my college years (from 1932 to 1936) marked a period of drought as well as depression in South Dakota. In high school, physics was the most exciting subject and Latin the least favorite; the high school had no chemistry course. We all learned some English, history, literature, speech, and math. I had taken up the trumpet in fifth grade, so I played in "Christy's" bands from fifth grade through college.

I always have been thankful for debate training in high school and college, as the experience was a great aid in teaching. Ruth Robinson, our high school debate coach, whipped us into line so that we made the semi-finals of the State debate tournament. My college debate team once got to the quarter-finals in national competition.

College Years

South Dakota State College (SDSC) offered a much greater choice of subjects than did my high school, and I worked in enough electives to get 5 years of credits in 4 years. I took most of the chemistry offerings and enjoyed them all. Physics was almost as exciting as chemistry, but math held me only through differential equations. Of the biology courses, bacteriology was outstanding, because Herb Batson was an excellent teacher. I participated in band and debate, but while taking extra classes and working 25 to 30 hours a week in the print shop, I did not have a lot of time for extracurricular activities. Irwin Gunsalus (of Cornell University and the University of Illinois) and I managed to run a Boy Scout troop for a while (we both had worked up to Eagle Scout as kids).

In my junior year I had an opportunity to work in the Experiment Station Chemistry Lab, so I abandoned the print shop and spent my spare time as a junior and senior in the chemistry lab. Dr. Kurt W. Franke was in charge of the lab. He did not teach in the chemistry department, but his lab and research operations occupied the bottom floor of the chemistry building. He recruited chemistry students for the lab, and a number of them did graduate level work there; the lab did not offer a PhD degree. It was a great place to get experience in chemistry, because the lab was well equipped and the research was challenging.

A disease had been reported many years before in cavalry horses stationed in Nebraska and South Dakota, and there were two diseases of unknown etiology in cattle: alkali disease (a chronic disease) and blind staggers (an acute disease). Franke and colleagues established that the diseases were caused by selenium. Pierre shale is the parent rock that yields seleniferous soils over large areas of Nebraska, Wyoming, and South Dakota. Many plant materials in the area have high enough levels of selenium to produce the chronic alkali disease, and a number of convertor plants (e.g. *Astragalus bisulcatus*) accumulate enough selenium to cause acute blind staggers.

Franke's group studied the effects of selenium in experimental animals, and its teratogenicity in chicks. One of my jobs was to develop a method of selenium analysis, which was the subject of my first published scientific paper. The lab hoods were not outstanding, and my manipulations included operating with Br_2, HBr, and SO_2; in comparison, manipulation of Se was relatively innocuous.

Van R. Potter had worked in the station lab as a graduate student, and we had taken German and physical chemistry courses together while I was an undergraduate. In early 1935 Van transferred to the University of Wisconsin to work on a PhD in biochemistry with C. A. Elvehjem. Van was the first of a succession of South Dakota State grads who came to Wisconsin for graduate work. He lived in the same dormitory with Wayne Umbreit, who was working with Perry Wilson on a PhD in bacteriology. Wayne was doing endless Kjeldahl analyses, and he told Van he was getting a bit tired of the routine. Van suggested that Wayne recruit me, because I had done Kjeldahl analyses for total N in the station chemistry lab. So Perry invited me to come to Wisconsin for the summer of 1935.

I was tremendously impressed with the beauty of the Wisconsin campus and Lake Mendota. I stayed in a dormitory on Lake Mendota, and the lab was a short walk from there. There were lots of nitrogen analyses to do, plus planting and harvesting soybeans and adding CO_2 to plants growing in closed bottles in the greenhouse. Before I returned to SDSC for my senior year, Perry offered me an assistantship for the following summer that would allow me to start graduate work in the Department of Bacteriology. I was delighted to accept the offer of $50 per month. So after finishing my BS in 1936, I moved to Madison, rented a room on Randall Court near the Agriculture campus, and started my graduate work.

Perry's office was in the basement of Ag Hall, where almost all of the Ag Bacteriology Department was housed, but our lab was in King Hall, the old Soils building, with an adjacent greenhouse. The King Hall lab was nothing special, but on the other hand, nobody cared if we altered it to fit our needs.

The Works Progress Administration (WPA) was in full swing to move the country out of the depression. It furnished three workmen to the lab: an ex-tailor, an ex-glazier, and O. A. Taliaferro, who had a BS in chemistry and ran the Kjeldahl analyses very competently. Later, "Auggie" Nelson, who had an MS in soils, joined the WPA crew and did most of the greenhouse work. The WPA helped the economy, but it took the war to pull us out of the depression.

Most of the first year, Perry was away in Cambridge, England, and in Helsinki, Finland, in A. I. Virtanen's lab. Perry and Virtanen had had a running controversy about probable intermediates in nitrogen fixation. Virtanen claimed that pea plants excreted aspartic acid, and he concluded that this supported his idea that hydroxylamine was the first demonstrable product of biological nitrogen fixation. He held that it combined with oxalacetic acid to yield an oxime that then was reduced to aspartic acid. Unfortunately, nobody could check this, because they couldn't get excretion from the roots of leguminous plants. Perry couldn't get excretion in Helsinki either, a result Virtanen blamed on the winter weather. Perry brought back samples of Virtanen's pea seeds, his inoculum, and even had a large box of soil shipped from Finland. None of these induced excretion in Madison. Since Virtanen implicated the seasonal changes in weather, a succession of graduate students had to run what we called the "marathon" series of experiments. Near the first of each month they set up a new experiment in the greenhouse, but no excretion was induced during any season in Madison. After leguminous plants have expended a great deal of energy to fix nitrogen, they seem very reluctant to give it up by excretion.

A major effort in the lab was to establish the physical constants for the process of biological nitrogen fixation. We grew red clover plants in 9 liter serum bottles under controlled gas atmospheres. The group established the pN_2 giving a half-maximum rate of nitrogen fixation, the effect of the pO_2, and demonstrated that H_2 was a specific competitive inhibitor of nitrogen fixation. When the clover plants grew to a reasonable size, CO_2 had to be supplied to their bottles very frequently on a bright day.

Upon his return from England, Perry Wilson bought a Warburg respirometer for the lab. He had mastered the apparatus during his stay at Cambridge University and thought it would be interesting to study the respiration of the root-nodule bacteria. I volunteered for this, as I was rather bogged down trying to show a difference in the amino acid distributions of effective and ineffective strains of the root-nodule bacteria. It was laborious to grow the bacteria in quantity and even more laborious to determine their amino acid composition. No suitable chromatographic method was available then, and one had to isolate crystalline derivatives of the amino acids from an acid hydrolysate; isolations were anything but quantitative.

I ran many manometric determinations and did comparative studies of the respiration of good and poor strains of species of most of the rhizobia on a variety of substrates. This work constituted the bulk of my PhD thesis. No real differences emerged in the properties of effective and ineffective strains. I was privileged to present the data at the 1939 Cold Spring Harbor Symposium—apparently Perry wasn't very anxious to attend. But it was a great experience for me as a graduate student to mix with W. Mansfield Clark, L. Michaelis, Eric Ball, Elmer Stotz, T. R. Hogness, C. G. King, Fred Stare, Fritz Lipmann, Carl Cori, Kurt Stern, Dave Goddard, Hans Gaffron, Dean Burk, and others.

There were other notable meetings. It probably was the summer of 1938 when Perry drove the lab group to the national meetings of the Society of American Bacteriology (SAB) in San Francisco. We stopped in Provo, Utah, and in Los Angeles to see Perry's relatives. We also paused at the north rim of the Grand Canyon where Perry announced, "fifteen minutes for the Grand Canyon"; at least we could say we had seen it. P. F. Clark of the University of Wisconsin was SAB president that year, and he had taken great care to prepare his presidential speech on "Alice in Virusland."

In 1939, Richard Tam (a graduate student from Hawaii in Perry's lab) drove the lab gang to the International Soil Science Congress at Rutgers University, in New Jersey. We were at the meeting when the Germans marched into Poland to start the war. The Germans attending the meeting extended an invitation for the next congress, then scheduled for Germany, but it never came off.

I also presented a paper at the Gibson Island Conference (precursor of the Gordon Conferences) in 1940. The only problem was that someone put my bag, complete with slides and paper, on the bus from Gibson Island back to Baltimore. So while others swam in the afternoons, I reconstructed a speech. A few days later I went into Baltimore and found the bag at the bus terminal—so there were slides for the talk after all.

Postdoctoral Training

Wayne Umbreit and I had talked about how useful it would be to have ^{15}N as a tracer in studies of N_2 fixation. Because nobody had exploited ^{15}N in this way, I wrote up a proposal in 1940 and submitted it for a National Research Council postdoctoral fellowship. The committee bought the idea and awarded me a fellowship. Harold Urey at Columbia University agreed to be the sponsor. I received my PhD in June 1940, and later that summer my mother and I settled in a Columbia University apartment on 125th Street near Broadway. My father had died in 1937, and my mother was recovering from a serious stroke. New York was hardly like Brookings, South Dakota, but to me New Yorkers seemed perhaps more provincial than South Dakotans. Some New Yorkers I met had managed to go as far west as Philadelphia, but beyond that lay nothing they perceived to be of interest.

Professor Urey was from Indiana and was educated in Montana, so this gave him a broader outlook. When I explained my projected work to him, he grasped the concepts immediately, although the work was far from his experience. He promptly suggested that I team up with C. E. Miller and test whether the nitrogenase enzyme system supported an exchange reaction that might invalidate its use as a tracer. He explained that if I mixed two equilibrium mixtures of N_2 species with different ^{15}N concentrations, it would produce a nonequilibrium mixture of $^{14}N^{14}N$, $^{14}N^{15}N$, and $^{15}N^{15}N$. When the nitrogenase enzyme system was exposed to this nonequilibrium mixture it would approach an equilibrium condition if an exchange reaction occurred, but otherwise the gas composition would remain unchanged. I learned a bit about high vacuum systems, did some glassblowing, generated N_2 from $^{15}NH_4^+$ salts that Professor Urey had given me, and exposed an actively fixing culture of *Azotobacter vinelandii* to the nonequilibrium N_2. Mass spectrometric analysis showed no equilibration and, hence, no invalidating exchange reaction.

Professor Urey was delighted with the result, because Hugh Taylor, the distinguished physical chemist from Princeton University, shortly before had asked Urey for some ^{15}N (Urey had a virtual corner on the supply of ^{15}N-enriched material, as he had developed the method for concentrating it; he had sent his exchange columns to Eastman and they were getting into the concentrating business). Taylor had plans for a postdoctoral student to study the exchange reaction postulated for the nitrogenase system. Urey met Taylor shortly thereafter at a meeting and had the fun of telling Taylor to seek a new postdoc problem, because the exchange reaction did not exist.

There was another rather simple but interesting experiment at Columbia that clarified a controversial point. In 1940, Ruben et al (8) at the University of California at Berkeley had produced ^{13}N, the radioisotope of nitrogen with a half-life of 10.1 min, and had exposed barley plants to ^{13}N-enriched N_2 for 20 min. The barley plants became radioactive. Lipman, the graduate dean at Berkeley, was a long-time supporter of controversial claims for nitrogen fixation by wheat and barley, so he encouraged Rubin and his colleagues to publish this work. Because we were skeptical, we exposed barley and inoculated clover in interconnecting chambers for 54 days to a gas mixture containing ^{15}N-enriched N_2. The clover was generously enriched in ^{15}N, and the barley had no detectable enrichment with ^{15}N (1).

It was a delightful year in New York. There were plays, concerts, and operas to attend, and a number of great parks to explore. The lab group in Chemistry was very supportive, and I learned a lot about glassblowing from my labmate Kelly. I also had a lab with Trelease's botany group, and attended their seminars and did any needed plant work there. Trelease had worked during the summers in Wyoming, studying the selenium convertor plants, so we had some common interests. The physical chemistry seminars were terrific, because not only were Urey's group and the other physical chemists there, but Fermi, Rabi, Maria Göppert Mayer, and Joe Mayer also attended.

In the spring and summer I spent most of my lab time at the College of Physicians and Surgeons with Schoenheimer and Rittenberg. Rittenberg had taken his PhD degree with Urey and hence was well acquainted with mass spectrometry and the stable isotopes. He and Schoenheimer exploited ^{15}N as a tracer in the study of animal metabolism. Before their studies, proteins were generally considered to be stable in the body. Their use of ^{15}N as a tracer demonstrated clearly that the proteins in the body were in a dynamic state. Schoenheimer's lab also was a lively place with people like Konrad Bloch there. Other Physician and Surgeons labs also had very distinguished crews. Schoenheimer, Rittenberg, and Sarah Ratner assisted in my separation of amino acid derivatives from a hydrolysate of *A. vinelandii* that I had exposed to ^{15}N-enriched N_2 for 45 min (2). One could not recover all the amino acids with the available methods, but among the amino acids recovered, glutamic acid carried the highest enrichment with ^{15}N. This was compatible with the concept that N_2 is reduced to NH_4^+ and that reductive amination of α-ketoglu-tarate yields glutamate that, in turn, transaminates to form the other amino acids. Virtanen's hydroxylamine hypothesis would give the highest ^{15}N en-richment in aspartic rather than glutamic acid.

These were war years. Harold Urey had discovered heavy water and was the acknowledged expert on concentrating rare isotopes. The Manhattan Pro-ject was under way, and there was a demand for D_2O for use as a moderator in nuclear reactors. Funds were virtually pushed onto Urey to get him to pursue war work, and he accepted this responsibility. I undoubtedly could have been a part of it, but I was not very interested. My draft board was in Harlem, and they gave me a deferment.

When I returned to Madison in the summer of 1941, I signed on as an instructor in the Department of Bacteriology and set up to use ^{15}N as a tracer in Perry Wilson's lab. Bacteriology was short-handed, as many of its staff spent much of their time during the war years at Camp Detrick or one of the other biological warfare stations. Perry was stationed in New York City, so I not only taught labs and seminar courses and gave some lectures, but I was very involved with Perry's graduate research students.

I didn't get into war work, and always did more than enough teaching to be deferred. I took night courses in electronics during the war, so that if drafted I'd have a good chance for the Signal Corps rather than the trenches. The Camp Detrick group wanted quantities of spores to test dispersal of microor-ganisms, so Wayne Umbreit and I cranked up the old yeast pilot plant and scoured the literature for information on growing spores in quantity. We couldn't find such data, so we devised media and culture conditions for sporulation and produced pounds of spores of *Bacillus niger* (black pigment) and *Bacillus globegii* (orange pigment). When these formed colonies on a petri dish, their pigments marked them clearly. Later Camp Detrick produced lots of *B. globegii* for testing, and I understand it became difficult there to avoid getting orange colonies on many of your petri dish cultures.

The group spent much of its time on the wartime penicillin project. Fleming had described the response of bacteria to penicillin some years earlier, but penicillin never had been used therapeutically. The group at Oxford University produced enough to test on patients and the curative results were spectacular. The Oxford group launched a program of production and invited cooperation in an attempt to bring production to a useful level. The cooperation was unprecedented. Data were shared from both sides of the Atlantic. In the United States, universities, national labs, and pharmaceutical firms shared all results openly and published bulletins regularly to enhance rapid distribution of information. Peterson's and Johnson's labs in the Department of Biochemistry were the focal point for local activities, but Botany, Bacteriology, and other departments on campus all pitched in.

Initially, penicillin was produced in bottles as surface cultures of *Penicillium chrysogenum,* but this was a labor-intensive process. Pfizer had the most knowledge about such cultures, based on their experience with citric acid fermentation. Commercial Solvents in Terre Haute, Indiana had had experience since World War I with the acetone-butanol submerged fermentation in tanks. But nobody could get reasonable production of penicillin from submerged cultures. We were interested in the physiology of the organism and how it might affect submerged production of penicillin. Henry Koffler (recently retired as president of the University of Arizona) and I worked together, and he wrote his thesis on the physiological changes in submerged penicillin cultures. We collaborated closely with the biochemists. Fritz Stauffer in Botany produced mutants that were tested. The Northern Regional Research Lab in Peoria, Illinois, contributed a good and cheap corn-steep liquor medium. Ken Raper and colleagues there isolated a greatly improved penicillium strain from a moldy muskmelon. This was mutagenized in an East Coast lab, tested at the University of Minnesota and at Wisconsin, and recognized as a winner for submerged culture. Initial yields had been about one Oxford unit of penicillin per ml, but this organism brought submerged yields to about 500 units per ml. Within a month every pharmaceutical firm in the country had this culture in its fermentation tanks. The cost of penicillin started at about $20/100,000 Oxford units, but by the end of the war it was 3¢/100,000 units wholesale. The penicillin project provided an excellent example of the progress that can be made when the scientific community tackles a project cooperatively and shares all data openly.

During the war we also continued work with ^{15}N as a tracer. There was no mass spectrometer on campus, but Professor Wahlin in the Physics Department volunteered to have a Nier-type unit built in the physics shops. Fred Eppling, an undergraduate, assisted in the project. The unit was battery operated, and readings were taken from a galvanometer image projected on a long scale (nonlinearity corrections were applied). When a filament burned out a new one had to be sealed in the glass envelope and precisely aligned with the filament slit; alignment could not be checked empirically until the instrument had been evacuated and baked for about a day. But the unit worked effectively, and we could gather data with adequate accuracy and precision. I spent so much time in the physics building that when a biophysics program was launched on campus, I was put on the committee because people thought I must be a physicist.

Biochemistry

Van Potter came into the picture again. As one of C. A. Elvehjem's most productive and favorite students, Van suggested to Elvehjem that I might be a prospect for a staff position in Biochemistry—although I had been associated with Bacteriology as a graduate student, I actually had completed a joint major with Biochemistry, so that made me reasonably respectable. I talked to C. A. Elvehjem and E. B. Hart about an appointment and was offered an assistant professorship to start tentatively on July 1, 1944. Life was simple then—no departmental committees or deans with whom to interview. In the spring, Professor W. E. Tottingham died suddenly. He was teaching Plant Biochemistry, a course I had taken with him, and they asked me to finish off the semester together with Fred Smith, one of K. P. Link's PhD students.

On July 1, I cut formal ties with Bacteriology, although they still were rather short-handed, and moved down to 150 Biochemistry, which was in the 1939 wing of the building and had a very adequate lab. Peterson and Johnson had a large group on the same floor and occupied part of our lab. Shortly after my transfer to Biochemistry, Mark Stahmann had a run-in with K. P. Link and moved down to the first floor to join us. At this point fellows were being discharged from the armed services and were returning to school. This furnished a group of unusually experienced students, and they were highly productive. Bob MacVicar and Carl Clagett had been at the University of Wisconsin before joining the armed services, and they came back to complete PhD degrees. Ed Tolbert had served in the Pacific theater. Henry Little had contracted a fungal disease in the South Pacific and received a medical discharge. Bob Stutz had picked up tuberculosis and was recovering after his discharge from the Navy. Al Krall was an ex-radar technician and was skilled in electronics. Oscar Olson was an old South Dakota State classmate who did his degree work jointly between our lab and C. A. Elvehjem's lab. It was a great group for a 30-year-old staff member to acquire, as they were not much younger than I was and were anxious to complete a degree and get going on their own.

We tackled a variety of problems—biological nitrogen fixation, photosynthesis, plant respiration, organic acid metabolism, and the mode of action of plant growth substances. Carl Clagett and Ed Tolbert got us going on glycolic acid oxidation, and after leaving, Tolbert cleared up many aspects of this field. Bob MacVicar did some work on boron and on plant growth substances, but about three years after leaving he became a departmental chairman and then a dean (I had predicted five years to a deanship), and then, in succession, a chancellor at the University of Illinois-Carbondale and president at Oregon State University. Al Krall followed ^{14}C products of photosynthesis in as short a time as 15 sec, but it was hard for graduate students to compete with Calvin's group (Ed Tolbert's brother was there), so we threw in the towel on photosynthesis. Bob Stutz worked out nice methods for separating organic acids on chromatographic columns and followed their fate in plants. The research continued to emphasize biological nitrogen fixation, and we worked in close collaboration with Perry Wilson.

To this point I haven't said anything about my family. All through high school and college I was scared to death of women and never dated them. In graduate school I encountered some female graduate students in the lab who didn't seem so terrifying after all. So I dated one of them occasionally and learned the jitterbug, which was the dance of the day. I even had two or three dates during my year at Columbia University in New York. In 1944 I met Katherine Brusse, who had an MS degree in Home Economics and had done research on biotin. She had had experience in the Food Relief Program in Chicago and as a nutrition agent traveling the state of Wisconsin. As a freshman she had been the youngest student at the University of Wisconsin. Not only was she incredibly bright, but she was highly competent in broad areas and a wonderfully nice person. It is obvious why I fell for her. My proposal was the smartest thing I ever have done. She accepted and we were married in Madison on September 12, 1945. Katherine served for many years in training dieticians at the University of Wisconsin and as dietician for the Medical School cafeteria. She proved to be a great mother and raised three great kids who never were any problem to us and who all have done very well on their own. They in turn have given us five granddaughters. I could go on at considerable length about the family, but this probably is not the place for it.

In 1954, I received a Guggenheim Fellowship. We boarded a ship in New York with the three kids—Jean, 7, John, 5, and Ellen, 2—and sailed to Sweden. Katherine wonders now how she ever kept three kids in line. We disembarked in Göteberg, located our 13 pieces of luggage strewn from one end of the pier to the other, and managed to make our train to Stockholm. We saw a bit of the city before boarding another boat for Helsinki. Helsinki was very impressive with its cleanliness and order. Dr. Kivimaa had been a visitor at the Forest Products Lab in Madison the previous year, and we had arranged to rent his apartment in Helsinki for seven weeks, while his family occupied their cabin in the country. We didn't know any Finnish, but by saying "*Linnenkosken katu vizi toista*" to a cabbie enough times he would get the idea of taking us home to that address. Jean, our 7-year-old, learned to swim in the Olympic pool (the Olympics had been in Helsinki a couple of years earlier; Britton Chance had won a sailing gold medal in the last event to give the US its last Olympic win over Russia). The cabbie would deliver the crew to the pool when they said "*Umastadium*." Of course all the people at the lab spoke English, and Virtanen knew eight or nine languages.

I took the bus to the lab early in the morning (it never got really dark) and managed to get quite a bit done. Virtanen and Sinnika Lundbom had published a short paper a year earlier that disagreed with our reported findings on nitrous oxide as a specific inhibitor of nitrogenase. Their paper was so short that one couldn't really tell how they had run the experiments. So I asked Sinnika for the details. She described the experiment in which they reported that nitrous oxide inhibited the use of nitrate rather than being specific for N_2. They had varied the oxygen concentrations.

I said, "*Azotobacter vinelandii* would have used all the oxygen in a day and you analyzed the cultures after six days. You were measuring a pO_2 effect, not an effect of nitrous oxide. Who designed this experiment?"

"The Professor did," she said. I replied, "Let's go talk to the Professor about it." Said Sinnika, "Oh no. My voice even changes when I go into the Professor's office."

Assuring her that my voice wouldn't change when confronting Finland's only Nobel laureate, I told Virtanen about the problem. We reran the experiment with three independent methods of measurement (I sent samples to Madison for ^{15}N analyses), and all methods verified our claims for the specificity of nitrous oxide as an inhibitor of nitrogen fixation. So we wrote up a paper, which I left with Virtanen. He submitted this retraction a few months later (7). But subsequently, Virtanen repeated the experiment in the original incorrect way, obtained the incorrect answer again, and retracted the retraction. I often have told this story to my students as a method to get three papers for one. Virtanen really was a very nice guy, and his AIV silage did wonders for the dairy industry of Finland. It was just that being the only Nobel Prize winner in the country put him on such a high pedestal that others were loath to challenge him, and even he may have come to believe in his infallibility.

After a pleasant and productive time in Helsinki, we moved on to Cambridge University for a stay of about five months, during which time I worked in Keilin's lab at the Molteno Institute. Keilin was a parasitologist by training but was known primarily as a great figure in biochemistry. He rediscovered cytochrome *c* in 1925 and studied its role in respiration. He was small, retiring, modest, and never in really good health. But he was a terrific scientist who was respected and loved by his colleagues. We didn't see a great deal of him in the lab, but Hartree was always there to help. Keilin was known as a great lecturer. I sat in on a fall course, and although attendance usually was about 5 students (class attendance at Cambridge is rather relaxed) for most of the distinguished staff lectures, attendance swelled to about 35 when Keilin took over with Hartree backing him up as demonstrator. His lectures were dramatic, filled with meaningful props, and the students ate up the lectures and were educated in the process.

Robin Hill was in the nearby biochemistry labs. There were lots of stories about Robin, but you really had to see his lab to believe the clutter. After seeing it you could believe the story about when a distinguished visitor was brought into the lab and discovered Robin writing a manuscript on the floor; that was the only bit of clear space in the lab. He and Keilin could discover a tremendous amount of information by looking through their microspectroscopes. Robin probably could tell you the concentration and the reduction level of the cytochromes ± 10% by eye. Biochemists might insist on isolating the cytochrome, lose 60% in the process, measure the extract to ± 1% in the spectrophotometer, and feel superior with their ± 60% result to Hill or Keilin with their ± 10% value.

Britton Chance had worked in Keilin's lab before I came, and he always examined Keilin's papers for any error. Finally, back in Philadelphia, with one of the best spectrophotometric setups in the world, Brit found a discrepancy of a few millimicrons between his measurements and those made by Keilin with a microspectroscope. However, his triumph was short-lived, because Brit found that Keilin had corrected the error when he had reported another microspectroscopic observation in a subsequent paper.

Kubo had reported (about 1939) that the red pigment in leguminous root nodules was hemoglobin. This assumed importance as a clue to how the aerobic rhizobia functioned without inactivation by oxygen (the high sensitivity of nitrogenase components to oxygen actually had not been demonstrated at this time). I went to the University of Chicago to share the expertise and the spectrophotometer of E. Haas for examining the oxygenation-deoxygenation of hemoglobin and leghemoglobin. We saw the classical spectral shifts with our control mammalian hemoglobin but not with nodule hemoglobin. This we reported (4), but we were wrong, as was pointed out later by Keilin and colleagues. Suitably humbled by this error, I wanted to learn techniques in Keilin's lab and to check there whether any of the free-living nitrogen fixers, or the rhizobia grown apart from the host plant, had hemoglobin. We found none, but we did verify Keilin's observations that we had missed the deoxygenation of leghemoglobin because it has a much higher affinity for oxygen than that of our mammalian hemoglobin control.

The hemoglobin study in other organisms quickly proved to be a blind alley, so I teamed up with Alfred Tissieres to study the cytochromes of *A. vinelandii*. We described cytochromes c_4 and c_5 in the organism. Tissieres was an avid mountain climber, and it was interesting to hear of his exploits in the Himalayas. I later showed pictures and described the time in Cambridge to my graduate students. Years later Brian Mudd had the pleasure of being on top of the east face of Long's Peak in Colorado when Tissieres emerged at the top after scaling the face. Brian recognized him from the pictures and was able to greet him with "Dr. Tissieres, I presume." We returned from England over the stormy North Atlantic by ship at Christmas-time 1954 and reestablished our research program in Madison.

The next jaunt of interest was to Central America with a United Fruit team. United Fruit was having trouble with Panama disease, a fungal disease in bananas, and they had contracted with the Stanford Research Institute to look into the problem. We toured United Fruit production and research facilities in most of the Central American countries, found out a lot about bananas, and ate bananas in quantities that amazed our hosts. We returned on a banana ship that departed from Costa Rico headed for San Francisco. On the ship there was plenty of time to ruminate on our findings and to suggest potential cures. The rather obvious thing we suggested was to develop and grow a banana resistant to Panama disease. United Fruit had been built by engineers who had drained the swamps, built railroads, controlled the mosquitoes, produced a fleet of ships, and had established a banana trade. Their engineering solution to Panama disease was to build dikes around the fields and to flood the fields for a couple of years. The anaerobic conditions killed the fungus. Small operators had no facilities to use this technique. But when United Fruit introduced a resistant banana (not as desirable to eat as the old Gros Michel) anyone could steal a chunk of tissue from the base of a stalk and propagate the bananas vegetatively. United Fruit lost its near-monopoly, and our successful solution to the disease problem depressed United Fruit stock drastically and rapidly. I still eat bananas regularly despite our disastrous advice to United Fruit and probably derive a high percentage of my potassium from bananas.

We had struggled to get consistent cell-free fixation of nitrogen. ^{15}N gave us a sensitive tracer, as we could detect 0.003 atom % excess above the normal abundance quite readily. We considered five times this, i.e. 0.015 atom % ^{15}N excess, as positive evidence of fixation, and we achieved this level in many experiments. At the Federation meetings in 1958 we reported fixation with cell-free preps from *Clostridium pasteurianum* of about 1.1 atom % ^{15}N excess; that was about 70 times our arbitrary level for positive fixation. But such fixation was not consistent. In 1960, Carnahan et al (5) reported consistent fixation by extracts from *C. pasteurianum* that had been dried in a rotary drier at 45°, extracted anaerobically, and supplied with generous levels of pyruvate as a substrate. We verified their fixation on the first trial using their technique and reported the verification along with the application to other organisms. They reported that ATP inhibited the cell-free nitrogen fixation.

Not content to accept ATP as an inhibitor (ADP is the inhibitor), Jack McNary, as a postdoc in our lab, reported that in fact it supported cell-free nitrogen fixation (6). At first, this was not accepted by the proponents of ATP as an inhibitor, but when they tried the proper experiment they verified McNary's results and published their findings. I was surprised to read a recent review that ignored McNary and attributed the discovery of the need for ATP for nitrogen fixation to those who had reported it as an inhibitor and had protested its need in nitrogen fixation.

Mortenson, a Wisconsin PhD, soon demonstrated that nitrogenase consisted of more than one protein (two or three were first suggested, but the number soon was proved to be two). Mortenson's work opened the way for a more rational approach to purification of the components. Everyone jumped into the fray, and I would be hard-pressed to say who should get credit for purifying the nitrogenase components. We had the most active preps for a time, but several labs contributed helpful methodology. One of the purified units turned out to be an iron-molybdenum protein and the other an iron protein. A reconstructed system also required an ATP-generating system and a reducing agent. Both protein components are very labile to oxygen, so great care is required to avoid inactivation by oxygen.

For years we had supported ammonium ion as the first demonstrable product of biological nitrogen fixation, but most live organisms converted it to amino acids so quickly that we had demonstrated ammonia directly only in *C. pasteurianum* (this anaerobe produced little α-ketoglutaric acid as an acceptor for ammonium, and under certain conditions ammonium ion was excreted with a high level of ^{15}N when $^{15}N_2$ was supplied). Cell-free preparations turned out ammonium unequivocally.

The MoFe and Fe proteins have characteristic EPR spectra at low temperatures. We teamed up with Bill Orme-Johnson, Perry Wilson, and Winston Brill's groups in studying the EPR changes when purified nitrogenase components were catalytically active. Clearly, the electron transfer was from an agent such as ferredoxin (a flavodoxin also can serve) to the Fe protein, to the MoFe protein, to N_2 (or an alternative substrate). ATP is used in the transfer of electrons from the Fe to the MoFe protein. The nitrogenase components have gone under a variety of names, but we prefer to designate them in a way that signifies their function, i.e. dinitrogenase reductase for the Fe protein, as its primary function is to transfer electrons to the MoFe protein (it does have some secondary functions), and dinitrogenase for the MoFe protein, as it binds and reduces dinitrogen. The complex of the two proteins we refer to as nitrogenase.

Shah, Brill, and their associates were able to dissociate a MoFe complex from dinitrogenase, and they named it FeMoco. Later, Shah, Ludden, and associates reactivated apodinitrogenase by adding back FeMoco under the proper conditions.

Nitrogenase reduces several substrates in addition to dinitrogen. The first of these demonstrated was nitrous oxide (one might claim protons came before nitrous oxide). Cyanide, methyl isocyanide, azide, cyclopropene, diazarine, cyanamide also are reduced, but they are mostly laboratory curiosities. However, acetylene as an alternative substrate has had a real impact on research on biological nitrogen fixation, and I have described its discovery as a substrate in some detail (3). In 1965, Schöllhorn, a postdoc in the lab, observed that acetylene and azide were reduced by the nitrogenase system. We reported this at an informal meeting at Sage Hen in 1965, and Schöllhorn reported it at the Federation meetings early in 1966. I wrote to Mike Dilworth about it at Christmas 1965, as he had been in our lab the year before. He wrote that they also had observed acetylene reduction independently. The method is simple and extremely sensitive, hence it has proved very useful in studies of biological nitrogen fixation. Schöllhorn and Dilworth never had met until I introduced them to each other at the International Conference on Biological Nitrogen Fixation in Cologne in 1988.

The era of genetics hit nitrogen fixation like it hit other research areas. Brill initiated an active program on campus, and Roberts, Ludden, and others have followed it up. Over 20 genes for the nitrogenase system have been described, and studies on these genes now dominate all conferences on biological nitrogen fixation. Not being much of a geneticist, I have collaborated on genetic studies with Gary Roberts of Bacteriology. There are many other aspects of our long research program on nitrogen fixation that I could detail, but I will restrain the urge.

Administration and Students

There was an interlude of administration in my career. In 1958, C. A. Elvehjem became president of the University of Wisconsin. He had been holding down the chairmanship of Biochemistry and the deanship of the Graduate School while supervising a vigorous research program, but the presidency was demanding enough to challenge even his organizational abilities. My colleagues asked me to chair Biochemistry. I have never had any ambition for administration and always had a ready no for offers to be put on a list for a potential deanship. But this was a request from my colleagues, who were a great bunch. So I chaired Biochemistry from 1958 until I resigned from the job in 1970. Some people claim that chairing a department is a full-time job, but I had an excellent role model in Connie Elvehjem, who managed research, the department, and the grad school all at once. I found the mornings adequate for the department and the afternoons for research and teaching. The only change was that I had to drop some research efforts and concentrate on biological nitrogen fixation, because one cannot easily cover the literature and progress in several fields at once these days. But the research continued in productive fashion, and I didn't get into too much trouble administratively, although our great staff did have some members generally considered to be mavericks.

My feeling was that a major task of a chairman was to protect his staff from nonsense from above; many messages from above that would eat into the productive time of the staff could be answered satisfactorily directly to the administrators. In this regard, E. B. Hart, who chaired the department from about 1906 to 1946, was a great role model. He diverted many directives from administration to his staff directly into the waste basket, and the department continued to be productive despite this aberration of Hart's. Anne Terrio, who could be termed chief secretary or administrative assistant, was a great asset. She had served under Hart and Elvehjem and took no guff from anyone. K. P. Link never was considered a push-over or an easy person to deal with, but I have heard Anne "tell him off" on a number of occasions. "So what if he is a professor; if he doesn't make sense I'll tell him so," was her attitude.

I had good colleagues on the staff and an excellent university tradition, but the greatest asset in research has been highly competent students. I often have wondered what impact I would have had at "Out-of-Contact-University." Biochemistry at Wisconsin had a distinguished reputation. Today there are many excellent departments of biochemistry, but when I joined the staff in 1944 there were just a few. People knew about Steenbock and vitamin D, Elvehjem and Strong and nicotinamide, Link and dicumarol, Peterson and Johnson and penicillin, Lardy and stabilized semen, and the others. It is hard to spoil good graduate students—give them a challenge and they will meet it. As director of a research group, people often gave me credit for the ideas that originated with my students. They were a terrific group, and I owe them a lot. They got me into the National Academy of Sciences and the American Philosophical Society, and they earned me the National Medal of Science and other recognition. They did most of the work on 329 published papers. One develops close ties with students who work around the lab, and they become almost like members of a family: a big family—my count shows about 30 MS and 75 PhD students and about 50 postdocs, plus a dozen undergraduates. At one time I chaired the NIH panel awarding training grants in biochemistry. During our visit to Cornell University I was being given the "pitch" by a staff member who said, "We get excellent grad students at Cornell; we don't get them from places like South Dakota." That was not quite the right pitch to make to a loyal South Dakotan like me, but it brings up the indefensible concept that you can judge people by their point of origin, race, religion, or gender. Nonsense! It's in the genes.

There were a number of assignments off campus. I was on an NSF panel directly after the creation of the NSF in 1950. I can remember how we deliberated over whether to grant Arthur Kornberg $8000 or $9000 on an outstanding research proposal. I'm glad to say we gave him $9000. I served on NIH panels and chaired the training grant panel in biochemistry in the early 1960s. At times I served on ONR, USDA, NASA, and EPA panels. Most were good, with EPA clearly the worst. The most fun, however, was on National Research Council panels. I served for its first six years on the Commission on Life Science panel, and that was a keen and distinguished group. The Board on Science and Technology for International Development was diverse and hence very interesting. It was well run, and the foreign grantees appreciated the attention and aid they received. NASA's Controlled Environment Life Support System program held interesting challenges on how to grow plants in space vehicles to sustain astronauts, recover carbon dioxide, regenerate oxygen, and recycle wastes in a closed system.

Travels for scientific tasks have taken me to a variety of places. Most stays were too transitory to give much of a feel for the country, but there were enough trips to Brazil to convey some appreciation for the country. In 1960 I spent six weeks in India evaluating biochemistry programs for the State Department, and I traveled quite widely there. I have returned to India three times since then. The most interesting trip was when a group of six of us, chaired by Allan Bromley (later science adviser to President Bush), set up the joint Indo-US Science Initiative that Indira Ghandi and President Reagan had formulated when Ghandi visited the United States. Bromley oversaw formulation of a program in New Delhi, and then we attended the signing ceremony with Ghandi.

I also have developed some feel for Australia from a couple of international conferences plus two expeditions to the Great Barrier Reef. The first expedition to the Reef was led by Ed Tolbert, and we were on the Alpha Helix anchored off Lizard Island for seven weeks. I studied nitrogen fixation on the Reef and the others studied photorespiration of marine plants. The second expedition was in the area of the Tropic of Capricorn, near Heron Island, and there we studied nitrogen metabolism of the zooxanthellae of hermatypic corals. My son, John, was chief scientist, and this had much more weight in getting me aboard than my expertise as a marine biologist. Of course, being on the Reef for five weeks then, and seeing only one fishing boat in that period, is not exactly a representative sampling of Australia.

At age 80 one has a tendency to think things were better in the past, but I'm not at all sure this is true. We were more self-sufficient than we are now. When I joined the biochemistry staff in 1944 with 4 years of postdoc experience, my salary was $3000 for 12 months, and E. B. Hart said I could have $200 credit in the stockroom to run my lab for the year. This was before NSF, NIH, and DOE funding, and sources of funds were scarce. I received a small grant from Red Dot potato chips, and Bob Stutz worked out a way to produce chips with uniform salt, reduced oil content, proper color, and good taste panel acceptance. The fatal flaw was that they were flat and didn't bulk up in the bag like the conventional chips. The first federal grant was a modest one from the ONR. It carried little money but gave access to some surplus equipment in storage in Pennsylvania.

Because of a lack of funding, we built much of our own equipment. Many of the staff were competent glassblowers. The shop turned out fraction collectors and, as indicated, we built a mass spectrometer in the Physics Department shop. Those days are gone, and perhaps students waste less time on equipment now. A few weeks ago a student in a colleague's lab was asked to do an experiment. Her reply was, "I'll do it if you get me the kit." Perhaps something has been lost in self-sufficiency.

One nice aspect of work at the University of Wisconsin has been the interaction with colleagues. There are very few fences here. At times I have collaborated in research with people from Bacteriology, Botany, Physiological Chemistry, Plant Pathology, Oceanography and Limnology, Forest Products Laboratory, Dairy Forage Research Center, Food Science, Horticulture, Agronomy, Biotron, Entomology, Engineering, Physics, Soil Science, Chemistry, Zoology, and obviously Biochemistry and the Enzyme Research Institute. Such collaboration has been very helpful to me and to my students.

Since retirement in 1984 my duties have tapered off and I no longer have the legal status to direct graduate students. Yaoping Zhang, as a postdoc, continues collaborative work with Gary Roberts and Paul Ludden on the genetics of the control system for nitrogenase. The department kindly provides an office and lab space for this continuing research.

Literature Cited

1. Burris RH, 1941. Failure of barley to fix molecular ^{15}N. *Science* 94:238–39
2. Burris RH. 1942. Distribution of isotopic nitrogen in *Azotobacter vinelandii. J. Biol. Chem.*143:509–17
3. Burris RH. 1975. The acetylene reduction techniques. In *Nitrogen Fixation by Free-Living Micro-organisms,* ed. WDP Stewart, pp. 249–57. Cambridge: Cambridge Univ. Press
4. Burris RH, Haas E. 1944. The red pigment of leguminous root nodules. *J. Biol. Chem.* 155:227–29
5. Carnahan JE, Mortenson LE, Mower HF,

Castle JE. 1960. Nitrogen fixation in cell-free extracts of *Clostridium pasteurianum. Biochim. Biophys. Acta* 38:188–89

6. McNary JE, Burris RH. 1962. Energy requirements for nitrogen fixation by cell-free preparations from *Clostridium pasteurianum. J. Bacteriol.* 84:598–99

7. Mozen MM, Burris RH. 1954. The incorporation of ^{15}N-labeled nitrous oxide by nitrogen fixing agents. *Biochim. Biophys. Acta* 14:577–78

8. Ruben S, Hassid WZ, Kamen MD. 1940. Radioactive nitrogen in the study of N_2 fixation by non-leguminous plants. *Science* 91:578–79

Annu. Rev. Plant Physiol. Plant Mol. Biol. 1995. 46:21–44

MOLECULAR GENETICS OF SEXUALITY IN *CHLAMYDOMONAS*

U. W. Goodenough, E. V. Armbrust, A. M. Campbell, and P. J. Ferris

Department of Biology, Washington University, St. Louis, Missouri 63130

KEY WORDS: mating type, chloroplast inheritance, green algae, evolution, sporulation

CONTENTS

ABSTRACT

Recent molecular and cellular data on the sexual cycle of *Chlamydomonas reinhardtii,* a unicellular green alga, is considered in the context of current theories on the origins and evolution of eukaryotic sex. The mating-type locus of *C. reinhardtii* controls gamete recognition and fusion, organelle inheritance, and sporulation, the three traits that characterize most sexual cycles in lower eukaryotes. The mating-type locus comprises approximately one megabase of DNA on linkage group VI, is highly rearranged in its central region, and contains identified genes that govern both recognition/fusion and chloroplast inheritance. Sporulation, a diploid program designed to negotiate changing environments, is analogous if not homologous to the somatic differentiation

21

program of multicellular plants and animals, and its expression requires the fusion of compatible gametes.

INTRODUCTION

At the dawn of the Cambrian, so our evolutionary histories now go, most of the basics of eukaryotic cell structure and function had evolved in protists, and the stage was set for an astonishing burst of eukaryotic phylogeny. One of the resultant lineages gave rise to the modern green algae and higher plants, while two other lineages also included photosynthetic organisms: The dinoflagellates, ciliates, and apicomplexans form one cluster; the chromophytes, diatoms, and oomycetes form another (81). The histories suggest, therefore, that plastid endosymbionts were acquired polyphyletically, although whether the same or different types of endosymbionts were involved is still an open question (35, 55, 79).

The modern descendants of the Cambrian-burst eukaryotes are almost all sexual, whereas sex in the modern descendants of the pioneer protists is either absent or episodic (3, 37). The argument can therefore be made that well-developed sexual cycles were important features of the organisms that gave rise to the Cambrian explosion.

This review focuses on the green algae, and particularly on the unicellular *C. reinhardtii,* in the context of these two central features of eukaryotic evolution: the acquisition and maintenance of organelles and the acquisition and maintenance of sexuality. Several evolutionary theorists (6, 16, 38, 41, 42, 51) have in fact argued that sex and organelles have much to do with one another, an argument supported by observations on the molecular biology of the *C. reinhardtii* life cycle. We offer a speculative account of the origins of organelle-possessing eukaryotes and sexual eukaryotes, followed by an evaluation of this scenario in terms of known features of the *C. reinhardtii* life cycle. Readers are referred to other accounts of eukaryotic evolution (65, 83, 97) and to earlier reviews of the *Chlamydomonas* life cycle (30, 31, 33, 68, 89).

EVOLUTIONARY SCENARIO

We begin with a primitive protist that had a DNA-based mitotic nuclear genome, a well-developed cytoskeleton, and the capacity for endocytosis, but no organelles and no sex. The capacity for endocytosis offered both an abundant food supply and the opportunity to "domesticate" (6) engulfed prokaryotes as organelles (56). An essential feature of the domestication program was the ability to regulate the number of organelle genomes per cell. Another important feature was the ability to resist the invasion of both undomesticated prokaryotes and any engulfed eukaryotes bearing organelles with dissimilar domestication programs, because they were likely to be pathogens. Digestion within autophagic vacuoles and segregation within parasitophorous vacuoles represented two solutions to the invasion challenge, but any prokaryotes or foreign organelles that breached these defenses and entered the cytoplasm would likely wreak havoc with the lineage, particularly invaders that were capable of infecting the resident organelles with their own genomes using an F-element-like transfer of DNA (43) or a plasmid-mediated fusion of organelle membranes (44). Moreover, if we adopt Sogin's postulate (81) that the nucleus derived from an endocytosed archebacterium, then an additional challenge would be the invasion of any foreign (proto)nuclear archebacteria, particularly any capable of infecting or fusing with the host nucleus.

An early response to these challenges, we propose, was the development of self-counting systems, which would monitor the copy number of nuclear and organelle genomes, assuring that there were neither too many nor too few. For nucleated cells, self-counting is straightforward in concept: A haploid or diploid genome is retained and replicated once per cell division (5). For cells with organelles, the self-counting concept must include the retention of multiple copies of a self-replicating genome. The most insidious problem is the presence of unregulated genomes that multiply freely and take over the lineage. Thus the system must be titrated so genome copy number neither drops below some threshold needed for energy production (2) nor rises above some toxicity threshold.

A second response to these challenges, we propose, was the development of immunity systems to deal with non-self genomes. Although different immunity mechanisms doubtless evolved for nuclei, chloroplasts, and mitochondria in different lineages, all would be expected to share two features common to all immunity systems: the capacity to recognize and eliminate foreign entities and the capacity to protect self-entities from such elimination.

A third capability of the pioneer protists, displayed by virtually all extant protists and many extant prokaryotes (14, 18, 37), was the ability to form spores or cysts in response to environmental stress. This is a highly adaptive trait that would be expected to be a strong focus of natural selection. Sporulation genes are not expressed during the clonal expansion of population size, but their expression allows a species to persist under conditions that would otherwise be lethal and to exploit a favorable niche when such a niche emerges. A spectacular array of spores can be found among the lower eukaryotes (14, 18), surrounded by mucilaginous coats and desiccation-resistant walls of all sorts, indicating that numerous spore-generating genetic programs evolved in response to a variety of environmental challenges.

Whereas the endocytic engulfment of non-self organisms would be expected to have stimulated the evolution of self-counting and immunity systems, any fusions with members of one's own species would have had a different consequence. Self-counting and immunity systems would be shared, and the opportunity would arise for diploidy and, hence, genetic complementation. Population theorists continue to wrestle with the question of whether diploidy is or is not adaptive over the long term (46, 64, 67). Because most lower eukaryotic taxa conduct most of their life cycles as haploids and engage in diploidy only as a prelude to sporulation (3, 18), we can evaluate that strategy in its own right. By conducting vegetative clonal reproduction as a haploid, any organism incurring a deleterious mutation is eliminated, thereby purging the species of mutational load at the trivial expense of occasional individuals. On the other hand, when the environmental pressure is on and latent sporulation programs are activated, many individuals in the population might harbor mutations in their sporulation genes, and there would be strong selection for organisms that could acquire diploid status under environmental stress and, hence, complement one another.

Another advantage accrued by a diploid spore is the potential for gene recombination, either via crossing-over or via the independent assortment of chromosomes during the (primitive) meiosis that would need to have evolved to control runaway ploidy and return the spore products to the haploid, mutation-purging, vegetative state. That sex is important for generating new combinations of genes is a biological truism, but it is difficult to argue that sex evolved for the purpose of recombination. As pointed out by Maynard-Smith and many others (e.g. 65, 83), the individual genome is the unit of selection, and the sexual scrambling of genomes works against the propagation of well-adapted genomes even though it may be of long-term advantage to the species to harbor genetic diversity. The magnitude of this dilemma is attenuated, however, when it is realized that for most of the early eukaryotes, sex is tightly coupled to sporulation (30, 97). Spores are induced to form only when the parental habitat has deteriorated and well-adapted genomes may no longer be optimal. Moreover, spores often remain dormant for years, and they are often transported to new locations. Recombinant genomes of the spore products are likely to encounter new microenvironments to which they may be better adapted than their parents. Hence, recombination in a spore could contribute to the survival of the emergent individual as well as the group.

If the acquisition of temporary diploidy at the time of sporulation occurs between members of a vegetative clone, then the postulated advantages of genetic complementation and recombination are not likely to be realized. Such adaptive consequences require that fusions occur between genetically dissimilar clones, that is, as a consequence of outcrossing. Here we encounter a major obstacle. If two lineages have diverged sufficiently to carry complementing mutations in sporulation genes, they are likely to have incurred mutations in their self-counting and immunity systems, which would render them incompatible as diploids.

These difficulties are resolved by the evolution of mating types. A species that develops mating-type determinants that dictate the display of complementary, species-specific recognition and fusion molecules on the cell surface assures that the resultant diploids will be intraspecific but not clonal and, hence, maximizes the possibility of a successful, recombinatory sporulation. Coupled to this cell-fusion function would be a second mating-type regulated activity, namely, the capacity to render the various self-counting and immunity systems compatible. This capacity pertains to both nuclei and organelles. Thus mating-type loci often control whether or not nuclei will fuse (72), and they control the fate of organelles in a variety of ways [e.g. initiating the wholesale destruction of one set of organelles (8, 62) or the destruction of one set of organelle genomes, as in *Chlamydomonas* (see below)]. The most widely practiced solution to the "organelle problem" has been the adoption of anisogamy, the production of one set of gametes (e.g. sperm or pollen) that carries no transmissible organelles.

The first well-developed eukaryotic sexual cycles, therefore, would be expected to be controlled by mating-type loci that, in response to environmental stress, promote the fusion of complementary haploid cell and nuclear types, ensure the transmission of only one set of organelle genomes, and dictate that sporulation programs are activated only in diploid cells that are heterozygous for mating type. Most modern sexual systems conform to this pattern. The spectacular diversity of strategies that have evolved to assure these common goals is striking.

Such an evolutionary scenario raises another question. By imposing the requirement that an organism find and fuse with an organism of opposite mating type before it can express its adaptive sporulation program, and imposing this requirement at a time of environmental crisis, has not a "cost of sex" been introduced that is difficult to offset by the putative advantages of diploid sporulation? Would it not be preferable to preserve both options, with stress yielding either asexual or sexual spores depending on the availability of a suitable partner? In fact, many taxa have retained such flexibility: The filamentous ascomycete fungi, the slime molds, and many algae produce either asexual or sexual spores, using separate genetic programs to do so (14, 18, 19a). Most taxa also retain the option of self-fertilization (e.g. hermaphroditism or homothallism) (17, 73, 88)—for example, 88% of all genera of flowering plants are hermaphrodites (7)—an option that offers less opportunity for variation but is clearly preferable to producing no spore (seed) at all. The evolution of homothallism is considered in the final section of this review.

The perceived costs of sex have stimulated several theorists to postulate that sex is the product of (ultra)selfish genetic elements that came to impose cell fusion on asexual organisms as a means to insure their horizontal spread within a population (40). The distinctiveness and the occasional transposability of the sex-regulating loci that have been described (6, 21, 39, 53, 85) indicate that these loci are indeed unusual genetic elements. Whether their apparently autonomous properties indicate that they once were, or continue to function as, selfish genes is one of the more interesting questions being asked by students of molecular evolution.

SEX IN *CHLAMYDOMONAS*

The unicellular biflagellate green alga *C. reinhardtii* is a common resident of soils and fresh-water habitats. It conducts most of its life cycle as a haploid mitotic vegetative cell and undergoes gametogenesis when starved for ammonium (60, 75). Gametogenesis (reviewed in 30, 31) entails an extensive degradation of preexisting molecules and the expression of gamete-specific genes, several of which have been cloned (94). Gametes are of two mating types, *plus* and *minus*, a trait governed by two loci, mt^+ and mt^-, that map to the left arm of linkage group VI (Figure 1). Recognition and fusion of gamete pairs yields heterokaryotic quadriflagellated zygotes. Zygote maturation, which requires about 24 hr under laboratory conditions, entails nuclear fusion, chloroplast fusion, the withdrawal of flagella, and the synthesis of a cell wall, which

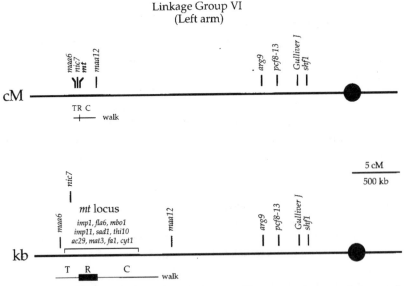

Figure 1   Genetic map of the left arm of linkage group VI. The upper map uses a centimorgan scale, the lower a kilobase scale, thereby emphasizing the effect of recombination suppression at the *mt* locus. The various genetic markers are described in References 21 and 36.

renders the resultant dormant spore resistant to dessication and temperature extremes (14, 99). Under favorable conditions, the spore undergoes meiosis and germination to yield four haploid products, two mt^+ and two mt^- (see 36 for details of the life cycle). Below, we consider this life cycle in the context of the foregoing evolutionary scenario, focusing on the regulatory role of mating type.

Chlamydomonas Mating Type (mt^+, mt^-) is a Complex Locus Under Recombinational Suppression

Because the mating-type locus of *C. reinhardtii* is the only sex-determining element to have been characterized in a green organism, it cannot yet be evaluated as typical or atypical in its molecular organization. It controls the same spectrum of traits as its analogs in other organisms and, therefore, has presumably responded to similar evolutionary constraints.

Figure 1 shows the location of *mt* on linkage group VI of the *C. reinhardtii* genome, 30 cM from its centromere. Gene mutations or cloned genes that fail to recombine with *mt* in sexual crosses include several (e.g. *imp1, imp11, sad1, mat3,* and *ezy1*) that affect mating or sporulation traits. The remaining loci linked to *mt* encode housekeeping functions concerned with metabolism, mitosis, or flagellar activity and have no known relevance to life-cycle transitions. These relationships generated the postulate (27) that the *mt* locus came to reside in an ordinary chromosome and to prevent recombination in flanking sequences.

The apparent molecular basis for this recombinational suppression was demonstrated by a chromosome walk through the *mt* region (21), which revealed three distinct domains of recombinational suppression: T (telomere-proximal), C (centromere-proximal), and R (rearranged) (Figure 1). The mt^+ and mt^- T and C domains fail to recombine during meiosis even though their DNA sequences are fully colinear and homologous. The central R domain, in contrast, is completely rearranged (Figure 2), with transpositions, inversions, deletions, and additions rendering the mt^+ and mt^- sectors (~ 200 kb) highly nonhomologous. Presumably these rearrangements preclude effective synapsis, with this effect spreading to the contiguous C (600 kb) and T (150 kb) domains. Three *mt*-linked housekeeping genes have been localized on the molecular map (Figure 2): *nic7* (nicotinamide-requiring) and *ac29* (chlorophyll-deficient) reside in the T domain, while *thi10* (thiamine-requiring) resides in the C domain (P Ferris, manuscript in preparation). The life-cycle-related *ezy1* (early zygote) gene cluster (see below) resides in the C domain, while the life-cycle-related *sfu* (sexual fusion) gene (see below) resides in the c region of the mt^+ R domain (Figure 2). The entire locus is likely to be a mosaic of housekeeping and life-cycle-related genes, with genes restricted to one or the other mating type possibly embedded within the R domain.

Two theories can explain how the molecular organization of the locus evolved. Both assume that it is essential to the specification of mating type that several key genes be inherited as a unit. The first theory suggests that the chromosomal rearrangements in the locus occurred by chance and were selected for their ability to preserve such linkages. The second theory postulates that the rearrangements were the work of genetic elements having the ability to transpose segments of the genome into a linked array. These alternatives, developed more fully in Reference 21, can be better evaluated once the molecular organization of mating-type loci in related sexual Volvocales has been analyzed.

As noted by many and stressed by Bell (6), there are two striking features of the various mating-type loci that have been cloned in the fungi (references in 21). First, they are distinctive: There exists no obvious sequence similarity, save the conservation of such DNA-binding motifs as homeodomains or HMG domains, between the mating-type genes of even closely related taxa. Given the highly conservative nature of eukaryotic gene evolution in general, the distinctiveness of sex-determining genes is most unusual. Second, the mating-type alternatives carried by a species, although occupying allelic positions on homologous chromosomes, are in no way allelic in the classical sense: They are very different in sequence from one another, leading Metzenberg & Glass (63) to suggest calling them idiomorphs. The *C. reinhardtii* R domain conforms to this pattern: Regions a, b, and c of the mt^+ locus have no counterparts in the mt^- locus, and regions d, e, and f in the mt^- locus have no counterparts in the mt^+ locus (Figure 2). Moreover, although the transposed and inverted sectors labeled 1–4 in Figure 2 are apparently homologous at the level of restriction mapping, they may have acquired distinctive patterns of expression as a consequence of their distinctive chromosomal positions. A possible explanation for these unusual features of mating-type loci is offered below.

Mating Type Controls Gamete Recognition and Fusion

Two independent recognition systems regulate the mating of *C. reinhardtii* gametes. Both are expressed only in cells starved for nitrogen, the known gametogenic stimulus under laboratory conditions (75). The first system is mediated by hydroxyproline-rich glycoproteins known as flagellar agglutinins, which associate with the flagellar surface in an extrinsic fashion and mediate the initial contact between motile gametes (reviewed in 1, 31). The *plus* and *minus* agglutinins are homologous proteins that belong to the same family as the proteins of the *C. reinhardtii* cell wall and the cell-wall hydroxyproline-rich proteins (e.g. the extensins) of higher plants (reviewed in 99). The instantaneous adhesion of agglutinins in mixtures of *plus* and *minus* gametes stimu-

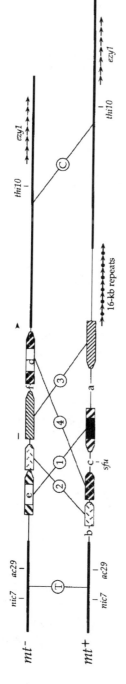

Figure 2 Schematic diagram of the *mt* loci of *C. reinhardtii*. The bullet-shaped units depict the four related segments of the two R domains of the *mt* loci, with their relative orientation indicated by the direction of their tapered ends. Regions a–c and the complete 16-kb repeat are unique to the *mt*⁺ locus; region c includes the *sfu* gene. Regions d–f are unique to the *mt*⁻ locus. The flanking T and C homologous domains, only partially shown, display the positions of the three localized housekeeping genes (*nic7*, *ac29*, and *thi10*) and the *ezy1* gene cluster.

lates an elevation in levels of intracellular cAMP (reviewed in 31, 68), which in turn activates the second recognition system.

The second recognition system is mediated by a fringe of complementary glycoproteins, unrelated to the flagellar agglutinins but not yet characterized molecularly. These glycoproteins coat the surface of small, plasma membrane-associated organelles known as *plus* and *minus* mating structures (32). After cAMP stimulation, the mating structures undergo an activation that in *plus* gametes entails the erection of an actin-filled microvillus called a fertilization tubule, which bears the *plus* fringe at its tip. When the fertilization tubule contacts the *minus* mating structure, the *plus* and *minus* fringes make adhesive contact (22). This is followed quickly by a fusion of the two mating-structure membranes and then by a total confluence of cytoplasm to form the binucleate zygote.

The flagellar agglutinins clearly function to bring the free-swimming gametes into contact and, hence, function in the same fashion as the pollen-stigma interactions in higher plants or the adhesion of sperm to the investment layers of animal eggs (52, 73). The fringe system, we propose, assures that only two cells fuse to form a zygote, analogous to the carefully controlled fusion of single pollen tubes and eggs in the plant ovary and to the various blocks to polyspermy that have evolved in the animals (52, 73).

Several gene loci (reviewed in 31, 33) affect these recognition systems: The *sag1* (sexual agglutination) locus (five alleles) and the unlinked *sag2* locus affect the production of active *plus* agglutinin, the *sad1* (sexual adhesion) locus (three alleles) affects the *minus* agglutinin, the *sfu* (sexual fusion) locus (three alleles) affects *plus* fringe display, and the *gam1* (gametogenesis) locus (three alleles) affects the membrane fusion that follows fringe adhesion. It is not clear why additional genes (i.e. affecting mating-structure formation) have not been identified in repeated mutant screens, but the the available loci define the three basic patterns of gametogenic gene expression:

1. Some genes are sex limited (23), affecting traits in one mating type but not the other even though they are unlinked to the *mt* locus. Thus *sag1* and *sag2* expression is limited to *plus* gametes, and *gam1* expression is limited to *minus* gametes: mt^- *sag1* meiotic products, for example, agglutinate normally as *minus* but transmit the nonagglutinating phenotype to their F2 mt^+ progeny. This pattern, analogous to the pattern of expression of various sterile mutations in yeast (39), is interpreted to indicate that expression of certain gametogenesis-related genes is regulated in *trans* by the *mt* loci. Care is necessary in the interpretation of particular mutations, however. For example, the *sag2* mutation has been shown to affect a general facet of protein glycosylation, which abolishes *plus* but not *minus* agglutinin activity (O Vallon, unpublished).

2. Some genes are sex linked (34). Thus the genes *sfu* and *sad1* are linked to the mt^+ and mt^- loci, respectively, and, hence, are always expressed in two of the four meiotic products. These genes may be regulated by the same *mt*-derived gene products that control the expression of the sex-limited genes, or they may be regulated in some other fashion.

3. Some genes are unlinked to *mt* and expressed in both mating types (34). This pattern is displayed by the mutations *imp3* (impotent) and *imp4*. The *imp3* and *imp4* mutations affect the activation of the flagellar-localized adenylate cyclase in response to flagellar adhesion (78; T Saito, unpublished), indicating that this signaling system is mating-type independent. A second signaling system, a blue-light requirement for gametogenesis, is controlled by two linked genes, *lrg1* (light requirement for gametogenesis) and *lrg2*, which are also expressed in both mating types (9a).

When mt^+/mt^- zygotes form sexually, most activate the sporulation-meiosis program described in the next section. A few, however, abort this program, for unknown reasons, and proceed to divide mitotically as vegetative diploids (19). When such cells are starved for nitrogen, they differentiate as *minus* gametes, indicating that mt^- is dominant to mt^+ with respect to mating functions.

A third recognition system operates within the zygote, namely, the fusion of *plus* and *minus* nuclei. Zygote-specific gene transcripts are translated in the young zygote, a translation inhibited by cycloheximide (see next section). If early zygotes are exposed to cycloheximide, they continue to be viable for several days but nuclei fail to fuse, even though they lie directly next to one another, and the sporulation program is arrested (U Goodenough, unpublished). When the cycloheximide is washed out, the nuclei fuse and the sporulation program resumes. Nothing is known about the genes involved in nuclear recognition, but heterozygosity at mating type is also required for nuclear fusion in the fungi (72).

The *sfu* gene, required for the display of fringe on the mt^+ mating structure, has been localized to region c of the mt^+ R domain (Figure 2) (P Ferris & U Goodenough, manuscript in preparation). This assignment was originally made by demonstrating that three mutations that yield a fringeless phenotype in *plus* gametes (*imp1*, *fus*, and *bs37*) all carry independent transposon insertions in the leftward portion of region c. When a wild-type clone carrying 4.7 kb of this region was used to transform the *imp1* mutant, transformants were recovered that fuse normally with *minus* partners and carry a normal display of fringe. Tetrad analysis of the resultant zygotes demonstrated that the introduced 4.7-kb DNA assorted independently of mt^+, indicating that it had inserted into a nonhomologous site [as is ordinarily the case during *C. reinhardtii* transformation (45)]. Northern analysis demonstrated that the 4.7-kb fragment hybridizes to a 3-kb message that is expressed uniquely in *plus* gametes. When the fragment was used to screen a *plus* gametic cDNA library, a positive clone was identified that is presently being analyzed. Data indicate that the *sfu* gene encodes either a component of the *plus* fringe system or a factor necessary for the expression of this system (see below). Presumably the cDNA sequence will distinguish between these possibilities.

Mating Type Controls Sporulation

The fusion of *C. reinhardtii* gametes into a binucleate zygote initiates zygote development. The earliest manifestation of zygote development is the expression of zygote-specific genes, whose transcripts are absent from gametes and vegetative cells (20, 87). We have sequenced eight genes whose transcripts appear only after the two cytoplasms achieve confluence (20, 61; J Woessner, unpublished), two of which encode zygote-specific cell wall glycoproteins (98, 99). Seven of these transcripts are expressed within 10 min of zygote formation, and their transcription occurs in the presence of cycloheximide; the eighth (Class V) is not transcribed until 90 min after cell fusion, and its expression is cycloheximide sensitive. We interpret these patterns to mean that the early genes are switched on as the consequence of an association (e.g. a dimerization) of preexisting transcription factor subunits that are sequestered in the *plus* and *minus* gametes; the late Class V gene, in contrast, is expressed only after some zygote-specific factor (encoded, presumably, by an early gene) is transcribed and translated. Additional zygote-specific genes are expressed at the time of spore germination (95). Such a cascade of differential gene expression describes the progression of all developmental systems, including the sporulation programs in prokaryotes (54).

The postulated transcription-factor subunits that are contributed to the zygote by the *plus* and *minus* gametes can be designated P and M, where the PM dimer is active in inducing the expression of the early zygote-specific genes (more complex circuitry may of course be involved; we are presenting here a minimalist model). Gene mutations that block P or M function have not been identified, but both the mt^+ and the mt^- gametogenesis programs must be activated if P and M are to be produced. Thus, artificially (PEG)-fused mt^+/mt^+ diploids, when starved, develop only into *plus* gametes, and starved mt^-/mt^- diploids develop only into *minus* gametes, which means that diploidy per se is not sufficient to switch on the zygote program (26, 57, 59). Moreover, when the above-mentioned mt^+/mt^- vegetative diploids are starved for nitrogen, they are only capable of *minus* gametic differentiation and fail to autosporulate, although sporulation ensues if they are mated to *plus* haploid or diploid gametes (19, 59). Hence, the mt^+ gametogenesis program must be allowed to proceed if P is to be generated, the mt^- program must proceed if M is to be generated, and the two can be said to be co-dominant in initiating the sporulation program. Whether P and M are also involved in regulating facets of their respective gametogenesis programs is unknown.

In considering why the interaction of two factors should be required to switch on a sporulation program, it may be relevant that this is the case for prokaryotes. In *Bacillus subtilis,* for example, a pair of transcriptional activators, members of the so-called two-component family of regulatory elements, are encoded in the haploid *B. subtilis* genome, and these proteins must interact to activate sporulation (54). This analogy suggests a speculation. If mating-type loci are crafted by elements that are able to transpose genes (21), then the fact that the regulatory genes in mating-type loci are idiomorphic may reflect their original status as independent genes that have subsequently been moved to allelic positions in chromosome homologs. Specifically, the postulated P and M genes in *C. reinhardtii* might originally have been members of a two-component sporulation program in a haploid genome that were subsequently moved to allelic positions on homologous chromosomes by the evolving *mt* loci. Such transpositions would assure that the fusion of opposite-type gametes is a prerequisite for the onset of sporulation.

It is unclear whether the postulated P gene is linked to the mt^+ locus. Hence, the foregoing speculation may not pertain to *C. reinhardtii* sporulation functions. It may, however, be a fruitful way to think about the evolution of idiomorphs in general, and it may be applicable to (some of the) nonhomologous regions in the *C. reinhardtii mt* loci (Figure 2). The speculation also predicts that mating-type genes may be very different from one lineage to the next as different sets of genes are recruited to the locus. As noted earlier, this is in fact what has been demonstrated for different groups of fungi.

Mating Type Controls Organelle Inheritance

Although it is well known that organelles are transmitted sexually via the large egg in the female line (13, 69), it is also the case that organelles are transmitted uniparentally, even when the gametes are equal in size (isogamy) and contribute equal numbers of organelle genomes to the zygote. The situation is particularly interesting in *C. reinhardtii* because meiotic products inherit chloroplast DNA from the *plus* parent only and mitochondrial DNA from the *minus* parent only (7a, 27, 28, 47, 57, 66, 70, 71, 74). Such reciprocity rules out simple models wherein one mating type simply took over the organelle problem and suggests far more interesting evolutionary dynamics. We focus here on plastid inheritance (see 70, 71 on the mitochondrial system).

Considerable information has been obtained on the uniparental transmission of chloroplast DNA in *C. reinhardtii*. A vegetative cell is estimated to carry roughly 150 copies of the chloroplast genome (a 200-kb circular chromosome) in its single chloroplast; a gamete carries about half that number (12,

86). The chromosomes are localized within discrete regions of the chloroplast stroma known as nucleoids. Within 1–2 hr of zygote formation, and well before the two chloroplasts are reported to fuse (11), the DNA within the *minus* nucleoids is degraded while the DNA in the *plus* nucleoids is left intact (15, 48). Crosses involving the mt^+/mt^- diploids described earlier have shown that mt^+ is dominant to mt^- in establishing this uniparental-plus (UP$^+$) pattern: The diploids, although *minus* with respect to gametogenesis functions, transmit chloroplast genomes as effectively as their *plus* partners, yielding a biparental (BP) inheritance of chloroplast genomes (57, 58).

Sager & Ramanis (76) discovered that if *plus* gametes, but not *minus* gametes, are briefly UV-irradiated just prior to mating, the system fails to operate properly, and the two sets of genomes are often inherited biparentally or, in extreme cases, from the *minus* parent only, a pattern called UP$^-$. The data indicate that UV irradiation affects a DNA target (77), but the target remains unidentified.

A candidate molecular participant in the control of chloroplast inheritance is the *ezy1* gene product (4). An *ezy1* gene cluster (7–8 tandem copies) is located in the C domain of both the mt^+ and the mt^- loci (Figure 2). Its zygote-specific transcription occurs within 10 min of mating and is selectively sensitive to prior UV-irradiation of the *plus* but not the *minus* gamete. The *ezy1* proteins enter both of the chloroplasts in the zygote and associate with all of the nucleoids, even though only the *minus* genomes are destroyed. They subsequently dissociate from the nucleoids and form large intrachloroplast aggregates that persist at least 10 hr into zygote development. Accompanying this dissociation is a change in the net charge of the proteins such that they migrate as more negative species in two-dimensional gels, a change that apparently is not the result of phosphorylation (EV Armbrust, unpublished). Three different *ezy1* cDNAs have been isolated, each encoding proteins with several amino-acid differences (4). It is not known whether these derive from one or both LG VI chromosomes nor whether the remaining 12 copies of the gene are expressed.

An additional set of observations on chloroplast DNA inheritance concerns the outcome of genetic crosses in which the input of *plus* chloroplast genomes has been reduced. In the first example, *plus* cells are grown in the presence of FUdR, a nucleotide analog that selectively blocks chloroplast DNA replication and, hence, generates cells with a low copy-number of plastid genomes. When such cells are mated as *plus* gametes, the resultant zygotes commonly display BP or UP$^-$ inheritance, as if the zygote is able to count input *plus* organelle genomes and adjust its destruction activities accordingly. In contrast, no effect on transmission patterns is observed when the input of *minus* chloroplast genomes is reduced using FUdR (58, 100). The second example involves crosses using the mutant strain *mat3*, whose mutant gene is linked to mt^+

(Figure 1) (29). The *mat3* mutants have a very small cell size, a very small chloroplast, and highly reduced levels of *plus* chloroplast DNA (EV Armbrust, unpublished), and they behave like FUdR-treated gametes, with BP and UP⁻ genome transmission commonly observed (29). The third example involves crosses between normal *minus* gametes and gametes that are chromosomally *minus* but engineered to undergo gametogenesis as *plus* (described in the following section). In this case, no mt^+-influenced chloroplast genomes enter the zygote, and meiotic progeny again receive chloroplast genetic markers from both parents (P Ferris, unpublished).

Investigators have proposed models to explain these various observations on chloroplast DNA inheritance in *C. reinhardtii*. The best-known model, proposed by Sager & colleagues (10, 77), suggests that restriction and modification enzymes might be involved, with an mt^+-encoded methylating enzyme providing protection to *plus* chloroplast DNA and a nuclease, activated in the zygote, destroying all unmethylated chloroplast genomes derived from the *minus* parent. In the context of this model, we have proposed (4) that the *ezy1* protein might be either the nuclease or an activator of the nuclease that degrades the *minus* genomes.

Evidence supporting or failing to support the Sager model has been reviewed (28, 36). A notable shortcoming of the model is that it does not obviously explain how the UP⁺ pattern is aborted when the input of *plus* genomes is low or absent. It would appear necessary to propose that a second system has evolved that first ascertains that the number of *plus* chloroplast genomes in the zygote is adequate and then activates the destruction of *minus* genomes. Nothing is currently known about this second system, but it may use facets of the self-counting system postulated earlier to monitor the number of organelle genomes in vegetative cells. Similarly, the destruction of *minus* chloroplast DNA may use facets of the immunity system postulated earlier to monitor the presence of non-self genomes; specifically, gene products derived from the *plus* parent and/or the zygote may perceive DNA in the *minus* chloroplast to be foreign. Since all four meiotic progeny inherit the same chloroplast genome, any molecular distinction between *plus* and *minus* chloroplast DNA must be conferred during subsequent vegetative growth or gametogenesis.

Finally, we note that the mt^+ locus carries a block of repeating 16-kb sequences (Figure 2), mirroring the repeating *ezy1* genes in the nearby mt^+ and mt^- C domains. The 16-kb repeat has been truncated to less than one unit in the mt^- locus, apparently during the course of rearrangement events (21). We are presently exploring whether the sequence is transcribed and, if so, whether it participates in UP⁺ inheritance.

The impl1 and iso1 Mutations Generate Fusion-Incompetent Pseudo-Plus Phenotypes

All strains of *C. reinhardtii* that have been isolated from the wild are heterothallic (dioecious), exhibiting no ability to mate intraclonally (36). Mutagenesis of an *minus* wild-type strain, however, yielded the mutant strain *impl1*, which has a phenotype we designate pseudo-*plus*: Its flagellar agglutinins and actin-filled fertilization tubules are of the *plus* type, but the *plus* mating-structures lack fringe and, therefore, when *impl1 mt⁻* gametes are mixed with *minus* gametes, all mating responses occur except cell fusion (32). By creating PEG-induced diploids carrying *impl1* and its wild-type allele, and analyzing the products of subsequent crosses, it was possible to demonstrate that *impl1* is recessive to its wild-type allele and is tightly linked to the *mt⁻* locus (25). These results indicate that the *impl1⁺* gene is responsible for the *mt⁻* dominance of mating functions, leading us to propose that the gene be designated *mid* (for minus dominance). The *impl1* mutation would then have two consequences: It would prevent *mt⁻*-limited genes from being expressed, and it would allow *mt⁺*-limited genes to be expressed. The only exception would be genes unique to the *mt⁺* locus because these would not be present in the *impl1 mt⁻* genome.

One such gene would be *sfu,* which, as noted earlier, is localized within the c region of the *mt⁺* R domain and specifies the formation of *plus*-specific fringe. We transformed the *impl1 mt⁻* mutant with the 4.7-kb genomic fragment carrying the *sfu* gene, a construct we henceforth designate as *sfu-T*. The transformants mated as normal *plus* gametes, fusing inefficiently but nonetheless effectively, and specifically, with *minus* gametes to form diploid zygotes (P Ferris, unpublished). Thus, the *sfu* gene is both necessary and sufficient to rescue the pseudo-*plus* phenotype.

A pseudo-*plus* phenotype is also generated by the *iso1* mutation, which was created by insertion of the transformation vector pARG7.8 into the nuclear genome of an *arg7 mt⁻* recipient (AM Campbell, HJ Rayala & UW Goodenough, manuscript submitted). The *iso1* strain has the intriguing phenotype that it can switch mating type. Thus, most of the cells in an *iso1* clone differentiate into *minus* gametes when starved for nitrogen, but a minority differentiate as pseudo-*plus*. The result is that the *iso1* gametes isoagglutinate but fail to consummate cell fusion. The *iso1* mutation is not linked to *mt⁻*, ruling out allelism to *mid*, but is sex limited in expression to *minus* gametes, with *iso1 mt⁺* meiotic products having a stable *plus* gametic phenotype. The simplest model proposes that *iso1* is required for the function of *mid* such that, for example, the *iso1* and *mid* products act as a dimer to switch on *mt⁻*-limited genes. The insertion of the transformation vector into the *iso1* gene would then result either in its low-level expression or in a product with compromised ability to interact with *mid*, generating a minority of cells that fail to express the *minus* program and instead express all resident *plus* genes. Insertional mutagenesis has the advantage that the mutant gene can be cloned and analyzed (84), a project that is now underway for *iso1*. We are also asking whether the pseudo-*plus* phenotype of the *iso1* mutant can be rescued by transformation with the *sfu-T* construct.

Matings Between mt⁻ Gametes and imp11 sfu-T mt⁻ Transformants

To our surprise, zygotes formed in matings between *mt⁻* gametes and *imp11 sfu-T mt⁻* transformants go on to sporulate and undergo meiosis even though they possess no *mt⁺* copy of linkage group VI (P Ferris, unpublished). We proposed earlier that the zygote program is switched on by an interaction between factors called P and M. The source of M in these matings is, of course, the *mt⁻* gamete, but the source of P is unclear. One possibility is that P and the *sfu* gene product are the same protein, a protein that first switches on the *plus* fringe program in gametes and then interacts with M in the zygote to switch on zygote-specific genes. The alternate possibility is that P is not encoded in the *mt⁺* locus but rather is a *plus*-limited gene whose expression is permitted in the *imp11* background.

When the above cross is performed with strains carrying chloroplast genetic markers, their transmission is biparental (P Ferris, unpublished). Thus the *mt⁺* version of linkage group VI is indeed necessary for UP⁺ inheritance, and the 4.7-kb *sfu-T* fragment does not encode this function. A search for the *mt⁺* gene(s) that specify the UP⁺ trait is currently under way in our laboratory: The strategy is that *minus* cells transformed with the appropriate constructs should display biparental inheritance when crossed with *plus* cells. As noted earlier, the 16-kb repeat in the *mt⁺* R domain is an attractive candidate for this function.

Our present understanding of mating-type determination in *C. reinhardtii* is summarized in Figure 3. We postulate that the *mid* gene product provides transcriptional activation, perhaps in conjunction with the *iso1* product, for the *minus*-specific genes. One of these genes is postulated to encode a repressor of *plus*-specific genes. Therefore, in the *imp11* mutant, *minus*-specific genes are not activated and *plus*-specific genes are not repressed. The *mt⁺* locus carries the *sfu* gene, which may or may not be the same gene that encodes P, and it also carries at least one gene involved in protecting *plus* chloroplast DNA from destruction in the zygote. If the *mid* gene product is the same as M, then the *mt* regulatory circuit will be elegantly simple in design. It will then become all the more compelling to learn how and why a small number of regulatory genes have become embedded in such a complex chromosomal locus.

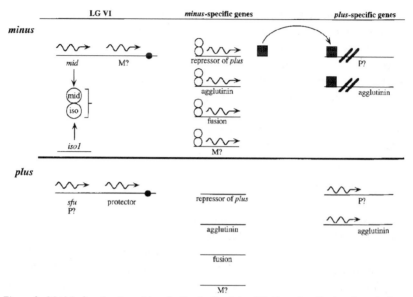

Figure 3 Model of mating-type determination in *C. reinhardtii*. Genes localized to the *mt* loci are shown at left; *minus*-specific genes are central; *plus*-specific genes are at right. The upper panel depicts *mt⁻*, the lower panel *mt⁺*. Transcription is indicated by the wavy arrows; proteins are circles (activators) or squares (repressors).

Recently, the opportunity to identify the *mid* gene has presented itself in a most fortuitous fashion (P Ferris, unpublished data). When the strain CC-421 (*nic7 ac29a spr-u-1-27-3 mt⁻*) is mated with a *plus* strain, the expected ratio of sexual phenotype (2 *plus*:2 *minus*) is often not observed among the meiotic progeny. Instead, many of the zygotes yield 1 *plus*:3 *minus* or 0 *plus*:4 *minus* progeny. When the extra *minus* products are examined more closely, they prove to be unstable. During mitotic growth, they generate stable *plus* clones as well as unstable *minus* clones that go on to throw off more stable *plus* clones. A culture of unstable *minus* gametes will isoagglutinate and isofuse to form zygotes that undergo meiosis to again yield unstable *minus* progeny. When the DNA from unstable *minus* strains is hybridized with *mt*-specific probes, the strains prove to be chromosomally *mt⁺*, meaning that they have picked up additional DNA that confers them, unstably, with a *minus* phenotype. The most obvious possibility is that, for unknown reasons, the CC-421 strain contains an extra copy of the *mid* gene that can be inherited, and subsequently lost, by *mt⁺* progeny. An intensive study of the sex-switching phenomenon in the unstable *minus* strains is currently underway in this laboratory.

HETEROTHALLISM VS HOMOTHALLISM

The evolutionary scenario presented at the beginning of this review assumes, as do most models of the origin of sex, that some benefit was accrued by outcrossing and that the first sexual species were heterothallic. A subsequent adoption of homothallism does not violate such reasoning because homothallism does not preclude outcrossing. It simply provides a useful option should an interclonal mate not materialize, and it would be particularly adaptive once the sporulation program had been co-opted by sexuality and required a PM kind of interaction.

The establishment of hermaphroditism, or monoecy, is readily conceptualized in higher plants by assuming that different kinds of germinative tissues elicit the expression of male or female germ-line options (96). For unicellular haploids, homothallism is a more challenging proposition. The best understood homothallic systems are those of *Saccharomyces cerevisiae* and *Schizosaccharomyces pombe*, in which an elaborate enzymology is responsible for recruiting copies of mating-type cassettes from nonexpressed to expressed positions in the genome (9, 24, 53, 85). A similar mechanism may operate in a homothallic basidiomycete (49). Of major interest is whether the homothallic species in the volvocine lineage (82) switch mating type in the fashion of the fungi or via some very different mechanism.

Until recently, the only tractable organism for the study of Chlorophyte homothallism has been *Chlamydomonas monoica*, a species in the same lineage as the heterothallic *Chlamydomonas eugametos*, a lineage that diverged from the *C. reinhardtii/Volvox* spp. line some 300 mY ago (50, 80). *C. monoica* and *C. eugametos* are similar in their cell structure and physiology (36). In contrast to the *C. reinhardtii* group, moreover, they fail to shed their walls during mating, fuse via apparently identical anterior papillae, and swim about as papilla-tethered pairs for 24 hr after mating rather than achieving immediate cytoplasmic confluence (89, 90). The *C. eugametos* gametes recognize one another via hydroxyproline-rich flagellar agglutinins, which are distinctly different in structure from those of their *C. reinhardtii* counterparts (31). A sexual version of *Chlamydomonas* spp. clearly evolved prior to the *eugametos-reinhardtii* divergence, and the *eugametos-monoica* divergence perhaps occurred quite recently. Because the *C. reinhardtii* lineage includes numerous homothallic haploids among the colonial Volvocales (82), *C. reinhardtii* is likely to carry "homothallic potential."

For *C. monoica,* clonal cells agglutinate and fuse as either *plus* or *minus* gametes. It is also possible to force a *C. monoica* clone to outcross by having it carry a recessive lethal mutation in a zygote-maturation (*zym*) gene (91a), underscoring the importance of outcrossing for the execution of mutant sporulation programs. *C. monoica* displays uniparental inheritance of chloroplast DNA: When a drug-resistant strain is outcrossed to a drug-sensitive strain, roughly half the zygotes yield four drug-resistant products and half yield four drug-sensitive products, the pattern expected if half the cells adopt the *plus* configuration and confer their chloroplast DNA with protected status (91).

A *C. monoica* mutation, *mtl1*, knocks out the ability of the strain to protect its *plus* chloroplast DNA such that *mtl1/mtl1* homozygotes, and half of *mtl1/mtl+* heterozygotes, die during zygote maturation (93). This lethality is suppressed by recessive mutations at the *sup1* locus, which prevent all chloroplast DNA degradation and, hence, generate biparental inheritance (92). The *sup1* gene, apparently expressed from both parental genomes, has been interpreted to encode a chloroplast-localized nuclease (92). If *sup1* is the *C. monoica* equivalent of *ezy1* (4), then the gene has become multi-copy in *C. reinhardtii*.

Available data indicate that a heterothallic ancestor of *C. eugametos*, transmitting its chloroplast DNA via the *plus* strain, acquired the ability to perform all such mating functions intraclonally and, hence, gave rise to *C. monoica*. A molecular analysis of the proposed conversion to homothallism has thus far been hampered by the fact that neither *C. eugametos* nor *C. monoica* has yielded to nuclear transformation.

ACKNOWLEDGMENTS

This review was written while UW Goodenough was a visiting scientist in the Dipartimento di Biologia Evolutiva at the University of Siena. Warm thanks are extended to Professor Romano Dallai and his colleagues for their hospitality. Research from this laboratory has been supported by grants from the NIH, NSF, and USDA.

Literature Cited

1. Adair WS. 1985. Characterization of the *Chlamydomonas* sexual agglutinins. *J. Cell Sci.* 2(Suppl.):233–60
2. Allen JF. 1993. Control of gene expression by redox potential and the requirement for chloroplast and mitochondrial genomes. *J. Theor. Biol.* 165:609–31
3. Anderson OR. 1988. *Comparative Protozoology. Ecology, Physiology, Life History*. Berlin: Springer Verlag. 482 pp.
4. Armbrust EV, Ferris PJ, Goodenough UW.

1993. A mating-type linked gene cluster expressed in Chlamydomonas zygotes participates in the uniparental inheritance of the chloroplast genome. *Cell* 74:801–11
5. Atherton-Fessler S, Hannig G, Piwnica-Worms H. 1993. Reversible tyrosine phosphorylation and cell cycle control. *Semin. Cell Biol.* 4:433–42
6. Bell G. 1993. The sexual nature of the eukaryotic genome. *J. Hered.* 84:351–59
7. Bierzychudek P. 1987. Patterns in plant

parthenogenesis. In *The Evolution of Sex and its Consequences,* ed. SC Stearn, pp. 197–217. Basel: Birkhauser Verlag

7a. Boynton JE, Harris EH, Burkhart BD, Lamerson PM, Gillham NW. 1987. Transmission of mitochondrial and chloroplast genomes in crosses of *Chlamydomonas. Proc. Natl. Acad. Sci. USA* 84:2391–95

8. Braten T. l973. Autoradiographic evidence for the rapid disintegration of one chloroplast in the zygote of the green alga *Ulva mutabilis. J. Cell Sci.* 12:385–89

9. Braunstein M, Rose AB, Holmes SG, Allis CD, Broach JR. 1993. Transcriptional silencing in yeast is associated with reduced nucleosome acetylation. *Genes Dev.* 7: 592–604

9a. Buerkle S, Gloeckner G, Beck CF. 1993. *Chlamydomonas* mutants affected in the light-dependent step of sexual differentiation. *Proc. Natl. Acad. Sci. USA* 90:6981–85

10. Burton WG, Graboway CT, Sager R. 1979. Role of methylation in the modification and restriction of chloroplast DNA in *Chlamydomonas. Proc. Natl. Acad. Sci. USA* 76: 1390–94

11. Cavalier-Smith T. 1976. Electron microscopy of zygospore formation in *Chlamydomonas reinhardtii. Protoplasma* 87:297–315

12. Chiang K-S, Sueoka N. 1967. Replication of chloroplast DNA in *Chlamydomonas reinhardtii* during vegetative cell cycle: its mode and regulation. *Proc. Natl. Acad. Sci. USA* 57:1506–13

13. Chiu W-L, Sears BB. 1993. Plastome-genome interactions affect plastid transmission in *Oenothera. Genetics* 133:989–97

14. Coleman AW. 1983. The roles of resting spores and akinetes in Chlorophyte survival. In *Survival Strategies of the Algae,* ed. GA Fryxell, pp. 1–21. Cambridge: Cambridge Univ. Press

15. Coleman AW. 1984. The fate of chloroplast DNA during cell fusion, zygote maturation and zygote germination in *Chlamydomonas reinhardtii* as revealed by DAPI staining. *Exp. Cell Res.* 152:528–40

16. Cosmides L, Tooby J. 1981. Cytoplasmic inheritance and intragenomic conflict. *J. Theor. Biol.* 89:83–129

17. Dallai R, ed. 1992. *Sex Origin and Evolution.* Modena, Italy: Mucchi Editore. 389 pp.

18. Dawes IW. 1981. Sporulation in evolution. In *Molecular and Cellular Aspects of Microbial Evolution,* ed. MJ Carlile, JF Collins, BEB Moseby, pp. 85–130. New York: Cambridge Univ. Press

19. Ebersold WT. 1967. *Chlamydomonas reinhardtii:* heterozygous diploid strains. *Science* 157:447–49

19a. Erdos GW, Raper KB, Vogen LK. 1973. Mating types and macrocyst formation in *Dictyostelium discoideum. Proc. Natl. Acad. Sci. USA* 70:1828–30

20. Ferris PJ, Goodenough UW. 1987. Transcription of novel genes, including a gene linked to the mating-type locus, induced by *Chlamydomonas* fertilization. *Mol. Cell. Biol.* 7:2360–66

21. Ferris PJ, Goodenough UW. 1994. The mating-type locus of Chlamydomonas reinhardtii contains highly rearranged DNA sequences. *Cell* 76:1135–45

22. Forest CL. 1987. Genetic control of plasma membrane adhesion and fusion in *Chlamydomonas* gametes. *J. Cell Sci.* 88:613–21

23. Forest CL, Togasaki RK. 1975. Selection for conditional gametogenesis in *Chlamydomonas reinhardtii. Proc. Natl. Acad. Sci. USA* 72:3652–55

24. Foss M, McNally F, Laurenson P, Rine J. 1993. Origin replication complex (ORC) in transcriptional silencing and DNA replication in *S. cerevisiae. Science* 262:1838–42

25. Galloway RE, Goodenough UW. 1985. Genetic analysis of mating locus linked mutations in *Chlamydomonas reinhardtii. Genetics* 111:447–61

26. Galloway RE, Holden LR. 1985. Transmission and recombination of chloroplast genes in asexual crosses of *Chlamydomonas reinhardtii.* II. Comparisons with observations in sexual diploids. *Curr. Genet.* 10: 221–28

27. Gillham NW. 1969. Uniparental inheritance in *Chlamydomonas reinhardtii. Am. Nat.* 103:355–88

28. Gillham NW, Boynton JE, Harris EH. 1991. Transmission of plastid genes. In *Cell Culture and Somatic Cell Genetics of Plants,* ed. L Bogorad, IK Vasil, 7A:55–92. San Diego: Academic

29. Gillham NW, Boynton JE, Johnson AM, Burkhart BD. 1987. Mating type linked mutations which disrupt the uniparental transmission of chloroplast genes in *Chlamydomonas. Genetics* 115:677–84

30. Goodenough UW. 1985. An essay on the origins and evolution of eukaryotic sex. In *The Origin and Evolution of Sex,* ed. HO Halvorson, A Monroy, pp. 123–40. New York: Liss

31. Goodenough UW. 1991. *Chlamydomonas* mating interactions. In *Microbial Cell-Cell Interactions,* ed. M Dworkin, pp. 71–112. Washington, DC: Am. Soc. Microbiol.

32. Goodenough UW, Detmers PA, Hwang C. 1982. Activation for cell fusion in *Chlamydomonas:* analysis of wild-type gametes and nonfusing mutants. *J. Cell Biol.* 92: 378–86

33. Goodenough UW, Ferris PJ. 1987. Genetic regulation of development in *Chlamydo-*

monas. In *Genetic Regulation of Development,* ed. W Loomis, pp. 171–89. New York: Liss

34. Goodenough UW, Hwang C, Martin H. 1976. Isolation and genetic analysis of mutant strains of *Chlamydomonas reinhardtii* defective in gametic differentiation. *Genetics* 82:169–86

35. Gray MW. 1991. Origin and evolution of plastid genomes and genes. In *Cell Culture and Somatic Cell Genetics of Plants,* ed. L Bogorad, IK Vasil, 7A:303–30. San Diego: Academic

36. Harris EH. 1989. *The* Chlamydomonas *Sourcebook.* San Diego: Academic

37. Harrison FW, Corliss JO, eds. 1991. *Microscopic Anatomy of Invertebrates.* Vol. 1: *Protozoa.* New York: Wiley-Liss. 493 pp.

38. Hastings IM. 1992. Population genetic aspects of deleterious cytoplasmic genomes and their effect on the evolution of sexual reproduction. *Genet. Res.* 59:215–25

39. Herskowitz I. 1988. Life cycle of the budding yeast *Saccharomyces cerevisiae. Microbiol. Rev.* 52:536–53

40. Hickey DA, Rose MR. 1988. The role of gene transfer in the evolution of sex. In *The Evolution of Sex: An Examination of Current Ideas,* ed. RF Michod, BR Levin, pp. 161–75. Sunderland, MA: Sinauer

41. Hurst LD. 1992. Intragenomic conflict as an evolutionary force. *Proc. R. Soc. London Ser. B* 248:135–40

42. Hurst LD, Hamilton WD. 1992. Cytoplasmic fusion and the nature of sexes. *Proc. R. Soc. London Ser. B* 247:189–94

43. Ippen-Ihler K, Skurray R. 1993. Genetic organization and transfer-related determinants on the sex factor F and related plasmids. In *Bacterial Conjugation,* ed. D Clewell, pp. 23–52. New York: Plenum

44. Kawano S, Takano H, Imai J, Mori K, Kuroiwa T. 1993. A genetic system controlling mitochondrial fusion in the slime mould, *Physarum polycephalum. Genetics* 133:213–24

45. Kindle KL.1990. High frequency nuclear transformation of *Chlamydomonas reinhardtii. Proc. Natl. Acad. Sci. USA* 87: 1228–32

46. Kondrashov AS, Crow JF. 1991. Haploidy or diploidy: Which is better? *Nature* 351: 314–15

47. Kuroiwa T. 1991. The replication, differentiation, and inheritance of plastids with emphasis on the concept of organelle nuclei. *Int. Rev. Cytol.* 128:1–62

48. Kuroiwa T, Kawano S, Nishibayashi S. 1982. Epifluorescent microscopic evidence for maternal inheritance of chloroplast DNA. *Nature* 298:481–83

49. Labarere J, Noel T. 1992. Mating type-switching in the tetrapolar basidiomycete *Agrocybe aegerita. Genetics* 131:307–19

50. Larson A, Kirk MM, Kirk DL. 1992. Molecular phylogeny of the volvocine flagellates. *Mol. Biol. Evol.* 9:85–105

51. Law R, Hutson V. 1992. Intracellular symbionts and the evolution of uniparental cytoplasmic inheritance. *Proc. R. Soc. London Ser. B* 248:69–77

52. Longo FJ. 1987. *Fertilization.* London: Chapman & Hall

53. Lorentz A, Heim L, Schmidt H. 1992. The switching gene *swi6* affects recombination and gene expression in the mating-type region of *Schizosaccharomyces pombe. Mol. Gen. Genet.* 233:436–42

54. Losick R, Stragier P. 1992. Crisscross regulation of cell-type-specific gene expression during development in *B. subtilis. Nature* 355:601–4

55. Manhart JR, Palmer JD. 1990. The gain of two chloroplast tRNA introns marks the green algal ancestors of land plants. *Nature* 345:268–70

56. Margulis L. 1981. *Symbiosis in Cell Evolution.* San Francisco: Freeman

57. Matagne RF. 1987. Chloroplast gene transmission in *Chlamydomonas reinhardtii.* A model for its control by the mating-type locus. *Curr. Genet.* 12:251–56

58. Matagne RF, Beckers M-C. 1983. Perturbation of chloroplast gene transmission in diploid and triploid zygotes of *Chlamydomonas reinhardtii* by 5-fluorodeoxyuridine. *Curr. Genet.* 7:335–38

59. Matagne RF, Mathieu D. 1983. Transmission of chloroplast genes in triploid and tetraploid zygospores of *Chlamydomonas reinhardtii:* roles of mating-type gene dosage and gametic chloroplast DNA content. *Proc. Natl. Acad. Sci. USA* 80:4780–83

60. Matsuda Y, Shimada T, Sakamoto Y. 1992. Ammonium ions control gametic differentiation and dedifferentiation in *Chlamydomonas reinhardtii. Plant Cell Physiol.* 33: 909–14

61. Matters GL, Goodenough UW. 1992. A gene/pseudogene tandem duplication encodes a cysteine-rich protein expressed during zygote development in *Chlamydomonas reinhardtii. Mol. Gen. Genet.* 232: 81–88

62. Meland S, Johansen S, Johansen T, Haugli K, Haugli F. 1991. Rapid disappearance of one parental mitochondrial genotype after isogamous mating in the myxomycete *Physarum polycephalum. Curr. Genet.* 19: 55–60

63. Metzenberg RL, Glass NL. 1990. Mating type and mating strategies in *Neurospora. BioEssays* 12:53–59

64. Michod RE. 1993. Genetic error, sex, and diploidy. *J. Hered.* 84:360–71

65. Michod RE, Levin BR. 1988. *The Evolution of Sex: An Examination of Current Ideas.* Sunderland, MA: Sinauer. 342 pp.

66. Munaut C, Dombrowicz D, Matagne RF. 1990. Detection of chloroplast DNA by using fluorescent monoclonal anti-bromodeoxyuridine antibody and analysis of its fate during zygote formation in *Chlamydomonas reinhardtii*. *Curr. Genet.* 18:259–63

67. Perrot V, Richerd S, Valero M. 1991. Transition from haploidy to diploidy. *Nature* 351:315–17

68. Quarmby LM. 1994. Signal transduction in the sexual life of *Chlamydomonas*. *Plant Mol. Biol.* In press

69. Reboud X, Zeyl C. 1994. Organelle inheritance in plants. *Heredity* 72:132–40

70. Remacle C, Bovie C, Michel-Wolwertz M-R, Loppes R, Matagne RF. 1990. Mitochondrial genome transmission in *Chlamydomonas* diploids obtained by sexual crosses and artificial fusions: role of the mating type and of a 1 kb intron. *Mol. Gen. Genet.* 223:180–84

71. Remacle C, Matagne RF. 1993. Transmission, recombination and conversion of mitochondrial markers in relation to the mobility of a group I intron in *Chlamydomonas*. *Curr. Genet.* 23:518–25

72. Rose MD. 1991. Nuclear fusion in yeast. *Annu. Rev. Microbiol.* 45:539–67

73. Russell SD, Dumas C, eds. 1992. Sexual reproduction in flowering plants. *Int. Rev. Cytol.*, Vol. 140. San Diego: Academic. 615 pp.

74. Sager R. 1954. Mendelian and non-Mendelian inheritance of streptomycin resistance in *Chlamydomonas reinhardtii*. *Proc. Natl. Acad. Sci. USA* 40:356–63

75. Sager R, Granick S. 1954. Nutritional control of sexuality in *Chlamydomonas reinhardtii*. *J. Gen. Physiol.* 37:729–42

76. Sager R, Ramanis Z. 1967. Biparental inheritance of nonchromosomal genes induced by ultraviolet irradiation. *Proc. Natl. Acad. Sci. USA* 58:931–35

77. Sager R, Ramanis Z. 1973. The mechanism of maternal inheritance in *Chlamydomonas*: biochemical and genetic studies. *Theor. Appl. Genet.* 43:101–8

78. Saito T, Small L, Goodenough UW. 1993. Activation of adenylyl cyclase in *Chlamydomonas reinhardtii* by adhesion and by heat. *J. Cell Biol.* 122:137–47

79. Scherer S, Lechner S, Boger P. 1993. psbD sequences of *Bumilleriopsis filiformis* (Heterokontophyta, Xanthophyceae) and *Porphyridium purpureum* (Rhodophyta, Bangiophycidae): evidence for polyphyletic origins of plastids. *Curr. Genet.* 24:437–42

80. Schmitt R, Fabry S, Kirk DL. 1992. In search of the molecular origins of cellular differentiation in *Volvox* and its relatives. *Int. Rev. Cytol.* 139:189–265

81. Sogin ML. 1991. Early evolution and the origin of eukaryotes. *Curr. Opin. Genet. Dev.* 1:457–63

82. Starr RC. 1968. Cellular differentiation in *Volvox. Proc. Natl. Acad. Sci. USA* 59:1082–88

83. Stearns SC. 1987. *The Evolution of Sex and its Consequences.* Basel: Birkhauser Verlag. 403 pp.

84. Tam L-W, Lefebvre PA. 1993. The use of DNA insertional mutagenesis to clone genes in *Chlamydomonas. Genetics* 135:375–84

85. Thon G, Klar AJS. 1993. Directionality of fission yeast mating-type interconversion is controlled by the location of the donor loci. *Genetics* 134:1045–54

86. Turmel M, Lemieux C, Lee RW. 1980. Net synthesis of chloroplast DNA throughout the synchronized vegetative cell-cycle of *Chlamydomonas. Curr. Genet.* 2:229–32

87. Uchida H, Kawano S, Sato N, Kuroiwa T. 1993. Isolation and characterization of novel genes which are expressed during the very early stage of zygote formation in *Chlamydomonas reinhardtii. Curr. Genet.* 24:296–300

88. Uyenoyama MK, Holsinger KE, Waller DM. 1993. Ecological and genetic factors directing the evolution of self-fertilization. *Oxford Surv. Evol. Biol.* 9:327–81

89. van den Ende H. 1992. Sexual signalling in *Chlamydomonas.* In *Perspectives in Plant Cell Recognition,* ed. JA Callow, JR Green, pp. 1–17. Cambridge: Cambridge Univ. Press

90. vanWinkle-Swift KP, Aliaga GR, Pommerville JC. 1987. Haploid spore formation following arrested cell fusion in *Chlamydomonas* (Chlorophyta). *J. Phycol.* 23:414–27

91. vanWinkle-Swift KP, Aubert B. 1983. Uniparental inheritance in a homothallic alga. *Nature* 303:167–69

91a. vanWinkle-Swift KP, Burrascano CG. 1983. Complementation and preliminary linkage analysis of zygote maturation mutants of the homothallic alga, *Chlamydomonas monoica. Genetics* 103:429–43

92. vanWinkle-Swift KP, Hoffman R, Shi L, Parker S. 1994. A suppressor of a mating-type limited zygotic lethal allele also suppresses uniparental chloroplast gene transmission in *Chlamydomonas monoica. Genetics* 136:867–77

93. vanWinkle-Swift KP, Salinger AP. 1988. Loss of mt$^+$-derived zygotic chloroplast DNA is associated with a lethal allele in *Chlamydomonas monoica. Curr. Genet.* 13:331–37

94. von Gromoff ED, Beck CF. 1993. Genes expressed during sexual differentiation of *Chlamydomonas reinhardtii. Mol. Gen. Genet.* 241:415–21

95. Wegener D, Beck CF. 1991. Identification

of novel genes specifically expressed in *Chlamydomonas reinhardtii* zygotes. *Plant Mol. Biol.* 16:937–46

96. Westergaard M. 1958. The mechanism of sex determination in dioecious flowering plants. *Adv. Genet.* 9:217–81

97. Williams GC. 1975. *Sex and Evolution.* Princeton, NJ: Princeton Univ. Press. 201 pp.

98. Woessner JP, Goodenough UW. 1989. Molecular characterization of a zygote wall protein: an extensin-like molecule in *Chla-mydomonas reinhardtii. Plant Cell* 1:901–11

99. Woessner JP, Goodenough UW. 1994. Volvocine cell walls and their constituent glycoproteins: an evolutionary perspective. *Protoplasma.* In press

100. Wurtz EA, Boynton JE, Gillham NW. 1977. Perturbation of chloroplast DNA amounts and chloroplast gene transmission in *Chlamydomonas reinhardtii* by 5-fluorodeoxyuridine. *Proc. Natl. Acad. Sci. USA* 74:4552–56

Annu. Rev. Plant Physiol. Plant Mol. Biol. 1995. 46:45–70

RESPIRATION DURING PHOTOSYNTHESIS

Silke Krömer

Lehrstuhl für Pflanzenphysiologie, University of Osnabrück, 49069 Osnabrück, Germany

KEY WORDS: mitochondrial substrate oxidation, nitrogen metabolism, oxidative phosphorylation, photorespiration, interorganellar redox transfer

CONTENTS

ABSTRACT

The respiratory activity of plants in the light, measured as CO_2 release from the TCA cycle or O_2 consumption by the respiratory chain, varies between 25 and 100% of the dark respiratory activity. This has been interpreted as evidence for an inhibition of respiration during photosynthesis. However, studies with specific respiratory inhibitiors have shown that oxidative phosphorylation occurs in the light and provides the cytosol with ATP, which is required for sucrose synthesis. Respiratory activity in the light might also be required to sustain a high photosynthetic capacity and might even prevent photoinhibition. Sources of redox equivalents for oxidative phosphorylation can be photosynthetically generated, externally oxidized NADPH; photorespiratory oxidation of glycine to serine; or a partial activity of the TCA cycle. Fifty to seventy-five percent of the redox equivalents produced in the mitochondria remain in the matrix and can function in ATP syntheses. The other 25–50% are exported via the mitochondrial malate-oxaloacetate shuttle and can function in nitrate reduction in the cytosol or hydroxypyruvate reduction in the peroxisomes.

INTRODUCTION

Mitochondrial respiratory activity in the dark includes the oxidation of carbon compounds in the TCA cycle and of NAD(P)H in the respiratory electron transport chain. The TCA cycle releases CO_2 and provides redox equivalents. During oxidation of these redox equivalents in the respiratory chain, O_2 is consumed and a proton gradient across the inner mitochondrial membrane is formed, which provides the energy for ATP synthesis in the mitochondria, a process termed oxidative phosphorylation. In the light, the oxidative decarboxylation of glycine to serine occurs in the mitochondrial matrix of photosynthetically active cells as a part of the photorespiratory cycle. Plant mitochondria have a malic enzyme that enables the operation of the TCA cycle without concomitant glycolysis. The respiratory chain of plant mitochondria has some unique features (17, 71): a cyanide-resistant nonphosphorylating electron transport pathway, a rotenone-insensitive oxidation site, and the ability to oxidize external NAD(P)H. The physiological importance of these mitochondrial features is not yet clear.

Chloroplastic photosynthetic electron transport uses light energy for the production of NADPH and ATP, which are required for the fixation of CO_2 in the Calvin cycle in the chloroplast stroma. In the light, dihydroxyacetone phosphate (DHAP), an intermediate of the Calvin cycle, is exported from the chloroplast in exchange with inorganic phosphate (P_i) or 3-phosphoglycerate (PGA) (25, 26) and functions in sucrose synthesis in the cytosol. In photosynthetic tissue the chloroplast is the site of most biosynthetic processes (e.g. CO_2 fixation, starch synthesis, nitrogen asssimilation, fatty acid synthesis, and the synthesis of many amino acids). These processes all require energy in form of ATP and NADPH.

Photosynthesis and dark respiration are metabolic pathways that produce redox equivalents and ATP to meet the cell's energy demands for growth and maintenance. The interactions between these pathways have been debated for years. Since the last reviews (36, 37, 97), our understanding of the interaction between respiration and photosynthesis has changed considerably. This review discusses the involvement of mitochondrial activities in photosynthetic metabolism, with emphasis on the function of mitochondrial ATP production, oxidation of redox equivalents, and supply of carbon skeletons. Dark respiration or light-enhanced dark respiration (for review see 77) are discussed only in relation to respiration in the light. This review concentrates on studies with C_3 plants (see 43 for mitochondrial functions in C_4 plants).

RESPIRATION DURING STEADY-STATE PHOTOSYNTHESIS

Respiratory Activity in the Light

The direct determination of respiratory O_2 uptake or CO_2 release in the light is a complicated task. Major components of net O_2 evolution are photosynthetic O_2 release, O_2 uptake by the Mehler reaction, respiratory O_2 uptake under nonphotorespiratory conditions, and additional O_2 consumption by the oxygenase activity of ribulose-1,5-bisphosphate carboxylase/oxygenase (Rubisco) and glycolate oxidase under photorespiratory conditions. Net CO_2 evolution involves CO_2 release from the TCA cycle and glycine decarboxylation as well as photosynthetic CO_2 fixation and phosphoenolpyruvate carboxylase (PEPC) activity. The use of high CO_2 concentrations to create nonphotorespiratory conditions impedes the detection of the relatively low respiratory CO_2 release. The use of low O_2 concentrations to prevent the occurrence of photorespiration can be problematic, because at O_2 concentrations required to suppress photorespiration, dark respiration is inhibited (96) and because photosynthetic O_2 evolution is considerably higher than respiratory O_2 uptake. Gas exchange measurements and the use of radiolabeled O_2 or CO_2 in combination with mass-spectrometric analyses have been used to differentiate O_2 uptake from O_2 evolution and CO_2 release from CO_2 fixation. Possible recycling of CO_2 and O_2 in illuminated cells cannot be determined with this technique; however, based on mathematical modeling, it has been suggested that recycling does not significantly contribute to gas exchange measurement (3).

The photosynthetic response to light often shows a break in the inital linearity that occurs at light intensities near the compensation point, a phenomenon termed the Kok effect (57). The O_2 uptake rate in *Anacystis nidulans* decreased progressively with increasing light intensities to about 50% of the dark respiratory rate, indicating a light-induced inhibition of dark respiration. A further increase in light intensity led to an increase in the O_2 uptake rate that exceeded the dark respiratory rate and indicated light-induced photorespiration (50). The contribution of either dark respiration or photorespiration to the Kok effect has been examined (52, 96), but controversial results and limited knowledge of the biochemistry behind the effect leave the cause of the phenomenon unclear.

The degree of dark respiration in the light has been estimated under saturating light intensities in intact plants. CO_2 release from the TCA cycle in the light, as determined by gas exchange measurements, varied between 25 and 100% of the respiratory rate in darkness (3, 13, 56, 79, 96). Some results from O_2 exchange measurements indicated that O_2 consumption is decreased in the light (6, 12), while other results implied that the rates of O_2 consumption in the light are comparable to those in the dark (33, 76, 106). At a low CO_2 concentration, a threefold increase in the O_2 uptake rate was found and shown to be largely of photorespiratory origin (81). It also has been suggested that the rate of light-enhanced dark respiratory O_2 uptake might represent the rate of respiratory electron transport of the preceding light period (4).

Inhibitors specifically blocking a reaction in a respiratory pathway can be used to determine the existence of respiratory activity during photosynthesis. Oligomycin inhibits F_0F_1-type ATP synthases found in the inner mitochondrial membrane, the thylakoid membrane, and the chloroplastic inner envelope membrane (63, 69); however, the sensitivity of these ATP synthases to oligomycin differs dramatically. Concentrations required to gain the same degree of inhibition are at least 400-fold lower in mitochondria than in chloroplasts (63). Oligomycin concentrations that inhibit oxidative phosphorylation in isolated mitochondria by about 80% (63) should, if respiratory electron transport occurs in the light, lead to an increase in net O_2 evolution when added to an intact photosynthesizing cell. Incubation of barley leaf protoplasts with oligomycin in the light, however, showed a decrease in the net O_2 evolution, which vanished when protoplasts were disrupted in a way that left chloroplasts and mitochondria intact (63). Thus, the decrease in photosynthetic activity is caused indirectly by the inhibition of mitochondrial ATP synthesis. From these results, three conclusions can be drawn: (*a*) oxidative phosphorylation occurs in the light, (*b*) photosynthetic activity depends on the activity of the mitochondrial electron transport chain, and (*c*) activity of the respiratory electron transport chain in the light is unlikely to result entirely from the alternative oxidase. Because the alternative pathway activity is linked at most to one phosphorylation site, it should only be affected slightly by oligomycin and thus could not explain the observed effect. Inhibitior studies do not permit quantification of the rate of respiratory O_2 uptake in the light or comparison to the rate of dark respiratory O_2 uptake. Therefore, it cannot be excluded that the rate in the light could be lower than in darkness.

Feedback Inhibition of Photosynthesis on Respiration

It has been suggested that light exerts an inhibitory effect on the respiratory electron transport chain via the cytosolic adenylate pool. The export of photosynthetically produced ATP would cause an increase in the cytosolic phosphorylation potential $[ATP(ADP \times P_i)^{-1}]$, leading to an inhibition of oxidative phosphorylation in the light that is controlled by the external ATP/ADP ratio (45). In fact, the cytosolic ATP/ADP ratio is lower in the light during steady-state photosynthesis at saturating CO_2 concentrations than in darkness (29, 39, 100); in contrast, it is higher in the light at CO_2 concentrations that limit photosynthesis (29). This increase, however, is unlikely to cause an inhibition of oxidative phosphorylation, because extramitochondrial ATP/ADP ratios higher than 20 are required to reduce respiratory O_2 uptake (20). Because of the large amounts of P_i in the vacuole [$\geq$ 50 mM (9)], it has been difficult to determine the cytosolic P_i concentration. In the dark, extravacuolar P_i concentrations range between 15 and 35 mM in pea, barley, wheat, and spinach in one report (101) and between 6 and 9 mM in spinach in another report (9). Chloroplastic P_i levels vary from 9 to 12 mM in leaves of beet root and spinach (87). Based on subcellular metabolite analysis with different species, it has been suggested that in the light the cytoplasmic level of P_i decreases by up to 10 mM under high CO_2 and increases by about 5 mM under low CO_2 in comparison to levels in the dark (34, 101). However, P_i-feeding experiments with spinach leaf pieces indicated that the vacuolar P_i pool might serve as a buffer to keep the cytoplasmic pool constant (9). Because of these variations in the results, it is not possible to determine the cytosolic phosphorylation potential, but it is unlikely that the increase in the phosphorylation potential would be significant enough to affect the respiratory O_2 uptake rate (20). Thus, the question remains of how the postulated light-induced inhibition of dark respiration could be obtained.

Sources of Cytosolic Adenylates

In the light, the cytosolic ATP pool can be sustained either by photosynthetic ATP synthesis or by oxidative phosphorylation. ATP produced in the chloroplast can be transferred to the cytosol by the ATP-ADP translocator (47) or in the form of DHAP via the phosphate translocator (25, 26). Both translocators are found in the inner envelope membrane. The activity and affinity of the ATP-ADP translocator is very low, which implies a minor function for the export of ATP (47). This view may be supported by the finding that a *Chlamydomonas reinhardtii* mutant deficient in chloroplast ATP synthase can grow photoautotrophically using ATP generated in the mitochondria to support CO_2 fixation by the Calvin cycle (66). In fact, the chloroplastic ATP-ADP translocator imports ATP into the stroma under conditions when the cytosolic ATP/ADP ratio is higher than the stromal ratio (45). In the light, DHAP exported from the chloroplast by the phosphate translocator can either be used for sucrose synthesis or be converted to PGA in the cytosol. The cytosolic conversion of DHAP to PGA can be catalyzed by the phosphorylating NAD-dependent glyceraldehyde 3-phosphate dehydrogenase/phosphoglycerate kinase (GAPDH/PGK) or by the nonphosphorylating irreversible NADP-dependent GAPDH/ PGK (Figure 1). The nonphosphorylating isoform has a K_m for DHAP and NADP of 20 μM and 3 μM, respectively (55). The concentrations of DHAP and NADP in the cytosol in the light in spinach are 0.6 mM and 38 μM, respectively (46, 107). Thus, in the light under atmospheric CO_2 concentrations the enzyme is substrate-saturated and can work at its maximal capacity. Although the phosphorylating NAD-dependent GAPDH has a higher activity than the nonphosphorylating NADP-dependent GAPDH when measured with saturating substrate concentrations, the relation of the activities is reversed when measured at substrate concentrations that are close to physiological conditions in the light (55). This might imply that the contribution of photophosphorylation to the cytosolic ATP pool is limited.

Plant mitochondria possess a very active ATP-ADP translocator that efficiently exports to the cytosol the ATP produced in the mitochondrial matrix (48). Specific inhibition of the mitochondrial ATP synthase by oligomycin (63) decreased the cytosolic ATP/ADP ratio in barley leaf protoplasts by about 60% under saturating as well as limiting light intensities and CO_2 concentrations (59, 62). Aminoacetonitrile (AAN), which specifically inhibits the pho-

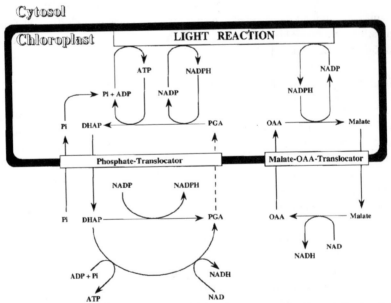

Figure 1   Transfer of ATP and redox equivalents across the chloroplast envelope via the phosphate-translocator or the malate-OAA translocator. The dashed line indicates the possible exchange of DHAP against PGA in the light.

torespiratory conversion of glycine to serine and thus the production of NADH for mitochondrial electron transport and oxidative phosphorylation (30), caused a decrease in the cytosolic ATP/ADP ratio by 25% and 45% at limiting and strictly limiting CO_2 concentrations, respectively (31). It can therefore be concluded that the cytosolic ATP pool is to a large extent maintained by oxidative phosphorylation.

Dependence of Photosynthetic Metabolism on Oxidative Phosphorylation

Sucrose synthesis occurring in the cytosol depends on the supply of UTP, the substrate for the cytosolic UDP-glucose pyrophosphorylase (Figure 2); 75% of the UTP and 70% of the UDP pools are located in the cytosol (15). The cytosolic UTP pool is linked to the ATP pool by a nucleoside-5'-diphosphate kinase. This enzyme is found in equal parts in the chloroplast and the cytosol (15). Its cytosolic activity is sufficient to catalyze the conversion of ATP to UTP. A drop in the cytosolic ATP level as a result of decreased mitochondrial oxidative phosphorylation (59, 62) is likely to lead to a decreased level of UTP. In a leaf, 90% of the UDP-glucose has been found in the cytosol at a concentration of 4.5 mM (15). Because the K_m of sucrose phosphate-synthase

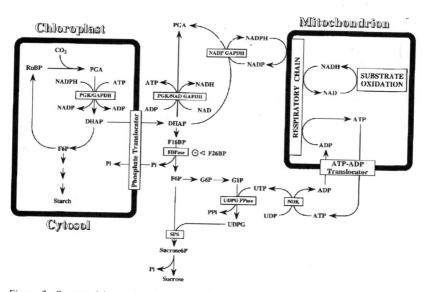

Figure 2 Suggested interactions between chloroplasts and mitochondria in the light. NDK: nucleoside-5'-diphosphate kinase; UDPG PPase: UDPGlucose pyrophosphorylase; FBPase: fructose-1,6-bisphosphatase; RuBP: ribulose-1,5-bisphospate; F6P: fructose-6-phosphate; G6P: glucose-6-phosphate; G1P: glucose-1-phosphate; F16BP: fructose-1,6-bisphosphate; F26BP: fructose-2,6-bisphosphate; UDPG: UDP-glucose; PPi: pyrophosphate.

(SPS) for UDP-glucose ranges from 4 to 8 mM (15), a decrease in the UDP-glucose level is likely to limit SPS activity. A decrease in SPS activity as well as in the activation state of SPS can be observed upon inhibition of oxidative phosphorylation (62). Limitation in the sucrose synthesis pathway leads to feedback regulation on photosynthesis (110), as indicated by an increase in the cytosolic glucose-6-phosphate (59, 62) and fructose-2,6-bisphosphate levels (59) upon inhibition of oxidative phosphorylation. The increase in the latter reduces cytosolic fructose-1,6-bisphosphatase activity (99), thus limiting the consumption of DHAP in the cytosol. Because the phosphate translocator in the inner chloroplast envelope is regulated by the equilibrium of the metabolite concentrations in the stroma and the cytosol (26), a limitation in the sucrose synthesis pathway can either lead to an elevated cytosolic level of DHAP or to a reduced efflux of DHAP from the chloroplast (62). In both cases, Calvin cycle activity will be reduced, as indicated by the decreased stromal PGA level (62).

The results described above show that the ATP provided to the cytosol by oxidative phosphorylation is required for sucrose synthesis. This conclusion is supported by the finding that inhibition of oxidative phosphorylation in the wild-type of *Nicotiana sylvestris*, a starch storer, did not affect the maximal photosynthetic activity, whereas a starchless mutant of the same species responded to such an inhibition by a decrease in photosynthetic activity (41). This result indicates that in the wild-type the cytosolic ATP level did not limit sucrose synthesis, whereas such a limitation did occur in the mutant, in which an increase in sucrose synthesis is required to compensate for the block in starch synthesis.

Dependence of Photosynthetic Capacity on Respiration

The stoichiometries of photosynthetic linear electron transport are still debated. H^+/e^--ratios of 2–3 (93) and H^+/P-ratios of 4 (10, 84) have recently been determined for photosynthetic linear electron transport. Based on these stoichiometries the ATP/NADPH ratio will range from 1.0 to 1.5. In the Calvin cycle, 1.5 ATP per NADPH are required to fix one carbon dioxide. The production of sufficient ATP for the Calvin cycle by linear electron transport alone would lead to an increase in the chloroplastic redox state, leading to feedback inhibition on the photosynthetic electron transport chain. To maintain a high photosynthetic activity, several processes are required to balance the ATP/NADPH ratio:

CYCLIC PHOTOPHOSPHORYLATION   This process enables the production of ATP from electron flow around photosystem I (PS I) without concomitant formation of NADPH. Rates of ATP synthesis linked to cyclic electron transport can amount to about 20% of the rates of linear electron transport (85). Cyclic photophosphorylation has been suggested to be necessary to support maximum rates of steady-state photosynthesis at high light and CO_2 concentration (28). Rapid cyclic electron transport has also been measured under O_2- or CO_2-free conditions but was largely suppressed under atmospheric conditons (42, 92). These findings bring into question the significance of a contribution of cyclic electron flow to ATP production in the chloroplast of C_3 plants under physiological conditions.

Q-CYCLE   Cyclic electron transport and probably also linear electron transport involve a proton-motive Q-cycle that contributes to the synthesis of ATP. In this case two protons are transported for every electron transferred to PS I. The Q-cycle might be an optional mechanism that operates only under low excitation light and simultanously with the ATP-synthesizing reaction pathways. This would imply that the H^+/e^--ratio is variable with the conditions applied (93) and, thus, with the contribution of the Q-cycle to photosynthetic ATP production.

MEHLER REACTION   This process describes the reduction of O_2 to H_2O_2 by electrons from the photosynthetic electron transport chain and the subsequent conversion of H_2O_2 to H_2O by an ascorbate-peroxidase pathway. ATP is produced in the same way and with the same stoichiometry as for linear electron transport. The in vivo rates of ferredoxin-dependent O_2 reduction are sufficient to cover the demand of ATP for the Calvin cycle (27). However, in the presence of OAA, H_2O_2 release from the Mehler reaction was suppressed, indicating that NADPH consumption by the chloroplastic NADP-MDH is favored over the Mehler reaction in order to balance the chloroplastic NADPH/ATP ratio (98). Inhibition of photorespiration under atmospheric conditions led to photoinactivation of the photosynthetic electron transport chain, although the Mehler reaction was unaffected by this treatment (111). Thus, photorespiration has been suggested to be more efficient in preventing an overreduction of the chloroplastic redox carriers than is the Mehler reaction. These results call into question the importance of the Mehler reaction for balancing the chloroplastic NADPH/ATP ratio under physiological conditions.

NADPH EXPORT   The stromal NADP-MDH is light activated and regulated by the redox state in the chloroplast stroma (91). The chloroplastic inner envelope membrane is impermeable to pyridine nucleotides, but an indirect transfer of reducing equivalents can proceed via a malate-OAA shuttle (44) (Figure 1), which has been suggested as a means to balance the chloroplastic NADPH/ATP ratio (90). Transfer of redox equivalents from the chloroplast can also procede via the phosphate translocator and the cytosolic nonphosphorylating NADP-dependent GAPDH/PGK system (Figure 1). In this way the transfer of NADPH could occur without simultaneous transfer of ATP and, thus, poise the stromal NADPH/ATP ratio.

Redox equivalents deriving from photosynthetic electron transport could be used for oxidation in the respiratory electron transport chain. Mitochondrial external oxidation of NAD(P)H is linked to two coupling sites, whereas substrate oxidation inside the mitochondria is linked to three sites (ignoring the possible contribution of alternative oxidase activity and rotenone-insensitive oxidative activity for simplicity's sake). Such a cooperation of photosynthetic linear electron transport and oxidative phosphorylation might be more efficient in balancing the chloroplastic NADPH/ATP ratio than other processes. Inhibition of mitochondrial ATP synthesis has been hypothesized to cause an increase in the cytosolic and chloroplastic redox levels (59, 62), indicating the utilization of redox equivalents from photosynthetic electron transport for oxidative phosphorylation. Inhibition of the mitochondrial ATP synthase is reported to enhance photoinhibition (88), which indicates that the mitochondrial electron transport chain reoxidizes part of the excess chloroplastic redox equivalents.

Thus, the activity of the mitochondrial electron transport chain in the light might be required to maintain a high photosynthetic capacity. These considerations do not imply that cyclic or pseudo-cyclic photophosphorylation do not occur in a plant cell in the light. They might, however, be of importance under more extreme enviromental conditions such as anoxia, cold stress, or water deficiency.

SOURCES OF REDOX EQUIVALENTS FOR OXIDATIVE PHOSPHORYLATION IN THE LIGHT

Glycine Oxidation

Under atmospheric conditions phosphoglycolate is produced by the oxygenase activity of Rubisco. To avoid the loss of carbon, phosphoglycolate is converted to phosphoglycerate in the photorespiratory pathway. The reactions of this pathway are spread over three different cellular compartments: the chloroplast, the mitochondria, and the peroxisomes. In the mitochondrial matrix, two molecules of glycine are converted to one molecule of serine. Simultaneously, CO_2 and NH_3 are released and NADH is produced. Thus, 75% of the carbon atoms in the phosphogycolate molecule are regained during photorespiration. The photorespiratory cycle accounts for 20–25% of the photosynthetic activity (68), which refers to a photorespiratory activity of 30–50 µmol CO_2 (mg chl h)$^{-1}$ (for reviews see 2, 51, 95).

NADH produced during glycine oxidation has to be reoxidized continously to sustain the flow through the photorespiratory pathway. Oxidation of glycine is coupled to three phosphorylation sites and occurs preferentially over other substrates such as malate or succinate (8, 18). If the mitochondrial matrix has a uniform NADH pool, it is hard to imagine how the dehydrogenases of the respiratory chain differentiate NADH from different substrates (16, 21). It has been suggested that groups of enzymes might be tied together in protein complexes (108), named metabolons, which may be located in the vicinity of the complexes of the respiratory chain or may even be attached to the membrane. In this way the reducing equivalents from the oxidation of some substrates would have facilitated access to the respiratory chain.

TCA Cycle Activity

In earlier studies TCA cycle activity in the light was determined by the incorporation of radiolabel into the cycle's intermediates, leading to the conclusion that the TCA cycle as well as glycolysis are active in the light, albeit with a decreased rate (for a summary see 36, 37).

The mitochondrial pyruvate dehydrogenase complex (mPDC) is the entry point of carbon into the TCA cycle. Its activity is sensitive to fine control through product feedback inhibition by NADH and acetyl-CoA and inactivation-reactivation by reversible phosphorylation (78 and references therein). Pea leaf mPDC kinase is inhibited by pyruvate, TPP, ADP, acetyl-CoA, NADH, and citrate, whereas K^+ and NH_4^+ activate it. Pea leaf mPDC phosphatase is stimulated by Mg^{2+} and inhibited by fluoride (78 and references therein). Determinations of in vivo mPDC activity from pea leaves showed a 40–60% inhibition of activity in the light under photorespiratory conditions (11, 32). The addition of inhibitors of the photorespiratory carbon flow prevented light-dependent inactivation of the mPDC. mPDC activity was 60–80% higher under nonphotorespiratory conditions in the light than under photorespiratory conditions. These results indicate that in pea, the oxidation of glycine suppresses mPDC activity (32). This decrease might be caused by NH_4^+ released during glycine oxidation stimulating the protein kinase that phosphorylates and inactives the mPDC (94). Light-dependent inactivation of the mPDC under photorespiratory conditions could not be observed in barley leaf protoplasts (61), indicating species-specific regulation of the mPDC. mPDC activity does not necessarily represent the operation of a complete TCA cycle. NADH, acetyl-CoA, and citrate inhibit the PDC kinase (78). An increase in these metabolite levels would signal a decreased demand for TCA cycle activity and should, thus, lead to stimulation of PDC kinase activity. Therefore, this pattern of regulation might indicate that products of mPDC activity (e.g. acetyl-CoA and citrate) are required for reactions outside the mitochondria. Citrate is indeed the main TCA cycle intermediate exported from the mitochondria (40). Acetyl-CoA can be converted to acetate by an acetyl-CoA hydrolase found in the mitochondrial matrix of pea (112). Acetate

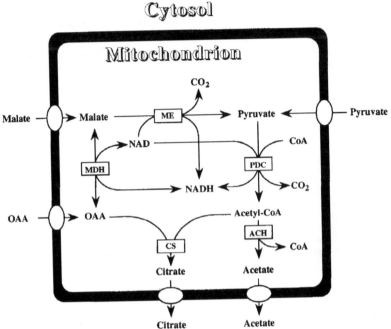

Figure 3   The flow of carbon through a partially active TCA cycle. ME: malic enzyme; ACH: acetyl-CoA hydrolase; CS: citrate synthase.

might be required for fatty acid synthesis in the chloroplasts. This would allow mPDC activity to occur without concomitant TCA cycle activity (Figure 3).

Substrates for mPDC activity in the light could be cytosolic pyruvate, malate, or OAA. Malate import by the mitochondria is supposed to be rather limited (113), whereas transport of OAA occurs with a high rate and affinity (22). In swelling experiments with isolated pea leaf mitochondria, the uptake of malate was decreased significantly by the presence of physiological OAA concentration (113), indicating that the uptake of OAA might be favored over malate uptake. However, malate can be generated from OAA inside the mitochondria. Malate oxidation inactivates PDC in isolated pea leaf mitochondria, unless the medium pH is 7 or lower, which leads to activation of malic enyzme and, thus, pyruvate production, which activates mPDC (78). It has been calculated from respiratory activities and metabolite export from isolated pea and spinach mitochondria that the operation of the complete TCA cycle during steady-state photosynthesis ranges from 10 to 20% (40). A partial activity of the cycle, involving MDH, malic enzyme, mPDC, and citrate synthase, has instead been suggested. In this way, reducing equivalents from substrate oxidation could contribute to mitochondrial ATP formation. The conversion of

OAA to malate in the mitochondrial matrix, as mentioned above, would consume one NADH and, thus, lower the yield of ATP.

Oxidation of External Redox Equivalents

EXTERNAL NADH   The oxidation of external NADH is pH dependent with a pH-optimum at 6.8–7.2 for most species (23, 72, 73). The pH of the cytosol is assumed to be between 7.1 and 7.6 (64), enabling oxidation of NADH with nearly optimal rates. Thus, small changes in the cytosolic pH should not affect the activity strongly.

NADH oxidation is stimulated in vitro by cations such as K^+, Mg^{2+}, or Ca^2. In the presence of high salts the V_{max} increases and the K_m decreases, whereas the removal of salts increases the K_m (23). Cations are thought to bind electrostatically to the negatively charged surface of the inner membrane and, thus, decrease the surface potential (23). Exogenous NADH oxidation specifically requires Ca^{2+} ions for activity (71 and references therein). The dependence on Ca^{2+} is strongly pH dependent and disappears at low pH. The oxidation of external NADH requires about 1 μM Ca^{2+} for activity (71). The cytosolic Ca^{2+} concentration is in the range of 0.1–1 μM (54). Τηυσ, α χηανγε ιν τηε χψτοσολιχ Χα^{2+} concentration could have a regulatory effect on the NADH oxidation rate. Polyamines are ubiquitous polycationic metabolites in eukaryotic cells. Used in the physiological concentration range they decrease the K_m and increase the V_{max} for the oxidation of external NADH by Jerusalem artichoke mitochondria suspended in a low-cation medium (82). Furthermore, they decrease the K_m for Ca^{2+} from about 1 μM to about 0.2 μM for external NADH oxidation in a high-cation medium. Thus, they may be important cytosolic factors affecting mitochondrial oxidation of external NADH in vivo.

External NADH is oxidized with a high rate and a K_m of 10–100 μM in many species and different plant tissues such as leaves and tubers (71).The concentration of NADH in spinach cytosol in the light is 0.6 μM (46). The oxidation rates of NADH at this concentration, calculated from the kinetic constants (58, 60), are shown in Table 1. Even though the V_{max} and the affinity toward NADH for the oxidation of external NADH are high (60, 71), the calculated in vivo rates are low because of the low level of NADH in the cytosol in the light. Because the K_m for NADH (≈ 1.4 μM) of nitrate reductase in spinach is low (86), the cytosolic NADH concentration might be kept well below the K_m for NADH (≈ 30 μM) of the external NADH dehydrogenase. The V_{max} for nitrate reduction in spinach is approximately 10 μmol (mg chl h)$^{-1}$ (53a). The cytosolic NADH concentration in spinach in the light would support an in vivo rate of about 3 μmol (mg chl h)$^{-1}$, which is at least twice the rate of the oxidation of external NADH by spinach mitochondria under the same condition. In an intact cell several substrates are oxidized simultaneously, and glycine and malate are oxidized preferentially over external NADH. Thus, the oxidation rate for external NADH is reduced in the presence of glycine and malate (Table 1). Furthermore, low $CaCl_2$ concentrations, which are closer to physiological conditions, lower the oxidation rates for NADH.

Even though the physiological oxidation rates of external NADH are low, they might increase significantly under conditions that increase the cytosolic redox state. Moreover, it remains to be elucidated how other, yet unknown, regulatory factors could modulate the in vivo respiratory activity of external NADH.

EXTERNAL NADPH   The oxidation of external NADPH is pH dependent with a pH-optimum at 6 for spinach (23) and tubers of Jerusalem artichokes (73) and at 6.6 for *Arum maculatum* spadices (73). However, NADPH oxidation with percoll-purified mitochondria of Jerusalem artichoke tubers is insensitive to pH variations between 5 and 7 (83). At the pH of the cytosol [7.1–7.6 (64)] the oxidation rate of external NADPH differs considerably depending on the species. The NADPH oxidation rates in spinach, Jerusalem artichoke, and *A. maculatum* at pH 7.2 amount to about 15% of the activity at the pH-optimum and to about 10% of the oxidation rate with NADH at the same pH (23, 73, 83). However, mitochondria from pea and sunflower leaves, etiolated corn and mung bean seedlings, and potato tuber showed oxidation rates of external NADPH at pH 7.2–7.4, amounting to 50–90% of those with external NADH (1, 60, 74). In species with a pH-optimum for the oxidation of external NADPH that is

Table 1 Calculated rates for the oxidation of external NADH and NADPH by spinach and pea leaf mitochondria in the presence and absence of glycine and/or malate and upon the addition of $CaCl_2$

Substrate	Addition	Oxidation rate [μmol NAD(P)H (mg chl h)$^{-1}$][a]			
		pea		spinach	
		2mM $CaCl_2$	no $CaCl_2$ added	2 mM $CaCl_2$	no $CaCl_2$ added
NADH	—	3.9	1.8	—	1.5
	glycine	4.1	1.3	—	0.8
	malate	3.8	1.4	—	1.8
	glycine + malate	—	0.8	—	0.6
NADPH	—	44	35	0	0
	glycine	35	32	0	0
	malate	38	26	0	0
	glycine + malate	—	25	0	0

[a]The data were evaluated from the K_m and V_{max} determined in (58,60) and the cytosolic concentrations of NAD(P)H estimated for the cytosol in spinach leaves in the steady-state of photosynthesis (46). Calculations are based on Michaelis-Menten Kinetics: $V = V_{max} \times S (S + K_m)^{-1}$.

distinctly different from the physiological cytosolic pH, a change in the latter can have a regulatory impact on the oxidation rate.

In spinach at the optimal pH range, the presence of cations increased the V_{max} of NADPH oxidation and decreased the K_m; however, removal of the cations did not lead to an increase in the K_m as it did for NADH (23). In Jerusalem artichoke mitochondria, oxidation of external NADPH was unaffected by the ionic strength but was dependent on Ca^{2+} (83). The requirement of external NADPH oxidation for free Ca^{2+} ions seems to be higher than for NADH (71). Oxidation of external NADPH by Jerusalem artichoke mitochondria suspended in a low-cation medium was stimulated by the addition of the polyamine spermidine (83), a response similar to that of external NADH oxidation.

In many species the oxidation rates and affinities for external NADPH are lower than for NADH (23, 60, 71, 73). The concentration of NADPH in spinach cytosol in the light is 150 μM (46). The oxidation rates for NADPH are 10–20 times higher than for NADH (see Table 1). Respiratory O_2 uptake in leaf slices and protoplasts in the dark is about 11–24 μatom O (mg chl h)$^{-1}$ (4, 13, 62, 81, 88), whereas intact wheat leaves show a higher respiratory activity of 36 μatom O (mg chl h)$^{-1}$ (13). The calculated physiological activity for the oxidation of NADPH is 25 μmol NADPH (mg chl h)$^{-1}$ (Table 1), which is in line with the respiratory O_2 uptake rates. The source for cytosolic NADPH could be photosynthetically generated NADPH (Figure 2). The physiological function of this pathway might be to poise the chloroplastic ATP/NADPH ratio as discussed in the preceding section. Cytosolic NADPH could also come from the pentose phosphate pathway. Unlike chloroplastic glucose-6-phosphate dehydrogenase, the cytosolic isoform is not inactivated in the light in pea leaves (24). Cytosolic glucose-6-phosphate dehydrogenase is regulated by the NADPH/NADP ratio, a decrease of which would lead to an increased activity of the enzyme. An alternative source for cytosolic NADPH might be the oxidation of isocitrate, an NADP-dependent reaction that is probably linked to the supply of carbon skeletons for amino acid synthesis in the light (14). Because not all species display oxidation of external NADPH at the physiological pH range of the cytosol, the general physiological importance of the oxidation of external NAD(P)H still has to be elucidated.

Redox Transfer Across the Mitochondrial Membrane

NADH formed from substrate oxidation in the mitochondrial matrix can be reoxidized in the respiratory chain or exported. Plant mitochondria have an active OAA translocator (Figure 4) (22, 75). The K_m for the import of OAA ranges from 3 to 7 μM in several plant tissues and the V_{max} is about 560 nmol (mg protein min)$^{-1}$ in green leaves of spinach and pea. The addition of OAA to isolated mitochondria oxidizing glycine in state 3 causes an inhibition of the

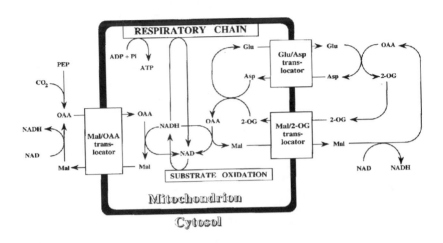

Figure 4   Transfer of redox equivalents across the mitochondrial inner membrane via the malate-OAA shuttle and the malate-aspartate shuttle. Mal: malate; Glu: glutamate; Asp: aspartate.

O_2 uptake rate (22), indicating that the oxidation of NADH by mitochondrial MDH is favored over the electron transport chain. In an intact photosynthesizing cell, cytosolic OAA can be provided from malate via cytosolic MDH. Although the equilibrium of cytosolic MDH is on the side of malate (103), the efficient removal of OAA by the mitochondrial OAA translocator would favor the reaction. OAA can also be generated from PEP by PEPC.

The cytosolic OAA concentration in the light in spinach is about 0.1 mM (46). At this concentration the OAA translocator should function close to its maximal activity, implying that the malate-OAA shuttle has the theoretical capacity to export all NADH produced in the mitochondria. The NADH/NAD ratio in the cytosol in the light is about 70 times lower than in the mitochondrial matrix during substrate oxidation in state 3 [cytosolic NADH/NAD ratio $\approx 10^{-3}$; mitochondrial NADH/NAD ratio ≈ 0.07 (46, 60)]. This gradient between the redox ratios in the cytosol and the mitochondrial matrix indicates that the malate-OAA shuttle is regulated in a way that avoids a depletion of NADH in the matrix. The degree of activity of the malate-OAA shuttle that depends on the malate/OAA ratio was determined as the extent of inhibition of respiratory O_2 uptake (40, 58, 60). At the physiological NADH/NAD ratio in the cytosol, about 25 and 50% of the NADH from substrate oxidation in spinach and pea mitochondria, respectively, was exported to the cytosol. Estimations of malate-OAA shuttle activity, based on malate efflux measurements, were found to be consistently lower than the OAA-dependent inhibition of glycine-dependent O_2 uptake (19), indicating that part of the malate remains in the mitochondria and might serve as a substrate for malic enzyme.

NADH can also be exported by a malate-aspartate shuttle found in plant mitochondria (53) (Figure 4). In pea and spinach the export of NADH via the malate-aspartate shuttle is 12–20% of that of the malate-OAA shuttle (40, 58, 60). Determinations of malate efflux from isolated pea leaf mitochondria showed that the malate-OAA shuttle activity exceeded malate-aspartate shuttle activity at OAA concentrations ≥ 15 μM (19). At OAA concentrations in the physiological range the malate-aspartate shuttle accounted for about 20% of the rate of NADH export as that of the malate-OAA shuttle.

Reoxidation of NADH exported from the mitochondria by the external NADH dehydrogenase is unlikely because the cytosolic NADH/NAD ratios under which the malate-OAA shuttle can operate are so low that hardly any oxidation of external NADH would occur. Export of NADH from the matrix seems to be strictly regulated to enable the use of a large proportion of redox equivalents produced in the mitochondria for the synthesis of ATP.

In Vivo Substrate Oxidation

Isolated mitochondria exposed to several substrates will oxidize these simultaneously until the maximal capacity of the respiratory chain is reached (60). In a cell the respiratory oxidation rate for each substrate depends on its concentration. Little is known about the concentrations of metabolites in the mitochondrial matrix. It cannot necessarily be assumed that the metabolite concentrations in the matrix and the cytosol are the same, because transport of the metabolites might be regulated in a way that increases or decreases their concentration in the matrix versus the cytosol. Thus, it is difficult to determine the oxidation rates for intramitochondrial substrates under physiological conditions. The maximal capacity of the respiratory chain in isolated mitochondria from green leaf tissue varies from 700 to 1000 nmol NADH (mg prot min)$^{-1}$ (60), which converts to 65–90 μatom O (mg chl h)$^{-1}$. Oxidation of glycine in the light in vivo occurs with a rate of 30–50 μatom O (mg chl h)$^{-1}$ depending on the tissue (68). Taking into account that 25–50% of the intramitochondrially produced redox equivalents are exported (40, 60), reoxidation of NADH in the respiratory chain would be 15–38 μatom O (mg chl h)$^{-1}$. These rates are consistent with reported dark respiratory O_2 uptake rates [11–36 μatom O (mg chl h)$^{-1}$] (see above) but are significantly lower than the maximal capacity of the respiratory chain. Thus, simultaneous oxidation of other substrates (e.g. external NADPH, malate, or pyruvate) would be allowed. This hypothesis concurs with the finding that oligomycin, which specifically inhibits mitochondrial ATP synthesis and consequently all mitochondrial substrate oxidation, causes a more distinct decrease in the cytosolic ATP/ADP ratio than does AAN, which specifically blocks the conversion of glycine to serine (31, 59, 62). It might even be speculated that the activity of the respiratory chain increases in the light in order to sustain the flux through the photorespiratory pathway, a response that might not be required under nonphotorespiratory conditions. An increase of dark respiratory O_2 uptake resulting from photorespiration has indeed been observed (81). In line with this hypothesis are the higher mitochondrial (3.6 vs 2.6) and cytosolic (7.3 vs 6.4) ATP/ADP ratios under photorespiratory conditions than under nonphotorespiratory conditions in the light (29, 100).

PEROXISOMAL REQUIREMENT FOR REDOX EQUIVALENTS

In the photorespiratory cycle, redox equivalents are required in the peroxisomal matrix for the reduction of hydroxypyruvate. The peroxisomal membrane has porins, which allow the diffusion of NADH. However, cytosolic NADH [0.6 µM in the light in spinach (46)] would only support 1% of the photorespiratory flow (80). An alternative source of NADH could be the oxidation of malate in the peroxisomes. Because the cytosolic malate concentration is about 10 times higher than OAA (46) and because no catalytic transport process is involved in the transfer of malate and OAA into the peroxisomes, the peroxisomal MDH reaction should be favored in the direction of NADH production. Moreover, peroxisomal MDH activity is sufficient to sustain the photorespiratory flow (49). In a photosynthezising leaf under photorespiratory conditions, the demands of the peroxisomal hydroxypyruvate reductase for reductant are met by oxidation of malate and not by transfer of NADH (80).

The requirement of the peroxisomal hydroxypyruvate reductase for NADH is equivalent to the production of NADH from glycine oxidation in the mitochondrial matrix. This observation has led to the proposal that NADH produced in the mitochondria is exported and used for peroxisomal reactions (53) (Figure 5). Because the export of only 25–50% of the redox equivalents produced in the mitochondrial matrix is not sufficient to sustain the reduction of hyroxypyruvate, another source for NADH is required. The activity of the malate-OAA shuttle in the chloroplastic envelope is high enough (44) to support peroxisomal hydroxypyruvate reductase activity. It is likely that both chloroplasts and mitochondria allocate NADH simultaneously to the peroxisomes; however, the contribution of the two sources has not been quantified.

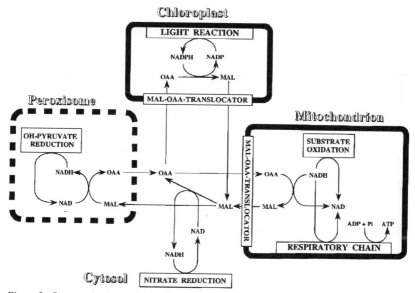

Figure 5   Suggested redox transfer by malate-OAA shuttles in a plant cell. Mal: malate.

INTERACTIONS BETWEEN NITRATE ASSIMILATION AND RESPIRATION IN THE LIGHT

SOURCES OF CARBON SKELETONS FOR AMINO ACID SYNTHESIS NO_3^- assimilation includes the reduction of NO_3^- to NO_2^- in the cytosol as well as the reduction of NO_2^- to NH_4^+ and the assimilation of the latter into amino acids via glutamine synthetase (GS) and glutamate synthase (GOGAT) in the chloroplast (95) (Figure 6). Chloroplasts have a limited ability to synthesize oxo-acids in the light, but they possess translocators that enable the import of 2-oxoglutarate (2-OG) and glutamine (26). It has been suggested that mitochondria synthesize and export 2-OG, which is then converted to glutamate via the chloroplastic GS/GOGAT system (70). Because plant mitochondria mainly export citrate (40) and because cytosolic isoenzymes for the conversion of citrate to 2-OG have been found (14), it has been hypothesized that this might be a route to provide carbon skeletons for NH_4^+ assimilation (Figure 6). This theory is supported by the regulatory pattern of mPDC, which enables the generation of acetyl-CoA for citrate production even when the complete TCA cycle is not functioning (see above).

The addition of NO_3^- or NH_4^+ to N-limited green algae in the light causes an increase in net production of amino acids and in respiratory CO_2 efflux, implicating the TCA cycle as a source of carbon skeletons for amino acid synthesis (105). The feeding of NO_3^- into N-limited wheat leaves causes a

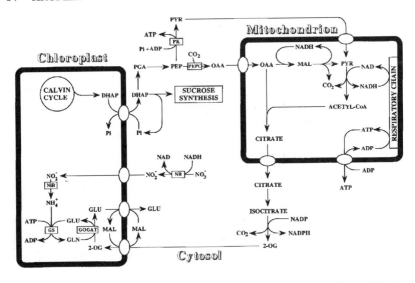

Figure 6   Suggested supply of carbon skeletons for photosynthetic nitrate assimilation. NR: nitrate reductase; NiR: nitrite reductase; pyr: pyruvate; mal: malate; glu: glutamate; gln: glutamine.

decrease in SPS activity and an increase in PEPC activity, which indicates a redirection of the carbon flow from sucrose synthesis to amino acid synthesis (102). PEP from the C_3-photosynthetic pathway can be converted to OAA by PEPC or to pyruvate by the cytosolic pruvate kinase (PK_c). Both metabolites, OAA and pyruvate, can serve as a substrate for the production of carbon skeletons within the mitochondria (Figure 6). Although PK_c was not affected by NO_3^- feeding (102), its activity increased upon NH_4^+ feeding (95). The regulatory pattern of PK_c from photosynthetic tissue [inhibition by glutamate (7, 67), activation by DHAP (67)] supports its function in providing carbon skeletons for amino acid synthesis. The finding that transgenic plants deficient in PK_c are not affected in their respiratory O_2 uptake rates or photosynthetic O_2 evolution rate (35) shows that there is no strict dependence of metabolism on PK_c activity. PEPC might provide an alternative pathway to bypass PK_c. PEPC activity in C_3 plants is regulated via protein phosphorylation (102, 104) as well as by metabolite effectors [inhibition by glutamate and aspartate; activation by glucose-6-P, fructose-6-P, DHAP, and PGA (S Krömer, P Gardeström & G Samuelsson, manuscript in preparation)]. This regulatory pattern supports the function of PEPC in supplying carbon skeleton for amino acid synthesis. This rather complex process to provide carbon skeletons for amino acid synthesis, as outlined in Figure 6, might be favorable because it would enable the synthesis of ATP in the mitochondria, which might be needed to support protein synthesis in the presence of high concentrations of amino acids in the cytosol.

SOURCES OF NADH FOR NITRATE REDUCTION   The addition of NH_4^+ or NO_3^- to N-limited alga cells stimulated respiratory O_2 uptake. This effect was more significant with NO_3^- than NH_4^+ (105, 106). NO_3^- reduction in the cytosol requires NADH, which can come from photosynthetic or respiratory-photorespiratory metabolism (89, 109) (Figure 5). The supply of mitochondrial NADH [15–38 µmol NADH (mg chl h)$^{-1}$; see above] is sufficient to support the rate of NO_3^- reduction [6–11 µmol NADH (mg chl h)$^{-1}$]. Nevertheless, NADH derived photosynthetically can also support cytosolic nitrate reduction.

CONCLUDING REMARKS

Considerable progress has been made in research on respiration in the light. Much work has focused on respiratory CO_2 release and O_2 uptake, but because results have varied greatly, the extent of respiratory activity in the light is still unknown. It has, however, become clear that the TCA cycle and the respiratory electron transport chain do function in the light, albeit with a lower activity than in darkness. Whether the complete TCA cycle or only a part of the cycle are active in the light is still debated, although evidence supports partial activity.

Little is known about the interrelation of respiratory activity in the light with other processes: Oxidative phosphorylation is required to support sucrose synthesis, oxidation of redox equivalents in the respiratory electron transport chain is likely to support the sustainment of a high photosynthetic capacity, and the activity of the TCA cycle provides carbon skeletons for nitrogen assimilation in the light. Respiratory activity in an illuminated cell also is likely to be important for other metabolic pathways: Mitochondria might be a source for acetyl-CoA or acetate for choroplastic fatty acid synthesis in the light; mitochondrial oxidative phosphorylation might also be required as an energy source for cytosolic biosynthetic reactions other than sucrose synthesis, e.g. maintaining interorganellar proton- and ion-gradients; and respiration in the light might be important for plant growth or plant development. In this context, studies on a maize mutant deficient in genes for proteins of the respiratory electron transport chain (38) showed disrupted developmental behavior. The use of mutants or transgenic plants will be a great help for this field of research in the future.

ACKNOWLEDGMENTS

I am very grateful to Dr. HW Heldt (University of Göttingen) in whose laboratories the major part of my work was performed. His interest and contribution to the work is highly appreciated. Furthermore, I thank Drs. P Gardeström and G Samuelsson (University of Umeå) for supporting my research in their laboratories and Dr. P. Gardeström for comments on the manuscript.

Any *Annual Review* chapter, as well as any article cited in an *Annual Review* chapter, may be purchased from the Annual Reviews Preprints and Reprints service. 1-800-347-8007; 415-259-5017; email: arpr@class.org

Literature Cited

1. Arron GP, Edwards G. 1979. Oxidation of reduced nicotinamide adenine dinucleotide phosphate by plant mitochondria. *Can. J. Biochem.* 57:1392–99
2. Artus NN, Somerville SC, Somerville CR. 1986. The biochemistry and cell biology of photorespiration. *CRC Crit. Rev. Plant Sci.* 4:121–47
3. Avelange M-H, Thiéry JM, Sarrey F, Gans P, Rébeillé F. 1991. Mass-spectrometric determination of O_2 and CO_2 gas exchange in illuminated higher-plant cells. Evidence for light-inhibition of substrate decarboxylations. *Planta* 183:150–57
4. Azcón-Bieto J, Lambers H, Day DA. 1983. Effect of photosynthesis and carbohydrate status on respiratory rates and the involvement of the alternative pathway in leaf respiration. *Plant Physiol.* 72:598–603
5. Baltscheffsky M, ed. 1990. *Current Research in Photosynthesis,* Vol. 2. Netherlands: Kluwer Academic
6. Bate GC, Sültemeyer DF, Fock HP. 1988. $^{16}O_2/^{18}O_2$ analysis of oxygen exchange in *Dunaliella tertiolecta.* Evidence for the inhibition of mitochondrial respiration in the light. *Photosynth. Res.* 16:219–31
7. Baysdorfer C, Bassham JA. 1984. Spinach pyruvate kinase isoforms. Partial purification and regulatory properties. *Plant Physiol.* 74:374–79
8. Bergman A, Ericson I. 1983. Effects of pH, NADH, succinate and malate on the oxidation of glycine in spinach leaf mitochondria. *Physiol. Plant.* 59:421–27
9. Bligny R, Gardeström P, Roby C, Douce R. 1990. ^{31}P-NMR studies of spinach leaves and their chloroplasts. *J. Biol. Chem.* 265:1319–26
10. Bogdanoff P, Gräber P. 1990. Proton efflux and phosphorylation in flash groups. See Ref. 5, pp. 217–20
11. Budde RJA, Randall DD. 1990. Pea leaf mitochondrial pyruvate dehydrogenase complex is inactivated in vivo in a light-dependent manner. *Proc. Natl. Acad. Sci. USA* 87:673–76
12. Canvin DT, Berry JA, Badger MR, Fock H,

Osmond CB. 1980. O_2 exchange in leaves in the light. *Plant Physiol.* 66:302–7
13. Cashin McBG, Cossins EA, Canvin DT. 1988. Dark respiration during photosynthesis in wheat leaf slices. *Plant Physiol.* 87:155–61
14. Chen R-C, Gadal P. 1990. Do the mitochondria provide the 2-oxoglutarate needed for glutamate synthesis in higher plant chloroplasts? *Plant Physiol. Biochem.* 28:141–45
15. Dancer J, Neuhaus HE, Stitt M. 1990. Subcellular compartmentation of uridine nucleotides and nucleoside-5'-diphosphate kinase in leaves. *Plant Physiol.* 92:637–41
16. Day DA, Neuburger M, Douce R. 1985. Interactions between glycine decarboxylase, the tricarboxylic acid cycle and the respiratory chain in pea leaf mitochondria. *Aust. J. Plant Physiol.* 12:119–30
17. Douce R, Neuburger M. 1989. The uniqueness of plant mitochondria. *Annu. Rev. Plant Physiol. Plant Mol. Biol.* 40:371–414
18. Dry IB, Day DA, Wiskich JT. 1983. Preferential oxidation of glycine by the respiratory chain of pea leaf mitochondria. *FEBS Lett.* 158:154–58
19. Dry IB, Dimitriadis E, Ward AD, Wiskich JT. 1987. The photorespiratory hydrogen shuttle. *Biochem. J.* 245:669–75
20. Dry IB, Wiskich JT. 1982. Role of the external adenosine triphosphate/adenosine diphosphate ratio in the control of plant mitochondrial respiration. *Arch. Biochem. Biophys.* 217:72–79
21. Dry IB, Wiskich JT. 1985. Characteristics of glycine and malate oxidation by pea leaf mitochondria: evidence of differential access to NAD and respiratory chains. *Aust. J. Plant Physiol.* 12:329–39
22. Ebbighausen H, Chen J, Heldt HW. 1985. Oxaloacetate translocator in plant mitochondria. *Biochim. Biophys. Acta* 810:184–99
23. Edman K, Ericson I, Møller IM. 1985. The regulation of exogenous NAD(P)H oxidation in spinach (*Spinacia oleracea*) leaf mitochondria by pH and cations. *Biochem. J.* 232:471–77
24. Fickenscher K, Scheibe R. 1986. Purifica-

tion and properties of the cytoplasmic glucose-6-phosphate dehydrogenase from pea leaves. *Arch. Biochem. Biophys.* 247:393–402

25. Fliege R, Flügge U-I, Werdan K, Heldt HW. 1978. Specific transport of inorganic phosphate, 3-phosphoglycerate and triosephosphates across the inner membrane of the envelope in spinach chloroplasts. *Biochim. Biophys. Acta* 502:232–47

26. Flügge U-I, Heldt HW. 1991. Metabolite translocators of the chloroplast envelope. *Annu. Rev. Plant Physiol. Plant Mol. Biol.* 42:129–44

27. Furbank RT, Badger MR. 1983. Oxygen exchange associated with the electron transport and photophosphorylation in spinach thylakoids. *Biochim. Biophys. Acta* 723:400–9

28. Furbank RT, Horton P. 1987. Regulation of photosynthesis in isolated barley protoplasts: the contribution of cyclic photophosphorylation. *Biochem. Biophys. Acta* 894:332–38

29. Gardeström P. 1987. Adenylate ratios in the cytosol, chloroplasts and mitochondria of barley leaf protoplasts during photosynthesis at different carbon dioxide concentrations. *FEBS Lett.* 212:114–18

30. Gardeström P, Bergman A, Ericson I. 1981. Inhibition of the conversion of glycine to serine in spinach leaf mitochondria. *Physiol Plant.* 53:439–44

31. Gardeström P, Wigge B. 1988. Influence of photorespiration on ATP/ADP ratios in the chloroplast, mitochondria and cytosol, studied by rapid fractionation of barley (*Hordeum vulgare*) protoplasts. *Plant Physiol.* 88:69–76

32. Gemel J, Randall DD. 1992. Light regulation of leaf mitochondrial pyruvate dehydrogenase complex. *Plant Physiol.* 100:908–14

33. Gerbaud A, Andre M. 1980. Effect of CO_2, O_2 and light on photosynthesis and photorespiration in wheat. *Plant Physiol.* 66:1032–36

34. Gerhardt R, Stitt M, Heldt HW. 1987. Subcellular metabolite levels in spinach leaves. Regulation of sucrose synthesis during diurnal alterations in photosynthetic partitioning. *Plant Physiol.* 83:399–407

35. Gottlob-McHugh S, Sangwan R, Blakeley S, Vanlerberghe G, Ko E, et al. 1992. Normal growth of transgenic tobacco plants in the absence of cytosolic pyruvate kinase. *Plant Physiol.* 100:820–25

36. Graham D. 1980. Effects of light on "dark" respiration. In *The Biochemistry of Plants*, ed. DD Davies, 2:525-79. New York: Academic

37. Graham D, Chapman EA. 1979. Interactions between photosynthesis and respiration in higher plants. In *Encyclopedia of Plant Physiology (NS)*, ed. M Gibbs, E Latzko, 6:150–62. Berlin/New York: Springer-Verlag

38. Gu J, Miles D, Newton KJ. 1993. Analysis of leaf sectors in the NCS6 mitochondrial mutant of maize. *Plant Cell* 5:963–71

39. Hampp R, Goller M, Ziegler H. 1982. Adenylate levels, energy charge, and phosphorylation potential during dark-light and light-dark transition in chloroplasts, mitochondria, and cytosol of mesophyll protoplasts from *Avena sativa* L. *Plant Physiol.* 69:448–55

40. Hanning I, Heldt HW. 1993. On the function of mitochondrial metabolism during photosynthesis in spinach leaves (*Spinacia oleracea* L.). Partitioning between respiration and export of redox equivalents and precursors for nitrate assimilation products. *Plant Physiol.* 103:1147–54

41. Hanson KR. 1992. Evidence for mitochondrial regulation of photosynthesis by a starchless mutant of *Nicotiana sylvestris*. *Plant Physiol.* 99:276–83

42. Harbinson J, Foyer C. 1991. Relationship between the efficiencies of photosystem I and II and stromal redox state in CO_2-free air. *Plant Physiol.* 97:41–49

43. Hatch MD, Carnal NW. 1992. The role of mitochondria in C_4 photosynthesis. See Ref. 65, pp. 135–48

44. Hatch MD, Dröscher L, Flügge U-I, Heldt HW. 1984. A specific translocator for oxaloacetate transport in chloroplasts. *FEBS Lett.* 178:15–19

45. Heber U, Heldt HW. 1981. The chloroplast envelope: structure, function and role in leaf metabolism. *Annu. Rev. Plant Physiol.* 32:139–68

46. Heineke D, Riens B, Grosse H, Hoferichter P, Peter U, et al. 1991. Redox transfer across the inner chloroplast envelope membrane. *Plant Physiol.* 95:1131–37

47. Heldt HW. 1969. Adenine nucleotide translocation in spinach chloroplasts. *FEBS Lett.* 5:11–14

48. Heldt HW, Flügge U-I. 1987. Subcellular transport of metabolites in plant cells. In *The Biochemistry of Plants*, ed. PK Stumpf, EE Conn, pp. 49–85. New York: Academic

49. Heupel R, Markgraf T, Robinson DG, Heldt HW. 1991. Compartmentation studies on spinach leaf peroxisomes. Evidence for channeling of photorespiratory metabolism in peroxisomes devoid of intact boundary membrane. *Plant Physiol.* 96:971–79

50. Hoch G, Owens OvH, Kok B. 1963. Photosynthesis and respiration. *Arch. Biochem. Biophys.* 101:171–80

51. Husic DW, Husic HD, Tolbert NE. 1987. The oxidative photosynthetic carbon cycle. *CRC Crit. Rev. Plant Sci.* 5:455–500

52. Ishii R, Murata Y. 1978. Further evidence of the Kok effects in C_3 plants and the effects of enviromental factors on it. *Jpn. J. Crop Sci.* 47:547–50

53. Journet E-P, Neuburger M, Douce R. 1981. Role of glutamate-oxaloacetate transaminase and malate dehydrogeanse in the regeneration of NAD^+ for glycine oxidation by spinach leaf mitochondria. *Plant Physiol.* 67:467–69

53a. Kaiser WM, Spill D. 1991. Rapid modulation of spinach leaf nitrate reductase by photosynthesis. *Plant Physiol.* 96:368–75

54. Kauss H. 1987. Some aspects of calcium-dependent regulation in plant metabolism. *Annu. Rev. Plant Physiol.* 38:47–72

55. Kelly GJ, Gibbs M. 1973. Nonreversible D-glyceraldehyde 3-phosphate dehydrogenase of plant tissues. *Plant Physiol.* 52:111–18

56. Kirschbaum MUF, Farquhar GD. 1987. Investigation of the CO_2 dependence of quantum yield and respiration in *Eucalyptus pauciflora. Plant Physiol.* 83:1032–36

57. Kok B. 1949. On the interrelation of respiration and photosynthesis in green plants. *Biochim. Biophys. Acta* 3:625–31

58. Krömer S, Hanning I, Heldt HW. 1992. On the source of redox equivalents for mitochondrial oxidative phosphorylation in the light. See Ref. 65, pp. 157–75

59. Krömer S, Heldt HW. 1991. On the role of mitochondrial oxidative phosphorylation in photosynthesis metabolism as studied by the effect of oligomycin on photosynthesis in protoplasts and leves of barley (*Hordeum vulgare*). *Plant Physiol.* 95:1270–76

60. Krömer S, Heldt HW. 1991. Respiration of pea leaf mitochondria and redox transfer between the mitochondrial and extramitochondrial compartment. *Biochim. Biophys. Acta* 1057:42–50

61. Krömer S, Lernmark U, Gardeström P. 1994. In vivo mitochondrial pyruvate dehydrogenase activity, studied by rapid fractionation of barley leaf protoplasts. *J. Plant Physiol.* In press

62. Krömer S, Malmberg G, Gardeström P. 1993. Mitochondrial contribution to photosynthetic metabolism. A study with barley leaf protoplasts at different light intensities and CO_2 concentrations. *Plant Physiol.* 192:947–55

63. Krömer S, Stitt M, Heldt HW. 1988. Mitochondrial oxidative phosphorylation participating in photosynthetic metabolism of a leaf cell. *FEBS Lett.* 226:352–56

64. Kurkdjian A, Guern J. 1989. Intracellular pH: Measurement and importance in cell activity. *Annu. Rev. Plant Physiol. Plant Mol. Biol.* 40:271–303

65. Lambers H, van der Plas LHW, eds. 1992. *Molecular, Biochemical and Physiological Aspects of Plant Respiration.* The Hague: SPB Academic

66. Lemaire C, Wollman F-A, Bennoun P. 1988. Restoration of phototrophic growth in a mutant of *Chlamydomonas reinhardtii* in which the chloroplast atpB gene of ATP synthase has a deletion: an example of mitochondria-dependent photosynthesis. *Proc. Natl. Acad. Sci. USA* 85:1344–48

67. Lin M, Turpin DH, Plaxton WC. 1989. Pyruvate kinase isozymes from the green alga, *Selenastrum minutum.* II. Kinetic and regulatory properties. *Arch. Biochem. Biophys.* 269:228–38

68. Lorimer GH, Andrews TJ. 1981. The C_2 chemo- and photorespiratory carbon oxidation cycle. In *The Biochemistry of Plants*, ed. PK Stumpf, EE Conn, pp. 329–74. New York: Academic

69. Maury WJ, Huber SC, Moreland DE. 1981. Effects of magnesium on intact chloroplasts. *Plant Physiol.* 68:1257–63

70. Miflin BJ, Lea PJ. 1982. Ammonia assimilation and amino acid metabolism. In *Encyclopedia of Plant Physiology*, ed. D Boulter, B Parthier, 14A:5–64. New York: Springer-Verlag

71. Møller IM, Lin W. 1986. Membrane-bound NAD(P)H dehydrogenases in higher plant cells. *Annu. Rev. Plant Physiol.* 37:309–34

72. Møller IM, Palmer JM. 1981. The inhibition of exogenous NAD(P)H oxidation in plant mitochondria by chelators and mersalyl as a function of the pH. *Physiol. Plant.* 53:413–20

73. Møller IM, Palmer JM. 1981. Properties of the oxidation of exogenous NADH and NADPH by plant mitochondria. *Biochim. Biophys. Acta* 638:225–33

74. Nash D, Wiskich JT. 1983. Properties of substantially chlorophyll-free pea leaf mitochondria prepared by sucrose density gradient separation. *Plant Physiol.* 71:627–34

75. Oliver DJ, Walker GH. 1984. Characterization of the transport of oxaloacetate by pea leaf mitochondria. *Plant Physiol.* 76:409–13

76. Peltier G, Thibault P. 1985. O_2 uptake in the light in Chlamydomonas: evidence for persistent mitochondrial respiration. *Plant Physiol.* 79:225–30

77. Raghavendra AS, Padmasree K, Saradedevi K. 1994. Interdependence of photosynthesis and respiration in plant cells: interactions between chloroplasts and mitochondria. *Plant Sci.* 97:1–14

78. Randall DD, Miernyk JA, David NR, Budde RJA, Schuller KA, et al. 1990. Phosphorylation of the leaf mitochondrial pyruvate dehydrogenase complex and inactivation of the complex in the light. *Curr. Top. Plant Biochem. Physiol.* 9:313–28

79. Rébeillé F, Gans P, Chagvardieff P, Pean

M, Tapie P, Thibault P. 1988. Mass spectrometric determination of the inorganic carbon species assimilated by photoautotrophic cells of *Euphorbis characias* L. *J. Biol. Chem.* 263:12373–77

80. Reumann S, Heupel R, Heldt HW. 1994. Compartmentation studies on spinach leaf peroxisomes. II. Evidence for the transfer of reductant from the cytosol to the peroxisomal compartment via malate-oxaloacetate shuttle. *Planta* 193:167–73

81. Rey P, Peltier G. 1989. Photorespiratory properties of mesophyll protoplasts of *Nicotiana plumbaginifolia*. *Plant Physiol.* 89: 762–67

82. Rugolo M, Antognoni F, Flamigni A, Zannoni D. 1991. Effects of polyamines on the oxidation of exogenous NADH by Jerusalem artichoke (*Helianthus tuberosus*) mitochondria. *Plant Physiol.* 95:157–63

83. Rugolo M, Zannoni D. 1992. Oxidation of external NAD(P)H by Jerusalem artichoke (*Helianthus tuberosus*) mitochondria. *Plant Physiol.* 99:1037–43

84. Rumberg B, Schubert K, Strelow F, Tran-Anh T. 1990. The H^+/ATP coupling ratio at the H^+-ATP-synthase of spinach chloroplasts is four. See Ref. 5, pp. 125–28

85. Rurainski HJ, Borchert S, Heupel R. 1987. Ferredoxin-dependent cyclic electron transport and photophosphorylation. In *Progress in Photosynthesis Research*, ed. J Biggins, 2:537–40.

86. Sanchez J, Heldt HW. 1989. Determination of the apparent K_m of nitrate reductase from spinach leaves for NADH by a coupled assay. *Bot. Acta* 102:186–88

87. Santarius KA, Heber U. 1965. Changes in the intracellular levels of ATP, ADP, AMP and P_i and regulatory function of the adenylate system in leaf cells during photosynthesis. *Biochim. Biophys. Acta* 102:39–54

88. Saradedevi K, Raghavendra AS. 1992. Dark respiration protects photosynthesis against photoinhibition in mesophyll protoplasts of pea (*Pisum sativum*). *Plant Physiol.* 99:1232–37

89. Sawhney SK, Naik MS, Nicholas DJD. 1978. Regulation of NADH supply for nitrate reduction in green plants via photosynthesis and mitochondrial respiration. *Biochem. Biophys. Res. Commun.* 81: 1209–16

90. Scheibe R. 1987. $NADP^+$-malate dehydrogenase in C_3-plants: regulation and the role of a light-activated enzyme. *Physiol. Plant.* 71:393–400

91. Scheibe R, Jacquot J-P. 1983. NADP regulates the light activation of NADP-dependent malate dehydrogenase. *Planta* 157: 548–53

92. Schreiber U, Neubauer C. 1990. O_2-dependent electron flow, membrane energization and the mechanism of non-photo-chemical quenching of chlorophyll fluorescence. *Photosynth. Res.* 25:279–93

93. Schubert K, Liese F, Rumberg B. 1990. Analysis of the variability of the H/e stoichiometry in spinach chloroplasts. See Ref. 5, pp. 279–82

94. Schuller KA, Randall DD. 1989. Regulation of pea mitochondrial pyruvate dehydrogenase complex: Does photorespiratory ammonium influence mitochondrial carbon metabolism? *Plant Physiol.* 89: 1207–12

95. Sechley KA, Yamaya T, Oaks A. 1992. Compartmentation of nitrogen assimilation in higher plants. *Int. Rev. Cytol.* 134:85–163

96. Sharp RE, Matthews MA, Boyer JS. 1984. Kok effect and the quantum yield of photosynthesis. *Plant Physiol.* 75:95–101

97. Singh P, Naik MS. 1984. Effect of photosynthesis on dark mitochondrial respiration in green cells. *FEBS Lett.* 165:145–50

98. Steiger H-M, Beck E. 1981. Formation of hydrogen peroxide and oxygen dependence of photosynthetic CO_2 assimilation by intact chloroplasts. *Plant Cell Physiol.* 22: 561–76

99. Stitt M. 1987. Fructose 2,6-bisphosphate as a regulatory molecule in plants. *Annu. Rev. Plant Physiol. Plant Mol. Biol.* 41:153–85

100. Stitt M, Lilley RM, Heldt HW. 1982. Adenine nucleotide levels in the cytosol, chloroplast and mitochondria of wheat leaf protoplasts. *Plant Physiol.* 70:971–77

101. Stitt M, Wirtz W, Gerhardt R, Heldt HW, Spencer C, et al. 1985. A comparative study of metabolite levels in plant leaf material in the dark. *Planta* 166:354–64

102. Van Quy L, Foyer C, Champigny M-L. 1991. Effect of light and NO_3- on wheat leaf phosphoenolpyruvate carboxylase activity leaves. *Plant Physiol.* 97:1476–82

103. Veech RL, Eggleston LV, Krebs HA. 1969. The redox state of free nicotinamide-adenine- dinucleotide phosphate in the cytoplam of rat liver. *Biochem. J.* 115:609–19

104. Wang Y-H, Chollet R. 1993. In vitro phosphorylation of purified tobacco-leaf phosphoenolpyruvate carboxylase. *FEBS Lett.* 328:215–18

105. Weger HG, Birch DG, Elrifi IR, Turpin DH. 1988. Ammonium assimilation requires mitochondrial respiration in the light: a study with the green alga *Selenastrum minutum*. *Plant Physiol.* 86:688–92

106. Weger HG, Turpin DH. 1989. Mitochondrial respiration can support NO_3- and NO_2- reduction during photosynthesis. *Plant Physiol.* 89:409–15

107. Winter H, Robinson DG, Heldt HW. 1994. Subcellular volumes and metabolite concentrations in spinach leaves. *Planta* 193: 530–35

108. Wiskich JT, Bryce JH, Day DA, Dry IB.

1990. Evidence for metabolic domains within the matrix compartment of pea leaf mitochondria. Implications for photorespiratory metabolism. *Plant Physiol.* 93:611–16

109. Woo KC, Jokinen M, Canvin DT. 1980. Nitrate reduction by a dicarboxylate shuttle in a reconstituted system from spinach leaves. *Aust. J. Plant Physiol.* 7:123–30

110. Woodrow IE, Berry JA. 1988. Enzymatic regulation of photosynthetic CO_2 fixation in C_3 plants. *Annu. Rev. Plant Physiol. Plant Mol. Biol.* 39:533–94

111. Wu J, Neimanis S, Heber U. 1991. Photorespiration is more effective than the Mehler reaction in protecting the photosynthetic apparatus against photoinhibition. *Bot. Acta* 104:283–91

112. Zeiher CA, Randall DD. 1990. Identification and characterization of mitochondrial acetyl-Coenzyme A hydrolase from *Pisum sativum* L. seedlings. *Plant Physiol.* 94:20–27

113. Zoglowek C, Krömer S, Heldt HW. 1988. Oxaloacetate and malate transport in plant mitochondria. *Plant Physiol.* 87:109–115

Annu. Rev. Plant Physiol. Plant Mol. Biol. 1995. 46:71–93
Copyright © 1995 by Annual Reviews Inc. All rights reserved

GENETIC CONTROL AND INTEGRATION OF MATURATION AND GERMINATION PATHWAYS IN SEED DEVELOPMENT

Donald R. McCarty

Program of Plant Molecular and Cellular Biology, Horticultural Sciences Department, University of Florida, Gainesville, Florida 32611

KEY WORDS: viviparous mutants, ABA-insensitive mutants, *VP1*, *ABI3*, mutants of maize, mutants of *Arabidopsis*

CONTENTS

71

ABSTRACT

The viviparous and germination mutants of maize and *Arabidopsis thaliana* illuminate the mechanism that integrates control of morphogenetic, maturation, dormancy, and germination pathways in seed development. Key elements of this mechanism include (*a*) developmental control of abscisic acid and gibberellin hormone synthesis and perception, (*b*) integration of maturation and anthocyanin pathways in the maize seed, (*c*) functions of the VP1 and ABI3 factors in abscisic acid–regulated gene expression, and (*d*) intrinsic developmental genes that couple seed maturation to the program of embryo morphogenesis. The scarcity of mutants that affect timing or tissue specificity of hormone synthesis in the seed is an important constraint to progress in understanding the role of hormone signals. The interactions among the abscisic acid–insensitive *abi1*, *abi2*, *abi3*, *abi4*, and *abi5* mutants of *A. thaliana* are consistent with multiple pathways of abscisic acid signal transduction in the seed. The maize *Vp1* and *A. thaliana Abi3* genes are functional homologs that mediate a seed-specific abscisic acid response necessary for maturation. The specific roles of these genes in controlling dormancy and anthocyanin synthesis in the seed have diverged since the evolutionary separation of maize and *A. thaliana*. The coupling of anthocyanin synthesis to maturation in maize may have resulted from changes in the downstream *c1* regulatory gene rather than a functional change in VP1. Functional analysis indicates that VP1 is a transcriptional activator of the Em and *C1* genes in maize, although its specific role in abscisic acid signal transduction remains poorly understood. The *lec1* and *fus3* mutants of *A. thaliana* and pleiotropic viviparous mutants of maize may identify intrinsic factors that couple the maturation pathway to embryo morphogenesis.

INTRODUCTION

A Brief Historical Perspective

The significance of maize mutants that cause vivipary, or precocious sprouting of the seed, was recognized almost immediately with the discovery of such mutants in the mid 1920s. The principle codiscoverers, Eyster (15) and Manglesdorf (39), however, gave very different interpretations to this phenotype. Eyster, who first identified the *vp1* and *vp2* genes, originally used the provocative gene name *primitive sporophyte* to reflect his view that vivipary might be a reversion to a more primitive state in plant evolution. In other words, mutants that block or delete seed maturation and the quiescent phase from the plant life cycle would result in a life history found in some primitive plant taxa. Eyster (16) later adopted the more conservative, if less interesting, designation of *vivipary,* evidently in recognition of the fact that vivipary occurs naturally in evolutionarily advanced plant species (e.g. mangrove) as well.

By concluding that viviparous mutants specifically blocked seed maturation, Eyster discounted the distinctive pleiotropic effects of the mutants he studied, largely because he had misinterpreted these traits (16). The absence of anthocyanins in the *vp1* mutant was erroneously attributed to tight linkage to the *r1* anthocyanin gene on chromosome 10L. *vp1* is in fact on chromosome 3 (61). Similarly, he concluded that the albino phenotype of the *vp2* mutant was also the result of a closely linked mutation that could be resolved from vivipary by recombination. Robertson (61) pointed out that the putative recombinants Eyster reported, which separated the pleiotropic characters of these mutants, were not confirmed by progeny tests and were almost certainly the result of heterofertilized seed in which the embryo and endosperm arose from sperm derived from different pollen grains and not the result of recombination. Despite (or possibly because of) these errors, Eyster clearly originated the concept that viviparous mutants are developmental mutants involved in specification of the maturation pathway during embryogeny.

Manglesdorf , in contrast, interpreted the phenotype from a physiological viewpoint (39, 39a). He regarded vivipary as accelerated or precocious germination of the seed rather than as a failure of the embryo to enter developmental arrest. Clearly, preharvest sprouting in maize and other cereals can have a variety of environmental or physiological causes. Manglesdorf suggested that the mutants were more susceptible to these factors. In line with this view he preferred the designation *ge* (for germination) instead of vivipary. Unlike Eyster, Manglesdorf recognized the pleiotropic nature of many of the viviparous mutants.

The discussion of whether viviparous mutants affect specification of the maturation pathway or affect developmental timing continues (28, 49, 50, 70). Both points of view have validity and both modes of action are represented among the available mutants. The challenge is to understand how the mechanisms underlying vivipary are related to the integrated control of the maturation, dormancy, and germination pathways in plant development.

Integration of Morphogenesis, Maturation, Dormancy, and Germination Pathways in Seed Development

Deletion, overlap, or uncoupling of one or more of the developmental phases of seed formation can occur in nature or experimentally. This suggests that the pathways of morphogenesis, maturation, and germination are loosely integrated and largely independent developmental programs. In *Brassica napus,* for example, the gene expression programs associated with late embryogenesis and germination overlap substantially in normal seed development (10). In culture, *B. napus* embryos can be induced to express aspects of embryogeny (cotyledon formation) during germination (18). In most species, somatic embryos and zygotic embryos grown in culture are viviparous, i.e. they complete morphogenesis and germinate into plantlets without entering an arrested or

Table 1 Viviparous mutants of maize

Gene	Chromosome location	Seed anthocyanins	ABA	Carotenoids	Plant phenotype	References
vp1	3L	deficient	insensitive	normal	normal	55, 61
vp2	5L	normal	deficient	deficient	albino	55, 61
vp5	1S	normal	deficient	deficient	albino	55, 61
ps1 (vp7)	5L	normal	deficient	deficient	albino/pink	55, 61
vp9	7S	normal	deficient	deficient	albino	55, 61
al1 (y3)	2S	normal		deficient	pale green	
y9	10S	normal		deficient	pale green	
w3	2L	normal		deficient	albino	
vp8	1L	normal	deficient	normal	small pointed leaves	55, 61
vp10		normal	deficient ?	normal	adherent leaves viable?	67
vp13[a]	10L	normal	?	normal	normal early, lethal at 5–6 leaf stage	DR McCarty, unpublished
dek33[b]	5S	normal-aleurone deteriorates	?	normal?	early seedling lethal	56
vp*-394[c]	unplaced	normal	?	normal	early seedling lethal	B Burr, unpublished
vp*-2274	unplaced	normal	?	normal	normal	DR McCarty, unpublished
vp*-2406	possible duplicate factor	normal	?	normal	normal embryo	DR McCarty, unpublished
emb*-8532	3L	normal	?	normal	precociously green, embryo lethal	9

[a]Resembles the nonextant vp6 described by Robertson (61).

[b]Variably abnormal embryo morphology—endosperm collapsed and floury.

[c]Endosperm size is markedly reduced—bears an interesting similarity to the nonextant ge2 mutant identified by Manglesdorf (39).

*Provisional gene names. Although these mutants have distinctive phenotypes, possible allelism to other mutants in the list has not been fully ruled out.

Table 2 Viviparous mutants of *Arabidopsis thaliana*

Gene	Chromosome location	Cotyledon phenotype	ABA	Plant phenotype	References
abi3[a]	3	normal	insensitive	normal	29
lec1		leaf-like trichomes and veination pattern	sensitive	normal	47, 49
fus3	3	like lec1	sensitive	normal	2, 28

[a]The *abi3-1, -2* mutant alleles are nondormant, but not viviparous.

desiccation-tolerant phase. Similarly, the viviparous mutants discussed in this review delete much of the maturation pathway and allow precocious expression of germination functions without altering the course of morphogenesis. Conversely, many of the embryo mutants of maize (65) and *Arabidopsis thaliana* (48) that are blocked at early stages of embryogeny do not become necrotic following desiccation, which indicates that some degree of maturation can occur in the absence of normal morphogenesis. In other species, seed maturation (e.g. in mangrove) and more commonly seed dormancy (as in many cultivated species) are dispensed with entirely.

Desiccation tolerance is typically a maturation-specific characteristic. However, resurrection plants induce a pathway closely related to seed maturation in vegetative tissues and thereby acquire desiccation tolerance throughout their life cycles (1, 5 and references therein). That this strategy of ectopic expression of desiccation tolerance has evolved independently in several higher plant groups suggests that a relatively small number of genetic steps may shuffle the ordering of the maturation and germination pathways. Nevertheless, much of the gene expression and the well-characterized abscisic acid (ABA) and gibberellic acid (GA) hormonal responses associated with maturation and germination are seed-specific in most species (for reviews see 66, 69).

This loosely integrated structure of overlapping pathways indicates that seeds can provide a useful and accessible stage for analyzing mechanisms that coordinate the component programs of plant development. The existing viviparous mutants and germination mutants of maize (Table 1), *A. thaliana* (Table 2), and barley illuminate important elements of this mechanism at several levels: (*a*) developmental regulation of hormone synthesis and perception, (*b*) integration of component pathways in the maturation program, (*c*) mechanisms of hormone-regulated gene expression, and (*d*) a possible higher-order coupling between morphogenesis and maturation. An attempt to incorporate these relationships into a unified framework is summarized in Figure 1.

A glance at Figure 1 should convince the reader that there are significant gaps in our knowledge: We lack informative mutants for several key steps, and a significant number of mutants are not firmly placed in our scheme. Moreover, the degree to which the pathways are conserved in monocots and dicots has yet to be established. This review assesses how far we can go in constructing a model for integration of pathways in seed development, with emphasis on areas where our understanding can be improved.

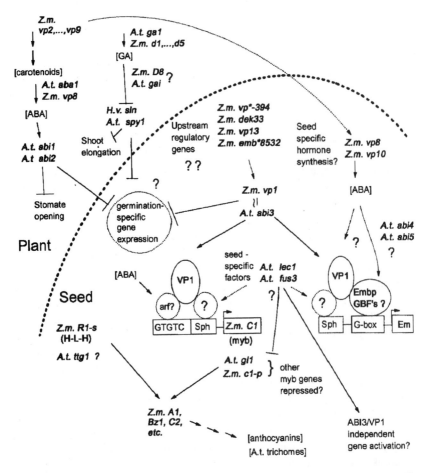

Figure 1 A working model for regulation of maturation and germination pathways in the seed. The dashed curve separates seed-specific processes from pathways that also function in vegetative organs of the plant. Gene activation or positive regulatory interactions are indicated by arrowheads. Negative regulation or inhibitory interactions are indicated by blunt end lines. The relevant species for each mutant or gene are indicated as follows: maize (*Zea mays*): Z.m.; *A. thaliana*: A.t.; barley (*Hordeum vulgare*): H.v. The other nongene designations include arf: abscisic acid regulated factor; GBF: G-box binding factor; H-L-H: helix-loop-helix DNA-binding protein; Embp: Em1a-binding-protein-1. Question marks and dotted lines indicate unknown or hypothetical factors or interactions. Detailed descriptions and references for the mutants and factors shown are cited in the tables and in text.

DEVELOPMENTAL CONTROL OF HORMONE SIGNALS

The key hormonal regulators of maturation, dormancy, and germination are abscisic acid and gibberellic acid. Many of the maize and *A. thaliana* mutants that are altered in synthesis and ability to perceive these hormones have maturation and germination phenotypes. From a developmental point of view we are most interested in what these mutants can tell us about (*a*) temporal and spatial regulation of hormone synthesis and (*b*) pathways of hormone perception and signal transduction.

ABA-Synthesis Mutants

ABA-deficient mutants have been particularly useful in determining the sources of ABA synthesis that are important for regulation of seed maturation and dormancy. From analysis of hormone levels and germination rates of F1 progeny seed derived from reciprocal crosses between *aba1* (ABA-deficient) mutants and wild-type parents, Karssen, Koornneef & coworkers (27, 29) concluded that although maternal tissues are a significant source of ABA present in the seed early in development, the endogenous hormone synthesized in the embryo is necessary and sufficient for seed dormancy in *A. thaliana*.

That maternal ABA has a significant role in regulating maturation of *A. thaliana* seed is indicated by an observation made by Koornneef et al (30) in constructing the *aba1, abi3-1* (ABA-insensitive) double mutant. As single mutants, *aba1* and *abi3-1* are nondormant but produce desiccation-tolerant seed, indicating that seed maturation is not strongly affected. Double mutant seed derived from selfing the double heterozygote are also nonviviparous. In contrast, the seed obtained in subsequent generations by selfing the resulting homozygous double mutant plants are viviparous and desiccation intolerant. The explanation offered for this difference is that in the first generation (F2) double mutant seed develop in the presence of maternal tissues that are heterozygous for *aba1* and thus capable of synthesizing ABA (30). The differential importance of maternal ABA on maturation and seed dormancy is consistent with evidence that maternally supplied ABA peaks relatively early in *A. thaliana* seed development, whereas seed dormancy is established late in seed formation well after the onset of maturation (27, 29).

Maternal ABA is evidently not sufficient to induce seed maturation in maize because viviparous seed clearly segregate on self-pollinated ears of plants that are heterozygous for various ABA-deficient mutants (54, 55). The endosperm can also be ruled out as a significant source of ABA in maize. Robertson (60, 61) used B-A translocation chromosomes that nondisjoin during microspore development to construct seed with nonidentical embryo and endosperm genotypes and thereby showed that *vp5* and *vp8* embryos are viviparous even in the presence of a genetically normal endosperm.

The suppressing effect of maternal ABA may be one reason why comparatively few viviparous mutants are known in *A. thaliana*. Another factor may be that many of the mutant screens specifically directed at finding ABA-response mutants in *A. thaliana* have been based on selection of M2 (second-generation plants) seed (31, 32, 53). This selection is biased strongly against recovery of desiccation-intolerant mutants because only homozygous mutants that can germinate following normal seed desiccation would be recovered.

Regulation of ABA Synthesis

The ABA-deficient mutants have, unfortunately, given us less insight into regulation of the biosynthetic pathway of hormone synthesis in the seed (71). Notably absent are mutants that affect organ specificity or timing of hormone synthesis. The single ABA-deficient mutant known in *A. thaliana* (*aba1*) affects both plant and seed hormone levels (31). In maize, most of the characterized mutants are also affected in carotenoid biosynthesis, which suggests that early steps in synthesis of ABA precursors are blocked (54). Much less is known about ABA synthesis in the class of viviparous mutants that have normal carotenoid pigmentation (Table 1). Candidates for mutants in late biosynthetic steps include *vp8* (54) and possibly *vp10* (67). Although both *vp8* and *vp10* affect vegetative development to some degree, neither mutant has the wilty plant phenotype that is characteristic of ABA mutants of *A. thaliana* (31). The possibility that these or other viviparous mutants of maize or *A. thaliana* affect hormone synthesis specifically in the seed is worthy of further study.

GA Synthesis in the Seed

The gibberellic acid–deficient (*ga1*) mutant of *A. thaliana* affects both plant and seed hormone levels. The requirement of *ga1* mutant seed for exogenous GA clearly establishes a role for the hormone in regulation of dormancy (29, 31). The suppression of the external GA requirement in the *ga1 aba1* double mutant supports the view that a balance of ABA and GA rather than absolute levels in the seed control dormancy in *A. thaliana*.

Although the role of GA as a regulator of hydrolase genes in the aleurone and scutellum of other cereals has been studied extensively, its function in germination of the maize seed remains elusive. In at least some strains of maize, hydrolase expression in aleurones of germinating seed occurs independently of exogenous GA (23). The observation that aleurones of GA-deficient dwarfs exhibit a somewhat greater GA-dependent alpha-amylase activity (23) has been taken as evidence that high endogenous levels of GA or greater sensitivity in the mature maize seed may obviate a need for new GA synthesis during germination. GA synthesis may be most critical for regulation of germination in species that have an imposed seed dormancy.

As in the case of the ABA-synthesis mutants, the GA dwarf mutants have illuminated the biochemical pathway (but less so the mechanisms) by which hormone synthesis is developmentally regulated. Although this situation should improve as greater effort is made to deliberately target regulatory steps in these pathways, one reason such mutants might remain elusive is duplication and functional redundancy of genes in the pathway. We have already discussed, for example, the partial overlap in the roles of maternal- and embryo-synthesized ABA in seed development. Gene duplication followed by functional diversification is emerging as a common theme in the evolution of developmentally regulated pathways in plants. Notably, the enzyme ACC synthase, which controls a key committed step in ethylene biosynthesis, is encoded by a complex differentially regulated gene family in plants (36). We should not be surprised if this also turns out to be the case for ABA and GA biosynthetic pathways. If there is substantial overlap in expression of family members, single mutant phenotypes may be difficult to detect. In this respect, it is interesting that in his early studies Manglesdorf (39a) reported viviparous mutants of maize with 15 to 1 and higher F2 segregation ratios indicative of duplicate and (less convincingly) triplicate factors. Atlhough the Manglesdorf collection has been lost, possibly similar mutants, including a potential duplicate factor, have been recovered recently (Table 1). Alternatively, the existence of duplicate factors in maize may be a vestige of its allotetraploid ancestry (26) and not necessarily indicative of a more broadly distributed gene duplication.

GA-Response Mutants

Two mutants that affect GA signal transduction pathways in the seed include the *slender* (*sln*) mutant of barley (7, 33) and the *spindly* (*spy*) mutant of *A. thaliana* (57). The *sln* mutant causes GA-independent shoot elongation as well as constitutive expression of GA-regulated hydrolase genes in the developing barley aleurone (7, 33). In at least some genetic backgrounds, the *sln* mutant has a viviparous phenotype (P Chandler, personal communication; U Hoecker, DR McCarty, unpublished data), which suggests that germination may be accelerated in this mutant. The antagonistic interaction between ABA and GA evidently occurs downstream of the SLN product, because ABA is still able to inhibit alpha-amylase expression in the *sln* mutant (33). The *spy* mutant of *A. thaliana* was recovered in a screen for seed able to germinate in the presence of the GA-synthesis inhibitor paclobutrazol (57). Like *sln*, *spy* mutant plants are highly elongated and have an appearance similar to plants treated with GA. Because the loss-of-function phenotype is activation of pathways that are normally induced by GA, *sln* and *spy* are suggested to encode negative regulatory factors that are common to both vegetative and seed GA response pathways. Olszewski & Jacobsen (57) suggest, however, that *sln* and *spy* may have different functions in the pathway beause unlike *sln*, *spy* mutants remain partially sensitive to exogenous GA.

ABA-Insensitive Mutants

The mutants that affect ABA sensitivity of the seed include *vp1* of maize (47, 63, 64) and *abi1*, *abi2*, and *abi3* of *A. thaliana* (32). The *A. thaliana* mutants are able to germinate on media containing inhibitory concentrations of ABA (32). The *abi1* and *abi2* mutants also affect ABA sensitivity and water relations of vegetative tissues, whereas *abi3* is seed specific. The *abi1* mutant is unique in being dominant over the wild-type allele. Although all three mutants cause reduced seed dormancy, only the null alleles of *abi3* have a pronounced affect on seed maturation and desiccation tolerance (19, 20, 30, 53).

The differential effects of these mutants on seed maturation may, however, be misleading because, as mentioned above, viviparous or desiccation-intolerant mutations are unlikely to have been recovered in the mutant screens used to isolate the original alleles of these genes. This bias is evident in the *abi3* allelic series. The original *abi3-1* allele, which was recovered as a homozygote from a screen of M2 seed is nonviviparous and desiccation tolerant (30, 32). As discussed above, *abi3-1* is only viviparous in combination with the *aba1* mutant. From this result, Koornneef et al (30) inferred that *abi3-1* is leaky and that null *abi3* mutations would be viviparous. This conjecture has subsequently been confirmed with the isolation of more severe *abi3* alleles that are viviparous as single mutants. The *abi3-3* mutation was recovered from a heterozygous M2 seed in a screen for mutants insensitive to a GA-synthesis inhibitor, uniconazole (53). The unexpected dominant, albeit weak, insensitivity of the *abi3-3* heterozygote to uniconazole implies that reduced dosage of the ABI3 product partially relieves a need for GA synthesis during germination of the *A. thaliana* seed (53). Two similar viviparous alleles, *abi3-4* (20) and *abi3-6* (52), have been analyzed at the DNA level and shown to have lesions that are likely to prevent expression of a functional ABI3 protein.

The *abi3* experience raises the question of whether the existing *abi1* and *abi2* alleles are null mutations, and if not, whether more severe alleles will affect maturation or have more broadly pleiotropic effects. Recent molecular and genetic studies indicate that the dominant *abi1-1* mutant is likely to express a functional protein. Triploid plants generated by pollination of a tetraploid line with *abi1* pollen are ABA insensitive, which indicates that two doses of a wild-type allele are insufficient to overcome the presence of a single dose of the mutant allele (35). This suggests that dominance results from a gain of function or dominant negative inhibitory product and is not a consequence of reduced dosage of the wild-type product in the heterozygote. The *abi1* gene has been cloned by chromosome walking and shown to encode a protein phosphatase 2C homolog that is potentially calcium regulated (35, 51).

The lesion in the *abi1-1* mutant is a point mutation that converts a glycine residue at position 180 of ABI1 to an aspartate. Although the glycine residue per se is not conserved in other protein phosphatase 2C enzymes, it occurs in a highly conserved region of the protein. One possibility is that the mutation induces a constitutive or improperly regulated phosphatase activity (51) that inhibits an ABA-stimulated protein phosphorylation cascade. If that is the case, ABA insensitivity may be conferred by a special class of protein phosphatase 2C alleles (35). By this line of reasoning, a loss-of-function allele could have a hormone-hypersensitive phenotype. An important challenge for researchers will be to establish the "knock out" phenotypes of *abi1* and *abi2* genes.

Double mutant combinations of these genes have been examined to determine whether the *abi1, abi2,* and *abi3* mutants are likely to function in a common pathway (17). The double mutant combination of *abi1* with the nonviviparous *abi3-1* allele has a roughly additive phenotype, which suggests that these genes may function in separate pathways. Epistatic interactions arising from a combination of a dominant and a leaky recessive mutation might, however, be difficult to interpret. The *abi1 abi2* double mutant is not distinguishable in ABA sensitivity from either single mutant parent. Attempts to recover the *abi2 abi3-1* double mutant have failed, possibly because of pollen lethality (17); it is unclear whether this reflects a novel or synergistic interaction between these mutants. Double mutant studies indicate that the recently identified *abi4* and *abi5* mutants function in the same pathway as *abi3* (17a).

DEVELOPMENTAL CONTROL OF ABA SENSITIVITY IN THE SEED

The seed-specific ABA-insensitive mutants of *A. thaliana* and maize, *abi3* and *vp1*, respectively, promise to be particularly informative in furthering our understanding of mechanisms that determine the developmental specificity of the ABA response.

Evidence that Vp1 and Abi3 Are Homologous Genes

The similarities in phenotype of the null *abi3-3* mutation and the *vp1* mutation of maize indicate that these genes might be functionally homologous (53). The structural similarities between the *Vp1* (47) and *Abi3* (20) genes lend support to this idea.The two genes have essentially the same structural organization, consisting of six exons with nearly identical intron placement. The largest contiguous block of conserved amino acid identity corresponds almost precisely to the region encoded by exons 2–5 (20). There is 95% amino acid identity within the 118-amino acid domain encoded by exons 2–5, and there are no gaps. In contrast, the large first exon is poorly conserved, with only about 25% amino acid identity overall and numerous amino-acid insertions or deletions. This exon, however, includes at least two isolated blocks of sequence identity. In addition, several regions with similar physical properties, including an extensive acidic domain near the N-terminus, can be discerned in the first exon.

The structural similarity does not rule out the possibility that VP1 and ABI3 are paralogous members of a gene family that may have duplicated and acquired specialized functions before the evolutionary separation of maize and *A. thaliana* (20). Evidence from sequence analysis of related genes cloned from rice (24) and barley (DR McCarty, unpublished results) by DNA hybridization is relevant to this point. In each case, a single *Vp1*-like gene was found. The pattern of sequence divergence among the grass genes is markedly similar to, though less extreme than, the ABI3 pattern. The first exons are approximately 45% identical in grasses, compared to 25% overall identity with ABI3 with clusters appearing in similar regions. As is true of ABI3, numerous gaps and/or insertions distinguish the three grass genes in the first exon. The divergence of the four genes is consistent with the evolutionary distance among these species. On that basis, we should not expect to find a second gene in *A. thaliana* that is substantially more closely related to VP1 than ABI3 or vice versa (24). If other more closely related genes exist, they should be detectable by DNA hybridization by virtue of the highly conserved C-terminal domain.

Divergent Functions of VP1 and ABI3

Although molecular evidence indicates that ABI3 and VP1 are orthologous (i.e. true homologs), the gene products may have diverged in function since the evolutionary separation of maize and *A. thaliana*. The *Abi3* gene functions in regulation of seed dormancy in addition to controlling maturation, whereas *Vp1* does not have a detectable postimbibition phenotype. The maize seed is, in any case, normally nondormant. Seed dormancy is present in closely related teosinte species as well as in many cereals (e.g. wheat and barley). Changes or selection at the *vp1* locus may have had a role in the loss of seed dormancy during the domestication of maize from teosinte. A shortening of the duration of VP1 expression during development, rather than a change in protein function, might be sufficient to reduce or eliminate seed dormancy. The second striking difference between *abi3* and *vp1* mutants is the pleiotropic effect of *vp1* on anthocyanin synthesis in seed tissues (61). Natural genetic variation in maize populations at the *c1* locus affects the coupling of anthocyanin regulation to VP1 (25, 43, 44; see below). These observations suggest that this novel integration of the anthocyanin and maturation pathways in maize arose through a change in the downstream regulator rather than a functional change in *Vp1* (25). Because both *A. thaliana* and maize have mutations in these genes and are transformable, there is an excellent opportunity for exploring the conservation of function between ABI3 and VP1 by making reciprocal exchanges.

Functional Analysis of VP1

The molecular analyses of VP1 and ABI3 show a novel protein structure that provides few direct clues to function. Studies showing that VP1 and ABI3 are required for maturation-related gene expression in maize (25, 46, 47, 59, 68) and *A. thaliana* embryos (17, 19, 30, 53), respectively, provide circumstantial evidence that these genes have a role in transcriptional regulation. Consistent with this idea, over-expression of VP1 in maize protoplasts causes a greater than 100-fold transcriptional activation of an Em promoter–GUS reporter gene construct (47). In addition, over-expressed VP1 causes a 5- to 10-fold synergistic enhancement of the ABA response of the Em promoter in maize protoplasts.

Both VP1 and ABI3 protein sequences include domains with physical characteristics typically found in transcriptional activators. These sequences include a large N-terminal acidic region that is present in both VP1 (47) and ABI3 and an amide-rich domain that is more prominent in ABI3 than in VP1 (20). Domain-swapping experiments show that the N-terminal region of VP1 has a potent transcriptional activation function (47). A 121-amino-acid segment of the acidic domain can functionally replace the activation domain of the yeast GAL4 transcription factor. Moreover, deletion of the N-terminal acidic domain of VP1 blocks *trans*-activation of Em-GUS in protoplasts. This *trans*-activation function is partially restored by substituting the core transcriptional activation sequence of the VP16 activator into the N-terminal sequence of VP1.

Although VP1 evidently functions as a transcriptional activator, there is little evidence that VP1 can bind directly to DNA (25, 47). At least three conserved, positively charged domains have been nominated as potential DNA-binding and/or nuclear-localization domains (20, 47). Similarities to known DNA-binding proteins are limited to general physical characteristics of these sequences. Moreover, recombinant VP1 does not bind detectably to the relevant regulatory DNA sequences identified in promoters of VP1-regulated genes (25, 47). McCarty et al (25, 43, 47) have proposed that VP1 is a coactivator protein that is targeted to *cis*-regulatory promoter elements primarily through physical interactions with other transcription factors. The tentative conclusion that VP1 cannot bind DNA by itself would not rule out the possibility that positively charged regions of VP1 or ABI3 contact the DNA helix. The possibility that protein-protein interactions have a principal role in targeting of the VP1-ABI3 activator to promoters of downstream genes suggests that the combinatorial interactions between VP1 and its DNA-binding partners could have important regulatory consequences (43).

A key to deciphering the role of VP1 and ABI3 in ABA signal transduction will be understanding the function of the highly conserved C-terminal domain. Although the level of sequence identity in this region rivals that of other highly conserved proteins (18), no homologies to this region have been found outside of the plant kingdom (4, 20, 24; DR McCarty, unpublished results). The *vp1* mutants that truncate this region in maize are desiccation tolerant (6, 45, 47), which indicates, surprisingly, that most of this domain is not absolutely required for ABA-regulated gene expression. If the essential transcriptional activation functions reside in the first exon, the C-terminal domain may contribute a regulatory function.

INTEGRATION OF MATURATION AND ANTHOCYANIN PATHWAYS IN MAIZE

The distinctive red and purple anthocyanin pigments that decorate the aleurone and scutellum of many strains of maize accumulate during seed maturation. Anthocyanin synthesis in these seed parts is blocked by the *vp1* mutant, suggesting a regulatory linkage between the maturation and anthocyanin pathways (43, 46). Does VP1 interact with *C1* and *R1*, the other key anthocyanin regulatory genes required for seed pigmentation? The C1 protein, which contains a myb-like DNA-binding domain and a transcriptional activation domain (21), is considered to interact directly with the R1 protein (38). R1 has a helix-loop-helix DNA-binding domain. Both proteins presumably must bind to upstream regulatory elements in the promoters of *a1, bz1,* and other structural genes in order for the pathway to be activated (for a review see 14).

The absence of *C1* mRNA in *vp1* mutant seed tissues indicates that VP1 may act upstream of one or both of the anthocyanin regulatory genes rather than acting directly on the structural genes in the pathway. If that is the case, then constitutive expression of the C1 and/or R1 factors in *vp1* mutant cells should be sufficient to activate the downstream structural genes without VP1. This test has been done by transiently expressing C1 and B1 in aleurone cells (the *B1* gene is functionally equivalent to *R1*). Using this approach, Hattori et al (25) showed that over-expression of the C1 protein in *vp1* mutant aleurone cells was both necessary and sufficient to induce the anthocyanin pathway in the complete absence of VP1. In contrast, over-expression of *R1* was not sufficient. Because the R1 protein is essential for expression of anthocyanin structural genes, Hattori et al concluded that the R1 product is probably already present in *vp1* mutant cells and that VP1 is, therefore, not likely to regulate *R1* expression. (Strictly speaking this conclusion should be applied only to the *R1-r* allele, which was present in the materials used in that experiment. Other diverse alleles of the *R1* and *B1* genes that also condition seed pigmentation may be regulated differently.) On the other hand, expression of the *C1* gene limits anthocyanin synthesis in *vp1* mutant cells. These results indicate that VP1 regulates the anthocyanin pathway principally by regulating expression of *C1* in the developing seed.

Consistent with VP1 regulation of *C1* transcription, the peak expression of *Vp1* mRNA precedes the appearance of *C1* transcripts and the onset of anthocyanin synthesis in developing seeds. Also, *C1* mRNA is absent in *vp1* mutant seed tissues (46). Hattori et al showed that over-expression of VP1 in maize protoplasts caused transcriptional activation of a GUS reporter gene driven by the *C1*-promoter region. In this assay, incubation of transformed protoplasts with ABA also caused induction of the reporter gene in the absence of VP1. In an interesting contrast to the response of Em-GUS discussed above, ABA and VP1 do not interact synergistically in activating *C1* in protoplasts.

Interaction of Light, ABA, and VP1

In addition to maturation signals, light and germination also play a role in anthocyanin regulation in the seed (8, 11, 13, 14). The interactions of light, ABA, and VP1 are illuminated by a series of *c1* alleles that differ in developmental regulation. In seeds that carry a dominant, functional *C1* allele, anthocyanin synthesis during seed formation requires an as yet poorly characterized light signal (11, 13). The null *c1-n* (not responsive to light) mutant allele produces a seed that is colorless (i.e. lacks anthocyanins) at maturity and remains colorless throughout germination. Seed that are homozygous for the recessive *c1-p* allele are also colorless at maturity but become pigmented during germination (in the dark) if they have received light exposure at some point in their development (8). In other words, regardless of when during development the light signal is perceived, anthocyanin induction is evidently delayed until after germination. This suggests that developing *c1-p* seed are competent to perceive the light signal but are not competent to respond until germination has begun.

In characterizing the *c1-p* phenotype, Chen & Coe (8) observed that the germination-dependent anthocyanin synthesis was not blocked in the *c1-p vp1* double mutant. In advanced stages of viviparous development, light induction of anthocyanins is also detected in single mutant *C1 vp1* seed, which suggests that the expression of *C1* and *c1-p* alleles during germination does not require VP1 (25, 44). These observations indicate that VP1 is required specifically for induction of *C1* alleles during seed maturation. From that point of view, *c1-p* has been interpreted as a regulatory variant in which regulation of the *c1* locus is uncoupled from VP1 and maturation (25, 44). Molecular evidence in support of this idea comes from the finding that the single DNA polymorphism (absence of a five-base repeat) in the proximal promoter region, which distinguishes *c1-p* from *C1* alleles, occurs within a region that is essential for ABA regulation and VP1 *trans*-activation of C1-sh-GUS in maize protoplasts (25). Direct functional comparisons of the *C1* and *c1-p* promoter-driven GUS reporter genes in the transient expression assay showed that ABA activation of *c1-p* is blocked, whereas *trans*-activation by VP1 is only partially inhibited (25).

Although the outcome of the transient expression analysis is consistent with the idea that a *cis*-element involved in coupling *c1* to a maturation signaling pathway is disrupted in *c1-p,* the basis for the delayed action of the light signal in developing *c1-p* seeds (8) remains a mystery. One possibility is that maturation-related factors repress *c1-p* during seed development. Alternatively, in the absence of VP1- and ABA-regulated activators, light induction may default to germination-specific activators. In either case, light regulation of the *c1* gene is strongly conditioned by maturation and germination signals. The mechanism that integrates these signals has yet to be uncovered.

CLUES TO THE ROLE OF VP1 IN ABA-REGULATED GENE EXPRESSION

The anthocyanin phenotype of *vp1* has long been intriguing because the ABA-deficient viviparous mutants of maize do not generally block anthocyanin pigmentation of the seed (61). Nevertheless, we have seen that the *C1* regulatory gene is activated by both ABA and VP1 in diploid maize protoplasts. The observation that there are no mutations in the *vp1* allelic series that produce the viviparous pigmented phenotype characteristic of ABA-deficient mutants indicates that reduced or altered expression of VP1 and reduced ABA dosage have different regulatory consequences (42). A distinction can be made between mutations that qualitatively alter VP1 structure and function and mutants that cause leaky expression of a normal protein. An important class of *vp1* alleles completely blocks anthocyanin synthesis without causing vivipary (6, 45, 62). Although these mutants usually complete maturation and acquire desiccation tolerance, the storage life of the dry seed is somewhat reduced relative to wild-type. Molecular studies indicate that mutants in this class truncate the highly conserved C-terminal domain of VP1 (6, 45, 47). Suprisingly, although the domain most highly conserved in ABI3 is disrupted, these mutants remain at least partially responsive to ABA (6). The *vp1-mum3* mutant, on the other hand, produces reduced levels of apparently unaltered VP1 protein (6). Unlike the anthocyaninless, nonviviparous alleles, *vp1-mum3* is variable in both pigment intensity and vivipary (6, 42). The pigment intensity appears to be correlated with the degree of developmental arrest.

One possible solution invokes multiple or branched pathways for ABA signal transduction that have different ABA dose-response characteristics. ABA activation of *C1* may have a much lower threshold than other maturation-related responses, in which case leaky synthesis or maternal sources of ABA might induce anthocyanin synthesis but not be sufficient to support maturation. If the low-threshold pathway that may be necessary but not sufficient for maturation specifically requires VP1, *vp1* mutants would block both anthocyanin synthesis and maturation; mutants that reduce ABA synthesis would block maturation but allow anthocyanin synthesis. It may be relevant, in this respect, that growth of *vp1* mutant embryos in culture is inhibited by high doses of ABA (63, 64). Consistent with the possibility of multiple receptors or signaling pathways, the dose-response curve for ABA activation of Em-GUS (pBM113kp) in rice (40, 41) and maize (47) protoplasts is nonsaturable over a broad range of ABA concentrations (10^{-7} to 10^{-3} M). In contrast, ABA activation of C1-sh-GUS in maize protoplasts reaches a maximum at 10^{-5} M ABA (25), which is consistent with the possibility that *C1* is coupled to a single high-affinity saturable system. As we have discussed, the genetics in *A. thaliana* are also consistent with multiple pathways for ABA perception (15, 17).

Although threshold differences can account for the essential features of anthocyanin regulation in viviparous mutants, the scenario does not explain the qualitative effects of C-terminal truncation mutants of *vp1* or the evidence that ABA and VP1 response elements in the *C1* promoter are partially separable (25). McCarty and coworkers have proposed that VP1 functions as a transcriptional coactivator (25, 43, 47). In this view, targeting of the VP1 activator to promoters of the *C1* and *Em* genes would depend on physical interactions with ABA-regulated transcription factors and possibly other DNA-binding factors (see Figure 1). Activation of the *C1* and *Em* genes may involve different DNA-binding partners, and interactions with these factors may require different domains of the VP1 protein (43, 47). Thus, the truncated mutants would retain the acidic activation function and a domain involved in interactions with a factor that binds to the *Em* promoter and other essential maturation genes but not to the *C1* promoter. The implication of multiple ABA-regulated transcription factors is consistent with idea of multiple ABA-signaling pathways. The Embp1 leucine zipper transcription factor, which specifically binds the Em1a (G-box) ABA-response element in the *Em* promoter (22), is a leading candidate for an ABA-regulated partner of VP1. Consistent with the above model, Em1a-like elements are evidently not involved in ABA or VP1 regulation of *C1* (25). This model predicts that a domain required for interaction with G-box binding factors will map to the N-terminal portion of the VP1 protein.

Differential regulation of *Em* and *Cl* is also evident in the synergistic interaction of VP1 and ABA. Over-expression of VP1 in protoplasts causes a synergistic enhancement of the ABA response of the *Em* promoter over the entire ABA concentration range (47). This result suggests that, if the multiple path model is correct, VP1 can participate in both high- and low-threshold signal transduction pathways. In an interesting contrast, there is no apparent synergistic effect on ABA regulation of *Cl* (25).

Finally, not all VP1 partners must be specifically related to ABA signal transduction. For example, the ABA-VP1 response sequence in the *Cl* promoter can be resolved into an ABA-specific domain and a sequence that is essential for VP1 *trans*-activation in the absence of exogenous ABA. The VP1 essential region (designated Sph) includes the RY sequence motif, which has been broadly implicated in seed-specific gene expression (3, 12, 34). This has led to speculation that through combinatorial interactions VP1 could participate in integrating hormonal and intrinsic developmental signals (25, 43), although the results based on expression of VP1 at higher than normal physiological levels could have other interpretations. Emerging (and potentially stronger) genetic evidence for involvement of hormone-independent factors in regulation of seed maturation in *A. thaliana* is described in the next section.

INTRINSIC DEVELOPMENTAL CONTROL OF MATURATION

Although the VP1-ABI3 factor is necessary for the ABA-regulated gene expression that is coupled to late embryogenesis, it is not clear that the ABA-response pathway(s) mediated by this factor are sufficient for developmental control of maturation. In the *lec1* (49) and *fus3* (2, 28) mutants of *A. thaliana*, a viviparous phenotype is associated with a transformation of the embryo cotyledons to more leaf-like structures. Specifically, the mutant cotyledons develop trichomes and a more leaf-like pattern of venation. In contrast to *abi3* and *vp1* mutants, *lec1* and *fus3* embryos remain sensitive to ABA when grown in culture (28, 49). The *lec1 abi3-3* (null) (48) and *fus3 abi3-3* (28) double mutants have roughly additive phenotypes, which suggests that these genes function in separate pathways. This result has been interpreted as evidence that *lec1* and *fus3* function in a hormone-independent regulatory cascade that couples key elements of maturation to embryogenesis (48). The more prosaic alternative that *lec1* and *fus3* function in a seed-specific ABA-synthesis pathway has not been directly ruled out, however. In this vein, the arrest of *fus3* and *lec1* embryo development by ABA in culture could be interpreted as rescue of the mutant phenotype by exogenous hormone. As demonstrated by *vp1* alleles (42), ABA deficiency and insensitivity do not necessarily produce equivalent phenotypes in developing seeds. More detailed analyses of gene expression patterns in the single and double mutant genotypes will determine whether or to what extent the *lec1* and *fus3* factors overlap with the *abi3*-mediated pathway.

The possibility of hormone-independent regulatory factors in *A. thaliana* is consistent with the model for VP1 regulation of *C1* in maize (43). In this model, factors such as LEC1 and FUS3 would be candidates for a Sph binding factor (Figure 1). Other intriguing parallels exist between the anthocyanin pathway of maize and the pathway for trichome development in *A. thaliana*. Both pathways are regulated by myb-type transcription factors (58). Moreover, the maize *R1* gene can complement the *ttg1* mutant of *A. thaliana*, which regulates both trichomes and anthocyanin synthesis (37). In this speculative vein, the repression of trichomes [via repression of the *gl1 myb* gene perhaps (58)] in developing *A. thaliana* cotyledons may be analogous to repression of light-activated *c1-p* during maize scutellum development (8).

In addition to the *A. thaliana* mutants, several profoundly pleiotropic viviparous mutants of maize point to a deeper coupling of maturation to embryo morphogenesis. At least two defective embryo mutants, *dek33* (56) and *emb*-8532* (9), exhibit elements of viviparous development. The *vp*-394* mutant causes a pleiotropic reduction in endosperm size. None of these mutants block anthocyanin synthesis (DR McCarty, unpublished results) and therefore are unlikely to act directly upstream of *Vp1*. A complete model for regulation of maturation will need to find a place for these genes.

Homeotics, Heterochronies, or Just Friends?

An echo of the Eyster-Manglesdorf debate over specification of maturation versus acceleration of development can be detected in modern discussions of the *lec1* and *fus3* mutants of *A. thaliana*. The conversion of cotyledons to more leaf-like structures has been interpreted alternatively as a homeotic transformation of organ identity (49) and as a heterochronic transition to a more advanced developmental time (28). Because of the sequential nature of organ appearance in plant development, it may always be possible to confound the concepts of homeosis and heterochrony as originally defined in animal models. If it is difficult to distinguish these concepts by experiment, should we discard these notions as semantic baggage? Perhaps the value of these ideas, like those of Eyster and Manglesdorf, is not their ability to really explain anything but rather that they have motivated good experiments, which will eventually uncover real mechanisms. For the time being, either point of view will serve.

CONCLUDING COMMENTS

Although viviparous and germination mutants have greatly expanded our understanding of the mechanisms responsible for integrating morphogenesis, maturation, and germination pathways in the developing seed, we are far from having a complete model. Progress is needed in several key areas: (*a*) Identification of genes involved in developmental control of hormone synthesis. Several recently discovered seed-specific viviparous mutants of maize and *A. thaliana* deserve closer scrutiny. (*b*) A better understanding of the ABA and GA signal transduction pathways. In particular, the possibility of multiple pathways for ABA signaling and the relationships between the *A. thaliana* ABA-insensitive mutants need to be resolved. With the cloning of *abi1*, rapid progress in this area can be expected in the near future. Although the structures of VP1 and ABI3 have been known for several years, these novel proteins still hold many mysteries, including the function of the highly conserved C-terminal domain and the identities of the hypothetical partners. (*c*) Identification of upstream regulators of *vp1* and *abi3* and intrinsic developmental factors involved in regulation of maturation. The pleiotropic developmental phenotypes associated with several of the viviparous mutants of maize and *A. thaliana* indicate that hormones may not be the whole story. Following in the long tradition of Eyster and Manglesdorf, it is probably better to work from a wrong or incomplete model than from no model at all.

ACKNOWLEDGMENTS

The author is supported by the Florida Agricultural Experiment Station and grants from the National Science Foundation and USDA. Journal Series No. N-00969.

Literature Cited

1. Bartels D, Hanke C, Schneider K, Michel D, Salamini F. 1992. A desiccation-related Elip-like gene from the resurrection plant *Craterostigma plantagineum* is regulated by light and ABA. *EMBO J.* 11:2771–78
2. Baumlein H, Misera S, Luersen H, Kolle K, Horstmann C, Wobus U, Muller AJ. 1994. The *FUS3* gene of *Arabidopsis thaliana* is a regulator of gene expression during late embryogenesis. *Plant J.* 3:379–87
3. Baumlein H, Nagy I, Villarroel R, Inze D, Wobus U. 1992. Cis-analysis of a seed protein gene promoter: the conservative RY repeat CATGCATG within the legumin box is essential for tissue-specific expression of a legumin gene. *Plant J.* 2:233–39
4. Berge SK, Bartholomew DM, Quatrano RS. 1989. Control of the expression of wheat embryo genes by abscisic acid. In *Molecular Basis of Plant Development,* ed. RB Goldberg, pp. 193–201. New York: Liss
5. Bewley JD.1979. Physiological aspects of

desiccation tolerance. *Annu. Rev. Plant Physiol.* 30:195–238

6. Carson CB. 1993. *Characterization of the structure and expression of the mutant alleles of the viviparous-1 gene of maize.* PhD thesis. Univ. Fla., Gainesville

7. Chandler PM. 1988. Hormonal regulation of gene expression in the "slender" mutant of barley (*Hordeum vulgare* L.). *Planta* 175:115–20

8. Chen S-M, Coe EH. 1977. Control of anthocyanin synthesis by the C1 locus in maize. *Biochem. Genet.* 15:333–46

9. Clark A, Sheridan WF. 1991. Isolation and characterization of 51 embryo specific mutations of maize. *Plant Cell* 3:935–51

10. Comai L, Harada JJ. 1990. Transcriptional activities in dry seed nuclei indicate that timing of the transition from embryogeny to germination. *Proc. Natl. Acad. Sci. USA* 87:2671–74

11. Cone KC, Cocciolone SA, Moehlenkamp CA, Weber T, Drummond BJ, et al. 1993. Role of the regulatory gene *pl* in the photocontrol of maize anthocyanin pigmentation. *Plant Cell* 5:1807–16

12. Dickinson CD, Evans RP, Nielsen NC. 1988. RY repeats are conserved in the 5′-flanking regions of legume seed-protein genes. *Nucleic Acids Res.* 16:371

13. Dooner HK, Ralston E. 1994. Light requirement for anthocyanin pigmentation of C aleurones. *Maize Genet. Coop. Newsl.* 68:74–75

14. Dooner HK, Robbins TP, Jorgensen RA. 1991. Genetic and developmental control of anthocyanin biosynthesis. *Annu. Rev. Genet.* 25:173–99

15. Eyster WH. 1924. A second factor for primitive sporophyte in maize. *Am. Nat.* 58:436–38

16. Eyster WH. 1931. Vivipary in maize. *Genetics* 16:575–90

17. Finkelstein RR. 1993. Abscisic acid-insensitive mutations provide evidence for stage-specific signal pathways regulating expression of an *Arabidopsis* late embryogenesis-abundant (lea) gene. *Mol. Gen. Genet.* 238:401–8

17a. Finkelstein RR. 1994. Mutations at two new *Arabidopsis* ABA response loci are similar to ABI3 mutations. *Plant J.* 5:765–71

18. Finkelstein RR, Crouch ML. 1984. Precociously germinating rapeseed embryos retain characteristics of embryogeny. *Planta* 162:125–31

19. Finkelstein RR, Somerville CR. 1990. Three classes of abscisic acid insensitive mutations of *Arabidopsis* define genes that control overlapping subsets of ABA responses. *Plant Physiol.* 94:1172–79

20. Giraudat J, Hauge BM, Valon C, Smalle J, Parcy F, Goodman HM. 1992. Isolation of the *Arabidopsis ABI3* gene by positional cloning. *Plant Cell* 4:1251–61

21. Goff S, Cone KC, Fromm ME. 1991. Identification of functional domains in the maize transcriptional activator C1: comparison of wild-type and dominant inhibitor proteins. *Genes Dev.* 5:298–309

22. Guiltinan MJ, Marcotte WR, Quatrano RS. 1990. A leucine zipper protein that recognizes an abscisic acid response element. *Science* 250:267–71

23. Harvey BMR, Oaks A. 1974. The role of gibberellic acid in the hydrolysis of endosperm reserves in *Zea mays*. *Planta* 121:67–74

24. Hattori T, Terada T, Hamasuna ST. 1994. Sequence and functional analyses of the rice gene homologous to the maize *Vp1* gene. *Plant Mol. Biol.* 24:805–10

25. Hattori T, Vasil V, Rosenkrans L, Hannah LC, McCarty DR, Vasil IK. 1992. The *viviparous1* gene and abscisic acid activate the *C1* regulatory gene for anthocyanin biosynthesis during seed maturation in maize. *Genes Dev.* 6:609–18

26. Helentjaris T, Weber DF, Wright S. 1988. Duplicate sequences in maize and identification of their genomic locations through restriction fragment length polymorphisms. *Genetics* 118:353–63

27. Karssen CM, Brinkhorst-van der Swan DLC, Breekland AE, Koornneef M. 1983. Induction of dormancy during seed development by endogenous abscisic acid: studies on abscisic acid deficient genotypes of *Arabidopsis thaliana* L. Heynh. *Planta* 157:158–65

28. Keith K, Kraml M, Dengler NG, McCourt P. 1994. *Fusca3:* a heterochronic mutation affecting late embryo development in *Arabidopsis*. *Plant Cell* 6:589–600

29. Koornneef M. 1986. Genetic aspects of abscisic acid. In *A Genetic Approach to Plant Biochemistry, Plant Gene Research,* ed. AD Blostein, PJ King, pp. 35–54. Vienna: Springer-Verlag

30. Koornneef M, Hanhart CJ, Hilhorst HWM, Karssen CM. 1989. *In vivo* inhibition of seed development and reserve protein accumulation in recombinants of abscisic acid biosynthesis and responsiveness mutants in *Arabidopsis thaliana*. *Plant Physiol.* 90:463–69

31. Koornneef M, Jorna ML, Brinkhorst-van der Swan DLC, Karssen CM. 1982. The isolation of abscisic acid (ABA) deficient mutants by selection of induced revertants in non-germinating gibberellin sensitive lines of *Arabidopsis thaliana* L. Heynh. *Theor. Appl. Genet.* 61:385–93

32. Koornneef M, Reuling G, Karssen CM. 1984. The isolation and characterization of abscisic acid-insensitive mutants of *Arabidopsis thaliana*. *Physiol. Plant.* 61:377–83

33. Lanahan MB, Ho TD. 1988. Slender barley: a constitutive gibberellin-response mutant. *Planta* 175:107–14

34. Lelievre J-M, Oliveira LO, Nielsen NC. 1992. 5′-CATGCAT-3′ elements modulate the expression of glycinin genes. *Plant Physiol.* 98:387–91

35. Leung J, Bouvier-Durand M, Morris P-C, Guerrier D, Chefdor F, Giraudat J. 1994. *Arabidopsis* ABA response gene *ABI1:* features of a calcium-modulated protein phosphatase. *Science* 264:1448–52

36. Liang XW, Abel S, Keller JA, Shen NF, Theologis A. 1992. The 1-aminocyclopropane-1-carboxylate synthase gene family of *Arabidopsis thaliana. Proc. Natl. Acad. Sci. USA* 89:11046–50

37. Lloyd AM, Walbot V, Davis R. 1992. *Arabidopsis* and *Nicotiana* anthocyanin production activated by maize regulator-R and regulator-C1. *Science* 258:1773–75

38. Ludwig SR, Habera LF, Dellaporta SL, Wessler SR. 1989. Lc, a member of the maize R1 gene family responsible for tissue-specific anthocyanin production, encodes a protein similar to transcriptional activators and contains the *myc* homology region. *Proc. Natl. Acad. Sci. USA* 86:7092–96

39. Manglesdorf PC. 1926. The genetics and morphology of some endosperm characters in maize. *Conn. Agric. Exp. Stn. Bull.* 279:513–614

39a. Manglesdorf PC. 1930. The inheritance of dormancy and premature germination in maize. *Genetics* 15:462–94

40. Marcotte WR, Bayley CC, Quatrano RS. 1988. Regulation of a wheat promoter by abscisic acid in rice protoplasts. *Nature* 335:454–57

41. Marcotte WR, Russell SH, Quatrano RS. 1989. Abscisic acid-responsive sequences from the Em gene of wheat. *Plant Cell* 1:969–76

42. McCarty DR. 1992. The role of VP1 in regulation of seed maturation in maize. *Biochem. Soc. Trans.* 20:89–92

43. McCarty DR. 1993. The role of the maize viviparous-1 gene in regulation of seed maturation. In *Cellular Communication in Plants,* ed. RM Amasino, pp. 27–36. New York: Plenum

44. McCarty DR, Carson CB. 1990. Molecular genetics of seed maturation in maize. *Plant Physiol.* 81:267–72

45. McCarty DR, Carson CB, Lazar M, Simonds SC. 1989. Transposable element induced mutations of the viviparous-1 gene of maize. *Dev. Genet.* 10:473–81

46. McCarty DR, Carson CB, Stinard PS, Robertson DS. 1989. Molecular analysis of viviparous-1: an abscisic acid insensitive mutant of maize. *Plant Cell* 1:523–32

47. McCarty DR, Hattori T, Carson CB, Vasil V, Vasil IK. 1991. The *viviparous1* developmental gene of maize encodes a novel transcriptional activator. *Cell* 66:895–905

48. Meinke DW. 1985. Embryo-lethal mutants of *Arabidopsis thaliana:* analysis of mutants with a wide range of lethal phases. *Theor. Appl. Genet.* 69:543–52

49. Meinke DW. 1992. A homeotic mutant of *Arabidopsis thaliana* with leafy cotyledons. *Science* 258:1647–50

50. Meinke DW, Franzmann LH, Nickle TC, Yeung EC. 1995. Leafy cotyledon mutants of *Arabidopsis. Plant Cell* 6:1049–64

51. Meyer K, Leube MP, Grill E. 1994. A protein phosphatase 2C involved in ABA signal transduction in *Arabidopsis thaliana. Science* 264:1452–55

52. Nambara E, Keith K, McCourt P, Naito S. 1994. An internal deletion mutant of the *Arabidopsis thaliana Abi3* gene. *Plant Cell Physiol.* 35:509–13

53. Nambara E, Naito S, McCourt P. 1992. A mutant of *Arabidopsis* which is defective in seed development and storage protein accumulation is a new *abi3* allele. *Plant J.* 2:435–41

54. Neill SJ, Horgan R, Parry AD. 1986. The carotenoid and abscisic acid content of viviparous kernels and seedlings of *Zea mays* L. *Planta* 169:87–96

55. Neill SJ, Horgan R, Rees AF. 1987. Seed development and vivipary in *Zea mays* L. *Planta* 171:358–64

56. Neuffer MG. 1992. Location and designation of four EMS induced kernel mutants. *Maize Coop. Genet. Newsl.* 66:39

57. Olszewski NE, Jacobsen SE. 1993. Mutations at the *SPINDLY* locus of *Arabidopsis* alter gibberellin signal transduction. *Plant Cell* 5:887–96

58. Oppenheimer DG, Herman PL, Sivakumaran S, Esch J, Marks MD. 1991. A *myb* gene required for leaf trichome differentiation in *Arabidopsis* is expressed in stipules. *Cell* 67:483–93

59. Paiva R, Kriz AL. 1994. Effect of abscisic acid on embryo-specific gene expression during normal and precocious germination in normal and viviparous maize embryos. *Planta* 192:332–39

60. Robertson DS. 1952. The genotype of the endosperm and embryo as it influences vivipary in maize. *Proc. Natl. Acad. Sci. USA* 38:580–83

61. Robertson DS. 1955. The genetics of vivipary in maize. *Genetics* 40:745–60

62. Robertson DS. 1965. A dormant allele of *vp1. Maize Genet. Coop. Newsl.* 39:104–5

63. Robichaud CS, Sussex IM. 1986. The response of *viviparous1* and wild-type embryos of *Zea mays* to culture in the presence of abscisic acid. *J. Plant Physiol.* 126:235–42

64. Robichaud CS, Wong J, Sussex I. 1980.

Control of in vitro growth of viviparous embryo mutants of maize by abscisic acid. *Dev. Genet.* 1:325–30

65. Sheridan WF, Clark JK. 1993. Mutational analysis of morphogenesis of the maize embryo. *Plant J.* 3:347–58

66. Skriver K, Mundy J. 1990. Gene expression in response to abscisic acid and osmotic stress. *Plant Cell* 2:503–12

67. Smith JD, Neuffer MG. 1992. *Viviparous10:* a new viviparous mutant in maize. *Maize Genet. Coop. Newsl.* 66:34

68. Thomann EB, Sollinger J, White C, Rivin CJ. 1992. Accumulation of group 3 late

embryogenesis abundant proteins in *Zea mays* embryos: roles of abscisic acid and the *Viviparous1* gene product. *Plant Physiol.* 99:607–14

69. Thomas TL. 1993. Gene expression during plant embryogenesis and germination: an overview. *Plant Cell* 5:1401–10

70. Walbot V. 1978. Control mechanisms for plant embryogeny. In *Dormancy and Developmental Arrest,* ed. ME Clutter, pp. 439–73. New York: Academic

71. Zeevart JAD, Creelman RA. 1988. Metabolism and physiology of abscisic acid. *Annu. Rev. Plant Physiol.* 39:439–73

Annu. Rev. Plant Physiol. Plant Mol. Biol. 1995. 46:95–122

CALCIUM REGULATION IN PLANT CELLS AND ITS ROLE IN SIGNALING

Douglas S. Bush

Department of Biological Sciences, Rutgers University, Newark, New Jersey 07102

KEY WORDS: calcium ATPase, cytosolic calcium, ion channel, gibberellic acid, abscisic acid, signal transduction, intracellular mediators

CONTENTS

ABSTRACT

Regulation of cellular Ca^{2+} is an essential cell function that is accomplished by a complex of processes collectively called the Ca^{2+} homeostat. This review discusses some of the progress that has been made in identifying the components of the Ca^{2+} homeostat in plants—pumps, secondary transporters, and ion channels—as well as progress in characterizing the operation of the homeostat in living cells during signal transduction. A current hypothesis for understanding how the Ca^{2+} homeostat influences cell function suggests that the spatial and temporal properties of changes in Ca^{2+} levels induced by stimuli are important. An examination of the Ca^{2+} changes induced by a number of stimuli in plants shows the Ca^{2+} homeostat to be capable of generating changes in Ca^{2+} that have characteristic spatial and temporal properties. Thus, the timing and location of Ca^{2+} changes in the cell, together with changes in

0066-4294/95/0601-0095$05.00

the activity of other cellular mediators, can explain the variety and specificity of cellular responses that are triggered by Ca^{2+}.

INTRODUCTION

Calcium plays a key role in plant growth and development because changes in cellular Ca^{2+}, acting through Ca^{2+}-modulated proteins and their targets, regulate an astonishing variety of cellular processes. The regulatory actions of Ca^{2+} range from control of ion transport to gene expression and are possible because of a homeostatic system that regulates Ca^{2+} levels. The development of this homeostatic system is evolutionarily ancient and probably reflects a biochemical necessity of maintaining low levels of Ca^{2+} in the phosphate-rich environment of the cytosol (101). Evolutionary invention, however, has turned this requirement into a complex system for regulating cellular function through controlled fluctuations in Ca^{2+} levels. Progress in understanding the Ca^{2+} homeostat and its significance for signaling in plant cells has been accelerated by direct measurements of ion channel activities and Ca^{2+} levels in living cells, and by molecular characterization of Ca^{2+} transport proteins. The convergence of these three kinds of information has shown that the cellular machinery for establishing Ca^{2+} homeostasis in plants, although used to diverse ends, is remarkably similar to that found in other eukaryotic cells. This review considers what has been learned recently about the components of the Ca^{2+} homeostat in plant cells with reference to its role in signaling. Because Ca^{2+} is involved in so many cellular functions, Ca^{2+} regulation has been the subject of several recent reviews (e.g. 64, 122, 133; see also 13, 43, 142).

OPERATION OF THE CALCIUM HOMEOSTAT

Distribution of Cellular Ca^{2+}

A contour map of the cell, drawn to show the electrochemical potential of Ca^{2+} ($\mu_{Ca^{2+}}$), would be a cartographer's nightmare. Steep electrochemical gradients for Ca^{2+} exist across the plasma membrane (PM), tonoplast (TN), and probably the endoplasmic reticulum (ER). Gradients across these organelles are normally relatively stable; however, changes may occur during signal transduction. Much smaller gradients are likely to exist between the cytoplasm, plastids, and mitochondria, but these gradients will change with the photosynthetic (119) and respiratory activities (141) of the cell. Even within the cytoplasm and nuclear space, usually considered to have continuous volumes, imaging of Ca^{2+} levels indicates that cytoplasmic Ca^{2+} (Ca^{2+}_{cyt}) is not the same everywhere in the cell (65). Gradients of Ca^{2+}_{cyt} are important for normal cellular function, not only during signal transduction events, but also for regulation of ongoing metabolic processes in the cytoplasm and organelles. Even when the gradients are relatively stable, they are established by a dynamic balance between influx and efflux of Ca^{2+} across each of the cellular membranes.

PLASMA MEMBRANE   The steepest gradient for Ca^{2+} normally exists across the PM. In unstimulated cells, Ca^{2+}_{cyt} levels are maintained at 10^{-7} M, whereas the levels of apoplastic Ca^{2+} for land plants are typically 10^{-4} to 10^{-3} M (11, 76). Electrical potentials across the PM (Ψ_{pm}) are -150 to -200 mV, and the resultant difference in $\mu_{Ca^{2+}}$ ($\Delta\mu_{Ca^{2+}}$), calculated from the Nernst equation (80), is 45–60 kJ · mol^{-1}. This value is comparable to the free energy change for ATP hydrolysis, which is approximately -44 kJ · mol^{-1} in living cells (120).

Cytosolic and extracellular Ca^{2+} are in equilibrium with a much larger pool of calcium that is complexed to polysaccharides, lipids, and proteins. Total exchangeable calcium levels in the cytosol of *Chara corallina* are estimated to be 50 mM (139), whereas proton-exchangeable calcium in the cell wall approaches 10^{-3} M (32). Stimuli that alter Ψ_{pm} or cell wall pH could mobilize cell wall Ca^{2+} and significantly increase $\Delta\mu_{Ca^{2+}}$. Light (6, 156, 158) and auxin (56), for example, both hyperpolarize the plasma membrane and lead to cell wall acidification that could increase $\Delta\mu_{Ca^{2+}}$ by 10–15 kJ · mol^{-1} and alter Ca^{2+} fluxes across the membrane. The magnitude of these fluxes is normally quite large in comparison to the level of Ca^{2+} in the cytosol. Estimates of calcium influx across the PM using $^{45}Ca^{2+}$ vary from approximately 6 to 500 μmol · l(intracellular water)$^{-1}$ · min^{-1} (120, 139). However, even the most conservative estimates indicate that small changes in influx or efflux could greatly alter Ca^{2+}_{cyt}.

VACUOLE   Recent studies have not changed the view of the vacuole as a storage compartment for Ca^{2+}; however, they have increased our appreciation of its role in maintaining Ca^{2+}_{cyt} (120, 139) and as a source of messenger Ca^{2+} during signal transduction (134). $\Delta\mu_{Ca^{2+}}$ across the TN may be calculated to be 28–34 kJ · mol^{-1}. The concentration of vacuolar Ca^{2+}, approximately 10^{-3} M, largely determines $\Delta\mu_{Ca^{2+}}$ (57). As a result, moderate depolarizations in membrane potential, such as those proposed to occur during signal transduction in guard cells (168), will not significantly alter $\Delta\mu_{Ca^{2+}}$. However, larger depolarizations of the TN potential induced by opening a specific Ca^{2+} conductance would effectively set $\Delta\mu_{Ca^{2+}}$ to zero and prevent the vacuole from acting as a releasable Ca^{2+} store (114). Because of the presumed importance of the vacuole as a stimulus-releasable Ca^{2+} store, more information is needed on changes in vacuolar membrane potential that occur during signaling. Large fluxes of Ca^{2+} occur across the TN membrane, similar in magnitude to those occuring across the PM, even in mature, unstimulated cells (e.g. 120, 139).

ENDOPLASMIC RETICULUM   The ER, like the vacuole, is thought to function as a Ca^{2+} store. It may soon be possible to measure directly the concentration of lumenal ER Ca^{2+} (96, 141), although uncertainties about the magnitude of the membrane potential across the ER make it impossible to reliably estimate $\Delta\mu_{Ca^{2+}}$. Several lines of evidence suggest that elevated levels of Ca^{2+} exist in the ER lumen. Ca^{2+} concentrations in isolated ER vesicles from aleurone cells have been measured to be 3–4 μM (28), and based on Ca^{2+}-flux measurements in isolated ER vesicles, it has been argued that lumenal Ca^{2+} does not exceed 50 μM (33). The Ca^{2+} level in the ER of living cells is generally assumed to be much higher (141). Total calcium levels can approach micromolar concentrations in the mammalian sarcoplasmic reticulum (SR), which contains an abundant Ca^{2+}-binding protein, calsequestrin (35). Calsequestrin and its homologs bind Ca^{2+} with a low affinity (Kd = 1 mM) (125). Plant cells contain homologs of these low-affinity Ca^{2+}-binding proteins (83, 91, 98, 118) that localize to the ER. Additional indirect evidence of high Ca^{2+} levels in the ER lumen comes from the observation that proper synthesis and targeting of lumenal ER proteins, a basic function of the ER shared by plant cells, is blocked in yeast (145) and mammalian cells (111) by ionophores or other agents that perturb ER Ca^{2+} levels.

Measurements of Ca^{2+} efflux from ER vesicles indicate that there is a rapid exchange of Ca^{2+} across the ER. $^{45}Ca^{2+}$ efflux from isolated ER membranes loaded with $^{45}Ca^{2+}$ shows an exponential decay with a half-time of approximately 12 min (29). This is similar to the time-course of Ca^{2+} efflux from permeabilized hepatocytes (160) and suggests that the ER level of Ca^{2+} is maintained by the balance between rapid influx and efflux processes. Efflux from the ER of animal cells is stimulated markedly by agents such as inositol tris phosphate (IP3) (8) or cyclic ADP-ribose (107), intracellular mediators involved in signaling pathways. This reflects the role of the ER, or a specialized portion of it (166), as the primary Ca^{2+} store in animal cells (165). Similar stimulation of efflux by IP3 has not been reported for plant cells, and the importance of the ER as a stimulus-releasable Ca^{2+} store in plants is not yet determined.

PLASTIDS AND MITOCHONDRIA   Although there is some uncertainty about the magnitude of $\Delta\mu_{Ca^{2+}}$ across the plastid and mitochondrial outer membranes, it is unlikely to be large in comparison with that across the PM, TN, and ER. Chloroplasts and mitochondria contain millimolar levels of total calcium, but most of this calcium is complexed with phosphate or membrane lipids. Ca^{2+} levels have been estimated to be at micromolar levels in plastids (99). Direct measurement of mitochondrial Ca^{2+} levels in bovine endothelial cells using the Ca^{2+}-sensitive photoprotein aequorin established an upper limit for Ca^{2+} of 200 nM, about twice the cytosolic Ca^{2+} level in these cells (141). Ca^{2+} elevation in plastids and mitochondria relative to the cytosol is attributed to balancing membrane potentials that arise as a result of photosynthesis (100) or respiration (47). Transient $\Delta\mu_{Ca}^{2+}$ may arise, therefore, at a light-dark transition or under hypoxic conditions (39). This transient $\Delta\mu_{Ca}^{2+}$ drives a large Ca^{2+} influx into chloroplasts in the light (100), where the increase in Ca^{2+} is thought to regulate a number of plastid processes, including gene expression (18, 123) and energy transduction (83). The early prediction (100) that light-generated $\Delta\mu_{Ca}^{2+}$ would change Ca_{cyt}^{2+} has been verified in algal cells, where a tide of Ca^{2+} flows in and out of the cytosol as a result of light-driven Ca^{2+} fluxes across the plastid envelope (119). This tide of Ca^{2+} has been suggested as a mechanism for coordinating activities between plastids and mitochondria through the mito-chondrial Ca^{2+}-sensitive NADH oxidase (119).

NUCLEUS   Many of the optical probes commonly used to measure Ca_{cyt}^{2+} are small enough to penetrate the nuclear envelope and report nuclear Ca^{2+} levels. Imaging of Ca^{2+} levels does not show any consistent difference between the nucleus and the surrounding cytosol (65, 157). This is not surprising because pores in the nuclear envelope are apparently large enough to permit diffusive movement of Ca^{2+}. Recent work in mammalian cells indicates that stimulus-in-duced changes in nuclear Ca^{2+} may be greater (10) or smaller (3) than changes in the surrounding cytoplasm. Some of the observed amplifications of Ca^{2+} changes in the nucleus may be an artifact (3); however, attenuation of nuclear Ca^{2+} changes has been observed regularly (7). Independent regulation of nuclear Ca^{2+} seems likely, therefore, and a mechanism based on calcium pumps and IP₃-operated Ca^{2+} channels on the nuclear envelope has been proposed (7). The existence of such a mechanism is important because there is increasing evidence that Ca^{2+} modulates many aspects of nuclear function in animal cells, including DNA transcription and repair (7).

Regulation of Cytosolic Ca^{2+}

A desire to understand the role of Ca^{2+} in signal transduction has oriented much of the recent research toward regulation of Ca_{cyt}^{2+}. A crucial step has been the development of techniques for measuring Ca_{cyt}^{2+} using fluorescent dyes, the photoprotein aequorin, and microelectrodes (138). The operation of the Ca_{cyt}^{2+} homeostat is seen clearly when cells are challenged with metabolic poisons, ionophores, or caged compounds. The caged compounds, which permit the controlled release of Ca^{2+} by photolysis, are the least likely to alter the intrin-sic properties of the homeostat (92, 139). Photolytic release of caged Ca^{2+} in guard cells (70) or in mesophyll protoplasts (55) produces a transient (5–10 min) increase in Ca_{cyt}^{2+}. Both the increase in Ca_{cyt}^{2+} and the return to resting levels are 10–100 times slower than those observed in animal cells (92). The apparently sluggish rise in Ca_{cyt}^{2+} is particularly intriguing because release of the caged Ca^{2+} should be essentially instantaneous. This finding raises the possibility of a positive feedback loop in which an increase in Ca^{2+}_{cyt} induces further Ca^{2+} release from intracellular stores. Calcium-induced calcium release (CICR) has been studied extensively in animal cells and is potentially important for amplifying changes in Ca^{2+}_{cyt} in plants (168).

Challenging cells with nonphysiological agents has shown the plant homeostat, even when apparently sluggish, to be remarkably robust. Cyanide, for example, at concentrations that strongly depolarize the PM, has no effect on Ca^{2+}_{cyt} in wheat aleurone cells (37) or *Neurospora crassa* (120) and causes only small changes in root hairs (57). Even the Ca^{2+} ionophore A23187, at low extracellular Ca^{2+} concentrations, has little effect on Ca^{2+}_{cyt} (59). Abrupt increases in extracellular Ca^{2+} from 0.1 to 1 mM result in a transient Ca^{2+}_{cyt} increase in root hairs followed by a steady-state decrease (57), indicating a negative feedback mechanism. The same treatment produces a modest 2–3-fold steady-state increase in aleurone cells (30, 37) and in guard cells (65).

None of these treatments indicates a clear relationship between Ca^{2+}_{cyt} and any other cellular parameter that has been measured. Depolarizations of Ψ_{pm}, an important regulator of Ca^{2+} homeostasis in animal cells, show no consistent correlation with Ca^{2+}_{cyt} in plants. In root hairs, agents that depolarize Ψ_{pm} cause both increases and decreases in Ca^{2+}_{cyt} (57). For example, increases in extracellular K^+, which depolarize Ψ_{pm}, transiently decrease Ca^{2+}_{cyt} in *Riccia fluitans* (58), as do weak acids, which hyperpolarize the membrane (57). Extracellular K^+ increases ^{45}Ca influx into moss protonema (152) and *C. corallina* (140), but its effects on Ψ_{pm} and Ca^{2+}_{cyt} have not been measured in these cells. Similarly, some evidence for Ψ_{pm} regulation of Ca^{2+}_{cyt} has been obtained in guard cells treated with abscisic acid (ABA). In this case, however, depolarization of Ψ_{pm} blocked increases in Ca^{2+}_{cyt} (150). It is possible that the lack of correlation between Ψ_{pm} and Ca^{2+}_{cyt} in plants is because the proper depolarizing stimuli have not been tested (58) or because other factors, in addition to depolarization, control Ca^{2+}_{cyt}. Nevertheless, the relationship between Ψ_{pm} and Ca^{2+}_{cyt}, if any, is unclear.

A stronger correlation appears to exist between Ca^{2+}_{cyt} and cytoplasmic pH (pH_{cyt}). Nonphysiological agents that acidify the cytosol (37, 57), alkalinize the extracellular solution (31, 37), or alkalinize the vacuole tend to increase Ca^{2+}_{cyt} (57). Natural stimuli that alter Ca^{2+}_{cyt} also frequently alter pH_{cyt} in the same way, although ABA is a notable exception (Table 1). These observations indicate a role for the proton-motive force (pmf) across the PM and TN and perhaps a direct role for vacuolar pH (pH_{vac}) in regulating Ca^{2+}_{cyt}. Such roles are plausible because transporters that mediate efflux of Ca^{2+} from the cytosol are driven wholly or in part by the pmf, as discussed below. In addition, Ca^{2+} channels identified at the TN mediate influx into the cytosol at elevated pH_{vac} (2). Whether the effects of pH on Ca^{2+}_{cyt} result from specific regulation or nonspecific, energetic limitation of transporters remains to be shown. It is also unclear whether changes in pH_{cyt} normally drive changes in Ca^{2+}_{cyt} or vice versa. Nevertheless, the interdependence of H^+ and Ca^{2+} regulation is undoubtedly important in plant cells.

Many natural stimuli induce changes in Ca^{2+}_{cyt}. Although these changes have been measured in relatively few types of cells, they show enormous variability in amplitude, kinetics, and spatial distribution of the change in Ca^{2+}_{cyt} (Table 1). It has been suggested that this variation allows cells to distinguish one kind of

Table 1 Stimulus-induced changes in cytosolic Ca^{2+}

Stimuli	Cell type	Character of Ca^{2+} change	Other cellular mediators	Reference
Touch	Tobacco epidermis	Rapid transient		95, 97
Low temperature	Tobacco epidermis	Rapid transient		96
Red light	Wheat mesophyll protoplasts	Transient increase, then decrease		157
	Oat mesophyll protoplast	Increase	pH_{cyt}	
Auxin	*Zea* coleoptile	Ramping up with oscillation period of 20–30 min	Ψpm, pH_{cyt}	56
	Sinapis root hairs	Steady-state increase/decrease	Depolarization Ψpm, increase in pH_{cyt}	164
	Coleoptile Epidermis	Increase	Decrease in pH_{cyt}	61
Elicitor	Tobacco epidermis	Rapid transients (1–2 min)		96
GA	Barley aleurone protoplast	Slow steady-state increase		68
	Wheat aleurone cells	Rapid steady-state increase		27
Cytokinins	*Funaria* caulonema	Steady-state increase		74
	Paphiopedilum guard cell	Rapid increase	Decrease in pH_{cyt}	88
ABA	*Commelina* guard cell	Repetitive increase, steady-state increase, and no change		65, 116
	Vicia guard cell	Repetitive increase	Ψm	150
	Paphiopedilum guard cell	Small increase	Increase in pH_{cyt}	85
	Barley aleurone protoplast	Rapid decrease	Increase in pH_{cyt}	167
	Zea, Petroselinum coleoptile	Rapid increase	Increase in pH_{cyt}	62
Hypoxia	*Zea* suspension cells	Steady-state increase		39
Gravity	*Zea* coleoptile	Increase	Decrease in pH_{cyt}	62

stimulus from another (8, 55, 97). This suggestion remains to be proven in plants, but the effect of natural stimuli on Ca^{2+}_{cyt} clearly shows the great flexibility of the Ca^{2+} homeostat.

Stimulus-induced changes in Ca^{2+}_{cyt} may be transient, sustained, or oscillatory, and the time required to see the full response varies from several seconds to several hours. Transient increases in Ca^{2+}_{cyt} are induced by stimuli as diverse as touch (i.e. mechanical stimuli), temperature shifts, ABA, auxin, red light, and fungal elicitors (Table 1). Mechanical stimulation (touch or wind) and cold shock result in a spike of Ca^{2+} lasting only a few seconds in whole plants transformed with aequorin (55, 96, 97). Almost equally rapid responses have been found in response to ABA in guard cells (116) and in barley aleurone protoplasts (167). The rapidity and simplicity of the kinetics of the change in Ca^{2+}_{cyt} for these stimuli indicate that the change occurs very early in the transduction pathway.

A second group of stimuli induce a slower, more complex Ca^{2+}_{cyt} response. Auxin, ABA, red light, cytokinin, gibberellic acid (GA), and a fungal elicitor induce increases in Ca^{2+}_{cyt} that develop over a period of several minutes to several hours (Table 1). The effect of red light and auxin on calcium homeostasis is particularly complex. In etiolated oat protoplasts, red light induces a transient elevation and then a decline in Ca^{2+}_{cyt} to levels that are significantly lower than before stimulation (55, 157). This complex pattern of change is not seen when Ca^{2+} levels are raised using caged compounds (55), indicating that more than one parameter of the homeostat is altered by red light. Auxin and ABA also induce complex changes in Ca^{2+}_{cyt}. In root hairs of *Sinapis alba,* auxin induces oscillations in Ca^{2+}_{cyt} with a period of 20–30 min superimposed on a rampwise increase (56). These oscillations in Ca^{2+} are accompanied by oscillations in pH_{cyt} and Ψ_{pm} (56). Oscillations with irregular periods are also induced by ABA in guard cell protoplasts (150). These complex, relatively slow kinetics indicate multiple intervening processes in Ca^{2+} regulation following stimulation. Oscillations in animal cells are often associated with the continued presence of stimuli that produce IP_3 (165). Coincident changes in Ψ_{pm}, pH_{cyt}, or potential second messengers have been measured or are suspected in plants in response to stimuli that induce altered Ca^{2+}_{cyt} (Table 1).

In addition to differences in kinetics of Ca^{2+}_{cyt} change, imaging techniques have shown that Ca^{2+} changes have differing spatial characteristics. Steady-state gradients have been demonstrated in pollen tubes (121), root hairs (59), and aleurone protoplasts (67). The formation of these gradients is possible because of low rates of Ca^{2+} diffusion in the cytosol (159), the fragmented nature of the plant cytoplasm, the nonhomogeneous distribution of Ca^{2+} stores, and Ca^{2+} flux throughout the cell (103). These factors increase the possibility of generating spatial variation in Ca^{2+}_{cyt} when cytosolic Ca^{2+} is perturbed by stimuli. Indeed, localized changes in Ca^{2+}_{cyt} may be required for the correct response to stimuli that control polarized growth (121), polarized secretion in the endosperm, and chloroplast movement (144).

Maps of localized changes in Ca^{2+}_{cyt} have been constructed for only a few stimuli. ABA-induced changes in Ca^{2+}_{cyt} in guard cells were greatest in the regions of the cytosol around the nucleus and vacuole (65), whereas GA-induced changes in Ca^{2+}_{cyt}, which required extracellular Ca^{2+}, were greatest at the periphery of the cell (67). In both cases, the spatial distribution of the Ca^{2+} was thought to reflect the localization of the mobilized store used to raise Ca^{2+}_{cyt}.

In summary, the Ca^{2+} homeostat exhibits complex behavior in response to a variety of stimuli. The parameters controlling Ca^{2+}_{cyt} are poorly understood, but a partial list includes Ψ_{pm}, pH_{cyt}, pH_{vac}, and other intracellular mediators, such as IP_3. There is evidence for negative feedback loops, which damp increases in Ca^{2+}_{cyt}, and for positive feedback loops, which amplify changes. Understanding the molecular basis of the homeostat requires identification of the principal transport proteins together with information on how they are regulated coordinately.

Mechanism of Ca^{2+} Regulation

Cellular Ca^{2+} homeostasis is maintained by an ensemble of Ca^{2+} transport proteins. Functionally, these transporters fall into two classes: those that mediate efflux from the cytoplasm (the Ca^{2+} ATPases and Ca^{2+}/nH^+ antiporters) and those that mediate influx (the Ca^{2+} channels). The use of isolated membrane vesicles and the patch-clamp technique have allowed characterization of the activities of these two classes of transporters in reasonably well-defined in vitro systems. Molecular characterization and understanding of functional significance have lagged behind somewhat. Several excellent reviews have been written on the biochemical and electrical properties of these transport proteins (e.g. 51, 52, 87, 151). The remainder of this section discusses the functional significance of these proteins in the control of cellular Ca^{2+}.

EFFLUX TRANSPORTERS

ATPases Plant cells contain a diverse group of primary ion pumps, the Ca^{2+} ATPases, that mediate Ca^{2+} efflux from the cytosol. These Ca^{2+} ATPases appear to belong to an ancient and evolutionarily diverse group of primary ion pumps, the P-type ATPases, that form a phosphorylated intermediate during transport (106). This phosphorylation, and thus transport activity, is inhibited by vanadate (106). Functional reconstitution and molecular cloning of the Ca^{2+} ATPases from animals and yeast show that they are catalytically active as monomeric peptides of molecular weights ranging from 100,000 to 140,000 (36, 117, 147). The catalytic functions of the Ca^{2+} ATPases are highly conserved, as are the structural motifs responsible for orientation of the protein in the membrane (36).

Two major classes of the Ca^{2+} ATPases have been identified in non-plant organisms. The first type, originally localized on the SR/ER of animal cells and therefore called the ER type, has a molecular mass of 100–120 kDa, is not stimulated by calmodulin (Cam), transports two Ca^{2+} for each ATP hydrolyzed, and like other P-type pumps, is prevented from forming a phosphorylated intermediate by La^{3+} (147). The second type, originally found on the PM of erythrocytes and therefore called the PM type, has a molecular mass of 134–140 kDa, is stimulated by Cam and acidic phospholipids, transports one Ca^{2+} per ATP hydrolyzed, and shows increased levels of the phosphoenzyme in the presence of La^{3+} (28).

In plants, Ca^{2+} transport in isolated membrane vesicles has been attributed to the P-type Ca^{2+} ATPases if it is strictly ATP dependent, inhibited by vanadate, insensitive to protonophores, and shows Ca^{2+}-dependent formation of a phosphorylated intermediate. Calcium transport activities with these characteristics have been found in virtually all membrane preparations from plants, including those from roots of monocots and dicots (50, 63, 127), leaves (72), cotyledons (137), cultured cells (26, 82, 110), coleoptiles (23), and aleurone cells (28, 68). In addition to the above characteristics that are indicative of P-type ATPases, plant Ca^{2+} ATPases have a high affinity for Ca^{2+} (52), are inhibited by low concentrations of erythrosin B (EB) (127, 167, 171), and have a low specificity for ATP as the substrate for formation of the phosphoenzyme and transport (40, 171). The most important transporters for regulating cellular Ca^{2+} are commonly thought to be those with high affinity for Ca^{2+}. Kinetic analysis of ATPases in many cell types has yielded estimates of K_ms for Ca^{2+} of 0.07 μM (137) to 12 μM (110). This range of Ca^{2+} affinities, although broad, encompasses the range of values normally observed in Ca^{2+}_{cyt}. A similar range has been shown for the Ca^{2+} ATPases in a single-cell type. A comparative study of the Ca^{2+} transporters in wheat aleurone cells identified three Ca^{2+} ATPases with K_ms for Ca^{2+} that ranged from 0.15 μM, for an ATPase associated with the TN, to 2 μM, for an ATPase associated with the PM (33).

There is increasing evidence that plant Ca^{2+} ATPases may be divided into groups that are roughly similar to the ER- and PM-types of Ca^{2+} pumps in non-plant cells. In plants, Cam-stimulated transport activity, a characteristic of the PM-type ATPase, has been found principally in the ER (40, 68, 82) and TN (60, 167) and, rarely, with the PM (24, 135). In general, Cam stimulates Ca^{2+} transport rates two- to threefold but only slightly decreases the K_m for Ca^{2+}, a significant difference from non-plant PM-type ATPases (36). Nevertheless, Cam-stimulated transport has many characteristics that are similar to the PM-type ATPases in animal cells. Cam-stimulated transport is associated with phosphoproteins of 120 kDa in carrots (40, 82), 115 kDa in cauliflower (5), and 140 kDa in maize; and the maize protein is recognized by antibodies to the erythrocyte PM-type ATPase (23). Phosphorylation of the 120-kDa and 140-

kDa proteins is increased by La^{3+} and by Cam (22, 40). Transport associated with the 120-kDa protein is also stimulated by acidic phospholipids and is insensitive to specific inhibitors of the mammalian ER-type ATPase (40). These data indicate that the Cam-stimulated Ca^{2+} ATPase is a PM-type, although it is not necessarily associated with the PM in plants (40, 82). Molecular evidence for the existence of intracellular PM-type ATPases has also been obtained. A cDNA, PEA1, with homology to the mammalian PM-type ATPase has been identified in *Arabidopsis thaliana* (82). The protein encoded by this cDNA has a predicted mass of 104 kDa, has 45% identity with mammalian PM-type Ca^{2+} ATPase, and localizes to the chloroplast envelope (83). The function of this putative Ca^{2+} ATPase is not known.

A second group of Ca^{2+} ATPases that are not stimulated by Cam, has also been found in plants. Although the inability of Cam to stimulate transport may be attributable to high levels of endogenous Cam that is tightly associated with the membrane (68, 135) or to proteolytic degradation of the ATPase (135), attempts to remove Cam by repeated washing of membranes does not, in some cases, significantly increase Cam stimulation (171). Based on the lack of Cam stimulation in a Ca^{2+} ATPase associated with the ER and the formation of a 100-kDa phosphoenzyme, similar in mass to the 100-kDa SR ATPase, Giannini et al proposed that ER-type transporters exist in plants (63). A strict homology with the ER-type ATPases, which transport two Ca^{2+} for each ATP hydrolyzed, would seem to be precluded at the PM by the large $\Delta\mu_{Ca^{2+}}$ (120). Nevertheless, Ca^{2+} ATPases that are independent of Cam are commonly found at the PM (27, 34, 72, 127, 136). Moreover, several lines of evidence suggest that the Ca^{2+} ATPase of the PM may function as a Ca^{2+}/nH^{+} transporter moving protons in the opposite direction to Ca^{2+}. This countertransport has been argued on energetic grounds as necessary for ATPases at the PM (120). Miller & Sanders argued for a minimum stoichiometry of two H^{+} for each Ca^{2+} transported, and evidence for this proposal has been obtained for PM Ca^{2+} transport in root hairs (59) and isolated membrane vesicles (136, 137).

Further characterization of the Cam-insensitive ATPases is required to determine if they are ER-type enzymes. Nevertheless, support for this idea has come from the discovery of cDNAs that code for proteins with homology to ER-type Ca^{2+} ATPases in tobacco (129) and tomato (174). The tobacco clone, pH27, appeared to encode the terminal third of an ER-type Ca^{2+} ATPase with 68% identity to an invertebrate Ca^{2+} ATPase (129). The tomato clone, LCA1, appeared to encode a full-length protein with a predicted mass of 116 kDa, eight predicted transmembrane domains, and 50% identity to mammalian SR/ER ATPases (174). Direct proof that these putative Ca^{2+} ATPases transport Ca^{2+} is still necessary.

The importance of Ca^{2+} ATPases in regulating Ca^{2+} in living cells has been inferred from their abundance, intracellular locations, and measured affinities for Ca^{2+}. However, relatively few studies have directly assessed the function of the ATPases in vivo. Currently, ATPase activity can only be modulated through the use of nonspecific inhibitors such as La^{3+}, vanadate, Cam inhibitors, and erythrosin B (EB). La^{3+}, for example, increases Ca^{2+}_{cyt} levels in guard cells (65) and decreases Ca^{2+}_{cyt} in root hairs (58). However, in a variety of experimental systems La^{3+} inhibits not only ATPases (72) but also Ca^{2+}/nH^+ antiporters (153) and may block voltage-operated Ca^{2+} channels (58). Similar difficulties arise in attributing the increase in Ca^{2+}_{cyt} induced by Cam inhibitors (66) and vanadate (37) to an effect on ATPases because Cam inhibitors may affect a variety of transport processes (72) and because vanadate is a general P-type ATPase inhibitor (106). EB is also a general inhibitor of ATPases, but plant Ca^{2+} ATPases at the PM (24) and ER (40) are inhibited at submicromolar levels, whereas other ATPases, such as the H^+ ATPases, require higher concentrations for inhibition (41). In *Sinapis alba* root hairs, EB (0.5 μM) by itself caused a small hyperpolarization of the membrane potential and had no effect on Ca^{2+}_{cyt} (59). However, in the presence of the Ca^{2+} ionophore A23187, EB blocked Ca^{2+} regulation and led to a large increase in Ca^{2+}_{cyt}. These data point to an important but not exclusive role for the ATPases in regulating Ca^{2+}_{cyt} (59).

During signal transduction the Ca^{2+} ATPases are presumed to function primarily to restore Ca^{2+}_{cyt} to prestimulus values. Because of difficulties in measuring ATPase activities in living cells, demonstrations of in vivo regulation of the ATPases are scarce. The rapid decrease in Ca^{2+}_{cyt} that is induced by ABA in aleurone protoplasts has been attributed to activation of the Ca^{2+} ATPases, but the mechanism of this putative activation is unknown (167). Membranes isolated from plants treated with auxin (IAA) (175), far-red light (46), GA (28, 68), and mechanical stimulation (17) have altered Ca^{2+} transport capacities when compared to untreated controls, and in the case of IAA (175), GA (68), or mechanical stimulation (17), the increase in transport has been related to activation by Cam.

The importance of the ATPases as a negative feedback system that resists changes in Ca^{2+}_{cyt} has been demonstrated in yeast by mutations in the *PMC1* gene, which encodes a PM-type ATPase that is found on the yeast TN membrane (44). *pmc1* mutants will grow on standard Ca^{2+}-containing media but show large reductions in the pool of sequestered Ca^{2+}. These mutants fail to grow, however, under conditions that increase Ca^{2+} influx into the cells. This growth inhibition results from activation of a Ca^{2+}-dependent protein phosphatase, calcineurin (44). Growth inhibition can be blocked by mutations in *CDM1*, a yeast Cam. These data indicate strongly that the TN ATPase in yeast is important for regulating Ca^{2+}_{cyt} under conditions of increased Ca^{2+} influx. If Ca^{2+} ATPases function in a similar way in higher plants, it may be difficult to obtain mutants in ATPase function because so many common stimuli, such as touch, wind, and stress, raise Ca_{cyt}^{2+}. The molecular genetic characterization of higher plant ATPases has great promise for unraveling their role in signal transduction.

Antiporters The second major class of efflux transporters found in plants are the Ca^{2+}/nH^+ antiporters. The antiporters are secondary transporters that do not directly require ATP for transport. Antiporters appear to be ubiquitous and highly active in plants (153–155). The antiporters can use either one of the components of a pmf, ΔpH (153) or $\Delta\Psi$ (12), to drive Ca^{2+} uptake. Accordingly, inhibition of Ca^{2+} transport by protonophores but not by vanadate has been commonly used to distinguish the antiporters from ATPases. Alternatively, antiporter activity has been estimated by measuring Ca^{2+}-induced dissipation of pH gradients (4, 15), although this indirect method may overestimate antiporter activity (50). The requirement for a pmf to drive Ca^{2+} transport by the antiporter would presumably restrict it to membranes with a primary H^+ pump. The PM, TN, and Golgi are possible locations for the Ca^{2+}/nH^+ antiporter, although antiporter activity is consistently associated only with the TN, which contains a highly active V-type primary proton pump. Antiporter activity rarely has been associated with the PM (93).

In vitro studies have demonstrated two properties of the antiporter that are important for understanding its role in Ca^{2+} regulation. These properties are a low affinity for Ca^{2+} and a stoichiometry of greater than two H^+ per Ca^{2+} transported. Kinetic analyses of the antiporter activity in membrane preparations from oat, wheat, carrots, barley, and red beet have estimated the K_m for Ca^{2+} to be 10 to 67 μM (12, 26, 33, 50, 153). These values are at least an order of magnitude higher than those for the Ca^{2+} ATPases. Evidence that the affinity of the antiporter for Ca^{2+} could be increased by Cam has been reported (4). In this case, direct measurements of Ca^{2+} transport were not made, and the effect of Cam could, therefore, be attributed to an increase in proton conductance not related directly to Ca^{2+} transport (50, 104). In a comparison of the efflux transporters in wheat aleurone cells, the K_m for Ca^{2+} of an antiporter associated with the TN was 100 times greater than that for a Ca^{2+} ATPase associated with the same membrane (33). However, the antiporter operated at a much higher maximal rate than did the ATPase, so that at 1 μM Ca^{2+}, both transported similar amounts of Ca^{2+} (33).

The precise stoichiometry of the number of H^+ exchanged per Ca^{2+} transported by the antiporter is relevant to the question of how large a $\Delta\mu_{Ca^{2+}}$ the antiporter could establish and whether or not the antiporter, by running in reverse, could serve as a Ca^{2+} influx pathway. Based on the apparent electrogenicity of transport, the stoichiometry of H^+ exchange for Ca^{2+} has been proposed to be one (153), two (15), or three (12). Assuming that the antiporter was solely responsible for Ca^{2+} accumulation by the vacuole, at least two H^+

would be required to account for the magnitude of the gradients normally established across the TN (12). Although this assumption may not be valid because Ca^{2+} ATPases are also associated with the TN (33), a stoichiometry of at least three is indicated by data showing that antiporter activity was stimulated by membrane potentials with the vacuolar side positive (12). This stoichiometry would also indicate that the antiporter would not normally run in reverse and serve as a pathway for Ca^{2+} influx into the cytosol.

In vitro characterizations of the antiporter have led to the view that it functions by damping large changes in Ca_{cyt}^{2+} and maintaining the Ca^{2+} store in the vacuole. Direct measurements of Ca_{cyt}^{2+} tend to support this view. Acidification of the cytosol by 0.3 pH units increased Ca_{cyt}^{2+}, and alkalization of the cytosol decreased Ca_{cyt}^{2+} in *Zea mays* root hairs (57). Perhaps more importantly, alkalization of the vacuolar pH by neutral red induced relatively large increases in Ca_{cyt}^{2+} (57). Evidence for a role of the antiporter in buffering Ca_{cyt}^{2+} has been obtained in *Neurospora crassa* under conditions where net Ca^{2+} influx occurs. In this case, the protonophore FCCP induced increases in Ca_{cyt}^{2+} whereas cyanide did not (120). Similar results have been obtained through the analysis of Ca^{2+}-sensitive mutants in yeast (126), which are deficient in the vacuolar H^+ ATPase. Analysis of these mutants points to the importance of the vacuolar proton gradient regulating Ca_{cyt}^{2+}. The effects of altering pH on Ca_{cyt}^{2+} are consistent with an important role for the antiporter in homeostasis, but they are not unequivocal. Other transporters, including the ATPase and Ca^{2+} channels (2), may also be affected by pH changes. A Ca^{2+}/H^+ ATPase that is inhibited by protonophores has been characterized in *Dictyostelium discoidecum* (143) and in TN membranes from wheat aleurone (167). Moreover, it has been argued that the principal substrate of the Ca^{2+}/nH^+ antiporter in living cells is Mg^{2+}, not Ca^{2+} (130). Direct and specific modulation of the antiporter is not yet possible because inhibitory compounds such as Ruthenium Red, La^{3+}, and DCCD (153), which inhibit the antiporter, also inhibit Ca^{2+} channels and ATPases.

INFLUX TRANSPORTERS   The steep electrochemical gradients for Ca^{2+} that occur within the cell require Ca^{2+} influx into the cytosol to be tightly regulated. Although Ca^{2+} influx could occur through a pump or an antiporter operating in reverse, these transporters appear to operate far away from thermodynamic equilibrium and are likely to function only in export of Ca^{2+}. The principal routes of Ca^{2+} entry are likely to be through Ca^{2+} channels.

The speed of Ca^{2+} flux through channels (80), together with sensors in the channel that determine the open or closed state, permit fine control over the kinetic and spatial properties of Ca^{2+} influx. Ca^{2+} channels in animal cells are extraordinarily diverse in intracellular location and in stimuli that lead to opening or closing of the pore (165). Ca^{2+} channels may be classified by their intracellular location: Those found on the PM are classified as influx channels; those found on an intracellular membrane are release channels. Another classification, based on the conditions that determine the open or closed state of the channel, recognizes five classes of Ca^{2+} channels (165). Three of these—voltage-operated, second-messenger-operated, and mechanically operated—have been identified in plants.

Voltage-operated channels   Some of the earliest evidence that Ca^{2+} was involved in regulation of metabolism and in signal transduction in plant cells was obtained through the use of pharmacological agents that, in animal cells, block or promote the opening of voltage-operated Ca^{2+} channels. These pharmacological agents include dihydropyridines such as nifedipine, phenylalkamines such as verapamil and bepridil, and benzothiazepines such as diltiazem. These agents block Ca^{2+} entry by binding to various regions of the $\alpha 1$ subunit of voltage-operated channels (165).

Verapamil was first shown to block cytokinin-induced, calcium-dependent nuclear migration and cell division in *Funaria hygrometrica* protonema (146). In a separate study, $^{45}Ca^{2+}$ influx into *Physcomiterella patens* protonema was increased by cytokinins and also by extracellular K^+, which presumably depolarized the membrane (152). Verapamil, nifedipine, and diltiazem all blocked the K^+-induced Ca^{2+} influx with half maximal inhibition between 0.1 and 0.5 μM (152). The ability of phenylalkamines to bind to membrane proteins has been used to identify components of putative Ca^{2+} channels (73, 77, 163). These studies have identified a 75-kDa protein that has channel activity in artificial lipid bilayers (163).

Studies with Ca^{2+}-channel blockers point to the existence of voltage-operated Ca^{2+} channels at the PM, but direct electrical measurements of channel activity are scarce. The need for ion channel measurements is accentuated by the finding that Ca^{2+}-channel blockers can block K^+ channels in plants (94, 161) and by the observation that Ca^{2+} can permeate K^+ channels (54, 150). Indeed, La^{3+} has been observed to block both physiological responses and Ca^{2+} influx in many processes, including spore germination (170), organogenesis (84), cold shock (95), and setting of circadian rhythms (112). Although La^{3+} may exert these effects through a Ca^{2+} channel, La3+ has been shown to block both inward-rectifying (162) and outward-rectifying (94) K^+ channels. In view of these findings, the possibility that significant Ca^{2+} entry occurs through K^+ channels or other nonselective channels cannot be ignored.

In contrast to influx Ca^{2+} channels, direct evidence has been obtained for two types of highly selective, voltage-operated release channels on the TN. Although both of these channels are voltage regulated, they appear to be gated open at different membrane potentials. A Ba^{2+}-permeable channel found on sugar-beet vacuoles was gated open at highly negative vacuolar potentials and conducted large Ba^{2+} currents into the vacuole (128). These channels were blocked by verapamil and by La^{3+} and had a single-channel conductance of 40 pS (in 100 mM Ba^{2+}) (128). A similar voltage-gated channel has been found on the tobacco vacuole (132). This Ca^{2+}-permeable channel was gated open at large negative vacuolar potentials. The function of these channels is not known. Negative potentials inside the vacuole greater than 100 mV could be generated by opening Ca^{2+} channels in the vacuolar membrane (151). The possibility that these channels actually represent a slow-vacuolar channel (79) has been suggested (168). The slow-vacuolar channel is regulated by Ca^{2+}, by Cam, and by positive (vacuolar negative) membrane potentials (9, 79). This channel was originally thought to be an anion channel but has since been shown to be selective for cations and, in guard cell vacuoles, to be permeable to Ca^{2+} (168). This discovery provides one possible mechanism for amplifying Ca^{2+} changes by CICR. In animal cells, studies of CICR are largely dependent on the plant alkaloid ryanodine, which inhibits the CICR channel. The characterization of the putative plant-CICR channel and identification of inhibitors of its activity are essential first steps to understanding the role of CICR in plant cells.

A second type of voltage-operated channel, also found on the TN of sugar beet, is gated open at potentials that are positive inside the vacuole (88). This channel has a conductance of 12 pS in 5–20 mM $CaCl_2$ and is inhibited by Gd^{3+} but not by other channel antagonists. The activity of the channel was increased by increases in vacuolar Ca^{2+} levels in the physiological range (88). A channel with similar voltage and conductance characteristics, described in broad bean vacuoles (2), has been shown to be activated by elevated pH_{vac} but is inactive at normal pH_{vac} (2). A channel with these characteristics provides an explanation of the observed effect of pH_{vac} on Ca^{2+}_{cyt}. Moreover, its sensitivity to Gd^{3+} further suggests that it could be important in generating the Ca^{2+} signals that are blocked by Gd^{3+}, such as that induced by cold shock in tomato (97).

Second-messenger-operated channels   The phosphoinositide pathway is an important, perhaps ubiquitous, second-messenger pathway leading to changes in cytosolic Ca^{2+} (8). One well-established mechanism for the release of internal stores of Ca^{2+} is through activation of IP_3-operated Ca^{2+} channels. Although the existence of physiologically meaningful levels of IP_3 in higher plants may still be questioned (43), the ability of IP_3 to alter cellular Ca^{2+} has been demonstrated. An IP_3-releasable pool of Ca^{2+} was first identified in isolated oat TN vesicles (154). Subsequently, IP_3 was shown to induce Ca^{2+} release from isolated vacuoles (134) and from unidentified organelles in intact cells (55, 70). Characterization of the IP_3-releasable pool in isolated membranes from red beet showed it to be sensitive to submicromolar levels of IP_3 and, like the mammalian

IP_3 channel, to be inhibited by heparin (25). A patch-clamp study has identified Ca^{2+}-permeable channels that are activated by 10^{-7} M IP_3 in isolated red-beet vacuoles (1). These channels had a single-channel conductance of 30 pS in 5 mM Ca^{2+} (vacuolar side), were active only at positive vacuolar membrane potentials, and were not specifically blocked by verapamil or TMB-8 (1). These single-channel properties agreed well with characterizations of the level of IP_3 required by isolated membrane vesicles for activation and sensitivity to channel blockers. It will be important to obtain similar single-channel data for guard cell and etiolated protoplasts that respond to IP_3 with elevations in Ca^{2+}_{cyt}. IP_3 metabolism and its potential role in signaling have been reviewed recently (43).

Stretch-operated channels Ca^{2+}-permeable channels that are activated by tension have now been identified on the PM of several plants (42, 49, 131). Stretch-operated channels have been proposed to be involved in turgor regulation (42), thigmotropic responses (19, 20, 42, 131), aluminum toxicity (48), and responses to temperature (49) and hormones (131). These stimuli have been shown to elevate Ca^{2+}_{cyt}, although like for many of the channels described to date, direct evidence of the role of stretch-operated channels in regulation of Ca^{2+}_{cyt} is lacking. The stretch-operated channels that have been described to date are not highly selective for Ca^{2+} but also allow K^+ to permeate (42). Complex behaviors of adaptation and linkage have been ascribed to these channels (131), which suggests multiple pathways of regulation.

Ca^{2+} AND SIGNALING

Ca^{2+} plays a key role in stimulus-response coupling in all eukaryotic organisms. In plants, Ca^{2+} regulates many cellular functions, including ionic balance (42, 78, 94, 104, 162, 168), motility (172), gene expression (18–20, 105, 123), carbohydrate metabolism (21), mitosis (109), and secretion (176). Although Ca^{2+} is not the only regulator of these processes, any stimulus that induces a complex cellular response will likely involve Ca^{2+} at some point in the transduction pathway. The converse is even more likely. Gross perturbation of Ca^{2+} homeostasis will interfere with cellular responses whether or not they are normally mediated by Ca^{2+}. Establishing the role of Ca^{2+} in a signal transduction pathway requires information about how a stimulus regulates Ca^{2+} levels and what target proteins are activated by the change in Ca^{2+}. A key element in understanding Ca^{2+}-mediated transduction has been the discovery of proteins that serve as the target of Ca^{2+} regulation. The linchpin of many Ca^{2+}-based signaling pathways in animal cells are Ca^{2+}/Cam-dependent kinases or phosphatases that control the phosphorylation state of key effector proteins. Ca^{2+}- (142), Ca^{2+}/Cam- (158, 169), and Ca^{2+}/phospholipid- (124) dependent kinases have been identified in plants, and evidence for a Ca^{2+}-dependent phosphatase has been obtained (107a, 113, 118a). In addition, Ca^{2+}/Cam-regulated ion transporters are essential components of the Ca^{2+}-mediated transduction pathways that have been described to date.

The simplest paradigm for stimulus-response coupling by Ca^{2+} is one in which the transduction chain is unbranched before the changes in Ca^{2+} that lead to the full response. This paradigm predicts that Ca^{2+} alone is necessary and sufficient for inducing a cellular response (86). Examples of stimulus-response couples that appear to fit this paradigm involve Ca^{2+} regulation of cytoskeletal properties such as cytoplasmic streaming (173), coordinate regulation of ion transporters that control cell volume (71, 157), and mechanical stimuli that induce the Cam-like genes (19, 20). The existence of multiple unbranched transduction pathways presents the dilemma of understanding the specificity of the Ca^{2+} response. Because many stimuli induce changes in Ca^{2+}_{cyt} in individual cells, an important question to address is, Why are stimuli all not equally effective at inducing the same responses? One possible solution is in the differences in spatial and temporal organization of the Ca^{2+} signals (8). Data to support this spatio-temporal model in plants are still meager, although differences in Ca^{2+}-dependent protein phosphorylation have been correlated to the nature of Ca^{2+} change (55) and the Ca^{2+} homeostat is clearly capable of generating responses that are highly characteristic of specific stimuli.

Another solution to the question of specificity is the existence of branched transduction chains that produce two or more mediators, both of which are necessary for the response. In a manner analogous to the use of nucleotides to compose codons, combinations of intracellular mediators can be used to encode the response. In this multiple-messenger model, Ca^{2+} acts in branched transduction pathways either in parallel or synergistically with other intracellular mediators. These pathways appear to be used by stimuli that induce complex metabolic or developmental shifts in cell function such as those induced by red light (18, 123) and plant hormones (14, 67).

Signaling Pathways

Much progress has been made in identifying the function of Ca^{2+} in the transduction of a variety of stimuli. Three stimulus-response couples demonstrate the involvement of Ca^{2+} in transduction pathways.

ABA AND CELL VOLUME   The ability of Ca^{2+} to coordinate cellular activities is most apparent in its effect on ionic balance. Many of the identified target proteins for Ca^{2+} and Cam are ion transporters involved in regulation of turgor pressure and cell volume. Because the vacuole contains much of the K^+ and Cl^- of the cell, coordinate regulation of these ions across the TN and PM is required for turgor and volume regulation. Studies of the guard cell have yielded much information on the role of Ca^{2+} in ABA-induced volume changes. Changes in cytosolic Ca^{2+} occur rapidly in guard cell protoplasts, and this change is sufficient to induce stomatal closure (70). The source of Ca^{2+} that leads to a rise in Ca_{cyt}^{2+} is not resolved; however, there is evidence that PM Ca^{2+} channels (150), IP₃-operated channels (70, 115), and voltage-operated slow-vacuolar channels (168) are involved. Changes in Ca_{cyt}^{2+} are intercepted by Ca^{2+}/Cam-regulated ion channels at the TN (168) and PM (113, 151, 168) of the guard cells. Elevated Ca_{cyt}^{2+} appears to coordinate K^+ efflux by inactivation of the inward K^+ channel at the PM (149) and by activation of an outward K^+ channel at the TN (168). The effects of Ca^{2+} on the inward K^+ channel at the PM may be mediated through a homolog of the Ca^{2+}-dependent protein phosphatase calcineurin (113) and may involve the action of heterotrimeric GTP-binding proteins (53). In addition to coordinating K^+ fluxes at the TN and PM, Ca^{2+} also activates two voltage-operated anion channels at the PM (78, 108). These channels, which differ in the kinetics of their inactivation, have been proposed to be responsible for depolarizing Ψ_{pm} (78, 108), thereby allowing K^+ efflux through a Ca^{2+}-insensitive but pH-sensitive outward K^+ channel (14). Although other factors are important for regulating guard cell volume, the role of Ca^{2+} as a messenger that coordinates ion fluxes at different membranes appears to be essential.

GIBBERELLIC ACID AND SECRETION   Gibberellins are involved in the control of many physiological processes in plants, from germination to flowering. At the cellular level, GA action has been studied most extensively in the cereal aleurone layer, where it promotes hydrolase production (90). Considerable progress has been made in understanding the role of Ca^{2+} in the transduction of the GA signal in aleurone cells. GA is perceived at the PM and therefore requires an intracellular mediator for its effects (69, 81). Induction of mRNA for α-amylase by GA does not require extracellular Ca^{2+}, which indicates that Ca^{2+} may not be involved in the earliest stages of GA action leading to expression of the α-amylase gene (45, 89). This view is confirmed by the observation that okadaic acid (OA), an inhibitor of Ca^{2+}-independent phosphatases (75), blocks all of the GA-induced responses, including α-amylase production, ion mobilization, and accelerated cell death (102).

In the wheat aleurone, GA induces a steady-state increase in Ca^{2+} of 300–500 nM (27). This increase is initiated immediately after GA addition and is complete within 30 min (27). A similar increase is observed in barley aleurone protoplasts (67). Potential target proteins for Ca^{2+} have been identified in aleurone cells. These include Cam (68) and ion transporters in the ER (28, 68) and vacuolar (9) membranes that together mediate the GA-induced synthesis of α-amylase, as well as the release of ions and enzymes. Membranes isolated from aleurone layers treated with GA show increased levels of Ca^{2+} transport activity associated with an ATPase on the ER membrane (28, 68). The mechanism of GA stimulation of the Ca^{2+} ATPase is not known, although Cam stimulation of the Ca^{2+} ATPase has been proposed (68). There is also evidence that the secretory apparatus is a target for Ca^{2+} in aleurone cells. Measurements of membrane surface area in barley protoplasts show Ca^{2+}-dependent increases that were attributed to exocytotic events (176). A third potential target protein for Ca^{2+} is the Ca^{2+}/Cam slow-vacuolar channel. This channel is found on the storage protein vacuole of the aleurone cell and is activated by increasing Ca^{2+}_{cyt} above 600 nM (9). The increase in channel activity is further stimulated by exogenous Cam. Because GA increases both Ca^{2+}_{cyt} and Cam in the barley aleurone cell, GA would also regulate the activity of this slow-vacuolar channel. These data indicate that Ca^{2+}-independent processes, associated with protein phosphatase activity, are important in the early events of GA action and that Ca^{2+}-dependent events are activated later or in parallel with them.

RED LIGHT AND GENE EXPRESSION   Calcium is involved in many of the responses of plants to red light that are mediated by phytochrome. These responses range from rapid changes in ion fluxes (16, 157) that control protoplast swelling and organelle movements to changes in gene expression that control organelle biogenesis and plant development (105, 123). An early model, developed to explain the effects of red light and Ca^{2+} on chloroplast movement and spore germination (144), proposed that phytochrome increased Ca^{2+}_{cyt}, which activated Cam-dependent kinases and phosphatases. Some of the essential features of this model have been verified in studies of phytochrome action in higher plant cells. These studies also have shown that the transduction pathway may be branched and may involve intracellular mediators other than Ca^{2+}.

Far-red reversible increases in Ca^{2+}_{cyt} are observed in cereal protoplasts immediately after exposure to red light (38, 157). Following simulation by red light, Ca^{2+}_{cyt} increases to micromolar levels over a period of 5–10 min, after which it declines below the prestimulus level (55, 157). This complex change in Ca^{2+}_{cyt} indicates that red light may affect more than one Ca^{2+} transporter and that mobilization of both intracellular (38) and extracellular (55) Ca^{2+} may be involved in the elevation of Ca^{2+}_{cyt}. Evidence indicates that there are intermediate steps between the formation of active phytochrome and changes in Ca^{2+}_{cyt} in several cell types. Red-light-induced germination of *Dryopteris paleacea* spores requires extracellular Ca^{2+}, but this can be supplied up to 40 h after illumination at a time when the spores have escaped far-red reversibility (148). In wheat protoplasts, GTPγS in the presence of Ca^{2+} could be substituted for the red light to induce protoplast swelling (16). Moreover, in phytochrome-deficient mutants of tomato, injection of GTPγS produces the full phytochrome-dependent red-light response of normal chloroplast development and anthocy-anin production. (18, 123). Changes in Ca^{2+} are not sufficient to induce the full phytochrome response because injection of Ca^{2+} or Cam leads to the induction of many of the enzymes associated with photosystem II, of ATPase subunits, and of partial chloroplast development, but not to the induction of anthocyanin development. Experiments with biochemical rescue of phyto-chrome-deficient mutants provide striking support for the hypothesis that phy-tochrome activates GTP-binding proteins that are in turn involved in regulat-ing Ca^{2+}_{cyt}. They also indicate that phytochrome initiates a branched transduc-tion pathway leading to the Ca^{2+}_{cyt} and another mediator that has been identified as cGMP (18). The existence of these multiple intracellular mediators in the transduction of phytochrome response may prove to be a common feature of Ca^{2+}-mediated signal transduction in plants.

PERSPECTIVE

The Ca^{2+} homeostat of plant cells is broadly similar to that of other eukaryotic cells. Only a few of the components of the Ca^{2+} homeostat—pumps, secon-dary transporters, and channels—have been characterized at the molecular level in plants. Nevertheless, the biochemical properties of Ca^{2+} transporters in isolated plant membranes indicate that Ca^{2+} efflux from the cytosol is mediated by the action of a multiplicity of P-type ATPases localized at each of the major cellular membranes. In addition, plant cells contain a Ca^{2+}/nH^+ anti-porter that is presumed to play a major role in Ca^{2+} sequestration in the vacuole, the main Ca^{2+} store in plants. Future studies are likely to show a redundancy in the mechanisms for regulating Ca^{2+} levels in each of the vari-ous compartments of the plant cell. This predicted redundancy is likely to complicate attempts to characterize the function of each transporter. Neverthe-less, the functions of transporters that have been identified at the molecular or biochemical level must be characterized in vivo in order to understand Ca^{2+} homeostasis.

Although the Ca^{2+} transporters in plants are biochemically similar to those in other organisms, the structural organization of the plant cell requires the homeostat to function in unique ways. For example, the vacuole appears to serve both as a primary source of stimulus-releasable Ca^{2+} and as a primary sink for Ca^{2+} in unstimulated cells. The discovery of an increasing diversity of Ca^{2+} channels and efflux transporters at the TN supports the hypothesis that it is a major regulator of cellular Ca^{2+}. Because the vacuole occupies a large fraction of the cell, spatial gradients in Ca^{2+}_{cyt} are likely to occur through localized release of Ca^{2+}. An important goal of future research will be to understand how properties of the vacuole, such as membrane potential and pH, regulate transporter activity at the TN to generate localized changes in Ca^{2+} in response to stimuli.

Just as the role of the TN in regulating cytosolic Ca^{2+} has become clearer, a role for the PM, long suspected by analogy with signaling in animal cells, is also becoming apparent. A number of stimuli that alter Ca_{cyt}^{2+}, including auxin, ABA, GA, and elicitors, appear to have receptors at the PM and require extracellular Ca^{2+} for stimulus-response coupling. This observation suggests that Ca^{2+} influx across the PM where $\Delta\mu_{Ca^{2+}}$ is largest, has an important role in Ca^{2+} signaling. The role of the PM in maintaining resting levels of Ca_{cyt}^{2+} is less clear. Until suitable methods are developed for measuring and manipulating the activity of the efflux transporters, either genetically or biochemically, their role in signaling and homeostasis can be understood only in general terms.

There is now overwhelming evidence for a role for calcium in regulating cellular functions in response to many stimuli in plants. Three intensively studied stimulus-response couples have been examined in this review. Although this is only a small portion of the responses that involve Ca^{2+} regulation (Table 1), these studies illustrate the diversity in the spatial and temporal organization of the Ca^{2+} change as well as its position, early or late, in the transduction pathway. Understanding how information is encoded in a Ca^{2+} change to dictate and specify a particular response is an important question for future work. There are two currently prevalent hypotheses for explaining specificity: a spatio-temporal model, in which the timing and location of the Ca^{2+} change encode for specificity, and a multiple-messenger model, in which specificity is encoded by the simultaneous action of several intercellular mediators. These models are not mutually exclusive, and evidence for both has been discussed in this review. Indeed, discovering other cellular messengers, such as the cyclic nucleotides, membrane fatty acids, and even pH, that function with Ca^{2+} to control the specificity of the response will be a major step toward understanding the role of Ca^{2+} in signaling in plants.

ACKNOWLEDGMENTS

The author gratefully acknowledges E Crump, SA Hodges, and M Cervantes-Cervantes for their help in preparing this manuscript, and the National Science Foundation for financial support (DCB-9206692).

Literature Cited

1. Alexandre J, Lassalles JP, Kado RT. 1990. Opening of Ca^{2+} channels in isolated red beet root vacuole membrane by inositol-1,4,5-trisphosphate. *Nature* 343:567–70

2. Allen GJ, Sanders D. 1994. Two voltage-

gated, calcium release channels coreside in the vacuolar membrane of broad bean guard cells. *Plant Cell* 6:685–94

3. Al-Mohanna FA, Caddy KWT, Bolsover SR. 1994. The nucleus is insulated from large cytosolic calcium ion changes. *Nature* 367:745–50

4. Andreev IM, Koren'kov V, Molotkovsky YG. 1990. Calmodulin stimulation of Ca^{2+}/nH^+ antiport across the vacuolar membrane of sugar beet taproot. *J. Plant Physiol.* 136:3–7

5. Askerlund P, Evans DE. 1992. Reconstitution and characterization of a calmodulin-stimulated Ca^{2+}-pumping ATPase purified from *Brassica oleracea*. *Plant Physiol.* 100:1670–81

6. Assmann SM, Simoncini L, Schroeder JI. 1985. Blue light activates electrogenic ion pumping in guard cell protoplasts of *Vicia faba* L. *Nature* 318:285–87

7. Bachs O, Agell N, Carafoli E. 1992. Calcium and calmodulin function in the cell nucleus. *Biochim. Biophys. Acta* 1113:259–70

8. Berridge MJ. 1993. Inositol trisphosphate and calcium signalling. *Nature* 361:315–25

9. Bethke PC, Jones RL. 1994. Ca^{2+}-calmodulin modulates ion channel activity in storage protein vacuoles of barley aleurone cells. *Plant Cell* 6:277–85

10. Birch B, Eng D, Kocsis J. 1992. Intranuclear Ca^{2+} transients during neurite regeneration of an adult mammalian neuron. *Proc. Natl. Acad. Sci. USA* 89: 7978–82

11. Bjorkman T, Cleland RE. 1991. The role of extracellular free-calcium gradients in gravitropic signalling in maize roots. *Planta* 185:379–84

12. Blackford S, Rea PA, Sanders D. 1990. Voltage sensitivity of H^+/Ca^{2+} antiport in higher plant tonoplast suggests a role in vacuolar calcium accumulation. *J. Biol. Chem.* 265:9617–20

13. Blatt MR, Thiel G. 1993. Hormonal control of ion channel gating. *Annu. Rev. Plant Physiol. Plant Mol. Biol.* 44:543–67

14. Blatt MR, Thiel G, Trentham DR. 1990. Reversible inactivation of K^+ channels of *Vicia* stomatal guard cells following the photolysis of caged insoitol 1,4,5-trisphosphate. *Nature* 346:766–69

15. Blumwald E, Poole RJ. 1986. Kinetics of Ca^{2+}/H^+ antiport in isolated tonoplast vesicles from storage tissue of *Beta vulgaris* L. *Plant Physiol.* 80:727–31

16. Bossen ME. 1990. *Plant protoplasts as a model system to study phytochrome-regulated changes in the plasma membrane.* PhD Diss. Landbouwuniv. Wageningen, The Netherlands

17. Bourgeade P, De Jaegher G, Boyer N. 1991.

18. Bowler C, Neuhaus G, Yamagata H, Chua N-H. 1994. Cyclic GMP and calcium mediate phytochrome phototransduction. *Cell* 77:73–81

19. Braam J. 1992. Regulated expression of the calmodulin-related TCH genes in cultured *Arabidopsis* cells: induction by calcium and heat-shock. *Proc. Natl. Acad. Sci. USA* 89:3213–16

20. Braam J, Davis RW. 1990. Rain-, wind-, and touch-induced expression of calmodulin and calmodulin-related genes in Arabidopsis. *Cell* 60:357–64

21. Brauer M, Sanders D, Stitt M. 1990. Regulation of photosynthetic sucrose synthesis: a role for calcium? *Planta* 182:236–43

22. Briars SA, Evans DE. 1989. The calmodulin-stimulated ATPase of maize coleoptiles forms a phosphorylated intermediate. *Biochim. Biophys. Acta* 159:185–91

23. Briars SA, Kessler F, Evans DE. 1988. The calmodulin-stimulated ATPase of maize coleoptiles is a 140,000-M_r polypeptide. *Planta* 176:283–85

24. Briskin DP. 1990. Ca^{2+}-translocating ATPase of the plant plasma membrane. *Plant Physiol.* 94:397–400

25. Brosnan JM, Sanders D. 1993. Inositol trisphosphate mediated Ca^{2+} release in beet microsomes is inhibited by heparin. *FEBS Lett.* 260:70–72

26. Bush DR, Sze H. 1986. Calcium transport in tonoplast and endoplasmic reticulum vesicles isolated from cultured carrot cells. *Plant Physiol.* 80:549–55

27. Bush DS. 1992. Regulation of cytosolic calcium by gibberellic acid and abscisic acid in the cereal aleurone. *Plant Physiol.* 99(Suppl.):38

28. Bush DS, Biswas AK, Jones RL. 1989. Gibberellic-acid-stimulated Ca^{2+} accumulation in endoplasmic reticulum of barley aleurone: Ca^{2+} transport and steady-state levels. *Planta* 178:411–20

29. Bush DS, Biswas AK, Jones RL. 1993. Hormonal regulation of Ca^{2+} transport in the endomembrane system of the barley aleurone. *Planta* 189:507–15

30. Bush DS, Jones RL. 1988. Cytoplasmic calcium and α-amylase secretion from barley aleurone protoplasts. *Eur. J. Cell Biol.* 46:466–69

31. Bush DS, Biswas AK, Jones RL. 1988. Measurement of cytoplasmic Ca^{2+} and H^+ in barley aleurone protoplasts. *Plant Cell Tissue Organ. Cult.* 12:159–62

32. Bush DS, McColl JG. 1987. Mass-action expressions of ion exchange applied to Ca^{2+}, H^+, K^+ and Mg^{2+} sorption on isolated

cell walls of *Brassica oleraceae*. *Plant Physiol.* 85:247–60

33. Bush DS, Wang T. 1994. Diversity of calcium transporters in wheat aleurone cells. *Plant Physiol.* 105(Suppl.):149

34. Butcher RD, Evans DE. 1987. Calcium transport by pea root membranes I. Purification of membranes and characteristics of uptake. *Planta* 172:265–72

35. Campbell KP, MacLennan DH, Jorgensen AO, Mintzer MC. 1983. Purification and characterization of calsequestrin from canine cardiac sarcoplasmic reticulum and identification of the 53,000 dalton glycoprotein. *J. Biol. Chem.* 258:1197–1204

36. Carafoli E. 1992. The Ca^{2+} pump of the plasma membrane. *J. Biol. Chem.* 267:2115–18

37. Cervantes-Cervantes M, Bush DS. 1994. Role of a Ca^{2+}/nH$^+$ antiporter in the regulation of cytoplasmic calcium in aleurone layers from wheat. *Plant Physiol.* 105 (Suppl.):149

38. Chae Q, Park HJ, Hong SD. 1990. Loading of quin2 into the oat protoplast and measurement of cytosolic calcium ion concentration changes by phytochrome action. *Biochim. Biophys. Acta* 1051:115–22

39. Chalivendra SC, Bush DS, Sachs MM. 1994. Elevation of cytosolic calcium precedes anoxic gene expression in maize suspension cultured cells. *Plant Cell.* In press

40. Chen F, Ratterman DM, Sze H. 1993. A plasma membrane-type Ca^{2+}-ATPase of 120 kilodaltons on the endoplasmic reticulum from carrot (*Daucus carota*) cells. *Plant Physiol.* 102:651–61

41. Coccuci MC. 1986. Inhibition of plasma membrane and tonoplast ATPases by erythrosin B. *Plant Sci.* 47:21–27

42. Cosgrove DJ, Hedrich R. 1991. Stretch-activated chloride, potassium, and calcium channels coexisting in plasma membranes of guard cells of *Vicia faba* L. *Planta* 186:143–53

43. Coté GG, Crain RC. 1993. Biochemistry of phosphoinositides. *Annu. Rev. Plant Physiol. Plant Mol. Biol.* 44:333–56

44. Cunningham KW, Fink GR. 1994. Calcineurin-dependent growth control in *Saccharomyces cerevisiae* mutants lacking PMC1, a homolog of plasma membrane Ca^{2+} ATPases. *J. Cell Biol.* 124:351–63

45. Deikman J, Jones RL. 1985. The control of α-amylase mRNA accumulation by gibberellic acid and calcium in barley aleurone layers. *Plant Physiol.* 78:192–98

46. Dieter P, Marmé D. 1981. Far-red light irradiation of intact corn seedlings affects mitochondrial and calmodulin-dependent microsomal Ca^{2+} transport. *Biochem. Biophys. Res. Commun.* 101:749–55

47. Dieter P, Marmé D. 1983. The effect of calmodulin and far-red light on the kinetic properties of the mitochondrial and microsomal calcium-ion transport system from corn. *Planta* 159:277–81

48. Ding JP, Badot P-M, Pickard BG. 1993. Aluminum and hydrogen ions inhibit a mechanosensory calcium-selective cation channel. *Aust. J. Plant Physiol.* 20:771–78

49. Ding JP, Pickard BG. 1993. Modulation of mechanosensitive calcium-selective cation channels by temperature. *Plant J.* 3:713–20

50. DuPont FM, Bush DS, Windle JJ, Jones RL. 1990. Calcium and proton transport in membrane vesicles from barley roots. *Plant Physiol.* 94:179–88

51. Evans DE. 1994. Calmodulin-stimulated calcium pumping ATPases located at higher plant intracellular membranes: a significant divergence from other eukaryotes? *Physiol. Plant.* 90:420–26

52. Evans DE, Briars S-A, Williams LE. 1991. Active calcium transport by plant cell membranes. *J. Exp. Bot.* 42:285–303

53. Fairley-Grenot K, Assman SM. 1991. Evidence for G-protein regulation of inward K$^+$ channel current in guard cells of fava bean. *Plant Cell* 3:1037–44

54. Fairley-Grenot KA, Assmann SM. 1992. Permeation of Ca^{2+} through K$^+$ channels in the plasma membrane of *Vicia faba* guard cells. *J. Membr. Biol.* 128:103–13

55. Fallon KM, Shacklock PS, Trewavas AJ. 1993. Detection in vivo of very rapid red light-induced calcium-sensitive protein phosphorylation in etiolated wheat (*Triticum aestivum*) leaf protoplasts. *Plant Physiol.* 101:1039–45

56. Felle H. 1988. Auxin causes oscillations of cytosolic free calcium and pH in *Zea mays* coleoptiles. *Planta* 174:495–99

57. Felle H. 1988. Cytoplasmic free calcium in *Riccia fluitans* L. and *Zea mays* L.: interaction of Ca^{2+} and pH? *Planta* 176:248–55

58. Felle HH. 1991. Aspects of Ca^{2+} homeostasis in *Riccia fluitans*: reactions to perturbations in cytosolic-free Ca^{2+}. *Plant Sci.* 74:27–33

59. Felle HH, Tretyn A, Wagner G. 1992. The role of the plasma-membrane Ca^{2+}-ATPase in Ca^{2+} homeostasis in *Sinapis alba* root hairs. *Planta* 188:306–13

60. Gavin O, Pilet P-E, Chanson A. 1993. Tonoplast localization of a calmodulin-stimulated Ca^{2+}-pump from maize roots. *Plant Sci.* 92:143–50

61. Gehring CA, Irving HR, Parish RW. 1990. Effects of auxin and abscisic acid on cytosolic calcium and pH in plant cells. *Proc. Natl. Acad. Sci. USA* 87:9645–49

62. Gehring CA, Williams DA, Cody SH, Parish RW. 1990. Phototropism and geotropism in maize coleoptiles are spatially correlated with increases in cytosolic free calcium. *Nature* 345:528–30

63. Giannini JL, Gildensoph LH, Reynolds-Niesmann I, Briskin DP. 1987. Calcium transport in sealed vesicles from red beet (*Beta vulgaris* L.) storage tissue. Characterization of a Ca^{2+} pumping ATPase associated with the endoplasmic reticulum. *Plant Physiol.* 85:1129–36

64. Gilroy S, Bethke PC, Jones RL. 1993. Calcium homeostasis in plants. *J. Cell. Sci.* 106:453–62

65. Gilroy S, Fricker MD, Read ND, Trewavas AJ. 1991. Role of calcium in signal transduction of *Commelina* guard cells. *Plant Cell* 3:333–44

66. Gilroy S, Hughes WA, Trewavas AJ. 1987. Calmodulin antagonists increase free cytosolic calcium levels in plant protoplasts in vivo. *FEBS Lett.* 212:133–37

67. Gilroy S, Jones RL. 1992. Gibberellic acid and abscisic acid coordinately regulate cytoplasmic calcium and secretory activity in barley aleurone protoplasts. *Proc. Natl. Acad. Sci. USA* 89:3591–95

68. Gilroy S, Jones RL. 1993. Calmodulin stimulation of unidirectional calcium uptake by the endoplasmic reticulum of barley. *Planta* 190:289–96

69. Gilroy S, Jones RL. 1994. Perception of gibberellin and abscisic acid at the external face of the plasma membrane of barley (*Hordeum vulgare* L.) aleurone protoplasts. *Plant Physiol.* 104:1185–92

70. Gilroy S, Read ND, Trewavas AJ. 1990. Elevation of cytoplasmic calcium by caged calcium or caged inositol trisphosphate initiates stomatal closure. *Nature* 346:769–71

71. Gilroy S, Trewavas A. 1990. Signal sensing and signal transduction across the plasma membrane. In *The Plant Plasma Membrane-Structure, Function and Molecular Biology,* ed. C Larsson, IM Møller, pp. 203–32. Berlin: Springer-Verlag

72. Gräf P, Weiler EW. 1989. ATP-driven Ca^{2+} transport in sealed plasma membrane vesicles prepared by aqueous two-phase partitioning from leaves of *Commelina communis. Physiol. Plant.* 75:469–78

73. Graziana A, Fosset M, Ranjeva R, Hetherington AM, Lazdunnski M. 1988. Ca^{2+} channel inhibitors that bind to plant cell membranes block Ca^{2+} entry into protoplasts. *Biochemistry* 27:764–68

74. Hahm SH, Saunders MJ. 1991. Cytokinin increases intracellular Ca^{2+} in *Funaria*: detection with indo-1. *Cell Calcium* 12:675–81

75. Hardie DG, Haystead TAJ, Sim ATR. 1991. Use of okadaic acid to inhibit protein phosphatases in intact cells. *Methods Enzymol.* 201:469–76

76. Harker FR, Venis MA. 1991. Measurement of intracellular and extracellular free calcium in apple fruit cells using calcium selective microelectrodes. *Plant Cell Environ.* 14:525–30

77. Harvey HJ, Venis MA, Trewavas AJ. 1989. Partial purification of a protein from maize (*Zea mays*) coleoptile membranes binding the Ca^{2+}-channel antagonist verapamil. *Biochem. J.* 257:95–100

78. Hedrich R, Busch H, Raschke K. 1990. Ca^{2+} and nucleotide dependent regulation of voltage dependent anion channels in the plasma membrane of guard cells. *EMBO J.* 9:3889–92

79. Hedrich R, Neher E. 1987. Cytoplasmic calcium regulates voltage-dependent ion channels in plant vacuoles. *Nature* 329: 833–36

80. Hille B. 1992. *Ionic Channels of Excitable Membranes.* Sunderland, MA: Sinauer Assoc. 607 pp.

81. Hooley R, Beale MH, Smith SJ. 1991. Gibberellin perception at the plasma membrane of *Avena fatua* aleurone protoplasts. *Planta* 183:274–80

82. Hsieh W-L, Pierce WS, Sze H. 1991. Calcium-pumping ATPases in vesicles from carrot cells. Stimulation by calmodulin or phosphatidyl serine, and formation of a 120 kilodalton phosphoenzyme. *Plant Physiol.* 97:1535–44

83. Huang L, Franklin AE, Hoffman NE. 1993. Primary structure and characterization of an *Arabidopsis thaliana* calnexin-like protein. *J. Biol. Chem.* 268:6560–66

84. Hush JM, Overall RL, Newman IA. 1991. A calcium influx precedes organogenesis in *Graptopetalum. Plant Cell Environ.* 14: 657–65

85. Irving HR, Gehring CA, Parish RW. 1992. Changes in cytosolic pH and calcium of guard cells precede stomatal movements. *Proc. Natl. Acad. Sci. USA* 89:1790–94

86. Jaffe LF. 1980. Calcium explosions as triggers of development. *Ann. NY Acad. Sci.* 339:86–101

87. Johannes E, Brosnan JM, Sanders D. 1991. Calcium channels and signal transduction in plant cells. *BioEssays* 13:331–36

88. Johannes E, Brosnan JM, Sanders D. 1992. Parallel pathways for intracellular Ca^{2+} release from the vacuole of higher plants. *Plant J.* 2:97–102

89. Jones RL. 1973. Gibberellic acid and ion release from barley aleurone tissue. Evidence for hormone-dependent ion transport capacity. *Plant Physiol.* 52:303–8

90. Jones RL. 1985. Protein synthesis and secretion by the barley aleurone: a perspective. *Isr. J. Bot.* 34:377–95

91. Jones RL, Bush DS. 1991. Gibberellic acid regulates the level of a BiP cognate in the endoplasmic reticulum of barley aleurone cells. *Plant Physiol.* 97:456–59

92. Kao JPY, Harootunian AT, Tsien RY. 1989. Photochemically generated cytosolic cal-

cium pulses and their detection by fluo-3. *J. Biol. Chem.* 264:8179–84

93. Kasai M, Muto S. 1990. Ca^{2+} pump and Ca^{2+}/H^+ antiporter in plasma membrane vesicles isolated by aqueous two-phase partitioning from corn leaves. *J. Membr. Biol.* 114:133–42

94. Ketchum KA, Poole RJ. 1991. Cytosolic calcium regulates a potassium current in corn (*Zea mays*) protoplasts. *J. Membr. Biol.* 119:277–88

95. Knight MR, Campbell AK, Smith SM, Trewavas AJ. 1991. Recombinant aequorin as a probe for cytosolic calcium in *Escherichia coli*. *FEBS Lett.* 282:405–8

96. Knight MR, Campbell AK, Smith SM, Trewavas AJ. 1991. Transgenic plant aequorin reports the effects of touch and cold-shock and elicitors on cytoplasmic calcium. *Nature* 352:524–26

97. Knight MR, Smith SM, Trewavas AJ. 1992. Wind-induced plant motion immediately increases cytoslic calcium. *Proc. Natl. Acad. Sci. USA* 89:4967–71

98. Krause K-H, Chou M, Mitchell AT, Sjolund RD, Campbell KP. 1989. Plant cells contain calsequestrin. *J. Biol. Chem.* 264: 4269–72

99. Kreimer G, Melkonian M, Holtum JAM, Latzko E. 1988. Stromal free calcium concentration and light-mediated activation of choloroplast fructose-1-6-bisphosphatase. *Plant Physiol.* 86:423–28

100. Kreimer G, Melkonian M, Latzko E. 1985. An electrogenic uniport mediates light-dependent Ca^{2+} influx into intact spinach chloroplasts. *FEBS Lett.* 180:253–58

101. Kretsinger RH. 1990. Why cells must export calcium. In *Intracellular Calcium Regulation,* ed. F Bronner, pp. 439–57. New York: Wiley-Liss

102. Kuo A, Sheng O, Bush DS. 1994. Effect of okadaic acid on GA and ABA action in wheat aleurone layers. *Plant Physiol.* 105 (Suppl.):72

103. Kütreiber W, Jaffe LF. 1990. Detection of extracellular calcium gradients with a calcium-specific vibrating electrode. *J. Cell Biol.* 110:1565–73

104. Ladror US, Zielinski RE. 1990. Effect of Ca^{2+} and calmodulin on ΔpH formation in tonoplast vesicles from corn roots. *Plant Physiol.* 92:850–54

105. Lam E, Benedyk M, Chua N-H. 1989. Characterization of phytochrome-regulated gene expression in a photoautotrophic cell suspension: possible role for calmodulin. *Mol. Cell Biol.* 9:4819–23

106. Läuger P. 1992. *Electrogenic Ion Pumps.* Distinguished Lect. Ser. Soc. Gen. Physiol. Sunderland, MA: Sinauer Assoc. 313 pp.

107. Lee HC, Walseth TF, Bratt GT, Hayes RN, Clapper DL. 1989. Structural determination of a cyclic metabolite of NAD^+ with intracellular Ca^{2+}-mobilizing activity. *J. Biol. Chem.* 264:1608–15

107a. Leung J, Bouvier-Durand M, Morris P-C, Guerrier D, Chefdor F, Giraudat J. 1994. *Arabidopsis* ABA response gne *ABI1*: features of a calcium-modulated protein phosphatase. *Science* 264:1488–52

108. Linder B, Raschke K. 1992. A slow anion channel in guard cells, activating at large hyperpolarization may be principal for stomatal closing. *FEBS Lett.* 313:27–30

109. Lino M, Endo M, Wada M. 1989. The occurrence of a Ca^{2+}-dependent period in the red light-induced late G1 phase of germinating *Adiantum* spores. *Plant Physiol.* 91: 610–16

110. Liss H, Siebers B, Weiler EW. 1991. Characterization, functional reconstitution and activation by fusicoccin of a Ca^{2+}-ATPase from *Corydalis sempervirens* Pers. cells suspension cultures. *Plant Cell Physiol.* 32: 1049–56

111. Lodish HF, Kong N. 1990. Perturbation of cellular calcium blocks exit of secretory proteins from the rough endoplasmic reticulum. *J. Biol. Chem.* 265:10893–99

112. Lonergan TA. 1990. Steps linking the photosynthetic light reactions to the biological clock require calcium. *Plant Physiol.* 93: 110–15

113. Luan S, Li W, Rusnak F, Assmann SM, Schreiber SL. 1993. Immunosuppressants implicate protein phosphatase regulation of K^+ channels in guard cells. *Proc. Natl. Acad. Sci. USA* 90:2202–6

114. Maathuis FJM, Prins HBA. 1991. Outward current conducting ion channels in tonoplasts of *Vigna unguiculata*. *J. Plant Physiol.* 139:63–69

115. MacRobbie EAC. 1993. Ca^{2+} and cell signalling in guard cells. *Semin. Cell Biol.* 4:113–22

116. McAinsh MR, Brownlee C, Hetherington AM. 1992. Visualizing changes in cytosolic-free Ca^{2+} during the response of stomatal guard cells to abscisic acid. *Plant Cell* 4:1113–22

117. McLennon DH, Holland PC. 1975. Calcium transport in sarcoplasmic reticulum. *Annu. Rev. Biophys. Bioeng.* 4:377–404

118. Menegazzi P, Guzzo F, Baldan B, Mariani P, Treves S. 1993. Purification of calreticulin-like protein(s) from spinach leaves. *Biochem. Biophys. Res. Commun.* 190:1130–35

118a. Meyer K, Leube MP, Grill E. 1994. A protein phosphatase 2C involved in ABA signal transduction in *Arabidopsis thaliana*. *Science* 264:1452–55

119. Miller AJ, Sanders D. 1987. Depletion of cytosolic free calcium induced by photosynthesis. *Nature* 326:397–400

120. Miller AJ, Vogg G, Sanders D. 1990. Cytosolic calcium homeostais in fungi: roles

of plasma membrane transport and intracellular sequestration of calcium. *Proc. Natl. Acad. Sci. USA* 87:9348–52

121. Miller DD, Callaham DA, Gross DJ, Hepler PK. 1992. Free Ca^{2+} gradient in growing pollen tubes of *Lilium. J. Cell Sci.* 101: 7–12

122. Muto S. 1992. Intracellular Ca^{2+} messenger system in plants. *Int. Rev. Cytol.* 142: 305–45

123. Neuhaus G, Bowler C, Kern R, Chua N-H. 1993. Calcium/calmodulin-dependent and -independent phytochrome signal transduction pathways. *Cell* 73:937–52

124. Nickel R, Schütte M, Hecker D, Scherer GFE. 1991. The phospholipid platelet-activating factor stimulates proton extrusion in cultured soybean cells and protein phosphorylation and ATPase activity in plasma membranes. *J. Plant Physiol.* 139:205–11

125. Oberdorf JA, Lebeche D, Head JF, Kaminer B. 1983. Identification of a calsequestrin-like protein from sea urchin eggs. *J. Biol. Chem.* 263:6806–9

126. Ohya Y, Umemoto N, Tanida I, Ohta A, Iida H, Anraku Y. 1991. Calcium-sensitive *cls* mutants of *Saccharomyces cerevisiae* showing a Pet⁻ phenotype are ascribable to defects of vacuolar membrane H^+-ATPase activity. *J. Biol. Chem.* 266:13971–77

127. Olbe M, Sommarin M. 1991. ATP-dependent Ca^{2+} transport in wheat root plasma membrane vesicles. *Physiol. Plant.* 83: 535–43

128. Pantoja O, Gelly A, Blumwald E. 1992. Voltage-dependent calcium channels in plant vacuoles. *Science* 255:1567–70

129. Pérez-Prat E, Narasimhan ML, Binzel ML, Botella MA, Chen Z, et al. 1992. Induction of a putative Ca^{2+}-ATPase mRNA in NaCl-adapted cells. *Plant Physiol.* 100:1471–78

130. Pfeiffer W, Hagar A. 1993. A Ca^{2+}-ATPase and a Mg^{2+}/H^+-antiporter are present on tonoplast membranes from roots of *Zea mays* L. *Planta* 191:377–85

131. Pickard BG, Ding JP. 1993. The mechanosensory calcium-selective ion channel: key component of a plasmalemmal control centre? *Aust. J. Plant Physiol.* 20:439–59

132. Ping Z, Yabe I, Muto S. 1992. Identification of K^+, Cl^-, and Ca^{2+} channels in the vacuolar membrane of tobacco cell suspension cultures. *Protoplasma* 171:7–18

133. Poovaiah BW, Reddy ASN. 1993. Calcium and signal transduction in plants. *Crit. Rev. Plant Sci.* 12:185–211

134. Ranjeva R, Carrasco A, Boudet AM. 1988. Inositol trisphosphate stimulates the release of calcium from intact vacuoles isolated from *Acer* cells. *FEBS Lett.* 230:137–41

135. Rasi-Caldogno F, Carnelli A, De Michelis MI. 1993. Controlled proteolysis activates the plasma membrane Ca^{2+} pump of higher plants. *Plant Physiol.* 103:385–90

136. Rasi-Caldogno F, Pugliarello MC, De Michelis MI. 1985. Electrogenic transport of protons driven by the plasma membrane ATPase in membrane vesicles from radish. Biochemical characterization. *Plant Physiol.* 77:200–5

137. Rasi-Caldogno F, Pugliarello MC, Olivari C, De Michelis MI. 1989. Identification and characterization of the Ca^{2+}-ATPase which drives active transport of Ca^{2+} at the plasma membrane of radish seedlings. *Plant Physiol.* 90:1429–34

138. Read ND, Allan WTG, Knight H, Knight MR, Malhó R, et al. 1992. Imaging and measurement of cytosolic free calcium in plant and fungal cells. *J. Microscopy* 166: 57–86

139. Reid RJ, Smith FA. 1992. Measurement of calcium fluxes in plants using ^{45}Ca. *Planta* 186:558–66

140. Reid RJ, Smith FA. 1992. Regulation of calcium influx in *Chara. Plant Physiol.* 100:637–43

141. Rizzuto R, Simpson AWM, Brini M, Pozzan T. 1992. Rapid changes of mitochondrial Ca^{2+} revealed by specifically targeted recombinant aequorin. *Nature* 358:325–27

142. Roberts DM, Harmon AC. 1992. Calcium-modulated proteins: targets of intracellular calcium signals in higher plants. *Annu. Rev. Plant Phyisol. Plant Mol. Biol.* 43:375–414

143. Rooney EK, Gross JD. 1992. ATP-driven Ca^{2+}/H^+ antiport in acid vesicles from *Dictyostelium. Proc. Natl. Acad. Sci. USA* 89: 8025–29

144. Roux SJ, Wayne RO, Datta N. 1986. Role of calcium ions in phytochrome responses: an update. *Physiol. Plant.* 66:344–48

145. Rudolph HK, Antebi A, Fink GR, Buckley CM, Dorman TE, et al. 1989. The yeast secretory pathway is perturbed by mutations in PMR1, a member of a Ca^{2+} ATPase family. *Cell* 58:133–45

146. Saunders MJ, Hepler PK. 1983. Calcium antagonists and calmodulin inhibitors block cytokinin-induced bud formation in *Funaria. Dev. Biol.* 99:41–49

147. Schatzmann HJ. 1989. The calcium pump of the surface membrane and of the sarcoplasmic reticulum. *Annu. Rev. Physiol.* 51: 473–85

148. Scheuerlein R, Wayne R, Roux SJ. 1989. Calcium requirement of phytochrome-mediated fern-spore germination: no direct phytochrome-calcium interaction in the phytochrome-initiated transduction chain. *Planta* 178:25–30

149. Schroeder JI, Hagiwara S. 1989. Cytosolic calcium regulates ion channels in the plasma membrane of *Vicia faba* guard cells. *Nature* 338:427–30

150. Schroeder JI, Hagiwara S. 1990. Repetitive

increases in cytosolic Ca^{2+} of guard cells by abscisic acid activation of nonselective Ca^{2+} permeable channels. *Proc. Natl. Acad. Sci. USA* 87:9305–9

151. Schroeder JI, Thuleau P. 1991. Ca^{2+} channels in higher plant cells. *Plant Cell* 3:555–59

152. Schumaker KS, Gizinski MJ. 1993. Cytokinin stimulates dihydropyridine-sensitive calcium uptake in moss protoplasts. *Proc. Natl. Acad. Sci. USA* 90:10937–41

153. Schumaker KS, Sze H. 1986. Calcium transport into the vacuole of oat roots. *J. Biol. Chem.* 261:12172–78

154. Schumaker KS, Sze H. 1987. Inositol 1,4,5-trisphosphate releases Ca^{2+} from vacuolar membrane vesicles of oat roots. *J. Biol. Chem.* 262:3944–46

155. Schumaker KS, Sze H. 1990. Solubilization and reconstitution of the oat root vacuolar H^+/Ca^{2+} exchanger. *Plant Physiol.* 92:340–45

156. Serrano EE, Zeiger E, Hagiwara S. 1988. Red light stimulates an electrogenic proton pump in *Vicia* guard cell protoplasts. *Proc. Natl. Acad. Sci. USA* 85:436–40

157. Shacklock PS, Read ND, Trewavas AJ. 1992. Cytosolic free calcium mediates red light-induced photomorphogenesis. *Nature* 358:753–55

158. Shimazaki K-I, Kinoshita T, Nishimura M. 1992. Involvement of calmodulin and calmodulin-dependent myosin light chain kinase in blue light-dependent H^+ pumping by guard cell protoplasts from *Vicia faba* L. *Plant Physiol.* 99:1416–21

159. Speksnijder JE, Miller AL, Weisenseel MH, Chen T-H, Jaffe LF. 1989. Calcium buffer injections block fucoid egg development by facilitating calcium diffusion. *Proc. Natl. Acad. Sci. USA* 86:6607–11

160. Taylor CW, Putney JW Jr. 1985. Size of the inositol 1,4,5-triphosphate-sensitive calcium pool in guinea-pig hepatocytes. *Biochem. J.* 232:435–38

161. Terry BR, Findlay GP, Tyerman SD. 1992. Direct effects of Ca^{2+}-channel blockers on plasma membrane cation channels of *Amaranthus tricolor* protoplasts. *J. Exp. Bot.* 43:1457–73

162. Tester M, MacRobbie EAC. 1990. Cytoplasmic calcium affects the gating of potassium channels in the plasma membrane of *Chara corallina*: a whole-cell study using calcium-channel effectors. *Planta* 180:569–81

163. Thuleau P, Graziana A, Ranjeva R, Schroeder JI. 1993. Solubilized proteins from carrot (*Daucus carota* L.) membranes bind calcium channel blockers and form

calcium-permeable ion channels. *Proc. Natl. Acad. Sci. USA* 90:765–69

164. Tretyn A, Wagner G, Felle HH. 1991. Signal transduction in *Sinapis alba* root hairs: auxins as external messengers. *J. Plant Physiol.* 139:187–93

165. Tsien RW, Tsien RY. 1990. Calcium channels, stores, and oscillations. *Annu. Rev. Cell Biol.* 6:715–60

166. Volpe P, Krause K-H, Hashimoto S, Zorzato F, Pozzan T, et al. 1988. "Calcisome," a cytoplasmic organelle: the inositol 1,4,5-trisphosphate-sensitive Ca^{2+} store of nonmuscle cells? *Proc. Natl. Acad. Sci. USA* 85:1091–95

167. Wang M, Duijn BV, Schram AW. 1991. Abscisic acid induces a cytosolic calcium decrease in barley aleurone protoplasts. *FEBS Lett.* 278:69–74

168. Ward JM, Schroeder JI. 1994. Calcium-activated K^+ channels and calcium-induced calcium release by slow vacuolar ion channels in guard cell vacuoles implicated in the control of stomatal closure. *Plant Cell* 6:669–83

169. Watillon B, Kettmann R, Boxus P, Burny A. 1993. A calcium/calmodulin-binding serine/threonine protein kinase homologous to the mammalian type II calcium/calmodulin-dependent protein kinase is expressed in plant cells. *Plant Physiol.* 101:1381–84

170. Wayne R, Hepler PK. 1985. Red light stimulates an increase in intracellular calcium in the spores of *Onoclea sensibilis*. *Plant Physiol.* 77:8–11

171. Williams LE, Schueler SB, Briskin DP. 1990. Further characterization of the red beet plasma membrane Ca^{2+}-ATPase using GTP as an alternative substrate. *Plant Physiol.* 92:747–54

172. Williamson JR, Monck JR. 1989. Hormone effects on cellular Ca^{2+} fluxes. *Annu. Rev. Physiol.* 51:107–24

173. Williamson RE. 1993. Organelle movements. *Annu. Rev. Plant Physiol. Plant Mol. Biol.* 44:181–202

174. Wimmers LE, Ewing NN, Bennett AB. 1992. Higher plant Ca^{2+}-ATPase: primary structure and regulation of mRNA abundance by salt. *Proc. Natl. Acad. Sci. USA* 89:9205–9

175. Zocchi G, Robotti G. 1993. Calcium transport in membrane vesicles isolated from maize coleoptiles. *Plant Physiol.* 101:135–39

176. Zorec R, Tester M. 1992. Cytoplasmic calcium stimulates exocytosis in a plant secretory cell. *Biophys. J.* 63:864–67

Annu. Rev. Plant Physiol. Plant Mol. Biol. 1995. 46:123–46

PEROXISOMES AND THEIR ASSEMBLY IN HIGHER PLANTS

Laura J. Olsen

Department of Biology, University of Michigan, Ann Arbor, Michigan 48109-1048

John J. Harada

Section of Plant Biology, Division of Biological Sciences, University of California, Davis, California 95616

KEY WORDS: glyoxysomes, leaf-type peroxisomes, microbodies, organelle biogenesis, protein targeting

CONTENTS

ABSTRACT

Peroxisomes in higher plants are nearly ubiquitous subcellular organelles that sequester enzymes involved in a variety of essential metabolic processes, including lipid mobilization, photorespiration, and nitrogen transport. Maintenance and proliferation of these nonautonomous organelles is thought to occur through enlargement of the organelle, resulting from the incorporation of nuclear-encoded proteins and lipids, and through fission of the peroxisomes.

0066-4294/95/0601-0123$05.00

An integrated understanding of the processes by which polypeptides are transported posttranslationally to peroxisomes is beginning to emerge through the determination of amino acid sequences that target proteins to the organelle and through the identification and analysis of genes required for peroxisome assembly. This review briefly discusses the characteristics of higher plant peroxisomes and focuses on the processes that underlie peroxisome formation.

INTRODUCTION

Peroxisomes are functionally diverse subcellular organelles that are nearly ubiquitous in eukaryotic cells. In higher plants, these organelles play critical roles in a variety of metabolic processes ranging from lipid mobilization to nitrogen transport. Over the past decade, significant progress has been made in understanding the mechanisms that underlie peroxisome biogenesis and functions. In this review, we focus on the advances made in understanding the assembly of higher plant peroxisomes. We also discuss information derived from studies of peroxisomes in other organisms, in part, to complement information available about the plant organelles. For additional details about peroxisomes, the reader is referred to other reviews (8, 20, 97, 110, 111, 118, 122, 148, 176, 183, 185, 187, 193).

Characteristics and Functions of Peroxisomes

Peroxisomes are a subclass of organelles designated as microbodies, which were originally identified ultrastructurally in animal (157, 160) and plant (63, 133) cells. All microbodies share a number of characteristics. First, they are bounded by a single membrane and have diameters generally ranging from 0.2 to 1.7 μm (62, 97). In some cells, they may be transiently or permanently interconnected through tubular extensions to form a reticulum (57, 79, 120, 215). Second, microbodies have a characteristically high equilibrium density of ~1.23 g/cm^3. Third, the microbody matrix is finely granular and sometimes contains paracrystalline or densely amorphous inclusions (see 41, 59, 62, 186). Fourth, microbodies are nonautonomous organelles in that they lack organellar DNA, ribosomes, and internal membrane systems (48, 104). Therefore, all microbody proteins must be encoded by the nuclear genome and synthesized elsewhere in the cell (see Models of Peroxisome Formation section). The characteristic that distinguishes peroxisomes from other microbodies is that they are respiratory organelles that contain H_2O_2-producing oxidases and catalase, which inactivates the reactive H_2O_2 (41, 60). All microbodies characterized in higher plants are peroxisomes (97).

A general characteristic of peroxisomes is their functional versatility. The involvement of these organelles in a variety of metabolic functions may reflect the fact that oxidases that are often sequestered in peroxisomes participate in a number of distinct cellular processes. For example, plant, fungal, and yeast peroxisomes contain β-oxidation enzymes as well as enzymes involved in the utilization of two-carbon compounds (9, 125, 128, 185, 193, 201); reactions involved in ureide metabolism occur in the peroxisomes of both plants and animals (185); and plasmalogens, an essential component of membranes, are synthesized, in part, in mammalian microbodies (88, 193). The specific function of a given class of peroxisomes appears to be determined by the environment or differentiated state of the cell type in which they are found.

Classes of Higher Plant Peroxisomes

Consistent with the diversity of functions displayed by peroxisomes in yeasts, fungi, and animals, at least four classes of peroxisomes have been identified in higher plants. Although they all share the general characteristics noted above, each class of peroxisomes participates in a distinct metabolic function, usually at a discrete stage of the plant life cycle.

GLYOXYSOMES   These specialized peroxisomes play a critical role in lipid mobilization. Glyoxysomes were designated as such because, with the exception of cytosolic aconitase (35, 38), they possess all of the glyoxylate-cycle enzymes and the β-oxidation enzymes (24, 34, 98). These enzymes catalyze the net conversion of fatty acids to succinate and are largely responsible for the ability of plants to utilize lipids as a carbon source (reviewed in 9, 188). Succinate generated by these reactions is generally converted to carbohydrates by gluconeogenic enzymes in the cytosol and mitochondria. The β-oxidation and glyoxylate-cycle enzymes are thought to be localized exclusively in peroxisomes (67, 186), although some reports suggest that β-oxidation may also occur in mitochondria (213). Glyoxysomes have been characterized most extensively during the postgerminative development of oilseed plants, where they are involved in mobilizing storage lipids to provide nutrients for growing seedlings. Functional glyoxysomes also occur in senescent organs, presumably in response to the mobilization of membrane lipids (17, 39, 40, 86). Additionally, the activities of two enzymes associated exclusively with glyoxysomes, isocitrate lyase and malate synthase, have been detected in developing seeds (33, 55, 191) and pollen (217), but it is unclear whether the other glyoxylate-cycle enzymes also are present at these stages of the life cycle.

LEAF-TYPE PEROXISOMES   Peroxisomes present in photosynthetically active tissues, such as green cotyledons and leaves, are typically appressed to chloroplasts and mitochondria (61), reflecting the cooperative roles of these organelles in photorespiration. Photorespiration is the oxygen- and light-dependent evolution of carbon dioxide that is associated with the reactions of photosynthesis, particularly in C3 plants (reviewed in 146, 184). The metabolic reactions involved in photorespiration are speculated either to dissipate excess photosynthetic capacity that cannot be used for carbon fixation or to recover carbon redirected from the photosynthetic carbon reduction cycle as a result of ribulose bisphosphate oxygenase activity. Reactions involved in photorespiration catalyze the net conversion of two molecules of phosphoglycolate, the oxidation product of ribulose bisphosphate carboxylase/oxygenase, into one molecule each of phosphoglycerate, carbon dioxide, and ammonia. Leaf-type peroxisomes contain the photorespiratory enzymes glycolate oxidase, serine-glyoxylate aminotransferase, and hydroxypyruvate reductase, which convert glycolate to glycine and serine to glycerate.

ROOT NODULE PEROXISOMES   Peroxisomes present in the root nodules of many nitrogen-fixing plants play a role in reactions involved in nitrogen transport (reviewed in 166). Many tropical legumes transport nitrogen in the form of the ureides allantoin and allantoic acid, which are derived from purine catabolism. Reactions involved in ureide biosynthesis are predicted to occur in several cellular compartments of two nodule cell types (cells infected with nitrogen-fixing rhizobia and uninfected cells). In infected cells, fixed ammonia is assimilated in the cytosol and the purine base xanthine is synthesized in the plastids. The subsequent conversion of xanthine to allantoin and allantoic acid is thought to occur in uninfected cells. One of the final steps in this pathway, the conversion of urate to allantoin, is catalyzed by urate oxidase and occurs in peroxisomes.

UNSPECIALIZED PEROXISOMES   Microbodies are also present in plant tissues that are not active photosynthetically and that lack storage lipids and waxes (reviewed in 97). They have been designated unspecialized peroxisomes because they contain catalase and, in most cases, H_2O_2-producing oxidases. Unspecialized peroxisomes are typically smaller than glyoxysomes or leaf-type peroxisomes, they occur less frequently than these other types of peroxisomes, and their buoyant density is lower than that of either glyoxysomes or leaf-type peroxisomes (96). Their designation as unspecialized probably reflects a lack of understanding of their specific role in cellular metabolism. For example, it has been postulated that these peroxisomes dissipate excess energy in the cell or protect cells from oxygen toxicity at high oxygen concentrations (97).

Origin of Specialized Classes of Peroxisomes

A thorough discussion of the processes that determine the specific function of a class of peroxisomes is beyond the scope of this review. However, we briefly consider two aspects of this discussion that are relevant to biogenesis because it is important to know whether each peroxisomal class undergoes a unique pathway of assembly or whether all peroxisomes are formed though a common mechanism. The evidence outlined below favors the latter hypothesis.

First, each functional class of peroxisomes contains a characteristic set of enzymes that confers upon the organelle its specialized function (148, 186). Thus, the metabolic role of a specific class of peroxisomes is determined by the mechanisms that regulate enzyme accumulation within the organelle. Although both transcriptional and posttranscriptional control mechanisms have been demonstrated to control the accumulation of individual peroxisomal enzymes (33, 55, 85, 140), the available evidence suggests that protein import does not play a determinative role in specifying organellar function. That is, specialized peroxisomes do not selectively import specific classes of peroxisomal proteins. In particular, glyoxysomal enzymes, which are not normally present in leaves and roots, are imported into leaf-type and root peroxisomes when expressed ectopically in transgenic plants (147, 149). Additionally, the peroxisomal protein firefly luciferase can be imported into peroxisomes from yeasts, insects, mammals, and plants (109). An implication of these results is that many, if not all, classes of peroxisomes are competent to import all peroxisomal proteins, which suggests the existence of common import pathways.

Second, peroxisomes of one functional class can be converted into peroxisomes with a different metabolic role. The best-characterized example is the conversion of glyoxysomes to leaf-type peroxisomes, which occurs as seedlings undergo the transition from heterotrophic to autotrophic growth. In cotyledons undergoing this transition, individual peroxisomes containing enzymes characteristic of both glyoxysomes and leaf-type peroxisomes have been identified (143, 163, 182). These results suggest that the conversion of peroxisomal function results from a replacement of glyoxysome-specific enzymes with those characteristic of leaf-type peroxisomes. Additional support for this premise comes from the finding that organelles containing enzymes characteristic of both types of peroxisomes are detected during cotyledon senescence when leaf-type peroxisomes are converted to glyoxysomes (142). Together, the data indicate that common mechanisms of assembly may be used in the biogenesis of peroxisomes with different functions.

MODELS OF PEROXISOME FORMATION

At least two models have been proposed to explain how peroxisomes are formed. An early model, based primarily on ultrastructural data, proposed that peroxisomes were derived by budding from the endoplasmic reticulum (ER). Currently, the more generally accepted theory is that new peroxisomes originate from the growth and division of preexisting peroxisomes.

The foundation of the ER vesiculation model of peroxisome formation was the interpretation that apparent direct membrane contacts between peroxisomes and the ER represent lumenal continuities (75, 97, 118, 144, 187). However, the presence of direct continuities between the budding membrane of nascent microbodies and the membrane of the presumed ER cisterna was questioned as early as 1975 (150). More recent higher-resolution micrographs have failed to convincingly demonstrate direct lumenal contacts between peroxisomes and the ER (118), although such contacts may be difficult to visualize if they do exist. Early micrographs have been reinterpreted as depicting a peroxisomal reticulum in which extensions, or "tails," between peroxisomes were mistaken for ER or as showing peroxisomes in close proximity to, but distinct from, the ER (118). In this regard, definite structural associations between peroxisomes and the ER are sometimes observed, especially in uninfected cells of developing root nodules, but clear lumenal continuities are not seen (106, 206, 216). A similar association between the outer mitochondrial membrane and the ER was also proposed more than 20 years ago (23, 58, 139), although it is commonly accepted now that such an association does not exist.

Other data do not support this model. First, no peroxisomal proteins have been shown to possess the necessary signal peptides that would be required for transit through the ER enroute to peroxisomes. Second, glycosylated forms of peroxisomal proteins are not routinely found; such modifications would be expected for ER-derived proteins (42, 118, 176). However, several groups have reported the presence of glycoproteins in glyoxysomes, leaving open the possibility that some glyoxysomal proteins may be of ER origin (10, 30, 46, 47, 73). Third, electrophoretic analyses have shown that ER, mitochondria, and peroxisomes contain different sets of membrane proteins (64, 114).

The general consensus is that peroxisomes arise from the enlargement and fission of preexisting peroxisomes (20, 42, 118, 151, 181). Organellar growth is accomplished through the addition of lipids and the posttranslational import of proteins. All peroxisomal matrix and membrane proteins analyzed to date are synthesized in the cytoplasm and transported posttranslationally directly into the organelle (19, 20, 65, 118, 121, 151). Demonstration that the increase in peroxisome number results from fission of the organelles in the yeast *Hansenula polymorpha* provides direct experimental support for this model (202). Using fusions between wild-type and peroxisome-deficient Chinese hamster ovary cells, Allen et al (2) found that a wild-type nucleus is necessary, but not sufficient, to restore normal peroxisome biogenesis. A cytoplasmic component of wild-type cells, most likely a normal peroxisome, is also required.

Assuming that peroxisomes do not bud from the ER and do not arise de novo, the origin of the first peroxisome remains unresolved. Do all classes of peroxisomes in plants, animals, and yeast have a single common ancestor? The widespread occurrence of peroxisomes in eukaryotic cells suggests that they arose early in evolution (28). Some reports suggest an endosymbiotic origin for peroxisomes (28, 42); others theorize that peroxisomes, like the ER, vacuole-lysosome, and nucleus, are derived from invaginations of plasma membranes (see 42). The answer to this question is needed to obtain a complete understanding of peroxisome biogenesis.

MECHANISMS OF PROTEIN TRANSPORT AND ORGANELLE ASSEMBLY

Proteins destined for peroxisomes are nuclear encoded, synthesized on free polyribosomes in the cytoplasm, and posttranslationally transported into the organelle (19, 20, 118, 121, 151, 158, 175, 176). In recent years, significant advances have been made toward elucidating the mechanisms and the components involved in protein import into peroxisomes.

Peroxisomal Protein Targeting Signals

Several recent articles provide excellent general reviews of the targeting signals that direct proteins to peroxisomes, with emphasis on mammalian and yeast systems (42, 151, 158, 174–176). In this section, we focus primarily on the targeting signals of higher plant peroxisomal proteins.

Most peroxisomal proteins are synthesized at their final size and are not detectably processed during or following import into the organelle (19, 20, 42, 118, 151, 158, 174–176). Therefore, the amino acid sequences within these proteins that target them to peroxisomes (the peroxisomal targeting signals) must reside within the mature polypeptide. A minority of peroxisomal proteins, including watermelon malate dehydrogenase (69, 70), pumpkin catalase (214), yeast carnitine acetyltransferase (192), and thiolases from cucumber (155), rat liver (18, 92, 152, 179), and humans (21), are synthesized with an amino-terminal peptide that is proteolytically cleaved after import in a process that has not been demonstrated to be tightly coupled to translocation (130). Other peroxisomal proteins apparently have neither a carboxyl-terminal nor an amino-terminal targeting signal. Thus, distinct signals have been localized to several different regions of peroxisomal proteins.

CARBOXYL-TERMINAL SIGNALS   The prototype peroxisomal targeting signal is a carboxyl-terminal tripeptide with a consensus amino acid sequence of serine-lysine-leucine (SKL). Subramani and co-workers found, using double immunofluorescence microscopy with anti-catalase antibodies as a peroxisomal marker, that the carboxyl-terminal SKL of firefly luciferase is necessary to target the protein to peroxisomes and is sufficient to direct a nonperoxisomal passenger protein to the organelles (80, 82, 83, 108). The SKL signal was also identified functionally in rat acyl-CoA oxidase, using import experiments with isolated peroxisomes (131). Substitution of each of the amino acids of the tripeptide signal has demonstrated that certain conservative changes can be tolerated without a loss of targeting activity (80). A number of plant peroxisomal proteins have related carboxyl-terminal tripeptides (148). To test the role of this tripeptide in a plant protein, the glyoxysomal protein isocitrate lyase, which includes the carboxyl-terminal amino acids serine-arginine-methionine, was shown to be both necessary and sufficient for import into peroxisomes of transgenic plants (147). Antibodies raised against a peptide containing the carboxyl-terminal SKL sequence recognized proteins from plants, mammals, yeast, fungi, and trypanosomes, indicating the extensive evolutionary conservation of this targeting signal (81, 84, 109).

Several peroxisomal proteins, including peroxisomal catalase, do not have carboxyl-terminal SKL motifs but possess apparently acceptable tripeptides upstream of the carboxyl terminus (42, 74, 148, 151, 158, 175). However, no definitive evidence indicates that a SKL-like tripeptide can function as a targeting signal at internal positions in a polypeptide. The addition of as few as one amino acid to the carboxyl terminus of the firefly luciferase SKL is sufficient to inactivate the targeting signal (42, 80, 131). In addition, a SKL-like tripeptide near but not at the carboxyl terminus of the peroxisomal protein amine oxidase from *H. polymorpha* has been shown not to be involved in targeting (25, 158). These results do not eliminate the possibility that internal SKL-like tripeptides have a role in targeting, because internal signals may function only within a specific sequence context or only in a protein conformation that permits it to be accessible. For example, the targeting of luciferase to peroxisomes was partially affected by a variety of mutations that did not eliminate the carboxyl-terminal targeting signal (82).

AMINO-TERMINAL SIGNALS   A second type of peroxisomal targeting signal is located in the amino-terminal region of a few peroxisomal proteins. In some cases, the signal is located within a propeptide that is cleaved during protein maturation. As few as 5–11 amino acids from the amino-terminal propeptide of rat thiolases A and B are necessary and sufficient for import into peroxisomes (93, 179). The propeptides of thiolases from cucumber, *Arabidopsis thaliana*, rat, and human are very similar to each other (155, 158, 174–176; LJ Olsen & JJ Harada, unpublished results). They also display some similarity to the propeptide from watermelon malate dehydrogenase and to the first 11 amino acids of peroxisomal amine oxidase from *H. polymorpha*, which suggests that they may share similar targeting roles (25, 42). A consensus sequence of RLxxxxxQ/HL has been suggested. The amino-terminal region of the microbody enzyme aldolase from *Trypanosoma bruceii* contains the consensus sequence, and that is sufficient to direct a cytosolic protein into microbodies (42,

71). Recently, the arginine and both leucines of the consensus sequence were shown to be critical for targeting thiolase to *Saccharomyces cerevisiae* peroxisomes, although the amino terminus of this thiolase is not cleaved upon import (51, 71). The amino-terminal targeting signal of peroxisomal proteins exhibits no structural similarities to those of chloroplast, mitochondrial, and ER polypeptides (158, 174, 179). Thus, this targeting signal of peroxisomal proteins is novel.

INTERNAL SIGNALS   Some peroxisomal proteins contain neither carboxyl-terminal SKL-like tripeptides nor cleavable amino-terminal signals, thus representing a third class of targeting signals, the exceptions (158). Experiments with acyl CoA oxidase from *Candida tropicalis* have identified two separate domains, one amino-terminal, but not cleaved, and another internal. Each domain was sufficient to target a passenger protein to peroxisomes using in vitro import assays and was itself imported into peroxisomes in transgenic *Candida maltosa*, but neither region includes a recognizable SKL-like tripeptide (105, 169). *S. cerevisiae* catalase A also contains two domains that are sufficient to direct a reporter protein to peroxisomes: a large amino-terminal domain and a small carboxyl-terminal region (113). Thus, in some cases, multiple signals are present in a single protein, each of which is capable of targeting the protein to peroxisomes. This functional redundancy may increase import efficiency (131).

Experiments with glyoxysomal isocitrate lyase provide another potential example of internal signals that function in targeting. Behari & Baker (11) showed that a series of modified isocitrate lyase polypeptides with carboxyl-terminal deletions that removed between 35 and 408 amino acids, including the SKL-like tripeptide, could be imported into glyoxysomes in vitro. However, another modified isocitrate lyase polypeptide lacking the carboxyl-terminal 37 residues was not detected in peroxisomes of transgenic plants (147). These seemingly conflicting results may indicate that the modified protein is unstable in peroxisomes in vivo and/or that the deletion derivative may assume a conformation that permits its import in vitro.

TARGETING SIGNALS FOR MEMBRANE PROTEINS   The targeting of peroxisomal membrane proteins from plants has not yet been investigated, but data from other organisms indicate that these polypeptides do not use the targeting signal(s) identified in matrix proteins (176). None of the characterized membrane proteins possess consensus sequences for amino-terminal or carboxyl-terminal targeting signals, and antibodies that recognize proteins containing the SKL tripeptide across many species appear to decorate matrix but not membrane proteins (109). In cells from patients with peroxisomal diseases such as Zellweger syndrome and in some of the peroxisome assembly mutants of yeast, visible membrane ghosts containing membrane proteins assemble, but they are unable to import matrix proteins (129, 162, 168, 176). Mutations of an internal SKL or the carboxyl-terminal AKE tripeptide of a 47-kDa peroxisomal membrane protein from *Candida boidinii* do not affect the localization of this protein, suggesting that some other signal directs its specific association with peroxisomal membranes (76).

Cellular Requirements

The principal role of peroxisomal targeting signals is to mediate recognition of the polypeptides by the factors required to effect specific transport into peroxisomes (203, 209). Such factors may include receptor components of the peroxisomal membrane and cytosolic components such as molecular chaperones. Clues about the identities and roles of these components have come from studies of mutants defective in peroxisome assembly in yeast (52, 53, 95, 121), Chinese hamster ovary cells (190, 218, 219), and humans (26, 159). All mutants identified so far have several common properties: (*a*) The peroxisomes do not appear normal at the ultrastructural level; (*b*) peroxisomal proteins may be present, but at least some are mislocalized to the cytosol; (*c*) the cells lack a complete complement of peroxisomal enzymatic functions; and (*d*) many, but not all, of the mutants possess membrane ghosts, although some mutants may lack peroxisomes entirely (52, 53, 117, 119). Analyses of these mutants provide important information that complements biochemical, physiological, and molecular investigations of the import process.

RECEPTORS   Biochemical and genetic evidence for the existence of proteinaceous receptors for peroxisomal import has accumulated, although definitive proof of receptor function is lacking. Wolins & Donaldson (212) recently demonstrated specific binding of a peptide with a carboxyl-terminal SKL to glyoxysomal membranes. The binding activity was saturable, protease-sensitive, and specific, as shown using competition experiments with excess peptide or matrix proteins. One interpretation of the results is that the binding activity represents a glyoxysomal membrane-bound receptor. Membrane translocation of proteins is receptor-mediated in all other transport systems studied. Specific receptor proteins have been identified for protein transport into chloroplasts (153, 164, 165) and mitochondria (154). Receptors will probably also be found for peroxisomal protein transport (118, 121, 176), and several promising candidates have been identified using the *peroxisome assembly* (*pas*) mutants, primarily in yeasts (52).

Two mutations of the yeasts *Pichia pastoris* and *S. cerevisiae*, *pas8* and *pas10*, respectively, cause defects in the import of peroxisomal matrix proteins possessing carboxyl-terminal SKL-like targeting signals but do not affect the import of thiolase, a protein targeted to peroxisomes via an amino-terminal propeptide (129, 194). The PAS8 and PAS10 proteins are homologous, and both appear to belong to the tetratricopeptide family (129, 194). Tetratricopeptide proteins are involved in transcription, chromosome segregation, cell-cycle control, RNA processing, and mitochondrial protein import (72, 167, 176). The PAS8 protein is tightly associated with peroxisomal membranes, has been localized to the cytosolic face of peroxisomes, and binds specifically to an SKL peptide (129, 176). Taken together, these properties make the proteins strong candidates for receptors that interact with peroxisomal proteins possessing a carboxyl-terminal SKL-like targeting signal. In addition, the proteins show some similarity to a mitochondrial import receptor (93).

pas7 mutants of *S. cerevisiae* exhibit the opposite peroxisomal import competence. Thiolase is not imported in these mutants, but proteins with SKL-like targeting signals are transported normally (52). Because the mutation does not reside in thiolase itself, the PAS7 protein has been suggested as a receptor for the import of proteins containing an amino-terminal propeptide (52, 176).

In addition to these potential receptors, other genes corresponding to peroxisome assembly mutations have been characterized. In some cases, the topologies of the corresponding proteins are consistent with their possible roles as components of the peroxisomal import machinery. The PAS3 protein from *S. cerevisiae* is an integral peroxisomal protein of 48 kDa, anchored in the membrane at its amino terminus, with the bulk of the protein exposed to the cytosol (95). The PAF1 protein, a 35-kDa integral membrane protein of peroxisomes, also has been proposed to be a part of the translocation apparatus by virtue of its ability to restore peroxisome assembly in a peroxisome-deficient mutant of Chinese hamster ovary cells (189). In other cases, identifying PAS2 protein homologs has provided little information about their role in peroxisome assembly. For example, the PAS2 protein of *S. cerevisiae* is related to ubiquitin-conjugating enzymes; its role in peroxisome assembly is not understood (210).

ENERGY REQUIREMENTS   The import of proteins into all other organelles requires energy in the form of ATP and/or a proton motive force across the organellar membrane (203, 209). Several groups have reported that ATP hydrolysis is required for protein import into peroxisomes (99, 156, 207). Neither GTP nor a proton motive force seem to play essential roles in peroxisomal import (99, 207). However, one group found that a proton ionophore inhibited import of peroxisomal proteins (13), possibly resulting from dissipation of a peroxisomal pH gradient or from depletion of intracellular ATP (42).

It is not clear whether a proton gradient exists across peroxisomal membranes. Proton gradients associated with peroxisomes have been detected in experiments using [^{31}P]-NMR and immunocytochemical methods (141, 205). However, it is difficult to imagine how a membrane potential could be maintained (42, 151) because peroxisomal membranes are permeable to small molecules (7, 197), and pore-like structures in peroxisomal membranes from animals allow passage of particles up to 800 Daltons (198). Glyoxysomal membranes may lack this pore protein (see Membrane Assembly section), which may explain why electron transport activity has sometimes been measured across these membranes (56, 91, 124, 134, 173). In addition, ATPase activities, all with different inhibitor sensitivities, have been measured in peroxisomes from yeast (49, 50, 208) and rat liver (37, 43). The PAS1 protein from *S. cerevisiae* and the PAY4 protein, which corresponds to a peroxisome assembly mutation from *Yarrowia lipolytica*, each possess a highly conserved consensus sequence for ATP binding, and they appear to belong to a novel family of putative ATPases (54, 145).

CYTOSOLIC FACTORS   Assuming that all new peroxisomes arise by budding from or fission of preexisting peroxisomes (see Models of Peroxisome Formation section), at least one peroxisome must be present in every cell as a basic cytosolic requirement for peroxisome biogenesis. As noted previously, Allen et al (2) suggest that complementation of a Chinese hamster ovary cell mutation that results in peroxisome deficiency requires a wild-type nucleus to provide the functional gene and a cytosolic factor that is presumably a normal peroxisome. Using a permeabilized cell system, Wendland & Subramani (207) demonstrated that peroxisomal import is cytosol dependent. Depletion of endogenous cytosol from permeabilized cells abolished their ability to import peroxisomal proteins; import competence was restored by the addition of exogenous cytosol. Thus, it is likely that all peroxisomes, including those from plants, require some cytosolic factor(s) for protein import.

Molecular chaperones are likely to be involved in peroxisomal protein import, although data to support this expectation are just beginning to appear. Protein import into most other organelles requires the participation of unfoldases, members of the HSP70 family, presumably to maintain the nascent polypeptide in a conformation appropriate for transport across a membrane (68, 90, 107). Because the majority of peroxisomal proteins are synthesized with a carboxyl-terminal targeting signal, the proteins may acquire substantial secondary structure before the targeting signals emerge from the ribosome (42). This suggests that peroxisomal proteins must be maintained in a loosely-folded or unfolded configuration in the cytoplasm to prevent the formation of translocation-incompetent polypeptides. Recently, a 53-kDa homolog of the bacterial DnaJ protein was identified from plants and localized to glyoxysomal membranes from cucumber cotyledons (155a), which provided the first indication that molecular chaperones may have a role in peroxisome biogenesis.

Membrane Assembly

Peroxisomes require a continuous supply of membrane components, both protein and lipid, to accommodate demands for increased mass during organelle enlargement and/or formation. The synthesis of peroxisomal membrane proteins precedes that of matrix components in regenerating rat liver (123). The exact mechanisms by which the protein and lipid particles assemble into functional peroxisomal membranes is unknown, but insights may be obtained from available information (reviewed in 199).

MEMBRANE PROTEINS   Substantial differences exist in the spectrum of membrane proteins present in various functional classes of peroxisomes. These differences are to be expected because, as with matrix proteins, each class of peroxisomes may require a distinct set of membrane proteins to fulfill its specialized metabolic role. For example, the assortment of proteins derived from glyoxysomal membranes of castor bean endosperm (10, 15, 16, 46, 89), cucumber cotyledons (112), and cottonseed cotyledons (30) are different from each other and from peroxisomal membrane proteins purified from potato tubers (173). Although the exact protein constituents of peroxisomal membranes have not been determined conclusively in any system, the functions of some of these polypeptides have been identified.

Despite the variations in peroxisomal membrane protein profiles, there are at least four membrane proteins from mammalian peroxisomes of known size and function. The largest is a 76-kDa long-chain acyl-CoA synthetase that is identical to the enzyme in microsomes and mitochondria (193). An approximately 70-kDa protein, whose cDNA has been analyzed, appears to be homologous to a superfamily of ATP-binding proteins, hinting at a possible role in ATP-dependent transport into peroxisomes (64, 103, 162, 168, 193). Evidence suggests that a 70-kDa protein in mouse liver peroxisomal membranes is enoyl-CoA hydratase bifunctional protein (32, 36), although this enzyme is generally thought to be a matrix protein (1, 193). A 35-kDa protein uniquely localized to peroxisome membranes is needed to complement a Chinese hamster ovary cell mutant lacking normal peroxisomes (189). This protein is involved in proper peroxisome assembly, but its exact role in this process is unknown. A 22-kDa membrane protein appears to be a pore-forming protein responsible for the permeability of mammalian peroxisomes to small molecules (64, 198). At least five other peroxisomal integral membrane proteins of unknown function have been identified in mammals. In addition, the activities of dihydroxyacetone phosphate acyltransferase, alkyl dihydroxyacetone phosphate synthase, and very-long-chain acyl-CoA synthetase have been localized to peroxisomal membranes but not assigned to specific polypeptides (193).

A number of membrane proteins, other than those that correspond to peroxisome assembly mutations, have been identified in yeasts (see Receptors section; 52, 95, 129, 176, 194). A 52-kDa proton-translocating ATPase may be associated with yeast peroxisomal membranes (49, 50). Numerous groups have studied the prevalent 47-kDa yeast peroxisomal membrane protein homolog of mitochondrial solute carrier proteins (100, 115, 126, 127, 177, 200). Although an abundant 20-kDa protein appears to be associated with peroxisomal membranes of *C. boidinii*, it contains no obvious membrane-spanning regions but does possess a carboxyl-terminal tripeptide that may target it to the matrix for peripheral association with the peroxisomal membrane (66). In addition, the protein is detected only during growth on methanol, which suggests that it may be involved in methanol metabolism (76–78).

Plant peroxisomal membranes appear to contain different proteins and activities than those present in yeast and mammalian peroxisomes. For example, peroxisomal membrane proteins characteristic of yeast and mammalian peroxisomes have yet to be identified in glyoxysomes. Consistent with these differences, there are several unique properties of glyoxysomal membranes. First, evidence for electron transport activity across glyoxysomal membranes has been reported (22, 56, 124). NADH-ferricyanide reductase was identified as a 32-kDa glyoxysomal membrane protein, which is distinct in size from the ER form (124). Potato tuber peroxisomes exhibit NADH-ferricyanide reductase activity that is specific for the β-hydrogen of NADH, whereas the ER reductase is α-specific (173). Interestingly, antibodies against ER-derived NADH-ferricyanide reductase identified a 52-kDa protein of potato tuber peroxisomal membranes (173). Second, no pore-forming protein, homologous to the abundant 22-kDa protein of mammalian peroxisomal membranes, has been found in glyoxysomes (10, 16, 30, 46). This is consistent with the glyoxysomal membranes' lack of permeability to small molecules (47). However, a 22-kDa protein of unknown function is detected among the membrane proteins of potato tuber peroxisomes (173). Third, there are several reports that some glyoxysomal membrane proteins are glycosylated, although this has been a controversial subject (114, 193). Of the 22 proteins observed by SDS-PAGE of castor bean endosperm glyoxysomal membranes, 12 were identified as glycoproteins (10). Chapman & Trelease (30) also identified at least one concanavalin A–reactive polypeptide in cottonseed cotyledon glyoxysomal membranes. At least seven proteins have been shown to be common to glyoxysomes and ER (46). Thus, although much remains to be learned about glyoxysomal membranes, it appears that they may have some unique proteins and properties, which is consistent with the specialized physiological role of the organelle.

MEMBRANE LIPIDS   Phosphatidylcholine and phosphatidylethanolamine are the principal phospholipids of castor bean endosperm glyoxysomal membranes (45). Lipid biosynthetic enzymes are located primarily in the ER of plants (8, 135) and animals (12). The only peroxisomal enzymes known to be involved in lipid biosynthesis are those that catalyze plasmalogen biosynthesis in mammalian liver and brain (6, 88). Thus, the ER appears to be the primary source of peroxisomal membrane lipids in yeast (27), animals (12), and plants (8, 44, 73, 102, 116, 122, 135). An exception is that lipid bodies appear to be the source of nonpolar lipids and phospholipids for membrane expansion of enlarging cottonseed glyoxysomes (29). Thus, cottonseed glyoxysomes and castor bean endosperm glyoxysomes differ in both membrane protein composition and source of membrane lipids (29, 30).

It is not known how these lipids travel from the ER or lipid bodies to peroxisomes. Diffusion of phospholipids in an aqueous phase is too slow and nonspecific to accommodate the demands of membrane synthesis (211). There is little evidence for vesicular shuttles or transfer by direct membrane contact between organelles, although these mechanisms may exist (136, 138). Lipid transfer has been postulated to involve phospholipid carrier proteins that catalyze the nonspecific exchange of lipids between membranes in vitro (3, 101, 211). An intriguing potential link between peroxisomes and nonspecific lipid transfer proteins is that mammalian cells that lack or have greatly reduced numbers of peroxisomes are deficient in these proteins (20, 119, 161, 178, 196). In plants, these small, basic lipid transfer proteins, which are probably structurally distinct from the nonspecific lipid transfer proteins of mammals (137), have been identified in a number of species and tissues (3, 4, 101, 132, 167, 171). They are synthesized with an amino-terminal signal sequence for cotranslational insertion into the ER, and thus are unlikely to appear in the cytoplasm and participate in bulk lipid transfer between intracellular membranes (3, 14, 31, 170, 180). By contrast, several reports specifically claim that these lipid transfer proteins are cytosolic (3,4). Regardless of their location, the lipid transfer activity by these proteins has only been demonstrated in vitro, and their role in vivo has not been established. For example, these proteins have also been proposed to function in plants in the transport of cutin monomers through the extracellular matrix to sites of cutin synthesis (172) and in plant defense against bacterial and fungal pathogens (132).

CONCLUDING REMARKS

Peroxisomes contribute to the overall regulation of metabolism in plant and other eukaryotic cells by compartmentalizing within the cell specialized sets of biochemical reactions that can play roles in a variety of metabolic processes. Thus, an understanding of peroxisomal function requires knowledge of the organelle's biogenesis. From the preceding discussion, it should be clear that peroxisome assembly is a complex process. Although the outlines of processes involved in peroxisome assembly are in place, much specific information remains to be learned.

Despite the complexity of these processes, a common theme of this discussion is that there are multiple pathways involved in peroxisome biogenesis. First, biogenesis of the organelle's contents and of the membrane proteins appear to proceed by different mechanisms. This conclusion is suggested by the analyses of peroxisome assembly mutants in which membrane ghosts have formed but the import of matrix proteins is defective (42, 161, 162, 176). Second, these mutants also provide genetic evidence that at least two distinct signal-dependent pathways exist for the import of proteins into the peroxisomal matrix. Proteins targeted to peroxisomes by a carboxyl-terminal tripeptide are not imported into peroxisomes of the *pas8* mutant of *P. pastoris* (129), the *pas10* mutant of *S. cerevisiae* (195), or some of the complementation groups of Zellweger syndrome fibroblasts (5, 204). However, other proteins with an amino-terminal presequence, such as thiolase, are correctly targeted to peroxisomes in these mutants. Conversely, the *pas7* mutant of *S. cerevisiae* is defective in the import of thiolase but is able to import matrix proteins possessing the carboxyl-terminal targeting tripeptide into peroxisomes (52, 94). Thus, it seems reasonable to expect that at least two different receptors are responsible for the recognition of the two peroxisomal targeting signals for matrix proteins (42, 176). Similar import mechanisms are likely to exist for protein transport into plant peroxisomes, and studies of these and other processes are needed to obtain a complete understanding of peroxisome assembly in higher plants.

ACKNOWLEDGMENTS

We thank Marilynn Etzler for her comments about the manuscript. Work cited from our laboratories was supported by funds from the University of Michigan and the USDA to LJO and from the NSF and the USDA to JJH.

Literature Cited

1. Alexson SEH, Fujiki Y, Shio H, Lazarow PB. 1985. Partial disassembly of peroxisomes. *J. Cell Biol.* 101:294–305
2. Allen L-AH, Morand OH, Raetz CRH. 1989. Cytoplasmic requirement for peroxisome biogenesis in Chinese hamster ovary cells. *Proc. Natl. Acad. Sci. USA* 86:7012–16
3. Arondel V, Kader J-C. 1990. Lipid transfer in plants. *Experientia* 46:579–85
4. Arondel V, Vergnolle C, Tchang F, Kader J-C. 1990. Bifunctional lipid-transfer: fatty

acid-binding proteins in plants. *Mol. Cell Biochem.* 98:49–56

5. Balfe A, Hoefler G, Chen WW, Watkins PA. 1990. Aberrant subcellular localization of peroxisomal 3-ketoacyl-CoA thiolase in the Zellweger syndrome and rhizomelic chondrodysplasia punctata. *Pediatr. Res.* 27:304–10

6. Ballas LM, Lazarow PB, Bell RM. 1984. Glycerolipid synthetic capacity of rat liver peroxisomes. *Biochim. Biophys. Acta* 795: 297–300

7. Beaufay H, Jacques P, Baudhuin P, Sellinger OZ, Berthet J, de Duve C. 1964. Tissue fractionation studies. 18. Resolution of mitochondrial fractions from rat liver into three distinct populations of cytoplasmic particles by means of density equlibration in various gradients. *Biochem. J.* 92:184–205

8. Beevers H. 1979. Microbodies in higher plants. *Annu. Rev. Plant Physiol.* 30:159–93

9. Beevers H. 1980. The role of the glyoxylate cycle. *Biochem. Plants* 4:117–30

10. Beevers H, Gonzalez E. 1987. Proteins and phospholipids of glyoxysomal membranes from castor bean. *Methods Enzymol.* 148: 526–32

11. Behari R, Baker A. 1993. The carboxyl terminus of isocitrate lyase is not essential for import into glyoxysomes in an in vitro system. *J. Biol. Chem.* 268:7315–22

12. Bell RM, Coleman RA. 1980. Enzymes of glycerolipid synthesis in eukaryotes. *Ann. Biochem.* 49:459–87

13. Bellion E, Goodman JM. 1987. Proton ionophores prevent assembly of a peroxisomal protein. *Cell* 48:165–73

14. Bernhard WR, Thoma S, Botella J, Somerville CR. 1991. Isolation of a cDNA clone for spinach lipid transfer protein and evidence that the protein is synthesized by the secretory pathway. *Plant Physiol.* 95: 164–70

15. Bieglmayer C, Graf J, Ruis H. 1973. Membranes of glyoxysomes from castor-bean endosperm. Enzymes bound to purified-membrane preparations. *Eur. J. Biochem.* 37:553–62

16. Bieglmayer C, Ruis H. 1974. Protein composition of the glyoxysomal membrane. *FEBS Lett.* 47:53–55

17. Birkhan R, Kindl H. 1990. Re-activation of the expression of glyoxysomal genes in green plant tissue. *Z. Naturforsch. Teil C* 45:107–11

18. Bodnar AG, Rachubinski RA. 1990. Cloning and sequence determination of cDNA encoding a second rat liver peroxisomal 3-ketoacyl-CoA thiolase. *Gene* 91: 193–99

19. Borst P. 1986. How proteins get into microbodies (peroxisomes, glyoxysomes, glycosomes). *Biochim. Biophys. Acta* 866: 179–203

20. Borst P. 1989. Peroxisome biogenesis revisited. *Biochim. Biophys. Acta* 1008:1–13

21. Bout A, Teunissen Y, Hashimoto T, Benne R, Tager JM. 1988. Nucleotide sequence of human peroxisomal 3-oxoacyl-CoA thiolase. *Nucleic Acids Res.* 16:10369

22. Bowditch MI, Donaldson RP. 1990. Ascorbate free-radical reduction by glyoxysomal membranes. *Plant Physiol.* 94:531–37

23. Bracker CE, Grove SN. 1971. Continuity between cytoplasmic endomembranes and outer mitochondrial membranes in fungi. *Protoplasma* 73:15–34

24. Breidenbach RW, Kahn A, Beevers H. 1967. Characterization of glyoxysomes from castor bean endosperm. *Plant Physiol.* 43:705–13

25. Bruinenberg PG, Evers M, Waterham HR, Kuipers J, Arnberg AC, Ab G. 1989. Cloning and sequencing of the peroxisomal amine oxidase gene from *Hansenula polymorpha*. *Biochim. Biophys. Acta* 1008: 157–67

26. Brul S, Westerveld A, Strijland A, Wanders RJA, Schram AW, et al. 1988. Genetic heterogeneity in the cerebrohepatorenal (Zellweger) syndrome and other inherited disorders with a generalized impairment of peroxisomal functions. *J. Clin. Invest.* 81: 1710–15

27. Carman GM, Henry SA. 1989. Phospholipid biosynthesis in yeast. *Annu. Rev. Biochem.* 58:636–69

28. Cavalier-Smith T. 1987. The simultaneous symbiotic origin of mitochondria, chloroplasts, and microbodies. *Ann. NY Acad. Sci.* 503:55–71

29. Chapman KD, Trelease RN. 1991. Acquisition of membrane lipids by differentiating glyoxysomes: role of lipid bodies. *J. Cell Biol.* 115:995–1007

30. Chapman KD, Trelease RN. 1992. Characterization of membrane proteins in enlarging cottonseed glyoxysomes. *Plant Physiol. Biochem.* 30:1–10

31. Chasan R. 1991. Lipid transfer proteins: moving molecules? *Plant Cell* 3:923–33

32. Chen N, Crane DI, Masters CJ. 1988. Analysis of the major integral membrane proteins of peroxisomes from mouse liver. *Biochim. Biophys. Acta* 945:135–44

33. Comai L, Dietrich RA, Maslyar DJ, Baden CS, Harada JJ. 1989. Coordinate expression of transcriptionally regulated isocitrate lyase and malate synthase genes in *Brassica napus* L. *Plant Cell* 1:293–300

34. Cooper TG, Beevers H. 1969. Beta-oxidation in glyoxysomes from castor bean endosperm. *J. Biol. Chem.* 244:3514–20

35. Courtois-Verniquet F, Douce R. 1993. Lack of aconitase in glyoxysomes and peroxisomes. *Biochem. J.* 294:103–7

36. Crane DI, Chen N, Masters CJ. 1988. Evidence that the enoyl-CoA hydratase bifunctional protein is identical with the 70,000 dalton peroxisomal membrane protein. *Biochem. Biophys. Res. Commun.* 160: 503–8

37. Cuezva JM, Santaren JF, Gonzalez P, Valcarce C, Luis AM, Izquierdo JM. 1990. Immunological detection of the mitochondrial F1-ATPase alpha subunit in the matrix of rat liver peroxisomes. A protein involved in organelle biogenesis? *FEBS Lett.* 270: 71–75

38. De Bellis L, Hayashi M, Biagi PP, Hara-Nishimura I, Alpi A, Nishimura M. 1994. Immunological analysis of aconitase in pumpkin cotyledons: the absence of aconitase in glyoxysomes. *Physiol. Plant.* 90: 757–62

39. De Bellis L, Nishimura M. 1991. Development of enzymes of the glyoxylate cycle during senescence of pumpkin cotyledons. *Plant Cell Physiol.* 32:555–61

40. De Bellis L, Picciarelli P, Pistelli L, Alpi A. 1990. Localization of glyoxylate-cycle marker enzymes in peroxisomes of senescent leaves and green cotyledons. *Planta* 180:435–39

41. de Duve C, Baudhuin P. 1966. Peroxisomes (microbodies and related particles). *Physiol. Rev.* 46:323–57

42. de Hoop MJ, Ab G. 1992. Import of proteins into peroxisomes and other microbodies. *Biochem. J.* 286:657–69

43. del Valle R, Soto U, Necochea C, Leighton F. 1988. Detection of an ATPase activity in rat liver peroxisomes. *Biochem. Biophys. Res. Commun.* 156:1353–59

44. Donaldson RP. 1976. Membrane lipid metabolism in germinating castor bean endosperm. *Plant Physiol.* 57:510–15

45. Donaldson RP, Beevers H. 1977. Lipid composition of organelles from germinating castor bean endosperm. *Plant Physiol.* 59:259–63

46. Donaldson RP, Gonzalez E. 1989. Glyoxysomal membrane proteins are present in the endoplasmic reticulum of castor bean endosperm. *Cell Biol. Int. Rep.* 13:87–94

47. Donaldson RP, Tully RE, Young OA, Beevers H. 1981. Organelle membranes from germinating castor bean endosperm. II. Enzymes, cytochromes, and permeability of the glyoxysome membrane. *Plant Physiol.* 67:21–25

48. Douglass SA, Criddle RS, Breidenbach RW. 1973. Characterization of deoxyribonucleic acid species from castor bean endosperm. Inability to detect a unique deoxyribonucleic acid species associated with glyxoysomes. *Plant Physiol.* 51:902–6

49. Douma AC, Veenhuis M, Sulter GJ, Harder W. 1987. A proton-translocating adenosine triphosphatase is associated with the peroxisomal membrane of yeasts. *Arch. Microbiol.* 147:42–47

50. Douma AC, Veenhuis M, Waterham HR, Harder W. 1990. Immunocytochemical demonstration of the peroxisomal ATPase of yeasts. *Yeast* 6:45–51

51. Einerhand AWC, Voorn-Brouwer TM, Erdmann R, Kunau W-H, Tabak HF. 1991. Regulation of transcription of the gene coding for peroxisomal 3-oxoacyl-CoA thiolase of *Saccharomyces cerevisiae*. *Eur. J. Biochem.* 200:113–22

52. Erdmann R, Kunau W-H. 1992. A genetic approach to the biogenesis of peroxisomes in the yeast *Saccharomyces cerevisiae*. *Cell Biochem. Funct.* 10:167–74

53. Erdmann R, Veenhuis M, Mertens D, Kunau W-H. 1989. Isolation of peroxisome-deficient mutants of *Saccharomyces cerevisiae*. *Proc. Natl. Acad. Sci. USA* 86:5419–23

54. Erdmann R, Wiebel FF, Flessau A, Rytka J, Beyer A, et al. 1991. *PAS1*, a yeast gene required for peroxisome biogenisis, encodes a member of a novel family of putative ATPases. *Cell* 64:499–510

55. Ettinger WF, Harada JJ. 1990. Translational or post-translational processes affect differentially the accumulation of isocitrate lyase and malate synthase proteins and enzyme activities in embryos and seedlings of *Brassica napus*. *Arch. Biochem. Biophys.* 281:139–43

56. Fang TK, Donaldson RP, Vigil EL. 1987. Electron transport in purified glyoxysomal membranes from castor-bean endosperm. *Planta* 172:1–13

57. Ferreira RMB, Bird B, Davies DD. 1989. The effect of light on the structure and organization of *Lemna* peroxisomes. *J. Exp. Bot.* 40:1029–35

58. Franke WW, Kartenbeck J. 1971. Outer mitochondrial membrane continuous with endoplasmic reticulum. *Protoplasma* 73: 35–41

59. Frederick SE, Gruber PJ, Newcomb EH. 1975. Plant microbodies. *Protoplasma* 84: 1–29

60. Frederick SE, Newcomb EH. 1969. Cytochemical localization of catalase in leaf microbodies (peroxisomes). *J. Cell Biol.* 43:343–52

61. Frederick SE, Newcomb EH. 1969. Microbody-like organelles in leaf cells. *Science* 163:1353–55

62. Frederick SE, Newcomb EH, Vigil EL, Wergin WP. 1968. Fine-structure characterization of plant microbodies. *Planta* 81: 229–52

63. Frey-Wyssling A, Grieshaber E, Muhlethaler K. 1963. Origin of spherosomes in plant cells. *J. Ultrastruct. Res.* 8:506–16

64. Fujiki Y, Fowler S, Shio H, Hubbard AL,

Lazarow PB. 1982. Polypeptide and phospholipid composition of the membrane of rat liver peroxisomes. Comparison with endoplasmic reticulum and mitochondrial membranes. *J. Cell Biol.* 93:103–10

65. Fujiki Y, Rachubinski RA, Lazarow PB. 1984. Synthesis of a major integral membrane polypeptide of rat liver peroxisomes on free polysomes. *Proc. Natl. Acad. Sci. USA* 81:7127–31

66. Garrard LJ, Goodman JM. 1989. Two genes encode the major membrane-associated proteins of methanol-induced peroxisomes from *Candida boidinii. J. Biol. Chem.* 264:13929–37

67. Gerhardt B. 1986. Basic metabolic function of higher plant peroxisomes. *Physiol. Veg.* 24:397–410

68. Gething M-J, Sambrook J. 1992. Protein folding in the cell. *Nature* 355:33–45

69. Gietl C. 1990. Glyoxysomal malate dehydrogenase from watermelon is synthesized with an amino-terminal transit peptide. *Proc. Natl. Acad. Sci. USA* 87:5773–77

70. Gietl C. 1992. Partitioning of malate dehydrogenase isoenzymes into glyoxysomes, mitochondria, and chloroplasts. *Plant Physiol.* 100:557–59

71. Glover JR, Andrews DW, Subramani S, Rachubinski RA. 1994. Mutagenesis of the amino targeting signal of *Saccharomyces cerevisiae* 3-ketoacyl-CoA thiolase reveals conserved amino acids required for import into peroxisomes *in vivo. J. Biol. Chem.* 269:7558–63

72. Goebl M, Yanagida M. 1991. The TPR snap helix: a novel protein repeat motif from mitosis to transcription. *Trends Biochem. Sci.* 16:173–77

73. Gonzalez E. 1986. Glycoproteins in the matrix of glyoxysomes in endosperm of castor bean seedlings. *Plant Physiol.* 80:950–55

74. Gonzalez E. 1991. The C-terminal domain of plant catalases. Implications for a glyoxysomal targeting sequence. *Eur. J. Biochem.* 199:211–15

75. Gonzalez E, Beevers H. 1976. Role of the endoplasmic reticulum in glyoxysome formation in castor bean endosperm. *Plant Physiol.* 57:406–9

76. Goodman JM, Garrard LJ, McCammon MT. 1992. Structure and assembly of peroxisomal membrane proteins. See Ref. 139a, pp. 221–29

77. Goodman JM, Maher J, Silver PA, Pacifico A, Sanders D. 1986. The membrane proteins of the methanol-induced peroxisome of *Candida boidinii.* Initial characterization and generation of monoclonal antibodies. *J. Biol. Chem.* 261:3464–68

78. Goodman JM, Trapp SB, Hwang H. 1990. Peroxisomes induced in *Candida boidinii* by methanol, oleic acid and p-alanine vary in metabolic function but share common integral membrane proteins. *J. Cell Sci.* 97:193–204

79. Gorgas K. 1985. Serial section analysis of mouse hepatic peroxisomes. *Anat. Embryol.* 172:21–32

80. Gould SJ, Keller G-A, Hosken N, Wilkinson J, Subramani S. 1989. A conserved tripeptide sorts proteins to peroxisomes. *J. Cell Biol.* 108:1657–64

81. Gould SJ, Keller G-A, Schneider M, Howell SH, Garrard LJ, et al. 1990. Peroxisomal protein import is conserved between yeast, plants, insects and mammals. *EMBO J.* 9:85–90

82. Gould SJ, Keller G-A, Subramani S. 1987. Identification of a peroxisomal targeting signal at the carboxy terminus of firefly luciferase. *J. Cell Biol.* 105:2923–31

83. Gould SJ, Keller G-A, Subramani S. 1988. Identification of peroxisomal targeting signals located at the carboxy terminus of four peroxisomal proteins. *J. Cell Biol.* 107:897–905

84. Gould SJ, Krisans S, Keller G-A, Subramani S. 1990. Antibodies directed against the peroxisomal targeting signal of firefly luciferase recognize multiple mammalian peroxisomal proteins. *J. Cell Biol.* 110:27–34

85. Graham IA, Smith LM, Leaver CJ, Smith SM. 1990. Developmental regulation of expression of the malate synthase gene in transgenic plants. *Plant Mol. Biol.* 15:539–49

86. Gut H, Matile P. 1988. Apparent induction of key enzymes of the glyoxylic acid cycle in senescent barley leaves. *Planta* 176:548–50

87. Deleted in proof

88. Hajra AK, Bishop JE. 1982. Glycerolipid biosynthesis in peroxisomes via the acyl dihydroxyacetone phosphate pathway. *Ann. NY Acad. Sci.* 386:170–82

89. Halpin C, Conder MJ, Lord JM. 1989. Different routes for integral protein insertion into *Ricinus communis* protein-body and glyoxysome membranes. *Planta* 179:331–39

90. Hartl F-U, Hlodan R, Langer T. 1994. Molecular chaperones in protein folding: the art of avoiding sticky situations. *Trends Biochem. Sci.* 19:20–25

91. Hicks DB, Donaldson RP. 1982. Electron transport in glyoxysomal membranes. *Arch. Biochem. Biophys.* 215:280–88

92. Hijikata M, Ishii N, Kagamiyama H, Osumi T, Hashimoto T. 1987. Structural analysis of cDNA for rat peroxisomal 3-ketoacyl-CoA thiolase. *J. Biol. Chem.* 262:8151–58

93. Hines V. 1992. The mitochondrial protein import machinery of *Saccharomyces cerevisiae.* See Ref. 139a, pp. 241–52

94. Hohfeld J, Mertens D, Wiebel FF, Kunau W-H. 1992. Defining components required for peroxisome assembly. See Ref. 139a, pp. 185–208

95. Hohfeld J, Veenhuis M, Kunau W-H. 1991. *PAS3*, a *Saccharomyces cerevisiae* gene encoding a peroxisomal integral membrane protein essential for peroxisome biogenesis. *J. Cell Biol.* 114:1167–78

96. Huang AHC, Beevers H. 1971. Isolation of microbodies from plant tissues. *Plant Physiol.* 48:637–41

97. Huang AHC, Trelease RN, Moore TS Jr. 1983. *Plant Peroxisomes*. New York: Academic

98. Hutton D, Stumpf PK. 1969. Fat metabolism in higher plants XXXVII. Characterization of the beta-oxidation systems from maturing and germinating castor bean seeds. *Plant Physiol.* 44:508–16

99. Imanaka T, Small GM, Lazarow PB. 1987. Translocation of acyl-CoA oxidase into peroxisomes requires ATP hydrolysis but not a membrane potential. *J. Cell Biol.* 105:2915–22

100. Jank B, Habermann B, Schweyen RJ, Link TA. 1993. PMP47, a peroxisomal homologue of mitochondrial solute carrier proteins. *Trends Biochem. Sci.* 18:427–28

101. Kader J-C. 1990. Intracellular transfer of phospholipids, galactolipids, and fatty acids in plant cells. In *Subcellular Biochemistry*, ed. HJ Hilderson, 16:69–111. New York: Plenum

102. Kagawa T, Lord JM, Beevers H. 1973. The origin and turnover of organelle membranes in castor bean endosperm. *Plant Physiol.* 51:61–65

103. Kamijo K, Taketani S, Yokota S, Osumi T, Hashimoto T. 1990. The 70-kDa peroxisomal membrane protein is a member of the Mdr (P-glycoprotein)-related ATP-binding protein superfamily. *J. Biol. Chem.* 265:4534–40

104. Kamiryo T, Abe M, Okazaki K, Kato S, Shimamoto N. 1982. Absence of DNA in peroxisomes of *Candida tropicalis*. *J. Bacteriol.* 152:269–74

105. Kamiryo T, Sakasegawa Y, Tan H. 1989. Expression and transport of *Candida tropicalis* peroxisomal acyl-coenzyme A oxidase in the yeast *Candida maltosa*. *Agric. Biol. Chem.* 53:179–86

106. Kaneko Y, Newcomb EH. 1987. Cytochemical localization of uricase and catalase in developing root nodules of soybean. *Protoplasma* 140:1–12

107. Keegstra K. 1989. Transport and routing of proteins into chloroplasts. *Cell* 56:247–53

108. Keller G-A, Gould SJ, de Luca M, Subramani S. 1987. Firefly luciferase is targeted to peroxisomes in mammalian cells. *Proc. Natl. Acad. Sci. USA* 84:3264–68

109. Keller G-A, Krisans S, Gould SJ, Sommer JM, Wang CC, et al. 1991. Evolutionary conservation of a microbody targeting signal that targets proteins to peroxisomes, glyoxysomes, and glycosomes. *J. Cell Biol.* 114:893–904

110. Kindl H. 1992. Plant peroxisomes: recent studies on function and biosynthesis. *Cell Biochem. Funct.* 10:153–58

111. Kindl H, Lazarow PB. 1982. *Peroxisomes and Glyoxysomes*, Vol. 386. New York: Ann. NY Acad. Sci. 550 pp.

112. Koller W, Kindl H. 1977. Glyoxylate cycle enzymes of the glyoxysomal membrane from cucumber cotyledons. *Arch. Biochem. Biophys.* 181:236–48

113. Kragler F, Langeder A, Raupachova J, Binder M, Hartig A. 1993. Two independent peroxisomal targeting signals in catalase A of *Saccharomyces cerevisiae*. *J. Cell Biol.* 120:665–73

114. Kruse C, Kindl H. 1982. Integral proteins of the glyoxysomal membranes. *Ann. NY Acad. Sci.* 386:499–501

115. Kuan J, Saier MH Jr. 1993. The mitochondrial carrier family of transport proteins: structural, functional, and evolutionary relationships. *Crit. Rev. Biochem. Mol. Biol.* 28:209–33

116. Kunce CM, Trelease RN, Doman DC. 1984. Ontogeny of glyoxysomes in maturing and germinated cotton seeds—a morphometric analysis. *Planta* 161:156–64

117. Lazarow PB. 1993. Genetic approaches to studying peroxisome biogenesis. *Trends Cell Biol.* 3:89–93

118. Lazarow PB, Fujiki Y. 1985. Biogenesis of peroxisomes. *Annu. Rev. Cell Biol.* 1:489–530

119. Lazarow PB, Moser HW. 1989. Disorders of peroxisome biogenesis. In *The Metabolic Basis of Inherited Disease*, ed. CR Scriver, AI Beaudet, WS Sly, D Valle, pp. 1479–1509. New York: McGraw-Hill

120. Lazarow PB, Shio H, Robbi M. 1980. Biogenesis of peroxisomes and the peroxisome reticulum hypothesis. In *31st Mosbach Colloq. Biol. Chem. Organelle Formation*, ed. R Bucher, W Sebald, H Weiss, pp. 187–206. New York: Springer-Verlag

121. Lazarow PB, Thieringer R, Cohen G. 1991. Protein import into peroxisomes in vitro. *Methods Cell Biol.* 34:303–26

122. Lord JM, Roberts LM. 1983. Formation of glyoxysomes. *Int. Rev. Cytol.* 15:115–156

123. Luers G, Beier K, Hashimoto T, Fahimi HD, Volkl A. 1990. Biogenesis of peroxisomes: sequential biosynthesis of the membrane and matrix proteins in the course of hepatic regeneration. *Eur. J. Cell Biol.* 52:175–84

124. Luster DG, Bowditch MI, Eldridge KM, Donaldson RP. 1988. Characterization of membrane-bound electron transport en-

zymes from castor bean glyoxysomes and endoplasmic reticulum. *Arch. Biochem. Biophys.* 265:50–61

125. Maxwell DP, Armentrout VN, Graves LB Jr. 1977. Microbodies in plant pathogenic fungi. *Annu. Rev. Phytopathol.* 15:119–34

126. McCammon MT, Dowds CA, Orth K, Moomaw CR, Slaughter CA, Goodman JM. 1990. Sorting of peroxisomal membrane protein PMP47 from *Candida boidinii* into peroxisomal membranes of *Saccharomyces cerevisiae. J. Biol. Chem.* 265: 20098–20105

127. McCammon MT, McNew JA, Willy PJ, Goodman JM. 1994. An internal region of the peroxisomal membrane protein PMP47 is essential for sorting to peroxisomes. *J. Cell Biol.* 124:915–25

128. McCammon MT, Veenhuis M, Trapp SB, Goodman JM. 1990. Association of glyoxylate and beta-oxidation enzymes with peroxisomes of *Saccharomyces cerevisiae. J. Bacteriol.* 172:5816–27

129. McCollum D, Monosov E, Subramani S. 1993. The *pas8* mutant of *Pichia pastoris* exhibits the peroxisomal protein import deficiencies of Zellweger syndrome cells—The PAS8 protein binds to the COOH-terminal tripeptide peroxisomal targeting signal, and is a member of the TPR protein family. *J. Cell Biol.* 121:761–74

130. Miura S, Mori M, Takiguchi M, Tatibana M, Furuta S, et al. 1984. Biosynthesis and intracellular transport of enzymes of peroxisomal β-oxidation. *J. Biol. Chem.* 259: 6397–6402

131. Miyazawa S, Osumi T, Hashimoto T, Ohno K, Miura S, Fujiki Y. 1989. Peroxisome targeting signal of rat liver acyl-CoA oxidase resides at the carboxy terminus. *Mol. Cell. Biol.* 9:83–91

132. Molina A, Segura A, Garcia-Olmedo F. 1993. Lipid transfer proteins (nsLTPs) from barley and maize leaves are potent inhibitors of bacterial fungal plant pathogens. *FEBS Lett.* 316:119–22

133. Mollenhauer HH, Morré DJ, Kelly AG. 1966. The widespread occurrence of plant cytosomes resembling animal microbodies. *Protoplasma* 62:44–52

134. Møller IM, Lin W. 1986. Membrane-bound NAD(P)H dehydrogenases in higher plant cells. *Annu. Rev. Plant Physiol.* 37:309–34

135. Moore TS Jr. 1982. Phospholipid biosynthesis. *Annu. Rev. Plant Physiol.* 33:235–59

136. Moreau P, Bertho P, Juguelin H, Lessire R. 1988. Intracellular transport of very long chain fatty acids in etiolated leek seedlings. *Plant Physiol. Biochem.* 26:173–78

137. Mori T, Tsukamoto T, Mori H, Tashiro Y, Fujiki Y. 1991. Molecular cloning and deduced amino acid sequence of nonspecific lipid transfer protein (sterol carrier protein 2) of rat liver: A higher molecular mass (60kDa) protein contains the primary sequence of nonspecific lipid transfer protein as its C-terminal part. *Proc. Natl. Acad. Sci. USA* 88:4338–42

138. Morré DJ. 1975. Membrane biogenesis. *Annu. Rev. Plant Physiol.* 26:441–81

139. Morré DJ, Merritt WD, Lambi CA. 1971. Connections between mitochondria and endoplasmic reticulum in rat liver and onion stem. *Protoplasma* 73:43–49

139a. Neupert W, Lill R, eds. 1992. *Membrane Biogenesis and Protein Targeting.* Amsterdam: Elsevier

140. Ni W, Trelease RN. 1991. Post-translational regulation of catalase isozyme expression in cotton seeds. *Plant Cell* 3:737–44

141. Nicolay K, Veenhuis M, Douma AC, Harder W. 1987. A ^{31}P NMR study of the internal pH of yeast peroxisomes. *Arch. Microbiol.* 147:37–41

142. Nishimura M, Takeuchi Y, De Bellis L, Hara-Nishimura I. 1993. Leaf peroxisomes are directly transformed to glyoxysomes during senescence of pumpkin cotyledons. *Protoplasma* 175:131–37

143. Nishimura M, Yamaguchi J, Mori H, Akazawa T, Yokota S. 1986. Immunocytochemical analysis shows that glyoxysomes are directly transformed to leaf peroxisomes during greening of pumpkin cotyledons. *Plant Physiol.* 80:313–16

144. Novikoff AB, Shin W-Y. 1964. The endoplasmic reticulum in the Golgi zone and its relations to microbodies, Golgi apparatus and autophagic vacuoles in rat liver cells. *J. Micros. (Oxford)* 3:187–206

145. Nuttley WM, Brade AM, Eitzen GA, Veenhuis M, Aitchison JD, et al. 1994. *PAY4*, a gene required for peroxisome assembly in the yeast *Yarrowia lipolytica*, encodes a novel member of a family of putative ATPases. *J. Biol. Chem.* 269:556–66

146. Ogren WL. 1984. Photorespiration: pathways, regulation, and modification. *Annu. Rev. Plant Physiol.* 35:415–42

147. Olsen LJ, Ettinger WF, Damsz B, Matsudaira K, Webb MA, Harada JJ. 1993. Targeting of glyoxysomal proteins to peroxisomes in leaves and roots of a higher plant. *Plant Cell* 5:941–52

148. Olsen LJ, Harada JJ. 1991. Biogenesis of peroxisomes in higher plants. In *Molecular Approaches to Compartmentation and Metabolic Regulation*, ed. AHC Huang, L Taiz, pp. 129–37. Rockville, MD: Am. Soc. Plant Physiol.

149. Onyeocha I, Behari R, Hill D, Baker A. 1993. Targeting of castor bean glyoxysomal isocitrate lyase to tobacco leaf peroxisomes. *Plant Mol. Biol.* 22:385–96

150. Osumi M, Fukuzumi F, Teranishi Y, Tanaka A, Fukui S. 1975. Development of mi-

crobodies in *Candida tropicalis* during incubation in a n-alkane medium. *Arch. Microbiol.* 103:1–11

151. Osumi T, Fujiki Y. 1990. Topogenesis of peroxisomal proteins. *BioEssays* 12:217–22

152. Osumi T, Tsukamoto T, Hata S, Yokota S, Miura S, et al. 1991. Amino-terminal presequence of the precursor of peroxisomal 3-ketoacyl-CoA thiolase is a cleavable signal peptide for peroxisomal targeting. *Biochem. Biophys. Res. Commun.* 181:947–54

153. Perry SE, Keegstra K. 1994. Envelope membrane proteins that interact with chloroplast precursor proteins. *Plant Cell* 6: 93–105

154. Pfanner N, Sollner T, Neupert W. 1991. Mitochondrial import receptors for precursor proteins. *Trends Biochem. Sci.* 16:63–67

155. Preisig-Muller R, Kindl H. 1993. Thiolase mRNA translated in vitro yields a peptide with a putative N-terminal presequence. *Plant Mol. Biol.* 22:59–66

155a. Preisig-Muller R, Muster G, Kindl H. 1994. Heat shock enhances the amount of prenylated DnaJ protein at membranes of glyoxysomes. *Eur. J. Biochem.* 219:57–63

156. Rapp S, Soto U, Just WW. 1993. Import of firefly luciferase into peroxisomes of permeabilized Chinese hamster ovary cells: a model system to study peroxisomal protein import in vitro. *Exp. Cell Res.* 205:59–65

157. Rhodin J. 1954. *Correlation of ultrastructural organization and function in normal and experimentally changed proximal convoluted tubule cells of the mouse kidney.* PhD thesis. *Aktiebolaget Godvil, Stockholm.* 76 pp.

158. Roggenkamp R. 1992. Targeting signals for protein import into peroxisomes. *Cell Biochem. Func.* 10:193–99

159. Roscher AA, Hoefler S, Hoefler G, Paschke E, Paltauf F, et al. 1989. Genetic and phenotypic heterogeneity in disorders of peroxisome biogenesis. A complementation study involving cell lines from 19 patients. *Pediatr. Res.* 26:67–72

160. Rouiller C, Bernhard W. 1956. "Microbodies" and the problem of mitochondrial regeneration in liver cells. *J. Biophys. Biochem. Cytol.* 2(Suppl.):355–59

161. Santos MJ, Hoefler S, Moser AB, Moser HW, Lazarow PB. 1992. Peroxisome assembly mutations in humans: structural heterogeneity in Zellweger syndrome. *J. Cell Physiol.* 151:103–12

162. Santos MJ, Imanaka T, Shio H, Small GM, Lazarow PB. 1988. Peroxisomal membrane ghosts in Zellweger syndrome-aberrant organelle assembly. *Science* 239: 1536–38

163. Sautter C. 1986. Microbody transition in greening watermelon cotyledons. Double

immunocytochemical labeling of isocitrate lyase and hydroxypyruvate reductase. *Planta* 167:491–503

164. Schnell DJ, Blobel G. 1993. Identification of intermediates in the pathway of protein import into chloroplasts and their localization to envelope contact sites. *J. Cell Biol.* 120:103–15

165. Schnell DJ, Blobel G, Pain D. 1990. The chloroplast import receptor is an integral membrane protein of chloroplast envelope contact sites. *J. Cell Biol.* 111:1825–38

166. Schubert KR. 1986. Products of biological nitrogen fixation in higher plants: synthesis, transport, and metabolism. *Annu. Rev. Plant Physiol.* 37:539–74

167. Sikorski RS, Boguski MS, Goebl M, Hieter P. 1990. A repeating amino acid motif in CDC23 defines a family of proteins and a new relationship among genes required for mitosis and RNA synthesis. *Cell* 60:307–17

168. Small GM, Santos MJ, Imanaka T, Poulos A, Danks DM, et al. 1988. Peroxisomal integral membrane proteins are present in livers of patients with Zellweger syndrome, infantile Refsum's disease and X-linked adrenoleukodystrophy. *J. Inher. Metab. Dis.* 11:358–71

169. Small GM, Szabo LJ, Lazarow PB. 1988. Acyl-CoA oxidase contains two targeting sequences each of which can mediate protein import into peroxisomes. *EMBO J.* 7: 1167–73

170. Somerville C, Browse J. 1991. Plant lipids: metabolism, mutants, and membranes. *Science* 252:80–87

171. Sossountzov L, Ruiz-Avila L, Vignois F, Jolliot A, Arondel V, et al. 1991. Spatial and temporal expression of a maize lipid transfer protein gene. *Plant Cell* 3:923–33

172. Sterk P, Booij H, Schellekens GA, Van Kammen A, De Vries SC. 1991. Cell-specific expression of the carrot EP2 lipid transfer gene. *Plant Cell* 3:907–21

173. Struglics A, Fredlund KM, Rasmusson AG, Møller IM. 1993. The presence of a short redox chain in the membrane of intact potato tuber peroxisomes and the association of malate dehydrogenase with the peroxisomal membrane. *Physiol. Plantarum* 88: 19–28

174. Subramani S. 1991. Peroxisomal targeting signals—the end and the beginning. *Curr. Sci.* 61:28–32

175. Subramani S. 1992. Targeting of proteins into the peroxisomal matrix. *J. Membr. Biol.* 125:99–106

176. Subramani S. 1993. Protein import into peroxisomes and biogenesis of the organelle. *Annu. Rev. Cell Biol.* 9:445–78

177. Sulter GJ, Waterham HR, Vrieling EG, Goodman JM, Harder W, Veenhuis M. 1993. Expression and targeting of a 47 kDa

integral peroxisomal membrane protein of *Candida boidinii* in wild type and a peroxisome-deficient mutant of *Hansenula polymorpha. FEBS Lett.* 315:211–16

178. Suzuki Y, Yamaguchi S, Orii T, Tsuneoka M, Tashiro Y. 1990. Non-specific lipid transfer protein (sterol carrier protein 2) defective in patients with deficient peroxisomes. *Cell Struct. Funct.* 15:301–8

179. Swinkels BW, Gould SJ, Bodnar AG, Rachubinski RA, Subramani S. 1991. A novel cleavable peroxisomal targeting signal at the amino-terminus of the rat 3-ketoacyl-CoA thiolase. *EMBO J.* 10:3255–62

180. Tchang F, This P, Stiefel V, Arondel V, Morch M-D, et al. 1988. Phospholipid tranfer protein: full-length cDNA and amino acid sequence in maize. *J. Biol. Chem.* 263:16849–55

181. Thieringer R, Shio H, Han Y, Cohen G, Lazarow PB. 1991. Peroxisomes in *Saccharomyces cerevisiae*: immunofluorescence analysis and import of catalase A into isolated peroxisomes. *Mol. Cell. Biol.* 11:510–22

182. Titus DE, Becker WM. 1985. Investigation of the glyoxysome-peroxisome transition in germinating cucumber cotyledons using double-label immunoelectron microscopy. *J. Cell Biol.* 101:1288–99

183. Tolbert NE. 1971. Microbodies—peroxisomes and glyoxysomes. *Annu. Rev. Plant Physiol.* 22:45–74

184. Tolbert NE. 1980. Photorespiration. *Biochem. Plants* 2:487–523

185. Tolbert NE. 1981. Metabolic pathways in peroxisomes and glyoxysomes. *Annu. Rev. Biochem.* 50:133–57

186. Tolbert NE, Essner E. 1981. Microbodies: peroxisomes and glyoxysomes. *J. Cell Biol.* 91:271S–283S

187. Trelease RN. 1984. Biogenesis of glyoxysomes. *Annu. Rev. Plant Physiol.* 35:321–47

188. Trelease RN, Doman DC. 1984. Mobilization of oil and wax reserves. In *Seed Physiology*, ed. DR Murray, pp. 202–45. Sydney: Academic

189. Tsukamoto T, Miura S, Fujiki Y. 1991. Restoration by a 35K membrane protein of peroxisome assembly in a peroxisome-deficient mammalian cell mutant. *Nature* 350: 77–81

190. Tsukamoto T, Yokota S, Fujiki Y. 1990. Isolation and characterization of Chinese hamster ovary cell mutants defective in assembly of peroxisomes. *J. Cell Biol.* 110: 651–60

191. Turley RB, Trelease RN. 1990. Development and regulation of three glyoxysomal enzymes during cotton seed maturation and growth. *Plant Mol. Biol.* 14:137–46

192. Ueda M, Tanaka A, Horikawa S, Numa S, Fukui S. 1984. Synthesis in vitro of precursor-type carnitine acetyltransferase with messenger RNA from *Candida tropicalis. Eur. J. Biochem.* 138:451–57

193. van den Bosch H, Schutgens RBH, Wanders RJA, Tager JM. 1992. Biochemistry of peroxisomes. *Annu. Rev. Biochem.* 61:157–97

194. Van der Leij I, Franse MM, Elgersma Y, Distel B, Tabak HF. 1993. PAS10 is a tetratricopeptide-repeat protein that is essential for the import of most matrix proteins into peroxisomes of *Saccharomyces cerevisiae. Proc. Natl. Acad. Sci. USA* 90:11782–86

195. Van der Leij I, Van den Berg M, Boot R, Franse M, Distel B, Tabak HF. 1992. Isolation of peroxisome assembly mutants from *Saccharomyces cerevisiae* with different morphologies using a novel positive selection procedure. *J. Cell Biol.* 119:153–62

196. Van Heusden GPH, Bos K, Raetz CRH, Wirtz KWA. 1990. Chinese hamster ovary cells deficient in peroxisomes lack the nonspecific lipid transfer protein (sterol carrier protein 2). *J. Biol. Chem.* 263:4105–10

197. van Veldhoven P, de Beer LJ, Mannaerts GP. 1983. Water- and solute-accessible spaces of purified peroxisomes. Evidence that peroxisomes are permeable to NAD^+. *Biochem. J.* 210:685–93

198. van Veldhoven PP, Just WW, Mannaerts GP. 1987. Permeability of the peroxisomal membrane to cofactors of β-oxidation. *J. Biol. Chem.* 262:4310–18

199. van Veldhoven PP, Mannaerts GP. 1994. Assembly of the peroxisomal membranes. In *Subcellular Biochemistry*, ed. JR Harris, AH Maddy, 22:231–61. New York: Plenum

200. Veenhuis M, Goodman JM. 1990. Peroxisomal assembly: membrane proliferation precedes the induction of the abundant matrix proteins in the methylotrophic yeast *Candida boidinii. J. Cell Sci.* 96:583–90

201. Veenhuis M, Harder W. 1989. Occurrence, proliferation, and metabolic function of yeast microbodies. *Yeast* 5:517–24

202. Veenhuis M, van Dijken JP, Pilon SAF, Harder W. 1978. Development of crystalline peroxisomes in methanol-grown cells of the yeast *Hansenula polymorpha* and its relation to environmental conditions. *Arch. Microbiol.* 117:153–63

203. Verner K, Schatz G. 1988. Protein translocation across membranes. *Science* 241: 1307–13

204. Walton PA, Gould SJ, Feramisco JR, Subramani S. 1992. Transport of microinjected proteins into peroxisomes of mammalian cells: inability of Zellweger cell lines to import proteins with the SKL tripeptide peroxisomal targeting signal. *Mol. Cell Biol.* 12:531–41

205. Waterham HR, Keizer-Gunnick I, Goodman JM, Harder W, Veenhuis M. 1990.

Immunocytochemical evidence for the acidic nature of peroxisomes in methylotrophic yeasts. *FEBS Lett.* 262:17–19

206. Webb MA, Newcomb EH. 1987. Cellular compartmentation of ureide biogenesis in root nodules of cowpea (*Vigna unguiculata* (L.) Walp.). *Planta* 172:162–75

207. Wendland M, Subramani S. 1993. Cytosol-dependent peroxisomal protein import in a permeabilized cell system. *J. Cell Biol.* 120:675–85

208. Whitney AB, Bellion E. 1991. ATPase activities in peroxisome-proliferating yeast. *Biochim. Biophys. Acta* 1058:345–55

209. Wickner WT, Lodish HF. 1985. Multiple mechanisms of protein insertion into and across membranes. *Science* 230:400–7

210. Wiebel FF, Kunau W-H. 1992. The Pas2 protein essential for peroxisome biogenesis for peroxisome biogenesis is related to ubiquitin-conjugating enzymes. *Nature* 359:73–76

211. Wirtz KWA. 1991. Phospholipid transfer proteins. *Annu. Rev. Biochem.* 60:73–99

212. Wolins NE, Donaldson RP. 1994. Specific binding of the peroxisomal protein targeting sequence to glyoxysomal membranes. *J. Biol. Chem.* 269:1149–53

213. Wood C, Jalil MNH, McLaren I, Yong BCS, Ariffin A, et al. 1984. Carnitine long-chain acyltransferase and oxidation of palmitate, palmitoyl coenzyme A and palmitoylcarnitine by pea mitochondria preparations. *Planta* 161:255–60

214. Yamaguchi J, Nishimura M, Akazawa T. 1984. Maturation of catalase precursor proceeds to a different extent in glyoxysomes and leaf peroxisomes of pumpkin cotyledons. *Proc. Natl. Acad. Sci. USA* 81:4809–13

215. Yamamoto K, Fahimi HD. 1987. Three-dimensional reconstuction of a peroxisomal reticulum in regenerating rat liver: evidence of interconnections between heterogeneous segments. *J. Cell Biol.* 105:713–22

216. Zaar K, Volkl A, Fahimi HD. 1987. Association of isolated bovine kidney cortex peroxisomes with endoplasmic reticulum. *Biochim. Biophys. Acta* 897:135–42

217. Zhang JZ, Laudencia-Chingcuanco DL, Comai L, Li M, Harada JJ. 1994. Isocitrate lyase and malate synthase genes from *Brassica napus* L. are active in pollen. *Plant Physiol.* 104:857–64

218. Zoeller RA, Allen L-A, Santos MJ, Lazarow PB, Hashimoto T, et al. 1989. Chinese hamster ovary cell mutants defective in peroxisome biogenesis. Comparison to Zellweger syndrome. *J. Biol. Chem.* 264:21872–78

219. Zoeller RA, Raetz CRH. 1986. Isolation of animal cell mutants deficient in plasmalogen biosynthesis and peroxisome assembly. *Proc. Natl. Acad. Sci. USA* 83:5170–74

Annu. Rev. Plant Physiol. Plant Mol. Biol. 1995. 46:147–66

REGULATION OF CHLOROPLAST GENE EXPRESSION

Stephen P. Mayfield, Christopher B. Yohn, Amybeth Cohen, and Avihai Danon

Department of Cell Biology, The Scripps Research Institute, La Jolla, California 92037

KEY WORDS: gene regulation, plant gene expression, plastid, organelle, translation

CONTENTS

0066-4294/95/0601-0147$05.00

ABSTRACT

Plastid gene expression during plant growth and development requires several independent processes. Although the plastid contains all the basic components for gene expression, it must rely on the import of nuclear-encoded proteins to carry out plastid biogenesis and photosynthesis. By understanding the individual roles of transcription, mRNA processing, mRNA stability, and mRNA translation in plastids, and how they depend on the nuclear genome, general trends can be identified in the complex mechanisms of chloroplast gene expression.

INTRODUCTION

Chloroplast gene expression involves both the activation of a set of plastid genes whose products are required in the chloroplast and the modulation of gene expression within the developed plastid in response to changing environmental conditions. Various aspects of chloroplast gene expression have been reviewed in detail elsewhere (18, 25, 31, 67, 74). This review attempts to reduce the vast literature on plastid gene expression to some general rules addressing plastid transcription, mRNA processing, mRNA stability, and mRNA translation as regulators of plastid gene expression. Although post-translational processes such as protein turnover are key to plastid gene expression, there is little current information from which clear conclusions can be drawn, so such processes are not covered here.

Chloroplast gene expression is linked with the expression of nuclear-encoded genes. These nuclear genes encode key structural and enzymatic elements required for plastid function, as well as the vast majority of regulatory factors identified to date. Understanding the regulation of chloroplast gene expression must therefore involve an examination of the role of nuclear-encoded regulatory factors. These nuclear-encoded factors are also addressed in this review.

TRANSCRIPTION AS A REGULATOR OF PLASTID GENE EXPRESSION

Examination of plastid mRNA accumulation in a number of species has shown that mRNA levels fluctuate in response to both developmental and environmental signals. Although both transcription rate and mRNA stability generally are agreed to be key components in determining plastid mRNA levels, there is some debate as to which of these mechanisms is dominant in regulating mRNA accumulation of specific plastid mRNAs (reviewed in 31, 67). Transcription rates change under a variety of developmental and environmental conditions, but these changes appear to affect most or all of the genes examined rather than specific individual genes.

Plastid Genes Contain Simple Prokaryotic-Like Promoters

Comparison of sequences from plastid promoter regions shows that plastid genes contain consensus bacterial-like -35 (TTGACA) and -10 (TATAAT) sequences. In vitro analysis of these promoters shows that the -35 region is not always essential for promoter activity (32, 33) and is not always identifiable in plastid genes (48). In vivo analysis of the *atpB* promoter shows that deletion of 5' sequences, up to -12 from the transcription start, still allows for measurable accumulation of these mRNAs (6, 48). Based on these analyses, plastid gene promoters appear to be simple elements that contain a -10 consensus, and in some cases a -35-like element, but that require few additional sequences for basal transcriptional activity.

Transcription Rates Vary Among Promoters

Although plastid promoters appear to contain only one or two consensus elements 5' of the transcription start, transcription rates still vary a great deal from one promoter to another. Transcription rates have been measured in at least a dozen plastid genes and range up to 50-fold or greater in a single plastid type (6, 19, 20, 46, 55, 70, 72, 80). Between plant species, relative transcription rates can differ for a single gene. For example, *psbA* transcription is about 30% that of the 16S rRNA in spinach (19), only 15% in *Chlamydomonas reinhardtii* (6), and 150% in barley (72). As a general rule, however, genes that are transcribed at reduced rates in one species are also transcribed at reduced rates in other species. If plastid promoters contain only a -35 and a -10 element, then variations in transcription from different promoters must be affected by either downstream sequences or by the position of the gene within the chloroplast chromosome. Transcription rates from chimeric genes, in which foreign mRNAs are transcribed from plastid promoters, show that these chimeric RNAs are transcribed at similar rates to the corresponding endogenous RNA and that this transcription appears to be independent of the position of the gene within the chromosome (6, 29, 78, 80, 81, 85). In most of these chimeric constructs, sequences downstream of the transcription start were included. In vivo analysis of chimeric constructs containing the *atpB* promoter shows that downstream sequences, up to +55 after transcription start, are required for high levels of transcription (48). These data indicate that sequences downstream of the -10 and -35 elements may have some role in determining the rate of transcription from individual promoters. Perhaps these elements act as transcription enhancers, as have been identified for the *psbA* gene in cyanobacteria (56). In addition, a nuclear mutation has been described that reduces transcription of the *rbcL* chloroplast gene in *C. reinhardtii* (39a). This report supports the possibility of gene-specific promoters.

Overall Transcription Rates Change During Plastid Development

Measurement of plastid gene transcription in developing barley seedlings indicates that overall transcription rates change during development and that a few specific genes are differentially transcribed during this period (49). Examination of photosynthetic gene expression during barley chloroplast development again shows a concomitant change in the transcription rates for many plastid genes, depending on the stage of plastid development, and shows that a subset of these genes appears to be differentially transcribed (3). The relative transcription rates of several photosynthetic mRNAs have been measured during chloroplast development (19) and in plastids of nonphotosynthetic spinach cells (20). Under these conditions, transcription was found to be essentially the same in all plastid types, suggesting that transcription is of minor importance in determining the level of plastid mRNA accumulation in these cells. Changes in transcription of chloroplast genes measured during cell cycles in *C. reinhardtii* show that transcription of plastid genes fluctuates during these cycles (55). However, these changes appear to affect most or all of the genes examined, suggesting that overall transcription rather than transcription of specific genes is regulated in this way.

From the available information to date it seems that transcription of plastid genes fluctuates in relation to plastid development, plastid type, and cell cycle, but the relative transcription of individual genes varies only slightly compared to the overall change in total transcriptional activity. This result suggests that transcription is a general regulator of plastid gene expression rather than the rate-limiting step for expression of individual genes.

Transcription Rates Change During Light and Dark Growth

The transcription of many plastid genes changes markedly during light and dark growth phases. In tomato fruit, transcription of both plastid and nuclear-encoded photosynthetic genes increases during the night and early morning (70). In *C. reinhardtii* cells grown under a 12-hour-light/12-hour-dark cycle, a 10-fold increase in transcription is observed during the dark phase, but this increase directly correlates with total UTP incorporation, rather than with transcription of a specific set of genes (55). This diurnal fluctuation of transcription in *C. reinhardtii* is the result of a light-entrained circadian cycle (81), which again appears to affect many genes in a similar manner. In barley, shifting five-day-old plants from dark into light has little effect on transcription (68). However, when barley seedlings are held eight days in the dark and then shifted into the light there is a marked, although transient, increase in plastid gene transcription, which affects some mRNAs to a greater extent than others (47). These data show that light-dark cycles influence the transcription of plastid genes, but in most cases this induction seems to affect most or all of the genes measured.

The best evidence to date for specific light-activated transcription comes from the *psbD-psbC* operon of barley (83). In barley, *psbD* is encoded in a polycistronic mRNA that also contains the *psbK*, *psbI*, *psbC*, *orf62*, and *trnG* mRNAs. Transcription of this polycistronic message is normally light independent. However, a subset of this gene cluster containing the *psbD* and *psbC* mRNAs is transcribed from an internal promoter in a light-dependent manner (83). The overall impact of this light-activated transcription is to raise the level of *psbD-psbC* mRNA, perhaps by 50% above dark conditions.

Gene Expression as a Function of Transcriptional Regulation

A generalization from the available data is that plastid promoters are very simple elements, perhaps requiring as little as a prokaryotic-like −10 element for basal activity. Additional sequences, downstream of the transcription start, may act as enhancer or repressor elements that influence the rate of transcription from individual promoters. These downstream elements may give rise to the differential transcription rates observed between individual plastid genes. Regulation of plastid transcription appears to influence the accumulation of the total population of plastid mRNAs. This transcriptional activation, which is responsive to light-dark cycles, development of the plastid, and plastid type, appears in most cases to affect many genes in unison. There are exceptions [e.g. the *psbD-psbC* operon of barley (12, 83)] in which some promoters appear to contain additional information that may result in specific gene expression.

RNA PROCESSING AS A REGULATOR OF PLASTID GENE EXPRESSION

RNA processing is not defined well enough to assign it a role in the regulation of chloroplast gene expression. Scattered examples and hints of regulation allow for the possibility of such a role, but no general rule can yet be established for chloroplast gene regulation utilizing RNA processing.

Transcript Processing Has a Limited Role in Gene Regulation

Although differential processing is a potential target for the regulation of plastid gene expression, little evidence has been identified to support this idea. An example of processing that affects gene expression is the methyl jasmonate–induced reduction in *rbcL* mRNA translation (73). Altered processing that produces an *rbcL* transcript with a longer 5′-untranslated region (UTR) is responsible for the reduction in translation. Many plastid genes, both polycis-tronic and monocistronic, give rise to multiple, overlapping RNAs through a variety of processing steps (4, 40, 62). These patterns of RNA accumulation can differ between developmental stages and environmental conditions (15, 89). This type of processing does not appear to be a general regulator of plastid gene expression, because processing is not required for translation of individ-ual coding regions within polycistronic mRNAs (1). One exception is the *crp1* mutant of maize (2a), in which lack of the monocistronic form of *petD* results in decreased synthesis of the *petD* gene product.

Intron Splicing Is Essential but Does not Regulate Gene Expression

Introns are prevalent in the chloroplast and in some cases are similar to nuclear pre-mRNA introns where they contribute to gene regulation. However, intron splicing has not been shown to be a regulator of chloroplast gene expression. Group I introns are found in bacteria, in rRNAs of both the chloroplast and nucleus, in protein-coding genes such as *psbA* in *C. reinhardtii*, and occasion-ally in higher plant chloroplasts (64; reviewed in 74). Group II introns are found almost exclusively in organelles and most of those in the chloroplast are degenerate Group II introns. Other more complex introns also have been identified. These are actually introns-within-introns, termed twintrons (13, 23; reviewed in 14). Several chloroplast introns contain open reading frames (ORFs) that encode endonucleases or RNA-splicing maturases (reviewed in 79). These complexities suggest that regulation must occur; however, no obvi-ous examples are available.

Group II and nuclear pre-mRNA introns have in common similar secondary RNA structures, similar mechanisms of splicing, and some specific base pair-ing interactions (reviewed in 63). The 5′ splice sites are similar, and mis-splic-ing in degenerate Group II introns and twintrons is similar to and may be a precursor to alternative splicing in nuclear genes. Although none of the regula-tory mechanisms present in nuclear pre-mRNA splicing have been identified in chloroplasts, their mechanistic similarities suggest that intron splicing may play some unknown role in chloroplast gene expression.

Many Nuclear Loci Are Dedicated to Trans-Splicing

Trans-splicing of plastid mRNAs has not yet been identified as a regulator of gene expression, although much of the data on *trans*-splicing suggests such a role. *Trans*-splicing is necessary to recombine gene segments that are discon-tinuous, and it has been identified in *rps12* and *psaA* mRNAs (reviewed in 74). Three scattered exons encode each of these genes. The exons and surrounding sequences form a structure characteristic of a Group II intron. In the case of *psaA*, a fourth chloroplast RNA, *tscA*, is necessary for efficient splicing, presumably because it completes the catalytic core of the intron (28). Fourteen nuclear loci have been identified that appear to be dedicated to *trans*-splicing of the chloroplast mRNAs of the three exons of *psaA* (29). The identification of so many nuclear loci dedicated to *trans*-splicing suggests that this process is a target for gene regulation that has yet to be identified.

RNA Editing Can Influence Gene Expression

Another important mechanism in chloroplast RNA processing is editing (reviewed in 30, 71). RNA editing, which involves changing specific bases of an RNA from those that are encoded in the DNA, plays an important role in producing a fully functional RNA molecule. An ACG start codon is changed to a functional AUG start codon in *rpl2* mRNA of maize and in *psbL* mRNA of tobacco and spinach (8, 39, 53). Editing of the spinach *psbF-psbL* transcript is differentially down-regulated in seeds and roots (8); the unedited *psbL* transcript lacks an initiator codon, and as such is untranslatable. In *ndhA* of maize, several codons are edited to restore conserved amino acids in the expressed protein (60). Some editing sites have also been discovered within introns (92), perhaps playing a role in maintaining intron structure and/or function. Editing precedes both splicing and cleavage to monocistronic mRNAs (77) and does not require chloroplast translation (94), indicating that protein components of the editing machinery must, by default, be encoded by the nucleus and imported into the chloroplast.

Regulation of mRNA Processing

The complex mechanisms used to accomplish RNA processing seem obvious targets for regulation of gene expression. However, few examples of regulated or differential processing have been elucidated. Alternative processing of the *rbcL* transcript has a role in *rbcL* gene expression (73), and observations of the *crp1* mutant (2a) suggest that monocistronic *petD* mRNA is translated more efficiently than is the polycistronic form. These examples seem to be exceptions and not the rule. Several chloroplast introns self-splice in vitro (37, 38, 52, 91), suggesting little or no regulation of this processing step. However, the requirement for nonphysiological conditions for this self-splicing implies that other components may be necessary in vivo. The nuclear factors involved in *trans*-splicing of *psaA* (29) and the chloroplast-localized ribonucleoproteins studied by Sugiura's group (57–59) are obvious candidates for regulating splicing. However, regulatory roles for any of these proteins have yet to be identified, and no direct evidence is available that splicing is used to regulate chloroplast gene expression. Regulation of editing, on the other hand, seems to play a direct role in plastid gene expression. Most editing sites identified in the chloroplast have functional significance, whether it is editing to introduce a functional start codon, or more subtle editing to restore conserved amino acids or intron sequences important in splicing. In the case of the spinach *psbF-psbL*, editing appears to play a direct role in the developmental regulation of plastid gene expression.

mRNA STABILITY AS A REGULATOR OF PLASTID GENE EXPRESSION

As discussed earlier, the differential accumulation of many plastid-encoded mRNAs appears to be regulated at the posttranscriptional level (19, 21, 44). Direct measurement of mRNA stability has shown that plastid mRNAs have dramatically different half-lives depending on the plastid developmental stage, type, and growth condition (reviewed in 31, 67). The UTRs of mRNAs, as well as the proteins that interact with them, have been examined as potential regulators of RNA stability.

mRNA Untranslated Regions Have a Role in RNA Stability

Investigation into the role UTRs and stem-loop structures may play in the regulation of mRNA accumulation has shown that both the 3'- and 5'-UTRs contribute to mRNA stability. Most chloroplast-encoded monocistronic and polycistronic mRNAs contain inverted repeats (IRs) within their 3'-UTR. IRs have the potential to form stem-loop structures, and by inference from other systems in which similar structures occur, it was proposed that these 3' stem-loops may play a role in RNA processing or as protective structures against nucleolytic degradation (reviewed in 31). Several chloroplast-encoded mRNAs also contain sequences within the 5'-UTR that have the potential to form stem-loop structures, again suggesting that these may play a role in mRNA stability (reviewed in 74).

3'-INVERTED REPEATS ACT AS PROCESSING ELEMENTS BUT NOT mRNA STABILITY ELEMENTS   Experiments using a spinach in vitro transcription system show that the 3'-IRs of several chloroplast-encoded mRNAs can function as RNA processing elements (86). RNAs containing IRs in their 3'-ends are also more stable in chloroplast extracts than are RNAs without IRs. However, deletions into the 3'-IR of the chloroplast-encoded *petD* RNA do not always result in reduced stability of the RNA in vitro (87). In addition, a two-base alteration in the 3'-IR of *petD*, which is predicted to form a stable stem-loop structure, results in unstable *petD* RNA in vitro. These conflicting data indicate that a 3'-IR element alone is not sufficient to confer RNA stability in vitro, but that sequences within the 3'-UTRs can have some impact on RNA accumulation.

In vivo experiments with chloroplast transformants of *C. reinhardtii* containing truncated *atpB* genes support the role of 3' stem-loop structures in correct 3'-end formation, but once again show a poor correlation between the requirement of an intact 3'-IR and mRNA stability (88). Fusion of the *petD* 3'-IR in both orientations 3' to the *atpB*-coding region maintains accurate processing and stability of the chimeric mRNA, indicating that orientation of the IR is not important in 3'-IR function. Deletions into this 3'-IR show that a complete IR is not a requirement for *atpB* mRNA accumulation in vivo (88), although large deletions into the IR region eventually result in a substantial reduction of *atpB* mRNA accumulation. These data show that a complete IR is not essential for mRNA accumulation in vivo, but that some portions of the 3'-UTR are required.

In vivo analysis of the *psaB* and *rbcL* 3'-IRs in *C. reinhardtii* supports IRs as necessary components for 3'-end formation in vivo but shows that IRs have limited importance in determining mRNA stability (7). *C. reinhardtii* cells transformed with a chimeric construct fused to either of these 3'-IR sequences accumulate chimeric transcripts of a single size. Transformants containing constructs without 3'-IRs accumulate chimeric mRNA near wild-type levels but of heterogeneous length. When multiple 3'-IRs are coupled to a reporter gene, transcripts accumulate that are terminated at the first IR in the forward orientation (7). These data show that the stem-loop structure formed by the 3'-IR, or adjacent sequences, act as recognition sites in vivo for endonucleolytic cleavage distal to each stem-loop. Exonuclease digestion in the 3' to 5' direction (86) cannot be involved in 3'-end processing of these mRNAs in vivo because this mechanism would have yielded transcripts of increased length when multiple IR sequences were added to the 3'-end of the chimeric gene. The rate of mRNA decay from these constructs in vivo is the same with and without the presence of a 3'-IR at the end of the mRNA (7). Furthermore, transformants containing a chimeric *rbcL* 5'-UTR fused to a reporter gene and two different 3'-ends produce mRNAs with identical half-lives, whereas chimerics containing different 5'-ends and the same *psaB* 3'-UTR produce mRNAs with very different half-lives. These data indicate that 3'-UTRs do not play a critical role in determining mRNA half-life in vivo, but 5'-UTRs may have a role in determining the half-life of a message (7).

Investigation of plastid-encoded mRNA 3'-IRs as protein-binding sites has demonstrated that most of the proteins that bind in vitro to these sequences are involved in RNA processing. Both endonuclease and exonuclease activities have been identified in spinach chloroplasts that act at the 3'-IR of *petD* pre-mRNA (11, 86). Whether these 3'-associated ribonuclease activities influence the stability of the *petD* mRNA in vivo is unknown.

THE 5'-UTR MAY CONTAIN ELEMENTS INVOLVED IN mRNA STABILITY   As described above, 5'-UTRs appear to have some role in mRNA stability. Whether this is a direct role or a pleiotropic effect resulting from some other function of these UTRs has not been determined. Sequential deletions into the *petD* 5'-UTR show that the entire 5'-UTR is required for stability of a chimeric transcript in vivo (78). Transformation of *C. reinhardtii* cells with the *rbcL* 5'-UTR fused to a reporter gene results in accumulation of a chimeric transcript in the dark that is rapidly degraded when cells are exposed to the light (80). Rapid turnover of the chimeric transcript, and not variability in the transcription rate, accounts for the light-induced degradation of the RNA. Adding part of the *rbcL*-coding region to this chimera stabilizes the transcript under both light and dark growth conditions. Transformation of *C. reinhardtii* cells with a *psbA* gene containing either a deletion of the ribosome binding site (RBS) or an alteration of bases resulting in pairing of the RBS caused a 90% or greater decrease in *psbA* mRNA accumulation (62a). Although these data support the 5'-UTR as an important component in mRNA stability, the mechanism of this stabilization is unknown. As discussed below, nuclear mutants that affect chloroplast mRNA stability also identify the 5'-UTR of chloroplast mRNAs as an important component of mRNA stability.

Nuclear Mutations Affect Chloroplast mRNA Stability

Several nuclear mutants of *C. reinhardtii* have been identified that fail to accumulate individual chloroplast-encoded mRNAs, despite having wild-type levels of transcription (reviewed in 74). Plastid-encoded *psbC* and *psbB* mRNAs are unstable in the nuclear mutants 6.2z5 and GE2.10, respectively (84). Accumulation of *psbB* mRNA is also greatly reduced in the nuclear mutant 222E (66), as is *psbD* mRNA in *nac2-26* (50). The *psbD* 3'-UTR has no influence on accumulation of a chimeric transcript in transformed *nac2-26* cells (69). A chimeric *psbD* 5'-UTR fused to a reporter gene accumulates the chimeric mRNA in wild-type but not in the *nac2-26* strain, suggesting that the *nac2-26* gene product interacts with the 5'-UTR of *psbD* to stabilize the mRNA. Rapid degradation of an in vitro–synthesized *psbD* 5'-UTR in *nac2-26* chloroplast lysates and not in wild-type lysates further demonstrates the influence of the 5'-UTR on message stability.

There are few examples of proteins that bind to the 5'-UTR of photosynthetic-related mRNAs to influence the stability of the transcript. The instability of the *psbD* mRNA in *nac2-26* lysates of *C. reinhardtii* has been reported to result from the loss of binding activity of a 47-kDa protein to the 5'-UTR of the mRNA (69). However, contradictory results showing that the 47-kDa protein is present and binds to the *psbD* 5'-UTR in *nac2-26* have also been obtained (CB Yohn, unpublished data). Therefore, despite the indications that nuclear-encoded factors confer stability to the *psbD* message by interacting with the 5'-UTR, identification of these proteins remains unresolved.

mRNA Stability Is Influenced by Ribosome Association

Analysis of translation-deficient mutants has shown that mRNA stability is not directly affected by mRNA translation but may be affected by ribosome association. Nuclear mutants deficient in the translation of the *psbD* mRNA (e.g. *nac1-11* and *nac1-18* in *C. reinhardtii*) accumulate levels of total and polysome-associated *psbD* mRNA greater than wild-type despite the absence of D2 synthesis (51; A Cohen, unpublished data). Association of the *psbD* mRNA with ribosomes in this mutant may lead to this increased stability. Barkan (2) has demonstrated that the *rbcL* mRNA is destabilized in several nuclear mutants of maize as a result of a decrease in mRNA-ribosome association. The *psbA* mRNA is less stable in mutants of *C. reinhardtii* containing an altered RBS (discussed earlier), possibly as a result of the inability of the mRNA to associate with ribosomes. A nuclear mutant of *C. reinhardtii*, F35, which fails to translate the *psbA* mRNA (26), also has reduced (25% of wild-type) accumulation of the *psbA* mRNA and diminished ribosome association (A Cohen, unpublished data). Other examples of mRNA stability resulting from association with ribosomes can be found in prokaryotes and eukaryotes (10). Perhaps the nuclear-encoded factors that affect mRNA stability are components of the translational apparatus that interacts with the 5'-UTR elements to regulate mRNA-ribosome association and, hence, mRNA stability.

Regulatory Components of mRNA Stability

Chloroplast mRNA turnover dramatically affects mRNA accumulation and, as a result, chloroplast gene expression. Analysis of the RNA elements and protein factors responsible for this differential mRNA accumulation suggests that 3'-UTRs can influence mRNA accumulation, but this effect does not appear to directly involve the 3'-IR elements. These IR elements appear to function as message-processing sites that are used to determine the 3'-end of the mRNA. RNA elements contained within the 5'-UTR appear to have a more profound affect on mRNA accumulation than do their 3' counterparts. To date, neither discrete RNA elements within the 5'-UTRs nor the protein components associated with these UTRs have been defined. Many of these protein factors may be involved in mRNA translation and only indirectly be involved with mRNA turnover, because a correlation between mRNA-ribosome interaction and mRNA stability clearly exists. The isolation and characterization of the nuclear factors involved in mRNA stability and the identification of the RNA elements with which these proteins interact should help identify the regulatory components of mRNA stability within the chloroplast.

TRANSLATION AS A REGULATOR OF CHLOROPLAST GENE EXPRESSION

The most dramatic impact on chloroplast protein accumulation is observed during the light-induced greening of the plastid. During this time the accumulation of some chloroplast proteins can increase more than 10,000-fold. Although transcription and mRNA turnover can influence this protein accumulation, the largest factor affecting this process is protein synthesis. Translation of specific chloroplast mRNAs can increase more than 100-fold during chloroplast biogenesis (reviewed in 25, 74).

Translation in chloroplasts has similarities to prokaryotic translation. Chloroplasts contain 70S ribosomes, fMet-initiator tRNA, and mRNAs that are not capped and do not carry a poly-A tail. A high degree of homology is found between rRNAs of chloroplasts and prokaryotes (24, 82), and sequences similar to Shine-Delgarno (SD) sequences (prokaryotic RBSs) can be found in many chloroplast mRNAs (9, 76). However, there are also several differences between translation in chloroplasts and prokaryotes. The spacing between the SD homologous sequences and the initiator codon is not as stringent in chloroplasts as in prokaryotes (9, 76). Chloroplast translation is not coupled to transcription, and chloroplast mRNAs can exist as stable ribonucleic protein (RNP) complexes with half-lives up to 40 h (44, 45). These differences indicate that translation in the chloroplast diverged from that of prokaryotes in both the components of translation and in the regulatory pathways that control it.

Nuclear-Encoded Factors Are Required for Chloroplast mRNA Translation

Genetic and molecular analyses of nuclear mutations have identified nuclear gene products as regulators of chloroplast mRNA translation (26, 42, 51, 75). These gene products affect the translation of specific individual mRNAs in the chloroplast and more than one factor may be required for the translation of any one chloroplast mRNA (reviewed in 25, 74). Each of these mutations identifies nuclear gene products as being functional components of chloroplast translation. A chloroplast suppressor of a nuclear mutant deficient in *psbC* translation identifies the site for interaction of the nuclear-encoded factor as the 5'-UTR of the *psbC* mRNA (75).

5'-UTRs Are Required for Translation

The effect of 5'-UTRs on chloroplast gene expression has been examined using chimeric genes containing the 5'-UTR of chloroplast mRNAs fused with coding sequences of reporter genes. Fusion of either the 3'- or 5'-UTR of *rps12* with the coding region of *atpB* indicates that the enhanced translation of *rps12* observed under conditions of reduced protein synthesis is mediated by *cis*-elements within the 5'-UTR, rather than the 3'-UTR (35). Chimeric genes containing a reporter gene fused to portions of the 5'-UTR of *atpB* and the 3'-UTR of *rbcL* have shown that the entire 5'-UTR of *atpB* is required for translation of the mRNA (78). Light-regulated translation of a chimeric mRNA containing the 5'- and 3'-flanking sequences of *psbA* is observed in tobacco chloroplasts, suggesting that either the 5'- or 3'-UTR of *psbA* is capable of conferring light-regulated translation to the chimeric coding region (85). A chimeric gene containing the *psbC* 5'-UTR has normal translation of the chimeric mRNA in wild-type cells but no translation in the nuclear *psbC* translation-deficient mutant F34. This observation shows that the 5'-UTR of the *psbC* mRNA is indeed the site for interaction with these nuclear-encoded factors (95). A chloroplast mutation that affects translation of the *psbC* mRNA localizes to the same region of the *psbC* 5'-UTR, again suggesting that the 5'-UTR of the chloroplast mRNA acts as a *cis*-element required for translation of the downstream coding region. A series of mutations, introduced into the 5'-UTR of the *psbA* mRNA, shows that the SD sequence is required for translation and that secondary structure around the RBS influences light-dependent *psbA* translation (62a). From the above data it seems clear that expression of chloroplast mRNAs requires nuclear-encoded factors that interact with the 5'-UTRs of mRNAs and that the 5'-UTRs contain *cis*-elements responsible for interacting with nuclear gene products to facilitate the translation of downstream coding regions.

5'-UTRs Are Associated with Specific RNA-Binding Proteins

Biochemical studies to characterize the regulatory proteins that bind to 5'-UTRs of various chloroplast mRNAs have identified a number of proteins (16, 35, 95). Comparison of proteins binding to the 5'-UTR of the *rps12* and *atpB* mRNAs in vitro identified a set of four proteins (38-, 45-, 54-, and 84-kDa) bound to both RNAs under normal or reduced protein synthesis (35). RNA affinity chromatography of proteins isolated from dark- and light-grown wild-type *C. reinhardtii* cells has identified a set of *psbA* RNA-binding proteins that includes a 38-, 47-, 55-, and 60-kDa protein (16). The 38- and 47-kDa proteins are bound at reduced levels in proteins isolated from dark-grown cells, whereas the levels of the 55- and 60-kDa proteins do not vary between protein isolates from dark- or light-grown cells. This result suggests that modulation of specific RNA-binding proteins could be used to regulate mRNA translation.

In the nuclear mutant *nac1*, which is deficient in *psbA* and *psbD* translation, the 47-kDa *psbA* mRNA-binding protein has altered mobility on polyacrylamide gels (CB Yohn, unpublished data). This observation suggests that the *nac1* mutation may result in altered 5'-UTR binding proteins, which leads to the absence of *psbA* and *psbD* translation. Proteins isolated from F64, a nuclear mutation that lacks translation of *psbC*, contain a 46-kDa protein that can be cross-linked to the 5′-UTR of *psbC* (95). This binding activity is missing in wild-type lysates, suggesting that the binding of the 46-kDa polypeptide to the *psbC* UTR might inhibit translation of that mRNA in F64 cells. The 46-kDa protein that cross-links to the *psbC* UTR in F64 does not appear to be related to the 47-kDa protein that binds to the *psbA* UTR (A Danon, unpublished data).

Although no conclusive data are available for the function of any of these 5′-UTR RNA-binding proteins, there is a correlation between the binding of these proteins and the level of translation of mRNAs in dark- and light-grown cells and in nuclear mutants that are deficient in translation (reviewed in 25). These 5′-UTR-binding proteins appear to be associated with the mRNAs as complexes. Some of the proteins are specific to individual mRNAs, and others are common to several mRNAs. The biochemical identification of these proteins is consistent with the genetic analysis in which the presence of specific translational enhancers for individual mRNAs has been predicted (reviewed in 25).

Binding Activity of 5′-UTR-Associated Proteins Is Regulated

Binding of proteins to the *psbA* 5′-UTR is regulated by altering the binding activity of proteins rather than by changing the protein content (16). Binding activity can be modulated in vitro by the addition of oxidizing and reducing agents to *C. reinhardtii* cell extracts (17a). This reduction-oxidation (redox) modulation of the RNA-binding activity may occur in vivo by the thioredoxin system. A threshold level of ADP (higher than 0.3 mM) was also found to inhibit the RNA binding of *psbA* translational activators by phosphorylation of one member of the protein complex (17). Levels of ADP high enough to activate this phosphorylation are found in chloroplasts following transfer into the dark (34, 36, 90). These results suggest that translation of *psbA* mRNA may be inhibited in the dark by inactivation of the *psbA* RNA-binding proteins resulting from a rise in stromal ADP concentrations (17). Upon illumination, translation may be activated by reduction of the RNA-binding proteins resulting from the production of reduced thioredoxin. The regulation of binding activity of other chloroplast mRNA-binding proteins has not been evaluated.

Translation Initiation Is a Rate-Limiting Step

Light induces a 50- to 100-fold enhancement of synthesis of specific chloroplast proteins (5, 41, 43, 54, 61) with no equivalent increase of their mRNAs. Analysis of light-induced ribosome association of the *psbA* mRNA has determined that only 30–50% of the available mRNA is found associated with polysomes (65; A Cohen, unpublished data) and that an excess of monosomes is detected in the stroma (65). These data indicate that factors other than mRNA levels or availability of free ribosomes may be the limiting step for translation. Recruitment of mRNA into polysomes is observed following illumination for both *rbcL* (5) and *psbA* (A Cohen, unpublished data), suggesting that translational initiation may play a role in light-activated translation.

The presence of SD sequences and their complementary regions in the 16S rRNA of the chloroplast (9, 77) indicates that initiation of translation in chloroplasts may follow internal binding of the 30S ribosomal subunits to the SD. Deletion of the SD sequence in *psbA* abolishes translation of the mRNA (62a) and excludes the mRNA from associating with ribosomes (A Cohen, unpublished data). Internal binding of ribosomes could explain why the processing of polycistronic mRNAs is not required for translation of internal open reading frames (1). Modulation of the degree of exposure of the RBS could regulate the translation initiation rate. Translation of specific chloroplast mRNAs could be regulated by altering the secondary structure surrounding the SD sequences or the initiator codon by the binding of specific mRNA-binding proteins in a fashion similar to that found in prokaryotes (22, 27, 93).

Mechanism of Translational Regulation

The above data support a model in which nuclear-encoded proteins are transported into the chloroplast to regulate the translation of specific mRNAs. The binding site for these translational activator proteins may consist of RNA structures located in the 5′-UTR of the regulated mRNA. Binding of these translational activators to their target mRNAs could result in alteration of the secondary structure surrounding the RBS to facilitate ribosome binding, or they could act as mRNA-specific initiation factors required for downstream events in translational initiation.

For light-activated translation, the binding of translational activators may be coordinated to the photosynthetic status of the chloroplast by two pathways. The first is the level of the redox potential, which is mediated by the production of reduced thioredoxin from photosynthesis. The higher the level of photosynthetic electron transport, the higher the level of reduced thioredoxin, resulting in the activation of RNA-binding proteins and ultimately increasing translation of plastid mRNAs. The second regulator of binding activity is induced by transfer of plants into the dark, resulting in increased levels of stromal ADP that cause a rapid inactivation of the RNA-binding proteins, leading to a cessation of translation. This rapid decrease in the translation of photosynthetic mRNAs would be required because the high energy cost for the translation of photosynthetic proteins that is observed in the light could not be supported in the dark.

CONCLUDING REMARKS

The regulatory events described in this chapter show the complex set of molecular mechanisms used in plastid gene expression. We have attempted to derive general rules for transcription, mRNA processing, mRNA stability, and mRNA translation as regulators of chloroplast gene expression. Overall, we have assigned transcription as a general regulator of plastid gene expression. In our view, rates of transcription vary dramatically depending on the plastid type, developmental stage, and environmental growth conditions, but in each case the rates of transcription of most or all of the genes appear to change in concert. Perhaps this activation involves the induction of a general RNA polymerase activity rather than the specific induction of transcription of individual genes. Message stability appears to affect the accumulation of individual mRNAs to a much greater degree than does transcription. The regulators of message stability, either the RNA or protein components, have not been identified. 3'-IR structures are not the prime RNA component involved in regulating mRNA stability, but whether other regions of the 3'-UTR are involved is unknown. Similarly, regions of the 5'-UTR influence mRNA stability, but whether these regions act directly as regulators of RNA stability or function in some other capacity and only pleiotropically affect mRNA stability has not been determined. Message processing remains an enigma. The number of nuclear loci dedicated to this process suggests that processing has to be a critical site for regulating chloroplast gene expression, yet few pieces of evidence exist to support processing as a primary regulator of plastid gene expression. Translational regulation has been shown both genetically and biochemically to be a regulator of plastid gene expression, and it is our prejudice that this form of gene regulation imparts the most dramatic affect on chloroplast gene expression. Obviously, without intact mRNAs, translational regulation is meaningless, so its impact is only of importance when the previous processes (transcription, mRNA processing, and stable accumulation of mRNAs) have all been performed.

These generalizations have been made based on indirect evidence of plastid gene regulation rather than on the direct identification of any individual regulatory factor. Many of these factors are encoded in the nuclear genome and are likely to be critical elements for both plastid gene expression and for coordinating the expression of plastid and nuclear genes. The isolation and characterization of these regulatory factors should clarify the role of the different processes in plastid gene expression. Many of these nuclear genes are now being cloned and their characterization in the coming years should provide important insights into the regulation of chloroplast gene expression.

Literature Cited

1. Barkan A. 1988. Proteins encoded by a complex chloroplast transcription unit are each translated from both monocistronic and polycistronic mRNAs. *EMBO J.* 7: 2637–44

2. Barkan A. 1993. Nuclear mutants of maize with defects in chloroplast polysome assembly have altered chloroplast RNA metabolism. *Plant Cell* 5:389–402

2a. Barkan A, Walker M, Nolasco M, Johnson D. 1994. A nuclear mutation in maize blocks the processing and translation of several chloroplast mRNAs and provides evidence for the differential translation of alternative mRNA forms. *EMBO J.* 13: 3170–81

3. Baumgartner BJ, Rapp JC, Mullet JE. 1993. Plastid genes encoding the transcription/translation apparatus are differentially transcribed early in barley (*Hordeum vulgare*) chloroplast development. Evidence for selective stabilization of *psbA* mRNA. *Plant Physiol.* 101:781–91

4. Berends T, Gamble PE, Mullet JE. 1987. Characterization of the barley chloroplast transcription units containing *psaA-psaB* and *psbD-psbC*. *Nucleic Acids Res.* 15: 5217–40

5. Berry JO, Breiding DE, Klessig DF. 1990. Light-mediated control of translational initiation of ribulose-1,5-bisphosphate carboxylase in amaranth cotyledons. *Plant Cell* 2:795–803

6. Blowers AD, Ellmore GS, Klein U, Bogorad L. 1990. Transcriptional analysis of endogenous and foreign genes in chloroplast transformants of *Chlamydomonas*. *Plant Cell* 2:1059–70

7. Blowers AD, Klein U, Ellmore GS, Bogorad L. 1993. Functional *in vivo* analyses of the 3′ flanking sequences of the *Chlamydomonas* chloroplast *rbcL* and *psaB* genes. *Mol. Gen. Genet.* 238:339–49

8. Bock R, Hagemann R, Kössel H, Kudla J. 1993. Tissue- and stage-specific modulation of RNA editing of the *psbF* and *psbL* transcript from spinach plastids—a new regulatory mechanism? *Mol. Gen. Genet.* 240:238–44

9. Bonham-Smith PC, Bourque DP. 1989. Translation of chloroplast-encoded mRNA: potential initiation and termination signals. *Nucleic Acids Res.* 17:2057–80

10. Brawerman G. 1989. mRNA decay: finding the right targets. *Cell* 57:9–10

11. Chen HC, Stern DB. 1991. Specific ribonuclease activities in spinach chloroplasts promote mRNA maturation and degradation. *J. Biol. Chem.* 266:24205–11

12. Christopher DA, Kim M, Mullet JE. 1992. A novel light-regulated promoter is conserved in cereal and dicot chloroplasts. *Plant Cell* 4:785–98

13. Copertino DW, Hallick RB. 1991. Group II twintron: an intron within an intron in a chloroplast cytochrome b-559 gene. *EMBO J.* 10:433–42

14. Copertino DW, Hallick RB. 1993. Group II and group III introns of twintrons: potential relationships with nuclear pre-mRNA introns. *Trends Biochem. Sci.* 18:467–71

15. Crossland LD, Rodermel SR, Bogorad L. 1984. Single gene for the large subunit of ribulosebisphosphate carboxylase in maize yields two differentially regulated mRNAs. *Proc. Natl. Acad. Sci. USA* 81:4060–64

16. Danon A, Mayfield SP. 1991. Light regulated translational activators: identification of chloroplast gene specific mRNA binding proteins. *EMBO J.* 10:3993–4001

17. Danon A, Mayfield SP. 1994. ADP-dependent phosphorylation regulates RNA-binding *in vitro*: implications in light-modulated translation. *EMBO J.* 13:2227–35

17a. Danon A, Mayfield SP. 1994. Light-regulated translation of chloroplast mRNAs through redox potential. *Science*. In press

18. Danon A, Yohn CB, Mayfield SP. 1993. Regulation of translation in plants. In *Genetic Engineering: Principles and Methods*, ed. JK Setlow, 15:41–55. New York: Plenum

19. Deng XW, Gruissem W. 1987. Control of plastid gene expression during development: the limited role of transcriptional regulation. *Cell* 49:379–87

20. Deng XW, Gruissem W. 1988. Constitutive transcription and regulation of gene expression in non-photosynthetic plastids of higher plants. *EMBO J.* 7:3301–8

21. Deng XW, Stern DB, Tonkyn JC, Gruissem W. 1987. Plastid run-on transcription. Application to determine the transcriptional regulation of spinach plastid genes. *J. Biol. Chem.* 262:9641–48

22. de Smit MH, van Duin J. 1994. Translational initiation on structured messengers. Another role for the Shine-Dalgarno interaction. *J. Mol. Biol.* 235:173–84

23. Drager RG, Hallick RB. 1993. A complex twintron is excised as four individual introns. *Nucleic Acids Res.* 21:2389–94

24. Edwards K, Kössel H. 1981. The rRNA operon from *Zea mays* chloroplasts: nucleotide sequence of 23S rDNA and its homology with *E. coli* 23S rDNA. *Nucleic Acids Res.* 9:2853–69

25. Gillham NW, Boynton JE, Hauser CR. 1994. Translational regulation of gene expression in chloroplasts and mitochondria. *Annu. Rev. Genet.* 28:71–93

26. Girard-Bascou J, Pierre Y, Drapier D. 1992. A nuclear mutation affects the synthesis of the chloroplast *psbA* gene production in *Chlamydomonas reinhardtii. Curr. Genet.* 22:47–52

27. Gold L, Hartz D. 1990. Initiation of protein synthesis in *E. coli*: the two crucial steps. In *Post-transcriptional Control of Gene Expression*, ed. JEG McCarthy, MF Tuite, 49:433–41. Berlin: Springer-Verlag

28. Goldschmidt-Clermont M, Choquet Y, Girard-Bascou J, Michel F, Schirmer-Rahire M, Rochaix JD. 1991. A small chloroplast RNA may be required for trans-splicing in *Chlamydomonas reinhardtii. Cell* 65:135–43

29. Goldschmidt-Clermont M, Girard-Bascou J, Choquet Y, Rochaix JD. 1990. Trans-splicing mutants of *Chlamydomonas reinhardtii. Mol. Gen. Genet.* 223:417–25

30. Gray MW, Covello PS. 1993. RNA editing in plant mitochondria and chloroplasts. *FASEB J.* 7:64–71

31. Gruissem W, Schuster G. 1993. Control of mRNA degradation in organelles. In *Control of Messenger RNA Stability*, ed. G Brawerman, J Belasco, pp. 329–65. Orlando, FL: Academic

32. Gruissem W, Zurawski G. 1985. Identification and mutational analysis of the promoter for a spinach chloroplast transfer RNA gene. *EMBO J.* 4:1637–44

33. Gruissem W, Zurawski G. 1985. Analysis of promoter regions for the spinach chloroplast *rbcL, atpB* and *psbA* genes. *EMBO J.* 4:3375–83

34. Hampp R, Goller M, Ziegler H. 1982. Adenylate levels, energy charge, and phosphorylation potential during dark-light and light-dark transition in chloroplasts, mitochondria and cytosol of mesophyll protoplasts from *Avena sativa* L. *Plant Physiol.* 69:448–55

35. Hauser CR, Randolph-Anderson BL, Hohl TM, Harris EH, Boynton JE, Gillham NW. 1993. Molecular genetics of chloroplast ribosomes in *Chlamydomonas reinhardtii.* In *The Translational Apparatus*, ed. KH

Nierhaus, F Fraceschi, AR Subramanian, VA Erdmann, B Wittmann-Liebold, pp. 545–54. New York: Plenum

36. Heber UW, Santarius KA. 1965. Compartmentation and reduction of pyridine nucleotides in relation to photosynthesis. *Biochim. Biophys. Acta* 109:390–408

37. Herrin DL, Bao Y, Thompson AJ, Chen YF. 1991. Self-splicing of the *Chlamydomonas* chloroplast *psbA* introns. *Plant Cell* 3:1095–107

38. Herrin DL, Chen YF, Schmidt GW. 1990. RNA splicing in *Chlamydomonas* chloroplasts. Self-splicing of 23 S preRNA. *J. Biol. Chem.* 265:21134–40

39. Hoch B, Maier RM, Appel K, Igloi GL, Kössel H. 1991. Editing of a chloroplast mRNA by creation of an initiation codon. *Nature* 353:178–80

39a. Hong S, Spreitzer RJ. 1994. Nuclear mutation inhibits expression of the chloroplast gene that encodes the large subunit of ribulose-1,5-bisphosphate   carboxylase/oxygenase. *Plant Physiol.* 106:In press

40. Hudson GS, Mason JG, Holton TA, Koller B, Cox GB, et al. 1987. A gene cluster in the spinach and pea chloroplast genomes encoding one CF1 and three CF0 subunits of the H$^+$-ATP synthase complex and the ribosomal protein S2. *J. Mol. Biol.* 196:283–98

41. Inamine G, Nash B, Weissbach H, Brot N. 1985. Light regulation of the synthesis of the large subunit of ribulose-1,5-bisphosphate carboxylase in peas: evidence for translational control. *Proc. Natl. Acad. Sci. USA* 82:5690–94

42. Jensen KH, Herrin DL, Plumley FG, Schmidt GW. 1986. Biogenesis of photosystem II complexes: transcriptional, translational, and posttranslational regulation. *J. Cell Biol.* 103:1315–25

43. Keller M, Chan RL, Tessier LH, Weil JH, Imbault P. 1991. Post-transcriptional regulation by light of the biosynthesis of *Euglena* ribulose-1,5-bisphosphate carboxylase/oxygenase small subunit. *Plant Mol. Biol.* 17:73–82

44. Kim M, Christopher DA, Mullet JE. 1993. Direct evidence for selective modulation of *psbA, rpoA, rbcL* and 16S RNA stability during barley chloroplast development. *Plant Mol. Biol.* 22:447–63

45. Klaff P, Gruissem W. 1991. Changes in chloroplast mRNA stability during leaf development. *Plant Cell* 3:517–29

46. Klein RR, Mullet JE. 1987. Control of gene expression during higher plant chloroplast biogenesis. *J. Biol. Chem.* 262:4341–48

47. Klein RR, Mullet JE. 1990. Light-induced transcription of chloroplast genes. *psbA* transcription is differentially enhanced in illuminated barley. *J. Biol. Chem.* 265:1895–902

48. Klein U, De Camp JD, Bogorad L. 1992. Two types of chloroplast gene promoters in *Chlamydomonas reinhardtii*. *Proc. Natl. Acad. Sci. USA* 89:3453–57

49. Krupinska K. 1992. Transcriptional control of plastid gene expression during development of primary foliage leaves of barley grown under a daily light-dark regime. *Planta* 186:294–303

50. Kuchka MR, Goldschmidt-Clermont M, van Dillewijn J, Rochaix JD. 1989. Mutation at the *Chlamydomonas* nuclear NAC2 locus specifically affects stability of the chloroplast *psbD* transcript encoding polypeptide D2 of PS II. *Cell* 58:869–76

51. Kuchka MR, Mayfield SP, Rochaix J-D. 1988. Nuclear mutations specifically affect the synthesis and/or degradation of the chloroplast-encoded D2 polypeptide of photosystem II in *Chlamydomonas reinhardtii*. *EMBO J.* 7:319–24

52. Kück U, Godehardt I, Schmidt U. 1990. A self-splicing group II intron in the mitochondrial large subunit rRNA (LSUrRNA) gene of the eukaryotic alga *Scenedesmus obliquus*. *Nucleic Acids Res.* 18: 2691–97

53. Kudla J, Igloi GL, Metzlaff M, Hagemann R, Kössel H. 1992. RNA editing in tobacco chloroplasts leads to the formation of a translatable *psbL* mRNA by a C to U substitution within the initiation codon. *EMBO J.* 11:1099–103

54. Laing W, Kreuz K, Apel K. 1988. Light-dependent, but phytochrome-independent, translational control of the accumulation of the P700 chlorophyll-a protein of photosystem I in barley (*Hordeum vulgare* L.). *Planta* 176:269–76

55. Leu S, White D, Michaels A. 1990. Cell cycle-dependent transcriptional and posttranscriptional regulation of chloroplast gene expression in *Chlamydomonas reinhardtii*. *Biochim. Biophys. Acta* 1049:311–17

56. Li R, Golden SS. 1993. Enhancer activity of light-responsive regulatory elements in the untranslated leader regions of cyanobacterial *psbA* genes. *Proc. Natl. Acad. Sci. USA* 90:11678–82

57. Li YQ, Nagayoshi S, Sugita M, Sugiura M. 1993. Structure and expression of the tobacco nuclear gene encoding the 33 kDa chloroplast ribonucleoprotein. *Mol. Gen. Genet.* 239:304–9

58. Li YQ, Sugiura M. 1990. Three distinct ribonucleoproteins from tobacco chloroplasts: each contains a unique amino terminal acidic domain and two ribonucleoprotein consensus motifs. *EMBO J.* 9: 3059–66

59. Li YQ, Ye LZ, Sugita M, Sugiura M. 1991. Tobacco nuclear gene for the 31 kd chloroplast ribonucleoprotein: genomic organiza-

tion, sequence analysis and expression. *Nucleic Acids Res.* 19:2987–91

60. Maier RM, Hoch B, Zeltz P, Kössel H. 1992. Internal editing of the maize chloroplast *ndhA* transcript restores codons for conserved amino acids. *Plant Cell* 4:609–16

61. Malnoè P, Mayfield SP, Rochaix JD. 1988. Comparative analysis of the biogenesis of photosystem II in the wild-type and *Y-1* mutant of *Chlamydomonas reinhardtii*. *J. Cell Biol.* 106:609–16

62. Matsubayashi T, Wakasugi T, Shinozaki K, Yamaguchi-Shinozaki K, Zaita N, et al. 1987. Six chloroplast genes (*ndhA-F*) homologous to human mitochondrial genes encoding components of the respiratory chain NADH dehydrogenase are actively expressed: determination of the splice sites in *ndhA* and *ndhB* pre-mRNAs. *Mol. Gen. Genet.* 210:385–93

62a. Mayfield SP, Cohen A, Danon A, Yohn CB. 1994. Translation of the *psbA* mRNA of *Chlamydomonas reinhardtii* requires a structured RNA element contained within the 5′ untranslated region. *J. Cell Biol.* 127:In press

63. Michel F, Umesono K, Ozeki H. 1989. Comparative and functional anatomy of group II catalytic introns—a review. *Gene* 82:5–30

64. Michel F, Westhof E. 1990. Modelling of the three-dimensional architecture of group I catalytic introns based on comparative sequence analysis. *J. Mol. Biol.* 216:585–610

65. Minami E-I, Shinohara K, Kawakami N, Watanabe A. 1988. Localization and properties of transcripts of *psbA* and *rbcL* genes in the stroma of spinach chloroplast. *Plant Cell Physiol.* 29:1303–9

66. Monod C, Goldschmidt-Clermont M, Rochaix JD. 1992. Accumulation of chloroplast *psbB* RNA requires a nuclear factor in *Chlamydomonas reinhardtii*. *Mol. Gen. Genet.* 231:449–59

67. Mullet JE. 1993. Dynamic regulation of chloroplast transcription. *Plant Physiol.* 103:309–13

68. Mullet JE, Klein RR. 1987. Transcription and RNA stability are important determinants of higher plant chloroplast RNA levels. *EMBO J.* 6:1571–79

69. Nickelson J, van Dillewijn J, Rahire M, Rochaix J-D. 1994. Determinants for stability of the chloroplast *psbD* RNA are located within its short leader region in *Chlamydomonas reinhardtii*. *EMBO J.* 13: 3182–91

70. Piechulla B, Gruissem W. 1987. Diurnal mRNA fluctuations of nuclear and plastid genes in developing tomato fruits. *EMBO J.* 6:3593–99

71. Pring D, Brennicke A, Schuster W. 1993.

RNA editing gives a new meaning to the genetic information in mitochondria and chloroplasts. *Plant Mol. Biol.* 21:1163–70

72. Rapp JC, Baumgartner BJ, Mullet J. 1992. Quantitative analysis of transcription and RNA levels of 15 barley chloroplast genes. Transcription rates and mRNA levels vary over 300-fold; predicted mRNA stabilities vary 30-fold. *J. Biol. Chem.* 267:21404–11

73. Reinbothe S, Reinbothe C, Heintzen C, Seidenbecher C, Parthier B. 1993. A methyl jasmonate-induced shift in the length of the 5′ untranslated region impairs translation of the plastid *rbcL* transcript in barley. *EMBO J.* 12:1505–12

74. Rochaix JD. 1992. Post-transcriptional steps in the expression of chloroplast genes. *Annu. Rev. Cell Biol.* 8:1–28

75. Rochaix JD, Kuchka M, Mayfield S, Schirmer-Rahire M, Girard-Bascou J, Bennoun P. 1989. Nuclear and chloroplast mutations affect the synthesis or stability of the chloroplast *psbC* gene product in *Chlamydomonas reinhardtii. EMBO J.* 8:1013–21

76. Ruf M, Kössel H. 1988. Occurrence and spacing of ribosome recognition sites in mRNAs of chloroplasts of higher plants. *FEBS Lett.* 240:41–44

77. Ruf S, Zeltz P, Kössel H. 1994. Complete RNA editing of unspliced and dicistronic transcripts of the intron-containing reading frame IRF170 from maize chloroplasts. *Proc. Natl. Acad. Sci. USA* 91:2295–99

78. Sakamoto W, Kindle KL, Stern DB. 1993. *In vivo* analysis of *Chlamydomonas* chloroplast *petD* gene expression using stable transformation of beta-glucuronidase translational fusions. *Proc. Natl. Acad. Sci. USA* 90:497–501

79. Saldanha R, Mohr G, Belfort M, Lambowitz AM. 1993. Group I and group II introns. *FASEB J.* 7:15–24

80. Salvador ML, Klein U, Bogorad L. 1993. 5′ sequences are important positive and negative determinants of the longevity of *Chlamydomonas* chloroplast gene transcripts. *Proc. Natl. Acad. Sci. USA* 90: 1556–60

81. Salvador ML, Klein U, Bogorad L. 1993. Light-regulated and endogenous fluctuations of chloroplast transcript levels in *Chlamydomonas*. Regulation by transcription and RNA degradation. *Plant J.* 3:213–19

82. Schwarz Z, Kössel H. 1980. The primary structure of 16S rDNA from *Zea mays* chloroplast is homologous to *E. coli* 16S rRNA. *Nature* 283:739–42

83. Sexton TB, Christopher DA, Mullet JE. 1990. Light-induced switch in barley *psbD-psbC* promoter utilization: a novel mechanism regulating chloroplast gene expres-

sion. *EMBO J.* 9:4485–94

84. Sieburth LE, Berry-Lowe S, Schmidt GW. 1991. Chloroplast RNA stability in *Chlamydomonas*: rapid degradation of *psbB* and *psbC* transcripts in two nuclear mutants. *Plant Cell* 3:175–89

85. Staub JM, Maliga P. 1993. Accumulation of D1 polypeptide in tobacco plastids is regulated via the untranslated region of the *psbA* mRNA. *EMBO J.* 12:601–6

86. Stern DB, Gruissem W. 1987. Control of plastid gene expression: 3′ inverted repeats act as mRNA processing and stabilizing elements, but do not terminate transcription. *Cell* 51:1145–57

87. Stern DB, Jones H, Gruissem W. 1989. Function of plastid mRNA 3′ inverted repeats. RNA stabilization and gene-specific protein binding. *J. Biol. Chem.* 264:18742–50

88. Stern DB, Radwanski ER, Kindle KL. 1991. A 3′ stem/loop structure of the *Chlamydomonas* chloroplast *atpB* gene regulates mRNA accumulation *in vivo. Plant Cell* 3:285–97

89. Stevenson JK, Hallick RB. 1994. The *psaA* operon pre-mRNA of the *Euglena gracilis* chloroplast is processed into photosystem I and II mRNAs that accumulate differentially depending on the conditions of cell growth. *Plant J.* 5:247–60

90. Stitt M, Wirtz W, Heldt HW. 1980. Metabolite levels during induction in the chloroplast and extrachloroplast compartments of spinach protoplasts. *Biochim. Biophys. Acta* 593:85–102

91. Winkler M, Kück U. 1991. The group IIB intron from the green alga *Scenedesmus obliquus* mitochondrion: molecular characterization of the *in vitro* splicing products. *Curr. Genet.* 20:495–502

92. Wissinger B, Schuster W, Brennicke A. 1991. Trans splicing in *Oenothera* mitochondria: *nad1* mRNAs are edited in exon and trans-splicing group II intron sequences. *Cell* 65:473–82

93. Wulczyn FG, Bolker M, Kahmann R. 1989. Translation of the bacteriophage Mu *mom* gene is positively regulated by the phage *com* gene product. *Cell* 57:1201–10

94. Zeltz P, Hess WR, Neckermann K, Borner T, Kössel H. 1993. Editing of the chloroplast *rpoB* transcript is independent of chloroplast translation and shows different patterns in barley and maize. *EMBO J.* 12:4291–96

95. Zerges W, Rochaix J-D. 1994. The 5′ leader of a chloroplast mRNA mediates the translational requirements for two nucleus-encoded functions in *Chlamydomonas reinhardtii. Mol. Cell. Biol.* 14:5268–77

Annu. Rev. Plant Physiol. Plant Mol. Biol. 1995. 46:167–88

FLORAL MERISTEMS TO FLORAL ORGANS: Genes Controlling Early Events in *Arabidopsis* Flower Development

Martin F. Yanofsky

Department of Biology and Center for Molecular Genetics, University of California at San Diego, La Jolla, California 92093-0116

KEY WORDS: flower development, *Arabidopsis*, inflorescence meristem, floral meristem, MADS-box

CONTENTS

ABSTRACT

Molecular and genetic studies show that the underlying mechanisms controlling flower development are largely conserved in distantly related dicotyledonous plant species. These studies have identified early-acting genes that promote the formation of floral meristems and later-acting genes that determine the fate of floral organ primordia. The floral meristem identity genes lie at the top of a regulatory hierarchy that leads to the activation of the organ identity

0066-4294/95/0601-0167$05.00

genes. Subsequent regulatory interactions between the organ identity genes restrict their domains of activity. The resulting spatial domains of gene activity allow cells within the floral meristem to assess their position and, hence, to differentiate into the appropriate organ type.

INTRODUCTION

The past five years have produced remarkable progress toward elucidating the molecular and genetic mechanisms controlling flower development in two model dicot species, *Arabidopsis thaliana* and *Antirrhinum majus* (8, 9, 11, 14, 15, 18, 25, 43, 48, 49, 54, 60, 70, 72, 86), hereafter referred to as *Arabidopsis* and *Antirrhinum*, respectively. More recently, these studies have been extended to other plant species by isolating homologs of the cloned *Arabidopsis* and *Antirrhinum* genes (80, 81). This review summarizes observations that detail events from the specification of floral meristem identity through the specification of floral organ identity. Particular emphasis is given to studies in *Arabidopsis*, but references to other plants are included when relevant.

During the vegetative phase of *Arabidopsis* growth, the shoot apical meristem produces leaf primordia on its flanks. Upon the transition to flowering, the apical meristem undergoes an abrupt transition (into an inflorescence meristem), after which flower primordia (floral meristems) are produced on its flanks in place of the leaf primordia. After several flower primordia are initiated at the apex, additional inflorescences, which had been initiated prior to the transition to flowering, begin to differentiate in the axils of the leaf primordia. These axillary inflorescences essentially repeat the pattern of the main flowering axis. Although flower primordia are initiated acropetally, the activation of axillary inflorescences occurs basipetally (27). The inflorescence meristem is indeterminate and can seemingly give rise to an endless array of lateral meristems in a spiral arrangement. In contrast, the floral meristem produces a determinate number of organ primordia in a whorled arrangement. The perianth organs, sepals and petals, occupy the first and second whorls, whereas the reproductive organs, stamens and carpels, occupy the third and innermost fourth whorl.

Based on morphological criteria, *Arabidopsis* flower development has been broken down into a number of distinct stages (75) (Figure 1*a*). During stage 1, the young flower primordium, or floral meristem, is first visible as a clump of cells on the flank of the inflorescence meristem. During stage 2, the flower primordium increases in size and becomes distinct from the inflorescence meristem. Stage 3 is defined by the appearance of the sepal primordia, which grow to overlie the flower primordium during stage 4. This review focuses on these early stages because they span the events from the formation of the floral meristem to the expression of the genes that specify floral organ fate. Each of

Figure 1   Photographs of wild-type and mutant flowers and inflorescences. *a–b*: wild-type; *c*: *lfy*; *d*: *ap1*; *e*: *ap1 cal*; *f*: *lfy ap1*. Numbers indicate the stage of flower development (7, 75)

the later steps in the development of flowers has been defined, leading up to anthesis at stage 13.

SPECIFICATION OF FLORAL MERISTEM IDENTITY

Genetic Studies of Meristem Identity

Rapid progress toward understanding the genetic basis for the specification of floral meristems has come from the isolation and characterization of mutants that prevent or alter meristem formation. In *Arabidopsis*, at least three loci have been extensively characterized that play major roles in this early step. Additional loci that play minor roles in specifying floral meristems have also been studied.

Mutations in the *Arabidopsis LEAFY* (*LFY*) gene lead to a partial conversion of flowers into inflorescence shoots. Only the early-arising flowers of *lfy* mutants develop as inflorescence shoots, while later-arising flowers have partial inflorescence characteristics, with floral organs often arising in an arrangement that is intermediate between spiral and whorled (30, 66, 67, 82). These later-arising flowers are variable in phenotype but often consist of an outer whorl of sepals, followed by inner whorls of sepaloid and carpelloid organs (Figure 1c). Thus, in addition to the role of *LFY* in specifying floral meristem identity, it also appears to be important in specifying petals and stamens (see below). Mutations in the *LFY* homolog from *Antirrhinum*, *FLORICAULA*, result in a complete conversion of flowers into inflorescence shoots (16). These studies illustrate one of the differences between these distantly related dicot plants, and indicate that at least in *Arabidopsis*, an additional gene can in part substitute for *LFY* in specifying floral meristems. Although flowers of most dicotyledonous plants, like those of *Antirrhinum*, are subtended by leaf-like bracts, these organs do not subtend *Arabidopsis* flowers. The abnormal flowers that develop in *lfy* mutants are subtended by leaf-like bracts, indicating that one of the functions of the wild-type *LFY* gene is to suppress bract formation.

The phenotype of *lfy* mutants is dramatically enhanced (30, 82) when combined with mutations in a second gene, *APETALA1* (*AP1*), such that essentially all the early- and late-arising floral primordia develop as inflorescence shoots (Figure 1f). Single *ap1* mutants (34) develop flowers that have partial inflorescence character in that secondary floral meristems arise in the axils of the first whorl organs (Figure 1d). These secondary meristems behave much like the primary floral meristem and give rise to tertiary and higher-order meristems in the axils of the subsequent first whorl organs. The *AP1* gene also plays a secondary role in specifying the identity of sepals and petals (see below).

Mutations in the *AP1* homolog from *Antirrhinum*, *SQUAMOSA* (*SQUA*), generally result in a much more severe phenotype than do corresponding mutations in the *Arabidopsis AP1* gene. Frequently, the flowers that would form in wild-type develop as inflorescence shoots in *squa* mutants. Occasionally, flowers form in *squa* mutants that range in phenotype from essentially wild-type to severely malformed (33).

In addition to the *LFY* and *AP1* genes, a third locus in *Arabidopsis* has been identified that greatly enhances the *ap1* mutant phenotype. This locus, designated *cal-1*, was first identified when the *ap1* mutation was introduced into the wild-type Wassilewskija (WS) ecotype, and more recently, several new alleles of *CAULIFLOWER* (*CAL*) have been identified (4; S Kempin, B Savidge & M Yanofsky, submitted). This locus derives its name from the resulting *cauliflower* phenotype, which is strikingly similar to the structures that develop in the dinner table variety of cauliflower (*Brassica oleracea* var. *botrytis*). The floral meristems that develop in wild-type plants behave as inflorescence meristems in the *ap1 cal* double mutant (Figure 1*e*). The end result is the formation of a massive proliferation of meristems in the positions that normally would be occupied by single flowers. Plants homozygous for the *cal-1* allele have no significant phenotype that distinguishes them from wild-type, indicating that the functions of *CAL* are encompassed by those of *AP1*. Because the floral meristems that form in *ap1* single mutants behave as inflorescence meristems in *ap1 cal* double mutants, *CAL* can largely substitute for *AP1* in specifying floral meristems. In contrast, because first and second whorl organs fail to develop normally in *ap1* single mutants (see below), *CAL* is largely unable to substitute for *AP1* in specifying organ identity. It is inferred from these genetic data that *CAL* and *AP1* encode partially functionally redundant activities. Eventually, *ap1 cal* double-mutant plants give rise to abnormal flowers that conform to the severe *ap1* mutant phenotype and, thus, lack first and second whorl organs.

Other genetic loci play at least minor roles in specifying floral meristem identity (Figure 2). For example, although *apetala2* (*ap2*) single-mutant flowers do not exhibit inflorescence characteristics, when *ap2* mutants are combined with either *ap1* or *lfy*, an underlying role for *AP2* in floral meristem specification is revealed. For example, *ap2-1 ap1-1* double mutants result in flowers developing as indeterminate inflorescences (4, 34). In addition, mutations in the *CLAVATA1* (*CLV1*) gene result in enlarged meristems, leading to a variety of phenotypes, including extra organs in the resulting flowers. In the *clv1 ap1* double mutant, flowers are initiated, but the center of each flower often develops as an indeterminate inflorescence (13). This result indicates that one of the functions of *CLV1* is to maintain floral meristem identity in the center of wild-type flowers. Mutations in another locus, designated *TERMINAL FLOWER* (*TFL*), produce phenotypes that are to a large extent opposite to those of mutations in the floral meristem identity genes. *tfl* mutants flower early, and the apical and lateral meristems develop as floral meristems. Thus, the *TFL* gene has been proposed to promote maintenance of the inflorescence

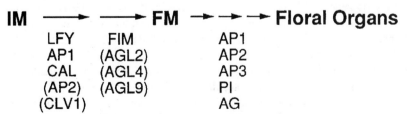

Figure 2   Overview of the genes involved in flower development. IM: inflorescence meristem; FM: floral meristem. All genes are from *Arabidopsis* except the *Antirrhinum FIM* gene.

meristem (1, 71, 72), although it also appears to function during the vegetative growth phase (67). *TFL* also has a role in negatively regulating *LFY* and *AP1* expression in apical and lateral inflorescence meristems, although this regulation may be indirect (4, 23, 82).

Molecular Characterization of Meristem Identity Genes

The *LEAFY* gene was isolated using a probe specific for the *Antirrhinum FLORICAULA* gene (82). Unlike many of the other recently cloned floral homeotic genes, *LFY* is not part of an extended gene family. Although the sequence of the LFY protein does not show any significant sequence similarity to previously characterized proteins, the existence of a proline-rich domain at the amino terminus and an acidic region in the middle of the protein suggests that LFY may function as a transcriptional activator (16). This conclusion is supported by immunolocalization experiments that demonstrate that the LFY protein is preferentially localized to plant nuclei (83). Consistent with the genetic data, *LFY* RNA first accumulates in the anlagen of floral primordia on the flanks of the inflorescence meristem (82), and *LFY* RNA is not detected in the inflorescence meristem. As floral meristems become morphologically distinct (stages 1 and 2), the expression of *LFY* RNA increases and is uniform over the young flower primordium. During stages 3 and 4, at which time the organ identity genes are first expressed (see below), *LFY* RNA largely disappears from the center of the incipient flower. *LFY* appears to be the earliest known molecular marker for cells destined to form flowers, because the two other meristem identity genes initiate their expression shortly after the onset of *LFY* expression (see below). Detailed molecular and genetic studies of the *Antirrhinum FLORICAULA* gene indicate that FLO activity in a subset of meristematic cell layers can activate the floral program in other cell layers (R Carpenter & E Coen, submitted; S Hantke, R Carpenter & E Coen, submitted). FLO activity therefore involves interdermal cell communication, perhaps mediated by a cell-transmissible signal. The FLO protein itself may be capable of moving between cells in a manner similar to that proposed for the maize KNOTTED protein (37).

The *APETALA1* gene was cloned as a new member of the MADS-box family of genes (46). The presence of a MADS-domain in the AP1 protein implies that AP1 binds DNA in a sequence-specific manner and functions as a transcriptional activator (see MADS-Box Genes section). *AP1* RNA is first detected in young flower primordia as soon as they are visible on the flanks of the inflorescence meristem (stage 1), and no *AP1* RNA is detected in the inflorescence meristem. The molecular and genetic data support the conclusion that *AP1* acts locally to specify the identity of floral meristems. In contrast to the uniform expression of *AP1* RNA throughout stage 1 and 2 flower primordia, *AP1* expression begins to decrease in the cells of the two inner whorls during stage 3, and at all later stages of flower development *AP1* RNA is localized to sepal and petal primordia. This raises the interesting question of what factor(s) negatively regulate *AP1* RNA accumulation in the inner whorls of the developing flower (see below). In addition, *AP1* RNA accumulates uniformly throughout the floral pedicel. This intriguing expression pattern is consistent with genetic data in that ectopic meristems frequently develop from pedicels of *ap1* mutant flowers. These results suggest that one of the functions for *AP1* is to define pedicel tissue as floral rather than inflorescence and, thus, to suppress meristem formation (4, 23).

The *CAULIFLOWER* gene was cloned by virtue of its close similarity to *AP1* (S Kempin, B Savidge & M Yanofsky, submitted). *CAL*, like *AP1*, encodes a protein with a MADS-domain and probably functions as a transcriptional activator. The CAL and AP1 protein sequences are 76% identical over their entire lengths. Within the MADS-domain, CAL and AP1 differ in only 5 of 56 residues, 4 of which represent conservative amino acid substitutions. These data suggest that the two proteins probably recognize similar DNA sequences and regulate similar target genes. Thus, the functional redundancy that was seen in genetic studies is reflected in the structural similarity of CAL and AP1. Although these data shed light on the underlying molecular basis for the functional redundancy of *CAL* and *AP1*, they fail to answer why *CAL* is only partially able to substitute for *AP1*. *CAL* RNA, like that of *AP1*, first begins to accumulate in stage 1 flower primordia and is later localized to the two outer whorls of organs. In contrast to *AP1* RNA, which accumulates at relatively high levels throughout sepal and petal development, *CAL* RNA is only detected at very low levels in these organ primordia. This result suggests that *CAL* is unable to substitute for *AP1* in specifying sepals and petals, at least in part because of the low levels of *CAL* RNA in these organ primordia.

What role does the *CAL* gene play in flower development? Genetic and molecular data address this question (4). The levels of *LFY* RNA are dramatically reduced in *ap1 cal* double mutants but are normal in the *ap1* single mutant. Thus, one role for *CAL* could be to boost the levels of *LFY* expression. This conclusion is supported by genetic data indicating that the phenotype of the *lfy ap1 cal* triple mutant is indistinguishable from the *lfy ap1* double mutant (4). Thus, if the *LFY* gene is inactivated, it makes no difference whether or not the *CAL* gene is active, because its proposed role is to modulate *LFY* gene activity.

Commercial cauliflower heads represent an example of a flower abnormality that has been recognized for hundreds of years (63, 88). Cauliflower (*Brassica oleracea* var. *botrytis*) is derived from wild-type *B. oleracea,* which also gave rise to many other crop plants such as broccoli, brussels sprouts, kale, and cabbage. *Brassica* and *Arabidopsis* spp. are members of the same family, the Brassicaceae, and their wild-type and *cauliflower* mutant inflorescences are strikingly similar in morphology. One of the interesting conclusions derived from studies of two distantly related dicot plant species, *Arabidopsis* and *Antirrhinum*, is that the underlying mechanisms controlling flower development are largely conserved. Indeed, counterparts of essentially all the cloned genes exist in both species, and corresponding expression patterns and mutant phenotypes are similar. Thus, it would not be surprising if the underlying molecular basis for the *cauliflower* phenotype in *Arabidopsis* and the closely related *B. oleracea* var. *botrytis* were related.

To determine if such a relation exists, the *CAULIFLOWER* gene was cloned from *B. oleracea* var. *botrytis* and compared to the corresponding genes from a wild-type *B. oleracea*, and to the *Arabidopsis* gene (S Kempin, B Savidge & M Yanofsky, submitted). The open reading frame of the *CAL* gene from *B. oleracea* var. *botrytis* is interrupted by a stop codon in the fifth exon, suggesting that only a truncated protein of 17 kDa, compared to a wild-type 30-kDa protein, could be produced. These data suggest strongly that the underlying molecular basis for the *cauliflower* phenotype in the dinner table *B. oleracea* var. *botrytis* is partly the result of a mutation in the *CAL* gene. Further studies are needed to demonstrate cosegregation of the *CAL* gene with the *cauliflower* phenotype and to determine the resulting phenotype when a wild-type copy of the *CAL* gene is introduced into *B. oleracea* var. *botrytis*. It also will be interesting to determine if the *AP1* gene is defective in *botrytis*.

SPECIFICATION OF FLORAL ORGAN IDENTITY

Genetic Studies of Organ Identity

The studies described above illustrate the progress toward understanding one of the earliest events in flower development, the specification of floral meristem identity. After floral meristems are generated, the later-acting organ

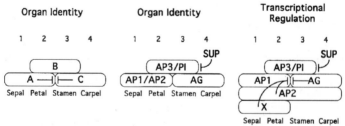

Figure 3 Models for activity and regulation of organ identity genes. One half of a flower primordium is represented, as shown by the position of the four whorls of organs. *Left* and *center*: genetic model showing how the indicated genes act alone or in combination to specify organ identity. *Right*: molecular model based on accumulation and regulation of RNAs specific for the indicated genes.

identity genes are activated. Genetic studies led to a model (Figure 3, *left*) in which three different homeotic activities, designated A, B, and C, specify the four different organ types (8, 15). Each of these activities functions in two adjacent whorls, with A in whorls 1 and 2, B in whorls 2 and 3, and C in whorls 3 and 4. Thus, activity A alone in whorl 1 specifies sepals, and C alone in whorl 4 specifies carpels. The combined activities of AB and BC specify petals and stamens, respectively, in whorls 2 and 3. This model further suggests that A and C are mutually antagonistic, such that A prevents the activity of C in whorls 1 and 2, and C prevents the activity of A in whorls 3 and 4.

The *Arabidopsis* genes (Figure 3, *middle*) initially proposed to be necessary for the A, B, and C functions are *APETALA2* (function A), *APETALA3* and *PISTILLATA* (function B), and *AGAMOUS* (function C). As additional data accumulate, the model must be revised, as has been the case recently with the addition of *APETALA1* as an A function gene (4, 23, 67). Thus, although the above-mentioned genes are necessary for the A, B, and C functions, it is unclear whether they are sufficient to specify each of the organ types. For example, as noted above, at least two genes are necessary for each of the A and B functions. Transgenic experiments (see below) address the question of sufficiency. The proposed model initially was based entirely on genetic data and, thus, deals specifically with the activities of gene products. The model does not imply that the genes are regulated at the transcriptional level, although this turns out to be the case for the majority of the genes thus far characterized.

Mutations in the A function *APETALA2* gene result in organ identity alterations in the outer two whorls, although the position and number of organs in all four whorls are affected (7, 8, 41, 42). The generalized phenotype of *ap2* mutants are first and fourth whorl carpels, and second and third whorl stamens. Strong *ap2* mutant alleles (*ap2-2*) produce flowers in which the medial first whorl organs are usually carpels, and the lateral first whorl organs either fail to develop or form leaf-like or carpelloid structures. Second whorl organs fail to develop in strong *ap2* mutant alleles because of a failure of these organ primordia to initiate, and although third whorl organs are generally absent, occasionally fertile stamens form. The fourth whorl carpels frequently fail to fuse properly and often are twisted. Finally, *ap2* mutant seeds show an altered seed coat morphology (38). The temperature-sensitive weak *ap2-1* allele generally displays phenotypes only in the two outer whorls; first whorl organs develop as leaves and second whorl organs are absent or form staminoid petals.

Mutations in the A-function *APETALA1* gene affect two distinct phases of plant development, floral meristem and organ identity. In addition to the meristem alterations discussed earlier, *ap1* mutants show organ defects in the two outer whorls. The first whorl organs of the strong *ap1-1* mutant allele either fail to develop or form leaf-like structures. The second whorl organs are generally absent because these organ primordia fail to initiate (4, 34, 67). Analysis of intermediate and weak *ap1* mutant alleles shows that when second whorl organs form, they often have characteristics of leaves, stamens, and/or petals. The genetic data demonstrate that *AP1* is involved not only in determining meristem and organ identity, but also in the intervening step of initiating organ primordia.

APETALA3 and *PISTILLATA* represent distinct B-function genes mapping to different chromosomes, and yet the phenotypic consequences of mutations in each of these genes are remarkably similar. In general, *ap3* and *pi* mutants show organ identity defects only in the second and third whorls, with these organs developing as sepals and carpels, respectively, rather than the wild-type petals and stamens (7, 8, 29, 35). In strong *ap3* and *pi* mutant alleles, the cells that would normally constitute the third whorl organs are largely incorporated into an enlarged central gynoecium.

Mutations in the C-function *AGAMOUS* gene result in organ identity alterations in only the two inner whorls (7, 8, 87). The third whorl organs of *ag* mutants develop as petals, and the fourth whorl is replaced by a new flower that reiterates the pattern of the primary flower. Thus, the overall phenotype of *ag* flowers is (sepal, petal, petal)$_n$. These data indicate that the wild-type *AG* gene is involved not only in specifying the identity of the reproductive organs, but also in suppressing the indeterminate growth of the floral meristem. In many ways the phenotypic alterations of *ag* and *ap2* mutants are opposite. In *ap2* mutants, the perianth organs are replaced by reproductive organs, whereas in *ag* mutants, the reproductive organs are replaced by perianth organs.

Because the genetic organ identity model was based largely on the single-gene mutations described above, it is not surprising that the phenotypes are mostly consistent with the model. The model makes specific predictions of the phenotypes of double-mutant strains, and such tests have largely confirmed its validity (8). A specific feature of the model—that the A and C functions are mutually antagonistic—is clearly demonstrated by double-mutant studies. For example, the C-function gene *AG* is normally active in only the two innermost whorls. However, the carpels and stamens that frequently develop in the two outer whorls of *ap2* single-mutant plants develop as leaves and staminoid petals, respectively, in the *ap2 ag* double mutant. This observation indicates that *AG* is active in the outer two whorls of *ap2* mutant flowers. Similarly, the petals and sepals that develop in the inner whorls of the *ag* single mutant develop as staminoid petals and leaves, respectively, in the *ag ap2* double mutant. This result indicates that *AP2* is active in specifying organ identity in the inner whorls of *ag* mutant flowers. The *ap2 ag* double mutant is particularly instructive because the model predicts that none of the organ identity functions will remain active in whorls 1 and 4. That the absence of A, B, and C activities leads to the formation of leaves suggests that these activities are largely responsible for converting leaves into the various floral organs. Indeed, the triple mutant, lacking all three activities, results in flowers in which all organs develop essentially as leaves (8). These recent genetic data are interesting from an historical perspective: In 1790, Goethe proposed that the four types of floral organs are all modified leaves (21). The *ap2 ag* double mutant also provides insight into the observation that most second- and third-whorl organ primordia fail to initiate in *ap2* single mutants. The fact that nearly wild-type numbers of organs form in the *ap2 ag* double mutant indicates that the ectopic AG activity in *ap2* mutants causes loss of organ initiation.

Mutations in the *SUPERMAN* (*SUP*) gene result in additional stamens developing at the expense of the fourth whorl carpels (6, 68). The number of extra stamens is variable, but some carpelloid tissue is usually present in the center of *sup* mutant flowers. The simplest interpretation of this phenotype, in the context of the organ identity model, is that *SUP* prevents the B-function genes from acting in the fourth whorl (Figure 3). This idea was confirmed by observing the phenotypes of *sup ap3* and *sup pi* double-mutant flowers, which are indistinguishable from the *ap3* and *pi* single-mutant flowers. Thus, because *SUP* suppresses B gene activity in the fourth whorl, in *b* mutant plants it does not matter if the *SUP* gene is active or not. Molecular data in support of the genetic experiments demonstrate that the expression domains of *AP3* and *PI* RNAs expand into the fourth whorl of *sup* mutant flowers (see below).

Molecular Studies of Organ Identity

The first floral homeotic gene to be isolated from *Arabidopsis* was the C-function gene *AGAMOUS*, using a T-DNA tagging approach (87). The *AG* gene product includes a MADS-domain indicating that *AG*, like all the other MADS-box genes, encodes a sequence-specific DNA-binding protein that functions as a transcription factor. Consistent with the genetic data, *AG* RNA specifically accumulates in the two inner whorls of organs (19, 87). *AG* RNA is first detected early in stage 3 of flower development, at which time the primordia for the sepals are first visible on the flanks of the floral meristem. Even at this early stage, *AG* RNA is specifically localized to the cells that give rise to stamens and carpels. *AG* RNA is uniformly expressed throughout stamen and carpel primordia until later in flower development when it becomes restricted to specific cell types. Within the stamens, *AG* RNA is excluded from the pollen grains as well as the sporogenous tissue that gives rise to pollen grains (5). Within the ovules, *AG* RNA is localized primarily to the endothelium, the cell layer that surrounds the embryo sac, and no expression is observed in either the embryo sac or in the integuments of the mature ovule.

The B-function gene *APETALA3* was isolated using a probe for its *Antirrhinum* homolog, *DEFICIENS A* (35, 76). Like *AG*, the *AP3* gene encodes a MADS-domain and, thus, probably functions as a transcription factor. *AP3* RNA is first detected early during stage 3, at about the same time that *AG* RNA is first detected. From stage 3 and beyond, *AP3* RNA accumulates in cells fated to form petals and stamens, the same two organ types that are affected in *ap3* mutants. *AP3* RNA is detected only in petals and stamens during much of flower development, but during stage 13, *AP3* RNA is detected in the outer integument of ovules (35, 51). The role, if any, for *AP3* expression in ovules is unclear because no ovule phenotype for *ap3* mutants has been observed. The other known B-function gene, *PISTILLATA*, was also cloned using a probe specific for its *Antirrhinum* homolog, *GLOBOSA* (22, 77). Like *AP3*, *PI* encodes a MADS-domain and is expressed preferentially throughout petal and stamen primordia. In contrast to *AP3*, the initial expression domain of *PI* includes cells of the fourth whorl. Studies in both *Antirrhinum* and *Arabidopsis* indicate that these B-function gene products likely interact as a heterodimer to affect downstream gene expression (22, 69, 77).

The A-function gene *APETALA1* was isolated as a new member of the MADS-box gene family (46). The genetic data indicating that AP1 functions in two phases of flower development are reflected in the biphasic expression pattern of *AP1* (see Molecular Characterization of Meristem Identity Genes section). Consistent with its role in specifying the identity of the two outer whorls of organs, *AP1* RNA accumulates in sepal and petal primordia throughout flower development. The A-function gene *APETALA2* was isolated by T-DNA tagging (38). AP2 does not show significant sequence similarity to functionally defined proteins, but its sequence suggests that it may be localized to the nucleus. One intriguing feature of the AP2 protein is that it contains an essential 68-amino-acid repeated motif, called the AP2-domain. Consistent with the *ap2* mutant phenotype, which shows alterations in all four floral whorls, *AP2* RNA accumulates in all whorls of developing flowers. *AP2* RNA also accumulates in stems and leaves, but no phenotype has been observed in vegetative development. Although *AP2* may not play a role in vegetative development, its activity in leaves and stems may be genetically redundant. Because *AP2* is a member of a small gene family in *Arabidopsis*, one of these related genes may provide such a function.

Activation and Regulation of Organ Identity Genes

The genetic organ identity model provides insights into the interactions between the organ identity genes, but does not allow one to determine if these genes are regulated at the transcriptional or posttranscriptional level. Furthermore, these data provide little insight into how these genes are initially activated. Good candidates for activators of the organ identity genes are the earlier-acting meristem identity genes. The expression of both B-function genes, *AP3* and *PI*, is severely reduced in *lfy* mutants, indicating that LFY can formally be considered a positive regulator of these genes (83). The class C gene, *AG*, is expressed at reduced levels in *lfy* single mutants, and its normal expression pattern is completely disrupted in *lfy ap1* double mutants. Furthermore, the onset of *AG* expression, which is unchanged in *ap1* single mutants, is significantly delayed in *ap1 cal* double mutants (B Savidge & M Yanofsky, unpublished data). These data indicate that *LFY*, *AP1*, and *CAL* encode partially redundant activities involved in the activation of *AG*. Thus the combined action of the meristem identity genes leads to the activation of the B- and C-function genes, although it is less clear how the uniformly expressed *LFY* and *AP1* genes lead to the region-specific expression domains of the later-acting genes. Additionally, the onset of expression of the meristem identity genes is much earlier than that of the organ identity genes, suggesting that the action of meristem identity genes in inducing the organ identity genes is probably indirect.

One candidate for a gene that mediates between these two classes of genes is the *Antirrhinum* gene *FIMBRIATA* (*FIM*), or its putative *Arabidopsis* counterpart *UNUSUAL FLORAL ORGANS* (74, 85). Both the expression and activity of *FIM* depend on the meristem identity genes, and the spatial expression domains of the organ identity genes depend on *FIM*. Thus *FIM* represents a new class of genes, and its protein product does not exhibit significant sequence similarity to previously characterized gene products. Other candidates for mediators between the meristem and organ identity genes are *AGL2*, *AGL4*, and *AGL9*, which are all MADS-box genes whose expression initiates after *LFY* and *AP1*, but before *AP3*, *PI*, and *AG* (20; A Mandel & M Yanofsky, unpublished data). In support of this conclusion, inactivation of the *AGL9* homologs from petunia (*FBP2*) and tomato (*TM5*) result in both meristem and organ defects (2, 57).

The genetic data indicate that AP2 prevents the activity of AG in the two outer whorls of wild-type flowers. To determine if this regulation occurs at the level of RNA accumulation, *AG* RNA was examined in the strong and weak *ap2* mutant backgrounds. In the strong *ap2-2* mutant allele, the expression domain of *AG* expands to include cells of all four whorls, demonstrating that AP2 is a negative spatial regulator of *AG* RNA accumulation (19). However, in the weak *ap2-1* allele, which produces leaves and stamens in place of the sepals and petals, *AG* RNA is not significantly expressed in the outermost whorl. These data suggest that the product of the *ap2-1* allele largely retains its ability to negatively regulate *AG* expression but has lost its ability to specify organ identity.

Because AP2 negatively regulates *AG* RNA accumulation in the first and second whorls of wild-type flowers, and the genetic model predicts that AP2 will expand its activity into the inner whorls of *ag* mutant flowers, it was anticipated that little or no *ag* RNA would accumulate in *ag* flowers. However, in situ hybridization with the *ag* mutant demonstrates that *ag* RNA accumulation is similar to that in wild-type flowers, following the same temporal and spatial pattern (23). These data suggest that either a factor (factor X) specific to the outer two whorls is necessary for the *AP2*-mediated negative regulation of *AG*, or that a factor specific to the inner two whorls prevents the *AP2*-mediated repression of *AG* (Figure 3, *right*). For example, factor X could allow *AP2* RNA to be translated only in the outer two whorls, allowing for specific regulation of *AG* in only those whorls in which AP2 protein is expressed. This seems unlikely because *ap2* mutants display distinct phenotypic effects in all four whorls, suggesting that AP2 is active in all four whorls of wild-type flowers. Another possibility is that factor X either posttranslationally modifies AP2 or directly interacts with AP2 to affect its negative regulation of *AG*. Alternatively, factor X could be active in the two inner whorls and have the opposite activity proposed above, such as preventing translation of *AP2* RNA in the inner whorls, posttranslationally modifying the AP2 protein, or directly interacting with AP2 to inhibit its ability to negatively regulate *AG*. *LEUNIG* is one candidate for a gene whose product interacts with AP2 to negatively regulate *AG* (84).

The *APETALA1* gene, which is needed for sepal and petal development, is uniformly expressed early in flower development but later is absent from the cells that give rise to stamens and carpels. Interestingly, the onset of *AG* expression in the two inner whorls is followed immediately by the loss of *AP1* expression in these cells (23). Furthermore, *AP1* is ectopically expressed in the inner whorls of *ag* single- and *ag ap2* double-mutant flowers, demonstrating that AG is a negative spatial regulator of *AP1* RNA accumulation in the two inner whorls of wild-type flowers. Thus *AP1*, which is involved in the activation of *AG*, is itself negatively regulated by AG. The timing of the onset of *AG* expression and the corresponding loss of *AP1* RNA accumulation suggests that this interaction may be direct, but further data are required to determine if AG represses *AP1* by binding to *cis*-acting regulatory sequences of the *AP1* gene.

The B-function genes, *AP3* and *PI*, which are activated initially by LFY (and perhaps by the *Arabidopsis* homolog of the *Antirrhinum FIM* gene), require each other's activity for maintenance of expression (22, 35, 36). As noted above, maintenance of expression and activation of the downstream target genes of *AP3* and *PI*, are probably mediated through direct interaction of these two proteins. Furthermore, the *SUP* gene product prevents B gene activity in the fourth whorl of wild-type flowers, but it is unclear why *AP3* and *PI* fail to be activated in the outer whorl.

Transgenic Studies

Genetic studies have identified many of the genes required for organ specification, but these studies cannot determine if the isolated genes are sufficient for their proposed activities. To determine if the *AG* gene is sufficient for the C function, transgenic plants were constructed in *Arabidopsis*, tobacco, tomato, and petunia in which the *AG* gene from each species was constitutively expressed by the CaMV 35S promoter (3, 53). The results of these studies were clear: Ectopic expression of the *AG* gene, within the context of the flower, is sufficient to convert sepals and petals into carpels and stamens, respectively (39, 45, 50, 58, 78). Furthermore, the *AG* gene is present in distantly related dicot plant species, as well as in monocots (65). However, it was not known if *AG* could function appropriately across species boundaries. This was demonstrated to be the case by showing that the *AG* gene from *Brassica napus* could produce the expected phenotype when ectopically expressed in transgenic tobacco. (45). This result indicates that the *B. napus* AG protein recognizes and regulates the promoters of its target genes across distant species boundaries. Furthermore, the transgenic tobacco plants had two novel phenotypes that could not have been predicted by the genetic data. Within the developing stamens, pollen fails to develop, and within the carpels, ovules are transformed into carpel-like structures. These data may explain the molecular data that had shown that late in stamen and carpel development, *AG* expression is specifically absent from certain cell types (5). That the loss of *AG* expression from certain cell types of the anther and ovules is functionally relevant is demonstrated by the aberrant development of these tissues when *AG* is expressed. Interestingly, a recently described *Arabidopsis* mutant, *bell1*, frequently develops carpels in place of ovules (51, 61, 62). Molecular data support the conclusion that the replacement of ovules by carpels results from ectopic *AG* expression.

Transgenic plants that constitutively express the B-function genes *AP3* and *PI* offer interesting insights into the roles of these genes in specifying petals and stamens (36, 84; B Krizek & EM Meyerowitz, unpublished information). Constitutive expression of *AP3* produces flowers that resemble the *superman* mutant, with extra stamens developing at the expense of carpels. Presumably, the constitutively expressed AP3 product is able to interact with PI, which is normally expressed only transiently in the fourth whorl, to produce the extra stamens. However, constitutive expression of *AP3* does not alter sepal development, presumably because *PI* is not expressed there. However, plants that constitutively express both *AP3* and *PI* show a conversion of sepals into petals as predicted by the genetic model. Constitutive expression of *AP3* also validates the idea that *AP3* is positively regulated by LFY, because introduction of the 35S-AP3 construct into *lfy* mutants partially restores the organ identity defects of *lfy* mutant plants.

MADS-BOX GENES

MADS-box genes play a central role in flower development, as evidenced by members of the A, B, and C organ identity functions being encoded by the MADS-box genes *AP1*, *AP3* and *PI*, and *AG*, respectively (22, 35, 46, 87). Furthermore, the two floral meristem identity genes *AP1* and *CAL* also are MADS-box genes, demonstrating that this gene family is important not only in the later step of specifying organ identity, but also for the early step of specifying floral meristem identity (46; S Kempin, B Savidge & M Yanofsky). Each of the above-mentioned MADS-box genes probably evolved from a common ancestral gene, with each gene acquiring a specific role in flower development. That cognate homologs of these genes exist in distantly related dicots (15, 70) and in monocots (65) suggests further that this family of genes has played an important role in flower development in many, if not all, flowering plants. Because the structure and function of MADS-box genes has been reviewed recently (18), it is mentioned here only briefly.

The term MADS-box arose from the first characterized genes that share this conserved region (*MCM1, AGAMOUS, DEFICIENS, SRF*) (70). The MADS-domain is present in the SRF and MCM1 transcription factors from humans and yeast, respectively, and is necessary for sequence-specific binding of these proteins to DNA (26, 28, 52, 55). Based on these studies, all MADS-box genes probably encode transcription factors that regulate various aspects of gene expression in distantly related organisms. This 56-amino-acid DNA-binding domain is localized to the amino terminus of all characterized plant gene products and generally begins immediately after the initiator methionine codon. The DNA sequences to which plant MADS-domain proteins bind [CC(A/T)$_6$GG)] have only recently begun to be determined and generally are related to the consensus binding sites of nonplant MADS-domain proteins (17, 31, 59, 73). The plant MADS-box genes all encode the K domain, a second weakly conserved region of roughly 70 amino acid residues. The K-domain shares structural similarity with the coiled coil domain of keratin, and has been suggested to mediate protein-protein interactions (44).

In addition to the above-mentioned MADS-box genes whose roles in flower development are well documented, low-stringency hybridizations identified additional members of this family in *Arabidopsis* and other plant species (10, 44, 46, 56, 70, 77, 79) (Figure 4). Six additional genes were isolated initially from *Arabidopsis* based on their similarity to *AGAMOUS*, and they were termed *AGL1-AGL6* (*AG*-like). More recently, 12 additional members of

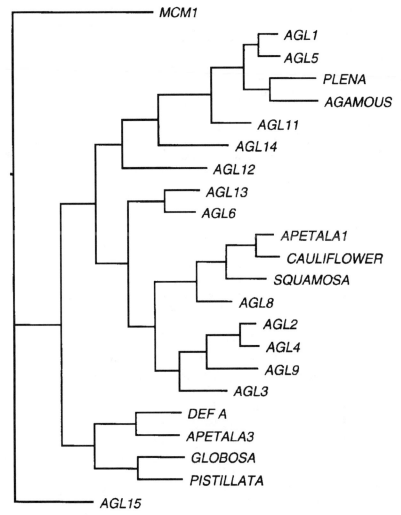

Figure 4   Tree diagram illustrating the evolutionary relationship between some of the cloned MADS-box genes.

this family have been isolated (S Kempin, B Savidge & M Yanofsky, submitted; S Rounsley, S Gold, S Liljegrin & M Yanofsky, unpublished information). RNA blotting and in situ hybridization analyses of these newly isolated genes suggest that MADS-box genes play diverse roles in *Arabidopsis* development. For example, the expression of *AGL2*, *AGL4*, and *AGL9* initiates during stage 2 of flower development, and they are thus candidate targets of the floral meristem identity genes and/or regulators of the organ identity genes.

AGL5, which requires AG for expression and whose RNA begins to accumulate shortly after that of *AG*, is a candidate target gene of AG (B Savidge & M Yanofsky, submitted). Several MADS-box genes are expressed at relatively high levels in developing ovules, suggesting a role for this family of genes in regulating ovule development. These genes may well play roles in addition to those already documented for flower development. For example, *AGL8* is expressed in the inflorescence meristem, but not in young flower primordia (A Mandel & M Yanofsky, manuscript in preparation); *AGL12* is preferentially expressed in roots; and *AGL15* is preferentially expressed during embryo development (S Rounsley & M Yanofsky, manuscript in preparation). The roles of MADS-box genes in plant development have only begun to be explored.

PERSPECTIVE

The recent cloning of the meristem identity genes has opened the door to understanding the molecular basis for the specification of floral meristem identity. Indeed, the genetic data that had implicated these loci in meristem identity is reflected in the temporal and spatial accumulation of RNAs specific for these genes. However, many questions remain unanswered. What factors initiate the expression of *LFY*, *AP1*, and *CAL* RNAs in young flower primordia? The *LFY*, *AP1*, and *CAL* genes should provide molecular tools that may allow for the cloning of genes whose products interact with the *cis*-acting regulatory sequences of these genes. This approach has led to the isolation of a novel family of DNA-binding proteins that recognize a specific element in the *SQUAMOSA* (the *AP1* homolog) gene promoter (32). Further analyses should determine if one or more members of this gene family play a role in regulating *SQUA* and *AP1* gene expression. In other experiments, RNA in situ hybridization showed that the onset of *AP1* expression is significantly delayed in *lfy* mutant plants (C Gustafson-Brown & M Yanofsky, unpublished information). This result suggests that *LFY*, which is not necessary for the later accumulation of *AP1* RNA, may play a role in activating *AP1* expression. Whether this activation of *AP1* by *LFY* is direct or indirect remains to be determined.

Do the LFY, AP1, and CAL proteins interact? Perhaps their activities are mediated in part through protein-protein interactions, a possibility because all three gene products probably accumulate in the same cells in young flower primordia. Are there additional proteins that interact with these gene products to determine meristem fate? Genetic experiments have already shown that the *AP2* and *CLV1* gene products play a role in meristem identity. What are the targets of these putative regulatory genes? Genetic and molecular data implicate these genes in the activation of the later-acting organ identity genes (83), and more recent data indicate that additional genes may mediate between the interactions of meristem and organ identity genes (74). Because *LFY*, *AP1*, and *CAL* RNAs are initially expressed in all cells of the floral meristem, what factor(s) prevents their accumulation in certain cell types of the flower later in development? This question has in part been answered for both *AP1* and *CAL*, which are negatively regulated by *AG* in the two inner whorls of wild-type flowers. Do these meristem identity genes exist and perform similar functions in all flowering plants, or do slight modifications in their expression patterns or activity play a role in the tremendous diversity of floral form and structure in distantly related plants?

The rapid progress made toward unraveling the underlying molecular mechanisms controlling flower development suggests that many of these questions will be answered in the next few years. But as these mysteries are solved, they will inevitably lead to new questions that will become the mysteries of tomorrow. Given the tremendous diversity of plants, the study of flower development will remain a fertile field for scientific investigation for many years to come.

ACKNOWLEDGMENTS

I am grateful to my colleagues for sending preprints of manuscripts and for sharing unpublished results, to Detlef Weigel for providing slides of the *lfy* and *lfy ap1* mutants, and to Steve Rounsley for preparing the MADS-box tree diagram. I thank members of my laboratory, Takashi Araki, Gary Ditta, Scott Gold, Cindy Gustafson-Brown, Sherry Kempin, Sarah Liljegrin, Ale Mandel, Steve Rounsley, and Beth Savidge, for their critical comments on this manuscript. Our *Arabidopsis thaliana* research is supported by grants from the National Science Foundation (DCB-9018749), the Arnold and Mabel Beckman Foundation, and by a David and Lucille Packard Fellowship in Science and Engineering.

Literature Cited

1. Alvarez J, Guli CL, Yu X-H, Smyth DR. 1992. *TERMINAL FLOWER*: a gene affecting inflorescence development in *Arabidopsis thaliana*. *Plant J.* 2:103–16
2. Angenent GC, Franken J, Busscher M, Weiss D, van Tunen AJ. 1994. Co-suppression of the petunia homeotic gene *fbp2* affects the identity of the generative meristem. *Plant J.* 5:233–44
3. Benfey PN, Chua N-H. 1990. The Cauliflower mosaic virus 35S promoter: combinatorial regulation of transcription in plants. *Science* 250:959–66
4. Bowman JL, Alvarez J, Weigel D, Meyerowitz EM, Smyth DR. 1993. Control of flower development in *Arabidopsis thaliana* by *APETALA1* and interacting genes. *Development* 119:721–43

5. Bowman JL, Drews GN, Meyerowitz EM. 1991. Expression of the *Arabidopsis* floral homeotic gene *AGAMOUS* is restricted to specific cell types late in flower development. *Plant Cell* 3:749–58

6. Bowman JL, Sakai H, Jack T, Weigel D, Mayer U, Meyerowitz EM. 1992. *SUPERMAN*, a regulator of floral homeotic genes in *Arabidopsis*. *Development* 114: 599–615

7. Bowman JL, Smyth DR, Meyerowitz EM. 1989. Genes directing flower development in *Arabidopsis*. *Plant Cell* 1:37–52

8. Bowman JL, Smyth DR, Meyerowitz EM. 1991. Genetic interactions among floral homeotic genes of *Arabidopsis*. *Development* 112:1–20

9. Bowman JL, Yanofsky MF, Meyerowitz EM. 1988. *Arabidopsis thaliana*: a review. In *Oxford Surveys of Plant Molecular and Cell Biology*, ed. BJ Miflin, 5:57-87. New York: Oxford Univ. Press

10. Bradley D, Carpenter R, Sommer H, Hartley N, Coen E. 1993. Complementary floral homeotic phenotypes result from opposite orientations of a transposon at the *plena* locus of *Antirrhinum*. *Cell* 72:85–95

11. Carpenter R, Coen ES. 1990. Floral homeotic mutations produced by transposon-mutagenesis in *Antirrhinum majus*. *Genes Dev.* 4:1483–93

12. Deleted in proof

13. Clark SE, Running MP, Meyerowitz EM. 1993. *CLAVATA1*, a regulator of meristem and flower development in *Arabidopsis*. *Development* 119:397–418

14. Coen ES, Carpenter R. 1993. The metamorphosis of flowers. *Plant Cell* 5:1175–81

15. Coen ES, Meyerowitz EM. 1991. The war of the whorls: genetic interactions controlling flower development. *Nature* 353: 31–37

16. Coen ES, Romero JM, Doyle S, Elliott R, Murphy G, Carpenter R. 1990. *FLORICAULA*: homeotic gene required for flower development in *Antirrhinum majus*. *Cell* 63:1311–22

17. Dalton S, Treisman R. 1992. Characterization of SAP-1, a protein recruited by serum response factor to the *c-fos* serum response element. *Cell* 68:597–612

18. Davies B, Schwarz-Sommer Z. 1994. Control of floral organ identity by homeotic MADS-box transcription factors. In *Results and Problems in Cell Differentiation*, ed. L Nover, pp. 235–58. Berlin: Springer-Verlag

19. Drews GN, Bowman JL, Meyerowitz EM. 1991. Negative regulation of the *Arabidopsis* homeotic gene *AGAMOUS* by the *APETALA2* product. *Cell* 65:991–1002

20. Flanagan CA, Ma H. 1994. Spatially and temporally regulated expression of the MADS-box gene *AGL2* in wild type and mutant *Arabidopsis* flowers. *Plant Mol. Biol.* In press

21. Goethe JW. 1790. Versuch die Metamorphose der Pflanzen zu erklaren. Gotha: CW Ettinger. Transl. A Arber, 1946. Goethe's botany. *Chron. Bot.* 10:63–126

22. Goto K, Meyerowitz EM. 1994. Function and regulation of the *Arabidopsis* floral homeotic gene *PISTILLATA*. *Genes Dev.* 8:1548–60

23. Gustafson-Brown C, Savidge B, Yanofsky MF. 1994. Regulation of the *Arabidopsis* floral homeotic gene *APETALA1*. *Cell* 76: 131–43

24. Deleted in proof

25. Haughn GW, Somerville CR. 1988. Genetic control of morphogenesis in *Arabidopsis*. *Dev. Genet.* 9:73–89

26. Hayes TE, Sengupta P, Cochran BH. 1988. The human *c-fos* serum response factor and the yeast factors GRM/PRTF have related DNA-binding specificities. *Genes Dev.* 2: 1713–22

27. Hempel FD, Feldman LJ. 1994. Bi-directional inflorescence development in *Arabidopsis thaliana*: acropetal initiation of flowers and basipetal initiation of paraclades. *Planta* 192:276–86

28. Herskowitz I. 1989 A regulatory hierarchy for cell specialization in yeast. *Nature* 342: 749–57

29. Hill JP, Lord EM. 1989. Floral development in *Arabidopsis thaliana*: a comparison of the wild type and the homeotic *pistillata* mutant. *Can. J. Bot.* 67:2922–36

30. Huala E, Sussex IM. 1992. *LEAFY* interacts with floral homeotic genes to regulate *Arabidopsis* floral development. *Plant Cell* 4:901–13

31. Huang H, Mizukami Y, Hu Y, Ma H. 1993. Isolation and characterization of the binding sequences for the product of the *Arabidopsis* floral homeotic gene *AGAMOUS*. *Nucleic Acids Res.* 21:4769–76

32. Huijser P, Cardon G, Cremer F, Hohmann S, Klein W-E, et al. 1994. Molecular and genetic analysis of early flower development in *Antirrhinum majus*. *4th Int. Congr. Plant Mol. Biol.*, No. 735 (Abstr.)

33. Huijser P, Klein J, Lönnig W-E, Meijer H, Saedler H, Sommer H. 1992. Bracteomania, an inflorescence anomaly, is caused by the loss of function of the MADS box gene *SQUAMOSA* in *Antirrhinum majus*. *EMBO J.* 11:1239–49

34. Irish VF, Sussex IM. 1990. Function of the *APETALA1* gene during *Arabidopsis* floral development. *Plant Cell* 2:741–53

35. Jack T, Brockman LL, Meyerowitz EM. 1992. The homeotic gene *APETALA3* of *Arabidopsis thaliana* encodes a MADS box and is expressed in petals and stamens. *Cell* 68:683–88

36. Jack T, Fox GL, Meyerowitz EM. 1994.

Ectopic expression of the *Arabidopsis* homeotic gene *APETALA3*: transcriptional and post-transcriptional regulation determine floral organ identity. *Cell* 76:703–16

37. Jackson D, Veit B, Hake S. 1994. Expression of maize *knotted1* related homeobox genes in the shoot apical meristem predicts patterns of morphogenesis in the vegetative shoot. *Development* 120:405–13

38. Jofuku KD, den Boer BGW, Van Montagu M, Okamuro JK. 1994. Control of *Arabidopsis* flower and seed development by the homeotic gene *APETALA2*. *Plant Cell.* 6:1211–25

39. Kempin SA, Mandel MA, Yanofsky MF. 1993. Ectopic expression of the tobacco floral homeotic gene *NAG1* converts perianth into reproductive organs. *Plant Physiol.* 103:1041–46

40. Deleted in proof

41. Komaki MK, Okada K, Nishino E, Shimura Y. 1988. Isolation and characterization of novel mutants of *Arabidopsis thaliana* defective in flower development. *Development* 104:195–203

42. Kunst L, Klenz JE, Martinez-Zapater J, Haughn GW. 1989. *AP2* gene determines the identity of perianth organs in flowers of *Arabidopsis thaliana*. *Plant Cell* 1:1195–1208

43. Ma H. 1994. The unfolding drama of flower development: recent results from genetic and molecular analyses. *Genes Dev.* 8:745–56

44. Ma H, Yanofsky MF, Meyerowitz EM. 1991. *AGL1-AGL6*, an *Arabidopsis* gene family with similarity to floral homeotic and transcription factor genes. *Genes Dev.* 5:484–95

45. Mandel MA, Bowman JL, Kempin SA, Ma H, Meyerowitz EM, Yanofsky MF. 1992. Manipulation of flower structure in transgenic tobacco. *Cell* 71:133–43

46. Mandel MA, Gustafson-Brown C, Savidge B, Yanofsky MF. 1992. Molecular characterization of the *Arabidopsis* floral homeotic gene *APETALA1*. *Nature* 360:273–77

47. Deleted in proof

48. Meyerowitz EM, Bowman JL, Brockman LL, Drews GN, Jack T, et al. 1991. A genetic and molecular model for flower development in *Arabidopsis thaliana*. *Development (Suppl.)* 1:157–67

49. Meyerowitz EM, Smyth DR, Bowman JL. 1989. Abnormal flowers and pattern formation in floral development. *Development* 106:209–17

50. Mizukami Y, Ma H. 1992. Ectopic expression of the floral homeotic gene *AGAMOUS* in transgenic *Arabidopsis* plants alters floral organ identity. *Cell* 71:119–31

51. Modrusan Z, Reiser L, Feldmann KA, Fischer RL, Haughn GW. 1994. Homeotic transformation of ovules into carpel-like structures in *Arabidopsis*. *Plant Cell* 6:333–49

52. Norman C, Runswick M, Pollock R, Treisman R. 1988. Isolation and properties of cDNA clones encoding SRF, a transcription factor that binds to the *c-fos* serum response element. *Cell* 55:989–1003

53. Odell JT, Nagy F, Chua N-H. 1985. Identification of DNA sequences required for activity of the cauliflower mosaic virus 35S promoter. *Nature* 313:810–12

54. Okamuro JK, den Boer BGW, Jofuku KD. 1993. Regulation of *Arabidopsis* flower development. *Plant Cell* 5:1183–93

55. Passmore S, Maine GT, Elble R, Christ C, Tye B-K. 1988. A *Saccharomyces cerevisiae* protein involved in plasmid maintenance is necessary for mating of MATa cells. *J. Mol. Biol.* 204:593–606

56. Pnueli L, Abu-Abeid M, Zamir D, Nacken W, Schwarz-Sommer Z, Lifschitz E. 1991. The MADS-box gene family in tomato: temporal expression during floral development, conserved secondary structures and homology with homeotic genes from *Antirrhinum* and *Arabidopsis*. *Plant J.* 1:255–66

57. Pnueli L, Hareven D, Broday L, Hurwitz C, Lifschitz E. 1994. The TM5 MADS box gene mediates organ differentiation in the three inner whorls of tomato flowers. *Plant Cell* 6:175–86

58. Pnueli L, Hareven D, Rounsley SD, Yanofsky MF, Lifschitz E. 1994. Isolation of the tomato *AGAMOUS* gene *TAG1* and analysis of its homeotic role in transgenic plants. *Plant Cell* 6:163–73

59. Pollock R, Treisman R. 1991. Human SRF-related proteins: DNA-binding properties and potential regulatory targets. *Genes Dev.* 5:2327–41

60. Pruitt RE, Chang C, Pang PP-Y, Meyerowitz EM. 1987. Molecular genetics and development of *Arabidopsis*. In *Genetic Regulation of Development, 45th Symp. Soc. Dev. Biol.*, ed. W Loomis, pp. 327–38. New York: Liss

61. Ray A, Robinson-Beers K, Ray S, Baker SC, Lang JD, et al. 1994. The *Arabidopsis* floral homeotic gene *BELL* controls ovule development through negative regulation of *AGAMOUS*. *Proc. Natl. Acad. Sci. USA* 91:5761–65

62. Robinson-Beers K, Pruitt RE, Gasser CS. 1992. Ovule development in wild type *Arabidopsis* and two female-sterile mutants. *Plant Cell* 4:1237–49

63. Sadik S. 1962. Morphology of the curd of cauliflower. *Am. J. Bot.* 49:290–97

64. Deleted in proof

65. Schmidt RJ, Veit B, Mandel MA, Mena M, Hake S, Yanofsky MF. 1993. Identification and molecular characterization of *ZAG1*, the maize homolog of the *Arabidopsis* flo-

ral homeotic gene *AGAMOUS. Plant Cell* 5:729–37

66. Schultz EA, Haughn GW. 1991. *LEAFY*, a homeotic gene that regulates inflorescence development in *Arabidopsis. Plant Cell* 3: 771–81

67. Schultz EA, Haughn GW. 1993. Genetic analysis of the floral initiation process (FLIP) in *Arabidopsis. Development* 110: 745–65

68. Schultz EA, Pickett FB, Haughn GW. 1991. The *FLO10* gene product regulates the expression domain of homeotic genes *AP3* and *PI* in *Arabidopsis* flowers. *Plant Cell* 3:1221–37

69. Schwarz-Sommer Z, Hue I, Huijser P, Flor PJ, Hansen R, et al. 1992. Characterization of the *Antirrhinum* floral homeotic MADS-box gene *deficiens*: evidence for DNA binding and autoregulation of its persistent expression throughout flower development. *EMBO J.* 11:251–63

70. Schwarz-Sommer Z, Huijser P, Nacken W, Saedler H, Sommer H. 1990. Genetic control of flower development: homeotic genes in *Antirrhinum majus. Science* 250: 931–36

71. Shannon S, Meeks-Wagner DR. 1991. A mutation in the *Arabidopsis TFL1* gene affects inflorescence meristem development. *Plant Cell* 3:877–92

72. Shannon S, Meeks-Wagner DR. 1993. Genetic interactions that regulate inflorescence development in *Arabidopsis. Plant Cell* 5:639–55

73. Shiraishi H, Okada K, Shimura Y. 1993. Nucleotide sequences recognized by the AGAMOUS MADS domain of *Arabidopsis thaliana in vitro. Plant J.* 4:385–98

74. Simon R, Carpenter R, Doyle S, Coen E. 1994. *FIMBRIATA* controls flower development by mediating between meristem and organ identity genes. *Cell* 78:99–107

75. Smyth DR, Bowman JL, Meyerowitz EM. 1990. Early flower development in *Arabidopsis. Plant Cell* 2:755–67

76. Sommer H, Beltran JP, Huijser P, Pape H, Lönnig W-E, et al. 1990. *DEFICIENS*, a homeotic gene involved in the control of flower morphogenesis in *Antirrhinum* ma-

jus: the protein shows homology to transcription factors. *EMBO J.* 9:605–13

77. Tröbner W, Ramirez L, Motte P, Hue I, Huijser P, et al. 1992. *GLOBOSA*: a homeotic gene which interacts with *DEFICIENS* in the control of *Antirrhinum* floral organogenesis. *EMBO J.* 11:4693–4704

78. Tsuchimoto S, van der Krol AR, Chua N-H. 1993. Ectopic expression of *pMADS3* in petunia phenocopies the petunia blind mutant. *Plant Cell* 5:843–53

79. van der Krol AR, Brunelle A, Tsuchimoto S, Chua N-H. 1993. Functional analysis of petunia floral homeotic MADS-box gene *pMADS1. Genes Dev.* 7:1214–28

80. van der Krol AR, Chua N-H. 1993. Flower development in petunia. *Plant Cell* 5:1195–1203

81. Veit B, Schmidt RJ, Hake S, Yanofsky MF. 1993. Maize floral development—new genes and old mutants. *Plant Cell* 5:1005–15

82. Weigel D, Alvarez J, Smyth DR, Yanofsky MF, Meyerowitz EM. 1992. *LEAFY* controls floral meristem identity in *Arabidopsis. Cell* 69:843–59

83. Weigel D, Meyerowitz EM. 1993. Activation of floral homeotic genes in *Arabidopsis. Science* 261:1723–26

84. Weigel D, Meyerowitz EM. 1994. The ABCs of floral homeotic genes. *Cell* 78: 203–9

85. Wilkinson MD, Haughn GW. 1994. The role of the *UNUSUAL FLORAL ORGANS* locus in determination of shoot type. *4th Int. Congr. Plant Mol. Biol.*, No. 742 (Abstr.)

86. Yanofsky MF, Araki T, Gustafson-Brown C, Kempin MA, Savidge B. 1993. Genes specifying floral meristem identity in *Arabidopsis.* In *NATO ASI Ser.*, ed. G Goruzzi, P Puigdomènech, 81:51–61. Berlin: Springer-Verlag

87. Yanofsky MF, Ma H, Bowman JL, Drews G, Feldmann K, Meyerowitz EM. 1990. The protein encoded by the *Arabidopsis* homeotic gene *AGAMOUS* resembles transcription factors. *Nature* 346:35–39

88. Yarnell SH. 1956. Cytogenetics of the vegetable crops. II. Crucifers. *Bot. Rev.* 22:81–166

Annu. Rev. Plant Physiol. Plant Mol. Biol. 1995. 46:189–214

CHEMOPERCEPTION OF MICROBIAL SIGNALS IN PLANT CELLS

Thomas Boller

Friedrich Miescher-Institut, P.O. Box 2543, CH-4002 Basel, Switzerland and Botanisches Institut, Universität Basel, Hebelstrasse 1, CH-4056 Basel, Switzerland

KEY WORDS: elicitors, Nod factors, resistance genes, signal perception, signal transduction

CONTENTS

ABSTRACT

Plants have exquisitely sensitive chemoperception systems for signal substances derived from microorganisms. Among the microbial substances that
plants can perceive at threshold concentrations of approximately 10^{-12}–10^{-10}
M are oligosaccharides and lipo-oligosaccharides, peptides and glycopeptides,
and lipophilic substances such as ergosterol. In many cases, the perception
systems of plants recognize "non-self" molecules characteristic of fungi and
bacteria with a high degree of specificity, and some of them show rapid
desensitization, leading to a refractory state reminiscent of smell perception.
Specific binding sites for microbial signal molecules have been found on intact
plant cells and on isolated plasma membranes; they probably function as
receptors. An intriguing question is whether the products of some of the
recently identified resistance genes are homologous to such receptors.
Chemoperception of microbial substances may not only play a role for the
plant in active defense and pathogen resistance, but also in mutualistic symbiosis and in the acquisition of basic information about microbes in the environment.

INTRODUCTION

For human beings, the impressions of the chemical senses, smell and taste, are
deeply rooted in our minds, particularly the scents perceived through our
olfactory receptors (58). Even the simplest motile organisms, such as bacteria,
paramecia, and slime molds, use similar olfactory systems. Thus, reviews and
textbooks often draw parallels between the chemoperception systems of microorganisms (4, 49, 162) and olfaction in animals (49, 58, 154). However,
common sense tells us that plants, as sedentary organisms, do not need olfactory systems because they would be unable to move when attracted by scents
or repelled by stenches. Accordingly, plants rarely are mentioned in overviews
on chemoperception (49), with the exception of the motile gametes of brown
algae (105).

This review summarizes evidence that, contrary to expectation, higher
plants have highly sensitive chemoperception systems that play a central role
in symbioses with bacteria and fungi [using symbiosis in its original, wide
sense of a close interaction, without consideration of benefits (147)]. These
systems allow plants to interact appropriately with mutualistic partners, neutral
endophytes, and antagonistic pathogens. Among the microbial substances perceived as signals by plants are the nodulation factors (Nod factors) of rhizobia,
the mutualistic nitrogen-fixing symbionts of legumes (41, 59), and the elicitors
derived from antagonistic, pathogenic fungi (1, 34, 50, 51). This review concentrates on the chemoperception process itself rather than on the responses
initiated by the various signals. The responses to microbial signals have been

reviewed elsewhere [e.g. the formation of root nodules induced by Nod factors in legumes (23, 41, 59, 61) or the defense reactions induced by elicitors in responsive plant tissues (47, 47a, 94)].

SURVEY OF MICROBIAL SIGNALS

At the heart of mutualistic and neutral as well as antagonistic symbioses is the transfer of assimilates from the autotrophic plant to the heterotrophic microbial partner; this massive chemical interaction at the trophic level is regulated and modulated by the exchange of less abundant chemical substances between the partners (147). Some of these substances, such as vitamins, antimetabolites, or toxins, act at the trophic level, directly affecting nutrient uptake or metabolism; others act at the informational level, being recognized as signals by appropriate chemoperception systems (49). Obviously, a given compound may act at both the trophic and the informational levels: Nutrients may act as attractants, or toxins as repellents (49). Theoretical considerations suggest that effective chemical signaling at the single-cell level requires threshold concentrations of 10^{-12} M (49). Several of the microbial signals perceived by plant cells are active at about this threshold, indicating that plants use the potential of chemical signaling up to its limits.

Bioassays are required to detect and characterize the plants' chemosensory systems. When the focus is on the responses to microbial signals, the biological reactions themselves, such as nodule formation or disease resistance, may serve as bioassays. However, these responses often can be recorded only hours, days, or even weeks after addition of the signals, making it difficult to draw conclusions about signal perception. Therefore, when the focus is on the chemoperception process, it is more appropriate to devise bioassays for the earliest detectable reactions, such as changes in ion fluxes and membrane potential, comparable to the receptor potentials monitored in animal olfaction (58, 154). Such bioassays have uncovered specific and highly sensitive perception systems for a vast variety of microbial substances. These are discussed below, with an emphasis on chemically well-characterized signal molecules.

Oligosaccharides

The prototype of microbial signals recognized by plant cells are the oligosaccharide elicitors, which induce the accumulation of phytoalexins (newly synthesized antimicrobial compounds of low molecular weight) (for reviews, see 34, 50, 51, 134, 173).

GLUCAN FRAGMENTS   In a classic study, a branched β-1,3-β-1,6-heptaglucoside has been isolated as the smallest elicitor-active compound from cell walls of *Phytophthora megasperma* f.sp. *glycinea*, a fungal pathogen of soybean, using phytoalexin accumulation in soybean cotyledons as a bioassay (144). This heptaglucoside is half-maximally active at a concentration of ~ 3×10^{-9} M; closely related heptaglucan isomers and smaller glucans are inactive, indicating that the chemoperception system is highly sensitive and selective (143). Larger glucan fragments from *P. megasperma* cell walls, which can be released by plant β-glucanases (75) and probably contain the heptaglucan unit, are equally active (51). Zoospores of *P. megasperma* release elicitor-active glucans without intervention of plant enzymes (168).

CHITIN OLIGOMERS   Chitin, the β-1,4-linked homopolymer of *N*-acetylglucosamine, is a major constituent of the cell walls of most higher fungi and of the exoskeleton of arthropods; it is not produced by plants and thus may suit them well for detection of fungi or, perhaps, invertebrates (55). Chitin, an insoluble macromolecule, induces lignification when injected into wheat leaves (9) and phytoalexin production when added to rice cell cultures (127). In these systems, plant chitinases probably release soluble chitin oligomers that can be perceived by the plant (16, 55). Indeed, chitin fragments are active as elicitors in rice cells (92). A highly sensitive perception system for chitin oligomers has been discovered in tomato cells, using alkalinization of the growth medium as a bioassay (55). In this system, chitin oligomers with a degree of polymerization (DP) ≥ 4 are half-maximally active at concentrations of $1-3 \times 10^{-10}$ M; the trimer is ~ 1000-fold and the dimer ~ 1,000,000-fold less active (55).

Chitosan (de-acetylated chitin) and its fragments elicit phytoalexin production in pea tissues (88) and cell cultures (29). However, this occurs only at comparatively high concentrations and might be the result of the polycationic nature of chitosan fragments, leading to membrane disturbances, rather than to a specific chemoperception process (71).

Lipo-chitooligosaccharides

Rhizobia produce nodulation factors (Nod factors), i.e. chemical signals essential for the formation of nodules, the symbiotic organs developing on the roots of their host plants (41, 59). These Nod factors, first isolated from *Rhizobium meliloti* on the basis of their ability to cause root hair deformation in alfalfa (101), were found to be lipo-chitooligosaccharides, i.e. derivatives of chitin oligomers with DP 4 and DP 5, carrying a *N*-fatty acyl group instead of an *N*-acetyl group on the glucosamine at the nonreducing end (41, 59). The *nodABC* genes, common to all rhizobia, are involved in synthesizing the lipo-chitooligosaccharide core, whereas other *nod* genes determine modifications important for host specificity, such as the addition of a sulfate, methylfucosyl, or arabinosyl group to the *N*-acetylglucosamine at the reducing end or the addition of an acetyl, methyl, or carbamoyl residue to the *N*-acetylglucosamine at the nonreducing end (40). Depending on these modifications and on the length of the chitooligosaccharide backbone, Nod factors are perceived differentially in different legumes (40, 41). In the most sensitive bioassays (e.g. root hair initiation or deformation, induction of a "thick and short root" phenotype, or induction of early nodulin genes), Nod factors are half-maximally active at concentrations of $10^{-10} - 10^{-9}$ M with a threshold of 10^{-12} M, whereas unmodified chitin fragments are inactive (41).

Surprisingly, Nod factors can also be perceived by tomato, a nonhost of rhizobia: Nod factors induce the same type of alkalinization of the growth medium in tomato cells as do chitin fragments (151). Perception of Nod factors in tomato is probably of little consequence biologically; however, it is tempting to speculate that the chemoperception systems for Nod factors in legumes, with their expected high selectivity, have evolved from a less specific perception system for chitin fragments such as the one observed in tomato (151).

Glycopeptides

Preparations of fungal glycopeptides and glycoproteins often act as elicitors in plants (1, 50). For example, glycopeptides derived from yeast extract (prepared from the yeast *Saccharomyces cerevisiae*, a saprophytic ascomycete) have been identified as elicitors in tomato cells, using induction of ethylene production as a bioassay (11). Chemically defined glycopeptide elicitors of this type have been obtained from purified yeast invertase, cleaved with α-chymotrypsin (10). The amino acid sequence of these glycopeptides is less important for activity than is the structure of their N-linked glycan side chains: Glycopeptides with side chains containing 8 mannosyl residues have little elicitor activity but the ones with side chains of 9–11 mannosyl residues are potent elicitors with half-maximal activities at concentrations of ~ 3×10^{-9} M. Interestingly, the yeast glycan side chains with 8 mannosyl residues represent a core structure occurring in all eukaryotes including plants, but the larger ones have their ninth mannosyl residue attached to this core at a position specific and unique for fungi; thus, the plant's chemoperception system is highly specific for "non-self" fungal glycopeptides (10).

Peptides and Proteins

Plants can also perceive substances other than oligosaccharide derivatives as elicitor signals. The first substance reported to induce phytoalexin accumulation, monilicolin A, was in fact an ~ 8-kDa polypeptide of unknown function (32), and a number of proteinaceous elicitors have since been characterized thoroughly.

PHYTOPHTHORA PROTEINS   Several types of proteins from various *Phytophthora* species act as elicitors in different plants.

Glycoprotein elicitor for parsley   The elicitor activity of a crude glucan preparation from *P. megasperma* in parsley cells is sensitive to proteinases rather than
to glucanases (119). The active molecule is a glycoprotein of $M_r \sim 42$ kDa (120);
a short nonglycosylated peptide generated by proteolytic cleavage is equally
active, inducing short-term ion fluxes as well as defense gene transcription and
phytoalexin accumulation half-maximally at a concentration of $\sim 2 \times 10^{-9}$ M
(118).

Elicitins   Culture fluids of most *Phytophthora* species contain proteins called
elicitins, which induce necrosis in tobacco (129) and in *Brassica* (83). Several
elicitins ($M_r \sim 10$ kDa) have been purified and sequenced (123, 129), and one
has been cloned (82). Purified elicitins display half-maximal activity in tobacco
cells at concentrations of $\sim 2 \times 10^{-9}$ M (15). Interestingly, *P. parasitica* var.
nicotianae strains, able to colonize tobacco, lack elicitins (130) and elicitin
mRNA (82). Nevertheless, culture fluids of *P. parasitica* var. *nicotianae* induce
phytoalexin synthesis in tobacco cells; this elicitor activity resides in a glycoprotein ($M_r = 46$ kDa) half-maximally active at a concentration of $\sim 5 \times 10^{-9}$ M
(53).

XYLANASE   A cellulase preparation from the saprophytic ascomycete
Trichoderma viride stimulates ethylene biosynthesis in tobacco leaves; the
active principle has been purified and identified as a xylanase (41). Xylanase
appears to be perceived directly, and not via its enzymatic products (142). In
tobacco cells, it induces ethylene biosynthesis half-maximally at a concentration
of $\sim 2 \times 10^{-9}$ M (6).

HARPINS   Phytopathogenic gram-negative bacteria frequently induce rapid localized cell death in non-host plants, a reaction called the hypersensitive
response (65). In various species of *Pseudomonas, Xanthomonas,* and *Erwinia,*
a cluster of conserved genes, the *hrp* cluster, is essential for induction of a
hypersensitive response in non-host plants as well as for pathogenesis in
susceptible host plants (65). A majority of the *hrp* genes are believed to encode
elements of a protein export system on the basis of their homology to genes
involved in protein secretion in two animal pathogens, *Yersinia* and *Shigella*
(57, 66). Specific proteins exported in a *hrp*-dependent fashion have been
discovered to induce a hypersensitive reaction, as summarized below.

Erwinia amylovora harpin   The product of the *hrpN* gene of *Erwinia amylovora,* a pathogen of Rosaceae, acts as a potent elicitor of the hypersensitive
response in tobacco leaves; this 44-kDa protein, called harpin, is resistant to
boiling but highly susceptible to proteases (171). Harpin induces within minutes

the production of active oxygen species in tobacco cells, acting half-maximally at a concentration of ~ 2×10^{-9} M (8).

Pseudomonas syringae harpin   In *Pseudomonas syringae* pv. *syringae*, a pathogen of bean, the *hrpZ* gene encodes an ~ 35-kDa protein, harpin$_{PSS}$, which induces a hypersensitive response in tobacco and potato at concentrations $>10^{-6}$ M (78). Harpin$_{PSS}$ resembles the *E. amylovora* harpin (171) in its heat resistance and sensitivity to proteases but differs completely in its sequence, except for a short stretch of 22 amino acids with 45% identity (78).

Pseudomonas solanacearum elicitor   PopA1 and PopA3, two heat-stable proteins (38 and 28 kDa, respectively) that are secreted in a *hrp*-dependent fashion by *Pseudomonas solanacearum*, a pathogen of tomato, elicit a hypersensitive response in tobacco and petunia but not in tomato (3). PopA3, identical to PopA1 but lacking 98 amino acids at the N-terminus, is most active, down to concentrations of $2-6 \times 10^{-7}$ M; it has no homology to the harpins (3).

PEPTIDES AS SPECIFIC ELICITORS   Some polypeptides produced by phytopathogenic fungi elicit a hypersensitive response only in specific cultivars of a plant.

Cladosporium AVR peptides   The interaction of the deuteromycete pathogen *Cladosporium fulvum* with its host plant, tomato, is an excellent model to study gene-for-gene relationships (43, 44), as discussed below. In this system, the first specific elicitor has been identified, a cysteine-rich 28-amino-acid polypeptide (~ 3 kDa) that elicits necrosis exclusively in tomato lines carrying the *Cf9* resistance gene (45). This elicitor, the so-called AVR9 peptide (139), is secreted as a precursor and then processed at the N-terminus by fungal and plant proteases (161). Fungal strains virulent on *Cf9* tomatoes lack the *Avr9* avirulence gene encoding the AVR9 precursor (163). Reintroduction of the gene leads to production of AVR9 and to an avirulent phenotype on *Cf9* tomatoes (160). Similarly, the *Avr4* avirulence gene of *C. fulvum* encodes the AVR4 protein, a larger cysteine-rich polypeptide without sequence homology to *Avr9* that specifically induces necrosis in tomato plants carrying the *Cf4* gene; races of *C. fulvum* virulent on *Cf4* tomatoes carry mutations in the *Avr4* gene (81).

Rhynchosporium secalis NIP1 peptide   The deuteromycete *Rhynchosporium secalis*, a pathogen of barley, produces at least three necrosis-inducing peptides (NIP) when colonizing host leaves (172). One of these peptides, an ~ 7-kDa protein called NIP1, induces defense reactions specifically in barley plants carrying the resistance gene *Rrs1*, at concentrations much below the ones required to induce necrosis (73). For the pathogen, NIP1 may function as a toxin mobilizing nutrients from its host plant; for the plant carrying the resistance gene *Rrs1*, it may serve as a chemical cue for an early induction of the defense response (73).

Microbial Enzymes Releasing Elicitor-Active Compounds from Plant Cells

Microorganisms feeding on living or dead plants secrete enzymes that digest plant cell wall components, such as pectolytic enzymes (28) or xylanases. Such enzymes are not necessarily recognized directly as polypeptides; rather, the products generated by the enzymes may be recognized through chemoperception. These products are derived from the plant's own cell wall and have been termed *endogenous elicitors* (33, 74). Endogenous elicitors may be considered as second messengers of the exogenous microbial enzymes (16, 50).

PECTOLYTIC ENZYMES   In a pioneering study, a polygalacturonase secreted by the saprophytic fungus *Rhizopus stolonifer* has been identified as an elicitor inducing casbene synthase (an enzyme producing the diterpene phytoalexin casbene) in *Ricinus communis* (97). The *Rhizopus* polygalacturonase is heat labile but liberates heat-stable elicitor-active compounds from plant cell walls or pectin (21). Pectic fragments of DP 10–15 are active as elicitors in various plant cells, inducing defense responses (22, 117, 134) and rapid changes in ion fluxes (108).

Plants contain proteinaceous inhibitors of pectic enzymes (28, 156). An obvious function of these inhibitors is to antagonize the pathogen's penetration effort (28). In addition, they may enhance elicitor signaling: In their presence, pectinases produce less total product but more of the larger pectic fragments, which are most active as elicitors (24).

Lipophilic Substances

Lipophilic substances may accumulate in membranes and affect cell metabolism without being perceived by a specific chemosensory mechanism. For example, nystatin, amphotericin B, and digitonin, at concentrations of $\sim 10^{-5}$ M, induce ion fluxes, phytoalexin production, and other defense reactions in parsley cells in a similar way as do elicitors (128). However, certain lipophilic substances appear to be perceived in a selective and specific manner, as discussed below.

ARACHIDONIC ACID   Arachidonic acid, a major fatty acid in the lipids of *Phytophthora infestans*, a pathogen of potato, elicits phytoalexin accumulation in potato tuber tissue (17). Possibly, the signal perceived by the plant is not arachidonic acid itself but a metabolite generated by lipoxygenase (18).

SYRINGOLIDES   Soybean cultivars carrying the resistance gene *Rpg4* react hypersensitively to *Pseudomonas syringae* pv. *glycinea* strains expressing the *avrD* gene of *P. syringae* pv. *tomato*, and, surprisingly, to *Escherichia coli* expressing *avrD* (86, 87). Culture fluids of these bacteria contain low–molecular weight host-specific elicitors (87), which have been purified and identified as syringolides, derived from the condensation of D-xylulose and a β-ketoalkanoic acid (113). Syringolides induce necrosis in soybean plants carrying *Rpg4* at concentrations in the micromolar range (113) but are inactive in soybean plants lacking *Rpg4* (86).

ERGOSTEROL   Tomato cells perceive ergosterol, the main sterol in most higher fungi, with exquisite sensitivity, as demonstrated using alkalinization of the growth medium as a bioassay (68). Ergosterol is half-maximally active at a concentration of 10^{-11} M with a threshold below 10^{-12} M; in contrast, a wide range of plant and animal sterols are completely inactive at concentrations of 10^{-6} M. The high selectivity and specificity of this perception system, apparently targeted to a non-self sterol, is reminiscent of steroid perception in animals (68).

PERCEPTION AND TRANSDUCTION OF MICROBIAL SIGNALS

Receptors

Selective perception of microbial signals implies the existence of specific receptors. At least for the hydrophilic signals, receptors are expected to be located at the outer surface of the plasma membrane. Binding sites with the expected selectivity and specificity for some of the microbial signals have been identified on the surface of plant cells and on their membranes; such binding sites are prime candidates for the chemoreceptors. They are generally present in very small amounts, rendering their purification difficult. Purification and/or cloning is a prerequisite, however, for proving that they function as receptors.

GLUCAN BINDING SITE   A high-affinity binding site for *P. megasperma* glucans has been identified in soybean plasma membranes (138). Its K_D of ~ 2 nM for the heptaglucoside elicitor correlates well with the heptaglucoside concentration required for half-maximal biological activity (27). Extensive studies with structural analogues have uncovered a close correlation between elicitor activity in vivo and the capacity to compete with elicitor binding in vitro (26). Thus, the binding site has the properties expected for the receptor. The binding site has been solubilized (25, 30) and labeled by photoaffinity techniques (31) but thus far only partially purified (62).

YEAST GLYCOPEPTIDE BINDING SITE   A high-affinity binding site for yeast glycopeptides has been identified on intact tomato cells and in tomato membranes; its K_D of ~3×10^{-9} M corresponds to the concentration required for half-maximal elicitor activity (12). Glycopeptides with N-linked side chains of 9–11 mannosyl residues have a high affinity for the binding site whereas the ones with side chains of 8 mannosyl residues have a low affinity, as expected for a binding site representing the biological receptor.

CHITIN OLIGOMER BINDING SITE   Tomato cells and membranes also possess a binding site for chitin oligomers. The highest affinity is for chitin oligomers with DP ≥ 4 and for a Nod factor (K_D of ~ $1–3 \times 10^{-9}$ M for intact tomato cells). The affinities for the trimer and dimer are ~ 300-fold and 300,000-fold lower, respectively (13). The relative affinities of these compounds are similar to their relative biological activities, and they distinguish the binding site from chitin-binding lectins (13). A high-affinity binding site for chitin oligomers has also been found in rice cells, although its selectivity has not been examined (145).

BINDING SITE FOR THE *PHYTOPHTHORA* ELICITOR PEPTIDE   A high-affinity binding site for an elicitor-active peptide has been found in parsley membranes (K_D of ~ 2×10^{-9} M); competition studies with similar peptides differing in biological activity demonstrate that this peptide has the selectivity and specificity expected for the biological receptor (118).

PLANT RESISTANCE GENES: ENCODING RECEPTORS FOR AVIRULENCE GENE PRODUCTS?   In the classic gene-for-gene interaction, single dominant resistance genes in the plant correspond to single dominant avirulence genes in the pathogen (60). This has led to the idea that the avirulence genes of the pathogen encode specific elicitors, and that the corresponding resistance genes in the plant encode the corresponding specific receptors (20, 43, 44, 86). The first resistance genes have recently been cloned, and the deduced amino acid sequences show intriguing homologies to proteins involved in signal perception and transduction. The *Pto* gene in tomato, conferring resistance to *Pseudomonas syringae* pv. *tomato* strains carrying the avirulence gene *avrPto*, encodes a serine-threonine protein kinase (107). It has high homology to the so-called receptor kinase family (169), among which is the *Brassica* SRK6 kinase involved in sporophytic self-incompatibility (153), but it lacks the putative extracellular domain of other family members thought to interact with (hitherto unknown) ligands (169). The *RPS2* gene of *Arabidopsis thaliana* (93), conferring resistance to *Pseudomonas syringae* carrying the avirulence gene *avrRpt2*, encodes a different putative transmembrane protein, with a nucleotide-binding domain (P loop) in the presumptive cytoplasmic part and a leucine-rich repeat structure in the presumptive extracellular part (14, 114a). The latter domain resembles ligand-interaction domains of animal receptors and thus might function as a binding site of a plant chemoreceptor; intriguingly, it also shares some homology with the bean pectinase inhibitor (156). The *N* gene of tobacco, specifying resistance to tobacco mosaic virus, has been cloned by transposon tagging; it is homologous to *RPS2* but lacks the membrane-spanning domain, which suggests an intracellular localization of the leucine-repeat structure (172a). A rust-resistance gene of flax, the classic object of Flor's study (60), has similarly been cloned by transposon tagging; it encodes a large protein with a P loop but no other apparent homologies (96). A gene called *Prf*, required for expression of *Pto*-mediated resistance against *P. syringae* pv. *tomato* as well as for sensitivity to the insecticide fenthion (135), has been cloned and shown to be a homolog of the *RPS2* gene (136). The protein kinase encoded by *Pto* may interact with the product of *Prf* in the functional recognition of *P. syringae* pv. *tomato*. In tomato, the resistance gene *Cf9*, mediating resistance to *C. fulvum* carrying the *Avr9* gene, has been cloned using an ingenious selection scheme with plants expressing the AVR9 peptide; *Cf9* encodes a protein with an extracellular domain homologous to the one found in receptor kinases but lacking the cytoplasmic protein kinase domain (80). It will be interesting to find out whether the *Cf9* product physically interacts with the AVR9 peptide. Binding studies with the labeled AVR9 peptide have demonstrated the same type of a high-affinity binding site in tomato lines with and without the *Cf9* gene, suggesting that the AVR9 elicitor also may be recognized in the absence of the resistance gene (91). Although the role of the AVR9 product as a specific elicitor has been demonstrated convincingly (160), *Cf9* may act downstream of chemoperception. Two additional genes implicated in *Cf9*-mediated resistance have been identified by mutagenesis (76).

In bacterial gene-for-gene interactions, the first avirulence genes were cloned more than ten years ago; however, except for *AvrD* (87), no race-specific elicitor has been attributed to any avirulence gene product (86). The syringolides formed in bacteria carrying *AvrD* genes behave as specific elicitors in the soybean lines carrying *Rpg4* but have a relatively low potency (113) and may act directly as membrane surfactants rather than by interaction with a specific receptor.

Second Messengers and Protein Phosphorylation

Recent studies indicate that transduction of microbial signals in plants proceeds along pathways known from animal and microbial chemoperception systems. The following summary combines results obtained with different signals in different plant systems. This may provide a broad overview, but it is important to note that different types of chemoreceptors may operate through different signal transduction chains.

CHANGES IN MEMBRANE POTENTIAL AND INTRACELLULAR ION CONCENTRA-TION   Nod factors induce depolarization of the plasma membrane in root hairs of host plants (but not in non-hosts) within minutes (52). Similarly, various elicitors can cause rapid membrane depolarization (108), as also indicated by the rapid alkalinization of the growth medium (55, 68) and the corresponding efflux of K^+ (108, 128) in cultured plant cells treated with microbial signals. Plasma membrane depolarization may lead to an enhanced calcium influx through voltage-dependent calcium channels (155). An elegant study with plants expressing aequorin has demonstrated elicitor-mediated increases in cytoplasmic calcium levels (90).

Are such increases in cytoplasmic calcium involved in the transduction of the elicitor signal? Experiments with calcium channel-blockers, chelators, and ionophores indicate a calcium requirement for induction of defense genes and phytoalexin accumulation (51, 128). A problem with such experiments is the incongruency of the time scales between the changes in calcium fluxes, occurring within minutes, and the responses examined, usually after hours or days.

PROTEIN KINASES AND PROTEIN PHOSPHATASES   In tomato cells, K-252a and staurosporine, two related inhibitors of protein kinases, block induction of ethylene biosynthesis and alkalinization of the growth medium mediated by various elicitors but do not affect basal metabolism and protein synthesis (54, 70). Derivatives of staurosporine differing in their potential to inhibit the protein kinase in tomato microsomes in vitro differ similarly in their effects on elicitor responses in vivo (54). The pattern of protein phosphorylation, visualized by an in vivo pulse-labeling technique, changes within minutes after application of microbial signals, and K-252a blocks the appearance of the newly labeled phosphoprotein bands (54, 55). Application of K-252a in mid-course of alkalinization leads to an arrest of elicitor responses and to a disappearance of label from the newly phosphorylated proteins within minutes, indicating rapid turnover of the phosphate groups in the phosphoproteins important in signaling (54). Elicitor-dependent release of active oxygen species (140, 166) is similarly blocked by staurosporine.

Calyculin A, a potent inhibitor of the two major protein phosphatases, PP1 and PP2A, mimics microbial stimuli in tomato cells at concentrations in the range of 10–100 nM: It induces alkalinization of the growth medium and an increase in the activity of 1-aminocyclopropane-1-carboxylate synthase, the key enzyme of ethylene biosynthesis, as well as the appearance of the newly phosphorylated proteins (56). This indicates that the critical phosphoproteins are continually phosphorylated and dephosphorylated in the nonelicited state, and that inhibition of the dephosphorylation is sufficient to initiate signal transduction. Okadaic acid, another protein phosphatase inhibitor, induces phenylalanine ammonia-lyase and phytoalexin accumulation in soybean cells and cotyledons (104).

ACTIVE OXYGEN PRODUCTION   Rapid production of active oxygen species has been found in various tissues in response to elicitor application (2, 19, 110). The oxidative burst leads to the cross-linking of cell wall proteins, rendering the cell walls more resistant to attack by fungal enzymes (19). In addition, active oxygen species may be toxic for pathogens (99, 110), or they may act as second messengers in the induction of defense genes (38, 110).

G PROTEINS, PHOSPHOLIPASE C, AND INOSITOL 1,4,5-TRISPHOSPHATE   When provided exogenously to cultured soybean cells, mastoparan, a peptide that activates G protein–dependent signaling in animal cells, induces an oxidative burst in a similar way as do elicitors (98). Pectic fragments induce a transient increase in inositol 1,4,5-trisphosphate; mastoparan is even more active in eliciting this effect, suggesting activation of phospholipase C in a process mediated by G proteins (99). Breakdown of phosphoinositides also has been observed in tobacco cells treated with pathogenic bacteria (5) and in elicitor-treated pea tissues and membranes (157).

LIPID OXIDATION   Rapid synthesis of *cis*-jasmonic acid (JA) occurs in elicitor-treated cells of various plant species (72, 115), and application of JA induces phytoalexin accumulation in these cells (72), suggesting that JA might mediate this elicitor response. JA is derived from oxidation of linolenic acid, which, along with an intermediate, 12-oxophytodienic acid, accumulates in elicitor-treated cells (115).

SYSTEMIC SIGNALS   Both in mutualistic and antagonistic symbioses, microbial signals are initially present locally, at the site of plant-microbe contact. However, in both cases, the plant may show a systemic response. In the interaction of legumes with rhizobia, the symbiotic response induced in one part of the root system leads to a suppression of symbiotic responses in other parts (23, 164). This suppression is based, at least in part, on a hypersensitive reaction against invading rhizobia (164).

In the hypersensitive response elicited locally by elicitors or infections causing necrosis, the whole plant may show systemic acquired resistance and systemic accumulation of pathogenesis-related proteins (133). Salicylic acid mimics this response, and its endogenous levels increase before the onset of resistance (106, 112, 126). Transgenic plants expressing salicylate hydroxylase fail to accumulate salicylic acid and to produce pathogenesis-related proteins and systemic acquired resistance upon local stimulation, implicating salicylic acid as a second messenger in these responses (64). However, the systemic signal appears to move out of a leaf before salicylic acid accumulates (126), and grafting of wild-type plants on plants expressing salicylate hydroxylase shows that systemic signaling is possible in the near absence of salicylic acid accumulation (165). Thus, the systemic signal for acquired resistance remains to be discovered.

Can microbial signals themselves move systemically? Pectic fragments appear to have some mobility in the phloem (131) and in the xylem (103). Xylanase (141) and elicitins (42) are translocated systemically under certain conditions, apparently by a hydraulic mechanism in the xylem (141). In general, however, systemic signaling appears to be based primarily on plant factors formed endogenously in response to localized microbial signals (23, 133).

Desensitization and Refractory State

Chemoperception systems in microbes (4, 162) and animals (58, 154) can be desensitized by increasing concentrations of a stimulus. This increases the dynamic range of the sensory system and allows detection of changes in stimulus concentration (49). Some chemoperception systems of plants similarly show rapid desensitization. For example, chitin oligomers induce alkalinization transiently; after about ten minutes, the cells enter a refractory state and can no longer be stimulated by chitin (55). This refractory state is maintained for several hours; the perception capacity returns slowly during the following days (55). Other stimuli, such as xylanase or ergosterol, induce a full response in cells rendered refractory for chitin oligomers, indicating that different chemoperception systems operate independently (55, 68).

INTERACTIONS OF MICROBIAL SIGNALS

Suppressors

In plant-pathogen interactions, it is in the best interest of the plant to recognize microbial elicitors; in contrast, it is in the best interest of the pathogen to camouflage these elicitors and to prevent their perception or action. Pathogen products that delay or prevent elicitor action are called suppressors. Although the importance of suppressors has been discussed frequently (79, 121), little is known about their chemical nature and mode of action. In the interaction of *Phytophthora infestans* with potato, soluble fungal glucans appear to act as suppressors against recognition of fungal cell wall components (48, 137). An extracellular glycoprotein of *P. megasperma* has been reported to block elicitor-induced phytoalexin accumulation in soybean (174). A glycoprotein fraction from the phytopathogenic ascomycete *Ascochyta rabiei* suppresses phytoalexin accumulation in its host, chickpea (89). A nonproteinaceous suppressor from *Cladosporium fulvum* prevents the action of nonspecific elicitors in tomato (102). Two short glycopeptides secreted by *Mycosphaerella pinodes*, an ascomycete pathogenic on pea, suppress elicitor-dependent induction of phenylalanine ammonia-lyase in pea tissue (146) and inhibit ATPase activity in pea plasma membranes (85). In all these cases, it is not known how the suppressor interferes with elicitor perception or signal transduction. In tomato cells, glycans released from yeast glycoproteins by endoglycosidase H act as suppressors and competitively inhibit the action of glycopeptide elicitors (10, 11). These glycan suppressors bind to the high-affinity binding site for the glycopeptide elicitors, indicating that they act as antagonists at the receptor (12). It remains to be seen whether such a system also operates in natural plant-microbe interactions.

A primary function of many microbial toxins may be suppression of pathogen recognition (20, 170). For example, *Helminthosporium carbonum*, a pathogen of maize, produces HC toxin, a cyclic peptide that allows the pathogen to spread in susceptible maize plants without an apparent defense response (170). Resistant maize lines detoxify the compound by a specific reductase (109); in these plants, pathogen growth is restricted.

Successful establishment of mutualistic symbioses may also depend on suppression of host defenses. In certain interactions of legumes with rhizobia, mutants of the bacterial symbionts defective in exopolysaccharides and lipopolysaccharides form small, ineffective nodules and elicit a defense response in their host plants (116, 124). Thus, the bacterial polysaccharides may function as suppressors, either by interfering with perception of an unknown elicitor (116) or by acting as specific symbiotic signals (124). Purified *Rhizobium meliloti* exopolysaccharides partially suppress the symbiotic deficiencies of the corresponding mutant when applied to alfalfa roots, suggesting that they are also perceived as signals (100).

Induction of certain elements of the defense response, followed by their suppression, occurs during colonization of roots by vesicular-arbuscular mycorrhiza, although nothing is known about the signals involved (77, 95, 150, 167).

Synergistic Effects

Elicitors may act synergistically, i.e. their combined effect may be greater than the sum of the individual effects. For example, endogenous oligogalacturonide elicitors induce phytoalexin synthesis synergistically with exogenous elicitors in soybean cotyledons and parsley cells (35, 36), and endogenous factors released by wounding, so-called competency factors, are necessary in soybean cotyledons to allow accumulation of phytoalexins in response to glucan elicitors (67).

SIGNIFICANCE OF MICROBIAL SIGNALS FOR THE PLANTS

Recognition of Non-Self: Information

Some chemoperception systems recognize compounds that occur widely in microorganisms. For example, tomato cells have sensitive perception systems for fragments of chitin (55), one of the most abundant biopolymers on earth, and ergosterol (68), the predominant sterol in most fungi. Thresholds for detection are at $\sim 10^{-12}$ M, corresponding to the dilution of the chitin or ergosterol content of a toothpick's tip of yeast in the volume of a swimming pool. Both compounds occur in mutualistic mycorrhizal fungi and neutral endophytes as well as in antagonistic pathogens. These molecules may signal the presence of a fungus, a non-self organism, but plants probably cannot qualify the nature of the organism (similar to the way in which a burnt smell may signal either a fireplace or a forest fire). The chemosensory capacity for such non-self molecules may confer basic information; however, additional cues are needed for the plant to recognize the non-self organism as a mutualistic symbiont or pathogen.

Recognition of Pathogens: Defense

The classic paradigm of plant chemoperception is the elicitor model, according to which elicitors from pathogens induce phytoalexin accumulation (34, 50, 51) or a hypersensitive response (65). Many of the chemoperception systems discussed above fit this model well. For example, tobacco has a sensitive perception system for elicitins, a class of proteins present in most *Phytophthora* strains except for the ones pathogenic on tobacco, and elicitin treatments induce resistance to the pathogen (130). Thus, elicitin perception allows tobacco to defend itself against *Phytophthora*, and *Phytophthora* can adapt to this situation by losing elicitin expression. The relationship between resistance and specific chemoperception of the peptides produced by *Cladosporium fulvum* and *Rhynchosporium secalis* is even more compelling (44). These peptides induce a hypersensitive reaction specifically in the host plants carrying a resistance gene, and the pathogen appears to reacquire virulence on the resistant host by getting rid of the corresponding signal, either by total loss of the relevant gene, as in the case of *Avr9* in *C. fulvum* (160), or by single point mutations, as in the case of *Avr4* in *C. fulvum* (81). As discussed above, these data are fully compatible with the model for gene-for-gene relationships in which the avirulence genes of the pathogen encode race-specific elicitors and the resistance genes of the plant encode the corresponding receptors (20, 43, 86).

In other cases, the relationship between elicitor recognition and resistance remains less clear. The glucan elicitor of *P. megasperma* (144) and the arachidonic acid in *P. infestans* (17) have been isolated from successful pathogens of soybean and potato, respectively. If these signals are present in the natural interaction, they apparently are not perceived effectively enough by the host plant to cause resistance, or the pathogen is insensitive to the defense mechanisms induced by the plant. Perhaps the pathogen produces suppressors that counteract elicitation, as suggested for *P. infestans* (48, 137). Glucans that are active as elicitors in soybean occur quite commonly in fungi (50) and even in the mutualistic partner of soybean, *Bradyrhizobium japonicum* (114). Thus, the perception system of soybean for glucans may function more generally in the acquisition of information about non-self organisms.

Recognition of Mutualistic Symbionts: Communication

In the interaction of legumes with rhizobia, perception of the Nod factor by the host plant is of central importance for the establishment of this symbiosis (41). Nod factors induce root hair curling and membrane depolarization (52) but also formation of preinfection threads (159), meristematic activity in roots (148), and, in alfalfa roots, even the formation of empty nodules (158). Specific modifications of the chitooligosaccharide backbone are decisive for host specificity (40, 41, 132). Host plants may have evolved highly selective chemoperception systems for specifically modified Nod factors in order to select the appropriate symbiont. In addition, specificity may be connected to differential stability of signals: The sulfated Nod factors are more stable than the nonsulfated homologs against hydrolysis by root chitinases (151, 152).

In mutually beneficial symbioses, chemical signaling typically does not occur in only one way but is based on chemical "cross-talk" (147). In the rhizobia, the whole set of *nod* genes involved in the production of the Nod factor is induced in response to flavonoids, secreted as signals by the plant roots; flavonoid secretion of the roots, in turn, is stimulated by Nod factors (41, 59). Antagonistic symbioses between plants and pathogens may well be based on similar cross-talk.

OUTLOOK

These are exciting times for students of plant-microbe interactions. Combinations of chemical and biochemical approaches have uncovered highly sensitive perception systems for various microbial signals, and studies of the receptors involved are well under way. At the same time, molecular genetic studies have led to the identification of plant resistance genes, and the analysis of their sequences indicates that some of them may function in signal perception. It will be interesting to see whether the biochemically defined binding sites for microbial signals and some of the genetically defined resistance gene products belong to the same class of receptor proteins.

Current studies on plant-microbe interactions tend to suggest that perception of a single chemical signal by the plant is all-important: Recognition of the Nod factor seems decisive for establishment of the mutualistic root nodule symbiosis in legumes, and the presence or absence of a single resistance gene, thought to represent the recognition system for a single avirulence gene product, appears to make all the difference between full resistance and full susceptibility to a given pathogen. However, these seemingly simple one-dimensional black-and-white systems are probably just the tip of an iceberg of chemical signaling: A given plant has an array of highly sensitive chemoperception systems both for molecules unique to a microbial species or genus and for molecules common and typical for a whole class of microbes. It would be surprising if the plant did not make use of this potential to combine and integrate microbial signals, allowing it to respond to an approaching microbe with a differentiated reaction on a gray scale between full acceptance and outright defense. This point may be illustrated with the recently characterized mutants of *A. thaliana* that spontaneously form necrotic lesions (46, 69). These mutants are more prone to undergo necrosis than is the wild-type, and they react with a hypersensitive defense response against pathogens that do not elicit a visible response in the wild-type, indicating that the plant has a potential to perceive these pathogens but does not normally react with the hypersensitive response (46, 69). Even more surprisingly, cytological observations have shown that in the interaction of potato with *Phytophthora infestans*, the presence or absence of a resistance gene leads to full resistance or susceptibility at the macroscopic scale but only to small, quantitative differences at the microscopic scale; most invading hyphae induce defense responses whether or not the resistance gene is present (63).

The perception and transduction of microbial signals can be seen as a paradigm of plant chemoperception in general. Some of the highly specific exogenous microbial signals may have endogenous analogs in the plant: Oligosaccharides influencing plant development may be produced not only by microbes or microbial enzymes but also by the plant itself, as suggested in the oligosaccharin concept (33, 134). For example, fruit ripening in tomato can be induced by pectic fragments resembling the endogenous elicitors (111) or by specific *N*-linked glycans from plants (125) resembling the suppressors derived from yeast glycopeptides (10). Nod factors specifically induce embryogenic development in a mutant carrot line (39), and the question has been raised whether analogous lipophilic chitin derivatives are produced by the plant (149). The elicitor-active peptides derived from *Phytophthora mega-*

sperma (118) can be likened to systemin, a peptide hormone involved in tomato plants in the systemic induction of proteinase inhibitors in response to wounding (122). Systemin induces a similar response in *Lycopersicon peruvianum* cells, with respect to medium alkalinization and ethylene biosynthesis, as do various microbial signals (53a). These data indicate that there may be endogenous signals comparable to the exogenous microbial ones. The nearest homologs of some of the recently cloned resistance genes (107) are the glycoproteins and receptor kinases involved in sporophytic self-incompatibility, i.e. in self-recognition (153). Thus, in evolutionary terms, the plant's chemoperception systems for self and non-self may be closely related, in fitting analogy to the olfactory system of animals, which functions both in the recognition of environmental signals and in intraspecific social relationships. It is a fascinating task for the future to find out how plants differentiate and integrate the self and non-self signals perceived through their chemical senses.

ACKNOWLEDGMENTS

I am grateful to Georg Felix and Ian Sanders for comments on the manuscript.

Literature Cited

1. Anderson AJ. 1989. The biology of glycoproteins as elicitors. In *Plant-Microbe Interactions. Molecular and Genetic Perspectives*, ed. T Kosuge, EW Nester, 3:87–130. New York: McGraw-Hill

2. Apostol I, Heinstein PF, Low PS. 1989. Rapid stimulation of an oxidative burst during elicitation of cultured plant cells. Role in defense and signal transduction. *Plant Physiol.* 90:109–16

3. Arlat M, Van Gijsegem F, Huet JC, Pernollet JC, Boucher CA. 1994. PopA1, a protein which induces a hypersensitivity-like response on specific *Petunia* genotypes, is secreted via the Hrp pathway of *Pseudomonas solanacearum. EMBO J.* 13: 543–53

4. Armitage JP. 1992. Behavioral responses in bacteria. *Annu. Rev. Physiol.* 54:683–714

5. Atkinson M, Bina J, Sequeira L. 1993. Phosphoinositide breakdown during the K^+/H^+ exchange response of tobacco to *Pseudomonas syringae* pv. *syringae. Mol. Plant-Microbe Interact.* 6:253–60

6. Bailey BA, Korcak RF, Anderson JD. 1992. Alterations in *Nicotiana tabacum* L. cv. *Xanthi* cell membrane function following treatment with an ethylene biosynthesis-inducing endoxylanase. *Plant Physiol.* 100: 749–55

7. Deleted in proof

8. Baker CJ, Orlandi EW, Mock NM. 1993. Harpin, an elicitor of the hypersensitive response in tobacco caused by *Erwinia amylovora*, elicits active oxygen production in suspension cells. *Plant Physiol.* 102: 1341–44

9. Barber MS, Bertram RE, Ride JP. 1989. Chitin oligosaccharides elicit lignification in wounded wheat leaves. *Physiol. Mol. Plant Pathol.* 34:3–12

10. Basse CW, Bock K, Boller T. 1992. Elicitors and suppressors of the defense response in tomato cells. Purification and characterization of glycopeptide elicitors and glycan suppressors generated by enzymatic cleavage of yeast invertase. *J. Biol. Chem.* 267:10258–65

11. Basse CW, Boller T. 1992. Glycopeptide elicitors of stress responses in tomato cells. *N*-linked glycans are essential for activity but act as suppressors of the same activity

when released from the glycopeptides. *Plant Physiol.* 98:1239–47

12. Basse CW, Fath A, Boller T. 1993. High affinity binding of glycopeptide elicitor to tomato cells and microsomal membranes and displacement by specific glycan suppressors. *J. Biol. Chem.* 268:14724–31

13. Baureithel K, Felix G, Boller T. 1994. Specific, high affinity binding of chitin fragments to tomato cells and membranes. Competitive inhibition of binding by derivatives of chitooligosaccharides and a Nod factor of *Rhizobium. J. Biol. Chem.* 269:17931–38

14. Bent AF, Kunkel BN, Dahlbeck D, Brown KL, Schmidt R, et al. 1994. *RPS2* of *Arabidopsis thaliana:* a leucine-rich repeat class of plant disease resistance genes. *Science* 265:1856–60

15. Blein J-P, Milat M-L, Ricci P. 1991. Responses of cultured tobacco cells to cryptogein, a proteinaceous elicitor from *Phytophthora cryptogea. Plant Physiol.* 95:486–91

16. Boller T. 1989. Primary signals and second messengers in the reaction of plants to pathogens. In *Second Messengers in Plant Growth and Development,* ed. DJ Morré, WF Boss, pp. 227–55. New York: Liss

16a. Boller T, Meins F Jr, eds. 1992. *Genes Involved in Plant Defense.* Vienna/New York: Springer-Verlag

17. Bostock RM, Kuć JA, Laine RA. 1981. Eicosapentaenoic and arachidonic acids from *Phytophthora infestans* elicit fungitoxic sesquiterpenes in the potato. *Science* 212:67–69

18. Bostock RM, Yamamoto H, Choi D, Ricker KE, Ward BL. 1992. Rapid stimulation of 5-lipoxygenase activity in potato by the fungal elicitor arachidonic acid. *Plant Physiol.* 100:1448–56

19. Bradley DJ, Kjellbom P, Lamb CJ. 1992. Elicitor-induced and wound-induced oxidative cross-linking of a proline-rich plant cell wall protein—a novel, rapid defense response. *Cell* 70:21–30

20. Briggs SP, Johal GS. 1994. Genetic patterns of plant host-parasite interactions. *Trends Genet.* 10:12–16

21. Bruce RJ, West CA. 1982. Elicitation of casbene synthetase activity in castor bean. The role of pectic fragments of the plant cell wall in elicitation by a fungal endopolygalacturonase. *Plant Physiol.* 69:1181–88

22. Bruce RJ, West CA. 1989. Elicitation of lignin biosynthesis and isoperoxidase activity by pectic fragments in suspension cultures of castor bean. *Plant Physiol.* 91:889–97

23. Caetano-Anollés G, Gresshoff PM. 1991. Plant genetic control of nodulation. *Annu. Rev. Microbiol.* 45:345–82

24. Cervone F, Albersheim P. 1989. Host-pathogen interactions. XXXIII. A plant protein converts a fungal pathogenesis factor into an elicitor of plant defense responses. *Plant Physiol.* 90:542–48

25. Cheong J-J, Alba R, Côté F, Enkerli J, Hahn MG. 1993. Solubilization of functional plasma membrane-localized hepta-β-glucoside elicitor-binding proteins from soybean. *Plant Physiol.* 103:1173–82

26. Cheong J-J, Birberg W, Fügedi P, Pilotti A, Garegg PJ, et al. 1991. Structure-activity relationships of oligo-β-glucoside elicitors of phytoalexin accumulation in soybean. *Plant Cell* 3:127–36

27. Cheong J-J, Hahn MG. 1991. A specific, high-affinity binding site for the hepta-β-glucoside elicitor exists in soybean membranes. *Plant Cell* 3:137–47

28. Collmer A, Keen NT. 1986. The role of pectic enzymes in plant pathogenesis. *Annu. Rev. Phytopathol.* 24:383–409

29. Conrath U, Domard A, Kauss H. 1989. Chitosan-elicited synthesis of callose and of coumarin derivatives in parsley cell suspension cultures. *Plant Cell Rep.* 8:152–55

30. Cosio EG, Frey T, Ebel J. 1990. Solubilization of soybean membrane binding sites for fungal β-glucans that elicit phytoalexin accumulation. *FEBS Lett.* 264:235–38

31. Cosio EG, Frey T, Ebel J. 1992. Identification of a high-affinity binding protein for a hepta-β-glucoside phytoalexin elicitor in soybean. *Eur. J. Biochem.* 204:1115–23

32. Cruickshank IAM, Perrin DR. 1968. The isolation and partial characterization of monilicolin A, a polypeptide with phaseollin-inducing activity from *Monilinia fructicola. Life Sci.* 7:449–58

32a. Daniels M, ed. 1994. *Abstracts of the Seventh International Symposium on Molecular Plant-Microbe Interactions.* Edinburgh: Int. Soc. Mol. Plant-Microbe Interact.

33. Darvill A, Augur C, Bergmann C, Carlson RW, Cheong JJ, et al. 1992. Oligosaccharins—oligosaccharides that regulate growth, development and defence responses in plants. *Glycobiology* 2:181–98

34. Darvill AG, Albersheim P. 1984. Phytoalexins and their elicitors—a defense against microbial infection of plants. *Annu. Rev. Plant Physiol.* 35:243–75

35. Davis KR, Darvill AG, Albersheim P. 1986. Host-pathogen interactions. XXXI. Several biotic and abiotic elicitors act synergistically in the induction of phytoalexin accumulation in soybean. *Plant Mol. Biol.* 6:23–32

36. Davis KR, Hahlbrock K. 1987. Induction of defense responses in cultured parsley cells by plant cell wall fragments. *Plant Physiol.* 85:1286–90

37. Dean JFD, Anderson JD. 1991. Ethylene

biosynthesis-inducing xylanase. II. Purification and physical characterization of the enzyme produced by *Trichoderma viride*. *Plant Physiol.* 95:316–23

38. Degousée N, Triantaphylidès C, Montillet J-L. 1994. Involvement of oxidative processes in the signaling mechanisms leading to the activation of glyceollin synthesis in soybean (*Glycine max*). *Plant Physiol.* 104: 945–52

39. De Jong AJ, Heidstra R, Spaink HP, Hartog MV, Meijer EA, et al. 1993. *Rhizobium* lipooligosaccharides rescue a carrot somatic embryo mutant. *Plant Cell* 5:615–20

40. Dénarié J, Cullimore J. 1993. Lipo-oligosaccharide nodulation factors: a new class of signaling molecules mediating recognition and morphogenesis. *Cell* 74: 951–54

41. Dénarié J, Debellé F, Rosenberg C. 1992. Signalling and host range variation in nodulation. *Annu. Rev. Microbiol.* 46:494–531

42. Devergne J-C, Bonnet P, Panabières F, Blein J-P, Ricci P. 1992. Migration of the fungal protein cryptogein within tobacco plants. *Plant Physiol.* 99:843–47

43. De Wit PJGM. 1992. Functional models to explain gene-for-gene relationships in plant-pathogen interactions. See Ref. 16a, pp. 25–47

44. De Wit PJGM. 1992. Molecular characterization of gene-for-gene systems in plant-fungus interactions and the application of avirulence genes in control of plant pathogens. *Annu. Rev. Phytopathol.* 30: 391–418

45. De Wit PJGM, Hofman JE, Velthuis GCM, Kuć JA. 1985. Isolation and characterization of an elicitor of necrosis isolated from intercellular fluids of compatible interactions of *Cladosporium fulvum* (syn. *Fulvia fulva*) and tomato. *Plant Physiol.* 77:642–47

46. Dietrich RA, Delaney TP, Uknes SJ, Ward ER, Ryals JA, Dangl JL. 1994. Arabidopsis mutants simulating disease resistance response. *Cell* 77:565–77

47. Dixon RA, Lamb CJ. 1990. Molecular communication in interactions between plants and microbial pathogens. *Annu. Rev. Plant Physiol. Plant Mol. Biol.* 41: 339–67

47a. Dixon RA, Harrison MJ, Lamb CJ. 1994. Early events in the activation of plant defense responses. *Annu. Rev. Phytopathol.* 32:479–501

48. Doke N, Garas NA, Kuć J. 1979. Partial characterization and aspects of the mode of action of a hypersensitivity-inducing factor (HIF) from *Phytophthora infestans*. *Physiol. Plant Pathol.* 15:127–40

49. Dusenbery DB. 1992. *Sensory Ecology. How Organisms Acquire and Respond to Information.* New York: Freeman. 558 pp.

50. Ebel J, Cosio EG. 1994. Elicitors of plant defense responses. *Int. Rev. Cytol.* 148:1–36

51. Ebel J, Scheel D. 1992. Elicitor recognition and signal transduction. See Ref. 16a, pp. 183–205

52. Ehrhardt DW, Atkinson EM, Long SR. 1992. Depolarization of alfalfa root hair membrane potential by *Rhizobium meliloti* Nod factors. *Science* 256:998–1000

53. Farmer EE, Helgeson JP. 1987. An extracellular protein from *Phytophthora parasitica* var. *nicotianae* is associated with stress metabolite accumulation in tobacco callus. *Plant Physiol.* 85:733–40

53a. Felix G, Boller T. 1995. Systemin induces rapid ion fluxes and ethylene biosynthesis in *Lycopersicon peruvianum* cells. *Plant J.* In press

54. Felix G, Grosskopf DG, Regenass M, Boller T. 1991. Rapid changes of protein phosphorylation are involved in transduction of the elicitor signal in plant cells. *Proc. Natl. Acad. Sci. USA* 88:8831–34

55. Felix G, Regenass M, Boller T. 1993. Specific perception of subnanomolar concentrations of chitin fragments by tomato cells. Induction of extracellular alkalinization, changes in protein phosphorylation, and establishment of a refractory state. *Plant J.* 4:307–16

56. Felix G, Regenass M, Spanu P, Boller T. 1994. The protein phosphatase inhibitor calyculin A mimics elicitor action in plant cells and induces rapid hyperphosphorylation of specific proteins as revealed by pulse-labelling with [^{33}P]phosphate. *Proc. Natl. Acad. Sci. USA* 91:952–56

57. Fenselau S, Balbo I, Bonas U. 1992. Determinants of pathogenicity in *Xanthomonas campestris* pv. *vesicatoria* are related to proteins involved in secretion in bacterial pathogens of animals. *Mol. Plant-Microbe Interact.* 5:390–96

58. Finger TE, ed. 1987. *Neurobiology of Taste and Smell*. New York: Wiley

59. Fisher RF, Long SR. 1992. *Rhizobium*-plant signal exchange. *Nature* 357:655–60

60. Flor HH. 1971. Current status of the gene-for-gene concept. *Annu. Rev. Phytopathol.* 9:275–96

61. Franssen HJ, Vijn I, Yang W-C, Bisseling T. 1992. Developmental aspects of the *Rhizobium*-legume symbiosis. *Plant Mol. Biol.* 19:89–107

62. Frey T, Cosio EG, Ebel J. 1993. Affinity purification and characterization of a binding protein for a hepta-β-glucoside phytoalexin elicitor in soybean. *Phytochemistry* 32:543–49

63. Freytag S, Arabatzis N, Hahlbrock K, Schmelzer E. 1994. Reversible cytoplasmic rearrangements precede wall apposition, hypersensitive cell death and defense-

related gene activation in potato/*Phytophthora infestans* interactions. *Planta* 194: 123–35

64. Gaffney T, Friedrich L, Vernooij B, Negrotto D, Nye G, et al. 1993. Requirement of salicylic acid for the induction of systemic acquired resistance. *Science* 261: 754–56

65. Goodman RN, Novacky AJ. 1994. *The Hypersensitive Reaction in Plants to Pathogens. A Resistance Phenomenon.* St. Paul, MN: Am. Phytopathol. Soc. 244 pp.

66. Gough CL, Genin S, Zischek C, Boucher CA. 1992. *hrp* genes of *Pseudomonas solanacearum* are homologous to pathogenicity determinants of animal pathogenic bacteria and are conserved among plant pathogenic bacteria. *Mol. Plant-Microbe Interact.* 5:384–89

67. Graham MY, Graham TL. 1994. Wound-associated competency factors are required for the proximal cell responses of soybean to the *Phytophthora sojae* wall glucan elicitor. *Plant Physiol.* 105:571–78

68. Granado J, Felix G, Boller T. 1995. Perception of fungal sterols in plants: subnanomolar concentrations of ergosterol elicit extracellular alkalinization in tomato cells. *Plant Physiol.* 107: In press

69. Greenberg JT, Guo A, Klessig DF, Ausubel FM. 1994. Programmed cell death in plants: a pathogen-triggered response activated coordinately with multiple defense functions. *Cell* 77:551–63

70. Grosskopf DG, Felix G, Boller T. 1990. K-252a inhibits the response of tomato cells to fungal elicitors *in vivo* and their microsomal protein kinase *in vitro*. *FEBS Lett.* 275:177–80

71. Grosskopf DG, Felix G, Boller T. 1991. A yeast-derived glycopeptide elicitor and chitosan or digitonin differentially induce ethylene biosynthesis, phenylalanine ammonia-lyase and callose formation in suspension-cultured tomato cells. *J. Plant Physiol.* 138:741–46

72. Gundlach H, Muller MJ, Kutchan TM, Zenk MH. 1992. Jasmonic acid is a signal transducer in elicitor-induced plant cell cultures. *Proc. Natl. Acad. Sci. USA* 89:2389–93

73. Hahn M, Jüngling S, Knogge W. 1993. Cultivar-specific elicitation of barley defense reactions by the phytotoxic peptide NIP1 from *Rhynchosporium secalis*. *Mol. Plant-Microbe Interact.* 6:745–54

74. Hahn MG, Darvill A, Albersheim P. 1981. Host-pathogen interactions. XIX. The endogenous elicitor, a fragment of a plant cell wall polysaccharide that elicits phytoalexin accumulation in soybeans. *Plant Physiol.* 68:1161–69

75. Ham KS, Kauffmann S, Albersheim P, Darvill AG. 1991. Host-pathogen interactions. XXXIX. A soybean pathogenesis-related protein with β-1,3-glucanase activity releases phytoalexin elicitor-active heat-stable fragments from fungal walls. *Mol. Plant-Microbe Interact.* 4:545–52

76. Hammond-Kosack KE, Jones DA, Jones JDG. 1994. Identification of two genes required in tomato for full *Cf-9*-dependent resistance to *Cladosporium fulvum*. *Plant Cell* 6:361–74

77. Harrison MJ, Dixon RA. 1993. Isoflavonoid accumulation and expression of defense gene transcripts during the establishment of vesicular-arbuscular mycorrhizal associations in roots of *Medicago truncatula*. *Mol. Plant-Microbe Interact.* 6:643–54

78. He SY, Huang H-C, Collmer A. 1993. *Pseudomonas syringae* pv. *syringae* harpin$_{Pss}$—a protein that is secreted via the hrp pathway and elicits the hypersensitive response in plants. *Cell* 73:1255–66

79. Jakobek JL, Smith JA, Lindgren PB. 1993. Suppression of bean defense responses by *Pseudomonas syringae*. *Plant Cell* 5:57–63

80. Jones DA, Thomas CM, Hammond-Kosack KE, Balint-Kurti PJ, Jones JDG. 1994. Isolation of the tomato *Cf9* gene for resistance to *Cladosporium fulvum* by transposon tagging. *Science* 266:789–93

81. Joosten MHAJ, Cozijnsen TJ, De Wit PJGM. 1994. Host resistance to a fungal tomato pathogen lost by a single base-pair change in an avirulence gene. *Nature* 367: 384–86

82. Kamoun S, Klucher KM, Coffey MD, Tyler BM. 1993. A gene encoding a host-specific elicitor protein of *Phytophthora parasitica*. *Mol. Plant-Microbe Interact.* 6: 573–81

83. Kamoun S, Young M, Glascock CB, Tyler BM. 1993. Extracellular protein elicitors from *Phytophthora*: host-specificity and induction of resistance to bacterial and fungal phytopathogens. *Mol. Plant-Microbe Interact.* 6:15–25

84. Deleted in proof

85. Kato T, Shiraishi T, Toyoda K, Saitoh K, Satoh Y, et al. 1993. Inhibition of ATPase activity in pea plasma membranes by fungal suppressors from *Mycosphaerella pinodes* and their peptide moieties. *Plant Cell Physiol.* 34:439–45

86. Keen NT, Dawson WO. 1992. Pathogen avirulence genes and elicitors of plant defense. See Ref. 16a, pp. 85–114

87. Keen NT, Tamaki S, Kobayashi DY, Gerhold D, Stayton M, et al. 1990. Bacteria expressing avirulence gene D produce a specific elicitor of the soybean hypersensitive reaction. *Mol. Plant-Microbe Interact.* 3:112–21

88. Kendra DF, Hadwiger LA. 1984. Characterization of the smallest chitosan oligomer

that is maximally antifungal to *Fusarium solani* and elicits pisatin formation in *Pisum sativum. Exp. Mycol.* 8:276–81

89. Kessmann H, Barz W. 1986. Elicitation and suppression of phytoalexin and isoflavone accumulation in cotyledons of *Cicer arietinum* L. as caused by wounding and by polymeric components from the fungus *Ascochyta rabiei. J. Phytopathol.* 117:321–35

90. Knight MR, Campbell AK, Trewavas AJ. 1991. Transgenic plant aequorin reports the effects of touch and cold-shock and elicitors on cytoplasmic calcium. *Nature* 352: 524–26

91. Kooman-Gersmann M, Honée G, de Wit PJGM. 1994. Binding of the AVR9 elicitor of the tomato pathogen *Cladosporium fulvum* to receptor sites in the plasma membrane. See Ref. 32a, p. 99

92. Kuchitsu K, Kikuyama M, Shibuya N. 1993. *N*-Acetylchitooligosaccharides, biotic elicitors for phytoalexin production, induce transient membrane depolarization in suspension-cultured rice cells. *Protoplasma* 174:79–81

93. Kunkel BN, Bent AF, Dahlbeck D, Innes RW, Staskawicz BJ. 1993. *RPS2*, an *Arabidopsis* disease resistance locus specifying recognition of *Pseudomonas syringae* strains expressing the avirulence gene *avrRpt2. Plant Cell* 5:865–75

94. Lamb CJ, Lawton MA, Dron M, Dixon RA. 1989. Signals and transduction mechanisms for activation of plant defenses against microbial attack. *Cell* 56:215–24

95. Lambais MR, Mehdy MC. 1993. Suppression of endochitinase, β-1,3-endoglucanase, and chalcone isomerase expression in bean vesicular-arbuscular mycorrhizal roots under different soil phosphate conditions. *Mol. Plant-Microbe Interact.* 6:75–83

96. Lawrence GJ, Ellis JG, Finnegan ES. 1994. Cloning a rust resistance gene in flax. See Ref. 32a, p. 86

97. Lee S-C, West CA. 1981. Properties of *Rhizopus stolonifer* polygalacturonase, an elicitor of casbene synthetase in castor bean (*Ricinus communis* L.) seedlings. *Plant Physiol.* 67:640–45

98. Legendre L, Heinstein PF, Low PS. 1992. Evidence for participation of GTP-binding proteins in elicitation of the rapid oxidative burst in cultured soybean cells. *J. Biol. Chem.* 267:20140–47

99. Legendre L, Yueh YG, Crain R, Haddock N, Heinstein PF, Low PS. 1993. Phospholipase-C activation during elicitation of the oxidative burst in cultured plant cells. *J. Biol. Chem.* 268:24559–63

100. Leigh JA, Walker GC. 1994. Exopolysaccharides of *Rhizobium*: synthesis, regulation and symbiotic function. *Trends Genet.* 10:63–67

101. Lerouge P, Roche P, Faucher C, Maillet F, Truchet G, et al. 1990. Symbiotic host-specificity of *Rhizobium meliloti* is determined by a sulphated and acylated glucosamine oligosaccharide signal. *Nature* 344:781–84

102. Lu H, Higgins VJ. 1993. Partial characterization of a non-proteinaceous suppressor of non-specific elicitors from *Cladosporium fulvum* (syn. *Fulvia fulva*). *Physiol. Mol. Plant Physiol.* 42:427–39

103. MacDougall AJ, Rigby NM, Needs PW, Selvendran RR. 1992. Movement and metabolism of oligogalacturonide elicitors in tomato shoots. *Planta* 188:566–74

104. MacKintosh C, Lyon GD, MacKintosh RW. 1994. Protein phosphatase inhibitors activate anti-fungal defence responses of soybean cotyledons and cell cultures. *Plant J.* 5:137–47

105. Maier I, Müller DG. 1986. Sexual pheromones in algae. *Biol. Bull.* 170:145–75

106. Malamy J, Carr JP, Klessig DF, Raskin I. 1990. Salicylic acid: a likely endogenous signal in the resistance response of tobacco to viral infection. *Science* 250:1002–4

107. Martin GB, Brommonschenkel SH, Chunwongse J, Frary A, Ganal MW, et al. 1993. Map-based cloning of a protein kinase gene conferring disease resistance in tomato. *Science* 262:1432–36

108. Mathieu Y, Kurkdjian A, Xia H, Guern J, Koller A, et al. 1991. Membrane responses induced by oligogalacturonides in suspension-cultured tobacco cells. *Plant J.* 1:333–43

109. Meeley RB, Johal GS, Briggs SP, Walton JD. 1992. A biochemical phenotype for a disease resistance gene in maize. *Plant Cell* 4:71–77

110. Mehdy MC. 1994. Active oxygen species in plant defense against pathogens. *Plant Physiol.* 105:467–72

111. Melotto E, Greve LC, Labavitch JM. 1994. Cell wall metabolism in ripening fruit. VII. Biologically active pectin oligomers in ripening tomato (*Lycopersicon esculentum* Mill.) fruit. *Plant Physiol.* 106: In press

112. Métraux J-P, Signer H, Ryals J, Ward E, Wyss-Benz M, et al. 1990. Increase in salicylic acid at the onset of systemic acquired resistance in cucumber. *Science* 250:1004–6

113. Midland SL, Keen NT, Sims JJ, Midland MM, Stayton MM, et al. 1993. The structures of syringolide-1 and syringolide-2, novel *C*-glycosidic elicitors from *Pseudomonas syringae* pv. *tomato. J. Organ. Chem.* 58:2940–45

114. Miller KJ, Hadley JA, Gustine DL. 1994. Cyclic β-1,6-1,3-glucans of *Bradyrhizobium japonicum* USDA 110 elicit isoflavonoid production in the soybean (*Glycine max*) host. *Plant Physiol.* 104:917–23

114a. Mindrinos M, Katagiri F, Yu G-L, Ausubel FM. 1994. The *A. thaliana* disease resistance gene *RPS2* encodes a protein containing a nucleotide-binding site and leucine-rich repeats. *Cell* 78:1089–99

115. Mueller MJ, Brodschelm W, Spannagl E, Zenk MH. 1993. Signaling in the elicitation process is mediated through the octadecanoid pathway leading to jasmonic acid. *Proc. Natl. Acad. Sci. USA* 90:7490–94

116. Niehaus K, Kapp D, Pühler A. 1993. Plant defence and delayed infection of alfalfa pseudonodules induced by an exopolysaccharide (EPS I)-deficient *Rhizobium meliloti* mutant. *Planta* 190:415–25

117. Nothnagel EA, McNeil M, Albersheim P, Dell A. 1983. Host-pathogen interactions. XXII. A galacturonic acid oligosaccharide from plant cell walls elicits phytoalexins. *Plant Physiol.* 71:916–26

118. Nürnberger T, Nennstiel D, Jabs D, Sacks WR, Hahlbrock K, Scheel D. 1994. High-affinity binding of a fungal oligopeptide elicitor to parsley plasma membranes triggers multiple defence responses. *Cell* 78: 449–60

119. Parker JE, Hahlbrock K, Scheel D. 1988. Different cell-wall components from *Phytophthora megasperma* f. sp. *glycinea* elicit phytoalexin production in soybean and parsley. *Planta* 176:75–82

120. Parker JE, Schulte W, Hahlbrock K, Scheel D. 1991. An extracellular glycoprotein from *Phytophthora megasperma* f. sp. *glycinea* elicits phytoalexin synthesis in cultured parsley cells and protoplasts. *Mol. Plant-Microbe Interact.* 4:19–27

121. Paxton JD, Groth J. 1994. Constraints on pathogens attacking plants. *Crit. Rev. Plant Sci.* 13:77–95

122. Pearce G, Strydom D, Johnson S, Ryan CA. 1991. A polypeptide from tomato leaves induces wound-inducible proteinase inhibitor proteins. *Science* 253:895–98

123. Pernollet J-C, Sallantin M, Sallé-Tourne M, Huet J-C. 1993. Elicitin isoforms from seven *Phytophthora* species: comparison of their physico-chemical properties and toxicity to tobacco and other plant species. *Physiol. Mol. Plant Pathol.* 42:53–67

124. Perotto S, Brewin NJ, Kannenberg EL. 1994. Cytological evidence for a host defense response that reduces cell and tissue invasion in pea nodules by lipopolysaccharide-defective mutants of *Rhizobium leguminosarum* strain 3841. *Mol. Plant-Microbe Interact.* 7:99–112

125. Priem B, Galli R, Bush CA, Gross KC. 1993. Structure of ten free *N*-glycans in ripening tomato fruit. Arabinose is a constituent of a plant *N*-glycan. *Plant Physiol.* 102:445–58

126. Rasmussen JB, Hammerschmidt R, Zook MN. 1991. Systemic induction of salicylic acid accumulation in cucumber after inoculation with *Pseudomonas syringae* pv. *syringae*. *Plant Physiol.* 97:1342–47

127. Ren YY, West CA. 1992. Elicitation of diterpene biosynthesis in rice (*Oryza sativa* L.) by chitin. *Plant Physiol.* 99:1169–78

128. Renelt A, Colling C, Hahlbrock K, Nürnberger T, Parker JE, et al. 1993. Studies on elicitor recognition and signal transduction in plant defence. *J. Exp. Bot.* 44(Suppl.): 257–68

129. Ricci P, Bonnet P, Huet J-C, Sallantin M, Beauvais-Cante F, et al. 1989. Structure and activity of proteins from pathogenic fungi *Phytophthora* eliciting necrosis and acquired resistance in tobacco. *Eur. J. Biochem.* 183:555–63

130. Ricci P, Trentin F, Bonnet P, Venard P, Mouton-Perronnet F, Bruneteau M. 1992. Differential production of parasiticein, an elicitor of necrosis and resistance in tobacco, by isolates of *Phytophthora parasitica*. *Plant Pathol.* 41:298–307

131. Rigby NM, MacDougall AJ, Needs PW, Selvendran RR. 1994. Phloem translocation of a reduced oligogalacturonide in *Ricinus communis* L. *Planta* 193:536–41

132. Roche P, Debellé F, Maillet F, Lerouge P, Faucher C, et al. 1991. Molecular basis of symbiotic host specificity in *Rhizobium meliloti*: *nodH* and *nodPQ* genes encode the sulfation of lipo-oligosaccharide signals. *Cell* 67:1131–43

133. Ryals J, Uknes S, Ward E. 1994. Systemic acquired resistance. *Plant Physiol.* 104: 1109–12

134. Ryan CA, Farmer EE. 1991. Oligosaccharide signals in plants: a current assessment. *Annu. Rev. Plant Physiol. Plant Mol. Biol.* 42:651–74

135. Salmeron JM, Barker SJ, Carland FM, Mehta AY, Staskawicz BJ. 1994. Tomato mutants altered in bacterial disease resistance provide evidence for a new locus controlling pathogen recognition. *Plant Cell* 6: 511–20

136. Salmeron JM, Rommens C, Barker SJ, Carland FM, Staskawicz BJ. 1994. Isolation of mutations in the *Pto* resistance locus of tomato and identification of a new locus, *Prf*, controlling pathogen recognition. See Ref. 32a, p. 86

137. Sanchez LM, Doke N, Kawakita K. 1993. Elicitor-induced chemiluminescence in cell suspension cultures of tomato, sweet pepper and tobacco plants and its inhibition by suppressors from *Phytophthora* spp. *Plant Sci.* 88:141–48

138. Schmidt WE, Ebel J. 1987. Specific binding of a fungal glucan phytoalexin elicitor to membrane fractions from soybean *Glycine max*. *Proc. Natl. Acad. Sci. USA* 84: 4117–21

139. Schottens-Toma IM, De Wit PJGM. 1988.

Purification and primary structure of a ne-crosis-inducing peptide from the apoplastic fluids of tomato infected with *Clado-sporium fulvum* (syn. *Fulvia fulva*). *Physiol. Mol. Plant Pathol.* 33:59–67

140. Schwacke R, Hager A. 1992. Fungal elici-tors induce a transient release of active oxygen species from cultured spruce cells that is dependent on Ca^{2+} and protein-ki-nase activity. *Planta* 187:136–41

141. Sharon A, Bailey BA, McMurtry JP, Taylor R, Anderson JD. 1992. Characteristics of ethylene biosynthesis-inducing xylanase movement in tobacco leaves. *Plant Physiol.* 100:2059–65

142. Sharon A, Fuchs Y, Anderson JD. 1993. The elicitation of ethylene biosynthesis by a *Trichoderma* xylanase is not related to the cell wall degradation activity of the en-zyme. *Plant Physiol.* 102:1325–29

143. Sharp JK, McNeil M, Albersheim P. 1984. The primary structure of one elicitor-active and seven elicitor-inactive hexa(β-D-glu-copyranosyl)-D-glucitols isolated from the mycelial walls of *Phytophthora megas-perma* f. sp. *glycinea. J. Biol. Chem.* 259: 11321–36

144. Sharp JK, Valent B, Albersheim P. 1984. Purification and partial characterization of a β-glucan fragment that elicits phytoalexin accumulation in soybean. *J. Biol. Chem.* 259:11312–20

145. Shibuya N, Kaku H, Kuchitsu K, Maliarik MJ. 1993. Identification of a novel high-af-finity binding site for *N*-acetylchitooligo-saccharide elicitor in the membrane frac-tion from suspension-cultured rice cells. *FEBS Lett.* 329:75–78

146. Shiraishi T, Saitoh K, Kim H-M, Kato T, Tahara M, et al. 1992. Two suppressors, supprescin A and B, secreted by a pea pathogen *Mycosphaerella pinodes*. *Plant Cell Physiol.* 33:663–67

147. Smith DC. 1994. *Symbiotic Interactions.* Oxford: Oxford Univ. Press. 148 pp.

148. Spaink HP, Sheeley DM, Van Brussel AAN, Glushka J, York WS, et al. 1991. A novel highly unsaturated fatty acid moiety of lipo-oligosaccharide signals determines host specificty of *Rhizobium. Nature* 354: 125–30

149. Spaink HP, Wijfjes AHM, van Vliet TB, Kijne JW, Lugtenberg BJJ. 1993. Rhizobial lipo-oligosaccharide signals and their role in plant morphogenesis; are analogous lipophilic chitin derivatives produced by the plant? *Aust. J. Plant Physiol.* 20:381–92

150. Spanu P, Boller T, Ludwig A, Wiemken A, Faccio A, Bonfante-Fasolo P. 1989. Chiti-nase in roots of mycorrhizal *Allium por-rum*: regulation and localization. *Planta* 177:447–55

151. Staehelin C, Granado J, Müller J, Wiemken A, Mellor RB, et al. 1994. Perception of *Rhizobium* nodulation factors by tomato cells and inactivation by root chitinases. *Proc. Natl. Acad. Sci. USA* 91:2196–200

152. Staehelin C, Schultze M, Kondorosi É, Mellor RB, Boller T, Kondorosi Á. 1994. Structural modifications in *Rhizobium meliloti* Nod factors influence their stabil-ity against hydrolysis by chitinases. *Plant J.* 5:329–40

153. Stein JC, Howlett B, Boyes DC, Nasrallah ME, Nasrallah JB. 1991. Molecular clon-ing of a putative receptor protein kinase gene encoded at the self-incompatibility locus of *Brassica oleracea. Proc. Natl. Acad. Sci. USA* 88:8816–20

154. Stengl M, Hatt H, Breer H. 1992. Periph-eral processes in insect olfaction. *Annu. Rev. Physiol.* 54:665–81

155. Thuleau P, Ward JM, Ranjeva R, Schroeder JI. 1994. Voltage-dependent calcium-per-meable channels in the plasma membrane of a higher plant cell. *EMBO J.* 13:2970–75

156. Toubart P, Desiderio A, Salvi G, Cervone F, Daroda L, et al. 1992. Cloning and charac-terization of the gene encoding the en-dopolygalacturonase-inhibiting protein (PGIP) of *Phaseolus vulgaris* L. *Plant J.* 2:367–73

157. Toyoda K, Shiraishi T, Yamada T, Ichinose Y, Oku H. 1993. Rapid changes in polyphosphoinositide metabolism in pea in response to fungal signals. *Plant Cell Physiol.* 34:729–35

158. Truchet G, Roche P, Lerouge P, Vasse J, Camut S, et al. 1991. Sulphated lipo-oli-gosaccharide signals of *Rhizobium meliloti* elicit root nodule organogenesis in alfalfa. *Nature* 351:670–73

159. Van Brussel AAN, Bakhuizen R, van Spronsen PC, Spaink HP, Tak T, et al. 1992. Induction of pre-infection thread structures in the leguminous host plant by mitogenic lipo-oligosaccharides of *Rhizobium. Sci-ence* 257:70–72

160. Van den Ackerveken GFJM, Van Kan JAL, De Wit PJGM. 1992. Molecular analysis of the avirulence gene *avr9* of the fungal to-mato pathogen *Cladosporium fulvum* fully supports the gene-for-gene hypothesis. *Plant J.* 2:359–66

161. Van den Ackerveken GFJM, Vossen P, De Wit PJGM. 1993. The AVR9 race-specific elicitor of *Cladosporium fulvum* is proc-essed by endogenous and plant proteases. *Plant Physiol.* 103:91–96

162. Van Houten J. 1992. Chemosensory transduction in eukaryotic microorgan-isms. *Annu. Rev. Physiol.* 54:639–63

163. Van Kan JAL, Van den Ackerveken GFJM, De Wit PJGM. 1991. Cloning and charac-terization of cDNA of avirulence gene *avr9* of the fungal pathogen *Cladosporium fulvum*, causal agent of tomato leaf mold. *Mol. Plant-Microbe Interact.* 4:52–59

164. Vasse J, de Billy F, Truchet G. 1993. Abortion of infection during the *Rhizobium meliloti*-alfalfa symbiotic interaction is accompanied by a hypersensitive reaction. *Plant J.* 4:555–66

165. Vernooij B, Friedrich L, Morse A, Reist R, Kolditz-Jawhar R, et al. 1994. Salicylic acid is not the translocated signal responsible for inducing systemic acquired resistance but is required in signal tranduction. *Plant Cell* 7:959–65

166. Viard M-P, Martin F, Pugin A, Ricci P, Blein J-P. 1994. Protein phosphorylation is induced in tobacco cells by the elicitor cryptogein. *Plant Physiol.* 104:1245–49

167. Volpin H, Elkind Y, Ojon Y, Kapulnik Y. 1994. A vesicular arbuscular mycorrhizal fungus (*Glomus intraradix*) induces a defense response in alfalfa roots. *Plant Physiol.* 104:683–89

168. Waldmüller T, Cosio EG, Grisebach H, Ebel J. 1992. Release of highly elicitor-active glucans by germinating zoospores of *Phytophthora megasperma* f. sp. *glycinea*. *Planta* 188:498–505

169. Walker JC. 1993. Receptor-like protein kinase genes of *Arabidopsis thaliana*. *Plant J.* 3:451–56

170. Walton JD, Panaccione DG. 1993. Host-selective toxins and disease specificity—perspectives and progress. *Annu. Rev. Phytopathol.* 31:275–303

171. Wei Z-M, Laby RJ, Zumoff CH, Bauer DW, He SY, et al. 1992. Harpin, elicitor of the hypersensitive response produced by the plant pathogen *Erwinia amylovora*. *Science* 257:85–88

172. Wevelsiep L, Rupping E, Knogge W. 1993. Stimulation of barley plasmalemma H$^+$-ATPase by phytotoxic peptides from the fungal pathogen *Rhynchosporium secalis*. *Plant Physiol.* 101:297–301

172a. Whitham S, Dinesh-Kumar SP, Choi D, Hehl R, Corr C, Baker B. 1994. The product of the tobacco mosaic virus resistance gene *N*: similarity to Toll and the interleukin-1 receptor. *Cell* 78:1101–15

173. Yoshikawa M, Yamaoka N, Takeuchi Y. 1993. Elicitors: their significance and primary modes of action in the induction of plant defense reactions. *Plant Cell Physiol.* 34:1163–73

174. Ziegler E, Pontzen R. 1982. Specific inhibition of glucan-elicited glyceollin accumulation in soybeans by an extracellular mannan-glycoprotein of *Phytophthora megasperma* f.sp. *glycinea*. *Physiol. Plant Pathol.* 20:321–31

Annu. Rev. Plant Physiol. Plant Mol. Biol. 1995. 46:215–36

APOPLASTIC WATER AND SOLUTE MOVEMENT: New Rules for an Old Space

M.J. Canny

Biology Department, Carleton University, Ottawa, Canada K1S 5B6

KEY WORDS: diffusivities in cell walls, osmotic pressure of xylem, solutes in intercellular spaces, solutes in xylem sap, water flow from vessels to symplast

CONTENTS

ABSTRACT

The history of the *apoplast,* both the word and the concept, is traced from its invention by Münch (1930), through its adoption by the translocation physiologists as a gateway by which to feed pesticides, and the independent evolution of the free-space concept by ion-uptake physiologists. Both usages

0066-4294/95/0601-0215$05.00

helped spread the idea that the cell walls were freely permeable to flowing solutions. Recent work has produced six contradictions to the prevailing notions of the rules that operate in the apoplast: (*a*) The flow of the apoplast is leaky, and the balance of the flow and leaks depends on vessel diameter. (*b*) Water leaves the vessels and enters the symplast faster than do some solutes, resulting in the accumulation of high concentrations of solutes at places called sumps. (*c*) Solutes diffuse away from sumps in the cell-wall apoplast at rates much slower than diffusion in water. (*d*) Ion concentrations in leaf vessels may be as high as 200 mM. (*e*) The intercellular-space apoplast of roots often contains solution with high concentrations of ions. (*f*) The threshold of cavitation of the flow-apoplast appears to be in the range of 1 to 2 bar below atmospheric, much less than the tensions required by the Cohesion Theory.

INTRODUCTION

The apoplast surrounds the living substance of a plant in the same way that the bureaucracy surrounds a government department. As physiologists we address all our questions to the apoplast, and they are translated by the bureaucracy into its own language and passed on, in the time frame habitual to the bureaucracy, to the decision chamber. The answers are returned to us through the bureaucracy, interpreted in its accustomed formal language. The more we can learn about the way the bureaucracy functions, and what rules regulate it, the more likely we are to phrase our questions appropriately and to interpret the answers correctly.[1] This review explores some of the simple general rules that govern the behavior of matter and energy in the apoplast. It suggests that at least six of these rules have been misunderstood and need to be replaced by new rules. Under the new rules, different procedures are appropriate for addressing questions to the apoplast, and different interpretations should be made from the answers. The review excludes consideration of the chemical and enzymic reactions of the apoplast, and of particular local obstacles in the apoplast such as the hydrophobic wall deposits of cuticle, Casparian strips, and suberized lamellae.

BRIEF HISTORY OF THE APOPLAST

The word and concept *apoplast* were invented by Münch (54) in his extended account of the mass-flow theory of translocation. After explaining the concept of the plant symplast, which he describes as well known, as a unit of living protoplasts surrounded by a single plasmalemma, Münch explains that the

---

[1]   See, for example, the manipulation of the bureaucratic process by a knowledge of what Parkinson calls the Standard Delay and by an alternative procedure (56).

symplast is separated from the "adjacent non-living space, particularly the water conducting tissues—the Hydrom.... These non-living parts of plant tissues we will include under the term Apoplast...." He refers to his famous figure of the mass-flow model—two osmotic cells connected by a tube (the whole constituting a symplast), immersed in a bath of water, the apoplast. The apoplast was the pathway in which water returned from the sink to the source. He then identified the apoplast with the water continuum of wet cell walls and vessel contents lying beside the phloem. For Münch, the apoplast was a watery space and intercellular spaces might be part of it if they contained water.

From its origin in translocation theory, the term apoplast spread into translocation physiology with the rise in economic importance of systemic herbicides and pesticides. The phloem-translocated herbicides, such as 2,4-D, were sprayed on plants and entered the apoplast. They were inserted into the phloem by getting them first into Münch's water-return path. Once disseminated there they could move into the symplast and be transported in the opposite direction by the sieve tubes. Crafts (28) seems to be the first to have appreciated and explicitly stated this bureaucratic function of the apoplast: "...the continuous wall phase, the apoplast, is a hydrated colloidal medium in dynamic equilibrium through imbibitional forces with water in the xylem,...it is evident that any source of water at atmospheric pressure that comes into physical contact with the apoplast will serve as a source for uptake and flow."

By analogy with the loading of herbicides from the leaf apoplast into the phloem (28), it was a short step to the hypothesis that the sucrose produced by mesophyll cells might move to the sieve tubes through the cell-wall apoplast. It would be released through the cell membranes of the mesophyll cells, travel in Münch's cell-wall apoplast return pathway (though here against the stream of transpiration water leaving the xylem) to the companion cells, and be accumulated to a high concentration by a sucrose-specific pump at the companion cell-membrane (35). The possibility of such apoplastic phloem loading was supported by experiments in which leaf disks, floated on labeled sucrose, accumulated label in the veins (71). The extent to which this pathway is significant in phloem loading is still debated (83).

For Münch, the important process in the xylem-vessel part of the apoplast pathway was the return of water from sink to source. The return from root sinks was easily identified as a combining of phloem-released water with the transpiration stream in the xylem vessels. Plant physiologists have considered the water moving from the roots in the transpiration stream as a large volume of dilute solution carrying nutrients and hormones that are lost to stem and leaf tissues on the way up (37, 48, 57). The return of water from fruit and meristematic sinks at the other end of the transpiration path was always something of a puzzle, because it demanded flow of water out of the fruit or away from the meristem in the direction opposite to transpiration (e.g. 24). Using

[3]H-labeled water, Pate et al (58) claimed to have shown such return movement in the xylem from cowpea pods, though the success of this demonstration implies a restriction of the mobility of water between the phloem and other tissues. Water is generally thought to exchange readily between cells or tissues.

In parallel with the translocation physiologists, another group was investigating the apoplast, although they had invented their own name for it, the apparent free space. This group studied the uptake of salts by pieces of storage tissue, especially disks of carrot, potato, and red beet (63). Briggs invented the term to describe the portion of the plant tissue into which substances in solution apparently move by free diffusion (12, 44). The cell-wall exchange space had already been given a name, the outer space, by Conway & Downey (25) working with yeast (31), but the term free space gained currency and was the term commonly used. The free space was subdivided into an electrically neutral fraction where positive and negative ions were present in equal concentrations (the water free-space), and a fraction with fixed negative charges where there were more positive than negative ions (the Donnan free-space) (12, 31). By about 1960 it became clear that both groups, those dealing with translocation and those with ion uptake, were studying approximately the same nonliving entity, although with different emphasis.

Sometime after 1970, and by a route I have not been able to trace, the word apoplast was adopted by plant anatomists apparently unaware of its special meaning. They began to use it to mean the walls and intercellullar spaces, even when these spaces contained gas, not liquid. Such spaces, when they contained liquid, were part of Münch's apoplast and of Briggs's apparent free space, but not when they contained gas. The earliest use of apoplast in this sense is in Gunning & Steer's atlas of plant fine structure (39). Anatomists' use of the word in this sense has become widespread (32, 50) and probably cannot be reversed. Once a scientific term has lost its precise meaning and begins to mean different things to different readers, it is probably best abandoned, or, if retained, requires qualification. This review asks for such qualification. The apoplast in Münch's sense is indeed a composite of spaces with quite different properties, and I propose that the word should not be used alone, but with qualifying adjectives.

The translocation group and the free-space group both held views of the space which cooperated (perhaps along with the name "free space") to spread the impression that solutions could move in the apoplast relatively unhindered. This was clearly true of the xylem vessel portion, and Münch's lumping of that portion with the wet cell walls implied that the walls also were freely permeable. Several lines of reasoning [all of them mistaken (see 15)] reinforced this impression:

1. The greatly respected authority of Sachs continued throughout his life (until 1897) to support his hypothesis [enunciated in 1865 (64)] that water moved in the vessel walls and not in the lumens of the vessels (the Imbibition Theory).

2. Strugger's experiments (74–77) following the spread of fluorescent dyes in the transpiration stream of living leaves showed the dyes confined to the walls of the mesophyll and epidermal cells. He interpreted this as evidence that the stream flowed out of the vessels and through the cell-wall apoplast in the plane of the walls (the Extended Cohesion Theory) (78).

3. Frey-Wyssling (34) coined the term holopermeable for cell walls through which solutions permeated freely. For evidence, he invoked the speed with which cells are plasmolyzed in a hypertonic solution, and argued that the plasmolyzing solution must flow easily through the wall.

4. Preston (61), after a lifetime's study of the physical properties of cell walls, said that the physiologists' proposal of flow in walls was "far from hopeless." He felt able to support the suggestion with a calculation in which the assumptions of 4-nm-wide capillary spaces in swollen walls were consistent with his experience.

DISTINCTIONS WITHIN THE APOPLAST[2]

This review is concerned with the rules that govern the behavior of substances in the apoplast, and from what has been said about the usage of the word, no one set of rules can apply to all of the apoplast. It is necessary to consider separately the xylem-lumen apoplast, the cell-wall apoplast, and the intercellular-space apoplast (i.e. considering the liquid and gaseous regions separately), settling on the rules appropriate in each. For those of us brought up on the body of facts and interpretations summarized above, the rules would be as follows:

1. In the xylem-lumen apoplast: There is Hagen-Poiseuille flow of a dilute solution along a gradient of pressure. During transpiration the pressure is negative (a tension). The widely accepted Cohesion Theory of the ascent of the transpiration stream requires a gradient of about 2 bar $(10m)^{-1}$ to lift water to the tops of tall trees. The lumen-apoplast is thought to be carefully constructed to withstand tensions in the water column of 60 bar or more.

2

Following the use of the term *protoplast* for the entity (i.e. the whole cell contents within the cell membrane), and the term *protoplasm* for the substance (i.e. the complex mixture of materials and organelles comprising the protoplast), the term *apoplast* would apply to the whole wet extracellular space, and *apoplasm* would apply to the substance of which it is made. However, because the apoplast consists of several quite different spaces with no common substance, apoplasm is not a useful concept or word.

2. In the water free-space (wet intercellular space + part of the cell-wall space): There is flow of solution in films or in a porous bed. Or, with no pressure gradient, there is diffusive exchange with contacting solutions, and the diffusivity for any solute is close to that in water (7).

3. In the Donnan free-space (another part of the wall space): The porous bed contains fixed negative charges, and the rules for charged solvent molecules would be modified by the electrical interactions (see 38a, 42, 49).

4. In the gaseous intercellular space: There is diffusion of gases, with diffusivity 10^4 times that in water. Under special circumstances there could be a wind generated in this space [e.g. in water lily petioles (4, 29)]. A gas phase in small spaces surrounded by water is a nonequilibrium state, and its preservation requires some still-undiscovered mechanism (87).

SAMPLING OF APOPLASTIC SPACES

Working from these assumptions, methods were devised to extract solution from the component apoplastic spaces. The xylem-lumen space is the easiest to sample, especially in roots or stems. A pooled average of xylem lumen contents can be collected as bleeding sap from cut roots or stems when positive pressure prevails there (8, 22). When the xylem does not bleed spontaneously, sap may often be forced from it by air pressure or centrifugation (3, 41, 53). Comparisons of bleeding sap with expressed sap from the same tissues have shown that they often differ in composition (see 9 for a critical review of these methods). These techniques usually cannot be applied to leaf xylem. Samples of fluid forced from the cut end of the xylem when a leaf is placed in a Scholander pressure chamber have often been assumed to represent leaf xylem lumen contents (e.g. 47). This assumption implies several other assumptions about the response of leaf cells and tissues to pressurization, none of which has been verified (19), and the reliability of this sampling is questionable. Solutes in the cell-wall (free-space) apoplast can be sampled by diffusive exchange with a solution brought into contact with them (59). Because access of the sampling solution is often restricted by anatomical barriers, the solution is injected by pressure differentials or centrifugation and then recovered by further application of forces after the exchanges are completed (52, 52a, 80). The time allowed for such exchanges is of course influenced by the experimenter's view of the permeability of the wall space. The perfusing or centrifugation of apoplastic solution, assumed to be lightly held in the holopermeable cell walls, has been widely practiced (e.g. 79). Cosgrove & Cleland (27) studied three methods of extracting solutes from the cell-wall apoplast: perfusion by water under pressure, infiltration and centrifugation, and expression by air pressure. They concluded that the solutes in the samples obtained were similar and representative of the apoplast contents. All samples showed substantial solute concentrations in the apoplast of young pea stems, sufficient to exert an osmotic pressure of 2.9 bar in the apical tissues and 1.8 bar in the basal tissues.

CONTRADICTIONS TO THE PREVAILING PARADIGM

This was the mental landscape in which I began experiments in the apoplast. I present below a summary of the changes of viewpoint forced upon me by unexpected contradicting results. Further details of the progress of this personal revolution may be found in References 14, 15, and 19.

Flow in Leaky Tubes

The first contradiction seemed a fairly slight and rational one, but in retrospect it held the clue to much of what followed. To verify Hagen-Poiseuille flow in wheat leaves (2), velocity was measured by the advance of dye solution in vessels of different sizes, and a plot was made of velocity vs (vessel diameter)^2, $(d)^2$. In accordance with the Hagen-Poiseuille law, this plot gave a straight line (Figure 6 in Ref. 2). However, the line intersected the d^2-axis, not at zero but at $d = 12$ μm. That is, there was no flow in vessels smaller than 12 μm diameter. This observation was interpreted as resulting from extraction of water (solution) from the vessels into the surrounding tissues. The significance of this interpretation was unappreciated until a later experiment in which the progress of dye solution was followed into the finest veins of a dicotyledon leaf (16). As the dye moved into smaller and smaller veins (vessels) it traveled slower and slower. In the smallest vessels (< 8 μm diameter) it took many minutes to move the last ~100 μm to the end. In these vessels, water extraction through the vessel walls (pits) was so high relative to axial flow that forward movement of the dye almost ceased.

The importance of vessel diameter to the balance between forward volume flow $\propto d^4$) and extraction radially (varies with surface, $\propto d$) explained some reasons why vascular strands contain vessels of different sizes (18). They have different functions. With a constant extraction rate per unit surface, wide vessels carry most of their contents longitudinally, and only a small proportion leaks out of them. In absolute terms, because their surface is large, the quantity leaked per vessel is also large. Small-diameter vessels leak much of their contents radially and achieve relatively little longitudinal flow. The absolute leakage from them is small individually, but because there are many more of them, most of the transpiring water passes out of them.

Sump Formation

The second contradiction was more fundamental. As the dye solution advanced through the vein network of a dicotyledon leaf into smaller and smaller vessels, the dye became more and more concentrated, until it passed the saturation limit and crystallized inside the vessels (16). This was shown at first

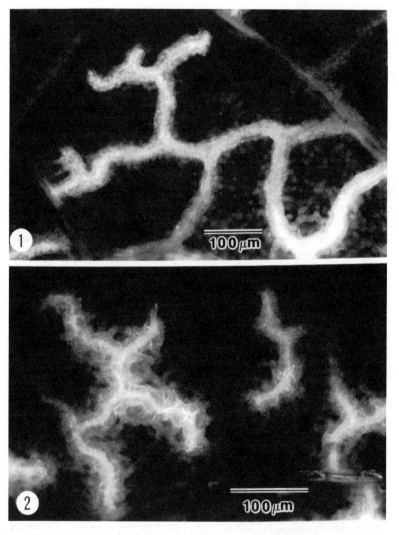

Figures 1 and 2 Whole mounts of freeze-substituted, anhydrously processed soybean leaves showing ultimate branches of fourth- and fifth-order veins after transpiration of fluorescent dye solutions, viewed by fluorescence optics. Figure 1 after 40 min transpiration in sulphorhodamine G (0.05%); Figure 2 after 40 min in pyrene trisulphonate (0.05%). Sumps of concentrated dye have formed in the terminal fifth-order veins. Note that the fourth-order veins supplying them contain much less dye. In both figures, the dye has started to diffuse from the sumps in the vessels through the cell-wall apoplast of the bundle sheath and mesophyll. The details of this diffusion are clearer at the higher resolution of Figure 2.

by a laborious technique (Figures 1 and 2) (16), but can now be demonstrated in an hour or so by anyone with access to a microscope (Figure 3) (55). Transpiration separates water from solutes, and it is no surprise that the dyes should accumulate somewhere in the leaf. But this accumulation was expected at the site of evaporation, i.e. the cell walls in the vicinity of the stomata. Separation of the dye from water within the vessels implied that the water extracted from the leaky vessels was entering a space impermeable to the dye, either the symplast and/or the cell wall. The weakest form of the contradiction was that the cell-wall apoplast in the plane of the wall was permeable to water flow but not to the dye. The stronger form was that the cell-wall apoplast was also impermeable to water flow along the wall, and that the water crossed the cell membrane into the symplast. For reasons that will become apparent later, the entry to the symplast is most likely.

To help with the economical discussion of these events some new terms were proposed (15). The places where the dye accumulates at high concentrations were called *sumps*. The site and flux of water crossing from the flow-apoplast to the symplast was called a *flume*. A sump is evidence of the presence of a flume. Solutes that cross cell membranes will not form sumps at flumes. Sulphorhodamine G is the test dye most useful for locating sumps, because it is nontoxic, negatively charged, and not very soluble in water.

The field of exploration of flumes by sump formation with a test dye is as large as the plant kingdom and has barely begun. A few generalizations are

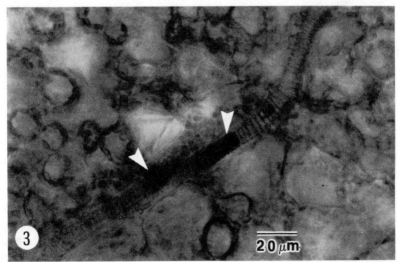

Figure 3 Fresh hand-section (paradermal) cut under oil of a leaf of tobacco after 90 min transpiration in sulphorhodamine solution (0.05%) showing crystals of dye (arrowheads) formed inside the vessel of a fifth-order vein. Bright-field optics. (Preparation and micrograph by NA O'Dowd.)

already apparent. Sumps are usually found in the smaller veins with smaller vessels and not in the larger veins that have large vessels. This observation is consistent with the relation of vessel diameter to the balance between flow and extraction discussed above. Sumps in grass leaves are usually not formed within vessels but away from the xylem inside the bundle sheath (55). The flumes in wheat leaves are at the parenchyma sheath cells outside the mestome sheath cells (14). Therefore, there must be flow of water and solution for a short distance through cell walls of a few of the mestome sheath cells, and these walls are one of the rare certain exceptions to the generalization that the cell-wall apoplast is not normally part of the flow space. In bamboo leaves, the sumps are formed, not near the xylem, but at the place where the parenchyma sheath adjoins the fusoid cells, the large clear cells that lie between two layers of chlorenchyma in the central plane of the lamina (55). These cells appear to distribute water from the veins to the mesophyll. Such water-conducting cells are probably widespread.

The combination of the two facts of leaky vessels and sump formation at flumes led me to predict that, in a wedge of contiguous xylem vessels of graded size such as is often seen in the metaxylem, sumps of solutes should form in the small vessel(s) at the apex of the wedge, in stems and leaves that were transpiring (18). Large vessels at the base of the wedge have predominantly axial flow and a small proportion of leakage, and small vessels at the apex of the wedge have mostly leaks and little or no forward flow. The water to supply the leaks of small vessels comes laterally from the larger vessels next to them. Thus there is radial flow of solution in the narrow part of the wedge, drawing upon the large vessels in the wide part. Solutes in the stream that do not enter the symplast will accumulate and move to the apex of the wedge, and may reach high concentrations there. This has been demonstrated for dyes (18), and to a lesser degree for ions (20), which more readily enter the symplast, though more slowly than does water (see below).

Diffusivity in the Cell-Wall Apoplast

The third contradiction related directly to the longitudinal permeability of the cell walls to the dyes. Strugger (74) demonstrated dye spreading to the leaf surface in anticlinal walls, and this is easily verified with a fluorescence microscope. The walls are indeed permeable to the dyes, but the time relations of their movement are not those of flow. The spread of the dye is not linear with time. The sump of dye forms quickly as transpiration water is separated from solution. But the spread of dye in the cell walls away from the sump is slow and gets slower as it proceeds. The distance of spread of the dye is proportional to $(time)^{1/2}$, which is characteristic of diffusion, not directly proportional to time as in flow. Furthermore, the rate of spread is independent of whether or not there is a flow of water in transpiration through the tissue. This result was shown long ago for roots (43, 46) and for leaves (68). The magnitude of the contradiction became apparent when the diffusivity of the dye in the cell walls was measured from the rate of its advance (18). In vein extensions and epidermal cell walls of wheat leaves, the diffusivity was from 1/100 to 1/10,000 of the diffusivity in water. Further, individual walls, and even parts of walls, were so structured as to have different diffusivities. For example, the diffusivity in the cell-wall apoplast downward from the vein and along the lower epidermis was twice the diffusivity to and along the upper epidermis.

The differentiation of wall diffusivity was very finely structured. In some walls the zone of high diffusivity was a narrow surface layer, so fine as to be beyond the resolution of the light microscope, but visible in the fluorescence microscope as a bright line of constant width. The dye diffused over the cell-wall surface in layers less than 200 nm thick, for which the term *nanopaths* was coined. Such nanopaths were seen also in maize roots (21), especially within the thick walls of the endodermal cells. In main and branch roots of maize, the values of diffusivity ranged from 1/50 to 1/1000 of the value in water (21).

What Strugger saw in his fluorescence microscope (74–77) and mistook for dye solution flowing in the cell walls was in fact dye diffusing in cell walls away from sumps that had formed in the veins beneath. It takes about 30 min for pyrene trisulphonate to diffuse from the sump in a vein of a soybean leaf to the epidermal surface (Figure 2). If the dye solution was flowing in these walls it would reach the surface in a few seconds. If Strugger's papers are reread with this interpretation in mind, the times he records for observations of the dyes are seen to be long and suggest the slow diffusion. Also, neither he nor anyone else has recorded the buildup of solute to a maximum concentration near the leaf surface, which would be expected if this was the terminus of the flowing solution.

What then has become of Frey-Wyssling's argument justifying the holopermeable cell wall by the observation of plasmolysis within a few minutes of placing a cell in a hypertonic solution? There are two elements to the answer: First, he is dealing with movement across the wall, not along it, and the permeability may be different in the two directions. Second, and more fundamentally, because distance of diffusion varies as $(time)^{1/2}$, diffusive equilibration through the thickness of a cell wall (e.g. 0.5 μm) is fast even if the diffusivity is low (see 17).

It can be objected that the molecule for which these diffusivities were measured is much larger than the ions of the xylem sap, and this is true. Sulphorhodamine has a molecular weight of 553. In mitigation, it can be said that the values quoted are relative to diffusion in water, and a similar proportional retarding effect may operate on small ions. Also, Strugger selected sulphorhodamine precisely because he found it did not bind to cell walls and was very mobile in filter paper and agar gels. Furthermore, the published estimates of the diffusivity of the cell-wall apoplast for the naturally occurring solutes in cell walls give values considerably less than the values in water (D_{aq}). Aikman et al (1) used an electrical method to determine diffusivities of ions in the apoplast of sugar beet tissue and found values around $D_{aq}/10$. Richter & Ehwald (62) found values for sucrose in the same tissue of $D_{aq}/5$ to $D_{aq}/10$. Briggs & Robertson (11) obtained a value in carrot disks of $D_{aq}/60$, and a similar value for the diffusivity of sodium in barley roots is quoted as unpublished by Pitman (60). The objection remains a valid one, and the measurements of diffusivity of the natural solutes in particular cell walls of a range of living tissues must be high on a list of priorities.

Ion Concentrations in Leaf Vessels

In redirecting the investigation away from dye tracers and toward seeking sumps of the natural solutes in leaves, I came upon the fourth contradiction. The technique chosen was X-ray microanalysis (EDX) in the scanning electron microscope (SEM) of quick-frozen, hydrated leaf vein xylem. This technique allows quantitative analyses of the elements present in single vessels.

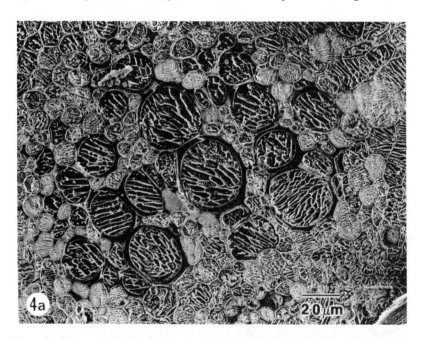

Figure 4a Frozen planed face of a second-order vein of a sunflower leaf viewed by SEM. The freezing sequesters solutes in the vacuoles of the cells into characteristic patterns. Analyses of the elements present in the solutions inside individual vessels were made by EDX-microanalysis and are given in Figure 4b.

The xylem sap is held in place by the freezing, and the frozen leaf is planed flat, etched very lightly to reveal the cell outlines, and coated with Al to make it conducting (45). X-ray spectra are collected from the chosen area of specimen (areas down to a few m^2), in which the elements present are recognized by the characteristic energies of their emitted X-rays. Because of absorption of X-rays by the ice, the method is not very sensitive. The threshold of detection of elements such as Na, K, Ca, Cl, S, and P is around 15 mM, higher than the concentrations characteristic of xylem sap leaving the roots in sunflower (37). If the ions present in the transpiration stream behaved like the dye tracers, they might become concentrated at sumps in the minor veins and reach levels measurable by the EDX/SEM. When the experiment was done, high and variable concentrations of K were found in individual vessels of sunflower leaves in the range of 25 to 200 mM, contradicting the traditional assumption that the xylem sap, even in leaf veins, is a weak solution of ions (Figure 4) (19). Balancing inorganic anions were often not apparent. Electrical balance could have been maintained by organic acids or nitrate, which the technique would not detect.

A more thorough investigation of ion concentrations in the veins of different orders in a single leaf (20) showed that K concentration increased through

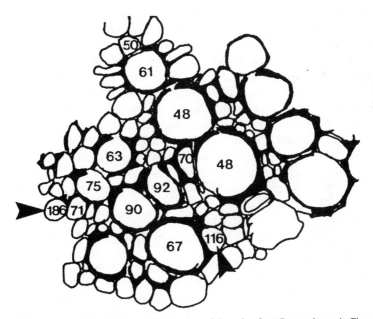

Figure 4b Drawing of the cells present in the face of the vein of sunflower shown in Figure 4a with the measured concentration of K (mM) marked in each. All the analyzed regions are vessels, except the two xylem parenchyma cells marked with arrowheads, which have higher concentrations of K.

the petiole, first-, second-, and third-order veins, but did not increase further in the smallest (fourth- and fifth-order) veins. Indeed, in contrast to the pattern of dye concentration, the K concentration fell to lower values in the fine veins. This observation was interpreted as evidence that K uptake into bundle sheath cells surrounding the vessels was more rapid in small veins than in large ones. The origin of the high K concentrations in leaf vessels was not simply attributable to the evaporative separation of water (as for the dyes) because substantial concentrations were often found entering the leaves in the vessels of the petiole. In seeking a source of K in stem xylem, I found a likely reservoir in the living, differentiating vessel elements of the secondary xylem. The K concentration was higher in these cells than in any other cells, occasionally reaching more than 0.5 M. Incidentally, these immature vessel elements, alone among the many cell types analyzed, contained high concentrations of Cl. The hypothesis was proposed that when a file of consecutive vessel elements differentiated to become a vessel, by loss of the end walls and protoplast, the contained K solution was released into the transpiration stream in the adjacent open vessels and, traveling with the stream through leaf traces to the leaf-vein system, could be detected as enhanced K concentrations in the veins downstream.

The further hypothesis was proposed (20) that the K in differentiating secondary vessel elements was part of a circulation of K between leaves, stems, and roots. It is well established that K is exported from leaves in the phloem and circulates back from root and stem in the xylem (48, 57). The hypothesis provides a route and mechanism by which K from the phloem would enter cambial cells and, moving to the cambial derivatives, accumulate at high concentration (and in large volume) in the vessel initials. The high K concentration would provide much of the osmotic force to expand the vessel elements to the size of mature vessels. Then, when the file of elements became a vessel, the released K would join the upward circulation back to the leaves. The recirculating of K probably also extends into roots, especially those with secondary thickening (51). According to this hypothesis, an appreciable fraction of the plant's complement of K would cycle between leaves, stems, and roots.

Solute Concentrations in Intercellular Spaces

The technique developed for studying ions in the vessels showed, incidentally, the fifth contradiction. In the frozen preparations of corn roots, many of the intercellular spaces between cells of the cortex, expected to contain gas and thus to be empty in the frozen specimen, contained frozen fluid. Such fluid-filled spaces had been reported previously (33, 73). More surprisingly, when the fluid in the spaces was analyzed, it was not just water, but contained high concentrations of K and Ca (21). On average, the mean ion concentrations in the spaces were not distinguishable from those in the vacuoles of the cortical cells surrounding the spaces. Not all the spaces contained fluid all the time. In some roots, a proportion of the spaces was dry and contained deposits that held the same elements as the fluid (K, Ca). The proportion of dry vs wet spaces appeared to depend on the age and water status of the root.

The suspicion that the high solute content of the intercellular-space apoplast in corn roots might be an artifact of the technique of preparation was allayed by the finding of a similar high solute apoplast in sugarcane, where the possibility of artifact was eliminated. In a search for the recently discovered N_2-fixing *Acetobacter diazotrophicus* (36), thought to be an endophyte that contributes substantial nitrogen to some varieties of sugarcane (23), the intercellular spaces of stem internodes were investigated (30). The frozen stem pieces were seen in the SEM to have fluid in the intercellular spaces, whose freezing pattern (and therefore probably solute content) was identical to those of the vacuoles of the stem parenchyma. The stem parenchyma is the main reservoir of sucrose in the plant. The fluid could be removed from the spaces by centrifugation, after which the SEM showed the spaces to be empty. The volume of the collected solution corresponded to that of the spaces. The fluid centrifuged from the stem pieces was a 10–12% solution of sucrose which, when collected under sterile conditions, contained *A. diazotrophicus*. In sugarcane, therefore, the intercellular-space apoplast acts as a culture broth to support the growth of an N_2-fixing endophyte. In this tissue the solution in the spaces could not have been produced by the preparation techniques for the SEM, because the solution can be extracted from the fresh tissue, leaving the spaces, as seen in the SEM, empty (30).

Casual observations of stem and root tissues of other plants in the cryo-SEM suggest that fluid-filled intercellular spaces may be fairly widespread there, but not in leaves. The two examples discussed above suggest the fluid-space apoplast may have a number of unexpected functions.

Sensitivity of the Flow-Apoplast to Cavitation

The high tensions in the flow-apoplast required by the Cohesion Theory would lead to breaking of the water columns under water stress, as a result of cavitation (production of water vapor) or air seeding (sucking air in through pit membranes) (82). The magnitude of the tension that can be sustained by water columns inside vessels has been much debated, and, to support the Cohesion Theory, emphasis has been placed on the maximum values measured for the cohesive strength of very pure water measured after high pressure has collapsed the small gas nuclei that initiate cavitation (13, 38). A recent study by Smith (70) has produced a sixth contradiction to another of the accepted rules for the apoplast: the sensitivity to cavitation. He emphasizes that the average threshold of cavitation is more relevant than the maximum, and that the adhesion of the water to the tube containing it is probably a graver limitation than is the cohesive strength of water to itself. Smith measured the cavitation threshold of filtered, deionized, degassed distilled water in contact with a range of tube linings from glass (completely wettable) to silicone grease (nonwettable). He showed that the cavitation threshold varied greatly, but even on the most wettable surface, on average, was around 3.5 bar below atmospheric pressure. In 84% of Smith's 354 trials, cavitation occurred between −1 and −8 bar. The degree to which nonwettable lignin surface is exposed on the inside of vessels becomes an important question, because cavitation in the vessel would occur at the lowest tension on the least wettable part of the wall. Smith's data suggest that the flow-apoplast cannot operate at the high tensions required by the theory.

Even ignoring the adhesion to the tube lining, there is growing evidence that water, unless specially treated, contains nuclei upon which bubbles form readily under tension. Yount et al (88) have investigated the nature of the nuclei on which bubbles are initiated. They identified a population of very small (1 nm to 1μm) gas bubbles "small enough to remain in solution and strong enough to resist collapse, their mechanical compression strength being provided by elastic skins...of surface-active molecules." Unless water is freed from these nuclei, it cavitates under quite low tensions.

Confidence in the existence of the high tensions in the flow-apoplast has been sustained by measurements with the Scholander pressure chamber, long believed to give indirectly an estimate of the tension inside xylem vessels. Zimmermann and his coworkers have used the xylem pressure probe to make direct measurements of the tension inside single vessels (6, 90). The tensions they measured mostly fall within the range where Smith found water to be practically safe from cavitation failure, and they found that when tensions approached the level near Smith's cavitation thresholds, the pressure probe recorded cavitation as a sudden jump back to the vapor pressure of water. It seems likely that the rule for the flow-apoplast may have to be changed to accept the much lower working tensions, not much more than 1 bar below atmospheric pressure.

IMPLICATIONS

The bureaucracy is more complicated than we thought, and its rules of operation are unfamiliar. We must reappraise our dialog with it, both the questions we address to it and the interpretations we make of the answers.

Supply and Collection of Substances via the Apoplast

There are two routes for feeding substances to the plant via the apoplast: flooding the wall-apoplast with the substance in solution, and feeding the solution into the flow-apoplast to be carried with the transpiration stream. Neither route can be relied on to deliver the substance to a target tissue at the supplied concentration without extensive investigation of the diffusivity of the substance in the walls between the source and the target, and of the concentrations of the substance in the various branches of the vascular system. The substance may be held up and its effect attenuated by low diffusivity, or it may be concentrated manyfold at sumps where its effect is greatly enhanced locally. Similar qualifications apply in reverse to the collection of material from the plant via the apoplast. In some regions of the plant, the exchange times are very long. Using some of the figures quoted above, a fiber with a wall thickness of 10 μm and a diffusivity of $D_{aq}/10,000$ will take 50 min to reach half-equilibration with a bathing solution; to reach 90% equilibration will take 85 h.

The Osmotic Xylem

Vessels are dead, have no cell membrane, and therefore in terms of the conventional paradigm would not exert their own osmotic pressure. If the walls of the cells surrounding the vessel were holopermeable ($D = D_{aq}$), they would have a reflection coefficient (σ) of zero, and solutes in the vessels still would not develop an osmotic pressure. But we have seen that the surrounding cell walls have low diffusivities ($D \leq D_{aq}/100$) and will therefore have a non-zero reflection coefficient, and solutes in the vessels will develop some osmotic pressure, say a third to a half of the full potential of the solutes. Moreover, quite high concentrations of ions have been found in some vessels.

The idea of an osmotic xylem is not new, but the rationale for it is new. The xylem saps of many trees contain sugars (66, 67, 89), and an osmotic pressure developed by the sugars has been considered as a driving force to explain the flow of maple sap from tapped xylem in spring (26) or to explain water uptake (10). Such a proposal is in conflict with the idea of holopermeable cell walls, but it becomes feasible if the cell walls have $\sigma \neq 0$. Cortes & Sinclair (26) rejected the proposal on several grounds, mainly because the concentration of the sap bleeding from the xylem of sugar maple did not correspond with the pressures measured simultaneously in the xylem. This objection is not conclusive. The pressure in a static tube is the same everywhere. If one part of a long tube ($\sigma \neq 0$) contains high solute concentration, the osmotic pressure in this part is transmitted to the whole tube contents. Bleeding from the tapped xylem may be driven by a pressure developed in a remote part of the xylem, and the concentration of the bled sap need not correspond with the measured pressure.

Positive pressures in the xylem in the absence of transpiration are known to be a common feature of roots, stems, and leaves (40, 69). The origins of the pressures are still debated (69), but most of the debate has involved the semipermeable membranes of living cells, ignoring the possibility of a solute-containing xylem surrounded by walls with $\sigma \neq 0$. Such positive pressures could serve important functions besides expelling guttation water. They could collapse and repair bubbles and embolisms formed by cavitation during periods of negative pressure. The positive pressures indicated by the bleeding saps of spring help refill the air embolisms formed by freezing (72). Nighttime positive pressures in summer help repair embolisms formed during the day's transpiration (81). Positive pressures produced by vessel solutes might counteract the tensions generated by a cohesion/surface-energy–driven transpiration, and lift the operating pressure range of such a system to tensions around 1 bar, in the range where the vessel contents are not very sensitive to cavitation (70).

The Structured Flow-Apoplast

The flow-apoplast is obviously structured to deliver solutes at particular sites at enhanced concentrations. These sites include the fine veins of dicotyledon leaves where water flux through flumes leaves some solutes in sumps within vessels (Figures 1–3). Because potassium and other ions are not found at higher concentrations in these vessels than in the vessels of larger veins (20), the cell membranes of bundle sheath and xylem parenchyma cells probably contain carrier systems to transfer them to the symplast. In grass leaves such carriers will be expected on the parenchyma sheath cells of the minor veins (14).

Sumps may be expected also in the smallest vessels of wedges of metaxylem (18), and the xylem parenchyma cells around these small vessels are likely to be especially active in accumulating the natural solutes of the xylem sap. For particular solutes of special interest in the transpiration stream, such as abscisic acid and cytokinins, studies will have to be initiated to see if they form sumps in the expected sites, and whether xylem parenchyma cells at these sites are active in transferring them to the symplast, or whether they remain to diffuse in the structured wall-apoplast.

The Structured Wall-Apoplast

Beyond the flow-apoplast, solutes not admitted to the symplast travel by diffusion in the wall-apoplast. Diffusivities in particular directions and layers of the cell walls vary over three or more orders of magnitude. A substance delivered by the flow-apoplast to a sump, and accumulated there at enhanced concentration, does not move evenly through the wall-apoplast in all directions. Accurate estimates of the effect of the substance on some remote target, as, for example, the effect of abscisic acid supplied in the xylem on guard cells in the epidermis, require knowledge of the concentration in the sump and of the diffusivities of the substance in the nanopaths that may connect the sump with the guard cells.

Root Uptake of Water and Solutes

The generalization that cell walls do not carry flowing solution, and that water movement through living tissue is through the symplast, has been applied to measure the sites and fluxes of water into roots (85, 86). A deduction from this generalization is that if a dye solution is supplied to the roots of a transpiring plant, the water will be transported through the symplast of the roots and the dye will stay outside. This experiment was done in 1871 by Baillon (5) and discussed by Sachs (65). I reasoned that the sites of water uptake would be shown as the sumps where the dye became concentrated, and the local rates of water uptake would be measured by the local rates of accumulation of the dye. The trick to permit sump formation, and prevent dispersal of the dye in the solution, is to feed the dye solution as a mist of fine drops. The rates measured over the root system of a mature corn plant (85) showed that more than 80% of the water entering the plant was taken up by the branch roots, though the flux per unit surface was the same into main roots and branches. In agreement with the generalization, the sumps of dye formed at the root surface (86), and only traces of dye penetrated the wall-apoplast by diffusion. On the branch roots, there seemed to be particularly vigorous fluxes of water over the grooves where the anticlinal walls met the surface.

The same generalization leads to the deduction that solute taken up by roots must be taken into the symplast at the outer layer of cells, and that in passing across the cortex to the xylem, the wall-apoplast is not a significant pathway. Thus, the specialized barriers in the apoplast in the hypodermis and endodermis may not affect substantially the movement of the solutes. However, the finding of concentrated ionic solutions in the intercellular-space apoplast of the maize root cortex (21) raises the question of what part this space may play in solute transport across the root. The root symplast in the cortex does not seem to be wholly distinct from the intercellular-space apoplast.

ENVOY

Anything that is alive is too complicated to be fully comprehended by the human mind. Even the dead part of a plant contains sufficient complications to exercise all our ingenuity. When we think that added to the simple, basic matters treated in this review there are the chemical and catalytic reactivities being found in the apoplast (84), we should humbly and thoughtfully plan our attempts to hold a dialog with the living plant.

ACKNOWLEDGMENTS

I thank the Natural Sciences and Engineering Research Council of Canada for an operating grant, Carleton University for the provision of space and facilities, Margaret McCully for criticism of the manuscript, Niamh O'Dowd for the preparation and micrograph of Figure 3, and Wayne England for making the plates.

Literature Cited

1. Aikman DP, Harmer R, Rust TS. 1980. Electrical resistance and ion movement through excised discs of sugar beet tissue. *Physiol. Plant.* 48:395–402

2. Altus DP, Canny MJ, Blackman DR. 1985. Water pathways in wheat leaves. II. Water-conducting capacities and vessel diameters of different vein types, and the behaviour of the integrated vein network. *Aust. J. Plant Physiol.* 12:183–99

3. Anderssen FG. 1929. Some seasonal changes in the tracheal sap of pear and apricot trees. *Plant Physiol.* 4:459–76

4. Armstrong J, Armstrong W. 1990. Light-enhanced convective throughflow increases oxygenation in rhizomes and rhizosphere of *Phragmites australis* (Cav.) Trin. ex Steud. *New Phytol.* 114:121–28

5. Baillon X. 1875. Expériences sur l'absorption par les racines du suc du *Phytolacca decandra*. *C. R. Acad. Sci.* 80:426–29

6. Balling A, Zimmermann U. 1990. Comparative measurements of the xylem pressure of *Nicotiana* plants by means of the pressure bomb and pressure probe. *Planta* 182:325–38

7. Bange GGJ. 1973. Diffusion and absorption of ions in plant tissue. III. The role of the root cortex in ion absorption. *Acta Bot. Neerl.* 22:529–42

8. Bollard EG. 1953. The use of tracheal sap in the study of apple-tree nutrition. *J. Exp. Bot.* 4:363–68

9. Bollard EG. 1960. Transport in the xylem. *Annu. Rev. Plant Physiol.* 11:141–66

10. Braun HJ. 1984. The significance of the accessory tissues of the hydrosystem for osmotic water shifting as the second principle of water ascent, with some thoughts concerning the evolution of trees. *IAWA Bull.* 5:275–94

11. Briggs GE, Robertson RN. 1948. Diffusion and absorption in disks of plant tissue. *New Phytol.* 47:265–83

12. Briggs GE, Robertson RN. 1957. Apparent free space. *Annu. Rev. Plant Physiol.* 8:11–30

13. Briggs LJ. 1950. Limiting negative pressure of water. *J. Appl. Phys.* 21:721–22

14. Canny MJ. 1988. Water pathways in wheat leaves. IV. The interpretation of images of a fluorescent apoplastic tracer. *Aust. J. Plant Physiol.* 15:541–55

15. Canny MJ. 1990. What becomes of the transpiration stream? *New Phytol.* 114: 341–68

16. Canny MJ. 1990. Fine veins of dicotyledon leaves as sites for enrichment of solutes of the xylem sap. *New Phytol.* 115:511–16

17. Canny MJ. 1990. Rates of apoplastic diffusion in wheat leaves. *New Phytol.* 116:263–68

18. Canny MJ. 1991. The xylem wedge as a functional unit—speculations on the consequences of flow in leaky tubes. *New Phytol.* 118:367–74

19. Canny MJ. 1993. The transpiration stream in the leaf apoplast: water and solutes. *Philos. Trans. R. Soc. London Ser. B* 341:87–100

20. Canny MJ. 1994. Potassium cycling in *Helianthus*: secondary vessel formation and the ions of the xylem sap. *Phil. Trans. R. Soc. London Ser. B.* In press

21. Canny MJ, Huang CX. 1994. Rates of diffusion into roots of maize. *New Phytol.* 126:11–19

22. Canny MJ, McCully ME. 1988. The xylem sap of maize roots: its collection, composition and formation. *Aust. J. Plant Physiol.* 15:557–66

23. Cavalcante VA, Döbereiner J. 1988. A new acid-tolerant nitrogen fixing bacterium associated with sugarcane. *Plant Soil* 108: 23–31

24. Clements HF. 1940. Movement of organic solutes in the sausage tree, *Kigelia africana*. *Plant Physiol.* 15:689–700

25. Conway EJ, Downey M. 1950. An outer metabolic region of the yeast cell. *Biochem. J.* 47:347–55

26. Cortes PM, Sinclair TR. 1985. The role of osmotic potential in spring sap flow of mature sugar maple trees (*Acer saccharum* Marsh). *J. Exp. Bot.* 36:12–24

27. Cosgrove DJ, Cleland RE. 1983. Solutes in

the free space of growing stem tissues. *Plant Physiol.* 72: 326–31

28. Crafts AS. 1956. The mechanism of translocation: methods of study with C^{14}-labelled 2,4-D. *Hilgardia* 26:287–334

29. Dacey JWH. 1980. Internal winds in water lilies: an adaptation for life in anaerobic sediments. *Science* 210:1017–19

30. Dong Z, Canny MJ, McCully ME, Roboredo MR, Cabadilla CF, et al. 1994. A nitrogen-fixing endophyte of sugarcane stems: a new role for the apoplast. *Plant Physiol.* 105:1139–47

31. Epstein E. 1972. *Mineral Nutrition of Plants*, p. 90. New York: Wiley. 412 pp.

32. Fahn A. 1982. *Plant Anatomy*, p. 37. New York: Pergamon. 544 pp. 3rd ed.

33. Frensch J, Stelzer R, Steudle E. 1992. NaCl uptake in roots of *Zea mays* seedlings: comparison of root pressure probe and EDX data. *Ann. Bot.* 70:543–50

34. Frey-Wyssling A. 1976. *The Plant Cell Wall*, p. 238. Berlin: Gebrüder Borntraeger. 294 pp. 3rd ed.

35. Geiger DR, Sovonick SA, Shock TL, Fellows RJ. 1974. Role of the free space in the translocation in sugar beet. *Plant Physiol.* 54:892–98

36. Gillis M, Kersters K, Hoste B, Janssens P, Kroppenstedt MP, et al. 1989. *Acetobacter diazotrophicus* sp. nov., a nitrogen-fixing acetic acid bacterium associated with sugarcane. *Int. J. Syst. Bacteriol.* 39:361–64

37. Gollan T, Schurr U, Schulze E-D. 1992. Stomatal response to drying soil in relation to changes in the xylem sap composition of *Helianthus annuus*. I. The concentration of cations, anions, amino acids in, and pH of the xylem sap. *Plant Cell Environ.* 15:551–59

38. Green JL, Durben DJ, Wolf GH, Angell CA. 1990. Water solutions at negative pressure: Raman spectroscopic study at −80 Megapascals. *Science* 249:649–52

38a. Grignon C, Sentenac H. 1991. pH and ionic conditions in the apoplast. *Annu. Rev. Plant Physiol. Plant Mol. Biol.* 42:103–28

39. Gunning BES, Steer MW. 1975. *Ultrastructure and the Biology of Plant Cells*, pp. 26–27. London: Arnold. 312 pp.

40. Haberlandt G. 1914. *Physiological Plant Anatomy.* Transl. M. Drummond, pp. 497–501. London: Macmillan. 777 pp.

41. Hardy PJ, Possingham JV. 1969. Studies on translocation of metabolites in the xylem of grapevine shoots. *J. Exp. Bot.* 20:325–35

42. Haynes RJ. 1980. Ion exchange properties of roots and ionic interactions within the root apoplasm: their role in ion accumulation by plants. *Bot. Rev.* 46:75–99

43. Höhn K. 1934. Die Bedeutung der Wurzelhaare für die Wasseraufnahme der Pflanzen. *Z. Bot.* 27:529–64

44. Hope AB, Stevens PG. 1952. Electric potential differences in bean roots and their relation to salt uptake. *Aust. J. Biol. Sci.* 5:335–43

45. Huang CX, Canny MJ, Oates K, McCully ME. 1994. Planing frozen hydrated specimens for SEM observation and EDX microanalysis. *Microsc. Res. Tech.* 28:67–74

46. Hülsbruch M. 1944. Fluoreszenzoptische Untersuchungen über den Wasserweg in der Wurzel. *Planta* 34:221–48

47. Jachetta JJ, Appleby AP, Boersma L. 1986. Use of the pressure vessel to measure concentrations of solutes in apoplastic and membrane-filtered symplastic sap in sunflower leaves. *Plant Physiol.* 82:995–99

48. Jeschke WD, Atkins CA, Pate JS. 1985. Ion circulation via phloem and xylem between root and shoot of nodulated white lupin. *J. Plant Physiol.* 117:319–30

49. Läuchli A. 1976. Apoplasmic transport in tissues. In *Encyclopaedia of Plant Physiology. Transport in Plants. II. B. Tissues and Organs,* ed. U Lüttge, MG Pitman, pp. 3–34. Berlin: Springer-Verlag

50. Mauseth JD. 1988. *Plant Anatomy*, p. 15. Menlo Park: Benjamin/Cummings. 560 pp.

51. McCully ME. 1994. The accumulation of high levels of potassium in the developing xylem elements of the roots of soybeans and some other dicotyledons. *Protoplasma.* In press

52. Mimura T, Yin Z-H, Wirth E, Dietz K-J. 1992. Phosphate transport and apoplastic phosphate homeostasis in barley leaves. *Plant Cell Physiol.* 33:563–68

52a. Moore PH, Cosgrove DJ. 1991. Developmental changes in cell and tissue water relations parameters in storage parenchyma of sugar cane. *Plant Physiol.* 96:794–801

53. Morrison TM. 1965. Xylem sap composition in woody plants. *Nature* 205:1027

54. Münch E. 1930. *Die Stoffbewegungen in der Pflanze*, p. 73. Jena: Fischer. 234 pp.

55. O'Dowd NA, Canny MJ. 1993. A simple method for locating the start of symplastic water flow (flumes) in leaves. *New Phytol.* 125:743–48

56. Parkinson CN. 1959. *Parkinson's Law or the Pursuit of Progress*, pp. 81–82. London: John Murray. 122 pp.

57. Pate JS. 1975. Exchange of solutes between phloem and xylem and circulation in the whole plant. In *Encyclopaedia of Plant Physiology, Transport in Plants. I. Phloem Transport*, ed. MH Zimmermann, JA Milburn, pp. 451–73. Berlin: Springer-Verlag. 535 pp.

58. Pate JS, Peoples MB, van Bel AJE, Kuo J, Atkins CA. 1985. Diurnal water balance of the cowpea fruit. *Plant Physiol.* 77: 148–56

59. Pfanz H, Dietz K-J. 1987. A fluorescence

236   CANNY

method for the determination of the apoplastic proton concentration in intact leaf tissues. *J. Plant Physiol.* 129:41–48

60. Pitman MG. 1965. Sodium and potassium uptake by seedlings of *Hordeum vulgare*. *Aust. J. Biol. Sci.* 18:10–25

61. Preston RD. 1974. *The Physical Biology of Plant Cell Walls*, pp. 381–82. London: Chapman & Hall. 491 pp.

62. Richter E, Ehwald R. 1983. Apoplastic mobility of sucrose in storage parenchyma of sugar beet. *Physiol. Plant.* 58:263–68

63. Robertson RN. 1952. Mechanism of absorption and transport of inorganic nutrients in plants. *Annu. Rev. Plant Physiol.* 3:1–25

64. Sachs J. 1865. *Handbuch der Experimental-Physiologie der Pflanzen*, pp. 210–21. Leipzig: Engelmann. 513 pp.

65. Sachs J. 1892. *Gesammelte Abhandlungen über Pflanzenphysiologie*, p. 476. Leipzig: Englemann. 1243 pp.

66. Sauter JJ. 1982. Efflux and re-absorption of sugars in the xylem. I. Seasonal changes in sucrose efflux in *Salix. Z. Pflanzenphysiol.* 106:325–36

67. Sauter JJ, Iten W, Zimmermann MH. 1973. Studies on the release of sugar into the vessels of sugar maple. *Can. J. Bot.* 51:1–8

68. Schlafke E. 1958. Kritische Untersuchungen zur Wanderung von Fluorochromen in Blättern. *Planta* 50:388–422

69. Schwenke H, Wagner E. 1992. A new concept of root exudation. *Plant Cell Environ.* 15:289–99

70. Smith AM. 1994. Xylem transport and the negative pressures sustainable by water. *Ann. Bot.* In press

71. Sovonick SA, Geiger DR, Fellows RJ. 1974. Evidence for active phloem loading in the minor veins of sugar beet. *Plant Physiol.* 54:886–91

72. Sperry J. 1991. Winter xylem embolism and spring recovery in *Betula cordifolia, Fagus grandifolia, Abies balsamea* and *Picea rubens*. In *Water Transport in Plants and Climatic Stress,* ed. A Raschi, M Borghetti, J Grace, pp. 86–98. Cambridge: Cambridge Univ. Press

73. Stelzer R, Kuo J, Koyro HW. 1988. Substitution of Na^+ by K^+ in tissues and root vacuoles of barley (*Hordeum vulgare* L. cv. Aramir). *J. Plant Physiol.* 132:671–77

74. Strugger S. 1938. Die lumineszenzmikroskopische Analyse des Transpirationsstromes in Parenchym. I. Die Methode und die ersten Beobachtungen. *Flora* 33:56–68

75. Strugger S. 1939. Die lumineszenzmikroskopische Analyse des Transpirationsstromes in Parenchym. II. Die Eigenschaften des Berberinsulfates und seine Spei-

76. Strugger S. 1939. Die lumineszenzmikroskopische Analyse des Transpirationsstromes in Parenchym. III. Untersuchungen an *Helxine soleirolii* Req. *Biol. Zentralbl.* 59:409–42

77. Strugger S. 1940. Studien über den Transpirationsstrom im Blatt von *Secale cereale* und *Triticum vulgare. Z. Bot.* 35:97–113

78. Strugger S. 1943. Der aufsteigende Saftstrom in der Pflanze. *Naturwissenschaften* 31:181–94

79. Terry ME, Bonner BA. 1980. An examination of centrifugation as a method of extracting an extracellular solution from peas, and its use for the study of indoleacetic acid-induced growth. *Plant Physiol.* 66: 321–25

80. Tetlow IJ, Farrar JF. 1993. Apoplastic sugar concentration and pH in barley leaves infected with brown rust. *J. Exp. Bot.* 44: 929–36

81. Tyree MT, Fiscus EL, Wullschleger SD, Dixon MA. 1986. Detection of xylem cavitation in corn under field conditions. *Plant Physiol.* 82:597–99

82. Tyree MT, Sperry JS. 1989. Vulnerability of xylem to cavitation and embolism. *Annu. Rev. Plant Physiol. Plant Mol. Biol.* 40:19–38

83. Van Bel AJE. 1993. Strategies of phloem loading. *Annu. Rev. Plant Physiol. Plant Mol. Biol.* 44:253–81

84. Varner J, Cassab GI. 1988. Cell wall proteins. *Annu. Rev. Plant Physiol. Plant Mol. Biol.* 39:321–53

85. Varney GT, Canny MJ. 1992. Rates of water uptake into the mature root system of maize plants. *New Phytol.* 123:775–86

86. Varney GT, McCully ME, Canny MJ. 1993. Sites of entry of water into the symplast of maize roots. *New Phytol.* 125:733–41

87. Wooley JT. 1983. Maintenance of air in intercellular spaces. *Plant Physiol.* 72:989–91

88. Yount DE, Gillary EW, Hoffman DC. 1984. A microscopic investigation of bubble formation nuclei. *J. Acoust. Soc. Am.* 76: 1511–21

89. Ziegler H. 1956. Untersuchungen über die Leitung und Sekretion der Assimilate. *Planta* 47:447–500

90. Zimmermann U, Haase A, Langbein D, Meinzer F. 1993. Mechanism of long-distance water transport in plants: a re-examination of some paradigms in the light of new evidence. *Philos. Trans. R. Soc. London Ser. B* 341:19–31

Annu. Rev. Plant Physiol. Plant Mol. Biol. 1995. 46:237–60

CELLULAR MECHANISMS OF ALUMINUM TOXICITY AND RESISTANCE IN PLANTS

Leon V. Kochian

US Plant, Soil and Nutrition Laboratory, USDA-ARS, Cornell University, Ithaca, New York 14853

KEY WORDS: aluminum resistance, metal stress, root, aluminum exclusion, aluminum tolerance

CONTENTS

ABSTRACT

Aluminum (Al) toxicity is the major factor limiting crop productivity on acid soils, which comprise up to 40% of the world's arable lands. Also, many native and crop plants exhibit genetic-based variability in Al sensitivity that has allowed plant breeders to develop Al-resistant crops. Considerable research has been directed to elucidating the mechanisms of Al toxicity and resistance in recent years. However, these processes are still poorly understood. This review examines our current understanding of cellular mechanisms of Al toxicity and resistance. The major symptom of Al toxicity is a rapid inhibition of root growth, which has been proposed to be caused by a number of different mechanisms, including Al interactions within the cell wall, the plasma membrane, or the root symplasm. Al resistance has been speculated to be the result of either exclusion of Al from the root apex or the tolerance of symplasmic Al. This review focuses on those mechanisms for which some evidence exists as well as on some controversial aspects of Al phytotoxicity and resistance.

INTRODUCTION

Native and cultivated plants exist in soil environments that can contain phyto-toxic levels of metals, including Al, Pb, Cd, Zn, Hg, Cu, Cr, Fe, Mn, and Ni. In recent years, plant biologists have become more aware of the importance of environmental issues in agriculture. This increased awareness has been re-flected in the interest expressed in plant responses to metal toxicities in the environment. Since the last review in this series on plant metal toxicities in 1978 (30), a significant portion of the research in this area has focused on the mechanisms of aluminum (Al) phytotoxicity and genetically based Al resis-tance. The intense research effort currently under way on Al toxicity indicates the agronomic importance of this problem: Al toxicity is the primary factor limiting crop productivity on acid soils, which comprise large areas of the world's lands, particularly in the tropics and subtropics (28, 30). Thus, it is an important factor limiting food production in many developing countries. To better develop Al-resistant crops via plant breeding and biotechnology, a con-siderable research effort is also under way to understand the fundamental processes that confer Al resistance in plants. Although research has been conducted on the cellular and physiological bases of Al toxicity and resistance, this research area is fraught with confusion and controversies. This review addresses our current understanding of these topics, with emphasis on the areas of controversy.

ALUMINUM TOXICITY

Aluminum is the most abundant metal and the third most common element in the earth's crust. At mildly acidic or neutral soil pH values, Al is primarily in the form of insoluble aluminosilicates or oxides. However, as soils become more acid, phytotoxic forms of Al are released into the soil solution to levels that affect root (and plant) growth. Thus, Al toxicity is the major growth-limit-ing factor for crop cultivation on acid soils (28, 29, 33). The initial and most dramatic symptom of Al toxicity is inhibition of root elongation (26, 108), which can occur within 1–2 hrs after exposure to Al (see e.g. 70, 114). In reviewing the literature, it is often difficult to separate primary or initial responses related to inhibition of root growth from secondary responses that arise as the result of a damaged root system (e.g. inhibition of mineral and water uptake). In recent years, a significant effort has been directed at identifying the initial processes disrupted by Al during the rapid onset of root growth inhibition, and this review focuses on the mechanistic basis for this acute Al toxicity. However, the performance of plants cultivated in acid soils is a function of the long-term health and growth of the root system (71), and there are examples of imperfect correlations between short-term Al-induced inhibition of root growth and long-term performance in the field (96). Although both short- and long-term aspects of Al toxicity contribute to plant responses to growth on acid soils, an overwhelming number of studies suggest that rapid inhibition of root growth is the primary Al toxicity symptom. For a more in-depth treatment of the different aspects of Al phytotoxicity, readers are directed to several recent reviews (26, 58, 82, 108).

What is the Toxic Al Species?

Because soluble Al can exist in many different ionic forms in aqueous solution, researchers attempting to elucidate mechanisms of Al phytotoxicity have been hampered by a lack of understanding and awareness of Al speciation. Most of the research on Al speciation in relation to Al phytotoxicity has focused justifiably on the Al chemistry of the solutions bathing the root, which are acidic soil solutions of fairly low ionic strength. However, recent evidence that Al can enter the symplasm of root cells fairly quickly (55, 56) indicates that Al speciation in a symplasmic solution with a pH above 7 and containing a mixture of inorganic and organic ions and macromolecules may also be an important aspect of Al phytotoxicity. If Al toxicity occurs as a result of Al interaction in the apoplasm, knowledge of Al speciation in the rhizosphere, or at least in the solution bathing the root, is of primary importance. However, if Al toxicity is the result of a symplasmic Al interaction, then the situation is more complex because it might be necessary to know which Al species is transported across the root-cell plasma membrane (PM) as well as which is the toxic Al species within the symplasm.

Al species that are relevant to phytotoxicity can be categorized into several different classes. With regard to the solution bathing the root, these classes include free or mononuclear forms of Al^{3+}, polynuclear Al, and Al as low molecular weight complexes. In the cellular cytoplasm, Al in either reversible or irreversible macromolecular complexes should also be considered (see 60–62). In acidic solutions (pH < 5.0), Al^{3+} exists as the octahedral hexahydrate, $Al(H_2O)_6^{3+}$, which by convention is usually called Al^{3+}. As the pH is increased, $Al(H_2O)_6^{3+}$ undergoes successive deprotonations to form $Al(OH)^{2+}$ and $Al(OH)_2^{+}$. At neutral pH the relatively insoluble $Al(OH)_3$ (gibbsite) forms, which limits the solubility of the other Al monomers (in the absence of other Al-binding ligands). As the solution pH is increased further to values commonly found in the cytoplasm ($\sim$ pH 7.4), the aluminate ion, $Al(OH)_4^-$, dominates Al speciation. When total Al activity is increased, particularly under conditions where the solution is partially neutralized, polynuclear forms of Al containing more than one Al atom can form, the most important of which is triskaideaaluminum $[AlO_4Al_{12}(OH)_{24}(H_2O)_{12}^{7+}]$, which is referred to as Al_{13} (15, 16, 73). Kinraide & Parker (48) have demonstrated that solutions remain mononuclear for many days when $\{Al^{3+}\}/\{H^+\}^3 \leq 10^{8.8}$, where braces denote ion activities. At values above $10^{8.8}$, significant formation of Al_{13} or precipitation of Al occurs.

Monomeric Al will also form low molecular weight complexes with a number of ligands, and Al^{3+} interacts most strongly with oxygen donor ligands such as carboxylate, phosphate, and sulfate groups. Thus, Al^{3+} forms low molecular weight complexes with organic acids, inorganic phosphate, polyphosphates (such as nucleotides), and sulfate. Equilibrium constants are available for many of the reactions involving Al hydrolysis and complexation, and these constants can be used to predict the concentrations and activities of these Al species, using computer speciation programs such as GEOCHEM-PC (74).

Aluminum can also bind reversibly to a number of macromolecules, including proteins, polynucleotides, and glycosides (60–62). It is often speculated that the major macromolecular interaction with Al^{3+} is binding of the polyphosphates in DNA, which could interfere with DNA replication. However, this interaction was dismissed by Martin (61, 62), who pointed out that Al^{3+} binds poorly to the weakly basic phosphate residues in DNA. Al^{3+} is more likely to bind tightly to some other molecule associated with DNA, such as phosphorylated proteins (e.g. histones).

Attempts have been made recently to demonstrate the toxicity of individual Al species in solution. This research has been hindered by a number of factors involving improper design of growth solutions, resulting in Al precipitation and formation of Al_{13}, and the lack of the proper information and tools to predict Al speciation. Another major hindrance has been the problem of collinearity between the activities of Al^{3+} and hydroxy-Al monomers in experiments where both Al concentration and pH were varied. It has been suggested that Al toxicity was better correlated with either the sum of all of the monomeric hydroxy-Al species or a combination of Al^{3+} and certain other monomeric hydroxy-Al species, instead of Al^{3+} alone (6, 18). It also has been suggested that for dicots, either $Al(OH)^{2+}$ or $Al(OH)_2^+$ were the phytotoxic species, and Al^{3+} was hypothesized to be much less toxic (6). For a more detailed review of this topic, see Reference 45.

Al^{3+} is the toxic species for wheat roots, based on an experimental design where solution Al^{3+} activities were increased while the activities of the hydroxy-Al species were decreased (45, 48, 72). In these experiments, care was taken to prevent the formation of Al_{13}, which appears to be considerably more toxic than Al^{3+} in the plant species that have been tested (72, 73, 97). It is unclear whether Al_{13} occurs naturally, and its contribution to soil toxicity is uncertain. Using NMR techniques, the presence of Al_{13} was reported in an acidic forest soil in Vermont (41). However, further research is needed to determine the relevance of Al_{13} to Al toxicity in an agricultural setting.

As mentioned above, Alva et al (6) have presented evidence that for dicotyledonous plants, $Al(OH)^{2+}$ or $Al(OH)_2^+$ were the primary toxic species, and Al^{3+} appears to be considerably less toxic. More recently, Kinraide & Parker investigated the apparent phytotoxicity of mononuclear hydroxy-Al to four different dicot species and also reanalyzed Alva et al's data. They also found that dicots may be more sensitive to $Al(OH)^{2+}$ and $Al(OH)_2^+$ than to Al^{3+} (49). This dramatic difference in Al response between dicots and monocots, which are sensitive to Al^{3+}, is puzzling, particularly since trivalent cations generally are more phytotoxic than are divalent and monovalent cations (19, 45). In these experiments, the researchers studied root toxicity as a function of increasing activities of hydroxy-Al monomers. These experiements used a series of growth solutions in which total Al concentration was held constant and the pH was increased incrementally in each solution of the series. As monomeric hydroxy-Al was increased, both Al^{3+} and H^+ activity were decreased concurrently. Kinraide & Parker have suggested that an alternative explanation for these results may be that an Al^{3+} toxicity response increases as pH rises, owing to a reduction in amelioration of Al^{3+} by H^+ (49).

It has been well documented that the addition of cations to growth solutions will ameliorate Al^{3+} toxicity, and this amelioration cannot be accounted for solely by ionic strength effects on Al^{3+} activity in solution. In particular, Ca^{2+} and Mg^{2+} are effective in ameliorating Al^{3+} toxicity (5, 47). A mechanistic explanation for the amelioration of Al^{3+} toxicity by other cations (including H^+) has been presented in models based on the interaction of Al^{3+} and other cations, either with the negatively charged surface of the root-cell plasma membrane or with the cell wall (32, 46, 50). Kinraide & coworkers have developed a model based on a modified Guoy-Chapman-Stern analysis of electrostatic interactions between Al^{3+}, other cations, and the negatively charged cell surface (46, 50). This model uses the assumptions that the negatively charged membrane surface will generate a strong attractive force for trivalent cations such as Al^{3+} and that other cations have the ability to reduce the negative surface charge of the PM, either through charge screening or charge neutralization. Based on this model, Al toxicity was best correlated with Al^{3+} activity (and not concentration) at the membrane surface and did not correlate well with bulk solution Al^{3+} activity. Furthermore, Al^{3+} toxicity was ameliorated in the following order of effectiveness: $H^+ \approx C^{3+} > C^{2+} > C^+$. This model indicates an important role for the negatively charged root-cell surface in toxicities not only to Al^{3+}, but also to other potentially toxic cations and anions such as La^{3+}, H^+, Na^+, and SeO_4^{2-} (46). The model also can be used to explain previously confusing observations such as the apparent phytotoxicity of $Al(OH)_2^+$ in dicots.

Nontoxic Al species appear to be formed by Al complexation with low molecular weight ligands such as sulfate, fluoride, and organic acids (45). This is particularly important for organic acids, such as citrate and malate, that have been implicated in mechanisms of Al resistance (11, 27, 66, 75, 90, 91). In a number of studies, the addition of citrate or malate either alleviates Al-induced inhibition of root growth (27, 70) or can alleviate potentially toxic effects of Al on membranes in vitro (107)

In summary, the demonstrated toxicities of Al^{3+} and Al_{13} suggest that Al toxicity might be limited to polyvalent Al cations with a charge greater than 2 (45), which is supported by observations that polyvalent cations (e.g. La^{3+}, Sc^{3+}, Ga^{3+}, $spermine^{4+}$), in general, are toxic (19, 45). The apparent toxicity of hydroxy-Al monomers should be viewed with caution in light of plausible alternative hypotheses.

The Root Apex is the Site of Al Toxicity

Direct evidence has recently demonstrated that the root apex is the primary site of Al-induced root growth inhibition. Ryan et al showed that only the terminal 2–3 mm of a maize root, which includes the meristem and root cap, needed to be exposed to Al to cause inhibition of root growth (92). If the entire root except the root apex was exposed to Al-containing solutions, root elongation rates were normal. Bennet and coworkers have suggested that the root cap is a site of perception of Al injury, based on anatomical studies of maize roots (12, 13). They observed rapid changes in the ultrastructure of root cap cells, including Al-induced alterations in the secretory pathway and suggested that Al could indirectly inhibit root growth via an unknown signal transduction pathway involving the root cap, apical meristem, hormones, and other putative signals. However, in the above-mentioned study by Ryan et al (92), Al inhibition of root growth was unaltered in decapped maize roots, which argues against a major role for the root cap in either Al toxicity or protection against Al toxicity.

Based on the above findings, research on mechanisms of Al toxicity and resistance should be directed to Al interactions within the root apex. Studies focusing on the entire root or root system probably will not provide the information necessary for elucidation of fundamental mechanisms of toxicity and resistance.

Mechanisms of Al Toxicity

Because Al can interact with a number of extracellular and intracellular structures, many different mechanisms of Al toxicity have been hypothesized.

These mechanisms include Al interactions within the root cell wall, Al disruption of the plasma membrane and PM transport processes, and Al interactions with symplasmic constituents such as calmodulin. No strong evidence exists in support of any one hypothesis. Several aspects of Al toxicity mechanisms have received considerable attention and/or are the subject of controversy; these are reviewed below.

CELL DIVISION VS CELL ELONGATION   The literature contains frequent reports that Al inhibition of root growth is the result of a reduction in cell division. Clarkson (19) was the first to show a correlation between Al inhibition of root elongation and cessation of mitotic figures in the root meristem. These results have been confirmed in other studies (68, 89, 114), and Al has been shown to inhibit DNA synthesis. In these studies Al inhibition of root growth either preceded or was concurrent with a decline in cell division. Because Al inhibition of cell division and DNA synthesis do not precede inhibition of root elongation, cell division cannot be the initial cause of root growth inhibition (114). The cell cycle in roots is approximately 24 hr (78), and a considerable lag between measurable inhibition of cell division and reduction in root growth must occur for cell division to be the cause. Root growth can be inhibited by Al within 1–2 hr. Because cell division is a relatively slow process and expansion of already divided cells is the major contributor to overall root elongation, during any 1–2 hr period cell division can only contribute 1–2% to the overall root extension (98). Thus, during the initial stages of Al inhibition of root growth, Al interactions with cell elongation must play a primary role. Because Al inhibits cell division, this inhibition could play a role in sustained root growth inhibition after the first 24 hr of Al exposure.

IS AL TOXICITY AN APOPLASMIC OR SYMPLASMIC PHENOMENON?   To understand the mechanistic basis for Al toxicity in roots, information about injury arising from apoplasmic vs symplasmic sites is important. There is no consensus on the cellular site of Al toxicity. Because Al^{3+} can bind and precipitate readily within the root cell wall and would be expected to cross the plasma membrane slowly, it is commonly thought that a majority of Al in the root is apoplasmic and resides in the wall or the mucigel surrounding the root apex (e.g. 20, 36). Because Al inhibits root growth rapidly and can alter certain physiological processes within minutes (e.g. Ca^{2+} influx; 37, 40), the apparent slow penetration of Al^{3+} into the root symplasm has led to hypotheses that focus on Al-induced damage within the wall (51) or on toxicity that is mediated through ion transporters or signal-transduction events initiated at the plasma membrane surface or root cap (12, 13, 37, 40, 81, 82). Techniques such as electron probe X-ray microanalysis have been used to investigate the relative distribution of Al across root sections, and the results have been equivocal. Some have found Al to be primarily extracellular, and others have found significant amounts of Al in cytoplasmic locations such as the nucleus (see e.g. 63, 65, 79). The problem with most of these studies is that Al exposures of many hours to days were often used, and high solution levels of Al (and P) could have resulted in Al precipitation and polymerization.

Recently, several labs have studied the absorption of Al into the apoplasm and symplasm of the wheat root apex in acidic simple salt solutions ($CaSO_4$ or $CaCl_2$). Taylor's lab was the first to attempt to quantify Al in the root symplasm and apoplasm by analysis of short-term Al uptake kinetics into the root apex of Al-sensitive and resistant wheat cultivars (118, 119). They found a rapid phase of Al uptake into the cell wall that could be desorbed with the Al-chelator, citrate. A slower phase of Al uptake appeared to include both uptake into the cell wall that was not easily desorbed and transport into the symplasm, although the symplasmic pool was identified indirectly, as a portion of the root Al that remained after desorption with citrate. It was estimated that 45–75% of the Al was apoplasmically located after a 3 hr exposure to Al. Interestingly, the addition of the uncoupler DNP increased Al uptake via this slower phase in the Al-resistant and not in the Al-sensitive wheat lines, which suggested that a metabolically coupled Al exclusion mechanism was operating in the resistant genotypes (118).

Tice et al (111) used a more direct method to quantify apoplasmic and symplasmic Al in Al-resistant and -sensitive wheat cultivars. Their method involved apoplasmic desorption of Al followed by freeze-thawing of the roots, which ruptured the plasma membrane. Subsequently, the roots were washed again to remove soluble symplasmic Al. The Al-fluorescent dye, morin, was used to determine the location of the residual Al in the root. Tice et al found that root tips from the Al-resistant cultivar always accumulated less Al in both the apoplasmic and symplasmic pools, when compared with the Al-sensitive genotype grown in the same solution. These results also suggested that an Al-exclusion mechanism operated in the resistant lines. Also, based on morin staining, 50–70% of the Al was estimated to be in the root apical symplasm after 2 days of growth in Al, which led the authors to speculate that a symplasmic Al-induced lesion was possible.

The strongest evidence for rapid Al uptake into the root apical symplasm comes from recent work on soybean roots (55, 56). A secondary-ion mass spectrometry (SIMS) system was used to analyze the spatial distribution of Al in cryosectioned and freeze-dried soybean root apices. The SIMS approach allows for the detection of low levels of tissue Al with a high degree of spatial resolution. After a 30 min exposure to a solution containing 38 μM Al^{3+} (activity), Al was found in the symplasm of the outer three layers of cells in the root apex. Intracellular Al concentrations for cells of the root periphery were estimated to be approximately 70 μM (55). After a 20 min or 2 hr exposure to

Al, an Al-sensitive cultivar accumulated significantly more Al in the entire root apex than did an Al-resistant cultivar (56). The entry of Al into the symplasm of the peripheral cells of the root apex occurs quickly enough for Al toxicity to involve a direct symplasmic interaction between Al and cellular components. Certainly more research should be conducted in this area.

AL TRANSPORT ACROSS THE ROOT-CELL PLASMA MEMBRANE   If we assume for strongly acid soils that Al^{3+} is the dominant transport species, Al^{3+} could cross the PM either via transport proteins that normally function in the absorption of other mineral ions or via fluid-phase or adsorption endocytosis. Al^{3+} has been shown to block ion channels in the root-cell PM (31, 39, 77). Ionic radius frequently is the most important parameter for predicting behavior of an ion in biology, and when Al^{3+} is compared with biologically important ions, it is closest in size to Fe^{3+} and Mg^{2+} (61, 62). Thus, Al^{3+} might cross the PM by sluggishly permeating a divalent cation channel normally functioning in Mg^{2+} uptake. Alternatively, at least in grasses, Al^{3+} might utilize the Fe(III)-phytosiderophore transport system to gain access to the cytoplasm. In response to Fe deficiency, grass roots release non-protein amino acids that not only bind Fe^{3+}, but can also complex other divalent and polyvalent cations fairly effectively. Because of the similarity in ionic radii for Fe^{3+} and Al^{3+}, phytosiderophores could chelate Al^{3+} and facilitate its absorption into the cytoplasm. For example, certain bacterial siderophores effectively chelate Al^{3+} (115). The other possible mechanism for Al^{3+} entry into the cytoplasm is endocytosis (2). Because Al^{3+} adsorbs tightly to the outer PM surface (3), this method would greatly enhance rates of Al^{3+} uptake via adsorption endocytosis.

The Al-citrate complex is another important Al species in terms of PM transport. Because citrate has been used to desorb Al from root cell walls and citrate release has been suggested as an important Al resistance mechanism in snapbean and maize (66, 75), the possibility should be considered that Al-citrate can cross the PM. At acidic pH values, the neutral Al-citrate complex dominates, and this neutral species may be able to pass through the lipid bilayer of the PM (61). However, Akeson & Munns showed that lipid bilayer permeation by the neutral Al-citrate complex was quite slow (1), and almost no uptake of this complex was measured in neuroblastoma cells (100). In addition, rates of Al-citrate uptake into maize roots were nearly undetectable (D Jones, personal communication). Therefore, significant Al-citrate transport is unlikely to occur across the root-cell PM.

THE ROLE OF AL INTERACTIONS WITH TRANSPORT PROCESSES IN AL TOXICITY
Considerable recent research and interest has been focused on Al interactions with the root-cell PM and PM transport processes. Electrophysiological investigations of the effects of Al on roots have indicated that Al toxicity is not the

result of overt membrane damage. In several different electrophysiological studies of severely Al-intoxicated wheat roots, root cells maintained reasonably negative transmembrane electrical potentials that were similar to those in control roots (40, 43, 67), which indicates a relatively intact PM. Also, both H^+ efflux via the PM H^+-ATPase and K^+ uptake appeared to function normally in roots exhibiting severe Al toxicity symptoms. Therefore, although Al induces dramatic alterations in root growth and morphology, Al does not cause significant damage to the PM. Al interactions with the PM must be relatively subtle and probably involve blockage of specific ion transport systems.

Given that the root apex is the primary site of Al toxicity, it is unfortunate that most of the studies on Al effects on root ion transport have focused on the mature root regions or the entire root system. Extracellular ion-selective microelectrodes have been used to map ion fluxes along roots (37, 40, 94) and have shown dramatic differences between the root apex and the mature root in terms of the important root ion transport systems. In general, a mature root has a smaller Ca^{2+} and Mg^{2+} influx than does the root apex. Also, a mature root has a large H^+ efflux and K^+ influx, whereas the root apex generally has a large H^+ influx and no K^+ uptake or a small K^+ efflux.

Investigations into Al inhibition of root apical ion transport processes that occur rapidly enough to be involved in Al toxicity have shown that only Ca^{2+} (and possibly Mg^{2+}) influx are inhibited rapidly at the root apex (37, 40, 94; JW Huang & LV Kochian, unpublished results). In studies comparing Al-resistant and -sensitive wheat cultivars, Huang et al have shown that Al inhibition of root apical Ca^{2+} influx correlated well with inhibition of root growth (37, 40). Based on the rapid inhibition and reversibility of the effects of Al^{3+} on Ca^{2+} influx, they suggested that Al^{3+} was blocking Ca^{2+} channels in the root-cell PM and this blockage could be involved in Al toxicity. The dramatic difference in sensitivity of the Ca^{2+} transport systems to Al^{3+} in the resistant and sensitive lines suggested that Ca^{2+} channels from these two cultivars might differ. Rengel & Elliott also found that Al^{3+} rapidly blocked Ca^{2+} influx into *Amaranthus tricolor* protoplasts, possibly via blockage of Ca^{2+} channels (83). More recently, both Huang et al (38, 39) and Pineros & Tester (77) have characterized a Ca^{2+} channel in the wheat root PM and have shown that micromolar activities of Al^{3+} effectively blocked this channel. Interestingly, Huang et al's work on Al^{3+} inhibition of Ca^{2+} influx at the root and membrane levels showed that although there was a differential inhibition of root apical Ca^{2+} influx in Al-sensitive vs -resistant cultivars, at the membrane level, Al^{3+} blocked Ca^{2+} channels equally well in membranes isolated from the Al-resistant and -sensitive cultivars (39). These results indicate that there are no major differences in the root PM Ca^{2+} channels from the two genotypes. The differential responses at the root level have been suggested to be the result of

Al-triggered organic acid release, resulting in Al exclusion from the root apex (39).

Although the correlation between rapid Al-induced inhibition of root apical Ca^{2+} influx and Al sensitivity to phytotoxicity provides an attractive (and speculative) framework for a possible Al toxicity mechanism, it is not clear how this blockage of Ca^{2+} channels would cause a rapid inhibition of root growth. Rengel has hypothesized that Al^{3+} blockage of Ca^{2+} channels could prevent the formation of important cytoplasmic Ca^{2+} transients needed for cell division to occur (81, 82). He has suggested that in the densely cytoplasmic cells of the root apex, the Ca^{2+} needed for these transients must come from the cell exterior, because the cells do not have a large central vacuole, which would be the alternative source for cytosolic Ca^{2+} in most plant cells. Although this is an imaginative hypothesis, there is no evidence that root apical cells cannot utilize Ca^{2+} stored in the ER or their numerous small vacuoles to generate these cytoplasmic Ca^{2+} transients.

Despite the attractiveness of the general hypothesis involving Al^{3+} blockage of Ca^{2+} influx in Al toxicity, it is based solely on correlative evidence. Two studies involving wheat roots and cells of the giant alga *Chara corallina* have demonstrated that it is possible to decouple Al^{3+} inhibition of root growth from Al^{3+} blockage of Ca^{2+} influx (80, 93). In Ryan et al's study, a growth solution containing a low Al^{3+} activity (1.5 μM) and a moderate Ca^{2+} activity (200 μM) stopped wheat root growth without inhibiting root apical Ca^{2+} influx. In addition, root growth was improved when other cations were added (e.g. Mg^{2+} and Na^+), and these cationic additions resulted in an inhibition of Ca^{2+} influx. Therefore, at least at moderate solution Ca^{2+} levels, the inhibition of root growth by low levels of Al^{3+} does not appear to be the result of inhibited Ca^{2+} influx. Since many acid soils can contain very low levels of Ca^{2+}, the general hypothesis might still hold at very low solution Ca^{2+} activities. This possibility awaits future verification.

Numerous reports have implicated Al inhibition of root K^+ uptake in Al toxicity (e.g. 31, 64, 67). However, most of these reports have focused on K^+ uptake into the mature root and not into the primary site of Al toxicity, the root apex. As reported above, a small K^+ efflux that is relatively insensitive to Al exposure is generally observed from the apex. Al^{3+} is probably an effective cation channel blocker, since it blocks both Ca^{2+} and K^+ channels in wheat root cells (31, 39, 77). Blockage of root K^+ channels in the long term could result in shoot K^+ deficiency symptoms; however, there is no evidence supporting the hypothesis that Al^{3+} inhibition of root apical K^+ influx is involved in root growth inhibition by Al^{3+}.

MECHANISMS OF SYMPLASMIC AL PHYTOTOXICITY   Since Lazof et al demonstrated that Al can enter the root cell symplasm after brief Al exposures (55, 56), the possibility must be considered more strongly that Al toxicity occurs because of Al interactions in the cytoplasm. The toxicity of cytosolic Al is often dismissed because in a cytoplasm with a pH of approximately 7.5, most of the monomeric Al would exist as $Al(OH)_4^-$, which appears to be nontoxic both in the apoplasm and symplasm (44, 61). Also, because of the abundance of potential ligands in the cytoplasm, a significant Al fraction could be tied up in nontoxic complexes. Thus, free cytoplasmic Al^{3+} activity in the cytoplasm is estimated to be in the picomolar to nanomolar range (61). Recent studies indicate that nanomolar or sub-nanomolar cytoplasmic Al^{3+} activities could be toxic as a result of Al interactions with sites regulated by Mg^{2+}. Because of ionic size similarities between Mg^{2+} and Al^{3+}, displacement of Mg^{2+} by Al^{3+} in biological systems is much more likely than are Al^{3+}-Ca^{2+} interactions (61, 62). For example, the complex of Al^{3+} with ATP^{4-} binds 1000 times more strongly to the enzyme hexokinase than does the Mg^{2+} complex, which is the proposed mechanism of Al^{3+} inhibition of this enzyme (61). The inhibition of Mg^{2+}-dependent processes by Al^{3+} could alter a number of cellular processes, including enzymes involved in phosphate transfer reactions, cytoskeletal interactions, and signal-transduction events.

Al interactions with the cytoskeleton   The changes in growth morphology induced by Al^{3+} in both roots and root hairs, which include a cessation of elongation and swelling of the root or root hair tip, indicate Al^{3+}-induced alterations in the cytoskeleton might occur. In humans and animals, Al^{3+}-induced neurological disorders are often associated with cytoskeletal defects (60). MacDonald et al found that Al^{3+} strongly promoted tubulin assembly into microtubules and inhibited subsequent Ca^{2+}-induced depolymerization of the microtubules (59, 60). The assembly of microtubules requires Mg^{2+}, which is believed to bind at GTP and GDP receptor sites. The association constant for Al^{3+} for this binding was 3×10^7 times that of Mg^{2+}, and an Al^{3+} activity of 4×10^{-10} M competed effectively against millimolar activities of Mg^{2+}. The rate of GTP hydrolysis by microtubules, which is important in the regulation of microtubule dynamics in vivo, was reduced dramatically by Al^{3+} binding. MacDonald et al suggest that the very low levels of Al^{3+} found in the cytoplasm of Al-challenged cells might be sufficient to disrupt the sensitive dynamics of microtubule formation and disassembly, which in turn could result in cellular disfunction.

Al INTERACTIONS WITH SIGNAL-TRANSDUCTION PROCESSES   Recent research from the medical field has indicated that cellular mechanisms of Al toxicity could involve interactions between Al^{3+} and components of the phosphoinositide signal transduction pathway that has been well characterized in animal cells (14) and is beginning to be understood in plant cells (21). For a complete account of this topic, readers are referred to Haug et al's comprehensive review of the evidence in favor of Al^{3+} interactions with this signal-transduction pathway (35). This signal-transduction pathway functions as follows: Interaction between an agonist and receptor at the PM surface causes signal transmission to a heterotrimeric G_p protein, which dissociates upon exchange of GTP for GDP. The G_α subunit then activates a PM-localized phospholipase C (PLC), which hydrolyzes the PM inositol lipid, phospatidylinositol-4,5-bisphosphate (PIP2), to produce two second messengers: inositol trisphosphate (IP_3), which is released into the cytoplasm, and diacyglycerol (DAG), which stays in the membrane and activates a protein kinase C (PKC). The primary role of IP_3 is to mobilize Ca^{2+} from endomembrane-localized stores via the opening of Ca^{2+} channels. This mobilization results in a transient rise in cytosolic Ca^{2+}, which can mediate a cellular response either indirectly, via binding and activation of calmodulin, or directly, through binding to specific enzymes. Al treatment interferes with this pathway by inhibiting the signal-induced rise in cytosolic IP_3 and, in some studies, by inhibiting the concomitant rise in cytosolic Ca^{2+} (95, 99, 101, 116). For example, Haug and coworkers have shown that exposure of neuroblastoma cells to micromolar levels of Al inhibited hormone-triggered IP_3 release and the concomitant rise in cytosolic Ca^{2+} (99, 101). When the results from a number of laboratories are considered together, it appears that Al^{3+} either binds a G_p protein (possibly at the Mg^{2+}-binding site) or interacts directly with phospholipase C. Al^{3+} does not appear to bind directly to and inactivate IP_3. Not all of the components of the phosphoinositide pathway have been identified in plants, but this signal-transduction pathway probably operates in plant cells, and Al^{3+} effects on the plant phosphoinositide pathway should be the subject of future research.

Al INTERACTIONS WITH CALMODULIN   Siegel & Haug first suggested that a primary target of Al phytotoxicity might be the Ca^{2+}-binding protein, calmodulin (102–104). This hypothesis has attracted considerable attention (26, 33, 34, 81, 108) and has spurred some research on the role of calmodulin and calmodulin-activated enzymes in Al toxicity and resistance (53, 105). The hypothesis is based on work from Haug's laboratory in which Al^{3+} effects on calmodulin structure were examined using several different techniques to investigate protein structure (e.g. fluorescence and circular dichroic spectroscopy, electron paramagnetic resonance), along with equilibrium dialysis measurements of Al^{3+} binding to calmodulin, and Al^{3+} effects on the activity of the calmodulin-activated enzyme, phosphodiesterase (102–104). It was reported that Al^{3+} binds stoichiometrically to the Ca^{2+}-binding sites in calmodulin with a 10-fold higher affinity than Ca^{2+} and elicits dramatic conformational changes in calmodulin that should inhibit its ability to activate other enzymes. In support of these observations, it was reported that Al^{3+} blocked the activity of calmodulin-activated phosphodiesterase. However, studies from other laboratories and reanalysis of the data used to develop this hypothesis cast serious doubt on the significance of Al^{3+}-calmodulin interactions in Al phytotoxicity. You & Nelson (117) repeated Siegel et al's studies (102) using EPR analysis of calmodulin spin-labeled at either methionine or tyrosine residues and found no effect of Al^{3+} on calmodulin structure at pH 6.5 or 5.5. They concluded that Al^{3+} does not interact with calmodulin at physiologically relevant pH values. Richardt et al (86) used flow dialysis techniques to investigate the effects of Al^{3+} on calmodulin-activated phosphodiesterase. They also found that Al^{3+} did not interact strongly with calmodulin, Al^{3+} did not effectively displace Ca^{2+} from calmodulin, and the observed inhibition of calmodulin-activated phosphodiesterase was the result of direct effects of Al^{3+} on phosphodiesterase and not on calmodulin. Finally, Martin (61) has criticized the equilibrium dialysis experiments used by Siegel and coworkers (102, 103) to calculate Al^{3+} binding constants for calmodulin because of errors in calculating the free Al^{3+} in their solutions. Martin's reanalysis of Siegel et al's data suggested that calmodulin would not tightly bind Al^{3+}. The obvious controversies in the literature concerning Al^{3+}-calmodulin interactions make it clear that considerations of calmodulin as a primary target for Al toxicity should be viewed with some skepticism.

ALUMINUM RESISTANCE

Several native and crop species exhibit significant genetically based variability in their responses to Al toxicity; this variability has been useful to plant breeders for the production of Al-resistant crops. I use the term Al resistance to denote plants that exhibit superior root growth, which ultimately results in enhanced plant vigor on acidic, Al-toxic soils or solutions. Al resistance can be divided into mechanisms facilitating Al exclusion from the root apex (Al exclusion) and mechanisms conferring the ability of plants to tolerate Al in the plant symplasm (Al tolerance). Many mechanisms of Al resistance have been proposed in the literature; most of these mechanisms are highly speculative, with little evidence supporting them. Specific mechanisms of Al resistance for which experimental evidence exists are examined below, and some controversial or confusing aspects of this topic are reconsidered. For a broader treatment of this topic, readers are referred to several reviews on Al resistance in plants (26, 109, 110).

The genetics of Al resistance has been looked at primarily in several important crop plants with a particular emphasis on wheat. In general, Al resistance is a dominant, multigeneic trait that can be controlled by one or more major genes and several minor genes (see e.g. 8, 9, 52, 84). Using ditelosomic lines of Chinese Spring wheat, Aniol & Gustafson found that Al resistance could be linked to seven different chromosome arms from the A, B, and D genomes (9), which indicates the potential genetic complexity of Al resistance. However, in some cases significant Al resistance can be conferred by single, dominant genes, and this has been used to produce near isogeneic lines that differed in Al resistance at a single locus (e.g. the *Alt1* locus; 24). Clearly, the potential exists for the operation of different cellular mechanisms that can confer Al resistance in plants, and it is unlikely that Al resistance is the result of a single mechanism in all plant species.

Al Exclusion Mechanisms

Several recent studies have shown that Al resistance in wheat involves exclusion of Al from the root apex. In both Al-resistant cultivars and near-isogeneic wheat lines, Al-sensitive genotypes accumulated three- to eight-fold more Al in the root apex (24, 88, 111), which is the critical site for Al toxicity (92). There was no difference in Al content of mature root tissues between resistant and sensitive genotypes (88). This mechanism appears to exclude Al from the entire root apex (both apoplasm and symplasm); therefore, the most likely mechanisms would involve either the release of Al-chelating ligands or root-induced increases in rhizosphere pH.

AL-INDUCED RELEASE OF ORGANIC ACIDS FROM THE ROOT APEX   One of the most interesting recent findings involves demonstrations of Al-induced release of root organic acids from Al-resistant genotypes. The first example of this response was reported in Al-resistant and -sensitive cultivars of snapbean (66). After growth in aseptic Al-containing solutions for 8 days, the root system of the Al-resistant cultivar released 70 times as much citrate as in the absence of Al and released 10 times as much citrate as did roots of the Al-sensitive cultivar +/− Al. The most complete recent work in this area comes from Delhaize and coworkers. Investigating Al resistance encoded by the *Alt1* locus, they found a good correlation between Al-triggered malate release, Al resistance, and Al exclusion from the root apex (24, 27). They have argued that the malate release protects the root apex by chelating Al^{3+}. Their conclusion is supported by the following observations: (*a*) Malate efflux is triggered only by Al^{3+} (and not Al_{13}, La^{3+}, Sc^{3+}, Mn^{2+}, or Zn^{2+}) (91); (*b*) malate efflux is localized solely to the root apex, the site of Al toxicity; (*c*) addition of malate to Al-toxic solutions protects Al-sensitive roots from toxicity; and (*d*) high rates of Al-triggered malate efflux cosegregates with Al resistance. Ryan and colleagues subsequently screened 36 different wheat cultivars for Al resistance and showed that Al-stimulated malate release correlated with Al resistance (90). Basu et al have shown similar differences in malate release from several other Al-resistant and -sensitive wheat cultivars (11). Thus, Al-stimulated malate release might be a general Al resistance mechanism in wheat. Also, as mentioned earlier, this mechanism appears to account for the differential inhibitory effects of Al on root apical Ca^{2+} influx in Al-resistant and -sensitive cultivars (39).

In Al-resistant and -sensitive maize cultivars, Pellet et al have shown that Al exposure triggered a rapid release of citrate in the Al-resistant genotype that was also localized within the root apex (75). The same low solution-activities of Al^{3+} (9 μM) that elicited the biggest difference in root growth between the Al-resistant and -sensitive lines also triggered the largest rates of root apical citrate release from the resistant line. Pellet et al also observed an Al-associated release of inorganic phosphate from the root apex of the resistant line, although it is not clear if this release was involved in Al exclusion.

An interesting aspect of this mechanism concerns the processes involved in Al perception and the associated organic acid release. Organic acids reside in the symplasm as organic anions, and release is presumed to be facilitated by anion channels. Ryan et al's recent work indicated that Al^{3+} exposure triggered an immediate release of malate in wheat that was inhibited by certain antagonists of anion channels (91). Thus, the primary mechanism probably involves an Al^{3+}-mediated opening of a malate-permeable channel (26). At this time, there is no information concerning the nature of the Al^{3+} trigger, which could involve a direct interaction with apoplasmic or symplasmic Al, or an indirect action via an Al-stimulated signal transduction event.

AL EXCLUSION VIA ALTERATIONS IN RHIZOSPHERE pH   The hypothesis that an Al^{3+} exclusion barrier could be created by plant-induced increases in rhizosphere pH has received much attention in the literature and conceptually is an attractive hypothesis (for review see 109, 110). However, no convincing evidence has been presented in support of this model. All the studies published in support of this model have measured changes in bulk solution pH, which is influenced primarily by the mature root regions and not the root apex, which is the site of Al toxicity. Even if this major problem is ignored, Taylor (110) points out that as many reports in the literature contradict this hypothesis as support it. More importantly, in the one study where spatial aspects of root apical rhizosphere pH were measured with pH microelectrodes in wheat cultivars that are often used as the standards for Al resistance (Atlas 66) and sensitivity (Scout 66), Miyasaka et al (67) found no difference in the rhizosphere pH at the root apex between the two cultivars in the absence of Al or during the first several hours of Al exposure.

THE ROLE OF THE CELL WALL IN AL EXCLUSION   It has been speculated that because of the fixed negative charges lining the water-filled pores within the cell wall, the root cell wall might be a site of Al^{3+} binding and immobilization, which would prevent Al^{3+} from associating with the PM or entering the symplasm. For this mechanism to be successful, cell-wall material would constantly have to be synthesized at the site of Al toxicity, which would be the case at the root apex. However, no experimental studies have tested this hypothesis, and it remains a purely speculative model. On the other hand, a number of soil-oriented scientists have hypothesized that Al resistance should be favored by low cell-wall cation-exchange capacity (CEC), which would result in the binding of lower amounts of Al within the cell wall (see e.g. 17, 69, 112). Inherent in this model is the assumption that Al binding within the wall is the first step leading to Al toxicity, presumably via Al uptake into the cell. However, based on studies in which the root cell wall CEC of the Al-resistant genotype was higher than that of the sensitive cultivar (4)—observations indicating that root CEC plays only a minor role in Al resistance (42)—and on studies in which root Al content greatly exceeded the root CEC (113), there is certainly no consensus among researchers that root CEC plays a role in Al exclusion. I also have technical and conceptual reservations concerning this hypothesis. First, in all of the studies where root CEC was measured, the entire root system and not the root apex was used for quantification of root CEC. Second, in many of the studies, root CEC was determined on intact roots after harsh treatments such as exposure of roots to HCl followed by immersion in plasmolyzing concentrations of KCl (1 molar). It appears that techniques originally designed to measure soil CEC have been used on roots, which should cause significant leakage of ions from the symplasm that could confound measurements of CEC. Finally, there is no real conceptual basis for the assumption that Al binding within the cell wall is a prerequisite for Al uptake. The actual transport site for uptake of ions across the PM would be the solution phase adjacent to the outer PM surface, and there is no reason to expect that the fairly tight Al binding onto the Donnan sites within the wall would have a large effect on this transport pool.

Al^{3+} EFFLUX ACROSS THE PLASMA MEMBRANE   Based on the observation that in some Al-resistant wheat cultivars, inhibition of metabolism causes an increase in root Al content, some researchers have hypothesized the existence of an active Al efflux at the root-cell PM (57, 110). Presumably this would be an Al^{3+}-translocating ATPase. Based on what we know about plant ion transporters, an Al^{3+}-ATPase would be an unlikely candidate for a transport system. First, the inwardly directed electrochemical gradient for Al^{3+} across the PM would be so large that the energy released via ATP hydrolysis would not be sufficient to drive the transport of Al^{3+} out of the cell. Second, given that the activity of Al^{3+} in the cytoplasm is probably in the sub-nanomolar range, a transport protein with a K_m for Al^{3+} of approximately 10^{-10} M is unlikely to exist. Based on what we now know about Al-stimulated release of organic acids as an Al exclusion mechanism in wheat, it is more likely that inhibition of metabolism causes a reduction in this process, with a concomitant rise in root Al levels.

Al Tolerance Mechanisms

Considerably more research has been conducted on Al exclusion mechanisms than on mechanisms of Al tolerance; hence, our understanding of mechanisms of Al tolerance are fragmentary at best.

IS AL TOLERANCE INDUCIBLE?   There is a considerable literature on the induction of low-molecular-weight binding peptides (e.g. the induction of phytochelatins by heavy metals such as Cd, Zn, and Pb) that might play a role in metal tolerance. In this case induction would mean the heavy metal activation of synthesis of metal-binding ligands, either at the gene or enzyme level. The possible induction of Al-binding compounds that confer Al tolerance has received some recent attention. Again, as in most other areas of Al resistance, little evidence exists in support of Al induction of tolerance. The most widely cited work in this area is that of Aniol (7), who reported that preincubation of wheat seedlings with sublethal levels of Al for 48 hr conferred substantial Al resistance to both Al-resistant and -sensitive wheat cultivars. Rincon & Gonzales repeated this same experiment more recently with different Al-resistant and -sensitive wheat cultivars and found no evidence for Al induction of Al tolerance (87). Several studies have showed that in the presence of moderate Al levels, root elongation declined initially followed by a partial recovery (23, 36, 71). These observations suggest induction of some type of Al resistance and warrant further study. However, mechanisms of Al resistance that are rapidly triggered by Al exposure, as in the case of organic acid release, could account for similar root growth responses to Al.

Several labs have shown that Al induces the synthesis of a number of different proteins in the root apex, but in most cases the proteins were found to be induced in both Al-resistant and -sensitive plants (22, 25, 76, 87). More recently, Basu et al identified, in the root apex of an Al-resistant cultivar, two 51-kDa microsomal proteins that were induced by 24–96 hr Al exposures (10). The same proteins were not found in the root apex of an Al-sensitive cultivar. Because differential Al resistance is observed in wheat roots within several hours, it will be necessary to demonstrate a rapid Al induction of a particular protein for it to play a role in Al tolerance.

Molecular investigations into Al tolerance are in their infancy. In Gardner's lab, differential hybridization screening of a root tip cDNA library from an Al-sensitive wheat genotype was used to clone seven cDNAs that were aluminum induced (85, 106). The putative proteins encoded by these cDNAs show homology to stress-related proteins in plants, including metallothionein-like proteins, phenylalanine ammonia lyase, proteinase inhibitors, and asparagine synthetase. Again, these genes were induced in both Al-resistant and -sensitive genotypes after 24–96 hr of Al exposure, which suggests that they are involved in Al-related stress responses and not tolerance. Larsen et al have taken a molecular genetic approach to studying Al resistance and have identified a number of Al-resistant and -sensitive mutants in *Arabidopsis thaliana* via screening of EMS-mutagenized populations of M2 seedlings (54). Genetic analysis indicates that Al tolerance is dominant or semidominant, whereas Al sensitivity is a recessive trait. The obvious goal of this approach is to identify the genes that confer Al resistance or sensitivity, via map-based cloning techniques.

CONCLUDING REMARKS

A recurrent theme in this review on Al toxicity and resistance in plants has been the lack of understanding of the fundamental processes underlying plant responses to Al. Nonetheless, significant progress has been made in these areas over the past 10 years, and there are many potentially exciting avenues for future research. In terms of Al phytotoxicity, I think it is safe to say that Al^{3+} is an important toxic species to plants, both in the rhizosphere and within the plant symplasm. The recent demonstration that Al can rapidly enter root cells forces researchers to consider more strongly the possibility of a symplasmic location for Al toxicity. Therefore, future research should be directed toward areas such as Al interactions with cytoskeletal components and signal-transduction processes, including the phosphoinositide pathway and control of cytosolic Ca^{2+}. It is heartening that we have begun to obtain solid evidence for a potentially important mechanism of Al resistance involving Al-stimulated release of organic acids from the root apex. The cellular mechanisms involved in the perception of the Al signal and the transduction of that signal to mediate organic acid release should provide for lively future research. Finally, molecular investigations of Al resistance and toxicity should provide important fundamental information and allow us to gain a deeper understanding of mechanisms of Al toxicity and resistance. The ultimate goal of this research will be the generation of crop species with greater Al resistance that farmers can use to cultivate the many acidic, Al-toxic soils throughout the world.

ACKNOWLEDGMENTS

I would like to thank Ms. Lisa Papernik and Drs. Jianwei Huang, David Jones, Thomas Kinraide, Didier Pellet, and Peter Ryan for providing helpful comments and thought-provoking discussions on this topic. I am also indebted to those colleagues that provided preprints and unpublished data. Financial support from the US Department of Agriculture-Agricultural Research Service, the USDA National Research Initiative Competitive Grant Program (#93-

37100-8874), and the Environmental Protection Agency (#R82-0001-010) is gratefully acknowledged.

Any *Annual Review* chapter, as well as any article cited in an *Annual Review* chapter, may be purchased from the Annual Reviews Preprints and Reprints service.
1-800-347-8007; 415-259-5017; email: arpr@class.org

Literature Cited

1. Akeson MA, Munns DN. 1989. Lipid bilayer permeation by neutral aluminum citrate and by three α-hydroxy carboxylic acids. *Biochim. Biophys. Acta* 984:200–6

2. Akeson MA, Munns DN. 1990. Uptake of aluminum into root cytoplasm: predicted rates for important solution complexes. *J. Plant Nutr.* 13:467–84

3. Akeson MA, Munns DN, Burau RG. 1989. Adsorption of Al^{3+} to phosphatidylcholine vesicles. *Biochim. Biophys. Acta* 986:33–40

4. Allan DL, Shann JR, Bertsch PM. 1990. Role of root cell walls in iron deficiency of soybean (*Glycine max*) and aluminium toxicity of wheat (*Triticum aestivum*). In *Plant Nutrition: Physiology and Applications*, ed. L van Beusichem, pp. 345–49. Dordrecht: Kluwer Academic

5. Alva AK, Asher CJ, Edwards DG. 1986. The role of calcium in alleviating aluminum toxicity. *Aust. J. Agric. Res.* 37:375–82

6. Alva AK, Edwards DG, Asher CJ, Blamey FP. 1986. Relationships between root length of soybean and calculated activities of aluminum monomers in nutrient solution. *Soil Sci. Soc. Am. J.* 50:959–62

7. Aniol A. 1984. Induction of aluminum tolerance in wheat seedlings by low doses of aluminum in the nutrient solution. *Plant Physiol.* 75:551–55

8. Aniol A. 1990. Genetics of tolerance to aluminum in wheat (*Triticum aestivum* L. Thell). *Plant Soil* 123:223–27

9. Aniol A, Gustafson JP. 1984. Chromosome location of genes controlling aluminum tolerance in wheat, rye, and triticale. *Can. J. Gen. Cytol.* 26:701–5

10. Basu A, Basu U, Taylor GJ. 1994. Induction of microsomal membrane proteins in roots of an aluminum-resistant cultivar of *Triticum aestivum* L. under conditions of aluminum stress. *Plant Physiol.* 104:1007–13

11. Basu U, Godbold D, Taylor GJ. 1994. Aluminum resistance in *Triticum aestivum* associated with enhanced exudation of malate. *J. Plant Physiol.* In press

12. Bennet RJ, Breen CM. 1991. The aluminium signal: new dimensions to mechanisms of aluminium tolerance. *Plant Soil* 134:153–66

13. Bennet RJ, Breen CM, Fey MV. 1987. The effects of aluminium on root cap function and root development in *Zea mays* L. *Environ. Exp. Bot.* 27:91–104

14. Berridge MJ. 1987. Inositol trisphosphate and diacylglycerol: two interacting second messengers. *Annu. Rev. Biochem.* 56:159–93

15. Bertsch PM. 1987. Conditions for Al_{13} polymer formation in partially neutralized aluminum solutions. *Soil Sci. Soc. Am. J.* 51:825–28

16. Bertsch PM, Thomas GW, Barnhisel RI. 1986. Characterization of hydroxy-aluminum solutions by aluminum-27 nuclear magnetic resonance spectroscopy. *Soil Sci. Soc. Am. J.* 50:825–30

17. Blamey FPC, Edmeades DC, Wheeler DM. 1990. Role of root cation-exchange capacity in differential aluminum tolerance of *Lotus* species. *J. Plant Nutr.* 13:729–44

18. Blamey FPC, Edwards DG, Asher CJ. 1983. Effects of aluminum, OH:Al and P:Al molar ratios, and ionic strength on soybean root elongation in solution culture. *Soil Sci.* 136:197–206

19. Clarkson DT. 1965. The effect of aluminium and some other trivalent metal cations on cell division in the root apices of *Allium cepa*. *Anal. Bot.* 29:309–15

20. Clarkson DT. 1967. Interactions between aluminium and phosphorus on root surfaces and cell wall material. *Plant Soil* 27:347–56

21. Coté GG, Crain RC. 1993. Biochemistry of phosphoinositides. *Annu. Rev. Plant Physiol. Plant. Mol. Biol.* 44:333–56

22. Cruz-Ortega R, Ownby JD. 1993. A protein similar to PR (pathogenesis-related) proteins is elicited by metal toxicity in wheat roots. *Physiol. Plant.* 89:211–19

23. Cumming JR, Cumming AB, Taylor GJ. 1992. Patterns of root respiration associated with the induction of aluminium tolerance in *Phaseolus vulgaris* L. *J. Exp. Bot.* 43:1075–81

24. Delhaize E, Craig S, Beaton CD, Bennet RJ, Jagadish VC, Randall PJ. 1993. Aluminum tolerance in wheat (*Triticum aestivum* L.) I. Uptake and distribution of aluminum in root apices. *Plant Physiol.* 103:685–93

25. Delhaize E, Higgins TJV, Randall PJ. 1991.

Aluminium tolerance in wheat: analysis of polypeptides in the root apices of tolerant and sensitive genotypes. In *Plant-Soil Interactions at Low pH,* ed. RJ Wright, VC Baligar, RP Murrmann, pp. 1071–80. Dordrecht: Kluwer Academic

26. Delhaize E, Ryan PR. 1994. Update: aluminum toxicity and tolerance in plants. *Plant Physiol.* In press

27. Delhaize E, Ryan PR, Randall PJ. 1993. Aluminum tolerance in wheat (*Triticum aestivum* L.) II. Aluminum-stimulated excretion of malic acid from root apices. *Plant Physiol.* 103:695–702

28. Foy CD. 1984. Physiological effects of hydrogen, aluminium, and manganese toxicities in acid soil. In *Soil Acidity and Liming,* ed. F Adams, pp. 57–97. Madison: Am. Soc. Agron.

29. Foy CD. 1988. Plant adaptation to acid, aluminium-toxic soils. *Commun. Soil Sci. Plant Anal.* 19:959–87

30. Foy CD, Chaney RL, White MC. 1978. The physiology of metal toxicity in plants. *Annu. Rev. Plant Physiol.* 29:511–66

31. Gassmann W, Schroeder JI. 1994. Inward-rectifying K^+ channels in root hairs of wheat. A mechanism for aluminum-sensitive low affinity K^+ uptake. *Plant Physiol.* 105:1399–1408

32. Grauer UE, Horst WJ. 1992. Modeling cation amelioration of aluminum phytotoxicity. *Soil Sci. Soc. Am. J.* 56:166–72

33. Haug A. 1984. Molecular aspects of aluminium toxicity. *Crit. Rev. Plant Sci.* 1:345–73

34. Haug A, Caldwell CR. 1985. Aluminum toxicity in plants: the role of the root plasma membrane and calmodulin. In *Frontiers of Membrane Research in Agriculture,* ed. JB St. John, PC Jackson, pp. 359–81. Totowa, NJ: Rowman & Allanheld

35. Haug A, Shi B, Vitorello V. 1994. Aluminum interaction with phosphoinositide-associated signal transduction. *Arch. Toxicol.* 68:1–7

36. Horst WJ, Wagner A, Marschner H. 1983. Effect of aluminium on root growth, cell-division rate and mineral element contents in roots of *Vigna unguiculata* genotypes. *Z. Pflanzenphysiol.* 109:95–103

37. Huang JW, Grunes DL, Kochian LV. 1992. Aluminum effects on the kinetics of calcium uptake into cells of the wheat root apex. Quantification of calcium fluxes using a calcium-selective vibrating microelectrode. *Planta* 188:414–21

38. Huang JW, Grunes DL, Kochian LV. 1994. Voltage-dependent Ca^{2+} influx into right-side-out plasma membrane vesicles isolated from wheat roots: characterization of a putative Ca^{2+} channel. *Proc. Natl. Acad. Sci. USA* 91:3473–77

39. Huang JW, Pellet DM, Papernik LA, Ko-

chian LV. 1994. Aluminum interactions with voltage-dependent calcium transport in plasma membrane vesicles isolated from roots of aluminum-sensitive and tolerant wheat cultivars. *Plant Physiol.* In press

40. Huang JW, Shaff JE, Grunes DL, Kochian LV. 1992. Aluminum effects on calcium fluxes at the root apex of aluminum-tolerant and aluminum-sensitive wheat cultivars. *Plant Physiol.* 98:230–37

41. Hunter D, Ross DS. 1991. Evidence for a phytotoxic hydroxy-aluminum polymer in organic soil horizons. *Science* 251:1056–58

42. Kennedy CW, Smith WC Jr, Ba MT. 1986. Root cation-exchange capacity of cotton cultivars in relation to aluminium toxicity. *J. Plant Nutr.* 9:1123–33

43. Kinraide TB. 1988. Proton extrusion by wheat roots exhibiting severe aluminum toxicity symptoms. *Plant Physiol.* 88:418–23

44. Kinraide TB. 1990. Assessing the rhizotoxicity of the aluminate ion, $Al(OH)_4^-$. *Plant Physiol.* 94:1620–25

45. Kinraide TB. 1991. Identity of the rhizotoxic aluminium species. *Plant Soil* 134:167–78

46. Kinraide TB. 1994. A Gouy-Chapman-Stern model for mineral rhizotoxicity. *Plant Physiol.* In press

47. Kinraide TB, Parker DR. 1987. Cation amelioration of aluminium toxicity in wheat. *Plant Physiol.* 83:546–51

48. Kinraide TB, Parker DR. 1989. Assessing the phytotoxicity of mononuclear hydroxy-aluminum. *Plant Cell Environ.* 12:479–87

49. Kinraide TB, Parker DR. 1990. Apparent phytotoxicity of mononuclear hydroxy-aluminum to four dicotyledenous species. *Physiol. Plant.* 79:283–88

50. Kinraide TB, Ryan PR, Kochian LV. 1992. Interactive effects of Al^{3+}, H^+, and other cations on root elongation considered in terms of cell-surface electrical potential. *Plant Physiol.* 99:1461–68

51. Klimashevskii EL, Dedov VM. 1975. Localization of the mechanism of growth-inhibiting action of Al^{3+} in elongating cell walls. *Soviet Plant Physiol.* 22:1040–46

52. Lafever HN, Campbell LG. 1978. Inheritance of aluminum tolerance in wheat. *Can. J. Gen. Cytol.* 20:355–64

53. Larkin PJ. 1987. Calmodulin levels are not responsible for aluminium tolerance in wheat. *Aust. J. Plant Physiol.* 14:377–85

54. Larsen PB, Tai CY, Howell SH, Kochian LV. 1994. Isolation and characterization of aluminum tolerant and sensitive *Arabidopsis* mutants. *Plant Physiol.* 105:S69

55. Lazof DB, Goldsmith JG, Rufty TW, Linton RW. 1994. Rapid uptake of aluminum into cells of intact soybean root tips: a

microanalytical study using secondary ion mass spectrometry. *Plant Physiol.* In press

56. Lazof DB, Rincon M, Rufty TW, MacKown CT, Carter TE. 1994. Aluminum accumulation and associated effects on $^{15}NO_3^-$ influx in roots of two soybean genotypes differing in Al tolerance. *Plant Soil.* In press

57. Lindberg S. 1990. Aluminium interactions with K^+ ($^{86}Rb^+$) and $^{45}Ca^{2+}$ fluxes in three cultivars of sugar beet (*Beta vulgaris*). *Physiol. Plant.* 79:275–82

58. Luttge U, Clarkson DT. 1992. Mineral nutrition: aluminium. *Prog. Bot.* 53:63–77

59. MacDonald TL, Humphreys WG, Martin RB. 1987. Promotion of tubulin assembly by aluminium ion *in vitro*. *Science* 236: 183–86

60. MacDonald TL, Martin RB. 1988. Aluminum ion in biological systems. *Trends Biol. Sci.* 13:15–19

61. Martin RB. 1988. Bioinorganic chemistry of aluminum. See Ref. 104a, pp. 2–57

62. Martin RB. 1992. Aluminium speciation in biology. In *Aluminium in Biology and Medicine*, ed. DJ Chadwick, J Whelan, pp. 5–25. New York: Wiley

63. Matsumoto H, Hirasawa E, Torikai H, Takahashi E. 1976. Localization of absorbed aluminium in pea root and its binding to nucleic acids. *Plant Cell Physiol.* 17:127–37

64. Matsumoto H, Yamaya T. 1986. Inhibition of potassium uptake and regulation of membrane-associated Mg^{2+}-ATPase activity of pea roots by aluminium. *Soil Sci. Plant Nutr.* 32:179–88

65. Millard MM, Foy CD, Coradetti CA, Reinsel MD. 1993. X-ray photoelectron spectroscopy surface analysis of aluminum ion stress in barley roots. *Plant Physiol.* 93: 578–83

66. Miyasaka SC, Buta JG, Howell RK, Foy CD. 1991. Mechanism of aluminum tolerance in snapbeans. Root exudation of citric acid. *Plant Physiol.* 96:737–43

67. Miyasaka SC, Kochian LV, Shaff JE, Foy CD. 1989. Mechanisms of aluminum tolerance in wheat. An investigation of genotypic differences in rhizosphere pH, K^+ and H^+ transport, and root-cell membrane potentials. *Plant Physiol.* 91:1188–96

68. Morimura S, Takahashi E, Matsumoto H. 1978. Association of aluminum with nuclei and inhibition of cell division in onion (*Allium cepa*) roots. *Z. Pflanzenernahr. Bodenk.* 88:395–401

69. Mugwira LM, Elgawhary SM. 1979. Aluminum accumulation and tolerance of triticale and wheat in relation to root cation exchange capacity. *Soil Sci. Soc. Am. J.* 43:736–40

70. Ownby JD, Popham HR. 1990. Citrate reverses the inhibition of wheat root growth caused by aluminum. *J. Plant Physiol.* 135: 588–91

71. Parker DR. 1994. Root growth analysis: an underutilized approach to understanding aluminum rhizotoxicity. *Plant Soil.* In press

72. Parker DR, Kinraide TB, Zelazny LW. 1988. Aluminum speciation and phytotoxicity in dilute hydroxy-aluminum solutions. *Soil Sci. Soc. Am. J.* 52:438–44

73. Parker DR, Kinraide TB, Zelazny LW. 1989. On the phytotoxicity of polynuclear hydroxy-aluminum complexes. *Soil Sci. Soc. Am. J.* 53:789–96

74. Parker DR, Zelazny LW, Kinraide TB. 1987. Improvements to the program Geochem. *Soil Sci. Soc. Am. J.* 51:488–91

75. Pellet DM, Grunes DL, Kochian LV. 1994. Organic acid exudation as an aluminum tolerance mechanism in maize (*Zea mays* L.). *Planta.* In press

76. Picton SJ, Richards KG, Gardner RC. 1991. Protein profiles in roottips of two wheat cultivars with differential tolerance to aluminium. In *Plant-Soil Interactions at Low pH*, ed. RJ Wright, VC Baligar, RP Murrmann, pp. 1063–10. Dordrecht: Kluwer Academic

77. Pineros M, Tester M. 1994. Characterization of a voltage dependent Ca^{2+}-selective channel from wheat roots. *Planta.* In press

78. Powell MJ, Davies MS, Francis D. 1986. The influence of zinc on the cell cycle in the root meristem of a zinc-tolerant and a non-tolerant cultivar of *Festuca rubra* L. *New Phytol.* 102:419–28

79. Rasmussen HP. 1968. Entry and distribution of aluminum in *Zea mays*. *Planta* 81: 28–37

80. Reid RJ, Tester MA, Smith FA. 1994. Calcium/aluminium interactions in the cell wall and plasma membrane of *Chara*. *Planta.* In press

81. Rengel Z. 1992. Disturbance of cell Ca^{2+} homeostasis as a primary trigger of Al toxicity syndrome. *Plant Cell Environ.* 15:931–38

82. Rengel Z. 1992. Role of calcium in aluminum toxicity. *New Phytol.* 121:499–513

83. Rengel Z, Elliott DC. 1992. Mechanism of aluminum inhibition of net $^{45}Ca^{2+}$ uptake by *Amaranthus* protoplasts. *Plant Physiol.* 98:632–38

84. Rhue RD, Grogan CO, Stockmeyer EW, Everett HL. 1978. Genetic control of aluminum tolerance in corn. *Crop Sci.* 18: 1063–67

85. Richards KD, Snowden KC, Gardner RC. 1994. *wali6* and *wali7*: genes induced by aluminum in wheat (*Triticum aestivum* L.) roots. *Plant Physiol.* 105:1455–56

86. Richardt G, Federolf G, Habermann E. 1985. The interaction of aluminum and other metal ions with calcium-calmodulin-

dependent phosphodiesterase. *Arch. Toxicol.* 57:257–59

87. Rincon M, Gonzales RA. 1991. Induction of protein synthesis by aluminium in wheat (*Triticum aestivum* L.) root tips. In *Plant-Soil Interactions at Low pH*, ed. RJ Wright, VC Baligar, RP Murrmann, pp. 851–58. Dordrecht: Kluwer Academic

88. Rincon M, Gonzales RA. 1992. Aluminum partitioning in intact roots of aluminum-tolerant and aluminum-sensitive wheat (*Triticum aestivum* L.) cultivars. *Plant Physiol.* 99:1021–28

89. Roy AK, Sharma A, Talukder G. 1989. A time-course study on effects of aluminium on mitotic cell division in *Allium sativum*. *Mutat. Res.* 227:221–26

90. Ryan PR, Delhaize E, Randall PJ. 1994. Malate efflux from root apices: evidence for a general mechanism of Al-tolerance in wheat. *Aust. J. Plant Physiol.* In press

91. Ryan PR, Delhaize E, Randall PJ. 1994. Characterisation of Al-stimulated efflux of malate from the apices of Al-tolerant wheat roots. *Planta*. In press

92. Ryan PR, DiTomaso JM, Kochian LV. 1993. Aluminium toxicity in roots: an investigation of spatial sensitivity and the role of the root cap. *J. Exp. Bot.* 44:437–46

93. Ryan PR, Kinraide TB, Kochian LV. 1994. Al^{3+}-Ca^{2+} interactions in aluminium rhizotoxicity. I. Inhibition of root growth is not caused by reduction of calcium uptake. *Planta* 192:98–103

94. Ryan PR, Shaff JE, Kochian LV. 1992. Aluminum toxicity in roots. Correlation among ionic currents, ion fluxes, and root elongation in aluminum-sensitive and aluminum-tolerant wheat cultivars. *Plant Physiol.* 99:1193–200

95. Schöfl C, Sanchez-Bueno A, Dixon CJ, Woods NM, Lee JAC, et al. 1990. Aluminium perturbs oscillatory phosphoinositide-mediated calcium signalling in hormone-stimulated hepatocytes. *Biochem. J.* 269:547–50

96. Scott BJ, Fisher JA. 1989. Selection of genotypes tolerant of aluminium and manganese. In *Soil Acidity and Plant Growth*, ed. AD Robson, pp. 167–203. Sydney: Academic

97. Shann JR, Bertsch PM. 1993. Differential cultivar response to polynuclear hydroxy-aluminum complexes. *Soil Sci. Soc. Am. J.* 57:116–20

98. Sharp RE, Silk WK, Hsiao TC. 1988. Growth of the maize primary root at low water potentials. *Plant Physiol.* 87:50–57

99. Shi B, Chou K, Haug A. 1993. Aluminum impacts elements of the phosphoinositide signalling pathway in neuroblastoma cells. *Mol. Cell. Biochem.* 121:109–18

100. Shi B, Haug A. 1990. Aluminum uptake by

neuroblastoma cells. *J. Neurochem.* 55:551–58

101. Shi B, Haug A. 1992. Aluminium interferes with signal transduction in neuroblastoma cells. *Pharmacol. Toxicol.* 71:308–13

102. Siegel N, Coughlin RT, Haug A. 1983. A thermodynamic and electron paramagnetic resonance study of structural changes in calmodulin induced by aluminum binding. *Biochem. Biophys. Res. Commun.* 115:512–17

103. Siegel N, Haug A. 1983. Aluminum interaction with calmodulin: evidence for altered structure and function from optical and enzymatic studies. *Biochem. Biophys. Acta* 744:36–45

104. Siegel N, Suhayda C, Haug A. 1982. Aluminum changes the conformation of calmodulin. *Physiol. Chem. Phys.* 14:165–67

104a. Sigel H, Sigel A, eds. 1988. *Metal Ions in Biological Systems: Aluminum and its Role in Biology*, Vol. 24. New York: Marcel Dekker

105. Slaski JJ. 1989. Effect of aluminum on calmodulin-dependent and calmodulin-independent NAD kinase activity in wheat (*Triticum aestivum* L.) root tips. *J. Plant Physiol.* 133:696–701

106. Snowden KC, Gardner RC. 1993. Five genes induced by aluminum in wheat (*Triticum aestivum* L.) roots. *Plant Physiol.* 103:855–61

107. Suhayda CG, Haug A. 1986. Organic acids reduce aluminum toxicity in maize root membranes. *Physiol. Plant.* 68:189–95

108. Taylor GJ. 1988. The physiology of aluminium phytotoxicity. See Ref. 104a, pp. 123–63

109. Taylor GJ. 1988. The physiology of aluminum tolerance. See Ref. 104a, pp. 165–98

110. Taylor GJ. 1991. Current views of the aluminum stress response: the physiological basis of tolerance. *Curr. Top. Plant Biochem. Physiol.* 10:57–93

111. Tice KR, Parker DR, DeMason DA. 1992. Operationally defined apoplastic and symplastic aluminum fractions in root tips of aluminum-intoxicated wheat. *Plant Physiol.* 100:309–18

112. Vose PB, Randall PJ. 1962. Resistance to aluminium and manganese toxicities in plants related to variety and cation-exchange capacity. *Nature* 196:85–86

113. Wagatsuma T, Ezoe Y. 1985. Effect of pH on ionic species of aluminum in medium and on aluminum toxicity under solution culture. *Soil Sci. Plant Nutr.* 31:547–61

114. Wallace SU, Anderson IC. 1984. Aluminium toxicity and DNA synthesis in wheat roots. *Agron. J.* 76:5–8

115. Winkler S, Ockels W, Budzikiewicz H, Korth H, Pulverer G. 1986. 2-hydroxy-4-methoxy-5-methylpyridine-N-oxide: an aluminum complexing metabolite from *Pseu-*

domonas cepacia. Z. Naturforsch. Teil C 41:807–8

116. Wood PC, Wojcikiewicz RJH, Burgess CM, Castleden CM, Nahorski SR. 1994. Aluminium inhibits muscarinic agonist-induced inositol 1,4,5-trisphosphate production and calcium mobilization in permeabilized SH-SY5Y human neuroblastoma cells. *J. Neurochem.* 62:2219–23

117. You G, Nelson DJ. 1991. Al^{3+} versus Ca^{2+} ion binding to methionine and tyrosine spin-labeled bovine brain calmodulin. *J.*

Inorg. Biochem. 41:283–91

118. Zhang GC, Taylor GJ. 1989. Kinetics of aluminum uptake by excised roots of aluminum-tolerant and aluminum-sensitive cultivars of *Triticum aestivum* L. *Plant Physiol.* 91:1094–99

119. Zhang GC, Taylor GJ. 1990. Kinetics of aluminum uptake in *Triticum aestivum* L. Identity of the linear phase of aluminum uptake by excised roots of aluminum-tolerant and aluminum-sensitive cultivars. *Plant Physiol.* 94:577–84

Annu. Rev. Plant Physiol. Plant Mol. Biol. 1995. 46:261–88
Copyright © 1995 by Annual Reviews Inc. All rights reserved

THE PLANT GOLGI APPARATUS:
Structure, Functional Organization and Trafficking Mechanisms

L. Andrew Staehelin and Ian Moore

University of Colorado, Department of MCD Biology, Boulder, Colorado 80309-0347

KEY WORDS: endoplasmic reticulum, Golgi stack, Golgi matrix, *trans*-Golgi network, partially coated reticulum, exocytosis, Rab proteins, SNAREs, clathrin

CONTENTS

ABSTRACT

This review focuses on three aspects of the Golgi apparatus: the structure of the Golgi stack–*trans* Golgi network (TGN) units, the biochemical and functional properties of the different cisternae, and the mechanism of vesicle-mediated transport. The plant Golgi apparatus consists of a set of dispersed, functionally independent Golgi-TGN units surrounded by a Golgi matrix. We postulate that this matrix controls vesicle transport and stabilizes the stacks as they are carried along by cytoplasmic streaming. The cell type–specific spatial organization of the N- and O-linked glycan and the complex polysaccharide synthesis pathways within Golgi stacks are reviewed. In two models we present an integrated overview of how ARF, Rab, vesicle coat, v- and t-SNARE, NSF, SNAP, GDI, and GDI-displacement-factor proteins bring about and regulate vesicular transport; where subtypes of these proteins operate in the secretory pathway; and which of these components have already been identified in plants.

INTRODUCTION

The Golgi apparatus is a carbohydrate assembly and processing organelle that occupies a central position in the secretory pathway. In plants, its principle functions include the biosynthesis of complex polysaccharides for the cell wall, the synthesis of glycolipids for the plasma membrane, and the assembly and processing of the oligosaccharide portions of cell wall, membrane, and storage glycoproteins. The enzymes that perform these functions are membrane bound and have their active sites facing the lumen of the flattened Golgi cisternae. The sugar nucleotides are transported from the cytosol to the enzymes by means of membrane transporters.

The functional unit of the plant Golgi apparatus appears to be the Golgi stack, its associated *trans*-Golgi network (TGN), and the Golgi matrix that surrounds both structures. Each Golgi stack consists of a series of cisternae that together contain a full set of enzymes and exhibit a distinct *cis*-to-*trans* polarity. Protein and lipid substrates are produced in the endoplasmic reticulum (ER) and delivered to the *cis*-Golgi cisterna in vesicles that bud from the transitional ER. Transport between the cisternae of each stack and to the TGN is mediated by nonselective carrier vesicles that appear to bud from the swollen rims of the cisternae and then fuse with the next. Upon arrival in the TGN, the completed products destined for the vacuole are collected into clathrin-coated vesicles by means of their sorting signals. The signal-less remaining molecules are packaged by default into a second set of vesicles that are targeted to the plasma membrane via the constitutive secretory pathway. Unlike the biosynthesis of glycoproteins, the biosynthesis of complex polysaccharides is confined exclusively to the Golgi stack–TGN units. All polysaccharides follow the default pathway to the cell surface, where they become integrated into the extracellular matrix.

Proper assembly and maintenance of the different compartments of the secretory pathway relies on a multitude of targeting and retention signals. Each protein contains one or more such signals (short peptide motifs or three-dimensional structural domains), which, in conjunction with specific receptors, help guide it to its proper destination. Retention signals are found on enzymatic and structural proteins that become part of the machinery associated with a given compartment of the secretory pathway. Vesicle-mediated transport between the different compartments is mediated by large, complex assemblies of proteins that drive the assembly and the budding of the transport vesicles and ensure that each vesicle fuses only with its target membrane. The regulation of these different processes involves numerous GTP-binding proteins of various classes.

This review focuses on three aspects of the Golgi apparatus of plants: (*a*) the structure of the Golgi stack–TGN units, (*b*) the biochemical and functional properties of the different cisternae, and (*c*) the machinery responsible for vesicle-mediated transport between the cisternae. Architectural aspects of the plant secretory pathway have been reviewed recently (41, 45, 106). Protein synthesis, protein trafficking, and protein targeting have been reviewed extensively elsewhere (9, 17, 27, 31, 60, 84, 121, 127), so they are only covered briefly here. See References 12, 38, and 58 for reviews of the literature on the synthesis of cell wall proteins, polysaccharides, and lignin; and see References 8 and 47 for reviews on plant exocytotic systems and vesicle transport.

THE GOLGI APPARATUS, TGN, AND PCR

The Plant Golgi Apparatus Consists of a Set of Dispersed, Functionally Independent Golgi Stack–TGN Units

The term Golgi apparatus refers to the complement of Golgi stacks (dictyosomes) and associated TGNs within a given cell. Each Golgi stack appears to contain a full set of Golgi enzymes and, together with its closely associated TGN that sorts and packages the products, corresponds to a functionally independent unit of the Golgi apparatus (80). Structurally, a Golgi stack is comprised of a series of flattened cisternae that exhibit a distinct polarity. This polarity reflects the underlying vectorial organization of the stacks, which receive protein products from the ER at the *cis* side and export them after processing and sorting from the *trans* side (100). Complex polysaccharides, which are synthesized exclusively in Golgi cisternae (80), are also assembled in a *cis*-to-*trans* direction (135).

In contrast to the clustered Golgi stacks of animal cells, those of plants are dispersed singly or in small clusters throughout the cytoplasm (61). Thin tubular connections between Golgi, ER, and TGN membranes have been reported in ZI-stained cotyledon cells of mung bean (46) but are absent in slime-secreting root-cap cells of maize and bean (61). Microtubule inhibitors have no major effects on the distribution or the structure of plant Golgi stacks (41), but treatment with cytochalasins, which disrupt actin filaments, causes an accumulation of secretory vesicles in the vicinity of the stacks (92). Whether this accumulation is caused by the disruption of direct actin-vesicle interactions, or results from the cessation of cytoplasmic streaming, which could be instrumental in dispersing the secretory vesicles, has yet to be determined.

The number of Golgi stacks per cell varies widely, depending on the plant, the stage of development of the cell, and the types and amounts of molecules produced by the endomembrane system. Interphase shoot apical meristem cells of *Epilobium hirsutum* contain on average 24 Golgi stacks (2) , whereas in onion root meristem cells there are approximately 400 per cell (37). An increase in Golgi stacks has also been observed during the maturation of root-cap cells (19). Unlike those of animal cells, plant Golgi stacks remain structurally intact and functionally active during mitosis to facilitate cell-plate formation. However, during the G2 phase of the cell cycle the stacks double in diameter and then divide in half by fission of the cisternae in a *cis*-to-*trans* direction (54).

A Golgi Matrix Encompasses Each Golgi Stack and its Trans-Golgi Network

As first noted by Sjöstrand & Hanzon (114), Golgi stacks are contained in a "ground substance," which subsequently became known as the Golgi zone of exclusion (79) because of its ability to exclude ribosomes from the vicinity of a Golgi stack. More recently, such zones have been observed in *Nicotiana sylvestris* and *Arabidopsis thaliana* root-tip cells preserved by ultrarapid freezing and shown to encompass both the actual Golgi stack and its TGN network (118). We propose that this zone of exclusion be named Golgi matrix (Figure 1) to emphasize its role as an integral and functional component of the Golgi apparatus.

Although no specific information is available on the composition of the Golgi matrix, several lines of indirect evidence suggest that this fine mesh network could contain actin- and spectrin-like molecules. Structural studies indicate that the Golgi matrix has a fine filamentous substructure (79, 118). Ribosome-excluding zones with similar ultrastructural features, such as the cortical actin networks or the zones around centrosomes of animal cells, have been shown to contain actin-related proteins (18). Furthermore, treatment of plant cells with actin filament–disrupting drugs such as cytochalasins causes Golgi stacks and secretory vesicles to aggregate into large ribosome-excluding clumps (79). Another candidate protein for the Golgi matrix is spectrin. Filamentous spectrin molecules are usually associated with plasma membranes,

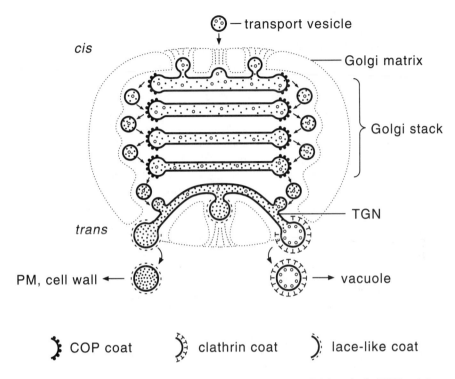

Figure 1 Diagram showing the spatial relationship between the Golgi stack, the TGN, and the Golgi matrix, as well as the distribution of COP, clathrin, and lace-like membrane coats. Lace-like coats have yet to be identified in plants.

but a homolog of erythrocyte β-spectrin has been shown to colocalize with Golgi markers in a variety of animal cell types (8a). More directed immunocyto-chemical and biochemical studies are needed to define with greater precision the chemical composition of this unexplored component of the Golgi appara-tus.

Although the functional significance of the Golgi matrix is still largely a matter of conjecture, the need for a protective and secretion-controlling Golgi covering is suggested by the following considerations: It is highly likely that the Golgi stacks of plant cells are dragged along and tumble around during cytoplasmic streaming. Such mobility could explain the frequent lack of a defined spatial relationship between ER cisternae and Golgi stacks (61, 110), as well as help to evenly distribute secretory vesicles to growing cell walls. Tumbling Golgi stacks carried around by cytoplasmic streaming may simply release secretory vesicles as they mature, and the strewn vesicles could find their way to fusion-competent plasma membrane or vacuolar domains by diffusion.

Because every Golgi stack–TGN unit appears to be contained within its own Golgi matrix (Figure 1), this network-like structure presumably affects the functioning of all Golgi stacks of a given cell. The Golgi matrix could serve (*a*) as a protective cocoon that prevents Golgi cisternae from shearing apart and the cisternae from becoming separated from the TGN during cytoplasmic streaming, (*b*) as a membrane-organizing network, (*c*) as a device for capturing ER-derived transitional vesicles, (*d*) as a vesicle-retaining network that prevents intercisternal transport vesicles from drifting away from the stack and becoming lost in the cytosol, and (*e*) as a quality-control structure for secretory vesicles, preventing immature, Golgi stack–derived vesicles, whose products have not been sorted and packaged in the TGN, from leaving the Golgi–TGN unit. Evidence in support of this last postulated function has come from studies of brefeldin A (BFA)–treated cultured sycamore maple cells under conditions (<10 µg/ml BFA) where the stacks persisted but the TGN appeared disrupted. In such cells, dense, *trans*-Golgi cisterna–derived, immature vesicles accumulate within the confines of the individual Golgi matrices, which causes them to expand and clump and fuse together. The large Golgi-dense vesicle aggregates created in this way appear to be held together by the enmeshed Golgi matrices (26). Even in cells in which Golgi cisternae have vesiculated in response to higher concentrations of BFA, the cisternal fragments do not disperse but instead accumulate in aggregates (105), presumably held together by Golgi matrix components. After BFA treatment the Golgi-derived vesicles of murine erythroleukemia cells also appear embedded in what might be the animal equivalent of a Golgi matrix (126).

Cryofixed Golgi Stacks: Differences in Cisternae and Stack Architecture

Golgi stacks of plant cells typically contain five to eight cisternae whose morphology and staining properties display a distinct *cis*-to-*trans* polarity (41). In cryofixed cells, the improved preservation of structural features has led to the distinction of three main morphological subtypes of Golgi cisternae: the *cis*, medial, and *trans* cisternae (51, 118, 135). Based on morphological criteria, most stacks appear to contain only one *cis*, but variable numbers of medial and *trans* cisternae (135).

The lightly stained *cis* cisterna occupies the end of the stack opposite to the most densely stained *trans* cisternae. Because of its fenestrated morphology and the presence of vesicular and tubular extensions toward the cytoplasm, the *cis* cisterna is often less readily discerned in electron micrographs. This variability in structure is consistent with its function as the site of entry of ER products into the Golgi stack and the site from which KDEL-, HDEL-, and RDEL-tagged ER proteins may be recycled back to the ER (23, 68, 85). The relative small size of *cis* cisternae compared to the extended *cis*-Golgi network of animal cells (94) probably reflects the reduced amount of protein trafficking in plant cells. Medial cisternae occupy the central region of the stack and possess a well-defined, flat central domain with parallel membranes and more fenestrated and bulbous margins. The contents of the relatively wide lumina exhibit a mottled staining pattern that often increases in density in a *cis*-to-*trans* direction. These variations in staining patterns probably reflect underlying differences in enzymatic composition of the membranes and of the products in the cisternae. The typifying features of *trans* cisternae are an osmotically collapsed lumen (118), where the membranes appear tightly pressed together, and darkly staining lumenal products. The contents of the highly fenestrated cisternal margins and the blebbing vesicles also stain heavily.

Several lines of incidental observations suggest that in the future it might be necessary to distinguish the terminal *trans* cisterna from the other *trans* cisternae because of some of its special properties. Indeed, the terminal *trans* cisterna may constitute a hybrid structure between *trans* cisternae and the TGN, at least in some cell types. For example, when cultured sycamore maple cells are exposed to the ionophore monensin, the first compartments to swell are the TGN and the transmost *trans* cisterna (134). Also, the transmost *trans* cisterna sometimes displays branched margins that appear more prominent in thin section images of Golgi stacks that lack a visible TGN (135). Electron microscopic data even indicate that the transmost *trans*-Golgi cisterna could substitute for the TGN, or even give rise to a replacement TGN when a TGN is used up, sloughed, or possibly converted to a partially coated reticulum (52, 91).

As discussed in greater detail below, vesicle transport between the different compartments of the secretory pathway is mediated and regulated by sets of proteins that form morphologically distinct coats on the budding vesicles. As evidenced in micrographs of cryofixed cells (118), COP types of coats (29) are typical of *cis*-, medial, and *trans*-Golgi cisternae, whereas clathrin coats (97) are characteristic of TGN membranes (Figure 1). The recently identified lace-like type of coated buds on the TGN of mammalian cells (66) will probably also be shown to be a typical component of exocytotic vesicles of plant cells, as depicted in Figure 1.

Intercisternal elements (78) are parallel fibrous elements (~10 nm repeat) that are sandwiched between Golgi cisternae. They are usually seen between *trans* cisternae of Golgi stacks of cells that are actively involved in slime synthesis, for example, in epidermal and peripheral cells of root tips or in cells destined to produce such molecules [e.g. columella cells (118)]. Golgi stacks of meristematic cells lack such structures. Their exact function is as yet unknown, but the alignment of freeze-fracture particles in *trans*-Golgi cisternae with such elements (118) suggests that they may anchor glycosyltransferase enzymes involved in slime synthesis and thereby prevent them from being dragged into forming secretory vesicles by the huge (10^3 kDa) slime molecules. This anchoring would also enhance the structural integrity of such Golgi stacks. No information is available on the nature of the molecular interactions that hold the cisternae of plant Golgi stacks together.

Aside from the common morphological features of Golgi stacks just described, major differences in appearance of Golgi stacks from different plant tissues have been reported (118). Thus, during the normal developmental progression of apical root meristem cells to columella and early and late peripheral cells, the Golgi stacks undergo a defined set of morphological changes suggestive of them being functionally retailored to meet the needs of each cell type. Support for the idea of a tissue-specific functional differentiation of Golgi stacks has come from immunocytochemical studies of root-cap cells of clover (71). In maize, the rate of mucilage production has been correlated with changes in soil impedance and with changes in the number of Golgi stacks and secretory vesicles (57). Changes in Golgi cisternal architecture, most notably increased fenestration, has been observed in response to heat-shock treatment (62).

TGN vs PCR: Two Functional Domains of the Same Organelle?

Structurally, the TGN (118) and the partially coated reticulum (PCR; 91) are two nearly indistinguishable organelles, which has led to considerable confusion in the plant literature (for review see 41, 70). Both are branched, tubular membrane systems with clathrin-coated terminal buds. The name PCR was initially given to all membrane networks of plant cells with clathrin-coated buds (91). However, many of the ramifying PCR networks are closely associated with the *trans* side of Golgi stacks and are included in the Golgi matrix (118). They appear to be functionally equivalent to the animal TGN, which serves as the sorting, packaging, and membrane-recycling compartment of the animal Golgi apparatus (42, 52). Consequently, it was proposed that the Golgi-associated PCR be called TGN (118). Support for this proposal has come from studies showing that the TGN contains both vacuolar and cell wall products and that isolated clathrin-coated vesicles contain seed storage proteins destined for the vacuole (9, 44, 98). Also, monensin-induced perturbation of the pH in the TGN leads to mistargeting of vacuolar proteins to the cell surface (20), presumably by affecting protein binding to putative sorting receptors (53).

In parallel with the just-described secretion studies, researchers investigating the endocytic uptake of cationized ferritin and heavy metal salts demonstrated that the PCR is an early station in the clathrin-coated vesicle–mediated endocytic pathway of plant cells (36, 95, 103). For example, in soybean protoplasts it takes plasma membrane–bound cationized ferritin approximately 2 min to arrive at the PCR, approximately 4 min to get to peripheral domains of some Golgi cisternae, approximately 6 min to enter multivesicular bodies, and approximately 1 h to accumulate in vacuoles (36). Interestingly, multivesicular bodies, Golgi cisternae, and vacuoles, but not PCR, stain for hydrolytic enzyme activity (95). These data are consistent with the idea that the PCR could serve as the functional equivalent of the early endosomal compartment of animal cells that sorts and recycles the components of endocytosed vesicles.

Based on our reading of the current status of the TGN vs PCR controversy, it appears that the two organelles serve different but overlapping functions in the exocytotic and endocytotic pathways of plants, respectively (Figure 2). In many instances they seem to be physically continuous structures (52), much like rough ER and smooth ER, and thus the segregation of their functional domains is not complete. For these reasons we propose that the TGN be defined as the clathrin-coated tubular network contained within the Golgi matrix of a given Golgi stack and whose primary function is to sort and package products of the secretory pathway. In contrast, the PCR would correspond to the parts of the clathrin-coated network that extend beyond the immediate boundaries of the Golgi matrix (52, 91) and function primarily in the endocytic pathway (36). The PCR, however, may still be embedded in a matrix connected to the Golgi matrix, and the TGN and the PCR will probably share functional activities.

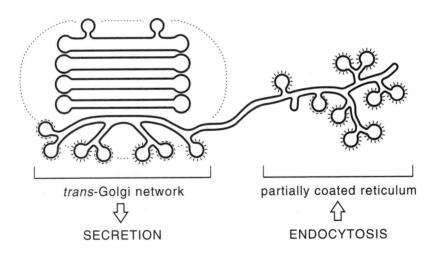

trans-Golgi network   partially coated reticulum

SECRETION   ENDOCYTOSIS

Figure 2 Model of the spatial relationship between TGN and PCR. The model postulates that TGN and PCR are two distinct functional domains of an interconnected membrane system much like the rough and smooth domains of the ER. Although the TGN serves primarily in the secretory pathway and the PCR in the endocytic pathway, the physical continuity between the two systems predicts that they should have overlapping functions and exchange components, which could be beneficial for membrane recycling. The model predicts that the PCR should have more clathrin-coated buds than does the TGN.

Lace-like coats (66) on exocytotic buds of the TGN (Figure 1) have not yet been reported in plants, but considering the need for a coating structure to produce vesicles of controlled size for transporting cell wall molecules by the default pathway (22) to the cell surface, their discovery seems imminent. Assuming that lace-like coated vesicles are involved strictly in the packaging of molecules destined for the plasma membrane and cell wall, our TGN-PCR model (Figure 2) predicts that in most cell types the TGN should have a balanced number of lace-like coated and clathrin-coated buds, whereas the PCR should have mostly clathrin-coated buds. Readers interested in the sorting of vacuolar proteins from the TGN are referred to some excellent reviews of the subject (9, 17, 84).

The Biosynthesis of N-Glycans Involves Enzymes in ER and Golgi Compartments

In plants, the majority of the enzymatic cell wall and membrane proteins are N-glycosylated. Assembly of the Glc_3Man_9 $(GlcNAc)_2$ glycan precursors occurs in the ER on lipid carriers known as dolichols (1). Upon completion, the 14-sugar oligosaccharide, which faces the ER lumen, is transferred *en bloc* and cotranslationally from the dolicholpyrophosphate carrier to an Asn in an Asn-Xaa-Ser/Thr sequence (Xaa being any amino acid other than proline) of the nascent polypeptide chain. The enzyme that catalyzes this reaction, oligosaccharyltransferase, is complexed in the ER membrane with two proteins, ribophorins I and II, the former having a dolichol-binding consensus sequence within its bilayer domain (63). Before the newly made glycoprotein leaves the ER, the three terminal glucose residues are removed efficiently by the hydrolytic enzymes glucosidase I and II (122). Considering the short half-life of these three glucose residues, one may ask, Why are they added in the first place? A possible explanation is that their presence signals the completion of the assembly of the 14-sugar oligosaccharide and promotes its binding to the oligosaccharide transferase.

Upon transfer of the glycoprotein to the Golgi, the enzyme mannosidase I trims the four α-1,2-linked Man residues (122). This action is needed for the addition of a GlcNAc residue by GlcNAc transferase I, which in turn is needed for the removal of two more Man residues by mannosidase II to yield $GlcNAcMan_3$ $(GlcNAc)_2$. Because all the steps up to this intermediate are common to plants and mammals, the enzymes performing these reactions in plants are probably localized to the *cis-* and medial-Golgi cisternae, as determined for mammalian cells (121).

N-glycans that are specific for either plants or animals arise from the enzymes associated with the second half of this pathway. Because the resulting molecules are more highly branched and contain additional sugars (in plants: fucose, xylose, and galactose, but not sialic acid), they are often called com-

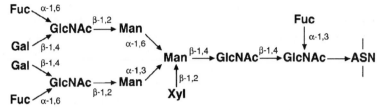

Figure 3   Structure of a typical, large, complex N-linked glycan of plant glycoproteins.

plex glycans (Figure 3), in contrast to the high-mannose earlier intermediates. There is some confusion in the literature concerning whether β-1,2-Xyl is added before or after α-1,3-Fuc. The preponderance of evidence, including immunocytochemical data, indicates that Xyl is added first (27), but there may be exceptions (121). In particular, the anti-β-Xyl and anti-α-Fuc labeling studies demonstrate that in sycamore maple cells the β-1,2-Xyl is added in medial- and the α-1,3-Fuc in *trans*-Golgi cisternae (67, 135, 34). Also, no plant glycoproteins containing an α-1,3-Fuc without a β-1,2-Xyl have been described, but the reverse situation is not uncommon, indicating that the xylosyltransferase does not require the presence of the α-1,3-Fuc residue for its activity. In contrast, when the GlcNAc transferase I enzyme is knocked out by the *cgl* mutation in *A. thaliana*, the conversion of high mannose to complex-type glycans is blocked at Man_5 (GlcNAc)$_2$. This block prevents the addition of β-1,2-Xyl by xylosyltransferase (128). Complementation of the *cgl* mutant with a human GlcNAc transferase I gene has been reported (39).

O-Glycan Biosynthesis is Initiated in Cis-Golgi Cisternae

The two major classes of O-linked glycoproteins of plants, the hydroxy-proline-rich glycoproteins (HRGPs) and the arabinogalactan proteins (AGPs; 113), serve structural and binding-recognition functions at the cell surface and in the cell wall. Most of these proteins are heavily glycosylated (50–98% of molecular mass), which protects them against proteolysis and often imposes on the peptide backbone an extended conformation. This latter property is responsible for the rod-like structure of extensin (119).

O-linked glycans are composed primarily of Ara and Gal, and appear to be less complex than N-glycans, but relatively few have been characterized to date. They are assembled by a series of glycosyltransferase enzymes that use sugar nucleotides to add sugar residues directly to Hyp and Ser residues and to the growing oligosaccharide chains. For example, in extensin, an HRGP, the Hyp residues are O-glycosylated with one to four Ara residues (123). In AGPs, the longer and sometimes branched side chains are O-linked via Gal to Hyp (4). Immunolabeling studies with antibodies that recognize the terminal residues of the arabinose-containing side chains of extensin (123) have demonstrated the presence of such epitopes in all Golgi cisternae, suggesting that Hyp arabinosylation starts in the *cis*-Golgi compartment (80). Prolyl-hydroxylation most likely occurs in the ER, but some of the biochemical fractionation data are also consistent with a Golgi location (27).

Complex Polysaccharides are Assembled in Golgi Stacks and the TGN

Complex polysaccharides constitute the bulk of the matrix molecules of primary cell walls and are the principle products of the Golgi of growing plant cells. The two major classes of complex polysaccharides are the neutral hemicelluloses and the acidic polysaccharides (5). Xyloglucan (XG), the most abundant hemicellulose of dicots, consists of a backbone of β-1,4 linked glycosyl residues and mostly Xyl or Xyl-Gal-Fuc side chains (48). Polygalacturonic acid/rhamnogalacturonan I (PGA/RGI), the dominant pectic polysaccharide of dicots, contains covalently linked blocks of PGA and RGI. During their assembly the PGA domains are methylesterified, only to be selectively deesterified in the cell wall. The backbone of RGI consists of alternating galacturonic acid and rhamnose residues, and half of the residues carry arabinan and arabinogalactan side chains (5).

The biosynthesis of complex cell wall polysaccharides has been studied using biochemical and immunocytochemical techniques. To date, the biochemical investigations have shown that these activities are associated with Golgi fractions and subfractions (38) and have led to the characterization of a number of glycosyltransferases, but none have been purified to homogeneity. For example, synthesis of the β-glucan backbone of XG also requires a simultaneously active XG-xylosyltransferase (40), which suggests that the two enzymes could be part of a multi-enzyme complex. In contrast, the addition of fucose and of galactose seems to be carried out by independent enzymes (30). In mammalian cells, the glycosyltransferases involved in the synthesis of terminal glycan sequences all appear to function as independent units and to have a lollipop configuration with a short cytoplasmic tail, one transmembrane helix, and a flexible stem or catalytic domain protruding into the Golgi lumen (85a, 90). Some progress has been made toward purifying the XG-fucosyltransferase (43) and the XG-glucosyltransferase (131) from pea, and an arabinosyl and a xylosyltransferase from bean (99). A Golgi membrane–bound nucleoside diphosphatase has been purified from rice (77).

Information on the compartmental organization of the biosynthetic pathways for PGA/RGI and for XG in Golgi stacks has come from recent immunocytochemical experiments with cryofixed sycamore cells (135) and root hairs of *Vicia faba* (111), using libraries of antibodies that recognize specific groups on these polysaccharides. The results are summarized in Figure 4. One of the

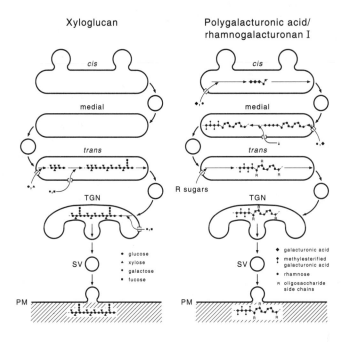

Figure 4   Models of the biosynthetic pathways of xyloglucan and polygalacturonic acid/rhamnogalacturonan I in the Golgi apparatus of sycamore maple suspension cultured cells as elucidated from immunolocalization studies. PM: plasma membrane; SV: secretory vesicles. Adapted from Reference 135.

most unexpected findings is that XG synthesis is confined to *trans*-Golgi cisternae and the TGN, with the backbone being synthesized in the former together with some side chains, and the fucosylation of additional side chains occurring in the latter. In contrast, the biosynthesis of PGA/RGI seems to encompass all Golgi cisternae. Thus, backbone assembly appears to start in *cis*- and to continue in medial-Golgi cisternae. Esterification of the PGA domains also seems to be a medial cisterna event, whereas the addition of side chains takes place in the *trans* cisternae. Immunocytochemical studies have also provided support for the concept of a tissue-specific retailoring of Golgi stacks in plants. Thus, when clover root-tip cells are labeled with anti-unesterified PGA/RGI antibodies, the gold particles are seen nearly exclusively over *cis*- and medial-Golgi cisternae in cortical cells (80) but mostly over *trans*-Golgi cisternae and the TGN in slime-secreting root-cap cells (71).

Golgi Membrane Proteins are Targeted by Signals in their Transmembrane Domains, and TGN Proteins by Signals in their Cytoplasmic Tails

As discussed in the preceding sections, *cis*-, medial-, and *trans*-Golgi cisternae contain different sets of processing enzymes. In particular, glycosyltransferases and glycosidases involved in the processing of glycoproteins, and complex polysaccharides appear to be enriched in specific Golgi subcompartments in the order in which they act. How are such integral membrane proteins targeted and retained? Mutational analysis of a number of cloned mammalian Golgi glycosyltransferases has shown that the targeting information is contained within the transmembrane domains and flanking sequences (87). However, there is no primary sequence homology within these domains. Instead, retention within a specific compartment appears to involve a degenerate signal defined by the three-dimensional structure and surface properties of the domains. In particular, resident Golgi proteins appear to be retained by the formation of oligomeric structures whose pH-dependent aggregation depends on polar residues that line one face of the transmembrane domains (72, 86). Golgi retention mechanisms in plants and animals are likely to be very similar, based on the demonstration that the *A. thaliana clg* (complex glycan) mutant can be complemented with a cDNA encoding the corresponding human GlcNAc transferase I gene (39). Targeting is never 100% accurate, however, and new enzymes are constantly on their way from their site of origin, the ER, to their site of action in the Golgi. Thus, one would expect that in localization studies the positioning of a specific enzyme to a specific type of cisterna will never be absolute.

In contrast to the transmembrane domain signals of Golgi proteins, the targeting signals of TGN proteins are associated with a tyrosine-containing sequence in the cytoplasmic tails (56). It is tempting to speculate that this sequence is a retrieval signal for TGN proteins much like the di-lysine retention motifs on transmembrane proteins of the ER (59). This idea is supported by the observation that the tyrosine-containing sequence also participates in the retrieval of TGN proteins from the plasma membrane (56).

VESICLE-MEDIATED TRANSPORT MECHANISMS

Transport Through the Golgi Stack: A Critical Evaluation of Models

Transport of newly synthesized membrane and soluble molecules through the Golgi stack occurs in a *cis*-to-*trans* direction. Three models have been proposed to account for this transport: the cisternal progression model (81), the tubular intercisternal transport model (3, 82), and the vesicle shuttle model (75, 101).

The cisternal progression model postulates that the transmost *trans* cisterna is used up during the packaging of secretory products and that to maintain a steady-state equilibrium of cisternae, new cisternae have to be assembled from ER-derived vesicles at the *cis* face. To maintain the functional integrity of the stacks during the turnover of the cisternae, either all of the enzymes are replaced every few minutes or they are conserved by a retrograde (perhaps vesicular) transport mechanism, while the products stay put and are processed during passage of the cisternae through the stack. The main evidence in support of this model has come from studies of cell wall scale-forming algae such as *Pleurochrysis scherffelii* (13). In such cells, large cellulose-containing scales are assembled in flat vesicles, which appear to be part of Golgi stacks, and then are discharged at the cell surface. Reexamination of the original micrographs, however, shows that these cells contain what appears to be a normal Golgi stack and that scale precursor assembly is initiated in a TGN-like compartment and completed in flattened secretory vesicles (see e.g. Figure 10 of Reference 13). The alignment of these flattened secretory vesicles with the Golgi stack may simply result from space constraints in this unicellular alga. Because most of the other supportive evidence is based on the kinetic interpretation of static micrographs (109), there seems to be little experimental evidence to support the cisternal progression product transport model. This discounting of the cisternal maturation model of Golgi transport does not preclude more long-term adjustments in numbers of cisternae per stack and interconversion of cisternal types when cisternal ratios change during development.

The recently formulated tubular intercisternal transport model (3, 82) postulates that transport between cisternal compartments is mediated by transient, fusogenic tubes, whose buds are stabilized by COP coats. Tubular connections between Golgi cisternae of adjacent stacks have been demonstrated in animal cells (see references in 3, 82). Furthermore, electron micrographs of Golgi stacks show very few coated vesicles compared to coated buds. Although we do not argue with these and other facts cited by the model's authors, we do not find their interpretations very convincing. In our opinion, if and when tubular connections form between successive cisternae of a stack, they may be more accidental in nature than main trafficking routes for secretory and membrane products. In particular, studies with the drug brefeldin A have shown that COP and clathrin coats prevent budding vesicles from becoming fusogenic tubes (65). The lack of free coated vesicles around Golgi stacks may simply reflect the kinetics of vesicle formation, fission, transport, and fusion. Thus, free vesicles may have a very short half-life compared to the budding structures, and chemical fixatives may preserve coated buds more efficiently than free vesicles.

The evidence in support of the vesicle shuttle model has been reviewed critically and in depth (75, 101). In our opinion, this is the only model that can account for nearly all transport data, in particular the in vitro transport findings and the selectivity of the anteriograde and retrograde transport mechanisms. A discussion of the protein players involved in vesicle-mediated transport is presented below.

The Vesicle Transport Machinery is Highly Conserved

Although the molecules that affect vesicle transport in the plant secretory pathway are still poorly characterized, they are probably similar to those of other eukaryotic cells whose vesicle transport mechanisms have been elucidated in detail. This expectation arises from two observations: (*a*) organisms as diverse as mammals and yeast use essentially the same mechanisms to regulate vesicle transport (11), and (*b*) key proteins of the yeast and mammalian vesicle transport machinery have close homologs in plants (24, 124).

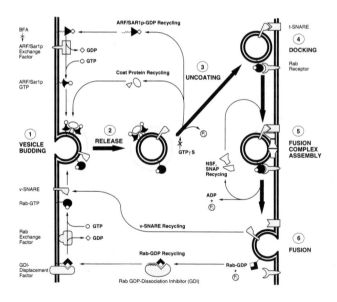

Figure 5   Model of vesicle-mediated transport between two membrane compartments. The model highlights where each of the main players (coat proteins, ARFs, Rabs, SNAPs, SNAREs, NSF, etc) act and how they interact to perform a specific function in the pathway. In the case of COP I–coated vesicles, docking may precede ARF GTP-hydrolysis and uncoating. The Rab receptor is hypothetical. See text for details. Compiled from References 32, 101, 108, 130, and 133.

The current understanding of vesicle transport mechanisms has come from the convergence of yeast genetics, neuronal-synapse biochemistry, and membrane transport assays in vivo and in permeabilized cells (32). This convergence has produced remarkable insights into the molecular mechanisms underlying these processes, as highlighted in Figure 5, which outlines the sites of action and function of the common molecular components responsible for the formation, targeting, and fusion of vesicles that mediate transport between

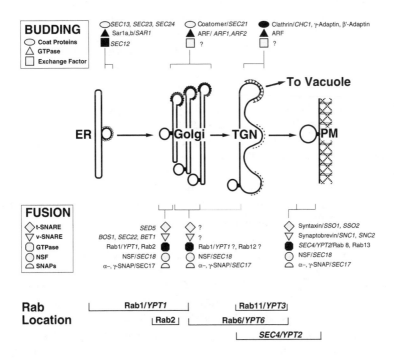

Figure 6 Diagram showing (*a*) where the different classes of vesicle budding and vesicle docking/fusion proteins illustrated in Figure 5 operate in the secretory pathway, and (*b*) how these classes are conserved across species boundaries. Note the repeated deployment of isoforms of these proteins along the pathway. The components in black are known to occur in plants. The mammalian factors and corresponding yeast genes (*italics*) are given where both have been characterized. Not shown is that *SEC23* has a mammalian counterpart and that yeast has a coatomer complex similar to mammalian coatomer (*SEC21* encodes yeast γ-COP). Question marks indicate proteins that have not been characterized, but whose presence is implied experimentally. The lower part of the figure shows where five classes of Rab protein known to occur in plants are likely to be found (based on the distribution and function of their mammalian and yeast counterparts). Plants also contain homologs of Rab5 and Rab7, which are associated with the endocytic pathway in other organisms. Compiled from References 6, 32, 33, and 108.

typical donor and acceptor compartments. Although the molecular analysis of plant secretory mechanisms still lags behind comparable studies of animal and yeast cells, significant progress has been made in identifying plant homologs of critical categories of trafficking-related proteins (Figure 6). A solid basis for future in vitro transport studies with plant membrane systems has also been created with the development of suitable in vitro transport assays (49) and the characterization of the first *clg* mutant of *A. thaliana* (128). With *clg*, which has a lesion in the GlcNAc transferase I gene, it should be possible to study intercisternal Golgi transport in vitro using methods similar to those pioneered by Rothman and coworkers (101) for mammalian Golgi.

COP and Clathrin Vesicle Coats are Composed of Many Proteins

Vesicle formation at a donor membrane involves the assembly of a protein coat around the budding vesicle from cytosolic precursors. Four types of vesicle coats have been recognized, one clathrin and three non-clathrin types known as COP I, COP II, and lace-like (108, 6, 66). Clathrin coats are associated with transport from the TGN to the vacuole or lysosome and with endocytosis at the plasma membrane. COP I coats mediate intra-Golgi, COP II, ER-to-Golgi, and lace-like coats, most probably TGN–to–plasma membrane transport. Electron microscope studies have confirmed these sites of action for clathrin and COP coats in plant cells (118).

Golgi COP I coats are composed of four major COP proteins (α, β, γ, δ; 160, 110, 97, and 61 kDa, respectively) and three smaller proteins (108). Prior to recruitment to the bud site, these proteins exist as a soluble cytosolic complex known as coatomer (108). ER COP II coats of yeast are formed from two multimeric cytosolic complexes, one consisting of Sec23p and Sec24p, and the other of Sec13p and a 150-kDa protein (6). No significant sequence similarities between COP I and COP II coat proteins have been reported to date, which suggests that these structures are analogous rather than homologous. Plant COPs have yet to be identified.

Clathrin coats contain two complexes: the clathrin triskelions, composed of three heavy and three light chains; and the assembly protein (AP) complexes. Two different AP complexes occur in mammals: AP1, comprising γ- and β'-adaptins plus two smaller subunits, is localized to the TGN; whereas AP2, comprising α- and β-adaptins and two smaller subunits, is confined to coated pits on the plasma membrane (108). β-COP is 17% identical to the β-type adaptins, so these proteins may perform similar functions in their respective coats (29). Plants are known to contain clathrin heavy and light chains (21, 69), and a homolog of the β/β'-adaptins has been detected immunologically and by southern hybridization in zucchini (55). Although this plant protein has been shown to be part of an AP1-type complex (55), it is also found in association with the plasma membrane (DG Robinson, personal communication). Thus, plants, like insects (14), may have only one form of β-adaptin.

Assembly of Vesicle Coats Involves Sar1p and ARF GTP-Binding Proteins

The assembly of a coat onto the membrane of a forming vesicle is regulated by small Ras-like GTP-binding proteins known as Sar1p and ARF, which become a major constituent of the coat (Figure 5, step 1). Sar1p-GTP is required for the assembly of COP II–type coats, and N-myristoylated ARF-GTP is required for COP I–type coats (6, 88). After vesicle release, hydrolysis of GTP bound to Sar1p or ARF converts each into its GDP-bound form, concomitant with a change in protein conformation. This leads to coat disassembly and solubilization of the coat proteins (Figure 5, step 3). GTP hydrolysis is catalyzed by GTPase activating proteins (GAPs). Sec23p is a GAP specific for Sar1p and is a constituent of COP II coats (6). Recycling of Sar1p/ARF-GDP onto the membrane involves replacement of the GDP with GTP, a reaction catalyzed by specific guanine-nucleotide exchange factors (GEFs) (Figure 5; 25, 50). Sec12p is a Sar1p-specific GEF in yeast (7), and *A. thaliana* contains homologs of Sec12p and Sar1p that are able to complement the corresponding yeast mutants (24). Upon recharging with GTP, Sar1p and ARF are ready for a new round of vesicle coat assembly. Heterotrimeric G-proteins also have been implicated in vesicle formation, but their exact roles are still unknown (74). A number of ARF homologs have been identified in plants but await functional characterization (76, 96).

Although the ARF-GEF activity has not been well characterized, it appears to be the direct or indirect target of BFA (25, 50). In the presence of BFA, neither ARF nor coatomer is recruited to the bud site, COP coats are not formed, and vesicles fail to pinch off. Instead, many buds seem to become fusogenic membrane tubes with the same targeting or fusion specificity as the normal vesicles (Figures 5 and 6). This causes Golgi cisternae to fuse with each other and with the ER (65). ARFs also have been implicated in transport between ER and Golgi cisternae, possibly in the retrograde Golgi-to-ER recycling pathway, in vesicle formation at the TGN, and in endosome function (108). Each of these systems is also sensitive to BFA.

Although no plant ARF-GEFs have been identified, Golgi stacks, TGN, and ER in plant cells are all affected by BFA to different extents (28, 73, 102, 105), suggesting that ARF-GEFs may also participate in vesicle trafficking in plants. The effects of BFA can differ considerably in cells from different plant species, which has led to some controversy. However, most differences can be traced to the different drug concentrations used (107).

Vesicle Targeting Involves Vesicle-Specific Ligand-Receptor Systems

Precision targeting of each vesicle is essential for the efficient transfer of products from one processing compartment to the next as well as for the retrograde recycling of lost proteins to their normal site of residence. Figure 5 (steps 4–6) presents an outline of our current understanding of the mechanism for discriminating vesicles of different origin and for promoting their fusion with the appropriate target membrane (32, 116, 117). Figure 6 shows how the same mechanism is adapted to promote vesicle formation at many sites, while retaining specificity of targeting.

Compartment-specific targeting is mediated primarily by integral membrane proteins called v-SNAREs and t-SNAREs (names derived from SNAP-receptors; see next section). v-SNAREs are thought to be recruited onto the nascent transport vesicles (Figure 5, step 1) and to act as the ligand that identifies the vesicle, whereas the t-SNAREs correspond to the target membrane receptors (Figure 5, step 4). Thus, each class of vesicle and its target membrane has its particular v-SNARE/t-SNARE pair (Figure 6). A second line of specificity, comparable to a verification system, may exist in the action of vesicle-specific GTP-binding proteins of the Rab family (133). Only when the specific targeting-docking systems are properly assembled are the conditions met for the recruitment of the membrane fusion proteins to the docking site.

The Vesicle Fusion Machinery is Shared by Different Transport Systems

The protein machinery that brings about fusion of a vesicle membrane with its target membrane apparently is shared by most if not all vesicle transport systems. The two main fusion complex factors are known as NSF (N-ethyl-maleimide Sensitive Factor) and SNAPs (Soluble NSF Attachment Proteins α and γ). The assembly of an NSF-SNAP fusion complex seems to depend on prior vesicle docking via the v/t-SNARE and the Rab/Rab-receptor pairs.

NSF is a soluble homotrimer of 76-kDa subunits, which exhibit ATPase activity but show no intrinsic affinity for membranes (132). NSF also does not associate with SNAPs when both are free in solution. However, when SNAPs become attached to a membrane by binding to v- and t-SNAREs they gain the ability to bind NSF (101, 117). Thus, the assembly of the fusion complex seems to be initiated by v/t-SNARE interaction, which is followed by the assembly of SNAPs and then NSF onto these pairs (Figure 5, step 5). Membrane fusion is facilitated, or perhaps precipitated, when NSF hydrolyzes ATP and is accompanied by disassembly of the particle (117).

In the model presented in Figure 5, the t-SNAREs reside permanently in the target membranes, possibly via retention signals in their transmembrane domains (130). How the v-SNAREs are recycled back to the donor membrane is unknown. Although no NSF, SNAP, or SNARE proteins have been demonstrated in plants, their occurrence in both yeast and animals suggests that it is only a matter of time until homologous proteins are discovered in plants.

Rab GTP-Binding Proteins may be Proofreaders of v/t-SNARE Pair Formation, and/or Activators/Deactivators of the Fusion Complex

The Rab subfamily of small Ras-like GTP-binding proteins appears to regulate the docking/fusion processes (133). Twenty-four classes of Rab proteins (Rab 1 to Rab 24) have been identified (33), and members of seven of these classes are known in plants (Figure 6). Each class analyzed to date appears to participate in the docking/fusion processes of a particular type of vesicle with its target membrane. This has been demonstrated by studying Rab mutants and by employing anti-Rab antibodies and Rab-derived peptides in in vitro assays (33, 133).

As shown in Figure 5, step 1, Rab proteins are recruited to the donor membrane in their GDP-bound form. Their binding to the membrane is mediated normally by two C20 geranylgeranyl tails attached to conserved cysteine residues at the C-terminus (133). Before insertion into the membrane, Rab-GDP exists in a soluble form complexed to the Rab-GDP dissociation inhibitor (GDI) protein (115, 125). The information that targets particular Rabs to particular membranes resides in the 30 to 40 C-terminal residues of the protein (15). Exchange of GDP for GTP by the Rab GEF proteins activates the Rabs in the donor membrane (115, 125), and Rab-GTP is incorporated into a vesicle as it forms (133) or possibly shortly thereafter (6). When the vesicle docks with the target membrane (Figure 5, step 4) Rab-GTP is thought to be bound by a receptor that may contribute to the specificity of docking. GTP hydrolysis by Rabs is stimulated by GAPs that are probably specific to each Rab class. Hydrolysis appears to occur just before (129) or just after (120) vesicle fusion, which sets the stage for Rab-GDP recycling (Figure 5, step 6). The Rab-GDI protein is able to remove a Rab-GDP from the target membrane and to inhibit GDP release from the Rab prior to return to the donor membrane.

What are the exact roles of Rabs and Rab-GTP hydrolysis? Two, nonexclusive models have been developed. The proofreading model proposes that the Rab-X/Rab-X-receptor interaction verifies the proper v/t-SNARE pair formation by completion of a v/t-SNARE-Rab/Rab receptor complex. In this context, Rab-GTP hydrolysis could trigger the fusion reaction (129), and in a properly assembled fusion complex, hydrolysis may be accelerated by a specific GAP. Recent data, however, suggest that fusion precedes GTP hydrolysis (120). This has led to the fusion-inactivation model, which postulates that Rab GTP associated with the v/t-SNARE-SNAP-NSF complex can promote membrane fusion without hydrolysis, and that subsequent Rab-GTP hydrolysis inactivates the fusion apparatus, thereby preventing additional, uncontrolled fusion activity (120).

As mentioned above, seven Rab classes have been identified in plants (reviewed in 10, 64, 83, 124), but their functions remain largely unexplored. They have been detected in tissues with high secretory activities such as meristems (35), pollen (I Moore, JS Schell & K Palme, unpublished data), and seedlings (89, 104), and some plant Rabs have been shown to complement corresponding yeast mutants (10, 16; I Moore, JS Schell & K Palme, unpublished data). The best evidence for a role in vesicle-mediated transport comes from the demonstration that antisense constructs against legume Rab 1 and Rab 7 prevent peribacteroid membrane formation in root nodules (16).

CONCLUDING REMARKS

The Future?

The goal of this review was to provide a conceptual overview of the structure and functional organization of the plant Golgi apparatus and of how vesicle-mediated transport between Golgi cisternae works at the molecular level. The plant Golgi apparatus has been the subject of numerous structural and biochemical studies since the mid 1950s, but insights into its molecular organization have come only during the past ten years. Cryofixation methods have provided reliable information on the structure and macromolecular organization of the Golgi stack–TGN units and their transport intermediates; biochemical and immunocytochemical investigations have led to new insights into the functional properties of the organelles and their subdomain organization; and genetic and molecular biological approaches have led to the identification of a number of proteins associated with trafficking and targeting functions. What will the future bring?

A taste of what is to come can be seen in the above sections devoted to vesicular trafficking. An important finding of that work is the high degree of conservation of components that bring about vesicle transport between the many different sets of membrane components within a given organism, as well as between highly divergent organisms such as mammals, yeast, and plants. This conservation is caused by the fact that each component of the transport machinery is an essential part of a larger complex; the discovery of one highly conserved component of such a complex suggests that others will be discovered too. Thus, plant equivalents for most of the components listed in Figure 6 are likely to be discovered in the next five years, but the functional characterization of these components will take much longer.

A glance at Figure 6 shows a preponderance of Ras-like GTP-binding proteins among the plant homologs currently known. This probably does not reflect the extent of the similarity between plant and other transport mechanisms, but rather the relative ease of isolation of cDNAs encoding these proteins owing to highly conserved elements within their structure. What are the prospects for isolating other components of the secretory machinery, especially those that are less well conserved? Functional complementation of yeast *sec* mutants is likely to be a useful way to isolate such cDNAs and to test the significance of sequence homologies observed in cDNAs isolated by other means (24). The random cDNA and the *A. thaliana* genome sequencing projects will soon be generating a large number of such clones.

Classical genetics has not yet been brought to bear on the plant secretory pathway. However, it has emerged that the *A. thaliana emb30* embryo lethal mutation is in a gene with homology to the yeast SEC7 locus (112). Loss of Sec7p in yeast leads to disruption of ER-to-Golgi transport and Golgi organization (93). In the *emb30* mutant cell division, expansion and wall deposition appear perturbed, consistent with the lesion affecting the secretory pathway and thereby plant growth and development. However, the functional similarity between Sec7p and *emb30* remains to be demonstrated.

Little is known at the molecular level about the Golgi apparatus and the TGN of plants. Few enzymes and transport proteins have been purified to homogeneity, and information on the genes that code for these proteins is similarly scarce. However, when such information becomes available it will be possible to address several questions. For example, how does a specific enzyme get targeted to a specific type of Golgi cisterna; can lost enzymes be recycled; how do Golgi stacks become retailored during development; what holds the cisternae of a stack together; and how is the TGN functionally related to the PCR? Finally, the Golgi matrix is a structure whose biochemical composition and molecular properties have yet to be explored in any system. Does this matrix really control vesicle transport as envisaged in Figure 3? These and many other questions should keep plant researchers challenged well into the next century.

ACKNOWLEDGMENTS

Thanks are due to Drs. Thomas Giddings and Dan La Flamme for reading the final draft of the manuscript, and to Janet Meehl for her untiring help in the preparation of the manuscript. Supported by NIH grant GM18639 to LAS.

Literature Cited

1. Abeijon C, Hirschberg C. 1992. Topography of glycosylation reactions in the endoplasmic reticulum. *Trends Biochem. Sci.* 17:32–36
2. Anton-Lamprecht I. 1967. Anzahl und Vermehrung der Zellorganellen im Scheitelmeristem von *Epilobium. Ber. Dtsch. Bot. Ges.*80:747–54
3. Ayala JS. 1994. Transport and internal organization of membranes: vesicles, membrane networks and GTP-binding proteins. *J. Cell Sci.* 107:753–63
4. Bacic A, Churms SC, Stephen AM, Cohen PB, Fincher GB. 1987. Fine structure of the arabinogalactan-protein from *Lolium multiflorum. Carbohydr. Res.* 162:85–94
5. Bacic A, Harris PJ, Stone BA. 1988. Isolation and characterization of plant cell walls and cell wall components. In *The Biochemistry of Plants,* ed. J Priess, 14:297–372. New York: Academic
6. Barlowe C, Orci L, Yeung T, Hosobuchi M, Hamamoto S, et al. 1994. COPII: a membrane coat formed by sec proteins that drive vesicle budding from the endoplasmic reticulum. *Cell* 77:895–907
7. Barlowe C, Schekman R. 1993. *SEC*12 encodes a guanine-nucleotide-exchange factor essential for transport vesicle budding from the ER. *Nature* 365:347–49
8. Battey NH, Blackbourn HD. 1993. The control of exocytosis in plant cells. *New Phytol.* 125:307–38
8a. Beck KA, Buchanan JA, Malhotra V, Nelson WJ. 1994. *J. Cell Biol.* 127:707–23
9. Bednarek SY, Raikhel NV. 1992. Intracellular trafficking of secretory proteins. *Plant Mol. Biol.* 20:133–50
10. Bednarek SY, Reynolds TL, Schroeder M, Grabowski R, Hengst L, et al. 1994. A small GTP-binding protein from *Arabidopsis thaliana* functionally complements the yeast *YPT6* null mutant. *Plant Physiol.* 104:591–96
11. Bennett MK, Scheller RH. 1993. The molecular machinery for secretion is conserved from yeast to neurons. *Proc. Natl. Acad. Sci. USA* 90:2559–63
12. Bolwell GP. 1988. Synthesis of cell wall components: aspects of control. *Phytochemistry* 27:1235–53
13. Brown RM Jr, Herth W, Franke WW, Romanovicz D. 1973. The role of the Golgi apparatus in the biosynthesis and secretion of a cellulosic glycoprotein in *Pleurochrysis*: a model system for the synthesis of structural polysaccharides. In *Biogenesis of Plant Cell Wall Polysaccharides,* ed. F Loewus, pp. 207–57. New York: Academic
14. Camidge DR, Pearse BMF. 1994. Cloning of Drosophila β-adaptin and its localization

on expression in mammalian cells. *J. Cell Sci.* 107:709–18
15. Chavrier P, Gorvel J-P, Stelzer E, Simons K, Gruenberg J, et al. 1991. Hypervariable C-terminal domain of Rab proteins acts as a targeting signal. *Nature* 353:769–72
16. Cheon C-I, Lee N-G, Siddique A-BM, Bal AK, Verma DPS. 1993. Roles of plant homologs of Rab 1p and Rab 7p in the biogenesis of the peribacteroid membrane, a subcellular compartment formed *de novo* during root nodule symbiosis. *EMBO J.* 12:4125–35
17. Chrispeels MJ. 1991. Sorting of proteins in the secretory system. *Annu. Rev. Plant Physiol. Mol. Biol.* 42:21–53
18. Clark SW, Meyer DI. 1993. Long lost cousins of actin. *Curr. Biol.* 3:54–55
19. Clowes FAL, Juniper BE. 1964. The fine stucture of the quiescent centre and neighbouring tissues in root meristems. *J. Exp. Bot.* 15:622–30
20. Craig S, Goodchild DJ. 1984. Golgi-mediated vicilin accumulation in pea cotyledon cells is re-directed by monensin and nigericin. *Protoplasma* 122:91–97
21. Demmer A, Holstein SEH, Giselbert H, Schauermann G, Robinson DG. 1993. Improved coated vesicle isolation allows better characterization of clathrin polypeptides. *J. Exp. Bot.* 44:23–33
22. Denecke J, Botterman J, Deblaere R. 1990. Protein secretion in plant cells can occur via a default pathway. *Plant Cell* 2:51–59
23. Denecke J, De Rycke R, Botterman J. 1992. Mammalian- and plant-sorting signals for protein retention in the endoplasmic reticulum form a closely related epitope. *EMBO J.* 11:2345–55
24. d'Enfert C, Gensse M, Gaillardin C. 1992. Fission yeast and a plant have functional homologues of the Sar1 and Sec12 proteins involved in ER to Golgi traffic in budding yeast. *EMBO J.* 11:4205–11
25. Donaldson J, Finazzi D, Klausner R. 1992. Brefeldin A inhibits Golgi membrane catalysed exchange of guanine nucleotide onto ARF protein. *Nature* 360:350–52
26. Driouich A, Faye L, Staehelin LA. 1993. The plant Golgi apparatus: a factory for complex polysaccharides and glycoproteins. *Trends Biochem. Sci.* 18:210–14
27. Driouich A, Levy S, Staehelin LA, Faye L. 1994. Structural and functional organization of the Golgi apparatus in plant cells. *Plant Physiol. Biochem.* 32:731–49
28. Driouich A, Zhang GF, Staehelin LA. 1993. Effect of brefeldin A on the structure of the Golgi apparatus and on the synthesis and secretion of proteins and polysaccharides in sycamore maple (*Acer pseudoplatanus*)

suspension-cultured cells. *Plant Physiol.* 101:1363–73

29. Duden R, Allan V, Kreis T. 1991. Involvement of β-COP in membrane traffic through the Golgi complex. *Trends Cell Biol.* 1:14–19

30. Farkas V, Maclachlan GA. 1988. Fucosylation of exogenous xyloglucans by pea microsomal membranes. *Arch. Biochem. Biophys.* 264:48–53

31. Faye L, Fitchette-Laine A-C, Gomord V, Chekkafi A, Delaunay A-M, et al. 1993. Detection, biosynthesis and some functions of glycans N-linked to plant secreted proteins. In *Soc. Exp. Biol. Semin. Ser.* 53: *Post-translational Modifications in Plants*, ed. NH Battey, HG Dickinson, AM Hetherington, pp. 213–42. Cambridge: Cambridge Univ. Press

32. Ferro-Novick S, Jahn R. 1994. Vesicle fusion from yeast to man. *Nature* 370:191–93

33. Fischer von Mollard G, Stahl B, Li C, Südhof T, Jahn R. 1994. Rab proteins in regulated secretion. *Trends Biochem. Sci.* 19:164–68

34. Fitchette-Lainé A-C, Gomord V, Chekkafi A, Faye L. 1994. Distribution of xylosylation and fucosylation in the plant Golgi apparatus. *Plant J.* 5:673–82

35. Fleming AJ, Mandel T, Roth I, Kuhlemeier C. 1993. The patterns of gene expression in the tomato shoot apical meristem. *Plant Cell* 5:297–309

36. Fowke LC, Tanchak MA, Galway ME. 1991. Ultrastructural cytology of the endocytotic pathway in plants. In *Endocytosis, Exocytosis and Vesicle Traffic in Plants*, ed. CR Hawes, JOD Coleman, DE Evans, pp. 15–40. Cambridge: Cambridge Univ. Press

37. Garcia-Herdugo G, Gonzalez-Reyes JA, Garcia-Navarro F, Navas P. 1988. Growth kinetics of the Golgi apparatus during the cell cycle in onion root meristems. *Planta* 175:305–12

38. Gibeaut DM, Carpita NC. 1994. Biosynthesis of plant cell-wall polysaccharides. *FASEB J.* 8:904–15

39. Gomez L, Chrispeels MJ. 1994. Complementation of an *Arabidopsis thaliana* mutant that lacks complex asparagine-linked glycans with the human cDNA encoding N-acetylglucosaminyltransferase I. *Proc. Natl. Acad. Sci. USA* 91:1829–33

40. Gordon R, MacLachlan GA. 1989. Incorporation of UDP-[^{14}C]glucose into xyloglucan by pea membranes. *Plant Physiol.* 91:373–78

41. Griffing LR. 1991. Comparisons of Golgi structure and dynamics in plant and animal cells. *J. Electron Microsc. Tech.* 17:179–99

42. Griffiths G, Simons K. 1986. The trans Golgi network: sorting at the exit site of the Golgi complex. *Science* 234:438–43

43. Hanna R, Brummell DA, Camirand A, Hensel A, Russell EF, et al. 1991. Solubilization and properties of GDP-fucosyltransferases from pea epicotyl membranes. *Arch. Biochem. Biophys.* 290:7–13

44. Harley SM, Beevers L. 1989. Coated vesicles are involved in the transport of storage proteins during seed development in *Pisum sativum*. *Plant Physiol.* 91:674–78

45. Harris N. 1986. Organization of the endomembrane system. *Annu. Rev. Plant Physiol.* 37:73–92

46. Harris N, Oparka KJ. 1983. Connections between dictyosomes, ER and GERL in cotyledons of mung bean (*Vigna radiata* L.). *Protoplasma* 114:93–102

47. Hawes C, Faye L, Satiat-Jeunemaitre B. 1995. The Golgi apparatus and pathways of vesicle trafficking. In *Membranes: Specialized Functions in Plants*, ed. M Smallwood, P Knox, D Bowles. Greenwich, CT: JAI. In press

48. Hayashi T. 1989. Xyloglucans in the primary cell wall. *Annu. Rev. Plant Physiol. Mol. Biol.* 40:139–68

49. Hellgren L, Morré DJ, Sellden G, Sandelius AS. 1993. Isolation of a putative vesicular intermediate in the cell-free transfer of membrane from transitional endoplasmic reticulum to the Golgi apparatus of etiolated seedlings of garden pea. *J. Exp. Bot.* 44:197-207

50. Helms J, Rothman J. 1992. Inhibition by Brefeldin A of a Golgi membrane enzyme that catalyses exchange of guanine nucleotide bound to ARF. *Nature* 360:352–54

51. Hess MW. 1993. Cell-wall development in freeze-fixed pollen: intine formation of *Ledebouria socialis* (Hyacinthaceae). *Planta* 189:139–49

52. Hillmer S, Fruendt H, Robinson DG. 1988. The partially coated reticulum and its relationship to the Golgi apparatus in higher plant cells. *Eur. J. Cell Biol.* 47:206–12

53. Hinz G, Hoh B, Robinson DG. 1993. Strategies in the recognition and isolation of storage protein receptors. *J. Exp. Bot.* 44:351–57

54. Hirose S, Komamine A. 1989. Changes in ultrastructure of Golgi apparatus during the cell cycle in a synchronous culture of *Catharanthus roseus*. *New. Phytol.* 111:599–605

55. Holstein SEH, Drucker M, Robinson DG. 1994. Identification of a β-type adaptin in plant clathrin-coated vesicles. *J. Cell Sci.* 107:945–53

56. Humphrey JS, Peters PJ, Yuan LC, Bonifacino JS. 1993. Localization of the TGN38 to the *trans* Golgi network: involvement of a cytoplasmic tyrosine-containing sequence. *J. Cell Biol.* 120:1123–35

57. Iijima M, Kono Y. 1992. Development of Golgi apparatus in the root cap cells of

maize (*Zea mays* L.) as affected by compacted soil. *Ann. Bot.* 70:207–12

58. Iiyama K, Lam TBT, Meikle PJ, Ng K, Rhodes DI, et al. 1993. Cell wall biosynthesis and its regulation. In *Forage Cell Wall Structure and Digestibility*, ASA-CSSA-SSSA, pp. 621–83. Madison: Univ. Wisc. Press

59. Jackson MR, Nilsson T, Peterson PA. 1993. Retrieval of transmembrane proteins to the endoplamsic reticulum. *J. Cell Biol.* 121:317–33

60. Jones RL, Jacobsen JV. 1991. Regulation of synthesis and transport of secreted proteins in cereal aleurone. *Int. Rev. Cytol.* 126:49–88

61. Juniper BE, Hawes CR, Horne JC. 1982. The relationship between the dictyosomes and the form of endoplasmic reticulum in plant cells with different export programs. *Bot. Gaz.* 143:1545–52

62. Kandasamy MK, Kristen U. 1989. Ultrastructural responses of tobacco pollen tubes to heat shock. *Protoplasma* 153:104–10

63. Kelleher DJ, Kreibich G, Gilmore R. 1992. Oligosaccharyl-transferase activity is associated with a protein complex composed of ribophorins I and II and a 48 kD protein. *Cell* 69:55–65

64. Kidou S, Anai T, Umeda M, Aotsuka S, Tsuge T, et al. 1993. Molecular structure of ras-related small GTP-binding protein genes of rice plants and GTPase activities of gene products in *Escherichia coli*. *FEBS Lett.* 332:282–86

65. Klausner RD, Donaldson JG, Lippencott-Schwartz J. 1992. Brefeldin A: insights into the control of membrane traffic and organelle structure. *J. Cell Biol.* 116:1071–80

66. Ladinsky MS, Kremer JR, Furcinitti PS, McIntosh JR, Howell KE. 1994. HVEM tomography of the trans-Golgi network: structural insights and identification of a lace-like vesicle coat. *J. Cell Biol.* 127:29–38

67. Lainé A-C, Gomord V, Faye L. 1991. Xylose-specific antibodies as markers of subcompartmentation of terminal glycosylation in the Golgi apparatus of sycamore cells. *FEBS Lett.* 295:179–84

68. Lee H-I, Gal S, Newman TC, Raikhel NV. 1993. The *Arabidopsis* endoplasmic reticulum retention receptor functions in yeast. *Proc. Natl. Acad. Sci. USA* 90:11433–37

69. Lin HB, Harley SM, Butler JM, Beevers L. 1992. Multiplicity of clathrin light-chain-like polypeptides from developing pea (*Pisum sativum* L.) cotyledons. *J. Cell Sci.* 103:1127–37

70. Low PS, Chandra S. 1994. Endocytosis in plants. *Annu. Rev. Plant Physiol. Plant Mol. Biol.* 45:609–31

71. Lynch MA, Staehelin LA. 1992. Domain-specific and cell type-specific localization of two types of cell wall matrix polysaccharides in the clover root tip. *J. Cell Biol.* 118:467–79

72. Machamer CE, Grin MG, Esquela A, Chung SW, Rolls M, et al. 1993. Retention of a *cis* Golgi protein requires polar residues on one face of a predicted α-helix in the transmembrane domain. *Mol. Cell Biol.* 4:695–704

73. Matsuoka K, Watanabe N, Nakamura K. 1994. Vacuolar targeting and post-translational processings of a precursor to sporamin expressed in tobacco cells. *Abstr. 4th Int. Congr. Plant Mol. Biol.*, #1248, Amsterdam

74. Melançon P. 1993. G whizz. *Curr. Biol.* 3:230–33

75. Mellman I, Simons K. 1992. The Golgi complex: in vitro veritas? *Cell* 68:829–40

76. Meman AR, Clark GB, Thompson GA Jr. 1993. Identification of an ARF type low molecular weight GTP binding protein. *Biochem. Biophys. Res. Commun.* 193:809–13

77. Mitsui T, Honma M, Kondo T, Hashimoto N, Kimura S, et al. 1994. Structure and function of the Golgi complex in rice cells 2. Purification and characterization of Golgi membrane-bound nucleoside diphosphatase. *Plant Physiol.* 106:119–25

78. Mollenhauer HH. 1965. An intercisternal structure in the Golgi apparatus. *J. Cell Biol.* 24:504–11

79. Mollenhauer HH, Morré DJ. 1980. The Golgi apparatus. In *The Biochemistry of Plants*, ed. P Stumpf, E Conn, pp. 437–88. New York: Academic

80. Moore PJ, Swords KMM, Lynch MA, Staehelin LA. 1991. Spatial organization of the assembly pathways of glycoproteins and complex polysaccharides in the Golgi apparatus of plants. *J. Cell Biol.* 112:589–602

81. Morré DJ. 1987. The Golgi apparatus. *Int. Rev. Cytol.* 17:211–53

82. Morré DJ, Keenan TW. 1994. Golgi apparatus buds—vesicles or coated ends of tubules? *Protoplasma* 179:1–4

83. Nagano Y, Murai N, Matsuno R, Sasaki Y. 1993. Isolation and characterization of cDNAs that endcode eleven small GTP-binding proteins from *Pisum sativum*. *Plant Cell Physiol.* 34:447–55

84. Nakamura K, Matsuoka K. 1993. Protein targeting to the vacuole in plant cells. *Plant Physiol.* 101:1–6

85. Napier RM, Fowke LC, Hawes C, Lewis M, Pelham HRB. 1992. Immunological evidence that plants use both HDEL and KDEL for targeting proteins to the endoplsmic reticulum. *J. Cell Sci.* 102:261–71

85a. Narimatsu H. 1994. Recent progress in molecular cloning of glycosyltransferase genes of eukaryotes. *Microbiol. Immunol.* 38:489–504

86. Nilsson T, Hoe MH, Sluarevicz P, Rabouille C, Watson R, et al. 1994. Kin recognition between medial Golgi enzymes in HeLa cells. *EMBO J.* 13:562–74

87. Nilsson T, Warren G. 1994. Retention and retrieval in the endoplasmic reticulum and the Golgi apparatus. *Curr. Opin. Cell Biol.* 6:517–21

88. Orci L, Palmer DJ, Amherdt M, Rothman JE. 1993. Coated vesicle assembly in the Golgi requires only coatomer and ARF proteins from the cytosol. *Nature* 364:732–34

89. Palme K, Diefenthal T, Vingron M, Sander C, Schell J. 1992. Molecular cloning and structural analysis of genes from *Zea mays* (L.) coding for members of the *ras*-related *ypt* gene family. *Proc. Natl. Acad. Sci. USA* 89:787–91

90. Paulson JC, Colley KJ. 1989. Glycosyltransferases. Structure, localization, and control of cell type-specific glycosylation. *J. Biol. Chem.* 264:17615–18

91. Pesacreta TC, Lucas WJ. 1985. Presence of a partially-coated reticulum and a plasma membrane coat in angiosperms. *Protoplasma* 1256:173–84

92. Phillips GD, Preshaw C, Steer MW. 1988. Dictyosome vesicle production and plasma membrane turnover in auxin-stimulated outer epidermal cells of coleoptile segments from *Avena sativa* (L.). *Protoplasma* 145:59–65

93. Pryer NK, Wuestehube LJ, Schekman R. 1992. Vesicle-mediated protein sorting. *Annu. Rev. Biochem.* 61:471–516

94. Rambourg A, Clermont Y. 1990. Three-dimensional electron microscopy: structure of the Golgi apparatus. *Eur. J. Cell Biol.* 51:189–200

95. Record RD, Griffing LR. 1988. Convergence of the endocytic and lysosomal pathways in soybean protoplasts. *Planta* 176:425–532

96. Regad R, Bardet C, Tremousayge D, Moisan A, Lescure B, et al. 1993. cDNA cloning and expression of an *Arabidopsis* GTP binding protein of the ARF family. *FEBS Lett.* 316:133–36

97. Robinson DG, Balusek K, Depta H, Hoh B, Holstein SEH. 1991. Isolation and characterisation of plant coated vesicles. In *Endocytosis, Exocytosis and Vesicle Traffic in Plants*, ed. CR Hawes, JOD Coleman, DE Evans, pp. 65–79. Cambridge: Cambridge Univ. Press

98. Robinson DG, Balusek K, Freundt H. 1989. Legumin antibodies recognize polypeptides in coated vesicles isolated from developing pea cotyledons. *Protoplasma* 150:79–82

99. Rodgers MW, Bolwell GP. 1992. Partial

purification of Golgi-bound arabinosyltransferase and two isoforms of xylosyltransferase from French bean (*Phaseolus vulgaris* L.). *Biochem. J.* 288:817–22

100. Rothman JE, Orci L. 1990. Movement of proteins through the Golgi stack: a molecular dissection of vesicular transport. *FASEB J.* 4:1460–68

101. Rothman JE, Orci L. 1992. Molecular dissection of the secretory pathway. *Nature* 355:409–15

102. Rutten TL, Knuiman B. 1993. Brefeldin A effects on tobacco pollen tubes. *Eur. J. Cell Biol.* 61:247–55

103. Samuels AL, Bisalputra T. 1990. Endocytosis in elongating root cells of *Lobelia erinus*. *J. Cell Sci.* 97:157–65

104. Sano H, Youssefian S. 1991. A novel *ras*-related *rpg1* gene encoding a GTP-binding protein has reduced expression in 5-azacytidine-induced dwarf rice. *Mol. Gen. Genet.* 228:227–32

105. Satiat-Jeunemaitre B, Hawes C. 1992. Redistribution of a Golgi glycoprotein in plant cells treated with brefeldin A. *J. Cell Sci.* 103:1153–66

106. Satiat-Jeunemaitre B, Hawes C. 1993. Insights into the secretory pathway and vesicular transport in plant cells. *Biol. Cell* 79:7–15

107. Satiat-Jeunemaitre B, Hawes C. 1994. G.A.T.T. (a general agreement on traffic and transport) and brefeldin A in plant cells. *Plant Cell* 6:463–67

108. Schmid SL. 1993. Biochemical requirements for the formation of clathrin- and COP-coated transport vesicles. *Curr. Opin. Cell Biol.* 5:621–27

109. Schnepf E. 1993. Golgi apparatus and slime secretion in plants: the early implications and recent models of membrane traffic. *Protoplasma* 172:3–11

110. Shannon TM, Steer MW. 1984. The root cap as a test system for the evaluation of Golgi inhibitors. II. Effect of potential inhibitors on slime droplet formation and structure of the secretory system. *J. Exp. Bot.* 35:1708–14

111. Sherrier DJ, VandenBosch KA. 1994. Secretion of cell wall polysaccharides in *Vicia* root hairs. *Plant J.* 5:185–95

112. Shevell DE, Leu W-M, Gillmor CS, Xia G, Feldmann KA, et al. 1994. EMB30 is essential for normal cell division, cell expansion, and cell adhesion in *Arabidopsis* and encodes a protein that has similarity to Sec7. *Cell* 77:1051–62

113. Showalter AM. 1993. Structure and function of plant cell wall proteins. *Plant Cell* 5:9–23

114. Sjöstrand FS, Hanzon V. 1954. Ultrastructure of Golgi apparatus of exocrine cells of mouse pancreas. *Exp. Cell Res.* 7:415–29

115. Soldati T, Shapiro A, Svejstrup ABD, Pfef-

fer SR. 1994. Membrane targeting of the small GTPase Rab9 is accompanied by nucleotide exchange. *Nature* 369:76–78

116. Söllner T, Bennett MK, Whitehart SW, Scheller RH, Rothman JE. 1993. A protein assembly-dissassembly pathway in vitro that may correspond to sequential steps of synaptic vesicle docking, activation, and fusion. *Cell* 75:409–18

117. Söllner T, Whitehart SW, Brunner M, Erdjument-Bromage H, Geromanos S, et al. 1993. SNAP receptors implicated in vesicle targeting and fusion. *Nature* 362:318–24

118. Staehelin LA, Giddings TH Jr, Kiss JZ, Sack FD. 1990. Macromolecular differentiation of Golgi stacks in root tips of *Arabidopsis* and *Nicotiana* seedlings as visualized in high pressure frozen and freeze-substituted samples. *Protoplasma* 157:75–91

119. Stafstrom JP, Staehelin LA. 1986. The role of carbohydrate in maintaining extensin in an extended conformation. *Plant Physiol.* 81:242–46

120. Stenmark H, Perton RG, Steele-Mortimer O, Lütke A, Gruenberg J, et al. 1994. Inhibition of Rab5 GTPase activity stimulates membrane fusion in endocytosis. *EMBO J.* 13:1287–96

121. Sturm A. 1994. N-glycosylation of plant proteins. In *Glycoproteins*, ed. J Montreuil, H Schachter, JFG Vliegenthart. Amsterdam: Elsevier Sci. In press

122. Sturm A, Johnson KD, Elbein AD, Szumilo T, Chrispeels MJ. 1987. Subcellular localization of glycosidases and glycosyltransferases involved in processing of N-linked glycans. *Plant Physiol.* 85:741–45

123. Swords KMM, Staehelin LA. 1993. Complementary immunolocalization patterns of cell wall hydroxyproline-rich glycoproteins studied with the use of antibodies directed against different carbohydrate epitopes. *Plant Physiol.* 102:891–901

124. Terryn N, Van Montagu M, Inzé D. 1993. GTP-binding proteins in plants. *Plant Mol. Biol.* 22:143–52

125. Ullrich O, Horiuchi H, Bucci C, Zerial M. 1994. Membrane association of Rab5 mediated by GDP-dissociation inhibitor and accompanied by GDP/GTP exchange. *Nature* 368:157–60

126. Ulmer JB, Palade GE. 1991. Effects of brefeldin A on the Golgi complex, endoplasmic reticulum and viral envelope glycoproteins in murine erythroleukemia cells. *Eur. J. Cell Biol.* 54:38–54

127. Vitale A, Chrispeels MJ. 1992. Sorting of proteins to the vacuoles of plant cells. *BioEssays* 14:151–60

128. von Schaewen A, Sturm A, O'Neill J, Chrispeels MJ. 1993. Isolation of a mutant *Aribidopsis* plant that lacks N-acetyl glucosaminyl transferase I and is unable to synthesize Golgi-modified complex N-linked glycans. *Plant Physiol.* 102:1109–18

129. Walworth NC, Brennwald P, Kabcenell AK, Garrett M, Novick P. 1992. Hydrolysis of GTP by Sec4 protein plays an important role in vesicular transport and is stimulated by a GTPase-activating protein in *Saccharomyces cerevisiae*. *Mol. Cell Biol.* 12:2017–28

130. Warren G. 1993. Bridging the gap. *Nature* 362:297–98

131. White AR, Xin Y, Pezeshk V. 1993. Xyloglucan synthase in Golgi membranes from *Pisum sativum* (pea). *Biochem. J.* 294:231–38

132. Whiteheart SW, Rossnagel K, Buhrow SA, Brunner M, Jaenicke R, et al. 1994. N-ethylmaleimide-sensitive fusion protein: a trimeric ATPase whose hydrolysis of ATP is required for membrane fusion. *J. Cell Biol.* 126:945–54

133. Zerial M, Stenmark H. 1993. Rab GTPases in vesicular transport. *Curr. Opin. Cell Biol.* 5:613–20

134. Zhang GF, Driouich A, Staehelin LA. 1993. Effect of monensin on plant Golgi: re-examination of the monensin-induced changes in cisternal architecture and functional activities of the Golgi apparatus of sycamore suspension-cultured cells. *J. Cell Sci.* 104:819–31

135. Zhang GF, Staehelin LA. 1992. Functional compartmentalization of the Golgi apparatus of plant cells. An immunocytochemical analysis of high pressure frozen and freeze-substituted sycamore maple suspension culture cells. *Plant Physiol.* 99:1070–83

Annu. Rev. Plant Physiol. Plant Mol. Biol. 1995. 46:289–315

PHYSIOLOGICAL AND ECOLOGICAL FUNCTION WITHIN THE PHYTOCHROME FAMILY

Harry Smith

Department of Botany, University of Leicester, Leicester LE1 7RH, United Kingdom

KEY WORDS: photomorphogenesis, ecology, physiology, mutants, transgenic plants, adaptive plasticity

CONTENTS

0066-4294/95/0601-0289$05.00

ABSTRACT

The phytochrome family of photoreceptors provides plants with a battery of sensors of the natural radiation environment. Each phytochrome can exist in two photoconvertible isomeric forms: The Pr form absorbs maximally in the red (R = 600–700 nm) and is photoconverted to the Pfr form, which absorbs maximally in the far-red (FR = 700–800 nm), thereby being converted to Pr. Investigations of mutant and transgenic plants with altered levels of expression of the genes for phytochromes A and B (i.e. the *PHYA* and *PHYB* genes) have begun to provide detailed information on the physiological functions of individual members of the family. Current evidence indicates that phytochrome A (phyA) mediates the far-red high-irradiance-response (FR-HIR) and functions differentially throughout the life cycle. Phytochrome B (phyB) is the principal phytochrome responsible for the classical red/far-red (R/FR) reversible responses, and for the responses of light-grown plants to the R:FR ratio. Other phytochromes have minor functions in these responses. Evidence is accumulating that the Pr form of phyB (PrB) has biological activity and that responses mediated by phyB may be the result of antagonistic actions of PrB and PfrB. For both phyA and phyB, response is related to gene dosage, indicating that tight control of *PHY* gene expression throughout the life cycle is required for the proper operation of the phytochrome family. The advent of molecular genetic techniques has begun to provide approaches for testing the fitness correlates of phytochrome-mediated responses, and thus to provide a comprehensive view of the adaptive value of photomophogenesis. The same approaches are also being used in attempts to improve crop plant performance through genetic engineering of the *PHY* genes.

INTRODUCTION

A profound reappraisal of the physiological and ecological significance of photomorphogenesis has begun since the discovery that plants contain not one but several genes that encode phytochrome photoreceptors (70) and the accompanying realization that each member of the extensive phytochrome family may be differentially active in regulating aspects of plant development (83). Powerful molecular and genetic techniques have provided means of allocating functional roles to individual members of the phytochrome family and of testing hypotheses of adaptive value and evolutionary significance. This review covers recent advances relating to the physiological functions of individual members of the phytochrome family, and attempts to provide some guidelines for extending these concepts to account for the behavior of plants in the natural environment. Coverage is limited strictly to those aspects that shed light on the ecophysiological functions of the phytochromes. Recent recommendations on phytochrome nomenclature (60) are followed throughout the review.

LIGHT AS AN ECOLOGICAL FACTOR

The dual nature of the significance of light for plants has been so often stated that it now reads as a trite observation of the obvious; nevertheless, the fundamental requirement for light as an energy source, and its importance as an informational signal, together constitute probably the most crucial factor in plant ecology. Plants use light energy for photosynthesis and use light signals to optimize their acquisition of light for photosynthesis. Light signals also are employed throughout the life cycle to synchronize development with seasonal changes, to ensure that resources are utilized effectively, to allow for appropriate reactions to competition from neighbors, and to initiate opportunistic advantage of environmental perturbations. Ecologically significant processes in which light signals are important include germination, etiolation, de-etiolation and seedling establishment, proximity perception and shade avoidance, photosynthetic acclimation to vegetation shade and to high irradiance, floral induction and the rate of flowering, induction of bud dormancy and tuberization, and tropic orientations. Investigations at the physiological level indicate that the phytochromes have roles in most, if not all, of these processes, although other photoreceptors, notably those detecting blue and UV radiation, are also involved in many processes. The challenge at the ecological and evolutionary level is to evaluate the importance of environmental signals perceived by the phytochromes as contributors to the survival of the individual plant until reproduction and thus, ultimately, to the survival of the species. Recent molecular genetic advances have the potential to enlarge our understanding of the ecophysiological functions of the phytochrome family, but the interpretations must be conditioned by a proper respect for the problems inherent in scaling-up concepts from the organism to the community.

THE PHYTOCHROME FAMILY

The PHY Genes

Sharrock & Quail (70) demonstrated the existence of five PHY genes (PHYA to PHYE) in Arabidopsis, and all five have now been sequenced and characterized (21). It appears, however, that the number of genes in the phytochrome family is not uniform in all higher plants. At least seven different PHY genes exist in tomato, including two presumably functional versions of the PHYB gene and a new PHY gene that does not have a counterpart in A. thaliana (59). The expression of the PHY genes is differentially regulated in A. thaliana (70) and tomato (35). Partial sequencing of the phytochrome-related sequences in a range of dicotyledonous species has revealed subfamilies of *PHY* genes representing the A, B, C, and E classes, with the *PHYD* gene (of *A. thaliana*) having close sequence similarity to *PHYB* (RA Sharrock, personal communication). One problem highlighted here is that of nomenclature; it may be more realistic to regard the *A. thaliana PHYB* and *PHYD* genes as analogous to the tomato *PHYB1* and *PHYB2* genes. More importantly, because research on the functions of members of the phytochrome family is largely concentrated on investigations of *A. thaliana*, it may be wise to retain a degree of caution on the universality of the conclusions derived.

The Phytochromes

The existence of several different phytochromes has been known for five years, but there is still very little information on the properties and characteristics of any phytochromes other than phytochrome A (phyA) from cereals and peas. Physiological distinction between the functions of the different phytochromes relies to some extent on matching the characteristics of photoresponses with those of the putative responsible phytochromes. The principal distinction that can be drawn is between the so-called Type I and Type II phytochromes: Type I phytochromes are regarded as being light-labile, whereas Type II phytochromes are light-stable. Light-labile phytochromes are those in which the Pfr form is unstable compared to the Pr form, so that in conditions under which the proportion of Pfr is maintained at a high level [i.e. during or after irradiation with red light (R), or with white light containing little far-red (FR)], Pfr is rapidly degraded, resulting in loss of total phytochrome. Pfr degradation, however, can occur in the dark, and therefore the term light-labile is factually incorrect, although it remains a convenient shorthand. This distinction allows, under certain restricted circumstances, the possibility of assigning specific functions to Type I and Type II phytochromes based on whether or not FR reversibility can be demonstrated after a prolonged period in darkness. The classic observation by Downs et al (26) that internode extension in light-grown plants can be accelerated by brief treatment with FR at the end of the daily light period, and reversed by R given several hours later, is now regarded as the first demonstration of light-stable phytochrome, and has led to the use of end-of-day FR (EOD-FR) as a standard test for the action of light-stable phytochromes. It is now widely accepted that Type I phytochromes are represented by phyA, and that the remaining members of the phytochrome family are Type II (see 29); detailed studies of the kinetics of Type II phytochromes, however, have shown that the light-stable epithet is not absolute, and that differences in the stability of Pr and Pfr can exist in Type II phytochromes (91, 92).

Phytochrome Mutants

The most important, and potentially the most powerful, advance in phyto-chrome research has been the selection of a range of photomorphogenic mu-tants (for reviews see 19, 42, 61, 96). Photomorphogenic mutants are of two classes: those in which the lesion is within a *PHY* gene, resulting in the complete absence of a specific phytochrome, and those in which the lesion affects a downstream component in one or more of the phytochrome transduc-tion pathways. Most photomorphogenic mutants have been selected in *A. thaliana*, but this plant, despite its advantages for molecular genetics, is not ideal for ecophysiological experiments. The original *hy* mutants of *A. thaliana* were selected for long hypocotyls when grown in white light (43) and include *hy1*, *hy2*, and *hy6*, chromophore mutants deficient in all phytochromes; *hy3*, which lacks phytochrome B (84) and of which several alleles have been shown to have lesions in the *PHYB* gene (63); *hy4*, a putative blue-light photoreceptor mutant (2); and *hy5*, a downstream transduction chain mutant (19). Rationali-zation of the nomenclature suggests that *hy3* should be termed *phyB* (63). The chromophore mutants *hy1*, *hy2*, and *hy6* should lack all phytochromes, but they must be regarded as leaky mutants because some undoubted phyto-chrome-mediated phenomena can be observed in the light-grown plants (99). Recently, *A. thaliana* mutants have been selected in which the FR-mediated inhibition of hypocotyl extension is absent. These mutants are variously known as *hy8* (58), *fre1* (52), and *fhy2* (97). Certain alleles of these selections have been shown to have lesions in the *PHYA* gene (22, 97) and thus should be termed *phyA*. No mutants affecting the levels or functions of any of the remaining *A. thaliana* phytochromes (C, D, or E) have been reported.

Photomorphogenic mutants have also been obtained in other species. In tomato, the *aurea* (*au*) mutant has long been regarded as a photomorphogenic mutant. Studies of phytochrome levels in *au* have shown that the plant con-tains no functional phyA, although phyA apoprotein is present at reduced levels (56). A photoreversible phyB-like molecule was found to be present at normal levels (69). It seems likely that *au* has a lesion that affects chromo-phore synthesis or assembly, and is not a phytochrome mutant per se (41). Selection for abnormal extension growth in R has revealed a class of temporar-ily red-light insensitive (*tri*) mutants (41). One of these mutants, *tri2*, lacks a protein that in the wild-type reacts to a phyB-specific antibody and a mRNA species that is recognized by a probe specific for the tomato *PHYB2* sequence (RE Kendrick, personal communication). The same selection procedure has also yielded tomato mutants that are probably lacking phyA (41).

Three other putative phytochrome mutants are important for ecophysiologi-cal studies. The long hypocotyl (*lh*) mutant of cucumber (45), the elongated internode (*ein*) mutant of *Brassica rapa* (25), and the maturity mutant (*ma3*^{R})

of *Sorghum bicolor* (17, 18, 28) lack a B-type phytochrome. In peas, the *lv* mutation, which phenotypically resembles the *lh* mutant of cucumber, is probably not a phytochrome mutant per se, but apparently results in impaired action of light-stable phytochrome (53, 94). Other mutants in which photomorphogenic development occurs in the absence of light [i.e. the *det* and *cop* mutants of *A. thaliana* (20, 24)] have been important in elaborating the sequence of regulatory events in the photomorphogenic transduction pathways, but as yet, these mutants have had little significance for studies of ecophysiological function.

Phytochrome Transgenics

The isolation of cDNA sequences for individual *PHY* genes has enabled the construction of transgenic plants that express introduced *PHY* sequences (see 78 for review). Several transgenic plants expressing *PHY* genes have been generated using tomato, tobacco, potato, or *A. thaliana* as host, and cereal or dicot *PHYA* or *PHYB* cDNAs as the transgenes, driven (in most cases) by the constitutive cauliflower mosaic virus (CaMV) 35S promoter (10, 39, 40, 90). In most cases the transgenes were expressed to produce phytochrome apoproteins that bound endogenously generated chromophore and became functionally active *in planta*. Large phenotypic effects of transgenic *PHY* gene expression have been observed. The relative lack of interest in antisense expression of specific *PHY* sequences has been somewhat surprising. Antisense expression may have been regarded as unimportant because of the existence of defined mutants lacking specific phytochromes. From the ecophysiological viewpoint, however, it would be useful to create transgenic plants in a range of species, for which mutation selection may be difficult, which carry antisense-mediated depression of the levels of specific phytochromes.

PHYSIOLOGICAL FUNCTIONS OF PHYTOCHROMES A AND B

The Phytochrome Response Modes

Three decades of physiological analysis have identified a number of different response modes of phytochrome action, based on criteria that include R/FR reversibility and its escape with time, dose-response relationships (both for fluence and for fluence rate), and whether or not the law of reciprocity is fulfilled (see 46 for detailed analysis). The distinct response modes provide clues to the different informational, or perceptive, functions of the phytochromes. Essentially, two classes of response mode exist: those in which induction of response can be saturated by brief treatments with R and subsequently reversed by brief FR, and those in which continuous irradiation for a significant period of time is necessary. In R/FR reversible response modes, reciprocity is usually fulfilled and response saturation occurs at relatively low fluences. These responses are exemplified by the low-fluence responses (LFR), such as the R/FR reversible regulation of germination and of extension growth in etiolated seedlings, but may also be at the basis of the R:FR ratio responses and EOD responses (e.g. 26) of light-grown plants (77). Complicating this picture are the so-called very low fluence responses (VLFR), which occur in imbibed seeds and dark-grown seedlings only and are saturated at such low fluences that even brief FR establishes sufficient Pfr for saturation. Phytochromes also function in photoperiod perception, but this is probably not a separate response mode, because both high irradiance responses (HIR) and R/FR reversible responses that fulfill reciprocity requirements have been identified.

For those responses that are maximized under continuous irradiation, the so-called high irradiance responses (HIR), response is a complex function of wavelength, fluence rate, and time of exposure to light (46). In etiolated seedlings of most dicotyledonous plants, continuous FR treatment causes strong inhibition of extension growth, often coupled with stimulation of anthocyanin synthesis, in a fluence-rate-dependent manner; this response is termed the FR-HIR. The wavelength maximum of the HIR in etiolated seedlings is variable, however, and some seedlings also show significant action in the R. The relative effect of continuous R and FR alters markedly as de-etiolation proceeds, with a changeover from maximum action in the FR to maximum in the R after 2–3 days in white light (50), indicating that the FR-HIR and the R-HIR are different response modes. For both the FR-HIR and the R-HIR, continuous light can be replaced by repeated pulses, but for FR the time in darkness between the pulses must be very short compared to the permissible dark periods for R pulses (46). Furthermore, repeated R pulses can be reversed by subsequent FR pulses, suggesting that the R-HIR may in fact be a manifestation of the classic R/FR reversible low-fluence response. For both the FR-HIR and the R-HIR, strong dependency on fluence rate is observed; mechanistic interpretations of fluence-rate dependency at constant photoequilibrium center on ideas of phytochrome cycling (i.e. repeated conversion between Pr and Pfr), but why such cycling should enhance phytochrome action remains a mystery (46).

PhyA Mediates the FR-HIR

The *phyA* mutants in *A. thaliana* were selected by their failure to show extension-growth inhibition under continuous FR, a strategy suggested by the physiological evidence that the FR-HIR is mediated by a light-labile Type I phytochrome. Hypocotyl growth in the *phyA* mutants, therefore, was insensitive to continuous FR, but responded as the wild-type (WT) to continuous white light or R (52, 58, 97). The FR-HIR is also often involved in the regulation of germination, and the *phyA* mutants correspondingly showed modified germination responses (62, 71, 97). The surprising finding was that the mutants, which entirely lack phyA, grew normally in white light, exhibiting virtually no detectable phenotype (97). This result indicates that phyA, which alone among the phytochromes accumulates to sufficient levels for it to be isolated and characterized, has only subtle roles in regulating development. This is also apparently true for the recently selected *phyA* mutants of tomato (41). Etiolated *A. thaliana phyA* mutants showed normal R/FR reversible responses and normal growth inhibition under continuous R; furthermore, light-grown seedlings showed normal (indeed, somewhat exaggerated) responses to changes in the R:FR ratio of white light (37, 97). That phyA is able to regulate extension growth in light-grown seedlings was also shown by studies of *phyA* mutants; WT seedlings grown in light-dark cycles, which presumably allow phyA levels to increase during the dark period, had shorter hypocotyls than did the mutants under the same conditions (37). A role for phyA in photoperiodic perception, at least in long-day plants, has also been proposed (3, 62). When seedlings were grown in short days (SD) with or without a 1 h white-light treatment in the middle of the night (light-break), the *phyA* mutant was less able to detect the light-break than were the corresponding WT or *phyB* mutants (62). Similarly, *phyA* seedlings grown in SD were much less capable of sensing a day extension of low-irradiance FR, or of white light of a low R:FR ratio, than were the corresponding WT, resulting in delayed flowering compared to the WT (3). These data fit with information on the flowering of long-day cereals, which indicated that at least a component of photoperiodic perception is mediated by a light-labile Type I phytochrome and has characteristics of an HIR (12, 13, 23, 86). An important corollary of these observations is that the light-labile phyA, which accumulates to relatively high concentrations in etiolated seedlings but is degraded to very low levels upon transfer to light, is clearly maintained at a level sufficient for action in the light-grown plant. The normal phenotype of light-grown *phyA* mutants contrasts markedly with the abnormal growth pattern of the other *phy* mutants; importantly, the *phyB* mutants show normal FR-HIR behavior. Evidence from mutants, therefore, is consistent with phyA being responsible for the FR-HIR-response mode. Through the FR-HIR, *phyA* is partially responsible for the regulation of germination, for the control of extension growth in etiolated seedlings and seedlings grown in light-dark cycles, and for a component of photoperiodic perception in long-day plants.

Photobiological investigations of transgenic *PHYA* expressers are generally consistent with the above conclusion but provide further insights. High levels of *PHYA* expression cause severe dwarfing in light-grown plants, seen most dramatically in tomato (10) and tobacco (40) expressing an oat *PHYA* driven by the CaMV 35S promoter. Detailed studies of the tobacco high-level expressers (designated 9A4), which have four- to ninefold WT levels of phytochrome in the etiolated seedlings (15), showed that the sensitivity of hypocotyl growth to inhibition by continuous FR was increased about fivefold, although the slope of the relationship between response and fluence rate was unchanged (47). When FR was given for several hours as a prolonged EOD treatment to light-grown seedlings, no inhibition was seen in the WT, whereas in transgenic seedlings, strong inhibition was observed that was dependent on fluence rate. In contrast, a pulse (15 min) of FR at the end of the light period caused a significant increase in extension in both WT and transgenic seedlings. These data suggest that the transgenic expression of an introduced *PHYA* cDNA leads to the persistence of the FR-HIR after de-etiolation. Similar results were later obtained with transgenic *A. thaliana* seedlings expressing an oat *PHYA* cDNA (98). Transgenic expression of an antisense *PHYA* construct in potato showed suppressed hook opening and early leaf development, especially in FR, although there seemed to be no persistence of a FR-HIR in sense transformants (B Thomas & C Gatz, personal communication). On the other hand, transformation of strawberry with a rice *PHYA* gene resulted in the predicted phenotype of light-grown plants being dwarfed, with reduced internode elongation and darker green leaves (B Thomas, personal communication).

Although these conclusions are consistent with those derived from the study of the *phyA* mutants, they pose certain problems for the interpretation of the FR-HIR. It is generally accepted that the FR-HIR must be mediated by a light-labile phytochrome (46). Hartmann's classic dichromatic experiments (34) showed that the FR wavelength maximum was a result of simultaneous excitation of both Pr and Pfr, establishing a photoequilibrium sufficiently low to prevent the rapid loss of total phytochrome, while maintaining sufficient Pfr available over a sufficient time period to induce the response. In the transgenic 9A4 tobacco seedlings, however, the rate of loss of the phyA encoded by the introduced oat *PHYA* cDNA was substantially lower ($t_{1/2} \sim 4.5$ h) than that of the endogenous tobacco phyA ($t_{1/2} \sim 40$ min) (51); in other words, the heterologous phyA was much less light-labile than the endogenous phyA, and yet it persisted in mediating a strong FR-HIR. Furthermore, because the transgene is driven by the CaMV 35S promoter, it is presumably not subject to the Pfr-mediated down-regulation that has been demonstrated for many WT plants. As a result, the concentration of the heterologous phyA is essentially uncontrolled, which does not fit well with the concepts that have up to now underpinned our incomplete understanding of the FR-HIR. It would be interesting to perform Hartmann-style dichromatic irradiation experiments with transgenic *PHYA* expressers to determine whether or not the critical photoequilibrium for the HIR is changed. Similarly, detailed studies of the wavelength maximum of the HIR in transgenic *PHYA* expressers may prove interesting. Other transgenic experiments have shown that the ability of phyA to mediate an HIR may be drastically modified by mutations within the *PHYA* cDNA sequence. The growth-inhibitory activity of transgenically expressed *PHYA* was strongly enhanced if a number of the N-terminal serines were replaced by alanines (85). In contrast, transformation with a *PHYA* cDNA from which the nucleotides encoding an N-terminal domain had been removed resulted in a dominant negative interaction in which the FR-HIR mediated by the endogenous phyA of the host plant was negated (9). These observations suggest that the FR-HIR may be a function that resides in a specific and unique domain of the phyA molecule, rather than simply resulting from the establishment of a critical photoequilibrium that minimizes the effect of rapid phyA proteolysis.

Another important insight gained from transgenic expression of *PHYA* genes that could not be deduced from the behavior of mutants is the capacity of a phytochrome to arrogate to itself functions that it may not normally mediate. An example is the regulation of hypocotyl extension by light pulses in etiolated *A. thaliana*; WT plants do not respond to a single pulse of R, but transgenic *PHYA* expressers exhibit strong R-mediated inhibition that is subsequently reversible by a pulse of FR (11). Thus, transgenically expressed phyA can mediate a R/FR reversible LFR in *A. thaliana* under conditions when none of the endogenous phytochromes are apparently able so to do. A similar example is the generation of the capacity for photoregulated gene expression in cell cultures of transgenic tobacco expressing a *PHYA* cDNA to high levels; such regulation is absent in WT cultures (48). These observations lead to the speculation that different members of the phytochrome family may be able to substitute for other members if and when necessary.

Response Modes Mediated by PhyB

The *A. thaliana phyB* mutants display a constitutively elongated phenotype in continuous white light or R but are short in FR, are early flowering, and their extension growth is virtually insensitive to EOD-FR treatments, or to daytime supplementation of white light with FR (i.e. low R:FR ratio) (19, 32, 99). In contrast, *A. thaliana phyA* mutants have normal responses to pulses of R or continuous R, to EOD-FR, and to daytime supplemental FR. Germination studies (71) have shown that although pulses of R induce, and subsequent pulses of FR reverse, the germination of WT and *phyA* seed, these treatments are completely without effect on *phyB* seed germination. On these grounds, phyB may be proposed to function in the classic R/FR reversible low-fluence responses, including the EOD response. In addition, phyB is clearly implicated in the R-HIR and the R:FR ratio responses seen in de-etiolated plants. Photoperiodic induction in short-day plants is associated with the presence of functional phyB. The ma_3^R maturity mutant of the short-day plant *S. bicolor* is virtually photoperiodically insensitive, flowering very early compared to WT (17, 18). The *lv* mutant of pea flowers early even in photoperiods that are noninductive for the WT plants (94). A similar role has been proposed for phyB in SD-induced tuberization in an Andean potato transformed with an antisense construct from the *PHYB* gene of cultivated potato (36). The transformed plants produced tubers under daylength conditions that would be noninducing for the WT. Most classical evidence indicates that the role of phytochromes in SD perception operates as a classical R/FR LFR. If the R-HIR and the R:FR ratio responses may be interpreted as manifestations of a basic R/FR reversible regulatory mechanism, as argued above, then the response-mode functions of phyB become simplified to R/FR reversible responses.

Although phyB may have primary responsibility, it is almost certainly not the only phytochrome capable of mediating the response modes outlined above. Even though phyB seems to be principally responsible for extension-growth regulation in R or white light, phyA still plays a role, if only in the background. The evidence cited above for the *A. thaliana phyA* mutants indicates a continuing role for phyA in light-grown plants (37). Also, in the *lv* mutant of pea, FR radiation inhibited extension growth even in plants de-etiolated in white light, indicating a continuing action of phyA in de-etiolated plants (94). In *A. thaliana phyA/phyB* double mutants, hypocotyl length under continuous R was found to be greater than in either of the single mutants or the WT; indeed, in the double mutant, R actually promoted hypocotyl extension compared to darkness (62). Thus, in the absence of phyB, the further absence of phyA reveals a photocontrol that is hidden in seedlings containing phyB; in other words, phyA is able to regulate hypocotyl extension under R, but this can only be seen when the overriding control by phyB is removed. Other aspects of the photophysiology of the double mutants indicate joint, or overlapping, control by phyA and phyB. In the double mutants, cotyledon development and *CAB* gene expression are only weakly regulated by R, whereas in both of the single mutants, control is similar to that in the WT (62). Thus, either phyA or phyB can mediate R effects on cotyledon development and *CAB* gene expression. Germination in *A. thaliana* is also under the dual control of phyA and phyB (62). The authors of this study pointed out that the separate actions of phyA and phyB in regulating development may be similar or opposite in direction.

Further examples that indicate action by phytochromes other than phyB include the retention of small, but significant, stem-extension responses to EOD-FR or reduced R:FR ratio in the putative *phyB* mutants of cucumber (*lh*) and *B. rapa* (*ein*) (82, 99). The responses of light-grown *A. thaliana* seedlings to low R:FR ratio are also not all mediated by phyB; the rates of flowering, radial stem expansion, and growth in leaf area, all aspects responsive to the R:FR ratio, behave normally in the *phyB* mutant (65). These data must mean that a phytochrome other than phyB is capable of perceiving R:FR ratio and of regulating aspects of development. The acceleration of flowering in *A. thaliana* by low R:FR ratio (which seems to be a separate phenomenon from the photoperiodic induction of flowering) has been further investigated by a genetic approach (33). The *phyB* mutant and the chromophore mutant *hy2* are both early flowering compared to WT, and both show slight accelerations of flowering under reduced R:FR ratio conditions. The double mutant (*phyB/hy2*) flowers even earlier than do either of the single mutants. When either *phyB* or *hy2* is placed in a late-flowering background (i.e. one of the *fca*, *fwa*, or *co* mutations), pronounced accelerations of flowering in response to low R:FR ratio are observed, although triple mutants of *phyB*, *hy2*, and one of the late-flowering mutations do not exhibit such a response. The rate of flowering in *A. thaliana* is evidently under the regulation of phyB and at least one other phytochrome. Another complication has come from recent mutant selection studies in tomato (41). The *tri* mutants temporarily elongate more rapidly than do WT in white light or R, but after a few days their growth rates become normal. It seems likely that in tomato, the functions of phyB may be shared by two phyB photoreceptors, encoded by the *PHYB1* and *PHYB2* genes, whose expression may be differentially regulated.

Is Pfr the Only Active Form?

Once a lively controversy, the idea that all phytochrome activity may not reside solely in the Pfr form, originally put forward by this reviewer essentially on theoretical grounds (73), has in general received little support. Subtle aspects of the phenotype of *phyB* mutants, however, are forcing a radical reappraisal of the central dogma that Pfr is the only active form. In particular, observations are accumulating of abnormalities in the developmental behavior of *phyB* mutants when grown in total darkness; such observations of differences between the mutant and WT in dark growth presumably must be related to the absence of the Pr form of phyB (i.e. PrB) in the mutant. Several lines of evidence point to the likelihood that, at least for phyB, both the Pr and the Pfr forms have biological activity, and that their activities may be antagonistic.

Evidence for a functional role for PrB comes from aberrant gravitropism behavior of the *A. thaliana phyB* mutant (44). Dark-grown WT seedlings exhibit pronounced negative gravitropism, but repeated brief pulses (1 min) of R lead to a randomization of growth direction, which can be reversed by brief FR given after each R; in other words, phytochrome mediates a loss of gravitropic orientation in *A. thaliana*. The gravitropic behavior of the *phyA* mutant is normal, but the *phyB* mutant is random in the dark; in effect, *phyB* behaves in the dark as if it had received R. The *hy2* chromophore mutant is negatively gravitropic both in the dark and after R treatment; that is, gravitropism in *hy2* is not randomized by R treatment. However, incubation of *hy2* seedlings in biliverdin, which allows chromophore synthesis and assembly (57), rescues the R-mediated randomization. Liscum & Hangarter (44) propose that PrB is responsible for the negative gravitropism observed in darkness, and that the removal of PrB upon R treatment, rather than the formation per se of PfrB, leads to growth randomization. The responses of the *hy2* mutant demonstrate that PrB must be active even in the absence of its chromophore. If this model is correct, one would expect that in transgenic plants expressing to high levels a *PHYB* cDNA, the presence of a large excess of PrB would cause gravitropism to be strongly negative even after R. However, the behavior of transgenic *PHYB* expressers is qualitatively similar to the WT (PRH Robson & H Smith, unpublished data). These data suggest an antagonism between opposing activities of PrB and PfrB; PrB promotes gravitropic orientation, and PfrB promotes randomization.

Root length in *A. thaliana phyB* mutants is substantially less than in the WT, and this difference is evident even in dark-grown seedlings (62). One of the interpretations offered for this result was that phyB may have some activity in the dark (presumably in the Pr form). Further evidence relates to the germination behavior of *phyA, phyB,* and *phyA/phyB* double mutants under continuous FR (62). Under these conditions, about 62% of the WT seed germinated, whereas essentially all the *phyB* seed germinated. The *phyA* seed all failed to germinate, but the *phyA/phyB* double mutant seed achieved about 33% germination. These data indicate that under continuous FR, phyA promotes, and phyB inhibits, germination. In other words, the *phyB* mutation compensates for the inhibitory effect of FR on germination, and the authors conclude that this could only result from direct action of PrB.

The concept that PrB has action that opposes that of PfrB may also explain the otherwise puzzling responses of transgenic plants that express introduced *PHYB* cDNAs to high levels (49). A range of *PHYB* transgenics expressing from about 2- to 30-fold WT levels were examined for their responses to treatments with continuous white light of a graded series of R:FR ratios. In WT *A. thaliana*, hypocotyl extension is very sensitive to reductions in R:FR ratio, and if the control of extension is entirely a function of Pfr concentration, then large increases in Pfr concentration, caused by transgene expression, would be expected to cancel out any possible effects of low R:FR ratio. Although the slopes of the relationships between calculated Pfr/P were lower in the overexpressers than in the WT, all the strains showed strong increases in extension with reductions in R:FR ratio. These relationships may be accounted for on trivial grounds, such as ectopic expression, but they also fit with the concept that PrB and PfrB regulate extension growth in an antagonistic manner so that, even in the presence of a high concentration of PfrB, the corresponding level of PrB still accelerates extension.

The idea that PrB and PfrB act antagonistically in regulating extension growth has also been proposed to explain the kinetics of extension growth in mustard seedlings exposed to rapid fluctuations in R:FR ratio at very high fluence rates (74). At high fluence rates, photoconversion intermediates between Pr and Pfr accumulate, and direct measurement of Pfr absorption at high fluence rates has demonstrated that at least 50% of the total measureable phytochrome (phyA in this case) would be present as photoconversion intermediates at daylight fluence rates (80). The question is therefore raised as to whether effects on extension growth mediated by the R:FR ratio will operate at very high fluence rates. At fluence rates tenfold greater than daylight values, changes in the R:FR ratio caused quite normal modulations of the extension rate of mustard seedlings (74). Changes in fluence rate of four orders of magnitude (i.e. 10,000-fold) caused temporary fluctuations in growth rate, but later steady-state growth rates were entirely determined by the R:FR ratio. Because the absolute concentrations of PrB and PfrB should be much lower at high than at low fluence rates, the normal responses to changes in R:FR ratio indicate that the ratio of PrB to PfrB, rather than the absolute concentration of PfrB, determines growth rate.

In summary, the concept that both PrB and PfrB have activity is gaining acceptance as a result of studies on mutant and transgenic plants, and these data fit with earlier observations on WT plants. If this is true for phyB, there is no a priori reason why it should not also be true for the other phytochromes. Reed et al (62) pointed out that there is precedence for alternate regulatory states of sensory molecules having opposite activities, rather than only one of two forms having activity. In *Escherichia coli*, NtrB, a member of the sensor class of two-component regulatory systems (55), acts as a kinase under conditions of nitrogen starvation and as a phosphatase under nitrogen excess.

The Importance of Phytochrome Concentration

An important conclusion from the responses of mutant and transgenic *PHY* expressers is that photomorphogenic response is tightly related to the level of *PHY* gene expression. The hypocotyls of light-grown seedlings heterozygous for the *phyB* mutation (i.e. *phyB/PHYB*) are intermediate in length between those of the WT and the homozygous mutants (*phyB/phyB*) (43). Similarly, the *phyA* mutations in *A. thaliana* exhibit partial dominance (97). In *S. bicolor*, gene dosage of the maturity alleles Ma_3, ma_3, and $ma_3{}^R$ regulates the levels of an immunochemically recognized 123-kDa protein (assumed to be phyB) and also regulates photoperiodic behavior in a manner consistent with partial dominance (28). This study concluded that two copies of the WT alleles Ma_3 or ma_3 are required to produce sufficient 123-kDa phytochrome to regulate growth and flowering correctly; interestingly, tillering was normally regulated in the heterozygotes, suggesting that the mechanisms mediated by phyB may vary for different phenotypic characters (28). These considerations indicate the existence of tight gene-dosage dependence, best explained on the basis of a close relationship between the phenotype and the abundance of the gene product. For the *PHYB* gene, this conclusion has been elegantly confirmed by Wester et al (95), who transformed the *A. thaliana phyB* mutant with a *PHYB* minigene driven by its own homologous promoter. This study demonstrated that the immunochemically detectable level of phyB varied linearly with *PHYB* gene copy number, and that photomorphogenic response was correlated with expression level. Thus, for normal photomorphogenesis to operate, it appears that expression of the genes for phyA and phyB must be tightly controlled.

Which Phytochrome Mediates the VLFR?

Although the VLFR appears to be an arcane response only detectable under rigorous experimental control, it may in fact have important ecological consequences, at least for seed germination (see below). Therefore, it is important to identify which of the phytochromes is responsible. As yet, little interest has been shown by those groups investigating the responses of mutants, although early studies on the tomato *aurea* mutant indicated that the VLFR was absent (1). Unfortunately, studies with *aurea* cannot conclusively identify the responsible phytochrome. Because phyA accumulates to relatively high concentrations in dark-grown seedlings, it has been suggested that phyA might operate as an antenna for small amounts of light, and evidence from transgenic plant studies indicates that phyA may perform such a role. Casal et al (14) analyzed the cotyledon-opening response of transgenic tobacco expressing an oat *PHYA* cDNA (the 9A4 strain) to single light pulses calculated to elicit LFRs and VLFRs. Overexpression of *PHYA* caused a marked increase in sensitivity to pulses that established very low Pfr/P levels, but only a slight increase in sensitivity at higher Pfr/P. Thus, enhancing the concentration of phyA increases the sensitivity to fluences within the VLFR range. These results are consistent with, but because of ectopic expression problems, not exclusive to the view that phyA normally mediates the VLFR. Detailed studies of VLFRs in *phyA* and *phyB* mutants are needed.

PRESUMED ROLES OF THE PHYTOCHROMES IN THE NATURAL ENVIRONMENT

Problems of Scaling-Up from Physiology to Ecology

Determining the ecological significance of a particular trait requires a quantitative assessment of its fitness attributes in a range of environments. Such assessments of single traits have been difficult to make, partly because of the experimental problems involved in comparing the performance of alternate phenotypes in a common environment, but more seriously because of the immense difficulty of obtaining by standard genetic methods precisely defined alternate phenotypes. A potentially productive approach is to carry out comparative investigations of species or, preferably, ecotypes adapted to habitats that differ in their light environments, but even in these cases, the precise definition of phenotype in terms of genotype is not feasible. The situation has been radically improved, however, by the advent of mutation selection and transgenic methodology, which allow single-gene variants to be generated and investigated in ecological experiments. It should now be possible to quantify the fitness attributes of traits mediated by specific phytochromes that are predicted on physiological considerations to be of ecological significance, by assessing the performance of mutant or transgenic plants in which those traits are absent or exaggerated. A few such investigations have begun, but until more detailed information is available, we are left with trying to identify ecological functions from correlative data. The coverage that follows concentrates on studies performed in the natural environment or those utilizing very close approximations of natural conditions.

Germination

Although large seeds with adequate resources often germinate well in darkness, a high proportion of plants produce seed that, depending on other environmental conditions, may require exposure to light before germination occurs. This requirement is common for those plants that produce large quantities of small seed, such as many herbs and pioneer tree species. Speculation on the ecological significance of light-controlled germination normally centers on a requirement to detect proximity to the soil surface, but other considerations are also important. Many seed that are imbibed in darkness become extremely sensitive to small amounts of light but can remain dormant in the soil for many years. These observations prompt the speculation that such seed are primed to germinate in response to very small light signals, such as may be perceived upon brief soil disturbance, for example by the activities of animals or the effects of wind or heavy rain. Seed of *Datura ferox*, an aggressive annual weed of temperate and subtropical South America, is produced in summer and autumn, but remains dormant if left on the soil surface or if buried; flushes of germination occur during the following spring or summer if the seed is brought back to the surface, which commonly occurs during soil cultivation. Studies of the photoresponses of buried seed indicated a shift of four orders of magnitude in the the fluence-response curve, equivalent to a shift from an LFR- to a VLFR-response mode (68). Calculations indicated that such sensitized seed would germinate after exposure to only a few milliseconds of normal sunlight. If, as speculated above, the VLFR is mediated by phyA, it would be of interest to measure the amounts of phyA in seed that has been sensitized by burial; freshly imbibed seed normally have very small quantities of phyA, and the antenna hypothesis of the VLFR would presumably require the synthesis of phyA in imbibed, but dormant, seed. The ecological significance of seed sensitization is evidently that such seed can take advantage of even the briefest soil disturbance.

A further role for phytochrome-mediated regulation of germination is in the colonization of canopy gaps. Forest soils commonly retain large banks of dormant seed of pioneer species, much of which is apparently present on the surface. When a gap of sufficiently large size appears in the canopy, these seed banks generate copious quantities of seedlings, which colonize the gap. The prevention of germination in surface seed is a response to the low R:FR ratio of the light environment within the canopy. The stimulation of germination within the gaps, however, is not a simple response to high R:FR ratio; it also involves detection of the size of the gap. Seed of *Piper auritum* and *Cecropia obtusifolia*, pioneer species prevalent in the tropical forests of Mexico, were shown to require sustained periods of high R:FR ratio irradiation for germination, consistent with the need to detect canopy gaps large enough for direct solar radiation to penetrate for several hours per day (87, 89). Subsequent comparative studies of tropical rainforest pioneer species of *Cecropia* and *Piper*, and of species of *Buddleja* and *Chenopodium* adapted to a high-altitude lava field, demonstrated that the latter seed responded to an instantaneous light stimulus, corresponding to the sudden exposure of buried seed upon soil disturbance, whereas the pioneer species required long periods of high R:FR ratio light (88). These two ecologically relevant controls on germination may be functions of two different phytochromes, with phyA perceiving the instantaneous light stimuli, and phyB responsible for the longer-term responses to R:FR ratio.

Etiolation and De-etiolation

For most seedlings, etiolation should be thought of as a last-resort strategy. Because of the controls on germination outlined above, the majority of small seeds presumably germinate at or very near the soil surface, and only those plants having large seeds with ample reserves are likely to be able to survive for long in the etiolated state. In consequence, de-etiolation and the assumption of photoautotrophy are critical requirements for survival. The processes that comprise de-etiolation are controlled by several photoreceptors, among which the phytochromes are relatively minor players, and the specific ecological significance of the phytochromes is therefore difficult to determine. Physiological studies indicate that the phytochromes interact with blue-light-absorbing photoreceptors, and with protochlorophyll, to regulate stem elongation, leaf development, and the synthesis and assembly of the photosynthetic apparatus. The *det* and *cop* mutants of *A. thaliana* (20, 24) have demonstrated that etiolation is a developmental pattern in which the normal growth pattern (i.e. de-etiolation) is repressed. Because etiolation is restricted to advanced orders of the plant kingdom, it seems clear that it confers substantial adaptive value, but to date, no studies have investigated its fitness consequences.

A related question is the ecological function of the phyA-mediated FR-HIR, most dramatically seen in the FR-mediated inhibition of hypocotyl extension in etiolated seedlings. The ecological relevance of this phenomenon is difficult to predict, because in theory it would mean that the extension growth of germinating seedlings would be strongly inhibited under vegetation canopies, or under leaf litter, where the radiation is predominantly FR. The adaptive advantage conferred by such a response is not immediately obvious, and it may be necessary to consider more subtle components of the FR-HIR. For example, the capacity of *A. thaliana* seedlings to synthesize chlorophyll and become photoautotrophic is markedly reduced after a sustained treatment with continuous FR, but this effect is absent in *phyA* mutants (H Smith, unpublished data; PH Quail, unpublished data). Perhaps the FR-HIR is responsible for delaying the commitment of scarce resources to leaf and plastid development in unpromising conditions. Consistent with this idea is the observation that the de-etiolation of *A. thaliana phyA* mutants, when compared with WT plants, was severely impaired if grown in deep canopy shade, leading to premature death (MJ Yanofsky, JJ Casal & GC Whitelam, unpublished data). It is conceivable, therefore, that the FR-HIR mediated by phyA could be important for seedling survival under dense canopies.

Proximity Perception and Shade Avoidance

As mentioned above, the transition from the heterotrophic etiolated seedling to the established photoautotrophic plant coincides with the assumption of the capacity to respond to R:FR ratio signals, which, in the natural environment, are associated with radiation reflected from, or filtered through, neighboring vegetation. The concepts of the shade-avoidance syndrome (72, 77), neighbor detection (4, 20), and proximity perception (79) are well established and their ecological significance seems obvious. Early research under controlled conditions (summarized in 72) demonstrated that stem-elongation rate was linearly related to the calculated proportions of Pr and Pfr established by the actinic radiation, and that the slope of the relationship was less for shade-tolerant species and greater for shade-avoiding species. These responses were later shown to be induced by direct irradiation of internodes with FR, and to be surprisingly rapid, with a lag phase of a few minutes (16). Elongation responses to reduced R:FR ratio occur even at sunlight fluence rates (74). Field experiments demonstrated that shade-avoidance responses occur in response to radiation reflected from neighbors before canopy closure and actual shading occurs (4, 38), and that the responses could be eliminated if reflected radiation was filtered through copper sulphate, which absorbs FR (6). In dense stands, elongation responses could be reduced by overhead cuvettes of copper sulphate that maintained a high R:FR ratio within the stand (7). The same group reported, somewhat surprisingly, that the allocation of resources into stem growth as a result of shade-avoidance reactions was not associated with complementary reductions in the allocation of resources to leaf development; indeed, it seems that productivity, measured in terms of plant biomass, was actually increased in conditions in which shade-avoidance reactions were induced (8). In contrast, Kasperbauer (38) has demonstrated marked reallocation of resources from leaf and root development into stem growth in tobacco plants grown at high densities.

Intriguingly, *Portulaca oleracea*, a recumbent weed, apparently detects reflected FR from neighboring vegetation and grows in the opposite direction (54). Similarly, cucumber seedlings grown in the field exhibit negative phototropism to neighboring vegetation and actively project new leaves into light gaps in patchy canopy environments (5). Shade-avoidance reactions are not restricted to herbaceous weeds. Studies of a range of tree species grown at different densities have shown that the growth dynamics of developing or regenerating canopies can be accounted for entirely on the basis of phytochrome-mediated proximity perception and the consequent shade-avoidance reactions (31). All of this evidence is consistent with the hypothesis that the primary function of the phytochromes is to detect fluctuations in the relative proportions of R and FR radiation, thereby providing the plant with information on the light environment (72).

In a thought-provoking essay, Schmitt & Wulff (67) recently drew together physiological observations on phytochrome-mediated shade avoidance and concepts developed by ecologists and population biologists on competition between plants in dense stands. Individuals in plant populations often vary greatly in size, and in dense stands a few dominant individuals may represent the majority of the biomass of that population. Size inequalities are of major significance in plant competition (30), and the concept of asymmetric competition (93) has become widely accepted as a principal basis of plant population structure. Asymmetric competition means that, in a given population, plants that can gain an initial size advantage have a greater chance of capitalizing on that advantage and of surviving to reproduction; in other words, a larger plant is a fitter plant. Because phytochrome-mediated proximity perception provides an early warning of imminent competition for the resource of light, Schmitt & Wulff (67) concluded that plasticity in response to the proximity of neighbors should be of adaptive value. Studies of ecotypes of *Impatiens capensis* (jewelweed) adapted to open and woodland habitats (27) showed that the open ecotypes were more sensitive to reflection signals (i.e. low R:FR ratio) than were the woodland ecotypes, indicating genetic variation in this trait. However, to demonstrate definitively that phytochrome-mediated responses to crowding are adaptive, it must be shown that elongation in dense stands increases relative fitness, while elongation in uncrowded conditions reduces relative fitness. These two predictions of the shade-avoidance hypothesis may now be tested through investigations of mutant and transgenic plants in which shade-avoidance responses are disabled.

To test the prediction that elongation in dense stands has adaptive value, Schmitt et al (66) analyzed the growth of mixed and pure stands of WT and transgenic tobacco that expressed an introduced oat *PHYA* cDNA to moderate levels. In the transgenic plants, the persistent FR-HIR mediated by the transgene prevented the normal phyB-mediated shade-avoidance response, but did not lead to severe dwarfing as is observed with high-level expressers. In dense stands, the transgenic plants, unable to respond to the proximity of neighbors by increased stem elongation, were shown by quantitative assessment of fitness correlates to be at a competitive disadvantage compared with the WT plants. In sparse stands, the fitness of the transgenic and WT plants were not different. Thus, plants in which phytochrome-mediated elongation responses to low R:FR ratio are disabled suffer serious competitive disadvantage when crowded, but not when grown in isolation. In a complementary test, the *ein* mutant of *B. rapa* was used to demonstrate that elongation in sparse stands is demonstrably disadvantageous (66). In dense stands, however, the mutant was much less disadvantaged in comparison with the WT plants. These tests provide critical evidence that proximity perception and shade avoidance, mediated principally by phyB, but involving the action of other phytochromes, represent components of facultative phenotypic plasticity that provide adaptive advantage.

Biotechnological Opportunities

If it is true that proximity perception and shade avoidance represent components of facultative phenotypic plasticity that provide adaptive advantage, then in theory there should be genetic variation in these traits that may be subject to selection within plant breeding programs. It has been argued that the disablement of shade avoidance in crop stands may be of significant value in terms of productivity, because resources committed to stem growth in dense stands may be reallocated to harvestable components (76). Although this view has been questioned in experiments in which the opportunity cost of shade avoidance was claimed to be very small (8), controlled-release field experiments with transgenic tobacco expressing an oat *PHYA* cDNA to moderate levels have demonstrated that the disablement of shade-avoidance responses leads to markedly different phenology in the field (H Smith, PRH Robson & AC McCormac, unpublished data). At low planting densities, the WT and transgenic plants had similar architecture, but as neighbor proximity increased, the WT plants allocated increasingly greater resources to stem growth at the expense of leaf development and thus became long and straggly. The transgenic *PHYA* expressers, on the other hand, became shorter at high densities, and the ratio of stem-to-leaf biomass did not increase as strongly as in the WT controls. These data demonstrate that transgenic manipulation of the relative levels of phyA and phyB can markedly modify the allocation of assimilates in crop plants in the field. For the majority of crop plants, the relationship between harvestable yield and total biomass is quite constant; if genetic engineering designed to disable the responses of plants to their neighbors results in improved ratios of yield to biomass, then increased productivity per unit of agricultural resource may be expected. There is great potential for applying genetic engineering of the *PHY* genes to the improvement of crop plants, especially because any improvements will not require changes in husbandry and, in principle, will therefore be ideal for application in areas of unsophisticated agriculture.

TOWARD A LIFE HISTORY OF PHOTOMORPHOGENESIS

Evidence derived from mutant and transgenic plants, when taken together with the vast body of information on the photobiology of wild-type plants, is reaching a stage when the overall significance of phytochrome-mediated photomorphogenesis throughout the plant's life history can be constructed. Using mainly *A. thaliana* as a paradigm but with additional information from other species, Figure 1 summarizes the contributions of the different phytochromes, as presently known, to the major developmental processes regulated by light; i.e. germination, etiolation/de-etiolation, shade avoidance, and photoperiodism. Both phyA and phyB contribute to photomorphogenic processes at all stages of plant development, from germination to flowering, but the relative importance of the two phytochromes changes throughout development.

Both phyA and phyB are capable of regulating seed germination; phyA may be responsible (via a VLFR) for the detection of very brief pulses of light during soil disturbance, and phyB may be responsible for graded responses to R:FR ratio under vegetation canopies. Additionally, FR-HIR regulation of germination is well known, and such a response may contribute, with the phyB-mediated R:FR ratio response, to the maintenance of dormancy under leaf litter.

The sharing of responsibility between phyA and phyB is most clearly seen in the regulation of cell extension. When an etiolated *A. thaliana* seedling is first exposed to light, its extension growth is inhibited; if the light to which it is exposed has a low R:FR ratio, then the inhibition, via the phyA-mediated FR-HIR, is very strong. As de-etiolation proceeds, the inhibitory action of low

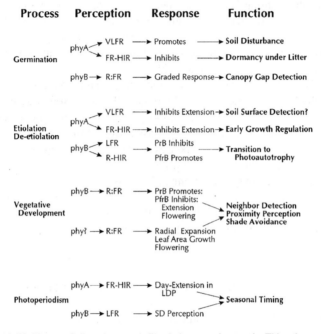

Figure 1 A life history of phytochrome-mediated photomorphogenesis. This scheme attempts to summarize the principal roles and functions of the phytochromes. In the left column, the life history of the plant is arbitrarily divided into the physiological phenomena of germination, etiolation/de-etiolation, vegetative development, and photoperiodism. Current knowledge on the regulation of these processes by individual phytochromes, operating through recognized response modes, is outlined in the center columns. The column on the right indicates the likely ecological significance of these physiological processes.

R:FR ratios is replaced by a strong promotion, mediated principally by phyB (49). This transition coincides with the degradation and loss of phyA down to low steady-state levels, and this removal of phyA upon de-etiolation is an essential prerequisite for the subsequent successful operation of the shade-avoidance syndrome. In the case of *A. thaliana*, the transition also coincides with the elongation phase of the hypocotyl, which may explain why the photophysiology of the hypocotyl is in many species substantially different from that of subsequent internodes. From a photomorphogenic viewpoint, the hypocotyl is a transitional organ. Subsequent growth and development is predominantly regulated by phyB, via R:FR ratio perception, but underlying this control, phyA still exerts an opposing effect. There is strong evidence that some separate components of shade avoidance are mediated by one or more phytochromes other than phyA or phyB. These phenomena include the radial expansion of stems and the growth in leaf area, leading to the speculation that phyA and phyB predominantly control polar cell-elongation, while another phytochrome is responsible for nondirectional cell enlargement (78). Another phenomenon partly under the control of a phytochrome other than phyA or phyB is the acceleration of flowering that is observed under low R:FR ratios. Flowering rate in facultative long-day plants such as *A. thaliana* may be a distinct phenomenon from that of floral induction, involving separate regulation by one or more phytochromes. The existence of phytochrome-mediated phenomena that are displayed in *phyA* and *phyB* mutants indicates that screens for mutants in other phytochromes might be devised.

The functions of the phytochromes in photoperiodic perception are the least well understood. A role for phyA in the perception of day extensions in long-day plants, via a FR-HIR, seems to be established, as does a role for phyB in the perception of short days, both for flowering and tuber induction.

Bearing in mind the demonstrated importance of phytochrome concentrations, it is obvious that fine control of individual *PHY* gene expression must be exerted throughout the life cycle. Variation in the regulation of *PHY* gene expression throughout the life cycle and between different species should be expected; indeed, such variation may be a necessary correlate of the adaptive plasticity conferred by the phytochrome family. One of the several advantages of gene families is functional flexibility (75). This flexibility is particularly evident within the phytochrome family, in those mutants in which other phytochromes partially substitute for an absent family member, and in those transgenic plants in which the member present in excess takes over the functions of its relatives, rather like some human families perhaps. In the natural state the flexibility conferred by plurality may be vital to the proper adaptation of the plant to its environment, requiring all members of the family to function both individually and in consort to take advantage of opportunities and resist constraints throughout the life cycle.

Literature Cited

1. Adamse P, Jaspers PAP, Bakker JA, Wesselius JC, Heeringa GH, et al. 1988. Photophysiology of a tomato mutant deficient in labile phytochrome. *J. Plant Physiol.* 133: 436–40
2. Ahmad M, Cashmore AR. 1993. *HY4* gene of *Arabidopsis thaliana* encodes a protein with characteristics of a blue-light photoreceptor. *Nature* 306:162–66
3. Bagnall D, King RW, Whitelam GC, Boylan MT, Wagner D, Quail PH. 1994. Flowering responses to altered expression of phytochrome in mutants and transgenic lines of *Arabidopsis thaliana*. *Abstr.4th Int. Congr. Plant Mol. Biol.*, p. 1060
4. Ballaré CL, Sánchez RA, Scopel AL, Ghersa CM. 1987. Early detection of neighbour plants by phytochrome perception of spectral changes in reflected sunlight. *Plant Cell Environ.* 10:551–57

5. Ballaré CL, Scopel AL, Radosevich SR, Kendrick RE. 1992. Phytochrome-mediated phototropism in de-etiolated seedlings. *Plant Physiol.* 100:170–77

6. Ballaré CL, Scopel AL, Sánchez RA. 1990. Far-red radiation reflected from adjacent leaves: an early signal of competition in plant canopies. *Science* 247:329–32

7. Ballaré CL, Scopel AL, Sánchez RA. 1991. Photocontrol of stem elongation in plant neighbourhoods: effects of photon fluence rate under natural conditions of radiation. *Plant Cell Environ.* 14: 57–65

8. Ballaré CL, Scopel AL, Sánchez RA. 1991. On the opportunity cost of the photosynthate invested in stem elongation reactions mediated by phytochrome. *Oecologia* 86: 561–67

9. Boylan MT, Douglas N, Quail PH. 1994. Dominant negative suppression of *Arabidopsis* photoresponses by mutant phytochrome A sequences identifies spatially discrete regulatory domains in the photoreceptor. *Plant Cell* 6:449–60

10. Boylan MT, Quail PH. 1989. Oat phytochrome is biologically active in transgenic tomatoes. *Plant Cell* 1:765–73

11. Boylan MT, Quail PH. 1991. Phytochrome A overexpression inhibits hypocotyl elongation in transgenic *Arabidopsis*. *Proc. Natl. Acad. Sci. USA* 88:10806–10

12. Carr-Smith HD, Johnson CB, Plumpton C, Butcher GW, Thomas B. 1994. The kinetics of type 1 phytochrome in green, light-grown wheat (*Triticum aestivum* L.). *Planta* 194:136–42

13. Carr-Smith HD, Johnson CB, Thomas B. 1989. Action spectrum for the effect of day-extensions on flowering and apex elongation in green light-grown wheat (*Triticum aestivum* L.). *Planta* 179:428–32

14. Casal JJ, Sánchez RA, Vierstra RD. 1994. *Avena* phytochrome A overexpressed in transgenic tobacco seedlings differentially affects red/far-red reversible and very-low-fluence-responses (cotyledon unfolding) during de-etiolation. *Planta* 192:306–9

15. Cherry JR, Hershey HP, Vierstra RD. 1991. Characterization of tobacco expressing functional oat phytochrome. *Plant Physiol.* 96:775–85

16. Child R, Smith H. 1987. Phytochrome action in light-grown mustard: kinetics fluence-rate compensation and ecological significance. *Planta* 172:219–29

17. Childs KL, Cordonnier-Pratt M-M, Pratt LH, Morgan PW. 1992. Genetic regulation of development in *Sorghum bicolor*. VII. ma_3^R flowering mutant lacks a phytochrome that predominates in green tissue. *Plant Physiol.* 99:765–70

18. Childs KL, Pratt LH, Morgan PW. 1991. Genetic regulation of development in *Sor-ghum bicolor*. V. The ma_3^R allele results in abnormal phytochrome physiology. *Plant Physiol.* 97:714–19

19. Chory J. 1992. A genetic model for light-regulated seedling development in *Arabidopsis*. *Development* 115:337–54

20. Chory J, Peto CA, Feinbaum R, Pratt LH, Ausubel F. 1989. *Arabidopsis thaliana* mutant that develops as a light-grown plant in the absence of light. *Cell* 58:991–99

21. Clack T, Mathews S, Sharrock RA. 1994. The phytochrome apoprotein family in *Arabidopsis* is encoded by five genes: the sequences and expression of *PHYD* and *PHYE*. *Plant Mol. Biol.* 25:413–27

22. Dehesh K, Franci C, Parks BM, Seeley KA, Short TW, et al. 1993. *Arabidopsis* HY8 locus encodes phytochrome A. *Plant Cell* 5:1081–88

23. Deitzer GF, Hayes R, Jabben M. 1979. Kinetics and time-dependence of the effect of far-red light on photoperiodic induction of flowering in winter barley. *Plant Physiol.* 64:1015–21

24. Deng XW, Quail PH. 1992. Genetic and phenotypic characterization of *cop-1* mutants of *Arabidopsis thaliana*. *Plant J.* 2:83–95

25. Devlin PF, Rood SB, Somers DE, Quail PH, Whitelam GC. 1992. Photophysiology of the elongated internode (*ein*) mutant of *Brassica-rapa ein* mutant lacks a detectable phytochrome b-like polypeptide. *Plant Physiol.* 100:1442–47

26. Downs RJ, Hendricks SB, Borthwick HA. 1957. Photoreversible control of elongation in pinto beans and other plants under normal conditions of growth. *Bot. Gaz.* 118: 199–208

27. Dudley SA, Schmitt J. 1994. Genetic differentiation between open and woodland *Impatiens capensis* populations in morphological responses to simulated foliage shade. *Funct. Ecol.* In press

28. Foster KR, Miller FR, Childs KL, Morgan PW. 1994. Genetic regulation of development in *Sorghum bicolor*. VIII. Shoot growth, tillering, flowering, gibberellin biosynthesis and phytochrome levels are differentially affected by dosage of the ma_3^R allele. *Plant Physiol.* 105:941–48

29. Furuya M. 1993. Phytochromes: their molecular species, gene families and functions. *Annu. Rev. Plant Physiol. Plant Mol. Biol.* 44:617–45

30. Geber M. 1989. Interplay of morphology and development on size inequality: a *Polygonum* greenhouse study. *Ecol. Monogr.* 59:267–88

31. Gilbert IR, Seavers GP, Jarvis PG, Smith H. 1995. Photomorphogenesis and canopy dynamics. Phytochrome-mediated proximity perception accounts for the growth dynamics of canopies of *Populus trichocarpa* x

deltoids 'Beaupré.' *Plant Cell Environ.* 18:In press

32. Goto N, Kumagai T, Koornneef M. 1991. Flowering responses to night-breaks in photomorphogenic mutants of *Arabidopsis thaliana*, a long-day plant. *Physiol. Plant* 83:209–16

33. Halliday KJ, Koornneef M, Whitelam GC. 1994. Phytochrome B and at least one other phytochrome mediate the accelerated flowering response of *Arabidopsis thaliana* to low red/far-red ratio. *Plant Physiol.* 104: 1311–14

34. Hartmann KM. 1966. A general hypothesis to interpret 'high energy phenomena' of photomorphogenesis on the basis of phytochrome. *Photochem. Photobiol.* 5:349–66

35. Hauser B, Cordonnier-Pratt M-M, Pratt LH. 1994. Differential expression of five phytochrome genes in tomato (*Lycopersicon esculentum* Mill.). *Plant Physiol.* 105 (Suppl.):72

36. Jackson SD, Heyer A, Prat S. 1994. Antisense potato plants with reduced levels of phytochrome B. *Abstr.4th Int. Congr. Plant Mol. Biol.,* p. 1067

37. Johnson E, Bradley M, Harberd NP, Whitelam GC. 1994. Photoresponses of light-grown *phyA* mutants of *Arabidopsis*: phytochrome A is required for the perception of daylength extensions. *Plant Physiol.* 105:141–49

38. Kasperbauer M. 1987. Far-red light reflection from green leaves and effects on phytochrome-mediated assimilate partitioning under field conditions. *Plant Physiol.* 85: 350–54

39. Kay SA, Nagatani A, Keith B, Deak M, Furuya M, Chua N-H. 1989. Rice phytochrome is biologically active in transgenic tobacco. *Plant Cell* 1:765–73

40. Keller JM, Shanklin J, Vierstra RD, Hershey HP. 1989. Expression of a functional monocotyledonous phytochrome in transgenic tobacco. *EMBO J.* 8:1005–12

41. Kendrick RE, Kerckhoffs LHJ, Pundsnes AS, van Tuinen A, Koornneef M, et al. 1994. Photomorphogenic mutants of tomato. *Euphytica.* In press

41a. Kendrick RE, Kronenberg GHM, eds. 1994. *Photomorphogenesis in Plants.* Dordrecht: Kluwer. 2nd ed.

42. Koornneef M, Kendrick RE. 1994. Photomorphogenic mutants of higher plants. See Ref. 41a, pp. 601–30

43. Koornneef M, Rolff E, Spruit CJP. 1980. Genetic control of light-inhibited hypocotyl elongation in *Arabidopsis thaliana* L. *Z.Pflanzenphysiol.* 100:147–60

44. Liscum E, Hangarter RP. 1993. Genetic evidence that the red-absorbing form of phytochrome B modulates gravitropism in *Arabidopsis thaliana. Plant Physiol.* 103: 15–19

45. Lopez-Juez E, Nagatani A, Tomizawa K-I, Deak M, Kern R, et al. 1992. The cucumber long hypocotyl mutant lacks a light-stable PHYB-like phytochrome. *Plant Cell* 4: 241–51

46. Mancinelli AL. 1994. The physiology of phytochrome action. See Ref. 41a, pp. 211–70

47. McCormac AC, Cherry JR, Hershey HP, Vierstra RD, Smith H. 1991. Photoresponses of transgenic tobacco plants expressing an oat phytochrome gene. *Planta* 185:162–70

48. McCormac AC, Smith H, Whitelam GC. 1993. Expression of oat-phyA-cDNA in a suspension cell culture of transgenic tobacco: a single-cell system for the study of phytochrome function. *J. Exp. Bot.* 44: 1095–1103

49. McCormac AC, Wagner D, Boylan MT, Quail PH, Smith H, Whitelam GC. 1993. Photoresponses of transgenic *Arabidopsis* seedlings expressing introduced phytochrome B-encoding cDNAs: evidence that phytochrome A and phytochrome B have distinct photoregulatory functions. *Plant J.* 4:19–27

50. McCormac AC, Whitelam GC, Boylan MT, Quail PH, Smith H. 1992. Contrasting responses of etiolated and light-adapted seedlings to red-far-red ratio—a comparison of wild-type mutant and transgenic plants has revealed differential functions of members of the phytochrome family. *J. Plant Physiol.* 140:707–14

51. McCormac AC, Whitelam GC, Smith H. 1992. Light-grown plants of transgenic tobacco expressing an introduced oat phytochrome A gene under the control of a constitutive viral promoter exhibit persistent growth inhibition by far-red light. *Planta* 188:173–81

52. Nagatani A, Reed JW, Chory J. 1993. Isolation and initial characterisation of Arabidopsis mutants that are deficient in phytochrome A. *Plant Physiol.* 102:269–77

53. Nagatani A, Reid JB, Ross JJ, Dunnewijk A, Furuya M. 1990. Internode length in *Pisum.* The response to light quality, and phytochrome type I and II levels in *lv* plants. *J. Plant Physiol.* 135:667–74

54. Novoplansky A, Cohen D, Sachs T. 1990. How *Portulaca* seedlings avoid their neighbors. *Oecologia* 82:490–93

55. Parkinson JS, Kofoid EC. 1992. Communication modules in bacterial signalling proteins. *Annu. Rev. Genet.* 26:71–112

56. Parks BM, Jones AM, Adamse P, Koornneef M, Kendrick RE, Quail PH. 1987. The *aurea* mutant of tomato is deficient in spectrophotometrically and immunochemically detectable phytochrome. *Plant Mol. Biol.* 9:97–107

57. Parks BM, Quail PH. 1991. Phytochrome-

deficient *hy1* and *hy2* mutants of *Arabidopsis* are defective in phytochrome chromophore biosynthesis. *Plant Cell* 5:39–48

58. Parks BM, Quail PH. 1993. *hy8*, a new class of *Arabidopsis* long hypocotyl mutants deficient in functional phytochrome A. *Plant Cell* 5:39–48

59. Pratt LH, Cordonnier-Pratt M-M, Hauser B, Kochert G, Caboche M. 1994. Phytochrome gene family in tomato (*Lycopersicon esculentum* Mill.) and sorghum (*Sorghum bicolor* [L.] Moench). *Abstr.4th Int. Congr. Plant Mol. Biol.*, pp. 1057

60. Quail PH, Briggs WR, Chory J, Hangarter RP, Harberd NP, et al. 1994. Spotlight on phytochrome nomenclature. *Plant Cell* 6: 468–71

61. Reed JW, Chory J. 1994. Mutational analyses of light-controlled seedling development in *Arabidopsis*. *Semin. Cell Biol.* 5:327–34

62. Reed JW, Nagatani A, Elich TD, Fagan M, Chory J. 1994. Phytochrome A and phytochrome B have overlapping but distinct functions in *Arabidopsis* development. *Plant Physiol.* 104:1139–49

63. Reed JW, Nagpal P, Poole DS, Furuya M, Chory J. 1993. Mutations in the gene for the red/far-red light receptor phytochrome B alter cell elongation and physiological responses throughout *Arabidopsis* development. *Plant Cell* 5:147–57

64. Deleted in proof

65. Robson PRH, Whitelam GC, Smith H. 1993. Selected components of the shade-avoidance syndrome are displayed in a normal manner in mutants of *Arabidopsis thaliana* and *Brassica rapa* deficient in phytochrome B. *Plant Physiol.* 102:1179–84

66. Schmitt J, McCormac AC, Smith H. 1995. A test of the adaptive plasticity hypothesis using transgenic and mutant plants disabled in phytochrome-mediated elongation responses to foliage shade. *Am. Nat.* In press

67. Schmitt J, Wulff RD. 1993. Light spectral quality phytochrome and plant competition *Trends Ecol. Evol.* 8:47–51

68. Scopel AL, Ballaré CL, Sánchez RA. 1991. Induction of extreme light sensitivity in buried weed seeds and its role in the perception of soil cultivations. *Plant Cell Environ.* 14:501–8

69. Sharma R, Lopez-Juez E, Nagatani A, Furuya M. 1993. Identification of photo-inactive phytochrome A in etiolated seedlings and photo-active phytochrome B in green leaves of the *aurea* mutant of tomato. *Plant J.* 4:1035–42

70. Sharrock RA, Quail PH. 1989. Novel phytochrome sequences in *Arabidopsis thaliana*: structure evolution and differential expression of a plant regulatory photoreceptor family. *Genes Dev.* 3:534–44

71. Shinomura T, Nagatani A, Elich TD, Fagan M, Chory J. 1994. The induction of seed germination in *Arabidopsis thaliana* is regulated principally by phytochrome B and secondarily by phytochrome A. *Plant Physiol.* 104:363–71

72. Smith H. 1982. Light quality, photoperception and plant strategy. *Annu. Rev. Plant Physiol.* 33:481–518

73. Smith H. 1983. Is Pfr the active form of phytochrome? *Philos. Trans. R. Soc. London* 303:443–52

74. Smith H. 1990. Phytochrome action at high photon fluence rates: rapid extension rate responses of light-grown mustard to variations in fluence rate and red:far-red ratio *Photochem. Photobiol.* 52:131–42

75. Smith H. 1990. Signal perception, differential expression within multigene families, and the molecular basis of phenotypic plasticity. *Plant Cell Environ.* 13:585–94

76. Smith H. 1992. Ecology of photomorphogenesis: clues to a transgenic programme of crop plant improvement *Photochem. Photobiol.* 56:815–22

77. Smith H. 1994. Sensing the light environment: the functions of the phytochrome family. See Ref. 41a, pp. 377–416

78. Smith H. 1994. Phytochrome transgenics: functional, ecological and biotechnological applications. *Semin. Cell Biol.* 5:315–25

79. Smith H, Casal JJ, Jackson GM. 1990. Reflection signals and the perception by phytochrome of the proximity of neighbouring vegetation. *Plant Cell Environ.* 13:73–78

80. Smith H, Fork DC. 1992. Direct measurement of phytochrome photoconversion intermediates at high photon fluence rates. *Photochem. Photobiol.* 56:599–606

81. Deleted in proof

82. Smith H, Turnbull M, Kendrick RE. 1992. Light-grown plants of the cucumber long hypocotyl mutant exhibit both long-term and rapid elongation responses to irradiation with supplementary far-red light. *Photochem. Photobiol.* 56:607–10

83. Smith H, Whitelam GC. 1990. Phytochrome, a family of photoreceptors with multiple physiological roles. *Plant Cell Environ.* 13:695–707

84. Somers DE, Sharrock RA, Tepperman JM, Quail PH. 1991. The *hy3* long hypocotyl mutant of *Arabidopsis* is deficient in phytochrome-b. *Plant Cell* 3:1263–74

85. Stockhaus J, Nagatani A, Halfer U, Kay S, Furuya M, Chua N-H. 1992. Serine-to-alanine substitutions at the amino-terminal region of phytochrome A result in an increase in biological activity. *Genes Dev.* 6:2364–72

86. Thomas B. 1991. Phytochrome and photoperiodic induction. *Physiol. Plant* 81: 571–77

87. Vázquez-Yanes C, Orozco-Segovia A.

1987. Light-gap detection by the photoblastic seeds of *Cecropia obtusifolia* and *Piper auritum*, two tropical rain forest trees. *Biol. Plant. (Praha).* 29:234–38

88. Vázquez-Yanes C, Orozco-Segovia A. 1990. Ecological significance of light-controlled seed germination in two contrasting tropical habitats. *Oecologia* 83:171–75

89. Vázquez-Yanes C, Smith H. 1982. Phytochrome control of seed germination in the tropical rain forest pioneer trees *Cecropia obtusifolia* and *Piper auritum* and its ecological significance. *New Phytol.* 92:477–85

90. Wagner D, Tepperman JM, Quail PH. 1991. Overexpression of phytochrome-B induces a short hypocotyl phenotype in transgenic *Arabidopsis. Plant Cell* 3:1275–88

91. Wang YC, Cordonnier-Pratt MM, Pratt LH. 1993. Temporal and light regulation of the expression of 3 phytochromes in germinating seeds and young seedlings of *Avena sativa* L. *Planta* 189:384–90

92. Wang YC, Cordonnier-Pratt MM, Pratt LH. 1993. Spatial distribution of 3 phytochromes in dark-grown and light-grown *Avena sativa* L. *Planta* 189:391–96

93. Weiner J. 1990. Asymmetric competition in plant populations. *Trends Ecol. Evol.* 5: 360–64

94. Weller JL, Reid JB. 1993. Photoperiodism and photocontrol of stem elongation in two photomorpogenic mutants of *Pisum sativum* L. *Planta* 189:15–23

95. Wester L, Somers DE, Clack T, Sharrock RA. 1994. Transgenic complementation of the *hy3* phytochrome B mutation and response to *PHYB* gene copy number in *Arabidopsis. Plant J.* 5:261–72

96. Whitelam GC, Harberd NP. 1994. Action and function of phytochrome family members revealed through the study of mutant and transgenic plants. *Plant Cell Environ.* 17:615–25

97. Whitelam GC, Johnson E, Peng J, Carol P, Anderson ML, et al. 1993. Phytochrome A null mutants of *Arabidopsis* display a wild-type phenotype in white light. *Plant Cell* 5:757–68

98. Whitelam GC, McCormac AC, Boylan MT, Quail PH. 1992. Photoresponses of *Arabidopsis* seedlings expressing an introduced oat phyA gene: persistence of etiolated plant type responses in light-grown plants. *Photochem. Photobiol.* 56:617–21

99. Whitelam GC, Smith H. 1991. Retention of phytochrome-mediated shade avoidance responses in phytochrome-deficient mutants of *Arabidopsis,* cucumber and tomato. *J. Plant Physiol.* 139:119–25

Annu. Rev. Plant Physiol. Plant Mol. Biol. 1995. 46:317–39

CELL CYCLE CONTROL

Thomas W. Jacobs

Department of Plant Biology, University of Illinois, Urbana, Illinois 61801

KEY WORDS: mitosis, cyclin, cyclin-dependent kinase, cell division, protein kinase

CONTENTS

ABSTRACT

Cell division in plants serves to subdivide space in the expanding organism rather than to supply cellular building blocks, as it does in animals. Despite these contrasting developmental contexts, transitions through and between stages of the mitotic cell cycle are controlled in both kingdoms by related families of cyclin-dependent kinases (CDKs) and their regulatory subunits (cyclins). Passage of animal cells through the G1 and S phases of the cycle is controlled by complexes of CDKs, cyclins, transcription factors, tumor supressor proteins, and inhibitor proteins. None of these complexes has been identified in plants. Most plant cell cycle regulators characterized to date resemble those active in the G2-M transition in yeast and animal cells, although bio-chemical demonstrations of their roles in plants are lacking. Also awaiting elucidation in plant systems are the essential connections between CDK-cyclin activities and hormonal or developmental signals.

INTRODUCTION

Plant growth and morphogenesis are the products of localized cell division and anisotropic cell expansion. Multicellularity and the absence of cell migration establish a unique developmental context in which cell division must be con-trolled in plants. This review provides, first, an overview of the cell cycle control model that has emerged in recent years and, second, an account of progress-to-date in applying this model to plant systems. Readers are also referred to other reviews on plant cell division (10, 11, 26–28, 52, 66, 94, 120) and to more thorough treatments of specific topics, including the eukaryotic cell cycle (85), yeast cell cycle control (25, 86), animal cell cycles (90), checkpoints (84, 109, 114), cyclin-dependent kinases (89, 101, 105, 116), cyclins (104), retinoblastoma protein (47), E2F/DRTF1 (62), and MAP kinase (53, 111).

Growth Control vs Checkpoint Control

Signal transduction pathways that guide cells into and out of the proliferative state constitute growth control in animal cell parlance. These pathways con-nect cellular physiology with the mitotic control engine per se. Growth con-trols in plants, then, are involved in seed germination, sprouting in bulbs, bud break, the wound response, lateral root initiation, induction of cell division by pathogens and rhizobia, and activation of shoot apical meristems during the floral transition. Cell cycle and checkpoint controls, on the other hand, oversee the passage of proliferating cells between the phases of the mitotic cycle—gap 1 (G1), DNA synthesis (S), gap 2 (G2), and mitosis (M) (Figure 1). Cancer research has clarified many features of growth control in animal systems, but far less is known on the botanical side.

Growing, multicellular eukaryotes consist entirely of cell lineages, the youngest members of which are likely to be in the proliferative state. In plants, these dividing cells are confined largely to meristems and cambia, having attained their proliferative epigenetic state by inheriting it somatically from embryonic forebears. The developmental timing of each lineage branch's ces-sation of division is an integral component of morphogenesis. Whether mitotic retirement in plants is triggered by the synthesis of a negative growth control signal or the depletion of a positive one, or both, is not known. Proliferative vs differentiating states can be mutually exclusive in plants (68). Our under-standing of the role of cell division in plant development will increase greatly

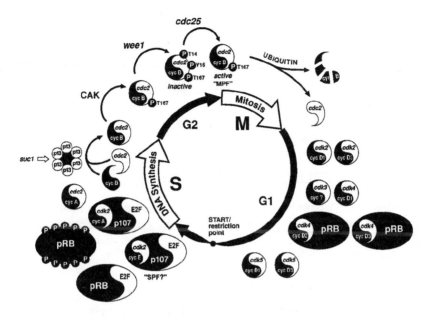

Figure 1 General scheme for eukaryotic cell cycle control. Protein complexes are shown at approximately the times they occur during cell cycle passage. The eight CDK-cyclins depicted in G1 phase are ordered by CDK number, not by the order in which they form or act, because the latter is not known in many cases. MPF = M-phase promoting factor, SPF = S-phase promoting factor. Other abbreviations are explained in the text.

as more information becomes available regarding the integration of growth vs cell cycle controls.

Regulators are needed to shepherd cells from one cell cycle phase to the next. The evolution of such controls arises out of the substantial biochemical and mechanical demands of mitosis and the requirement that each cell receives an intact genome after each division. Immediately following its birth at cytokinesis, the cell passes through G1, the first gap phase. Late in this phase, it choses among four fates: divide, arrest, differentiate, or senesce. This decision point is called START in yeast and R, the restriction point, in animal cells. It is in the best interests of the cell's somatic progeny, and thus the organism of which it is a part, that the cell not proceed into S without confirming that conditions are auspicious for another mitosis. For example, has the last mitosis been completed? Has the cell enlarged sufficiently since the last cytokinesis? Will ambient conditions support a complete genome replication? Likewise, the cell must not enter M-phase from G2 unless DNA synthesis is complete and damaged DNA repaired. At the end of mitosis, septation must not occur unless all chromosomes have condensed, congressed, and migrated to the spindle poles. These decision gates of the cell cycle are called checkpoints and were defined in yeast genes whose products arrest cycling cells until an essential condition is satisfied (12, 36, 84). The best-characterized checkpoints occur at G1-S and G2-M.

A tangle of details separates the yeast, mammalian, amphibian, and marine invertebrate model systems from which the consensus molecular paradigm has arisen. The following generalized account disregards many such complications in the interest of pedagogic clarity.

THE CELL CYCLE CONTROL PARADIGM

Cyclin-dependent protein kinases (CDKs) guard the checkpoints of the eukaryotic cell cycle (Figure 1). These enzymes are regulated primarily at the posttranslational level, by phosphorylation and by association with cyclin regulatory subunits (99).

$p34^{cdc2}$ Kinase and its Regulation

CDKs catalyze the transfer of a phosphate from ATP to specific serine or threonine residues on regulatory and structural proteins, the aggregate modification of which drives cells through cell cycle checkpoints (89). The prototype CDK is $p34^{cdc2}$, a 34-kDa protein kinase and product of the *Schizosaccharomyces pombe* (fission yeast) *cdc2* gene. $p34^{cdc2}$ holds the honorary designation of CDK1, although this identifier is rarely used. Mutant yeast carrying conditional null *cdc2* alleles arrest at either START (G1-S) or at G2-M (92), depending on their position in the cycle at the time the restrictive condition is imposed. This pleiotropy demonstrates that fission yeast require a single gene product to effect two very different cellular transitions—the initiation of DNA synthesis (G1-S) and entry into mitosis (G2-M). Budding yeast, *Saccharomyces cerevisiae,* rely on a highly homologous CDK, the CDC28 gene product, for the same purposes (106).

The consensus phosphorylation target for $p34^{cdc2}$ is S/T-P-X-Z [S = serine, T = threonine, P = proline, X = polar amino acid, Z = basic amino acid (81)]. $p34^{cdc2}$ is simply a serine-threonine protein kinase catalytic domain. Its invariant signature motifs include the GEGTYG variation of the GxGxxG (x = any amino acid) element found in the amino-terminal ATP-binding subdomain (I) of all serine-threonine protein kinases, the WYRAPE version of xxAPE in subdomain VIII, the HRDLKPQN version of HxDxxxN in subdomain VI, and the highly conserved EGVPSTAIREISLLKE (so-called PSTAIRE) motif in subdomain III (34). Although all functional $p34^{cdc2}$ homologs have a PSTAIRE domain, its presence does not assure a cell cycle role for every protein kinase that carries it (54).

The activation of yeast p34^{cdc2} kinase at G1-S and G2-M initiates DNA synthesis and mitosis, respectively. The same basic mechanism operates in animal and, presumably, plant cells, except G1-S regulation is more complex and involves additional CDKs in these cells. p34^{cdc2} is regulated at the levels of transcription, protein abundance, posttranslational modification (phosphorylation), and association with regulatory subunits (cyclins). Although activation of p34^{cdc2} at G1-S and G2-M is primarily posttranslational, plant studies have focused on the abundance and distribution of its transcript and polypeptide. The disproportionately heavy treatment of regulation at this level is prompted by trends in the extant botanical literature.

Variation in p34^{cdc2} mRNA and protein abundance has been examined in yeast and animal cells: (*a*) from one phase of the cycle to the next, (*b*) during shifts between proliferative and nonproliferative states, and (*c*) during cell differentiation. Alterations in the level of *cdc2* mRNA and protein play an important, albeit secondary, role in cell cycle transitions. The abundance of *cdc2* mRNA in yeast cells remains constant throughout the mitotic cycle (128). In animal cells, both the steady-state level of *cdc2* mRNA and its rate of synthesis vary in a cell cycle–dependent manner: *cdc2* transcript abundance and synthesis rate are low throughout G1 and rise five- and tenfold, respectively, between S and M (9, 73, 127). Consistent with observed kinetics, the human *cdc2* gene promoter contains elements subject to activation by the E2F and *c-myb* transcription factors, both known to act primarily at G1-S (9, 60). We return below to the role of E2F.

Despite phase-to-phase fluctuations in *cdc2* mRNA abundance, the steady-state level of p34^{cdc2} protein remains constant with respect to total cellular protein in yeast or animal cells (13, 56, 61, 127). p34^{cdc2} protein homeostasis is reflected in the inability of *cdc2* transcript overproduction to alter yeast cell cycle kinetic parameters (17). Yet this stability belies underlying variation in synthesis and degradation. In animal cells, the half-life of p34^{cdc2} protein drops twofold at G1-S, concomitant with its transcriptional activation (127). The p34^{cdc2} pool is thereby 75–85% turned over in late G1. Because p34^{cdc2} kinase is regulated predominantly by phosphorylation (see below), refreshing its pool at G1-S may obviate the need for a specific phosphatase to return it to its unphosphorylated ground state after mitosis. In summary, then, the transient appearance of p34^{cdc2} kinase activity at the G1-S and G2-M transitions is not attributable to spikes in the enzyme's abundance within the cell.

Do changes in *cdc2* expression mediate shifts into or out of the proliferative state? Fission yeast cells pass into stationary phase with no decline in *cdc2* transcript level (17). Similarly, a variety of growth-inhibitory conditions do not affect *cdc2* mRNA levels in animal cells (108). During serum or mitogen stimulation, *cdc2* mRNA and protein increase dramatically, although not equivalently, in various animal cell systems (14, 29, 63, 108). Antisense inhibition of *cdc2* expression during stimulation of cultured cells reduces DNA synthesis but does not affect other molecular events that occur during reentry into the cycle through G1 (29). These observations suggest that *cdc2* expression merely reflects, but does not determine single-handedly, the proliferative state of animal cells.

How do *cdc2* message and protein levels behave during cell differentiation? Overexpression of *cdc2* in cultured muscle cells fails to inhibit differentiation, suggesting that a decline in *cdc2* expression is not necessary for differentiation (1). p34^{cdc2} protein declines in cultured animal cells only after in vitro differentiation is well under way (30, 40, 56). Rough correlations have been drawn between *cdc2* gene product level and the proliferative state of embryonic chicken tissues (59). Thus, in animal cells, p34^{cdc2} pool size is relatively insensitive to growth control signals and only declines as a consequence of differentiation.

Senescence is one path of terminal differentiation. Senescent fibroblasts fail to initiate S-phase, divide, or accumulate *cdc2* transcripts or protein upon serum or growth factor stimulation (108, 121, 125). Introduction of additional, expressed copies of the *cdc2* gene into these cells, by nuclear microinjection (121) or transfection (108), fails to overcome the developmental block imposed by the senescence program. Thus, *cdc2* expression alone is not sufficient to drive developmentally arrested cells back into the proliferative state.

Other Cyclin-Dependent Protein Kinases

cdc2 and *CDC28* are sufficient to control the principal checkpoints in fission and budding yeast cell cycles, respectively. However, cells of metazoans must respond to a greater complexity of signals, particularly during G1. To support their developmental flexibility, animal cells have evolved families of CDKs that act exclusively during G1 and S. This repertiore includes CDK2, CDK3, CDK4, CDK5, and CDK6 (99). The human genome carries at least 11 p34^{cdc2}-related kinases (76). Only three of these (*cdc2*, CDK2, and CDK3) rescue yeast *CDC28* mutants. It is unknown whether these CDKs represent sentries at bona fide checkpoints or functionally redundant enzymes whose diversity affords the cell more signal-response options.

CDK2 (originally identified as p33 or Eg1) closely resembles *cdc2* in primary structure (110) and participates sequentially in at least three complexes during G1 and S (64). Early in G1, CDK2 associates with a D-type cyclin (see below), but the role of this complex is unclear. At G1-S, it changes partners to cyclin E and triggers the initiation of DNA synthesis (123). The large complex consisting of CDK2, cyclin E, and other proteins is therefore likely to be the S-phase promoting factor (SPF) (Figure 1) (57). Later, during

S-phase, CDK2's association with cyclin A is required for sustained DNA synthesis (Figure 1) (20).

CDK3 is weakly expressed in human cells and displays 66% and 76% amino acid sequence identities with $p34^{cdc2}$ and CDK2, respectively (75). CDK3 appears to play a role in regulating G1 progression (126). Human CDK4 is active exclusively in G1, displays 50% amino acid identity with $p34^{cdc2}$, and carries a PISTVRE variation of PSTAIRE. CDK4 phosphorylates the retinoblastoma tumor suppressor (pRB) and related proteins in vitro but not the standard CDK substrate, histone H1 (70). pRB phosphorylation is a key step in the G1-S transition (see below and Figure 1). CDK4's regulatory subunit appears to be a D-type cyclin. Some antagonistic growth factors arrest animal cell divisions by down-regulating CDK4 synthesis (19). CDK5, known as PSSALRE prior to the discovery of its association with a D-type cyclin, is the only CDK whose activity is detected exclusively in nondividing cells (124, 130). Finally, CDK6 (formerly PLSTIRE) binds D-type cyclins and is activated by mitogens in G1 (76).

Cyclin Regulatory Subunits

Merely adjusting the intracellular concentration of $p34^{cdc2}$ does not provide a mechanism for checkpoint controls that is sufficiently responsive or versatile to be properly exercised at G1-S and G2-M. If CDKs are the brawn of the checkpoint guardians, then their cyclin subunits are the brains (100, 115). Because monomeric CDKs have little or no activity, each CDK catalytic subunit must associate with one, or sequentially more than one, cyclin partner that regulates the timing, substrate specificity, and localization of its activity (97, 102). Cyclins are 30–55-kDa proteins that carry a central conserved domain of approximately 100 amino acids (the so-called cyclin box).

Cyclins are classified as either mitotic or G1, depending on the point or period in the cell cycle during which they activate their CDK partners. The mitotic cyclins include only A and B types and are conserved sufficiently to be cloned from any species by PCR. Cyclin A is predominantly nuclear and activates at least two CDKs. At S-phase onset, cyclin A associates with CDK2 and other proteins (see below and Figure 1) and plays a role in initiating DNA replication. During S- and M-phases, cyclin A also associates with $p34^{cdc2}$, which it activates just before nuclear envelope breakdown (95). Cyclin B accumulates in the cytoplasm during interphase as an inactive dimer with $p34^{cdc2}$ but appears in an activated, nuclear complex at G2-M (102). Completion and irreversibility of M-phase is assured, at least in part, by the ubiquitin-mediated degradation of cyclin B and other, as yet unidentified proteins during anaphase (48). Budding yeast has four mitotic cyclin genes designated CLB1–4, but, oddly enough, fewer have turned up in animal cells so far.

The G1 or START cyclins are related distantly to the mitotic cyclins in that both carry the cyclin box. However, this motif is amino terminal in G1 cyclins: Their carboxy ends have proline-, glutamic acid–, serine-, and threonine-containing PEST elements that mediate their rapid turnover. Budding yeast G1 cyclin genes include CLN1, CLN2, CLN3, HCS26, and ORFD, the first three of which are functionally redundant. G1 cyclins of higher eukaryotes have been isolated by screening for genes that rescue CLN1/CLN2/CLN3 triple mutants when overexpressed (65). Although A- and B-type cyclins were unexpectedly isolated in these screens, three classes of G1 cyclins were also recovered and classified by primary sequence as C, D (including D1, D2, and D3), and E types (115). The level of cyclin C oscillates only twofold during the cell cycle, peaking during G1. Its role and partners are unknown.

Because D-type cyclins are growth factor inducible, they link growth control–signaling cascades with the CDK-driven cell cycle engine (71). Aberrant D-type cyclin expression is highly correlated with several human cancers (82). Cyclin D1 is the product of a previously identified human proto-oncogene locus (83) and activates CDK2, CDK4, CDK5, and CDK6 kinases during G1 (129). The three human D cyclins are not functionally redundant, but they behave differently among cell types. They are also the only cyclins whose overexpression accelerates a cell cycle phase (G1) (103). Finally, D-type cyclins carry an amino-terminal motif (LxCxE) that mediates their direct binding to pRB and related proteins (18), thus conferring their participation in multimeric complexes that guide G1 passage (Figure 1).

Cyclin E associates with pRB, p107, and CDK2 during late G1. This putative SPF complex may catalyze G1-S passage by phosphorylating pRB, from which transcription factors are released whose activities contribute to DNA synthesis initiation (16, 115). The cyclin E–CDK2 complex is a target for negative regulators of mammalian G1 phase transit (58), and aberrations in cyclin E are also associated with some human cancers (69).

The Essential Role of Phosphorylation

Posttranslational control of CDK activity is achieved by cyclin association and phosphorylation. As patriarch of the family, $p34^{cdc2}$ has provided a model for control of CDK activity by covalent modification. $p34^{cdc2}$ is regulated by an activating phosphorylation during G2 at Thr-167 and by inhibitory phosphorylations at Thr-14 (absent in some species) and Tyr-15 soon thereafter (Figure 1). The enzyme complex responsible for the activating phosphorylation at Thr-167 is a serine-threonine kinase called CAK (*cdc2*-activating kinase) (117). Its 37-kDa catalytic subunit is a relative of the CDKs. CDK2 and CDK4 are also activated by CAK or a related complex (8, 55), as undoubtedly are other CDKs (116).

Once activated in G2 by CAK, p34^{cdc2} is immediately inactivated by phosphorylations at Tyr-15 and, in some cases, Thr-14, by the *wee1* kinase (72). This bifunctional threonine-tyrosine kinase is 50 and 107 kDa in human and yeast cells, respectively, and appears to be the seat of checkpoint control at the G2-M boundary, remaining activated until the "DNA replication and repair complete" signal is received. Other CDKs are probably controlled by similar negative regulatory mechanisms (33).

The *cdc25* phosphatase removes p34^{cdc2}'s inhibitory phosphorylation(s) on Tyr-15 (and Thr-14 in some species) and triggers M-phase entry (Figure 1) (78). The eukaryotic cell's abrupt G2-M transition is effected by a positive feedback loop wherein the *cdc25* phosphatase is activated by the product of its own catalysis, the activated p34^{cdc2} kinase (122). Other CDKs are probably regulated by removal of inhibitory phosphorylations (113).

Other Proteins in the Network

Additional proteins that regulate animal and yeast cell cycles are worth noting, because analogous functions, if not strict homologs, will certainly be found in plant cell cycle regulatory networks.

The yeast gene product p13^{suc1} is the prototype for a class of small proteins that bind many CDKs tightly enough to be used as nonspecific, CDK affinity purification reagents (5). Structural analysis suggests that the human p13^{suc1} homolog forms a hexamer that acts as a workbench upon which CDK-cyclin complexes are assembled (Figure 1) (96).

Proteins of 16–40 kDa that specifically inhibit CDKs have been identified in yeast and animal systems. These cyclin kinase inhibitors (CKIs) specifically target CDK-cyclin complexes and can be either inducible or constitutive. Such proteins provide additional opportunities for cellular conditions or deleterious mutations to promote or forestall progress around the cycle (50, 87).

Transcriptional regulation, although not central to control of p34^{cdc2} activity at G2-M, plays a central role in the G1-S transition, when a pervasive metabolic shift occurs as the cell begins to replicate the genome. Families of heterodimeric transcription factors, SWI4/SWI6 in yeast and E2F (or DRTF1) in mammalian cells, regulate genes whose products are essential for S-phase initiation (62, 74). pRB or related proteins (e.g. p107) sequester E2F prior to G1-S and thereby prevent premature transactivations. So-called pocket proteins, such as pRB and p107 (named for their E2F-binding "pockets"), lose E2F-binding activity at G1-S as a result of becoming heavily phosphorylated by CDK4-cyclin D, CDK2-cyclin E, and/or CDK2-cyclin A (88).

PLANT CDKS AND CYCLINS

Neither the model yeast nor animal cell systems remotely represents the developmental condition of a meristematic plant cell. Nor can we expect that every mitotic regulatory motif uncovered in these model systems will have a strict counterpart in plants. Thus, our task is twofold: to survey and characterize plants with respect to homologs of yeast and animal cell cycle regulators, but also to manipulate botanical systems to uncover controls that uniquely serve plant developmental strategies.

Plant CDK Structure

All plants examined to date have cyclin-dependent kinases. Given both the size of animal CDK gene families (75) and the botanical tendency toward large gene families, it is unlikely that any plant species' CDK coding content has been accounted for in its entirety. Plant CDKs are generally 294 amino acids in length, weigh in at 33–35 kDa, and display 60–65% primary sequence identities with yeast *cdc2* or *cdc28* (Table 1). Many plant CDKs rescue yeast *cdc2/CDC28* mutants. It remains to be determined whether the occasional negative results in these experiments (37, 46) imply functional divergence among plant CDKs that drive the G2-M transition, or a kinase that acts elsewhere in the plant cell cycle.

Based on primary sequence criteria, the majority of cloned plant CDKs encode p34^{cdc2} kinases or very close relatives. However, five CDKs stray significantly from p34^{cdc2}'s canonical 16-residue PSTAIRE domain. The *Antirrhinum majus cdc2c* and *cdc2d* polypeptides have unusual PPTALRE and PPTTLRE motifs, respectively (Table1) (24). The *Arabidopsis thaliana cdc2b* gene product is slightly long, at 309 amino acids, and carries the same PPTALRE domain as *A. majus cdc2c,* diverging at 5 of PSTAIRE's 16 conserved residues. Pea *cdkPs3* is an intriguing variant, bearing a 224–amino acid (23 kDa), serine- and proline-rich carboxy-terminal extension, as well as one smaller addition at the amino terminus and another within the coding region. Its PITAIRE variation on PSTAIRE differs from the consensus at five positions. CDKs with such large carboxy-terminal extensions have been reported

---

[a]Predicted number of amino acids in derived polypeptide; [b]Predicted molecular mass of derived polypeptide; [c]cDNA, PCR, or genomic clone; [d]Composition of "PSTAIRE" domain (canonical sequence shown at top of column); [e]Percent identity with *Saccharomyces cerevisiae* (Sc), CDC28 and *Schizosaccharomyces pombe* (Sp), *cdc2* gene products at the amino acid sequence level; [f]Rescue of temperature-sensitive (ts) *cdc28* mutant of *S. cerevisiae* or ts *cdc2* mutant of *S. pombe*; [g]GenBank/EMBL accession number; [h]cdc2MsB fails to rescue a *cdc2* fission yeast mutant or a *cdc28* budding yeast mutant allele that arrests at G2-M, but can rescue a *cdc28* budding yeast mutant that arrests at G1-S; [i]Only 7 of 16 residues of PSTAIRE domain reported; [j]Data unavailable due to truncated clone; [k] ME Prewett, HS Feiler, H Chen, JJ Poole, TW Jacobs, unpublished data; [l]Failed to rescue *S. cerevisiae cdc28* mutant; NR, Not reported

Table 1 Cloned cyclin-dependent kinases from plants

Species	Gene name	#aa[a]	MW[b]	Clone[c]	"PSTAIRE"[d]	%ID[e] Sc	Sp	Rescue[f]	Acc#[g]	Reference
alfalfa	*cdc2MsA*	294	33,886	cDNA	EGVPSTAIREISLLKE	65%	64%	*cdc2*	M58365	
	cdc2MsB	294	33,864	cDNA	EGVPSTAIREISLLKE	55–65%		*cdc28*[h]	X70707	46
Antirrhinum	*cdc2a*	NR	NR	cDNA	EGVPSTAIREISLLKE	NR	NR	NR	NR	45
	cdc2b	NR	NR	cDNA	EGVPSTAIREISLLKE	NR	NR	NR	NR	24
	cdc2c	NR	NR	cDNA	PPTALRE[i]	NR	NR	NR	NR	24
	cdc2d	NR	NR	cDNA	PPTTLRE[i]	NR	NR	NR	NR	24
Arabidopsis	*CDC2a*	294	34,008	cDNA	EGVPSTAIREISLLKE	61%	62%	*cdc28*	S45387	24
								cdc2	X57839	22
				genomic					D10850	43
	CDC2b	*j*	*j*	genomic	*j*	*j*	*j*	*j*	X57840	51
		309	35,295	genomic	EGIPPTALREISLLQM	NR	NR	NR	D10851	43
maize	*cdc2ZmA*	294	33,812	cDNA	EGVPSTAIREISLLKE	63%	63%	*cdc28*	M60526	51
	cdc2ZmB	*j*	*j*	cDNA	EGVPSTAIREISLLKE	*j*	*j*	*j*	NR	7
mothbean	NR	294	33,961	cDNA	EGVPSTAIREISLLKE	NR	NR	NR	M99497	7
pea	*cdkPs1*	294	33,864	cDNA	EGVPSTAIREISLLKE	NR	NR	*cdc28*	X56554	49
	cdkPs2	294	33,900	cDNA	EGVPSTAIREISLLKE	NR	NR	*cdc28*	NR	21[k]
	cdkPs3	518	57,224	cDNA	EGFPITAIREIKILKKL	NR	43%	NR	NR	k
petunia	*cdc2Pet*	*j*	*j*	PCR	*j*	*j*		*j*	X64321	k
rice	*cdc2Os-1*	294	34,049	cDNA	EGVPSTAIREISLLKE	62%	NR	*cdc28*	X60374	4
	cdc2Os-2	292	33,671	cDNA	EGVPSTAIREISLLKE	62%	NR	*l*	X60375	37
	R2	424	47,582	cDNA	EGVNFTALREIKLLKE	45%	47%	*l*	X58194	37
soybean	*cdc2-S5*	294	33,940	cDNA	EGVPSTAIREISLLKE	61–65%		*cdc28*	M93139	38
	cdc2-S6	294	33,950	cDNA	EGVPSTAIREISLLKE	61–65%		*cdc28*	M93140	77

in animal cells, and like the calmodulin domain protein kinases (35), this class of CDK may represent collinear catalytic and regulatory subunits. Cyclin-independent kinase activity has yet to be reported in any of these proteins in vitro.

The large rice protein, R2 (NFTALRE; 47.6 kDa) (38), is 56% identical (over 314 amino acid residues) to MO15, the catalytic subunit of *Xenopus laevis* CAK, the *cdc2*-activating kinase (117). Whether R2 is in fact the first plant CAK to be identified remains to be seen, because no enzyme activity has been reported for R2's gene product.

Despite the presence of Tyr-15 in essentially all reported plant p34^{cdc2} homologs, rigorous biochemical evidence for regulatory phosphorylation at this site in plants has not been demonstrated. This most likely results from the low cellular abundance of these enzymes, the brevity of Tyr-15 phosphorylation during the cycle in vivo, and the poor efficiency of metabolic labeling of plant cells.

Plant CDKs and the Cell Cycle

In the absence of conditional mutants or antisense experiments, we must derive support for hypotheses of plant CDK function from correlative data only. Protein levels of specific plant CDKs have yet to be assayed over the cell division cycle. RNA transcript abundance of one *cdc2* homolog remains unchanged throughout the alfalfa cell cycle (67), but others appear to increase from G1 to G2 in *Petunia hybrida* cells and *A. majus* plants (4, 24). The profound differences among the experimental systems from which these conflicting results were obtained preclude any general conclusions about plant CDK mRNA accumulation profiles over the cell cycle.

Although p34^{cdc2} kinase displays peaks of activity at both G1-S and G2-M in yeast, its activity is confined to G2-M in metazoans, where the G1-S peak of Ca^{+2}- and cAMP-independent histone H1 kinase (H1K) activity (CDK hallmarks) is attributable to the CDK2-cyclin A complex (Figure 1). In extracts from synchronous pea root tips and alfalfa cell cultures, H1K assays define a minor peak at G2-M and a much stronger one at G1-S (67) (HS Feiler & TW Jacobs, unpublished observations). These results may be attributable to the use of the nonspecific p13^{suc1}-Sepharose affinity matrix to semipurify the H1K activity. This matrix binds both CDK2-cyclin A and p34^{cdc2}-cyclin B (76). Two temporally separable S-phase H1K peaks could be resolved by precipitating the activity with anti-human cyclin antibodies (early S peak) and p13^{suc1} (later S peak) in alfalfa cells partially synchronized by double phosphate starvation (G1-S arrest) (67). Although these results, along with differential yeast complementation (46; see also note *j* in Table 1) may implicate a plant CDK2-like function as responsible for one or both of the S-phase peaks, no plant H1K activity has been rigorously linked to a CDK specifically identified by prior cloning.

P. hybrida leaf mesophyll cells can be induced to divide with partial synchrony by removing their cell walls and incubating the recovered protoplasts under appropriate conditions. In this system, an S-phase peak of H1K activity was not detected when H1K was purified by precipitation with a resin to which the human p13^{suc1} homolog (p9^{CKShs1}) was covalently linked. Perhaps this matrix's selectivity differs from that of p13^{suc1}, or the mimosine used to block (in S) and further synchronize the cells had unforeseen effects on CDK regulation (98).

Plant CDKs and the Proliferative State

CDK transcript levels are often higher in proliferating than in nonproliferating plant cells (7, 22, 46, 68). But it must be reiterated that induction of and withdrawal from the proliferative state is an issue of growth control, not cell cycle control. Therefore, CDK expression and activity should be viewed as a downstream consequence of the action of growth-control pathways, the activation states of which determine whether or not a cell divides. Indeed, plant CDK mRNA or protein has often been reported in nonmitotic cells (4, 7, 22, 32, 68, 77; ME Prewett & TW Jacobs, unpublished observations). Therefore, mere CDK presence is not sufficient for division, although it most assuredly is necessary.

It has been justifiably proposed that the presence of CDKs in nondividing plant cells confers upon them mitotic competence (42). This important conclusion reflects the well-documented developmental plasticity of plants and suggests that an intermediate, developmentally metastable state (G0, perhaps), between proliferation and truly terminal differentiation, may exist in the natural history of many, if not all, vegetative plant cells. However, the oversimplified interpretation of competence, that the presence of a single CDK in a plant cell somehow poises it at the brink of division, belies the necessity of the constellation of other proteins (e.g. CDKs, cyclins, and their attendant regulators) required for successful negotiation of the mitotic pathway.

For greater insights, these issues await experiments using ectopic expression of a variety of cell cycle–regulatory genes driven by inducible promoters in transgenic plants. A tentative first step in this direction has been taken in the construction of tobacco plants constitutively expressing the fission yeast *cdc25* gene (encoding the Tyr-15-dephosphorylating, p34^{cdc2}-activating phosphatase; see Figure 1) (3). These plants displayed a variety of developmental abnormalities, perhaps most significantly including cells of reduced size in the root apical meristem. This observation suggests that unregulated expression of the ultimate p34^{cdc2} activator, a *cdc25* phosphatase, drove cells into mitosis before they had attained their normal minimum size. Interpretation of this experiment is problematic, given the absence of direct evidence to date for the existence of either p34^{cdc2} Tyr-15 phosphorylation or a *cdc25* homolog in higher plants.

Plant CDKs and Phytohormones

Plant cell division is a ubiquitous response to auxin or to auxin plus cytokinin. Not surprisingly, considerable interest is focused on the role of CDKs in proliferative responses to phytohormones. But in the animal cell-growth control model, the CDK-cyclin-driven cell-cycle engine is linked to the downstream ends of signaling cascades, the upstream components of which interact directly with growth factors. Membrane or soluble cytoplasmic receptors bind ligands that activate the downstream pathways, ultimately resulting in transcriptional induction of a rate-limiting mitotic control element such as a G1 cyclin (2, 103). Although plant cell cycle and phytohormone work may often implicitly ignore the essential intervening pathways, hormonal effects on plant CDK expression have been documented in several cases.

The steady-state levels of plant *cdc2* transcripts usually increase in response to auxins but not always to cytokinins. Cultured alfalfa cells have displayed the most rapid response to date, with a transient peak in the steady-state level of an inexplicably large 2.5-kb *cdc2MsA* transcript 1–3 h after a pulse of 2,4-dichlorophenoxyacetic acid (2,4-D). A shorter *cdc2MsA* transcript showed a more complex response under these conditions (46). Intact plant organs respond more slowly to auxin, with rises in steady-state *cdc2* mRNA [or protein visualized with anti-PSTAIR antibodies (32)] detectable at least 24 h after treatment with naphthaleneacetic acid (NAA) (77), 2,4-D (32, 67), or indole-3-acetic acid (IAA) (42). Although cytokinin treatment alone has been reported less frequently to alter *cdc2* transcript levels, one of the G1 cyclins of *A. thaliana* (H3; see below) appears to be cytokinin-inducible. *A. thaliana* cyclins H1 and H2 are not hormone inducible, but H2 transcript levels rise in response to carbon source resupply in cultured cells (118). 6-Benzylaminopurine (BAP) and NAA were synergistic in their *cdc2A* induction effect on *A. thaliana* roots (42).

Rhizobium spp. bacteria induce mitotic responses in the root cortices of legumes via a signaling pathway that may or may not directly involve perturbations in endogenous auxin and cytokinin activities. Measured molecular responses of cell-cycle regulators to *Rhizobium* spp. induction in vivo have been slow. Elevation in steady-state levels of soybean cdc2-S5 (and to a lesser extent cdc2-S6) became detectable by a sensitive PCR assay only after roots had been exposed to *Rhizobium* spp. for 24 h (77). Even alfalfa microcallus suspensions require 15 h of treatment with purified cell-free nodulation factors from *Rhizobium meliloti* before an elevation in p13^{suc1}-precipitable H1K activity is detectable (112). One of several possible interpretations of these long lags is that nodulation factors impinge on one (or more) of the host cell's mitotic signaling cascades at a far-upstream end. Clearly, further work is needed to characterize the signal transduction pathways that link plant mitogen reception and molecular cell cycle responses.

Substrates for Plant CDKs

Yeast and animal CDKs phosphorylate cytoskeletal elements, transcription factors, regulatory enzymes, and other cellular proteins (89). Although plant CDKs probably interact with many of these same substrates, some unique targets are likely to be found in plant cells as well. To date, the most exciting results in this regard derive from efforts to localize $p34^{cdc2}$ protein within dividing plant cells by indirect immunocytochemistry (7, 79). Detected with specific antisera, the maize *cdc2ZmA* gene product is localized predominantly in the nucleus during interphase. In contrast, anti-PSTAIRE peptide antisera stain diffusely throughout the cell, indicating that diverse PSTAIRE-containing proteins are distributed among a variety of cellular compartments. Late in the G2-M transition, *cdc2ZmA* protein colocalizes with the preprophase band (PPB) of microtubules that presages the cell's ultimate division site. Blockage of PPB formation with the antimicrotubule drug oryzalin abolishes this *cdc2*ZmA localization, indicating that the PPB directs *cdc2*ZmA to the cortical ring and not vice versa (6). A CDK may imprint (presumably by phosphorylation) sites to which the phragmoplast perimeter will be directed, even attracted, on the inner plasma membrane. But it is also possible that $p34^{cdc2ZmA}$ kinase activity merely catalyzes the dissolution of the PPB or mediates the latter's separation from its cortical attachment points. Subcellular localization studies with plant cyclin probes have yet to be reported.

Plant Cyclins

Plant cyclin research is in its infancy. Mitotic cyclins have been cloned by polymerase chain reaction from several plant species (Table 2). Again, none of these cloning efforts can be considered exhaustive. Like their animal counterparts, reported clones encode proteins of approximately 50 kDa. Complementation of yeast mitotic cyclin mutants is not a reliable test for cyclin function; the accepted assay is microinjection of putative cyclin mRNAs into *X. laevis* oocytes and observation of meiotic maturation. Such demonstrations have been reported for cyclins cloned from maize, soybean, and *A. thaliana* (39, 41, 107). In no case has the gene product of a plant cyclin clone been rigorously shown to associate with or to activate a plant CDK.

Plant mitotic cyclin sequences do not fall unequivocally into the A- and B-type categories of animal cyclins (see 107 for discussion). The classical criterion, that A-cyclins peak earlier in the cycle than B-cyclins, has not yet been applied to plant mitotic cyclins [although the two alfalfa mitotic cyclins, cycMs1 and cycMs2, display transcript accumulation kinetics reminiscent of A- and B-type protein profiles, respectively (44)]. Classification by primary sequence alone is clearly problematic, because in the absence of detailed information relating primary sequence to tertiary structure and function, no conclusions regarding cellular roles can be drawn from amino acid sequences alone. And no data have been reported pertinent to plant cyclin turnover, apart from the observation that essentially all plant mitotic cyclin clones encode plausible 9-residue destruction boxes [i.e. ubiquitination targets, variations on RxALGxIxN (31, 91)] within 50 amino acids of their amino termini (Table 2).

The first candidates for plant G1 cyclins were recovered by selecting *A. thaliana* cDNAs capable of rescuing a budding yeast strain deficient in the three redundant G1 cyclins (*CLN1, CLN2,* and *CLN3*) (118). *A. thaliana* cyclins δ1, δ2, and δ3, as they have been provisionally named, are around 40 kDa and most closely resemble, at the primary sequence level, animal G1 cyclins, particularly those of the human D-type. Like their animal counterparts, the deduced polypeptides of these new clones lack the destruction box but carry PEST sequences at their amino- and carboxy-termini, indicative of rapid turnover. Of special significance is the presence of the pRB-binding consensus motif, LxCxE, at the amino terminus of each of the derived δ-cyclin polypeptide sequences. Although pRB homologs have not been identified in plants to date, this finding suggests not only that they might be present but that the availability of these new G1 cyclin clones might provide a means to identify and ultimately clone a plant pRB homolog. That plant cells might harbor a factor that, like pRB, sequesters and releases transcription factors needed to drive the G1-S transition is supported by the observation that RNA synthesis inhibitors block the plant cell cycle if administered during G1 but not during S phase (93).

Mitotic cyclins appear to be excellent macromolecular markers for the proliferative state, owing to their narrowly defined cellular function and rapid turnover (39, 41, 44, 107, 118). However, a dividing plant cell, when analyzed by in situ hybridization, can still be seen to lack cyclin mRNA if, at the moment of its fixation, it was not in a phase of the cell cycle during which the probe's cognate transcripts had accumulated to detectable levels. Recent results suggest that the cellular abundance of mitotic cyclins is sufficiently high

---

[a]Predicted number of amino acids in derived polypeptide; [b]Predicted molecular mass of derived polypeptide; [c]cDNA, PCR, or genomic clone; [d]Composition of destruction box for targeted, ubiquitin-mediated proteolysis (animal consensus provided at top of column); [e]GenBank EMBL accession number; [f]Uncertain due to truncated clone; [g]Cloned by rescue of *S. cerevisiae* G1 cyclin mutants; [h]Sequence unrelated to H1, H2, or H3; [i]Lacks destruction box but contains proline (P), glutamate(E), serine (S), threonine (E) domains at amino- and carboxy-termini, indicative of rapid proteolytic turnover; [j]JAH Murray, unpublished data; [k]H Chen, P Wakeley, J Dunphy, C Roush, H Feiler & TW Jacobs, unpublished data; NR, Not reported

Table 2 Cloned cyclins from plants

Species	Gene name	Type	aa[a]	MW[b]	Clone[c]	Dest Box[d] RxALGxIxN	Acc#[e]	Reference
alfalfa	cycMs1	mitotic	213[f]	f	cDNA	RRALGVINQ	X68740	44
	cycMs2	mitotic	328f	f	cDNA	f	X68741	44
Antirrhinum	cyclin1	mitotic	473	52,704	cDNA	RRALGDIGN	X76122	24
	cyclin2	mitotic	441	49,205	cDNA	RRALGDIGN	X76123	24
Arabidopsis	cyc1At	mitotic	428	48,415	cDNA	RQVLGDIGN	M80190	41
	cyc2aAt	mitotic	421	48,377	cDNA	RRVLRVINQ	Z31400	23
	cyc2bAt	mitotic	429	49,735	cDNA	RRALGVINH	Z31401	23
	cyc3aAt	mitotic	443	50,336	genomic	AKALGVSNS	Z31589	23
	cyc3bAt	mitotic	286[f]	32,371	cDNA	RAVLKDVSN	Z31402	23
	H1	G1	334	37,688	cDNA[g]	PEST[i]	NR	118
	H2	G1	383	42,900	cDNA[g]	PEST[i]	NR	118
	H3	G1	376	42,805	cDNA[gh]	PEST[i]	NR	118
	J18	G1	234	27,231	cDNA[gh]	none	NR	j
carrot	C13-1	mitotic	341[f]	f	cDNA	RVVLGEISN	X62819	39
maize	cyclaZm	mitotic	420[f]	47,482[f]	cDNA	RAPLGDIGN	U10079	107
	cycIZm	mitotic	456[f]	50,896[f]	cDNA	RASVGSLGN	U10077	107
	cycIIzm	mitotic	424	47,745	cDNA	RRALSDIKN	U10076	107
	cycbZm	mitotic	445	48,252	cDNA	RRALGDIGN	U10078	107
pea	cycPs1	mitotic	472	51,950	cDNA	RAALHDIGN	NR	k
	cycPs2	mitotic	449	51,162	cDNA	RAGLTDVTN	NR	k
	cycPs3	mitotic	466	52,714	cDNA	RAILHDVTN	NR	k
soybean	S13-6	mitotic	454	50,063	cDNA	RKALGDIGN	X62820	39
	S13-7	mitotic	f	f	cDNA	f	X62303	39

for detection by in situ hybridization only during G2 and M in meristematic cells of *A. majus* (24). The same may apply to detection of *A. thaliana* G1 cyclins, which are expressed at various periods during G1 and S (118). Enumeration of cyclin-positive cells by in situ hybridization therefore suffers from the same analytical limitation as does the mitotic index parameter (cells in mitosis/total number of cells). That is, the probability of a cell's being cyclin-positive by an in situ assay is dependent on its being in the proliferative state as well as on the fraction of the total cell cycle period during which the cognate cyclin is detectable. When antisera directed against specific plant cyclins become available for in situ detection, their quantitative application will be subject to the same caveats.

THE NEXT PHASE

The lure of applying these new tools to old questions in plant cell physiology and development is strong. But we cannot afford to bypass basic enzymatic analyses of the newly identified cell cycle–regulatory proteins. Many of the inherent difficulties in working with plant material must be circumvented if the promise of satisfying insights is to be fulfilled. Cloning of plant homologs of *suc1*, CAK, *wee1*, many G1 cyclins, most other CDKs, countless phosphatases, CDK inhibitors, pRB, and various transcription factors will require further PCR excursions (e.g. differential display), yeast complementation, interaction cloning, and surveys of expressed sequence tags. Of equal if not more importance, we must begin to identify genes and proteins unique to cell-cycle control in plants, perhaps by devising appropriate *A. thaliana* mutant screens. Ectopic expression from inducible promoters in transgenic plants has barely begun. Major developmental insights will come from the linking of cell-cycle–regulatory biochemistry with plant growth–control pathways. The recent identification of MAP kinases in alfalfa, pea, and *A. thaliana* (15, 53, 80, 119) is a promising advance in this direction and provides additional yeast and animal models from which an appropriately botanical analog can be fashioned. As the interfaces between such pathways and the cell cycle engine are elaborated in model systems, plant cell biologists will once again be offered a possible map of the path ahead.

ACKNOWLEDGMENTS

I thank Tim Hunt, James Murray, Venkatesan Sundaresan, John Doonan, and Dennis Francis for generously providing unpublished data, manuscripts, and thoughtful opinions; and past and present members of my laboratory for discussions, editorial assistance, and unpublished data. Work in my laboratory is supported by the National Science Foundation (Cell Biology Program) and the Department of Energy (Energy Biosciences Program).

Literature Cited

1. Akhurst RJ, Flavin NB, Worden J, Lee MG. 1989. Intracellular localization and expression of mammalian CDC2 protein during myogenic differentiation. *Differentiation* 40:36–41

2. Albers MW, Williams RT, Brown EJ, Tanaka A, Hall FL, et al. 1993. FKBP-rapamycin inhibits a cyclin-dependent kinase activity and a cyclin D1-Cdk association in early G1 of an osteosarcoma cell line. *J. Biol. Chem.* 268:22825–29

3. Bell MH, Halford NG, Ormrod JC, Francis D. 1993. Tobacco plants transformed with *cdc25,* a mitotic inducer gene from fission yeast. *Plant Mol. Biol.* 23:445–51

4. Bergounioux C, Perennes C, Hemerly AS, Qin LX, Sarda C, et al. 1992. A *cdc2* gene of *Petunia hybrida* is differentially expressed in leaves, protoplasts and during various cell cycle phases. *Plant Mol. Biol.* 20:1121–30

5. Brizuela L, Draetta G, Beach D. 1987. p13^{suc1} acts in the fission yeast cell division cycle as a component of the p34^{cdc2} protein kinase. *EMBO J.* 6:3507–14

6. Colasanti J, Cho SO, Wick S, Sundaresan V. 1993. Localization of the functional p34^{cdc2} homolog of maize in root tip and stomatal complex cells—association with predicted division sites. *Plant Cell* 5:1101–11

7. Colasanti J, Tyers M, Sundaresan V. 1991. Isolation and characterization of cDNA clones encoding a functional p34^{cdc2} homologue from *Zea mays. Proc. Natl. Acad. Sci. USA* 88:3377–81

8. Connell-Crowley L, Solomon MJ, Wei N, Harper JW. 1993. Phosphorylation independent activation of human cyclin-dependent kinase 2 by cyclin A in vitro. *Mol. Biol. Cell* 4:79–92

9. Dalton S. 1992. Cell cycle regulation of the human *cdc2* gene. *EMBO J.* 11:1797–804

10. Davidson D. 1991. Cell division. In *Plant Physiology: A Treatise,* ed. F Steward, pp. 341–436. New York: Academic

11. Doerner P. 1994. Cell cycle regulation in plants. *Plant Physiol.* In press

12. Downes CS, Wilkins AS. 1994. Cell cycle checkpoints, DNA repair and DNA replication strategies. *BioEssays* 16:75–79

13. Draetta G, Beach D. 1988. Activation of cdc2 protein kinase during mitosis in human cells: cell cycle-dependent phosphorylation and subunit rearrangement. *Cell* 54:17–26

14. Draetta G, Beach D, Moran E. 1988. Synthesis of p34, the mammalian homolog of the yeast cdc2/CDC28 protein kinase, is stimulated during adenovirus-induced proliferation of primary baby rat kidney cells. *Oncogene* 2:553–57

15. Duerr B, Gawienowski M, Ropp T, Jacobs TW. 1993. MsERK1: a mitogen-activated protein kinase from a flowering plant. *Plant Cell* 5:87–96

16. Dulic V, Drullinger LF, Lees E, Reed SI, Stein GH. 1993. Altered regulation of G1 cyclins in senescent human diploid fibroblasts: accumulation of inactive cyclin E-Cdk2 and cyclin D1-Cdk2 complexes. *Proc. Natl. Acad. Sci. USA* 90:11034–38

17. Durkacz B, Carr A, Nurse P. 1986. Transcription of the *cdc2* cell cycle control gene of the fission yeast *Schizosaccharomyces pombe. EMBO J.* 5:369–73

18. Ewen ME, Sluss HK, Sherr CJ, Matsushime H, Kato JY, et al. 1993. Functional interactions of the retinoblastoma protein with mammalian D-type cyclins. *Cell* 73:487–97

19. Ewen ME, Sluss HK, Whitehouse LL, Livingston DM. 1993. TGF-beta inhibition of cdk4 synthesis is linked to cell cycle arrest. *Cell* 74:1009–20

20. Fang F, Newport JW. 1993. Distinct roles of cdk2 and cdc2 in RP-A phosphorylation during the cell cycle. *J. Cell Sci.* 106:983–94

21. Feiler HS, Jacobs TW. 1990. Cell division in higher plants—a *cdc2* gene, its 34-kDa product, and histone-H1 kinase activity in pea. *Proc. Natl. Acad. Sci. USA* 87:5397–401

22. Ferreira PCG, Hemerly AS, Villarroel R, Van Montagu M, Inze D. 1991. The *Arabidopsis* functional homolog of the p34^{cdc2} protein kinase. *Plant Cell* 3:531–40

23. Ferreira PCG, Hemerly AS, De Almeida Engler J, Bergounioux C, Burssens S, et al. 1994. Three discrete classes of *Arabidopsis* cyclins are expressed during different intervals of the cell cycle. *Proc. Natl. Acad. Sci. USA.* In press

24. Fobert PR, Coen ES, Murphy GJP, Doonan JH. 1994. Patterns of cell division revealed

by transcriptional regulation of genes during the cell cycle in plants. *EMBO J.* 13: 616–24

25. Forsburg SL, Nurse P. 1991. Cell cycle regulation in the yeasts *Saccharomyces cerevisiae* and *Schizosaccharomyces pombe*. *Annu. Rev. Cell Biol.* 7:227–56

26. Fosket D. 1990. Cell division in plant development. *Semin. Dev. Biol.* 1:357–66

27. Francis D. 1992. The cell cycle in plant development. *New Phytol.* 122:1–20

28. Francis D. 1995. The cell cycle—minireview. *Physiol. Plant.* In press

29. Furukawa Y, Piwnica-Worms H, Ernst TJ, Kanakura Y, Griffin JD. 1990. *cdc2* gene expression at the G1 to S transition in human T-lymphocytes. *Science* 250:805–8

30. Gaetano C, Matsumoto K, Thiele C. 1991. Retinoic acid negatively regulates p34^{cdc2} expression during human neuroblastoma differentiation. *Cell Growth Differ.* 2:487–93

31. Glotzer M, Murray AW, Kirschner MW. 1991. Cyclin is degraded by the ubiquitin pathway. *Nature* 349:132–38

32. Gorst JR, John PCL, Sek FJ. 1991. Levels of p34^{cdc2}-like protein in dividing, differentiating and dedifferentiating cells of carrot. *Planta* 185:304–10

33. Gu Y, Rosenblatt J, Morgan DO. 1992. Cell cycle regulation of CDK2 activity by phosphorylation of thr160 and tyr15. *EMBO J.* 11:3995–4005

34. Hanks S, Quinn A, Hunter T. 1988. The protein kinase family: conserved features and deduced phylogeny of the catalytic domains. *Science* 241:42–52

35. Harper JF, Sussman MR, Schaller GE, Putman-Evans C, Charbonneau H, et al. 1991. A calcium-dependent protein kinase with a regulatory domain similar to calmodulin. *Science* 252:951–54

36. Hartwell L, Weinert T. 1989. Checkpoints: controls that ensure the order of cell cycle events. *Science* 246:629–34

37. Hashimoto J, Hirabayashi T, Hayano Y, Hata S, Ohashi Y, et al. 1992. Isolation and characterization of cDNA clones encoding *cdc2* homologues from *Oryza sativa*: a functional homologue and cognate variants. *Mol. Gen. Genet.* 233:10–16

38. Hata S. 1991. cDNA cloning of a novel *cdc2*$^{+}$/CDC28-related protein kinase from rice. *FEBS Lett.* 279:149–52

39. Hata S, Kouchi H, Suzuka I, Ishii T. 1991. Isolation and characterization of cDNA clones for plant cyclins. *EMBO J.* 10:2681–88

40. Hayes TE, Valtz NLM, McKay RDG. 1991. Down-regulation of CDC2 upon terminal differentiation of neurons. *New Biol.* 3:259–69

41. Hemerly A, Bergounioux C, Van Montagu M, Inze D, Ferreira P. 1992. Genes regulat-

ing the plant cell cycle: isolation of a mitotic-like cyclin from *Arabidopsis thaliana*. *Proc. Natl. Acad. Sci. USA* 89:3295–99

42. Hemerly AS, Ferreira P, Engler JD, Van Montagu M, Engler G, et al. 1993. *Cdc2A* expression in *Arabidopsis* is linked with competence for cell division. *Plant Cell* 5:1711–23

43. Hirayama T, Imajuku Y, Anai T, Matsui M, Oka A. 1991. Identification of two cell-cycle-controlling *cdc2* gene homologs in *Arabidopsis thaliana*. *Gene* 105:159–65

44. Hirt H, Mink M, Pfosser M, Bogre L, Gyorgyey J, et al. 1992. Alfalfa cyclins—differential expression during the cell cycle and in plant organs. *Plant Cell* 4:1531–38

45. Hirt H, Pay A, Bogre L, Meskiene I, Heberle-Bors E. 1993. cdc2MsB, a cognate cdc2 gene from alfalfa, complements the G1/S but not the G2/M transition of budding yeast cdc28 mutants. *Plant J.* 4:61–69

46. Hirt H, Pay A, Gyorgyey J, Bako L, Nemeth K, et al. 1991. Complementation of a yeast cell cycle mutant by an alfalfa cDNA encoding a protein kinase homologous to p34^{cdc2}. *Proc. Natl. Acad. Sci. USA* 88: 1636–40

47. Hollingsworth R, Chen P-L, Lee W-W. 1993. Integration of cell cycle control with transcriptional regulation by the retinoblastoma protein. *Curr. Opin. Cell Biol.* 5:194–200

48. Holloway SL, Glotzer M, King RW, Murray AW. 1993. Anaphase is initiated by proteolysis rather than by the inactivation of maturation-promoting factor. *Cell* 73: 1393–402

49. Hong ZG, Miao GH, Verma DPS. 1993. p34^{cdc2} protein kinase homolog from mothbean (*Vigna aconitifolia*). *Plant Physiol.* 101:1399–1400

50. Hunter T. 1993. Braking the cycle. *Cell* 75:839–41

51. Imajuku Y, Hirayama T, Endoh H, Oka A. 1992. Exon-intron organization of the *Arabidopsis thaliana* protein kinase genes cdc2A and cdc2B. *FEBS Lett.* 304:73–77

52. Jacobs T. 1992. Control of the cell cycle. *Dev. Biol.* 153:1–15

53. Jonak C, Heberle-Bors E, Hirt H. 1994. Map kinases—universal multi-purpose signaling tools. *Plant Mol. Biol.* 24:407–16

54. Kaffman A, Herskowitz I, Tjian R, Oshea EK. 1994. Phosphorylation of the transcription factor-PHO4 by a cyclin-CDK complex, PHO80-PHO85. *Science* 263: 1153–56

55. Kato JY, Matsuoka M, Strom DK, Sherr CJ. 1994. Regulation of cyclin D-dependent kinase 4 (cdk4) by cdk4-activating kinase. *Mol. Cell. Biol.* 14:2713–21

56. Kiyokawa H, Ngo L, Kurosaki T, Rifkind R, Marks P. 1992. Changes in p34^{cdc2} kinase activity and cyclin A during induced

differentiation of murine erythroleukemia cells. *Cell Growth Differ.* 3:377–83

57. Koff A, Giordano A, Desai D, Yamashita K, Harper JW, et al. 1992. Formation and activation of a cyclin E-cdk2 complex during the G1 phase of the human cell cycle. *Science* 257:1689–94

58. Koff A, Ohtsuki M, Polyak K, Roberts JM, Massague J. 1993. Negative regulation of G1 in mammalian cells: inhibition of cyclin E-dependent kinase by TGF-beta. *Science* 260:536–39

59. Krek W, Nigg EA. 1989. Structure and developmental expression of the chicken CDC2 kinase. *EMBO J.* 8:3071–78

60. Ku DH, Wen SC, Engelhard A, Nicolaides NC, Lipson KE, et al. 1993. c-myb transactivates *cdc2* expression via myb binding sites in the 5′-flanking region of the human *cdc2* gene. *J. Biol. Chem.* 268:2255–59

61. Labbe JC, Lee MG, Nurse P, Picard A, Doree M. 1988. Activation at M-phase of a protein kinase encoded by a starfish homologue of the cell cycle control gene cdc2+. *Nature* 335:251–54

62. La Thangue NB. 1994. DRTF1/E2F—an expanding family of heterodimeric transcription factors implicated in cell-cycle control. *Trends Biochem. Sci.* 19:108–14

63. Lee MG, Norbury CJ, Spurr NK, Nurse P. 1988. Regulated expression and phosphorylation of a possible mammalian cell-cycle control protein. *Nature* 333:676–79

64. Lees E, Faha B, Dulic V, Reed SI, Harlow E. 1992. Cyclin E/cdk and cyclin A/cdk2 kinases associate with p107 and E2F in a temporally distinct manner. *Genes Dev.* 6: 1874–85

65. Lew DJ, Dulic V, Reed SI. 1991. Isolation of three novel human cyclins by rescue of G1 cyclin (Cln) function in yeast. *Cell* 66: 1197–1206

66. Lloyd C, ed. 1991. *The Cytoskeletal Basis of Plant Growth and Form.* New York: Academic

67. Magyar Z, Bako L, Bogre L, Dedeoglu D, Kapros T, et al. 1993. Active *cdc2* genes and cell cycle phase-specific cdc2-related kinase complexes in hormone-stimulated alfalfa cells. *Plant J.* 4:151–61

68. Martinez MC, Jorgensen J-E, Lawton MA, Lamb CJ, Doerner PW. 1992. Spatial pattern of *cdc2* expression in relation to meristem activity and cell proliferation during plant development. *Proc. Natl. Acad. Sci. USA* 89:7360–64

69. Marx J. 1994. How cells cycle toward cancer. *Science* 263:319–21

70. Matsushime H, Ewen ME, Strom DK, Kato JY, Hanks SK, et al. 1992. Identification and properties of an atypical catalytic subunit (p34^{PSK-J3}/cdk4) for mammalian-D type-G1 cyclins. *Cell* 71:323–34

71. Matsushime H, Quelle DE, Shurtleff SA, Shibuya M, Sherr CJ, et al. 1994. D-type cyclin-dependent kinase activity in mammalian cells. *Mol. Cell. Biol.* 14: 2066–76

72. McGowan CH, Russell P. 1993. Human *wee1* kinase inhibits cell division by phosphorylating p34^{cdc2} exclusively on tyr15. *EMBO J.* 12:75–85

73. McGowan CH, Russell P, Reed SI. 1990. Periodic biosynthesis of the human M-phase promoting factor catalytic subunit p34 during the cell cycle. *Mol. Cell. Biol.* 10:3847–51

74. McKinney J, Cross F. 1992. A switch-hitter at the start of the cell cycle. *Curr. Biol.* 2:421–23

75. Meyerson M, Enders G, Wu C-L, Su L-K, Gorka C, et al. 1992. A family of human cdc2-related protein kinases. *EMBO J.* 11: 2909–21

76. Meyerson M, Harlow E. 1994. Identification of G1 kinase activity for cdk6, a novel cyclin D partner. *Mol. Cell. Biol.* 14:2077–86

77. Miao GH, Hong ZL, Verma DPS. 1993. Two functional soybean genes encoding p34^{cdc2} protein kinases are regulated by different plant developmental pathways. *Proc. Natl. Acad. Sci. USA* 90:943–47

78. Millar JBA, Russell P. 1992. The cdc25 M-phase inducer: an unconventional protein phosphatase. *Cell* 68:407–10

79. Mineyuki Y, Yamashita M, Nagahama Y. 1991. p34^{cdc2} kinase homologue in the preprophase band. *Protoplasma* 162:182–86

80. Mizoguchi T, Hayashida N, Yamaguchi-Shinozaki K, Kamada H, Shinozaki K. 1993. Atmpks—a gene family of plant map kinases in *Arabidopsis thaliana. FEBS Lett.* 336:440–44

81. Moreno S, Nurse P. 1990. Substrates for p34^{cdc2}: *in vivo* veritas? *Cell* 61:549–51

82. Motokura T, Arnold A. 1993. Cyclin D and oncogenesis. *Curr. Opin. Genet. Dev.* 3:5–10

83. Motokura T, Bloom T, Kim HG, Juppner H, Ruderman JV, et al. 1991. A novel cyclin encoded by a *bcl1*-linked candidate oncogene. *Nature* 350:512–15

84. Murray AW. 1992. Creative blocks: cell-cycle checkpoints and feedback controls. *Nature* 359:599–604

85. Murray AW, Hunt T, eds. 1993. *The Cell Cycle: An Introduction.* San Francisco: Freeman

86. Nasmyth K. 1993. Control of the yeast cell cycle by the Cdc28 protein kinase. *Curr. Opin. Cell Biol.* 5:166–79

87. Nasmyth K, Hunt T. 1993. Dams and sluices. *Nature* 366:634–35

88. Nevins JR. 1992. E2F—a link between the rb tumor suppressor protein and viral oncoproteins. *Science* 258:424–29

89. Nigg EA. 1993. Targets of cyclin-depend-

ent protein kinases. *Curr. Opin. Cell Biol.*
5:187–93

90. Norbury C, Nurse P. 1992. Animal cell cycles and their control. *Annu. Rev. Biochem.* 61:441–70

91. Nugent JHA, Alfa CE, Young T, Hyams JS. 1991. Conserved structural motifs in cyclins identified by sequence analysis. *J. Cell Sci.* 99:669–74

92. Nurse P, Bissett Y. 1981. Gene required in G1 for commitment to cell cycle and in G2 for control of mitosis in fission yeast. *Nature* 292:558–60

93. Ohnishi N, Kodama H, Ando S, Komamine A. 1990. Synthesis of protein and mRNA is necessary for transition of suspension-cultured *Catharanthus roseus* cells from the G1 to the S phase of the cell cycle. *Physiol. Plant.* 80:95–101

94. Ormrod J, Francis D, eds. 1993. *Molecular and Cell Biology of the Plant Cell Cycle.* Dordrecht: Kluwer Academic

95. Pagano M, Pepperkok R, Verde F, Ansorge W, Draetta G. 1992. Cyclin-A is required at two points in the human cell cycle. *EMBO J.* 11:961–71

96. Parge HE, Arvai AS, Murtari DJ, Reed SI, Tainer JA. 1993. Human cksHs2 atomic structure: a role for its hexameric assembly in cell cycle control. *Science* 262:387–95

97. Peeper DS, Parker LL, Ewen ME, Toebes M, Hall FL, et al. 1993. A- and B-type cyclins differentially modulate substrate specificity of cyclin-cdk complexes. *EMBO J.* 12:1947–54

98. Perennes C, Qin LX, Glab N, Bergounioux C. 1993. Petunia p34^{cdc2} protein kinase activity in G2/M cells obtained with a reversible cell cycle inhibitor, mimosine. *FEBS Lett.* 333:141–45

99. Pines J. 1993. Cyclins and cyclin-dependent kinases—take your partners. *Trends Biochem. Sci.* 18:195–97

100. Pines J. 1993. Cyclins and their associated cyclin-dependent kinases in the human cell cycle. *Biochem. Soc. Trans.* 21:921–25

101. Pines J, Hunter T. 1991. Cyclin-dependent kinases: a new cell cycle motif? *Trends Cell Biol.* 1:117–212

102. Pines J, Hunter T. 1991. Human cyclin-A and cyclin-B1 are differentially located in the cell and undergo cell cycle-dependent nuclear transport. *J. Cell Biol.* 115:1–17

103. Quelle DE, Ashmun RA, Shurtleff SA, Kato JY, Barsagi D, et al. 1993. Overexpression of mouse D-Type cyclins accelerates g(1) phase in rodent fibroblasts. *Genes Dev.* 7:1559–71

104. Reed SI. 1991. G1-specific cyclins: in search of an S-phase-promoting factor. *Trends Genet.* 7:95–99

105. Reed SI. 1992. The role of p34 kinases in the g1 to S-Phase transition. *Annu. Rev. Cell Biol.* 8:529–61

106. Reed SI, Wittenberg C. 1990. Mitotic role for the Cdc28 protein kinase of *Saccharomyces cerevisiae. Proc. Natl. Acad. Sci. USA* 87:5697–701

107. Renaudin J-P, Colasanti J, Rime H, Yan Z, Sundaresan V. 1994. Cloning of four cyclins from maize indicates that higher plants have three structurally distinct groups of mitotic cyclins. *Proc. Natl. Acad. Sci. USA* 91:7375–79

108. Richter KH, Afshari CA, Annab LA, Burkhart BA, Owen RD, et al. 1991. Downregulation of *cdc2* in senescent human and hamster cells. *Cancer Res.* 51:6010–13

109. Roberge M. 1992. Checkpoint controls that couple mitosis to completion of DNA replication. *Trends Cell Biol.* 2:277–81

110. Rosenblatt J, Debondt H, Jancarik J, Morgan DO, Kim SH. 1993. Purification and crystallization of human cyclin-dependent kinase-2. *J. Mol. Biol.* 230:1317–19

111. Ruderman J. 1993. MAP kinase and the activation of quiescent cells. *Curr. Opin. Cell Biol.* 5:207–13

112. Savoure A, Magyar Z, Pierre M, Brown S, Schultze M, et al. 1994. Activation of the cell cycle machinery and the isoflavonoid biosynthesis pathway by active *Rhizobium meliloti* nod signal molecules in *Medicago* microcallus suspensions. *EMBO J.* 13: 1093–102

113. Sebastian B, Kakizuka A, Hunter T. 1993. Cdc25M2 activation of cyclin-dependent kinases by dephosphorylation of threonine-14 and tyrosine-15. *Proc. Natl. Acad. Sci. USA* 90:3521–24

114. Sheldrick KS, Carr AM. 1993. Feedback controls and g2 checkpoints—fission yeast as a model system. *BioEssays* 15:775–82

115. Sherr CJ. 1993. Mammalian G1 cyclins. *Cell* 73:1059–65

116. Solomon MJ. 1993. Activation of the various cyclin/cdc2 protein kinases. *Curr. Opin. Cell Biol.* 5:180–86

117. Solomon MJ, Harper JW, Shuttleworth J. 1993. CAK, the p34^{cdc2} activating kinase, contains a protein identical or closely related to p40^{MO15}. *EMBO J.* 12: 3133–42

118. Soni R, Charmichael JP, Shah ZH, Murray JAH. 1995. A family of cyclin D homologs from plants differentially controlled by growth regulators and containing the conserved retinoblastoma protein interaction motif. *Plant Cell.* In press

119. Stafstrom J, Altschuler M, Anderson D. 1993. Molecular cloning and expression of a MAP kinase homologue from pea. *Plant Mol. Biol.* 22:83–90

120. Staiger C, Doonan J. 1993. Cell division in plants. *Curr. Opin. Cell Biol.* 5:226–31

121. Stein GH, Drullinger LF, Robetorye RS, Pereira Smith OM, Smith JR. 1991. Senescent cells fail to express *cdc2, cycA,* and

cycB in response to mitogen stimulation. *Proc. Natl. Acad. Sci. USA* 88:11012–16

122. Strausfeld U, Fernandez A, Capony JP, Girard F, Lautredou N, et al. 1994. Activation of p34^{cdc2} protein kinase by microinjection of human *cdc25C* into mammalian cells—requirement for prior phosphorylation of *cdc25C* by p34^{cdc2} at mitosis on sites phosphorylated. *J. Biol. Chem.* 269:5989–6000

123. Tsai LH, Lees E, Faha B, Harlow E, Riabowol K. 1993. The cdk2 kinase is required for the G1-to-S transition in mammalian cells. *Oncogene* 8:1593–602

124. Tsai LH, Takahashi T, Caviness VS, Harlow E. 1993. Activity and expression pattern of cyclin-dependent kinase-5 in the embryonic mouse nervous system. *Development* 119:1029–40

125. Tsuji Y, Ninomiyatsuji J, Torti SV, Torti FM. 1993. Selective loss of *cdc2* and *cdk2* induction by tumor necrosis factor-alpha in senescent human diploid fibroblasts. *Exp. Cell Res.* 209:175–82

126. Van den Heuvel S, Harlow E. 1993. Distinct roles for cyclin-dependent kinases in cell cycle control. *Science* 262:2050–54

127. Welch PJ, Wang JYJ. 1992. Coordinated synthesis and degradation of *cdc2* in the mammalian cell cycle. *Proc. Natl. Acad. Sci. USA* 89:3093–97

128. Wittenberg C, Reed SI. 1988. Control of the yeast cell cycle is associated with assembly/disassembly of the CDC28 protein kinase complex. *Cell* 54:1061–72

129. Xiong Y, Zhang H, Beach D. 1992. D-type cyclins associate with multiple protein kinases and the DNA replication and repair factor PCNA. *Cell* 71:505–14

130. Yue X, Connolly T, Futcher B, Beach D. 1991. Human D-type cyclin. *Cell* 65:691–99

Annu. Rev. Plant Physiol. Plant Mol. Biol. 1995. 46:341–68
Copyright © 1995 by Annual Reviews Inc. All rights reserved

REGULATION OF METABOLISM IN TRANSGENIC PLANTS

Mark Stitt

Institute of Botany, University of Heidelberg, 69120 Heidelberg, Germany

Uwe Sonnewald

Institute for Plant Genetics and Crop Plant Research, Correnstrasse 3, 06466 Gatersleben, Germany

KEY WORDS: transgenic plants, carbohydrates, photosynthesis, sucrose, starch, bioengineering

CONTENTS

ABSTRACT

This review discusses how genetically manipulated plants are being used to study the regulation of metabolism in plants, using carbohydrate metabolism as an example. The molecular tools required are introduced, including the history of *Agrobacterium tumefaciens*–mediated gene transfer and other transformation techniques, the availability of promoters to achieve a specific or induced expression, strategies to target proteins to subcellular compartments of the cell, and the use of antisense or cosuppression to inhibit expression of endogenous genes. A discussion then follows of how such plants can be used in biochemical and physiological experiments to identify and quantify the importance of enzymes and processes that control metabolic fluxes, storage, and growth. These results are leading to a reassessment of ideas about metabolic regulation and have consequences for design of bioengineering strategies in plants. Emerging commercial applications are also surveyed.

INTRODUCTION

Since the early 1980s when regeneration of the first transgenic plants was reported, the use of such plants has opened new and fascinating fields in plant biology and crop improvement (37, 172). We use carbohydrate metabolism as an example to discuss how genetically manipulated plants can be used to study regulation in vivo. In this mature research area reverse genetics can be used to test conventional ideas about regulation and to identify which elements of primary metabolism are of key importance for plant growth. Background information is available in reviews about photosynthetic (32, 62, 82, 111, 112, 144, 145, 148, 149, 178) and nonphotosynthetic (9, 55, 77, 144, 145) metabolism, and the manipulation of sink-source relations (140, 146, 147).

PRODUCTION OF TRANSGENIC PLANTS FOR METABOLIC DESIGN

Because of the complex organization and metabolic compartmentation of higher plants, many molecular tools have to be combined to succesfully manipulate their metabolic processes. Directed changes require a transformation system, a suitable gene, suitable promoter sequences for cell- and tissue-specific expression, and effective targeting signals to direct protein to its final destination within the cell. Metabolic design can involve antisense-inhibition or heterologous overexpression of plant-related or alien genes.

TRANSFORMATION   Table 1 summarizes the history of *Agrobacterium tumefaciens*–mediated gene transfer (for a recent review see 58). The observation (181) that large quantities of the T-DNA could be deleted and replaced by unrelated sequences indicated that any DNA of interest could be introduced into plant cells. The use of dominant selectable markers (52; summarized in 179), the removal of phytohormone biosynthetic genes from the T-DNA region (181), and the introduction of the leaf-disk transformation system (60) led to an explosive development of plant biology.

Table 1 History of *Agrobacterium tumefaciens*–mediated plant transformation

Year	Finding	Reference
1907	Identification of *Agrobacterium tumefaciens*, the causative agent of crown gall disease.	137
1974	Discovery of a large extrachromosomal element, involved in crown gall induction.	180
1975	Acquisition of tumor-inducing ability by nononcogenic agrobacteria as a result of plasmid transfer. The plasmid was called tumor inducing plasmid.	166
1977	Transfer and stable maintenance of T-DNA [a segment of the Ti (tumor-inducing) plasmid] in plant cells	13
1980	Use of the Ti-plasmid for the introduction and stable maintenance of foreign genes in higher plant cells.	51
1981	Mendelian transmission of genes introduced into plants by the Ti-plasmid.	104
1983	Construction of chimeric genes as dominant selectable markers in plant cells.	52
	Design of a Ti-plasmid vector for the introduction of DNA into plant cells without alteration of their normal regeneration capacity.	181
1984	Regeneration of kanamycin-resistant plants and inheritance of foreign genes in a Mendelian manner.	59
1985	Establishment of the leaf-disk transformation-regeneration method.	60
1988	First report of down-regulation of an authentic plant gene by antisense gene.	164

Agrobacterium tumefaciens gene transfer is the most frequently used strategy for generating transgenic plants. Limitations of the host range of *A. tumefaciens* restricts its use in many cases, particularly when dealing with monocots (6a, 133). One reason for host-range limitations is the different wound responses of different plants and tissues (109). In most cereals, a proper wound response is missing. Instead of dedifferentiation of wound-adjacent cells, they accumulate phenolic compounds and die (109), making a successful transformation difficult. Extensive efforts have been made to develop other systems for the delivery of free DNA into these plants, including particle bombardment (15), viral vectors (67), protoplast transformation via polyethylene glycol or electroporation (for review see 109), microinjection of DNA into mesophyll protoplasts (17), and macroinjection of DNA into young floral tillers of cereal plants (20). The most successful method to date is particle bombardment of embryogenic calli or developing embryos (133, 167).

AVAILABLE PLANT PROMOTERS   To efficiently manipulate a metabolic process, the gene must be expressed in a predictable and suitable manner. Depending on the process under examination, this may require organ- and/or cell-specific expression, and appropiate developmental and environmental control. The most commonly used promoter has been the constitutive 35S-CaMV promoter (34). This promoter has been used successfully to engineer herbicide- and pathogen-resistant plants (see Crop Improvement section) but is of limited use for studies of metabolic pathways. For example, synthesis and allocation of photoassimilates involves an interaction between the exporting leaves (source), the transport route, and the growing, importing organs (sinks). Organ-specific promoters have been reported for storage sink tissues such as seeds, potato tubers, and fruits (26), and for metabolic sink tissues such as meristems (78). The most extensively studied leaf-specific genes encode the small subunit of ribulose-1,5-bisphosphate carboxylase-oxygenase (33), the chloroplast a/b binding protein (134), and the 10-kDa protein of the oxygen-evolving complex of photosystem II (153). Within an organ, cell specifity may also be required. Expressing the GUS-reporter gene under control of the leaf-specific promoter sequences in transgenic plants resulted in activity in mesophyll cells, guard cells, and companion cells (26). These three cell types fulfill different and specialized functions in leaves: Carbon fixation occurs mainly in leaf mesophyll cells; guard cell metabolism regulates the opening and closure of stomata; and companion cells are responsible for the phloem loading. Recently, more specific promoter sequences have been described, allowing the guard cell–specific (91) and the mesophyll-specific (152) expression of reporter genes. In addition, photoassimilate transport may be manipulated by using either the phloem-specific sucrose synthase promoter (84) or the rolC-promoter (126) from *Agrobacterium rhizogenes*.

Inducible promoters provide an ideal tool to express heterologous genes. Several plant promoters described are inducible by endogenous signal metabolites. The best studied are abscisic acid–inducible (83) and salicylic acid–inducible (157) genes. However, use of these promoters is limited because (*a*) the naturally occurring level of signal metabolites and the responsiveness of the target tissue or cell may vary, (*b*) the signal metabolites alter the expression of many endogenous genes, and (*c*) the inducible genes described so far are under complex control, including environmental and developmental factors. Synthetic promoters that respond to chemical inducers would circumvent the above-mentioned problems (reviewed in 174). Chimeric promoter systems have been described using either gene activation (38, 177) or gene repression (176) to control gene expression, but they have have not been used successfully for physiological studies because the chemical inducers are not easily absorbed by plants.

TARGETING OF FOREIGN PROTEINS   Eukaryotic cells carry out metabolic processes in discrete subcellular compartments. With the exception of a few plastidic and mitochondrial proteins, the polypeptides present in organelles are encoded by nuclear DNA, and transport of the polypeptides to their appropriate destinations depends on the presence of specific targeting signals. Heterologous proteins must therefore be expressed as fusion proteins containing a suitable targeting sequence. Targeting of nuclear-encoded polypeptides into the cytosol does not require additional sorting information. Secreted or vacuolar proteins contain a N-terminal hydrophobic signal peptide, which directs entry of the nascent polypeptide into the lumen of the endoplasmic reticulum (reviewed in 14). Whereas secretion of proteins into the apoplast does not need further information in higher plants (21, 171), vacuolar targeting or retention in the endoplasmic reticulum requires additional targeting signals (139, 173; reviewed in 95). Nuclear-encoded mitochondrial and plastidic polypeptides are synthesized as precursor proteins in the cytosol containing N-terminal transit peptides. Fusion of mitochondrial (8) and plastidic (162) transit peptides to foreign proteins enables targeting of the polypeptides to the respective organelles.

ANTISENSE INHIBITION OF ENDOGENOUS PLANT GENES   Alteration in the expression of endogenous plant genes allows investigation of their contribution to plant growth. Classical mutants of specific genes can be used, if the phenotype allows screening of the resulting mutant. It is not possible to obtain mutants for specific genes when they are members of multigene families, when the phenotype cannot be predicted, or when the mutation is lethal. Antisense technology provides a useful alternative in such cases (first reported for plants in 164; reviewed in 89). The antisense inhibition of plant genes allows variable reduction in the amount of the target protein, including enzymes encoded by multigene families (122). In principle, antisense allows a tissue- and cell-specific inhibition of target-enzymes. However, several precautions have to be taken: (*a*) With standard transformation techniques, the site of integration of the antisense gene in the genome is random, and gene disruption may lead to unrelated phenotypes. To exclude artifacts caused by insertion-mutagenesis and somaclonal variation, several different transformants should be investigated. (*b*) Environmental and developmental differences in the efficiency of the antisense inhibition can occur (75, 117, 164). These differences are likely if heterologous promoter sequences are used to control expression of the antisense gene, because these do not allow coregulation of sense- and antisense-transcripts. The degree of inhibition should therefore always be measured in each of the samples used for analysis. (*c*) Variability of the organ-specific gene expression of reporter genes (163) complicates the analysis of antisense effects. The effect on expression should therefore be checked in each transformant.

COSUPPRESSION   Endogenous genes can also be down-regulated by expression of homologous sense-transcripts [referred to as cosuppression (96)]. Expression of a chimeric gene construct encoding partial sense–transcripts of the polygalac-turonase and pectinesterase gene under the control of the 35S-CaMV promoter led to the simultaneous down-regulation of both genes in transgenic tomatoes (128). Although cosuppression is more efficient than antisense in some cases, the effects can be variable, and the underlying mechanisms are unclear (see 7a and references therein). Several different mechanisms have been proposed, including DNA methylation, ectopic pairing of homologous DNA sequences and subsequent heterochromatin formation, and an undefined posttranscrip-tional regulation of mRNA levels.

OVEREXPRESSION   Another strategy for altering metabolism is ectopic expres-sion of heterologous enzymes. These enzymes can be plant-derived or from unrelated organisms. Overexpression of plant-derived enzymes does not always lead to an increased enzymatic activity in vivo, if the enzyme is subject to metabolic or posttranslational control (75). To overcome endogenous regulatory mechanisms, mutated enzymes with altered regulatory properties can be ex-pressed (142). Expression of novel polypeptides that are not found in higher plants has the advantage that such inhibitory regulation is likely to be absent.

APPLICATION TO THE STUDY OF CARBOHYDRATE METABOLISM REGULATION

Several criteria have been used traditionally to identify regulatory sites: (a) The responsible enzyme should catalyze an irreversible reaction; (b) it should possess regulatory properties; and (c) there should be characteristic reciprocal changes of flux and substrate concentration in vivo (100, 123, 143). Applica-tion of these criteria has produced a vast amount of detailed information about individual reactions, but it also has shown the limitations of this approach (146): Individual enzymes are modulated by multiple effectors that act in a highly interactive manner (63, 144, 147, 149); metabolic pathways often con-tain more than one regulatory enzyme (69, 98, 147, 178); the activity of the various enzymes in a pathway is highly coordinated (22, 69, 98, 178); and sometimes there are alternative enzymes that could substitute for each other (7, 55, 77, 145). An analogous problem exists at the level of whole plant physiol-ogy, where correlative studies rarely provide decisive information about the significance of a particular process for the growth and performance of a plant.

The limitations of the traditional approach are highlighted by the distinction (57) between the "regulatability" of an enzyme (i.e. whether mechanisms exist to alter its activity, for example, allosteric effectors, posttranslational modifi-cation, changed expression) and its "regulatory capacity" (i.e. whether a change in the activity of the enzyme will actually lead to a change in flux through the pathway). Much of what we loosely term regulation maintains a functional balance within a pathway (e.g. to allow an enzyme to respond to a change initiated elsewhere in the path, to coordinate the activities of the various enzymes in the pathway, and to maintain the concentrations of substrates and cofactors in a range suitable for the operation of all the other enzymes in the pathway). Such enzymes have high regulatability but a low regulatory capacity. New system-oriented approaches are needed to identify the sites and properties that are central for the control of flux and growth.

Contribution of Rubisco to the Control of Photosynthetic Rate

Rubisco represents 30–40% of leaf protein (27) and is thought to play a central role in the carbon, nitrogen, and water economy of the plant (28, 82, 130, 178). The contribution of an enzyme to the control of a process can be tested directly, by constructing a series of plants with a progressive decrease in the amount of one enzyme (leaving all the others unchanged) and investigating the impact on pathway flux (29, 68, 146, 148). For the simplest case of a linear pathway, a small (e.g. 25%) decrease in the amount of a given enzyme will (*a*) lead to proportional (25%) inhibition of flux through the pathway if the enzyme is solely responsible for control of flux (limiting), (*b*) have no effect if the enzyme exerts no control, or (*c*) lead to a significant but nonproportional (< 25%) inhibition of pathway flux if the enzyme contributes to control but is not the only site controlling flux through the pathway. This thought exercise can be formalized (68) by calculating the flux control coefficient (see 29, 68, 117, 148). A flux control coefficient, $C_{E_i}^{J_i}$, describes the contribution that an enzyme, E_i, makes to the control of flux through a pathway, J_i. It is formally defined as

$$C_{E_i}^{J_i} = \frac{dJ/J}{dE/E},$$

1.

where J and E are the magnitude of the flux and the amount of activity of the enzyme in the original state (e.g. in the wild-type), dE is a small change to the amount or activity of the enyzme (e.g. the difference between the wild-type and a mutant or antisense plant), and dJ is the resulting change of flux. Precautions needed in measuring and calculating control coefficients have been reviewed (1a, 135, 151). The most important precautions are that the change in expression must be relatively small and specific and that the changes of enzyme and flux must be measured accurately.

Two sets of antisense *rbc*S plants (64, 122) have been used to investigate the interaction between Rubisco content, photosynthetic rates, and growth. The contribution of Rubisco to the control of photosynthesis depends on the short-term and long-term conditions (reviewed in 151). When tobacco was grown at a high nitrogen supply and moderate irradiance (300 µmol m^{-2} s^{-1}), the flux control coefficient of Rubisco for photosynthesis ($C_{Rubisco}^A$) was low (0.1–0.2; i.e. a 25 % decrease in the amount of Rubisco led to a 2.5–5% inhibition of photosynthesis) in ambient growth conditions (117). It increased to 0.7–0.8 when the CO_2 was decreased suddenly or the irradiance was increased suddenly to 1000 µmol m^{-2} s^{-1} (79, 116, 150). This is broadly in agreement with the predictions of models of photosynthesis (28, 178). However, $C_{Rubisco}^A$ was only 0.2–0.3 at an irradiance of 1000 µmol m^{-2} s^{-1}, when the tobacco had previously grown at this high irradiance (79). Leaves obviously can adjust to avoid a one-sided limitation of photosynthesis by Rubisco over a wide range of growth conditions. In more extreme growth conditions, photosynthesis is strongly limited by Rubisco. When tobacco plants are grown under very high (up to 2000 µmol m^{-2} s^{-1}) natural irradiance, $C_{Rubisco}^A$ rises to over 0.80 (calculated from results in 64, 73). Even more interestingly, when the nitrogen supply was decreased, $C_{Rubisco}^A$ rose to 0.6 even at moderate growth irradiance (116). Clearly, an enzyme's contribution to control is not fixed. This has obvious implications for the design and interpretation of experiments to identify control sites.

Operation of Rubisco in Vivo and the Calvin Cycle

Efficient use of the large investment of nitrogen in Rubisco requires (*a*) efficient diffusion of CO_2 from the air to the active sites of Rubisco, (*b*) a saturating concentration of the CO_2 acceptor ribulose-1,5-bisphosphate (Ru1,5bisP), and (*c*) protection of Rubisco against various tight-binding inhibitors.

CO2-ENTRY   Transgenic plants are providing novel information about the significance of putative steps and regulation mechanisms during CO_2 entry into the leaf. In *rbc*S antisense plants (64, 149) and plants with decreased expression of glyceraldehyde-3-phosphate dehydrogenase (114) the inhibition of photosynthesis is not accompanied by parallel decline of stomatal conductance, and the CO_2 concentration in the air space within the leaf (c_i) increases. This shows that c_i does not play a major role in coordinating stomatal aperture and photosynthesis in tobacco in normal conditions. Recently, antisense was used to decrease carbonic anhydrase to 2% of wild-type levels (115). δ^{13}C-discrimination was significantly lower than in the wild-type (0.85), corresponding to a 15 bar reduction in the CO_2 partial pressure at the sites of carboxylation, from which a 4% inhibition of photosynthesis was calculated. These elegant studies confirm that carbonic anhydrase facilitates diffusion of inorganic carbon in the aqueous phase of the leaf but also show that this diffusion is of marginal importance for photosynthetic rate.

RUBISCO ACTIVASE   Biochemical evidence indicates that activase protects Rubisco against a series of inactivating events. These events include binding of Ru1,5bisP to the inactive decarbamylated enzyme (Ru1,5bisP binds to the decarbamylated form with a higher affinity than the carbamylated form; see 108), binding of inhibitors such as XubP, which are formed at a low frequency during catalysis (25), and binding of an inhibitor (Ca1P), which accumulates in the dark in some species (108). Activase provides an example of how mutants and transgenic plants can be used to discover a previously unknown protein and investigate its role and significance. Activase was discovered in investigations of the *Arabidopsis thaliana* mutant *rca*, which lacks the protein and cannot maintain Rubisco in the carbamylated (active) form at air levels of CO_2. Two sets of antisense tobacco plants have been investigated (66, 86). The decarbamylation of Rubisco and the inhibition of photosynthesis increased with length of illumination, as expected if activase protects Rubisco against inhibitory events, rather than catalyzing carbamylation per se (86). Ca1P (which binds to Rubisco in the dark and inhibits it) is released more slowly after illumination in activase antisense plants (86). Studies of plants with a small decrease in activase expression showed that activase could be reduced two- to threefold without inhibition of photosynthesis in growth conditions (66). In saturating irradiance there was an inhibition of the maximum light-saturated rate of photosynthesis, which correlated with a small decrease in Rubisco activation (66). Flux control coefficients of activase of zero and 0.28 for ambient and light-saturated photosynthesis, respectively, were estimated.

Ru1,5bisP REGENERATION BY THE CALVIN CYCLE   A Rubisco has a high affinity in vitro for Ru1,5bisP ($K_m = 18\mu M$), and there is a high concentration of active sites in the plastid (4 mM) (see 178), indicating that a 1:1 ratio of Ru1,5bisP:Rubisco sites should saturate Rubisco in vivo. However, empirical measurements (130, 170) and theoretical arguments (130, 170, 178) indicate that more Ru1,5bisP may be required. This problem has been investigated directly by decreasing the expression of enzymes involved in the regeneration of Ru1,5bisP and testing at what point a falling Ru1,5bisP/Rubisco active site ratio leads to an inhibition of photosynthesis. When NADP-GAPDH was decreased, photosynthesis was inhibited when the Ru1,5bisP/Rubisco active site ratio fell below about 2 (114). This demonstrates that one (or more) factor(s) operate in vivo to weaken the interaction between Ru1,5bisP and its binding site.

The contribution of individual Calvin cycle enzymes to the control of photosynthesis has been investigated in plants with decreased expression of NADP-GAPDH (114), plastid aldolase (140), plastid fructose-1,6-bisphosphatase (pFBPase) (72), and phosphoribulokinase (PRK) (JC Gray & M Paul, personal communication). Ambient photosynthesis was unaffected when NADP-GAPDH was decreased two-fold, or PRK was decreased 10- to

20-fold, and was inhibited by only 12% when pFBPase was decreased three-fold. A 60% decrease of p-aldolase led to a significant inhibition of photosynthesis (V Haake, unpublished data). These results show that the irreversible enzymic steps involved in the regeneration of Ru1,5bisP do not exert major control over the rate of photosynthesis. pFBPase and PRK are highly regulated (82, 148, 178), but this regulation maintains function of the Calvin cycle, rather than controlling photosynthetic rate (i.e. these enzymes have high regulatability but a low regulatory capacity). Unexpectedly, flux is quite sensitive to decreased expression of aldolase, which catalyzes a reversible reaction.

Control of Partitioning to Sucrose, Starch, and Amino Acids

Photosynthate can be converted to sucrose and exported, or stored as starch, or used for amino acid synthesis. Flux through these pathways must be regulated for two reasons (144, 145, 149). First, the overall rate of end-product synthesis must be coordinated with the rate of CO_2 fixation to avoid inhibition of photosynthesis resulting from depletion of the Calvin cycle intermediates or accumulation of phosphorylated intermediates and depletion of inorganic phosphate. The 3PGA/P_i ratio plays a key role in signaling the balance between photosynthetic rate and end-product synthesis (112, 145, 148, 178). Second, the distribution of fixed carbon between the products must be regulated.

PHOTOSYNTHETIC STARCH SYNTHESIS   Biochemical studies indicate that starch synthesis is regulated by ADPglucose-pyrophosphorylase (AGPase). This enzyme is strongly activated by 3PGA and inhibited by P_i (111, 112). *A. thaliana* plants heterozygous for a mutation in the structural gene of AGPase (99) and antisense AGPase potato plants (92) have been used to investigate the relation between enzyme amount and the rate of photosynthetic starch synthesis. A flux control coefficient for starch synthesis of 0.3–0.5 was estimated (99), directly confirming the central role of AGPase.

The contribution of plastid phosphoglucose isomerase (pPGI), plastid phosphoglucose mutase (pPGM), and branching enzyme (BE) have been investigated in heterozygotes of *Clarkia xantiana*, *A. thaliana*, and pea mutants (76, 97, 99, 136). A 50% reduction in enzyme amount did not affect flux in limiting light; however, it did decrease the rate of starch synthesis in saturating light and CO_2. The estimated flux control coefficients of pPGI and pPGM for starch synthesis in saturating light and CO_2 were 0.35 and 0.21, respectively (as compared to 0.64 for AGPase in these conditions). These results illustrate several general conclusions. First, the distribution of control in a pathway depends on the conditions. Second, control can be shared between several steps. Third, reactions such as pPGM and pPGI, which catalyze readily reversible reactions in vivo, can exert control over flux in some conditions. In this particular example, these "uninteresting" enzymes start to exert control when the flux to starch increases about twofold above that in growth conditions. Such enzymes obviously are not present in large excess.

Regulation often involves changes in kinetic properties rather than in enzyme amount. The importance of the kinetic properties of AGPase has been investigated in mutants and transgenic plants. Loss of the 3PGA activation of AGPase results in decreased starch synthesis in a *Clamydomonas reinhardtii* mutant (2). Overexpression of a mutated AGPase from *Escherichia coli* (glgC-16), which is active in the absence of allosteric activators (113), resulted in a large accumulation of starch in the leaves of heterotrophically growing tomato plantlets (142), whereas overexpression of the non-mutated (regulatable) enzyme did not. Control analysis has not considered the effect of changed properties in enough detail. In principle, a control coefficient could be calculated for a change in the properties of an enzyme. Preliminary theoretical analysis indicates that changes in kinetic properties may sometimes have a larger effect on flux than does a change in the amount (D Fell, personal communication).

Constitutive or leaf-specific overexpression of glgC-16 in potato or tomato prevented normal autotrophic growth (142). Overproduction of starch presumably inhibits photosynthetic sucrose production and export. This illustrates another general point: Regulation in vivo often has a dual function. In this example, (*a*) activation of AGPase by a rising $3PGA/P_i$ ratio mediates a stimulation of starch synthesis in response to a rising supply of photosynthate, and (*b*) inhibition of AGPase by a falling $3PGA/P_i$ ratio protects against an excessive drain of triose phosphates into starch.

SUCROSE SYNTHESIS   Biochemical studies indicate that sucrose synthesis is modulated at the first committed step [cytosolic fructose-1,6-bisphosphatase (cFBPase)] and the penultimate reaction [sucrose-phosphate synthase (SPS)]. cFBPase is regulated by an interaction between substrate availability and the signal metabolite fructose-2,6-bisphosphate (Fru2,6bisP) (144, 149). SPS activity is modulated by allosteric effectors [e.g. glucose-6-phosphate (Glc6P) activates and P_i inhibits)] and protein phosphorylation (62, 63). Indirect evidence indicates that control may be shared between these enzymes (98). Genetically manipulated plants allow direct assessment of their contribution as well as the contributions of other enzymes in the path.

The first step is the reversible counter-exchange of P_i, triose-phosphates, and 3PGA between the stroma and cytosol, catalyzed by the triose-phosphate translocator (TPT) (32). Antisense potato plants with a 20–30% reduction in TPT expression showed a 27–50% inhibition of sucrose synthesis during the photoperiod and a large accumulation of starch (46, 119). Changed expression of the TPT in tobacco also alters the starch/soluble sugars ratio (3). The importance of metabolite transporters for regulation may have been underestimated in the past.

C. *xantiana* mutants with decreased expression of the cytosolic phosphoglucose isomerase (cPGI) have been used to analyze the contribution of this reversible reaction to the control of sucrose synthesis (76, 97). Decreasing activity by 36% or 64% had no effect on flux, but plants with 18% of wild-type cPGI had lower sucrose and higher starch synthesis. The inhibition of sucrose synthesis was larger in limiting irradiance (21%) than in saturating irradiance and CO_2 (10%). Sucrose synthesis is inhibited because accumulation of substrate (Fru6P) leads to an increase of Fru2,6bisP (see 144, 145), which inhibits cFBPase. These plants have been used to evaluate how Fru2,6bisP and triose-phosphate supply interact in vivo to regulate cFBPase (97, 143, 148). Recently, Fru2,6bisP has been increased by overexpressing rat liver Fru6P,2-kinase in tobacco and has been demonstrated to inhibit sucrose synthesis (127).

UDP-glucose pyrophosphorylase (UGPase) converts glucose-1-phosphate (Glc1P) to UDP-glucose (UDPGlc). This step is irreversible in most organisms, because of hydrolysis of pyrophosphate (PP_i) by pyrophosphatase. It is reversible in the cytosol of plant cells, because this compartment contains a significant level of PP_i (144). Antisense UGPase plants with 5% of the wild-type UGPase activity showed no changes in photosynthate partitioning (183). To test whether sucrose synthesis can be accelarated by removal of PP_i, the *ppa* gene for pyrophosphatase from *E. coli* was introduced into tobacco and potato under control of the constitutive 35S-CaMV promoter (138). There was a small shift of partitioning in favor of sucrose (66), which reduced starch and increased sucrose levels in potato leaves and in young leaves of tobacco plants (138). A large stimulation of sucrose synthesis was not achieved because the steady-state levels of hexose-phosphates (Fru6P, Glc6P, and Glc1P) decreased (65), which restricts SPS activity. There were dramatic effects on growth and carbohydrates accumulated in the older leaves of tobacco plants, because constitutive expression of PPase had dramatic effects on phloem transport and sink metabolism (see below).

Maize SPS has been overexpressed three- to sevenfold in tomato (10, 36); spinach SPS has been overexpressed twofold in tobacco (75); and SPS expression has been decreased by 60% in potato leaves (75). Overexpression of heterologous SPS resulted in a small stimulation of sucrose synthesis (N Galthier, C Foyer, & WP Quick, personal communication) and a two- to threefold increase of sucrose content and the sucrose/starch ratio in tomato leaves (36) but not tobacco (75) leaves. A 60–70% reduction in SPS expression in potato leaves resulted in a 50% inhibition of sucrose synthesis and increased starch and amino acid synthesis (75). A flux control coefficient of SPS for sucrose synthesis of 0.3–0.45 was estimated for potato leaves (75).

These results show that SPS contributes to the control of sucrose synthesis in leaves but is not the only site of control.

Unpublished results indicate that changes in SPS expression are counteracted by posttranslational modification in the plants discussed above. Huber and colleagues have shown that SPS is regulated via reversible protein phosphorylation in vivo and have identified the kinases and phosphatases as well as a putative phosphorylation site (62, 63, 87). The production of genetically modified protein with altered properties, or the removal of the phosphorylation site, could lead to a more efficient manipulation of sucrose synthesis.

Export and Storage of Sugar

PROOF OF APOPLASTIC LOADING   Sucrose is exported via an apoplastic route in many species, involving passage through the cell wall and active uptake into the phloem via a H^+-sucrose cotransporter (159). Direct proof of this process has been obtained by expressing yeast invertase in the cell wall of tobacco, *A. thaliana*, tomato, and potato plants (21, 47, 171). These plants had strongly inhibited growth and accumulated carbohydrate in their leaves. The H^+-sucrose cotransporter has been identified by cloning and functional expression in yeast (35, 120). Antisense inhibition resulted in a massive accumulation of carbohydrate in the leaves and severe stunting of potato plants (121).

COMPARTMENTATION AND METABOLISM OF SUGARS IN THE LEAF   Transgenic plants are providing new information about the metabolism and compartmentation of sugars in leaves. Expression of invertase in the apoplast or vacuole leads to accumulation of sucrose as well as hexoses (47, 48, 139, 171). This observation implies that sucrose is being resynthesized and stored in a compartment that is inaccessible to the introduced invertases. In wild-type tobacco leaves, sucrose is located in the cytosol and free hexoses are located in the vacuole (48). The transformants accumulate large amounts of sucrose in the cytosol and accumulate hexoses in their vacuoles and amino acids in the stroma as counter-osmotica (48). This illustrates that plant cells use specific organic solutes to balance the osmotic potential in their subcellular compartments. It also indicates that the release and transport of hexoses across external and internal membranes are likely to be key sites in the regulation of their use.

Expression of invertase in the cytosol led to a decrease of sucrose (48, 139). Surprisingly, glucose remained low and fructose accumulated (139) in the vacuole (48). The rapid use of glucose can be explained by the general lack of feedback regulation of glucose-phosphorylating hexokinases. The accumulation of fructose may be the result of substrate-inhibition of fructokinase by fructose (5, 118).

FEEDBACK INHIBITION OF PHOTOSYNTHESIS   Studies with transgenic plants have demonstrated that photosynthesis is inhibited when carbohydrate accumulates in source leaves as a consequence of an inhibition of phloem transport (47, 48, 121, 139, 171). These plants also provide a novel system to investigate the signals and mechanisms that are involved in this feedback or "sink" regulation of phyotosynthesis. Sucrose and free hexoses behave differently when the balance between sucrose synthesis and export is changed. In antisense *rbc*S, pFBPase, TPT, and SPS plants, the decreased rate of sucrose synthesis is accompanied by a much larger decline of free hexoses than of sucrose (30, 46, 72, 75). In antisense H^+-sucrose transporter plants, there is a larger rise of free hexoses than of sucrose (121). Correlative evidence indicates that an increase of hexose may play an important role in the sink regulation of photosynthesis (41). In tobacco, an accumulation of sugars led to decreased levels of transcripts and a decline of proteins, such as Rubisco, that are required for photosynthesis (74, 148). In potato, photosynthesis was inhibited as a result of osmotic problems (48).

Impact of Altered Photosynthate Export on Allocation and Plant Growth

As discussed above, transgenic plants can be used to study regulation within a pathway. A decisive advantage of transgenic plants is that they also reflect the importance of a particular pathway or process for whole plant growth.

DIRECT INHIBITION OF EXPORT   Phloem processes are essential for whole plant function. Plant growth is severely inhibited when phloem export is inhibited directly by expression of invertase (21, 171) or antisense against the phloem H^+-sucrose transporter (121). Root growth is particularly strongly decreased and tuber yield drops sharply. The effect on sink tissue composition depends on how phloem transport has been inhibited. The amino acid/sucrose ratio increases in tubers of potato plants expressing invertase in the apoplast (47). In contrast, transport of amino acids is also reduced when phloem function (rather than sucrose loading) is inhibited by expression of pyrophosphatase (removal of PPi inhibits sucrose mobilization and energy metabolism because sucrose is degraded via sucrose synthase in the phloem; J Lerchl, P Geigenberger, U Sonnewald & M Stitt, unpublished results). It may also be possible to manipulate sink composition by altering phloem loading in the source leaves. Amino acid and sugar transport proteins are reviewed in another chapter of this volume (35).

ALTERED PARTITIONING OF PHOTOSYNTHATE.   Leaf starch is only essential for growth in some conditions. Starchless mutants of *A. thaliana* (12, 125) and tobacco (61) grow poorly in short-day conditions. However, they grow as well as the wild-type in long days (12) or on limiting nitrogen (125). Starchless mutants increase their root/shoot ratio in response to low nitrogen (125), showing that starch metabolism is not causally involved in the modulation of whole plant allocation by this environmental parameter. However, these mutants have a consistently higher root/shoot ratio (125) and an altered pattern of leaf expansion (61), indicating that root and shoot growth depends to a different extent on starch metabolism. Low-starch potato plants also flower earlier and have a large number of small tubers (92). Tissue-specific promoters are needed to distinguish which of these changes in allocation are the result of leaf starch metabolism and which are the result of starch metabolism in sink organs.

Partial inhibition of photosynthetic sucrose synthesis does not lead to a consistent inhibition of growth. Antisense TPT potato plants do not have decreased biomass (46). The lower rate of sucrose export during the day is compensated for by increased degradation of starch and export during the night. In *A. thaliana*, starch degradation leads to the formation of glucose, which is exported via the glucose transporter (156) and therefore bypasses the TPT. Partitioning may be important only in some conditions or developmental stages. Antisense TPT plant growth is inhibited immediately after establishing an explant (119), but not later. It will be important to compare plant growth under photoperiod and temperature regimes designed to alter the overall rate of growth, or to compare the relative importance of daytime and nighttime export.

ALTERATION OF PHOTOSYNTHETIC RATE   Agronomists have long realized that yield is not related to maximal photosynthetic rate (A_{max}) (40). A_{max} is unrelated to biomass production in antisense FBPase potato plants (72) and antisense *rbc*S plants (compare 30, 116, 151; see also 85). A closer relationship might be expected between ambient photosynthesis and growth. However, ambient photosynthesis was decreased without an immediate impact on tuber yield in antisense FBPase plants (72). In antisense *rbc*S tobacco, ambient photosynthesis was changed twofold without any effect on the rate of growth in nitrogen-limited plants (30, 151). At higher nitrogen supply, decreased photosynthesis led to decreased biomass production (30, 85, 151), but the impact was smaller than expected. This was the result of adaptations in storage and allocation strategies. Antisense *rbc*S plants had a larger leaf area on a whole plant basis and less nonproductive accumulation of starch in their leaves (31, 151). Analogous differences are found between wild species with varying growth rates (40, 107). The growth difference between wild-type and antisense *rbc*S plants is also restricted to the first two weeks (85). A similar development-dependent effect occurs in antisense-FBPase plants (72) and is reminiscent of a large literature finding that plants switch from source- to sink-limitation early in their life cycle (reviewed in 175).These results emphasize the flexibility of leaf metabolism and export strategies, and point to the importance of allocation and use in the importing sink organs for plant growth and yield.

The Import and Use of Sucrose in Sinks

CARBOHYDRATE ENTRY AND METABOLISM IN SINKS IS CHARACTERIZED BY THE PRESENCE OF MULTIPLE ROUTES   Phloem unloading can involve (a) apoplastic unloading of sucrose into the cell wall, followed by hydrolysis to glucose and fructose and their uptake into the growing cells or (b) symplastic unloading via plasmodesmata (159). In the cytosol, sucrose can be degraded via sucrose synthase (SuSy) or alkaline invertase (77).

SYMPLASTIC UNLOADING   Growing potato tubers are being used as a model for a symplastic sink. Reduction of SuSy to 5–30% of wild-type levels decreased starch synthesis and tuber dry weight (140, 182). The UDPGlc and fructose produced by SuSy are metabolized by UGPase and fructokinase. Antisense tubers with 5% of wild-type UGPase activity showed normal growth and composition (183). On the other hand, when UGPase activity was restricted by overexpressing *E. coli* pyrophosphatase to remove PPi, there was a decrease of starch and sugars accumulated in the tubers, tuber size was strongly reduced, and more tubers were induced per plant (65, 138). Mobilization of sucrose apparently is more sensitive to the availability of PPi than to the expression of UGPase per se. Because SuSy catalyzes a reversible reaction, the rate of sucrose breakdown responds to the use of carbohydrate in the sink. When yeast invertase was overexpressed in the cytosol, starch content declined and abnormally long tubers with a large number of lenticels developed (140). Thus normal tuber growth is disrupted when the reversible SuSy pathway is short-circuited in the cytosol.

APOPLASTIC UNLOADING   Developing seeds have no plasmodesmatal connections to the maternal tissue. In maize *minature-1* mutants (88), which lack a seed-encoded cell-wall invertase, seed development is impaired strongly in homozygotes but not in heterozygotes. This result implies that cell-wall invertase is essential but not rate-limiting for unloading in this apoplastic sink. The role of cell-wall invertase in other tissues is less well defined. The maize inbred line Oh-43 lacks invertase in its primary roots but grows normally except for an inability to use exogenously supplied sucrose (23). Other enzymes apparently can substitute for cell-wall invertase in these roots. Apoplastic unloading also requires hexose uptake systems in the sink cells. A large family of H^+-hexose cotransporters has been cloned in *A. thaliana* (124).

CREATION OF A NOVEL APOPLASTIC SINK   To convert a symplastic sink into an apoplastic sink, yeast invertase was introduced into potato, fusing it to the proteinase II signal peptide to direct it to the apoplast, and using the patatin promoter to obtain a tuber-specific expression (L Willmitzer, personal communication). The tubers had a normal starch content, a threefold decline of sucrose, and increased hexose and water content. Individual tuber size increased and even though tuber number per plant declined, tuber yield per plant increased by up to 30% on a fresh weight basis in greenhouse experiments (140) and in the field (L Willmitzer, personal communication).

GLYCOLYSIS AND RESPIRATION   Phosphofructokinase (PFK) is considered to be an important regulation site in mammals and microbes. In plants, this step can also be catalyzed by PFP (144). The importance of PFK in regulating plant respiration has been investigated using potato tubers with 20-fold overexpression of *E. coli* PFK, or 50-fold lower expression of PFP (11, 44). In both cases, there were large changes in the concentrations of glycolytic intermediates but no change in the rate of respiration or growth. The next site for regulation is thought to be at pyruvate kinase and PEP carboxylase. Tobacco plants with fivefold lower pyruvate kinase activity contained greatly elevated PEP, but respiration and growth were identical to the wild-type (42). These studies show that fine regulation can compensate for large changes in gene expression at both sites. This fine regulation allows glycolytic flux to respond automatically to the demand for ATP and carbon skeletons.

STARCH SYNTHESIS   Hexose phosphates enter the amyloplast via a modified TPT (49) and are then converted to starch. AGPases from storage tissues are also activated by a rising 3PGA/Pi ratio (111, 112). The relation between the amount of AGPase and starch accumulation has been investigated in pea seed mutants (K Denyer & AM Smith, personal communication) and in antisense AGPase potato tubers (92). A large reduction of AGPase led to a near-complete inhibition of starch synthesis, demonstrating that AGPase is essential for starch synthesis (92). However, a small (20–50%) decrease of AGPase did not result in a proportional decrease in starch content. Flux control coefficients of 0.1 and 0.2–0.3 can be estimated for pea seeds and potato tubers, respectively.

   AGPase presumably is down-regulated in vivo, to coordinate starch synthesis with other processes needed for growth. The importance of the kinetic properties of AGPase for storage starch accumulation is demonstrated directly in studies with 10–15-fold tuber-specific overexpression of glgC-16 (encoding a AGPase that is active in the absence of allosteric activation). This overexpression led to a 30% increase of starch content in Russet Burbank potatos, whereas overexpression of wild-type *E. coli* AGPase had no significant effect (142). These results show that AGPase contributes to the control of starch synthesis and that its kinetic properties are important. Nevertheless, much of the control is located at other sites. These sites could be located within, but could also lay outside, the pathway of starch synthesis.

Reevaluation of Metabolic Regulation

The results obtained in the first four to five years of research with genetically manipulated plants are leading to a revision of ideas about the regulation of metabolism: (a) Control is usually shared between several proteins. When one is changed, the impact on pathway flux is attenuated by compensation elsewhere in the system. (b) Many enzymes that catalyze irreversible reactions and possess regulatory properties have suprisingly low flux control coefficients (e.g. pFBPase, PRK, PFK, pyruvate kinase; see above). These enzymes operate at a fraction of their catalytic activity. In the wild-type, this fine regulation allows their activity to respond to flux changes that have been initiated elsewhere in the path. It also means they can compensate readily for decreased expression. The association of regulatory properties with enzymes that catalyze irreversible reactions may often just reflect a thermodynamic problem. If these highly exergonic reactions were allowed to run to equilibrium, there would be differences in the concentrations of metabolites of the order of 10^3–10^5, which is probably incompatible with the effective operation of a pathway. The activity of these enzymes therefore must be restricted by fine regulation of their catalytic activity, or by having only a small amount of the enzyme present. The latter strategy is not suitable for enzymes in central pathways that undergo rapid change of flux. (c) Enzymes that catalyze readily reversible reactions are not always present in large excess (e.g. pPGM, pPGI, pAldolase, cPGI, NADP-GAPDH, SuSy, TPT) and impose a ceiling on the flux that can be attained after manipulation of more obvious candidates. This observation contradicts earlier ideas about regulation (100, 123). However, with hindsight, it is difficult to envision how natural selection could maintain a lot of enzymes in large excess (76). Pathway flux is susceptible to decreased expression of these proteins because they lack regulatory properties (140). The only way to compensate for this susceptibility is via an increased substrate/product ratio; pathway flux will be inhibited once this increased ratio starts to restrict some other enzyme. (d) It would be impossible to coordinate fluxes into the various pathways if each were controlled by a quasi-independent enzyme with a very high flux control coefficient and no regulatory interactions with the other pathways. Even obviously important sites for regulation (e.g. AGPase and SPS) have flux control coefficients below 0.5 in most conditions investigated so far (75, 92, 99). Enzymes are tied into a pathway via the effectors and reactants that they share with other enzymes in the pathway. The theory of control analysis (29, 57, 68, 143, 148) explains how flux control coefficients emerge from the interactions between the kinetic properties of enzymes in a pathway. (*e*) Plant metabolism and growth are very flexible because of several factors, including the complexity of regulation, built-in redundancy owing to alternative enzymes and pathways for many processes, and the buffering effect of storage and allocation. As a result, the importance of an enzyme or process should be tested directly; indirect evidence should not be relied upon. It is striking that enzymes that generally have been considered essential and important can be almost totally removed without leading to any obvious phenotype (42, 44) and that intensively researched processes such as photosynthate partitioning can be altered without drastic effects on growth in many conditions (36, 46, 75, 125). The flexibility of plant metabolism also has implications for biotechnology. It is encouraging that plants are able to cope with large switches in their primary metabolism, including novel unloading routes (140), suppression of starch metabolism (12, 92), and massive accumulation of sugars (92, 121, 140). However, this flexibility also implies that the probability of obtaining a desired change will diminish as the number of steps between the manipulation and the goal increases.

CROP IMPROVEMENT

PLANT PROTECTION   Advances in basic research and the ability to efficiently introduce foreign genes into agronomically important crop plants has already led to several applications of genetic engineering in modern plant-breeding. Herbicide-tolerant plants can be engineered by altering the sensitivity of the target enzyme (45, 129) or by introducing a new enzyme to detoxify the herbicide (18, 141). Pathogen-resistant plants have been produced by several strategies. Expression of a modified toxin from the gram-positive bacterium *Bacillus thuringiensis* (158) or a trypsin inhibitor from cowpea (54) results in insect-resistant transgenic plants. The most commonly used approach for virus resistance is the expression of chimeric genes encoding viral coat protein (1, 6). Other promising strategies include expression of viral *cis*-acting elements, movement proteins, and defective replicase (4). Transgenic plants expressing a phaseolotoxin-resistant ornithyl transcarbamylase are insensitive to this toxin and are less susceptible to infection by the bacterial pathogen *Pseudomonas syringae* pv. *phaselicola* (19). Resistance to abiotic stresses has been improved: Resistance to high salt has been improved by overproduction of mannitol (154), improved resistance to chilling has been reported as a result of increased fatty acid desaturation (93), and ozone resistance has been induced by overexpressing superoxide dismutase (160).

QUALITY IMPROVEMENT   Major advances using mutants and transgenic plants in studies of starch structure (90, 112), lipid composition (9, 39, 70, 71, 103), and nitrate and amino acid metabolism (56, 80) allow several commercial goals to be approached: (*a*) The shelf life and/or solids content of tomato fruits has been altered by inhibiting ethylene production (43, 102) or by decreasing polygalacturonase activity (43, 131). (*b*) Starch content of potato tubers has been increased via the heterologous expression of mutated ADP-glucose pyrophosphorylase gene from *E. coli* (142). (*c*) Starch is composed of two polymers, a linear unbranched chain (amylose) and a branched chain (amylopectin). Expression of an antisense-RNA that inhibits the granule-bound starch synthase has yielded potato composed of amylopectin only (168). (*d*) Fatty acid composition has been altered by expressing acyl-ACP thioesterases in transgenic plants (169) or by altering desaturase expression (39). (*e*) Expression of heterologous storage proteins in seeds (94) or the stimulation of lysine biosynthesis (80) could improve the nutritional value of harvested seeds. (*f*) New products have been synthesized in transgenic plants (71). Expression of a heterologous levan-sucrase from *Bacillus subtilis* in transgenic tobacco (24) and potato (165) plants has resulted in fructan-storing transgenic plants, indicating that those plants can be manipulated to store a foreign carbohydrate. A chimeric gene encoding glycogen synthase from *E. coli* has been used to create potato plants with a novel starch in their tubers (132), and introduction of a cyclodextrin-glucosyl transferase into tobacco plants has resulted in production of small amounts of cyclodextrin (101). Another example for the synthesis of novel polymers in transgenic plants is the production of polyhydroxybutyrate (106).

Plants can also be used as bioreactors to produce proteins of interest. Because of low production costs and the possibility of easily scaling up, plants can be used for production of technical enzymes, as has shown by the expression of a chimeric xylanase gene from *Clostridium thermocellum* in tobacco (50). The neuropeptide Leu-enkephalin has been expressed as a translational fusion protein in transgenic *A. thaliana* and *Brassica* plants (161). Recent developments have demonstrated that even complex molecules such as antibodies (53, 81) or single-chain Fv antibodies (105, 155) can be synthesized in transgenic plants.

ACKNOWLEDGMENTS

Work of the authors has been supported by the Bundesministerium für Forschung und Technologie and the Deutsche Forschungsgemeinschaft. We are grateful to numerous colleagues for making unpublished articles available to us, to our coworkers for their support and enthusiasm, and to Frau Edda Lammer for her patient production of the final manuscript.

Literature Cited

1. Abel PP, Nelson RS, De B, Hoffmann N, Rogers SG, et al. 1986. Delay of disease development in transgenic plants that express the tobacco mosaic virus coat protein gene. *Science* 232:738–43

1a. ap Rees T, Hill SA. 1994. Metabolic control analysis of plant metabolism. *Plant Cell Environ.* 17:587–600

2. Ball S, Marianne T, Durick L, Fresnoy M, Belrue B, Desq A. 1992. A *Chlamydomonas reinhardtii* low-starch mutant is defective for 3-phosphoglycerate activation and orthophosphate inhibition of ADP-glucose pyrophosphorylase. *Planta* 185:17–26

3. Barnes SA, Knight JS, Gray JC. 1995. Alteration of amount of the chloroplast phosphate translocator in transgenic tobacco affects the partitioning of assimilate between starch and sucrose. *Plant Physiol.* In press

4. Baulcomb D. 1994. Novel strategies for engineering virus resistance in plants. *Curr. Opin. Biotechnol.* 5:117–24

5. Baysdorfer C, Kremer DF, Sicher RC. 1989. Partial purification and characterisation of fructokinase activity from barley. *J. Plant Physiol.* 134:56–61

6. Beachy RN, Loesch-Fries S, Tumer NE. 1990. Coat protein-mediated resistance against virus infection. *Annu. Rev. Phytopathol.* 28:451–74

6a. Binns AN. 1990. Agrobacterium mediated gene delivery and the biology of host range limitations. *Physiol. Plant.* 79:135–39

7. Black CC, Mustardy L, Sung SS, Komanik PP, Xu DP, Paz N. 1987. Regulation and roles for alternative pathways of hexose metabolism in plants. *Physiol. Plant.* 69:387–94

7a. Boerjan W, Bauw G, Van Montagu M, Inzé D. 1994. Distinct phenotypes generated by overexpression and suppression of γ-adenosyl-L-methione synthetase reveals developmental patterns of gene silencing in tobacco. *Plant Cell* 6:1401–14

8. Boutry M, Nagy F, Poulson C, Aoyagi K, Chua N-H. 1987. Targeting of bacterial chloramphenicol acetyltransferase to mitochondria in transgenic plants. *Nature* 328:340–42

9. Browse J, Somerville CR. 1992. Glycerolipid synthesis: biochemistry and regulation. *Annu. Rev. Plant Physiol. Plant Mol. Biol.* 42:467–506

10. Bruneau J-M, Worrell AC, Cambou B, Lando D, Voelker TA. 1991. Sucrose phosphate synthase, a key enzyme for sucrose biosynthesis in plants. *Plant Physiol.* 96:473–78

11. Burrell MM, Mooney PJ, Blundy M, Carter D, Wilson F, et al. 1994. Genetic manipulation of 6-phosphofructokinase in potato tubers. *Planta* 194:95–101

12. Caspar T, Huber SC, Somerville CR. 1986. Alterations in growth, photosynthesis and respiration in a starchless mutant of *Arabidopsis thaliana* deficient in chloroplast phosphoglucomutase activity. *Plant Physiol.* 79:1–7

13. Chilton M-D, Drummond MH, Merlo DJ, Sciaky D, Montoya AL, et al. 1977. Stable incorporation of plasmid DNA into higher plant cells: the molecular basis of crown gall tumorigenesis. *Cell* 11:263–71

14. Chrispeels MJ. 1991. Sorting of proteins in the secretory system. *Annu. Rev. Plant Physiol. Plant Mol. Biol.* 42:21–53

15. Christou P. 1992. Genetic transformation of crop plants using microprojectile bombardment. *Plant J.* 2:275–81

16. Deleted in proof

17. Crossway A, Oakes JV, Irvine JM, Ward B, Knauf VC, Shewmaker CK. 1986. Integration of foreign DNA following microinjection of tobacco mesophyll protoplasts. *Mol. Gen. Genet.* 202:79–85

18. De Block M, Vandewiele M, Dockx J, Thoen C, Gossele V, et al. 1987. Engineering herbicide resistance in plants by expression of a detoxifying enzyme. *EMBO J.* 6:2513–18

19. de la Fuente-Martinez JM, Mosqueda-Cano G, Alvarez-Morales A, Herrera-Estrella L. 1992. Expression of bacterial phaseolotoxin-resistant ornithyl transcarbamylase in transgenic tobacco confers resistance to *Pseudomonas syringae* pv. *Phaseolicola. Bio-Technology* 10:905–9

20. de la Pena A, Lörz H, Schell J. 1987. Transgenic rye plants obtained by injecting DNA into young floral tillers. *Nature* 325:274–76

21. Dickinson S, Altabella T, Chrispeels M. 1991. Slow growth phenotype of transgenic tomato expressing apoplastic invertase. *Plant Physiol.* 95:420–25

22. Dietz K-J, Heber U. 1986. Light and CO_2 limitation of photosynthesis and states of reactions regenerating ribulose-1,5-bisphosphate or reducing glycerate-3-phosphate. *Biochim. Biophys. Acta* 848:392–401

23. Duke SR, McCarty DR, Koch KE. 1991. Organ-specific invertase deficiency in the primary root of an inbred maize line. *Plant Physiol.* 97:523–27

24. Ebskamp MJM, Van der Meer IM, Spronk BA, Weisbeek PJ, Smeekens SCM. 1994. Accumulation of fructose polymers in transgenic tobacco. *Bio-Technology* 12:272–75

25. Edmondson DL, Badger MR, Andrews TJ.

1990. Slow inactivation of ribulose bis-phosphate carboxylase during catalysis is caused by accumulation of a slow tight binding inhibitor at the catalytic site. *Plant Physiol.* 93:1390–97

26. Edwards JW, Coruzzi GM. 1990. Cell-specific gene expression in plants. *Annu. Rev. Genet.* 24:275–303

27. Evans JR. 1989. Photosynthesis and nitrogen relationships in leaves of C3 plants. *Oecologia* 78:9–19

28. Farquhar GD, von Caemmerer S. 1982. Modelling of photosynthetic response to environmental conditions. In *Encyclopedia of Plant Physiology,* ed. OL Lange, PS Nobel, CB Osmond, H Ziegler, 12B:549–87. Heidelberg: Springer-Verlag

29. Fell DA. 1992. Metabolic control analysis: a survey of its theoretical and experimental development. *Biochem. J.* 286:313–30

30. Fichtner K, Quick WP, Schulze E-D, Mooney HA, Rodermel SR, et al. 1993. Decreased ribulose-1,5-bisphosphate carboxylase-oxygenase in transgenic tobacco transformed with antisense rbcS. V. Relationship between photosynthetic rate, storage strategy, biomass allocation and vegetative plant growth at three different nitrogen supplies. *Planta* 190:1–9

31. Fichtner K, Schulze E-D. 1992. The effect of nitrogen nutrition on annuals originating from habitats of different nitrogen availability. *Oecologia* 92:236–41

32. Flügge U-I, Heldt HW. 1991. Metabolite translocators of the chloroplast envelope. *Annu. Rev. Plant Physiol. Plant Mol. Biol.* 42:129–44

33. Fluhr R, Kuhlemeier C, Nagy F, Chua N-H. 1986. Organ-specific and light-induced expression of plant genes. *Science* 232:1106–12

34. Franck A, Guilley H, Jonard G, Richards K, Hirth L. 1980. Nucleotide sequence of Cauliflower Mosaic virus DNA. *Cell* 21:285–94

35. Frommer WB, Ninnemann O. 1995. Heterologous expresion of genes in bacterial, fungal, animal, and plant cells. *Annu. Rev. Plant Physiol. Plant Mol. Biol.* 46:419–44

36. Galthier N, Foyer CH, Huber JLA, Voelker TA, Huber SC. 1993. Effects of elevated sucrose phosphate synthase activity on photosynthesis, assimilate partitioning and growth in tomato (*Lycopersicon esculentum* var UC 82 B). *Plant Physiol.* 101:535–43

37. Gasser CS, Fraley RT. 1989. Genetically engineering plants for crop improvment. *Science* 244:1293–99

38. Gatz C, Kaiser A, Wendenburg R. 1991. Regulation of a modified CaMV 35S promoter by the Tn10-encoded Tet repressor in transgenic tobacco. *Mol. Gen. Genet.* 227:229–37

39. Gibson S, Falcone DL, Browse J, Somerville CR. 1994. Use of transgenic plants and mutants to study the regulation and function of lipid composition. *Plant Cell Environ.* 17:627–38

40. Gifford RM, Evans RT. 1981. Photosynthesis, carbon partitioning and yield. *Annu. Rev. Plant Physiol.* 32:485–509

41. Goldschmidt EE, Huber SC. 1992. Regulation of photosynthesis by end-product accumulation in leaves storing starch, sucrose and hexose sugars. *Plant Physiol.* 99:1443–48

42. Gottlob-McHugh SG, Sangwan RS, Blakeley SD, Vanlerberghe GC, Ko K, et al. 1992. Normal growth of transgenic tobacco plants in the absence of cytosolic pyruvate kinase. *Plant Physiol.* 100:820–25

43. Gray JE, Picton S, Giovannoni JJ, Grierson D. 1994. The use of transgenic plants and naturally occurring mutants to understand and manipulate tomato fruit ripening. *Plant Cell Environ.* 17:557–72

44. Hajirezaei M, Sonnewald U, Viola R, Carlisle S, Dennis DT, Stitt M. 1994. Transgenic potato plants with decreased expression of pyrophosphate:fructose-6-phosphate phosphotransferase show no visible phenotype and only minor changes in metabolic fluxes in tubers. *Planta* 192:16–30

45. Haughn GW, Smith J, Mazur B, Somerville C. 1988. Transformation with a mutant *Arabidosis* acetolactate synthase gene renders tobacco resistant to sulfonylurea herbicides. *Mol. Gen. Genet.* 211:266–71

46. Heineke D, Kruse A, Flügge UI, Frommer WB, Riesmeier J, et al. 1994. Effect of "antisense" repression of the chloroplast triose phosphate translocator on photosynthesis metabolism in transgenic potato plants. *Planta* 193:174–80

47. Heineke D, Sonnewald U, Bussis D, Günter G, Leidreiter K, et al. 1992. Apoplastic expression of yeast-derived invertase in potato. Effects on photosynthesis, leaf solute composition, water relations, and tuber composition. *Plant Physiol.* 100:301–8

48. Heineke D, Wildenberger K, Sonnewald U, Willmitzer L, Heldt HW. 1994. Accumulation of hexoses in leaf vacuoles: studies with transgenic tobacco plants expressing yeast-derived invertase in the cytosol, vacuole or apoplast. *Planta* 194:29–33

49. Heldt HW, Flügge UI. 1992. Metabolite transport in plant cells. In *Soc. Exp. Biol. Semin. Ser.* 50: *Plant Organelles,* ed. AK Tobin, pp. 21–47. Cambridge: Cambridge Univ. Press

50. Herbers K, Wilke I, Sonnewald U. 1995. Heat treatment purification of a thermo-

stable xylanase from *Clostridium thermocellum* expressed at high levels in the apoplast of transgenic tobacco plants. *Bio-Technology*. In press

51. Hernalsteens J-P, Van Vliet F, De Beuckeleer M, Depicker A, Engler G, et al. 1980. The *Agrobacterium tumefaciens* Ti plasmid as a host vector system for introducing foreign DNA in plant cells. *Nature* 287:654–56

52. Herrera-Estrella L, De Block M, Messens E, Hernalsteens J-P, Van Montagu M, Schell J. 1983. Chimeric genes as dominant selectable markers in plant cells. *EMBO J.* 2:987–95

53. Hiatt A, Cafferkey R, Bowdish K. 1989. Production of antibodies in transgenic plants. *Nature* 342:76–78

54. Hilder VA, Gatehouse AMR, Sheerman SE, Barker RF, Boulter D. 1987. A novel mechanism of insect resistance engineered into tobacco. *Nature* 300:160–63

55. Ho LC. 1988. Metabolism and compartmentation of imported sugars in sink organs in relation to sink strength. *Annu. Rev. Plant Physiol.* 39:355–78

56. Hoff T, Truong H-N, Caboche M. 1994. The use of mutants and transgenic plants to study nitrate metabolism. *Plant Cell Environ.* 17:489–506

57. Hofmeyer JH, Cornish-Bowden A. 1991. Quantitative assessment of regulation in metabolite systems. *Eur. J. Biochem.* 200: 223–36

58. Hooykaas PJJ, Schilperoort RA. 1992. Agrobacterium and plant genetic engineering. *Plant Mol. Biol.* 19:15–38

59. Horsch RB, Fraley RT, Rogers SG, Sanders PR, Lloyd A, Hoffmann N. 1984. Inheritance of functional foreign genes in plants. *Science* 223:496–98

60. Horsch RB, Fry JE, Hoffmann NL, Eichholtz D, Rogers SG, Fraley RT. 1985. A simple and general method for transferring genes into plants. *Science* 227:1229–31

61. Huber SC, Hanson JR. 1992.Carbon partitioning and growth of starchless mutant of *Nicotiana sylvestris*. *Plant Physiol.* 99: 1449–54

62. Huber SC, Huber JLA. 1992. Role of sucrose phosphate synthase in sucrose metabolism in leaves. *Plant Physiol.* 99:1275–78

63. Huber SC, Huber JLA, McMichael RW Jr. 1991. The regulation of sucrose synthesis in leaves. In *Carbon Partitioning Within and Between Organisms*, ed. CT Pollock, JF Farrar, AJ Gordon, pp. 1-26. Oxford: Bios Scientific

64. Hudson GS, Evans JR, von Caemmerer S, Arvidsson YBC, Andrews TJ. 1992. Reduction of ribulose-1,5-bisphosphate carboxylase/oxygenase content by "antisense" RNA reduces photosynthesis in transgenic tobacco plants. *Plant Physiol.* 98:294–302

65. Jelitto T, Sonnewald U, Willmitzer L, Hajirezaei M, Stitt M. 1992. Inorganic pyrophosphate content and metabolites in potato and tobacco plants expressing *E. coli* pyrophosphatase in their cytosol. *Planta* 188:238–44

66. Jiang C-Z, Quick WP, Alred R, Kliebenstein D, Rodermel SR. 1994. Antisense inhibiton of Rubisco activase expression. *Plant J.* 5:787–98

67. Joshi RL, Joshi V. 1991. Strategies for expression of foreign genes in plants. *FEBS Lett.* 281:1–8

68. Kacser H, Porteous J. 1987. Control of metabolism, what do we have to measure. *Trends Biochem. Sci.* 12:5–14

69. Kerr PS, Huber SC. 1987. Coordinate control of sucrose formation in soybean leaves by sucrose-phosphate synthase and fructose-2,6-bisphosphate. *Planta* 170:197–204

70. Kinney AJ. 1994. Genetic modification of the storage lipids of plants. *Curr. Opin. Bio-Technol.* 5:144–51

71. Kishore GM, Somerville CR. 1993. Genetic engineering of commercially useful biosynthetic pathways in transgenic plants. *Curr. Opin. Bio-Technol.* 4:152–57

72. Kossmann J, Sonnewald U, Willmitzer L. 1995. Reduction of the choroplast fructose-1,6-bisphosphatase in transgenic potato plants impairs photosynthesis and plant growth. *Plant J.* 6:637–50

73. Krapp A, Chaves MM, David MM, Rodriques ML, Pereira JS, Stitt M. 1994. Decreased ribulose-1,5-bisphosphate carboxylase/oxygenase in transgenic tobacco transformed with "antisense" rbcS. VII. Impact on photosynthesis and growth in tobacco growing under extreme high irradiance and high temperatures. *Plant Cell Environ.* 17:945–49

74. Krapp A, Hofman B, Schäfer C, Stitt M. 1993. Regulation of the expression of rbcS and other photosynthetic genes by carbohydrates: a mechanism for the "sink-regulation" of photosynthesis? *Plant J.* 3:817–28

75. Krause K-P. 1994. *Zur Regulation von Saccharosephosphatsynthase*. PhD diss., Univ. Bayreuth

76. Kruckeberg A, Neuhaus HE, Feil R, Gottlieb L, Stitt M. 1989. Dosage mutants of phosphoglucose isomerase in cytosol and chloroplast of *Clarkia xantiana*. I. Impact on mass action ratios and fluxes to sucrose and starch. *Biochem. J.* 261:457–67

77. Kruger N. 1990. Carbohydrate synthesis and degradation. In *Plant Physiology, Biochemistry and Molecular Biology*, ed. DT Dennis, DH Turpin, pp. 59–76. Singapore: Longmans

78. Ito M, Sato T, Fukuda H, Komamine A.

1994. Meristem-specific gene expression directed by the promoter of the S-phase-specific gene, cyc07, in transgenic *Arabidopsis*. *Plant Mol. Biol.* 24:863–78

79. Lauerer M, Saftic D, Quick WP, Fichtner K, Schulze ED, et al. 1993. Decreased ribulose-1,5-bisphosphate in transgenic tobacco transformed with "antisense" rbcS. VI. Effects on photosynthesis in plants grown at different irradiance. *Planta* 190: 332–45

80. Lea PJ, Forde BG. 1994. The use of mutants and transgenic plants to study amino acid metabolism. *Plant Cell Environ.* 17: 541–57

81. Ma JK-C, Lehner T, Stabila P, Fux CI, Hiatt A. 1994. Assembly of monoclonal antibodies with IgG1 and IgA heavy chain domains in transgenic tobacco plant. *Eur. J. Immunol.* 24:131–38

82. MacDonald FD, Buchanan BB. 1990. In *Plant Physiology, Biochemistry and Molecular Biology*, ed. DT Dennis, DH Turpin, pp. 249–62. Singapore: Longmans

83. Marcotte WR Jr, Russell SH, Quatrano RS. 1989. Absisic acid-responsive sequence from Em gene of wheat. *Plant Cell* 1:969–76

84. Martin T, Frommer WB, Salanoubat M, Willmitzer L. 1993. Expression of an *Arabidopsis* sucrose synthase gene indicates a role in metabolization of sucrose both during phloem loading and sink organs. *Plant J.* 4:367–77

85. Masle J, Hudson GS, Badger MR. 1993. Effects of ambient CO_2 concentrations on growth and nitrogen use in tobacco plants transformed with an "antisense" gene to the small subunit of ribulose-1,5-bisphosphate carboxylase/oxygenase. *Plant Physiol.* 103:1075–88

86. Mate CJ, Hudson GS, von Caemmerer S, Evans JR, Andrews TJ. 1993. Reduction of ribulose-1,5-bisphosphate carboxylase activase levels in tobacco by "antisense" RNA reduces ribulose bisphosphate carboxylase carbamylation and impairs photosynthesis. *Plant Physiol.* 102:1119–28

87. McMichael RW, Klein RR, Salvucci ME, Huber SC. 1993. Identification of the major regulatory phosphorylation sites in sucrose-phosphate synthase. *Arch. Biochem. Biophys.* 307:248–52

88. Miller ME, Chourey PS. 1992. The maize invertase-deficient minature-1 seed mutation is associated with aberrant pedicel and endosperm development. *Plant Cell* 4: 297–305

89. Mol J, Van der Krol A, Van Tunen A, Van Blokland R, de Lange P, Stuitje A. 1990. Regulation of plant gene expression by antisense RNA. *FEBS Lett.* 268:427–30

90. Müller-Röber B, Kossmann J. 1994. Approaches to influence starch quantity and starch quality in transgenic plants. *Plant Cell Environ.* 17:601–14

91. Müller-Röber B, la Cognata U, Sonnewald U, Willmitzer L. 1994. A truncated version of an ADP-glucose pyrophosphorylase promoter from potato specifies guard cell-selective expression in transgenic plants. *Plant Cell* 6:601–12

92. Müller-Röber BT, Sonnewald U, Willmitzer L. 1992. Inhibition of ADP-glucose pyrophosphorylase in transgenic potatoes leads to sugar-storing tubers and influences tuber formation and expression of tuber storage protein genes. *EMBO J.* 11:1229–38

93. Murata N, Ishizaki-Nishizawa O, Higashi S, Hayashi H, Tasaka Y, Nishida I. 1992. Genetically engineered alteration in the chilling sensitivity of plants. *Nature* 356: 710–13

94. Naito S, Dube PH, Beachy RN. 1988. Differential expression of conglycinin alpha and beta subunit genes in transgenic plants. *Plant Mol. Biol.* 11:109–23

95. Nakamura K, Matsuoka K. 1993. Protein targetting to the vacuole in plant cells. *Plant Physiol.* 101:1–5

96. Napoli C, Lemieux C, Jorgensen R. 1990. Introduction of chimeric chalcone synthase gene into *Petunia* results in reversible co-suppression of homologous genes in *trans*. *Plant Cell* 2:279–89

97. Neuhaus HE, Kruckeberg AL, Feil R, Gottlieb L, Stitt M. 1989. Dosage mutants of phosphoglucose isomerase in the cytosol and chloroplasts of *Clarkia xantiana*. II. Study of the mechanisms which regulate photosynthate partitioning. *Planta* 178: 110–22

98. Neuhaus HE, Quick WP, Siegl G, Stitt M. 1990. Control of photosynthate partioning in spinach leaves. Analysis of the interaction between feedforward and feedback regulation of sucrose synthesis. *Planta* 181:583–92

99. Neuhaus HE, Stitt M. 1990. Control analysis of photosynthate partitioning: impact of reduced activity of ADP-glucose pyrophosphorylase or plastid phosphoglucomutase on the fluxes to starch and sucrose in *Arabidopsis thaliana* L. Heynh. *Planta* 182:445–54

100. Newsholme EA, Start C. 1973. *Regulation in Metabolism*. London: Wiley

101. Oakes JV, Shewmaker CK, Stalker DM. 1991. Production of cyclodextrins, a novel carbohydrate, in the tubers of transgenic potato plants. *Bio-Technology* 9:982–86

102. Oeller PW, Wong LM, Taylor LP, Pike DA, Theologis A. 1991. Reversible inhibition of tomato fruit senescence by antisense RNA. *Science* 254:437–39

103. Ohlrogge JB. 1994. Design of new plant

products: engineering of fatty acid metabolism. *Plant Physiol.* 104:821–26

104. Otten L, De Greve H, Hernalsteens JP, van Montagu M, Schieder O, et al. 1981. Mendelian transmission of genes introduced into plants by the Ti plasmids of *Agrobacterium tumefaciens. Mol. Gen. Genet.* 183: 209–13

105. Owen MR, Gandecha A, Cockburn W, Whitelam GC. 1992. Synthesis of a functional anti-phytochrome single-chain Fv protein in transgenic tobacco. *Bio-Technology* 10:790–94

106. Poirier Y, Dennis DE, Klomparens K, Somerville C. 1992. Polyhydroxybutyrate, a biodegradable thermoplastic, produced in transgenic plants. *Science* 256:520–23

107. Poorter H, Remkes C. 1990. Leaf area ratio and net assimilation rate of 24 wild species differing in relative growth rate. *Oecologia* 83:553–59

108. Portis AR Jr. 1992. Regulation of ribulose-1,5-bisphosphate carboxylase/oxygenase activity. *Annu. Rev. Plant Physiol. Plant Mol. Biol.* 43:415–37

109. Potyrkus I. 1991. Gene transfer to plants: assessment of published approaches and results. *Annu. Rev. Plant Physiol. Plant Mol. Biol.* 42:205–25

110. Deleted in proof

111. Preiss J. 1988. Biosynthesis of starch and its degradation. In *Biochemistry of Plants,* ed. J Preiss, 13:181–254. San Diego: Academic

112. Preiss J. 1991. Biology and molecular biology of starch synthesis and its regulation. In *Oxford Surveys of Plant Molecular and Cell Biology,* ed. B Miflin, 7:59–114. Oxford: Oxford Univ. Press

113. Preiss J, Romeo T. 1994. Molecular biology and regulatory aspects of glycogen biosynthesis in bacteria. In *Progress in Nucleic Acid Research and Molecular Biology,* ed. WE Cohan, KE Moldance, Vol. 47. San Diego: Academic

114. Price GD, Evans JR, von Caemmerer S, Yu J-W, Badger MR. 1994. Specific reduction of chloroplast glyceraldehyde-3-phosphate dehydrogenase activity by antisense RNA reduces CO_2 assimilation via a reduction in RUBP regeneration in transgenic tobacco plants. *Planta* 193:331–40

115. Price GD, von Caemmerer S, Evans JR, Yu J-W, Lloyd J, et al. 1994. Specific reduction of chloroplast carbonic anhydrase activity by "antisense" RNA in transgenic tobacco has a minor effect on photosynthetic CO_2 assimilation. *Planta* 193:331–40

116. Quick WP, Fichtner K, Schulze E-D, Wendler R, Leegood RC, et al. 1992. Decreased ribulose-1,5-bisphosphate carboxylase/oxygenase in transgenic tobacco transformed with "antisense" *rbc*S. IV. Impact on photosynthesis and plant growth altered nitrogen supply. *Planta* 188:522–31

117. Quick WP, Schurr U, Scheibe R, Schulze E-D, Rodermel SR, et al. 1991. Decreased Rubisco in tobacco transformed with "antisense" *rbc*S. I. Impact on photosynthesis in ambient growth conditions. *Planta* 183: 542–54

118. Renz A, Stitt M. 1993. Substrate specificity and product inhibition of different forms of fructokinase and hexokinase in developing potato tubers. *Planta* 190:166–75

119. Riesmeier JW, Flügge UI, Schulz B, Heineke D, Heldt HW, et al. 1993. Antisense repression of the chloroplast triose phosphate translocator affects carbon partitioning in transgenic potato plants. *Proc. Natl. Acad. Sci. USA* 90:6160–64

120. Riesmeier JW, Willmitzer L, Frommer WB. 1992. Isolation and characterisation of a sucrose carrier cDNA from spinach by functional expression in yeast. *EMBO J.* 11:4705–13

121. Riesmeier JW, Willmitzer L, Frommer WB. 1994. Evidence for an essential role of the sucrose transporter in phloem loading and assimilate partitioning. *EMBO J.* 13:1–7

122. Rodermel SR, Abbott MS, Bogorad L. 1988. Nuclear-organelle interactions: nuclear antisense gene inhibits ribulose bisphosphate carboxylase enzyme levels in transformed tobacco plants. *Cell* 55:673–81

123. Rolleston FS. 1972. A theoretical background to the use of measured intermediates in the study of the control of intermediary metabolism. *Curr. Top. Cell Regul.* 5:47–75

124. Sauer N, Tanner W. 1993. Molecular biology of sugar transporters in plants. *Bot. Acta* 106:277–86

125. Schulze W, Stitt M, Schulze E-D, Neuhaus HE, Fichtner K. 1991. A quantification of the significance of assimilatory starch for growth of *Arabidopsis thaliana* L. Heynh. *Plant Physiol.* 95:890–95

126. Schmülling T, Schell J, Spena A. 1989. Promoters of the rolA, B, and C genes of *Agrobacterium rhizogenes* are differentially regulated in plants. *Plant Cell* 1:665–70

127. Scott P, Lange A, Pilkis SJ, Kruger NJ. 1995. Carbon metabolism in leaves of transgenic tobacco containing elevated fructose-2,6-bisphosphate levels. *Plant J.* In press

128. Seymour GB, Fray RG, Hill P, Tucker GA. 1993. Down-regulation of two non-homologous endogenous tomato genes with a single chimaeric sense gene construct. *Plant Mol. Biol.* 23:1–9

129. Shah DM, Horsch RB, Klee HJ, Kishore

GM, Winter JA, et al. 1986. Engineering herbicide tolerance in transgenic plants. *Science* 233:478–81

130. Sharkey TD. 1989. Evaluating the role of Rubisco regulation in photosynthesis in C_3 plants. *Philos. Trans. R. Soc. London Ser. B* 323:435–48

131. Sheey RE, Kramer M, Hiatt WR. 1988. Reduction of polygalacturonase activity in tomato fruit by antisense RNA. *Proc. Natl. Acad. Sci. USA* 85:8805–9

132. Shewmaker CK, Boyer CD, Wiesenborn DP, Thompson DB, Boersig MR, et al. 1994. Expression of *Escherichia coli* glycogen synthase in the tuber of transgenic potatoes (*Solanum tuberosum*) results in a highly branched starch. *Plant Physiol.* 104:1159–66

133. Shimamoto K. 1994. Gene expression in transgenic monocots. *Curr. Opin. Biotechnol.* 5:158–62

134. Simpson J, Timko MP, Cashmore AR, Schell J, Van Montagu M, Herrera-Estrella L. 1985. Light-inducible and tissue-specific expression of a chimaeric gene under control of the 5′ flanking sequence of a pea chlorophyll a/b-binding protein gene. *EMBO J.* 4:2723–29

135. Small JR, Kascer H. 1993. Responses of metabolic systems to large changes in enzyme activities and effectors. II. The linear treatment of branched pathways and metabolite concentrations. Assessment of the general non-linear case. *Eur. J. Biochem.* 213:625–40

136. Smith AM, Neuhaus HE, Stitt M. 1990. The impact of decreased starch branching enzyme activity on photosynthetic starch synthesis in leaves of wrinkled-seeded pea. *Planta* 181:310–15

137. Smith EF, Townsend CO. 1907. A plant tumor of bacterial origin. *Science* 25:671–73

138. Sonnewald U. 1992. Expression of *E. coli* pyrophosphatase in transgenic plants alters photoassimilate partitioning. *Plant J.* 2:571–81

139. Sonnewald U, Brauer M, von Schaewen A, Stitt M, Willmitzer L. 1991. Transgenic tobacco plants expressing yeast-derived invertase in either the cytosol, vacuole or apoplast: a powerful tool for studying sucrose metabolism and sink/source interactions. *Plant J.* 1:95–106

140. Sonnewald U, Lerchl J, Zrenner R, Frommer W. 1994. Manipulation of sink-source relations in transgenic plants. *Plant Cell Environ.* 17:649–58

141. Stalker DM, McBride KE, Malyj LD. 1988. Herbicide resistance in transgenic plants expressing a bacterial detoxification gene. *Science* 242:419–23

142. Stark DM, Timmerman KP, Barry GF, Preiss J, Kishore GM. 1992. Regulation of the amount of starch in plant tissues by ADP-glucose pyrophosphorylase. *Science* 258:287–92

143. Stitt M. 1989. Control of sucrose synthesis. Estimation of free energy charges, investigations of the contribution of equilibrium and non-equilibrium reactions and estimation of elasticities and flux control coefficients. In *Techniques and New Developments in Photosynthetic Research,* ed. J Barber, pp. 365–92. London: Plenum

144. Stitt M. 1990. Fructose 2,6-bisphosphate as regulatory metabolite in plants. *Annu. Rev. Plant Physiol. Mol. Biol.* 41:153–85

145. Stitt M. 1990. The flux of carbon between the chloroplast and the cytoplasm. In *Plant Physiology, Biochemistry and Molecular Biology,* ed. DT Dennis, DH Turpin, pp. 319–40. Singapore: Longmans

146. Stitt M. 1993. Control of photosynthetic carbon fixation and partitioning. How can use of genetically manipulated plants improve the nature and quality of information about regulation? *Philos. Trans. R. Soc. London Ser. B* 340:225–33

147. Stitt M. 1994. Manipulation of carbohydrate partitioning. *Curr. Opin. Bio-Technol.* 5:137–43

148. Stitt M. 1995. Metabolic regulation of photosynthesis. In *Photosynthesis and Stress,* ed. N Baker. San Diego: Academic. In press

149. Stitt M, Huber SC, Kerr P. 1987. Control of photosynthetic sucrose synthesis. In *Biochemistry of Plants,* ed. MD Hatch, NK Boardman, 10:327–407. New York: Academic

150. Stitt M, Quick WP, Schurr U, Schulze E-D, Rodermel SR, Bogorad L. 1991. Decreased Rubisco in tobacco transformed with "antisense" *rbs*S. II. Flux control coefficients for photosynthesis in varying light, CO_2 and air humidity. *Planta* 183:555–65

151. Stitt M, Schulze E-D. 1994. Does Rubisco control the rate of photosynthesis and plant growth? An exercise in molecular ecophysiology. *Plant Cell Environ.* 17:465–88

152. Stockhaus J, Poetsch W, Steinmüller K, Westhoff P. 1994. Evolution of C_4 phosphoenolpyruvate carboxylase promoter of the C_4 dicot *Flaveria trinervia*: an expression analysis in the C_3 plant tobacco. *Mol. Gen. Genet.* In press

153. Stockhaus J, Schell J, Willmitzer L. 1989. Identification of enhancer elements in the upstream region of the nuclear photosynthetic gene ST-LS1. *Plant Cell* 1:805–13

154. Tarczynski MC, Jensen RG, Bohnert HJ. 1993. Stress protection of transgenic tobacco by production of the osmolyte mannitol. *Science* 259:508–10

155. Tavladoraki P, Benvenuto E, Trinca S, Martinis DD, Cattaneo A, Galeffi P. 1993. Transgenic plants expressing a functional

single-chain Fv antibody are specifically protected from virus attack. *Nature* 366: 469–72

156. Trethewey RN, ap Rees T. 1994. A mutant of *Arabidopsis* lacking the ability to transport glucose across the chloroplast envelope. *Biochem. J.* 301:449–54

157. Uknes S, Dincher S, Friedrich L, Negrotto D, Williams S, et al. 1993. Regulation of pathogensis-related protein-1a gene expression in tobacco. *Plant Cell* 5:159–69

158. Vaek M, Reynaerts A, Höfte H, Jansens S, De Beuckeleer M, et al. 1987. Transgenic plants protected from insect attack. *Nature* 328:33–37

159. Van Bel AJE. 1993. Strategies of phloem loading. *Annu. Rev. Plant Physiol. Plant Mol. Biol.* 44:253–83

160. Van Camp W, Willekens H, Bowler C, Van Montagu M, Inze D, et al. 1994. Elevated levels of superoxide dismutase protect transgenic plants against ozone damage. *Bio-Technology* 12:165–68

161. Vandekerckhove J, Damme JF, Lijsebettens M, Botterman J, De Block M. 1989. Enkephalins produced in transgenic plants using modified 2S seed storage proteins. *Bio-Technology* 7:929–32

162. Van den Broeck G, Timko MP, Kausch AP, Cashmore AR, Van Montagu M, Herrera-Estrella L. 1985. Targeting of foreign proteins to chloroplasts by fusion to the transit peptide from the small subunit of ribulose-1,5-bisphosphate carboxylase. *Nature* 313: 358–63

163. Van der Hoeven C, Dietz A, Landsmann J. 1994. Variability of organ-specific gene expression in transgenic tobacco plants. *Transgenic Res.* 3:159–65

164. Van der Krol AR, Lenting PE, Veestra J, van der Meer IM, Koes RE. 1988. An antisense chalcone synthase gene in transgenic plants inhibits flower pigmentation. *Nature* 333:866–69

165. Van der Meer IM, Ebskamp MJM, Visser RGF, Weisbeek PJ, Smeekens SCM. 1994. Fructan as a new carbohydrate sink in transgenic potato plants. *Plant Cell* 6:561–70

166. Van Larcbeke N, Genetello C, Schell J, Schilperoort RA, Hermans AK, et al. 1975. Acquisition of tumour-inducing ability by non-oncogenic agrobacteria as a result of plasmid transfer. *Nature* 255:742–43

167. Vasil V, Castillo AM, Fromm ME, Vasil IK. 1992. Herbicide resistant fertile transgenic wheat plants obtained by microprojectile bombardment of regenerable embryonic callus. *Bio-Technology* 10:667–74

168. Visser RGF, Somhorst I, Kuipers GJ, Ruys NJ, Feenstra WJ, Jacobsen E. 1991. Inhibition of the expression of the gene for granule-bound starch synthase in potato by an-

tisense constructs. *Mol. Gen. Genet.* 225: 289–96

169. Voelker TA, Worrell AC, Anderson L, Bleibaum J, Fan C, et al. 1992. Fatty acid biosynthesis redirected to medium chains in transgenic oilseed plants. *Science* 257: 72–74

170. von Caemmerer S, Edmondson DL. 1986. Relationship between steady-state gas exchange, *in vitro* ribulose bisphosphate carboxylase activity and some carbon reduction cycle intermeditates in *Raphanus sativus. Aust. J. Plant Physiol.* 13:669–88

171. Von Schaewen A, Stitt M, Schmidt R, Sonnewald U, Willmitzer L. 1991. Expression of a yeast-derived invertase in the cell wall of tobacco and *Arabidopsis* plants leads to accumulation of carbohydrate, inhibition of photosynthesis and strongly influences growth and phenotype of transgenic tobacco plants. *EMBO J.* 9:3033–44

172. Walden R, Schell J. 1990. Techniques in plant molecular biology—progress and problems. *Eur. J. Biochem.* 192:563–76

173. Wandelt CI, Khan MRI, Craig S, Schroeder HE, Spencer D, Higgins TJV. 1992. Vicilin with carboxy-terminal KDEL is retained in the endoplasmic reticulum and accumulates to high levels in leaves of transgenic plants. *Plant J.* 2:181–92

174. Ward ER, Ryals JA, Miflin BJ. 1993. Chemical regulation of transgene expression in plants. *Plant Mol. Biol.* 22:361–66

175. Wardlaw IF. 1990. The control of carbon partitioning in plants. *New Phytol.* 116: 341–81

176. Wilde RJ, Cooke SE, Brammar WJ, Schuch W. 1994. Control of gene expression in plant cells using a 434:VP16 chimeric protein. *Plant Mol. Biol.* 24:381–88

177. Wilde RJ, Shufflebottom D, Cooke S, Jasinska I, Merryweather A, et al. 1992. Control of gene expression in tobacco cells using a bacterial operator-repressor system. *EMBO J.* 11:1251–59

178. Woodrow IE, Berry JA. 1988. Enzymatic regulation of photosynthetic carbon dioxide fixation. *Annu. Rev. Plant. Physiol. Mol. Biol.* 39:533–94

179. Yoder JI, Goldsbrough AP. 1994. Transformation systems for generating marker-free transgenic plants. *Bio-Technology* 12:263–67

180. Zaenen I, Van Larcbeke N, Van Montagu M, Schell J. 1974. Supercoiled circular DNA in crown-gall inducing agrobacterium strains. *J. Mol. Biol.* 86:109–27

181. Zambryski P, Joos H, Genetello C, Leemans J, Van Montagu M, Schell J. 1983. Ti plasmid vector for the introduction of DNA into plant cells without alteration of their normal regeneration capacity. *EMBO J.* 2: 2143–50

182. Zrenner R, Salanoubat M, Willmitzer L,

Sonnewald U. 1995. Evidence of the crucial role of sucrose synthase for sink strength using transgenic potato plants (*Solanum tuberosum* L.). *Plant J.* In press

183. Zrenner R, Willmitzer L, Sonnewald U. 1993. Analysis of the expression of potato uridine phosphate glucose pyrophosphorylase and its inhibition by "antisense" RNA. *Planta* 190:247–52

Annu. Rev. Plant Physiol. Plant Mol. Biol. 1995. 46:369–94

MOLECULAR GENETICS OF PLANT EMBRYOGENESIS

David W. Meinke

Department of Botany, Oklahoma State University, Stillwater, Oklahoma 74078

KEY WORDS: *Arabidopsis,* embryo-defective mutants, maize, mutant analysis, pattern formation, seed development

CONTENTS

ABSTRACT

Embryogenesis is a complex developmental pathway that plays a central role in the life cycle of higher plants. Recent advances in the application of genetics and molecular biology to the study of developmental processes in animals have led to a renewed interest in the analysis of plant development and the identification of genes with important functions during plant embryogenesis. The most extensive studies have dealt with maize and *Arabidopsis,* two organisms amenable to both genetic and molecular analysis. When combined with the results of large-scale genome projects, which should ultimately allow the identification of every gene with an essential function during plant growth and development, these studies of developmental mutants, cell lineages, gene expression patterns, and structural changes associated with seed development should eventually provide a detailed view of plant embryogenesis at the molecular level.

INTRODUCTION

Solving complex problems in the biological sciences often requires an integrated, multidisciplinary approach. Significant advances become possible only when different experimental strategies and perspectives are combined with appropriate biological systems. The field of plant developmental and molecular genetics provides a good example of this principle. The renewed interest in plant embryogenesis and the recent explosion of data on gene expression during seed development have resulted from the integration of descriptive, molecular and genetic approaches to the study of this important developmental pathway. Recent advances in the developmental and molecular genetics of plant embryogenesis are summarized in this review. For additional details, the reader is referred to other books and reviews dealing with plant development (20, 54, 154), classical plant embryology (70, 135), plant developmental genetics (106, 108, 146), somatic embryogenesis (172, 175), apomixis (77), embryogenesis (32, 55a, 83, 170), endosperm development (87), embryo-defective mutants (14, 55a, 104, 105, 109), and gene expression during seed development (55, 157).

Embryogenesis is a critical stage in the life cycle of higher plants. Large numbers of genes must be expressed in a highly coordinated manner to ensure that the single-celled zygote develops into an organized, multicellular structure capable of surviving desiccation and germinating to produce a viable seedling. Several unifying themes have emerged from studies of embryogenesis in different plant families: (*a*) Embryogenesis can usually be separated into an initial morphogenetic phase characterized by cell division and the onset of cell differentiation, followed by a maturation phase that involves accumulation of major storage products and preparation for seed desiccation, dormancy, and germination. (*b*) The zygote is not a unique cell in angiosperms; a wide range of somatic cells can be induced to produce embryos either as part of the normal life cycle or through experimental manipulation. (*c*) Plant embryo development normally occurs within the environment of the seed and fruit, which contain specialized tissues with different origins, functions, genotypes, and numbers of chromosomes. Developmental interactions between these structures must be considered in studies of plant embryogenesis.

Developmental geneticists and classical embryologists traditionally have viewed plant diversity from different perspectives. To the plant embryologist interested in characterizing different patterns of embryo development throughout the angiosperms, information must be obtained from large numbers of organisms representing different families and developmental strategies. The tendency has been to spread a small number of researchers over a large number of organisms. In contrast, developmental geneticists need to establish a large community of scientists dedicated to the analysis of a single organism with appropriate features. In this way, mutants and other genetic stocks required for detailed analysis of developmental pathways can be contributed by many different individuals within the community. Molecular biologists often fall in between these two extremes, with individuals choosing a limited number of plant species for a variety of technical reasons. Molecular geneticists tend to focus on the same organisms used by developmental geneticists because the basic research tools are already available. In recent years, two plants (maize and *Arabidopsis thaliana*) have emerged as model systems for developmental and molecular genetics in general and embryogenesis in particular (25, 35, 113). This review focuses on recent advances in our understanding of the genetics and molecular biology of embryogenesis in these model organisms.

DESCRIPTIVE STUDIES OF PLANT EMBRYOGENESIS

In most angiosperms, the zygote divides to form an embryo composed of two parts, the embryo proper and the suspensor. Cell-division patterns early in development have been characterized for a wide range of plant species (91, 134). The embryo proper usually differentiates an epidermis (protoderm) by the globular stage and then undergoes a transition from radial to bilateral symmetry with the initiation of cotyledons at the heart stage. The globular-heart transition is also marked by cell specialization within the embryo proper. The suspensor is typically a filamentous structure that functions early in development to support growth of the embryo proper. The suspensor subsequently undergoes programmed cell death and is not usually present at maturity (174). The endosperm tissue is produced from a second fertilization event and is triploid except in unusual cases. This nutritive tissue frequently begins as a syncytium of free nuclei that later become surrounded by cell walls. The endosperm is often present at maturity in monocots, particularly in grasses, whereas in many dicots it is absorbed during seed development.

Descriptive studies of normal development provide an essential foundation for subsequent genetic and molecular studies. Several reviews have been published dealing with morphological features of embryo development in *A. thaliana* (93, 97, 170). Seeds in most crucifers are located in fruits (siliques) that are arranged in a developmental progression along the length of the stem.

The small size of *A. thaliana* seeds actually facilitates descriptive studies of embryo development because relatively few sections are required to reconstruct the entire seed. Recent studies have focused on megagametogenesis and ovule development (52, 137), embryonic pattern formation (72, 169), the microtubular cytoskeleton early in development (167, 168), storage-product accumulation during embryonic maturation (94), and the origin of root (37, 138a) and shoot (3) apical meristems during embryogenesis. The use of interference microscopy to examine large numbers of cleared seeds in a short period of time has also become a powerful tool in the analysis of embryo development (97, 162, 170).

The morphology of embryogenesis in maize has also been examined in detail (1, 136). Development in maize and *A. thaliana* differs in several important respects: (*a*) the maize embryo is much larger at all stages of development, (*b*) the shoot apical meristem in maize begins to produce leaves during embryo development, whereas in *A. thaliana* the first leaves do not appear until after germination, (*c*) the endosperm is a massive tissue present at seed maturity in maize, and (*d*) the maize embryo follows a different pattern of differentiation and morphogenesis. Maize is an attractive system for studying embryo-endosperm interactions throughout development because translocations involving accessory (B) chromosomes (7) can be used to construct discordant kernels in which the embryo and endosperm differ in genotype. Early stages of embryogenesis are more difficult to examine in maize because the zygote is more deeply embedded in maternal tissues and is therefore less amenable to Nomarski optics. The recent establishment of in vitro fertilization techniques (40, 43) and the application of fluorescence microscopy to isolated megagametophytes (63) represent promising alternative strategies for studying early stages of embryogenesis in maize.

MOLECULAR ANALYSIS OF PLANT EMBRYOGENESIS

Molecular studies of plant embryogenesis have undergone an interesting evolution over the past 15 years. Initial studies were generally designed to (*a*) estimate the number of different RNAs present in developing seeds, (*b*) examine the spatial and temporal distribution of different RNA species, (*c*) isolate and characterize the genes that code for abundant seed proteins, particularly the seed-storage proteins, and (*d*) identify the *cis*-acting regulatory sequences and *trans*-acting DNA-binding proteins that regulate expression of seed-specific genes. The results of these extensive studies have already been reviewed in detail (55, 157). In recent years, there has been a shift toward combining the tools of molecular biology with the power of genetics. One problem with pursuing molecular biology in the absence of genetics is that the developmental significance of a particular gene may be difficult to establish. Future research in plant molecular biology must focus not simply on isolating and characterizing large numbers of genes expressed during plant embryo development, but also on determining the biological significance of these genes by demonstrating what happens when their function is disrupted. This can be accomplished either by creating transgenic plants that express an antisense construct or by working with genes that have already been disrupted through loss-of-function mutations. Either strategy could enhance the many significant advances made in recent years with other molecular approaches, some of which are described below.

Zygotic Embryogenesis

Large numbers of genes are expressed during seed development in higher plants. Nucleic-acid hybridization studies indicate that perhaps as many as 20,000 different RNAs are present in developing embryos of some angiosperms (55, 73). The corresponding genes can be grouped according to temporal patterns of expression. These genes may be transcribed at low levels, either throughout the plant or at specific stages of seed development, or expressed at high levels in specific seed tissues. Some genes are transcribed during both late embryogenesis and early germination. Genetic studies support the conclusion that many genes expressed in seeds perform duplicated functions that can be eliminated without disrupting development (51).

Several approaches have been pursued in the molecular analysis of zygotic embryogenesis. Expression of storage-protein genes has been examined in detail by characterizing conserved sequence motifs in upstream regulatory regions and their associated DNA-binding proteins (27, 48, 67, 74, 81, 140, 161). These elements often continue to function when transformed into other plant species (4, 55, 119), consistent with the view that at least some regulatory circuits active during embryogenesis have been conserved during plant evolution. Analysis of storage-protein synthesis in maize has been facilitated by the availability of endosperm mutants that fail to activate normal patterns of gene expression because they lack essential transcriptional regulators (87, 126, 140). The role of ABA in modulating gene expression during seed development has also been examined in detail (57, 79, 163). Extensive studies on gene-expression patterns in embryos cultured at different stages of development in the presence and absence of ABA have provided further details on gene function during germination and seed development (64, 69). Late embryogenesis has also been studied by characterizing the expression of molecular markers in ABA-insensitive mutants of A. thaliana (49) and in viviparous mutants of maize (128).

Localization of gene products through in situ hybridization (47, 71, 155) and immunocytochemistry (2) continues to provide insights into plant development. For example, localization of Knotted protein to the shoot apical meristem of wild-type maize plants (152) first indicated that this regulatory gene might normally function in the establishment of a shoot apex during embryogenesis. Homologs of important regulatory genes have been identified by screening cDNA libraries prepared from reproductive tissues. Recent examples include homologs of serine and threonine protein kinases isolated from *Petunia hybrida* ovules (29), proteins involved in DNA replication isolated from zygotes of maize (39, 39a), a MADS-box homolog expressed in zygotic *Brassica napus* embryos (61), and a *Knotted* homolog expressed in somatic embryos of soybean (90).

Several laboratories have used PCR methods such as differential mRNA display (82) to prepare libraries from small amounts of tissue and thereby facilitate the identification of genes active at early stages of development (39, 61, 122). This powerful approach could lead to significant advances in our understanding of gene expression during early embryogenesis. Another development has been the massive amount of sequence data generated by random partial sequencing of *A. thaliana* cDNA clones in both the United States and France (35, 62). As a result of this large-scale effort, it will soon be possible to assign names and likely functions, based on sequence homology, to many genes expressed during seed development. The challenge for the future will be to use this information to address basic questions in plant development.

Somatic Embryogenesis

Somatic embryogenesis continues to be an effective model for studying gene expression at different stages of plant development (175). A number of genes expressed in somatic embryos have been examined in detail (157, 175). Several interesting stories have emerged from recent studies of somatic embryogenesis. One story involves the role of auxin transport in plant embryogenesis, as discussed later in this review. Although auxin was once thought to play a minor role in embryogenesis, it now appears that auxin transport and perception may be required for the establishment of bilateral symmetry during early stages of development. Other stories have dealt with the possible roles of calmodulin (127) and Lea proteins (76, 164, 173) in somatic embryogenesis. Several genes expressed in carrot somatic embryos code for secreted extracellular proteins. One gene that produces a lipid transfer protein (EP2) has been particularly useful as a marker for epidermal cell differentiation during embryogenesis (155). Another gene product (EP1) with homology to *Brassica* S-locus glycoproteins is present in nonembryogenic callus but not in somatic embryos themselves (160). The precise role of these extracellular proteins remains to be established, but they may be involved in the regulation of cell expansion and the maintenance of biophysical features required for morphogenesis (160, 175).

Perhaps the most unexpected finding involves a secreted glycoprotein (EP3) that rescues a temperature-sensitive mutant of carrot (*ts11*) that fails to complete the transition from a globular to heart stage of somatic embryogenesis. Surprisingly, this glycoprotein is an endochitinase (30). Although the normal substrate remains to be identified, recent studies have demonstrated that two lipo-oligosaccharides (pentamers of β-1,4-linked N-acetylglucosamines) that function as *Rhizobium* nodulation factors can also rescue *ts11* embryos in culture (31). This raises the possibility that endochitinases may catalyze the release of an important signal molecule from a minor component of the plant cell wall during embryogenesis and that bacteria may have evolved the capacity to use this system of communication by producing signal molecules with similar biological effects (18, 31).

GENETIC ANALYSIS OF PLANT EMBRYOGENESIS

Principles of Genetic Analysis

The basic principles of plant developmental genetics have been reviewed elsewhere (106, 108). The general strategy in genetic analysis is to use mutants either as markers of cell lineages (see 65, 66, 133) or as vehicles for the identification of essential genes. Genetic analysis has played an increasingly important role in recent studies of plant development. However, not all genes can be identified readily by recessive loss-of-function mutations. Genes that are duplicated in the genome, genes that are required for early stages of gametogenesis, and genes that perform functions that are redundant, nonessential, or detectable only in unique circumstances often escape detection in mutant screens. The power of genetics is not that it leads to the identification of every transcriptional unit within the genome, but rather that it allows one to focus on genes that must be expressed in order for growth and development to proceed in a normal manner.

Three important principles of developmental genetics frequently have been overlooked in the rush to determine the molecular basis of mutant phenotypes. The first principle concerns the value of isolating multiple alleles at a given locus. This task is particularly difficult when dealing with mutants defective in embryo development because the number of target genes is quite large and consequently the chance that any two mutants are defective in the same gene is low (51). The availability of multiple alleles with different levels of normal gene function often facilitates attempts to clone the gene in question and determine its developmental significance. The second principle involves the importance of mapping a mutant gene of interest. In addition to being a prerequisite for map-based cloning, mapping studies often lead to the identification of additional mutant alleles, and with recent advances in mapping of random cDNA sequences, mapping may even help to identify putative wild-type clones in the same region of the chromosome. The third principle is that alleles may have different phenotypes depending on the severity of the mutation. Some mutants known primarily for their defects in root or shoot development may correspond to weak alleles of embryonic defectives. Other mutations may be pleiotropic in nature, affecting both embryo development and vegetative morphology. As a result, laboratories examining the same line may focus on different aspects of the mutant phenotype.

Diversity of Available Mutants

Several different classes of embryonic mutants have been identified in higher plants (103, 104). Many mutants are defective in early stages of cell division and morphogenesis. Others fail to accumulate pigments and storage materials during embryonic maturation. Still others are disrupted in the preparation for dormancy and germination. Many of these mutants are likely to be defective in genes with housekeeping functions that first become essential during embryo development. Although this information may be discouraging to developmental biologists interested in finding genes that play a direct role in the regulation of plant embryogenesis, it should be good news for biochemists, physiologists, and cell biologists, many of whom could use mutants defective in basic cellular processes.

Embryonic mutants also differ in the initial site of gene action. The primary defect in some embryonic mutants is limited to the endosperm tissue (87). Altered development of the embryo in these mutants is often an indirect consequence of endosperm failure (11, 103). The primary defect in other mutants appears to be restricted to the embryo proper. A few mutants are defective in the development of surrounding maternal tissues (46, 80). Although the most extensive studies of embryonic mutants have dealt with maize and A. thaliana, related mutants have also been described in barley (10, 46), carrot (88, 141), rice (118), and peas (9, 165). Most of these recessive loss-of-function mutations were induced with chemical mutagens, X rays, transposable elements, or T-DNA from Agrobacterium tumefaciens. Embryonic mutations typically are maintained as heterozygotes that produce 25% defective seeds following self-pollination. Mutant seeds usually appear normal early in development but then become more abnormal as development proceeds. Mutants with defects in surrounding maternal tissues are maintained as homozygotes that produce 100% abnormal seeds after self-pollination.

EMBRYO AND ENDOSPERM MUTANTS OF MAIZE

Hundreds of mutants defective in kernel development have been isolated and characterized in maize over the past 75 years (for reviews, see 103, 104, 146).

Three common types of seed mutants are recognized in maize: (*a*) defective-endosperm mutants, in which the initial defect occurs in the endosperm tissue and mutant embryos can produce normal plants, (*b*) defective-kernel mutants, in which development of both the embryo and endosperm are significantly altered, and (*c*) embryo-specific or germless mutants, in which development of the embryo is profoundly altered without disrupting growth of the endosperm. Mutants of the second and third types are particularly relevant to studies of plant embryogenesis.

One defective-endosperm mutant (*miniature seed*) characterized in the 1940s as defective in nutrient transport between the endosperm and adjacent maternal tissues (89) has recently been shown to lack invertase activity (16, 114). This result indicates that invertase activity in the endosperm is required for normal development of both the endosperm and surrounding maternal (pedicel) tissues. This example illustrates how the isolation and characterization of a developmental mutant of maize can unexpectedly provide information relevant to plant physiologists interested in nutrient transport. This type of information exchange between geneticists and physiologists should occur with increasing frequency as more mutants defective in genes with basic cellular functions are uncovered.

Large numbers of defective-kernel mutants have been isolated in maize following pollen mutagenesis (121, 147) and transposon tagging (129, 138). A significant number of these genes have been assigned to chromosome arms (121, 138). Mutant embryos have been examined for their general morphology (22, 23, 38, 150), response in culture (149), and ability to be rescued by a wild-type endosperm in discordant kernels (121). The recent isolation and characterization of 63 *dek* mutations from *Mutator* stocks (138) represents a significant addition to existing collections produced by chemical mutagenesis and should greatly facilitate further analysis of this mutant class at the molecular level.

Embryo-specific (*emb*) mutants are perhaps the most valuable for studies of early embryogenesis in maize because they are most likely to identify genes with an essential role in morphogenesis. Recently, an impressive collection of 51 *emb* mutants isolated from *Mutator* stocks has been described (24, 148). Included in these reports were analyses of segregation data, detailed phenotypic descriptions, and linkage assignments to chromosome arms. These mutants in combination with the tagged *dek* mutants described above represent a valuable resource for molecular studies of embryo and endosperm development in maize.

Another tagged mutant recently identified in maize (*dks8*) produces an embryo that appears normal except for the complete absence of a shoot apical meristem (153). This unique pattern of development has not been observed previously in maize, although similar mutants have been described in *A.*

thaliana (3). The analysis of shoot-meristemless mutants of maize and *A. thaliana* demonstrates how individuals working with different organisms can benefit from each other's efforts. This example begins with the *Knotted-1* (*Kn1*) gene of maize, the first homeobox gene characterized in plants, which causes aberrant cell divisions near the leaf surface when ectopically expressed as a result of a dominant mutation (56, 58). After in situ hybridization and immunocytochemical studies localized the *Kn* gene product in wild-type plants to cells of the shoot apical meristem, it appeared that recessive loss-of-function mutations might result in a *dek* or *emb* phenotype. Although none of the mutants examined to date appear to be altered in *Knotted*, the shootless mutant of maize has a phenotype similar to that predicted for a loss-of-function allele of *Knotted* and may define a related class of regulatory proteins. Molecular studies already have identified other members of the *Kn1* gene family with similar patterns of expression during plant development (68).

The relationship of the work described above to recent efforts with *A. thaliana* lies in the discovery that a recessive shoot-meristemless (*shm*) mutant of *A. thaliana* (3) appears to be defective in a *Knotted* homolog (MK Barton, personal communication). This result is exciting not only because it may provide additional information on *Knotted* function and meristem formation during embryogenesis, but also because it suggests that information obtained from molecular characterization of developmental mutants in maize and *A. thaliana* may be synergistic in nature. This conclusion, which is further supported by recent studies with *vp1* mutants of maize (59, 99) and *abi3* mutants of *A. thaliana* (53), could greatly facilitate the isolation and characterization of genes with important regulatory functions during plant embryogenesis. Further information on viviparous mutants of maize is presented in another review in this volume (98).

EMBRYO-DEFECTIVE MUTANTS OF *ARABIDOPSIS*

Diversity of Available Mutants

Several hundred mutants defective in embryogenesis have been isolated and characterized in *A. thaliana* following seed mutagenesis (72, 97, 102, 109, 112) and *A. tumefaciens*–mediated seed transformation (13, 42, 55a, 109). Included in these collections are large numbers of embryonic lethals, defectives, and pigment mutants. The distinction between these overlapping classes has been difficult to follow in the literature, partly because mutants with identical phenotypes have been given different names in different laboratories. There has also been a gradual change in nomenclature within my own laboratory, from embryonic lethals to embryonic defectives. This change reflects a growing understanding that many mutant embryos are capable of continued growth and development under appropriate conditions (6, 50, 109). Two complementary approaches have been used to isolate recessive *emb* mutants defective in embryo development. One approach involves screening immature siliques for the presence of 25% defective seeds following self-pollination. Mutant seeds can be recognized by their unusual size, shape, color, or embryo morphology prior to desiccation (42, 109, 111). The other approach involves screening germinated seedlings on agar plates for evidence of defects in embryogenesis (72, 97). This approach represents a more efficient method for mutant isolation but fails to provide detailed information on mutant embryos that cannot survive desiccation.

The high frequency of *emb* mutants identified following seed mutagenesis in *A. thaliana* is consistent with the presence of at least 500 target genes with essential functions at this critical stage of the life cycle. Because so many genes are involved, mutants with similar phenotypes are rarely defective in the same gene. As a result, the identification of multiple *emb* alleles presents a significant challenge to plant biologists. Several years ago, we began a large-scale mapping project designed to saturate the map with informative mutations and to facilitate the recovery of duplicate alleles (130). To date, we have obtained recombination data for 169 mutants defective in embryo development and placed 110 of these genes on the map (51). Embryo-defective mutations have consequently become the most common type of visible marker on the genetic map of *A. thaliana*. Nineteen examples of duplicate alleles have been found, consistent with our estimate that approximately 500 genes readily mutate to give an embryo-defective phenotype in *A. thaliana* (51). With continued progress, it may be possible to approach saturation for this important class of mutations.

Several different strategies have been pursued in the analysis of embryo-defective mutants. These have included studies designed to characterize the diversity of mutant phenotypes and patterns of abnormal development (13, 102, 130); examine the response of mutant embryos in culture (6, 50), search for evidence of gametophytic gene expression by examining the distribution of mutant seeds in heterozygous siliques (101, 102), and describe the ultrastructure (131, 143) and extent of cellular differentiation (60, 110, 166) in the mutant embryo proper and suspensor. Emphasis in recent years has been placed on the analysis of tagged mutants identified following *A. tumefaciens*–mediated seed transformation (45). Several laboratories have been involved in extensive screens for T-DNA insertional mutants defective in embryogenesis (see 13, 36, 42, 55a, 151, 166, 170, 171). Some groups are screening for embryo-defective mutations caused by insertion of maize transposable elements (e.g. 44, 159). Others are exploring the use of promoter trapping to identify genes expressed during embryogenesis (84, 156, 158, 158a). As a result of these combined efforts, there will soon be many tagged *emb* genes amenable to molecular analysis.

Housekeeping vs Regulatory Functions

Many embryo-defective mutants are likely to be altered in basic housekeeping functions that first become essential during early stages of development. The best known example is a biotin auxotroph of *A. thaliana* that produces mutant seeds unable to complete normal embryogenesis in the absence of supplemental biotin (142, 144, 145). Mutant embryos contain reduced levels of biotin and can be rescued in culture by intermediates of biotin synthesis in bacteria (desthiobiotin and 7,8-diaminopelargonic acid) but not by their immediate precursor (7-keto-8-aminopelargonic acid). Rescued plants are phenotypically normal when grown in the presence of biotin. Mutant embryos produced by heterozygous plants exhibit a wide range of phenotypes, presumably because they receive slightly different amounts of biotin from surrounding maternal tissues. The recent identification of a tagged *bio1* allele following *A. tumefaciens*–mediated seed transformation (123) should facilitate detailed analysis of this locus at the molecular level.

Other mutants are likely to be defective in genes that play a more direct role in the regulation of plant growth and development. Several candidate genes that appear to play important functions during embryogenesis in *A. thaliana* are described below. It may not be possible to make a clear distinction between housekeeping and regulatory functions during plant embryogenesis because many genes are likely to perform cellular functions that are directly related to both growth and morphogenesis. The challenge for the future is not only to identify genes that may regulate early embryogenesis directly, but even more importantly to determine how large numbers of genes interact to influence both morphogenesis and cellular differentiation throughout development.

Fusca Mutants

The *fusca* phenotype of *A. thaliana* is characterized by inappropriate accumulation of anthocyanin in cotyledons of developing embryos (116, 117). Mutant seeds often germinate but usually fail to develop beyond the seedling stage. Accumulation of anthocyanin is an indirect consequence of the mutation and not the primary cause of seedling lethality because double mutant embryos blocked in anthocyanin biosynthesis still fail to complete normal development (15). Although the change in pigmentation observed in *fusca* mutants might appear unrelated to the regulation of growth and development, the *fusca* phenotype has allowed the identification of an interesting class of regulatory genes that appear to function in the transduction of a wide range of developmental and environmental signals during embryogenesis and seedling development (15, 19).

Twelve complementation groups of recessive *fusca* mutations have been found by screening immature siliques and germinated seedlings for the presence of anthocyanin in cotyledons (115). In light of the large number of duplicate *fusca* alleles identified to date, it seems unlikely that the total number of *fusca* genes will increase significantly above this number. The *fusca* phenotype provides a good example of how a single class of mutants can be viewed from totally different perspectives in laboratories with contrasting interests. A surprising feature of the *fusca* phenotype is that some mutant seedlings exhibit a deetiolated or constitutive photomorphogenic response in the dark (15, 19, 21). As a result, several *fusca* mutants were identified through independent screens for defects in photomorphogenesis in the dark.

Three *FUSCA* genes have recently been cloned and sequenced: *FUS1/ COP1* (34, 100), *FUS2/DET1* (132), and *FUS6* (15). The *COP1* protein may play a role in gene regulation by interacting with DNA through an N-terminal zinc-binding domain and binding associated proteins through an internal coiled-coil helix structure and a C-terminal domain with homology to the β subunit of trimeric G proteins (34). The *COP1* protein also shares homology with a subunit of the *Drosophila melanogaster* TFIID transcription complex (41). The *DET1* protein is not homologous to any known protein in existing databases but appears to contain a nuclear localization signal and may play an important role in signal transduction (132). *FUS6* also encodes a novel protein that is hydrophilic, α-helical, and contains potential protein kinase C phosphorylation sites but no homology to known transcription factors (15).

Two contrasting models have been presented to explain the *fusca* phenotype. The first model states that *fusca* mutants are altered primarily in light response pathways and that other features of the mutant phenotype are an indirect consequence of this initial defect (33). The second model states that *FUSCA* genes play overlapping and complementary roles in transduction of a wide range of environmental and developmental signals during late embryogenesis and early seedling development (15). According to this model, some *FUSCA* genes are involved primarily in light response whereas others play a greater role in transduction of hormonal and nutritional signals. Regardless of the specific roles of *FUSCA* genes in plant growth and development, the *fusca* phenotype has clearly demonstrated that mutations with minor effects on seed morphology can still lead to the identification of genes with important regulatory functions.

Leafy Cotyledon Mutants

A more traditional approach to the identification of regulatory genes has been to look for mutants with striking abnormalities indicative of a dramatic change in developmental timing or spatial organization. Such heterochronic and homeotic mutants have been used extensively to dissect the regulation of development in a variety of model genetic organisms (92). A similar approach has also been used to study flower development in angiosperms (26). The *leafy cotyledon* phenotype of *A. thaliana* provides an example of a dramatic change in embryogenesis that appears to result from disruption of genes with critical regulatory functions. The first mutant examined in detail (*lec1*) exhibited striking defects in embryonic maturation and produced viviparous embryos with cotyledons that were partially transformed into leaves (107). Mutant seeds were desiccation intolerant at maturity because they failed to activate embryo-specific programs late in development. Immature embryos prematurely entered a germination pathway after the torpedo stage of development and then acquired characteristics normally restricted to vegetative parts of the plant. Another striking feature of this mutant was the production of trichomes on embryonic cotyledons. These surface hairs are normally restricted to leaves, stems, and sepals of wild-type plants.

Three additional mutants have been described that produce trichomes on cotyledons following precocious germination in culture (5, 75, 110, 171). One mutant represents a new allele of *lec1* and appears to be tagged with T-DNA (55a, 110, 171). The phenotype of this mutant is indistinguishable from that of the original allele. The second mutant (*fus3*) was classified originally as a *fusca* (115), but further examination showed that its pattern of development was similar to *lec1* and clearly different from other *fusca* mutants (5, 75, 110). Although the original *fusca* nomenclature for this gene has been retained, it actually belongs to the *leafy cotyledon* class. The third mutant (*lec2*) defines a new locus (110). Mutant embryos from these lines differ in morphology, desiccation tolerance, pattern of anthocyanin accumulation, presence of storage materials, size and frequency of trichomes on cotyledons, and timing of precocious germination in culture (110). In contrast to ABA-insensitive *abi3* embryos, which also exhibit striking changes in embryonic maturation (78, 120, 125), mutant embryos from the *leafy cotyledon* class are sensitive to ABA in culture. The *ABI3* and *LEC* genes therefore appear to function in different pathways (110).

The *leafy cotyledon* phenotype is consistent with a model in which the basal developmental state of embryonic cotyledons is leaflike (107, 110), much like the basal developmental state of floral organs appears to be leaflike (26). In the absence of normal *LEC* function and embryonic maturation programs, mutant cotyledons revert to this primitive state and acquire features normally restricted to vegetative leaves. This model is consistent with the view that cotyledons originally evolved as modified leaves through gradual imposition of embryo-specific programs. The alternative model that *LEC* genes function to repress precocious activation of vegetative programs during embryogenesis (75) cannot be excluded based on available data but seems less plausible from an evolutionary perspective. Regardless of the precise function of these novel genes, the *leafy cotyledon* phenotype has clearly facilitated the discovery of a fascinating and previously unknown network of regulatory genes with overlapping functions during late embryogenesis.

Abnormal Suspensor Mutants

Abnormal growth of the suspensor is a characteristic feature of many embryodefective mutants of *A. thaliana* (174). The suspensor produced during normal embryogenesis consists of an enlarged basal cell and a single column of six to eight additional cells that attach the embryo proper to surrounding maternal tissues. The suspensor stops dividing at the globular stage and then undergoes programmed cell death beginning at the heart stage of development. Several examples of mutants with abnormal suspensors have been analyzed in detail (17, 55a, 95, 143). The pattern of abnormal development observed in these mutants is consistent with a model in which continued growth of the suspensor during normal development is inhibited by the embryo proper. The developmental potential of suspensor cells becomes apparent only when this inhibitory effect is removed. This model is further supported by experimental studies demonstrating that destruction of the embryo proper through exposure to chemicals or radiation is often followed by abnormal growth of the suspensor (174).

Several recent studies have provided additional clues to the underlying causes of abnormal suspensor growth and have extended our understanding of normal suspensor development. The first study involves a fascinating mutant called *twin,* in which viable secondary embryos are occasionally produced from the suspensor of the primary embryo (162). This mutant provides the first documented example of suspensor polyembryony in plants caused by mutation at a single genetic locus. The pattern of abnormal development observed in *twin* elegantly demonstrates that cells of the suspensor have the potential to duplicate the entire spectrum of developmental programs normally restricted to the embryo proper. The presence of developmental abnormalities in the mutant embryo proper indicates that in wild-type plants, *TWIN* serves a dual function in promoting normal development of the embryo proper as well as maintaining cell identity and suppressing embryogenic potential in the suspensor.

Another recent study of abnormal suspensor (*sus*) mutants with particularly large suspensors (143) has provided additional information on the developmental significance of this mutant phenotype. Characterization of these mutants has been facilitated by the isolation of three alleles each of two different genes (*sus1* and *sus2*). Examination of immature seeds with Nomarski optics showed that abnormal growth of the mutant suspensor was always immediately preceded by a disruption of morphogenesis in the globular embryo proper. This observation is consistent with the initial model that suspensor defects are an indirect consequence of altered development in the embryo proper. The mutant suspensor also acquired characteristics normally restricted to the embryo proper late in development, specifically the accumulation of storage protein and lipid bodies. Further analysis of abnormal development associated with different members of this allelic series suggested that growth of the mutant suspensor ultimately was limited by the same genetic defect that initially disrupted the embryo proper.

Two models for the role of *SUS* genes in plant development have been presented (143). According to the first model, these genes are required primarily for continued morphogenesis in the embryo proper. In the absence of normal gene function, the embryo proper is unable to transmit an inhibitory signal to the suspensor, which then begins dividing and acquiring features characteristic of the embryo proper. The mutant suspensor is unable to revert completely to its basal embryogenic state because it encounters the same defect that originally prevented continued morphogenesis in the embryo proper. The second model assumes that *SUS* genes perform a more direct role in producing a signal with a dual function, promoting morphogenesis in the embryo proper and inhibiting further growth of the suspensor.

Further analysis of abnormal suspensor mutants will require molecular isolation of the disrupted genes. My laboratory has recently cloned and sequenced genomic regions flanking the T-DNA insert in a tagged allele of *sus2* (B Schwartz & D Meinke, unpublished data). Although molecular complementation of mutant seeds with wild-type *SUS2* sequences has not yet been completed, it seems likely that the gene interrupted by T-DNA in this tagged allele is indeed responsible for the mutant phenotype. *SUS2* appears to encode a large protein that throughout its length exhibits striking homology to the yeast *PRP8* gene and its putative homolog from *Caenorhabditis elegans*. Extensive genetic and molecular studies with yeast have shown that *PRP8* functions as a critical spliceosome assembly factor during RNA processing (12). We therefore believe that *sus2* may be defective in RNA processing during early embryo development. This discovery is consistent with many aspects of the models of *SUS* function noted above and provides further support for our view that embryo-defective mutants represent a valuable resource of genes with essential cellular functions during plant growth and development. If these initial results are confirmed, *sus2* will become the first example, in a multicellular eukaryote, of a mutant that is known to be defective in this particular spliceosome assembly factor.

Embryonic Pattern Mutants

Several years ago, Jürgens and colleagues performed a comprehensive screen to identify mutant seedlings that appeared to be defective in embryonic pattern formation (72). Several classes of mutants identified in this screen have subsequently been examined in detail (8, 96, 97, 158b). Particular emphasis has been placed on mutants altered in apical-basal organization. Multiple mutant alleles have been identified for several of these genes. The strategy used in the isolation and characterization of these mutants has been similar to that used with great success in the genetic dissection of embryonic pattern formation in *D. melanogaster*. Whether this approach will be equally successful in elucidating the genetic control of early embryogenesis in plants remains to be determined. Although polarity and pattern formation during plant embryogenesis may be controlled through the direct action of a few specialized gene products, it seems equally likely that these important events are regulated through complex interactions of signal transduction networks similar to those known to operate during vegetative development in higher plants. The analysis of mutants that appear to be defective in embryonic pattern formation nevertheless provides a unique opportunity to study the regulation of early embryogenesis from a genetic perspective.

Two pattern mutants (*gnom* and *monopteros*) have been examined in the most detail (97). *Monopteros* seedlings lack the hypocotyl, radicle, and root meristem regions normally associated with the basal portion of the plant. This defect has been attributed to changes in cell-division patterns at the octant stage of embryo development (8). The appearance of abnormalities early in development is consistent with the proposed role of this gene in establishing the basal region of developing embryos. The observation that mutant tissues occasionally produce adventitious roots in culture indicates that *MONOPTEROS* is not required for development of a functional root apex.

Analysis of *gnom* has an interesting history that clearly illustrates two points raised in this review: (*a*) One mutant can be viewed from different perspectives in different laboratories, and (*b*) solving fundamental questions in plant development often requires an interdisciplinary approach that combines descriptive, genetic, and molecular studies. The first *gnom* allele was identified in my laboratory following seed mutagenesis and screening of immature siliques for the presence of defective seeds (102). This mutant was originally classified as an embryonic lethal following the tradition of Müller (116), even though as noted previously, some of the mutations identified in this screen did not actually become lethal during embryogenesis. This mutant was first named 112A-2A (6, 102) and then renamed *emb30* when a different system of nomenclature was adopted for embryo-defective mutants (50). Although early studies in my laboratory dealt with the ultrastructure and extent of cellular differentiation in mutant embryos (60, 131), the most striking features noted in publications were the fused appearance of mutant cotyledons and the failure of mutant embryos to produce roots under a variety of culture conditions (6, 102).

Many additional alleles of *emb30/gnom* were subsequently identified in the Jürgens laboratory by screening at the seedling stage for defects in embryonic pattern formation (72). The rootless phenotype was then interpreted within the context of embryonic pattern formation, and subsequent analysis shifted to a detailed characterization of alterations in cell-division patterns during early stages of embryogenesis (97). This important feature of the mutant phenotype was not addressed originally. The Jürgens laboratory therefore appropriately focused attention on describing the first signs of developmental abnormalities in mutant embryos.

Experimental studies of plant embryogenesis provided a different perspective on the fused-cotyledon phenotype when this same abnormality appeared following exposure of carrot somatic embryos to inhibitors of auxin transport (28, 139). Several years later, these inhibitors were found to elicit a similar response from zygotic embryos of *Brassica juncea* cultured early in development (85, 86). Fused cotyledons were also present in seeds produced by *A. thaliana* plants homozygous for the *pin* mutation (86), which is known to disrupt auxin transport during vegetative development (124). Thus a connection seemed to exist between defects in auxin transport and the initiation of cotyledons with altered patterns of symmetry. Nevertheless, the relevance of these observations to the proposed role of *GNOM/EMB30* in apical-basal pattern formation remained obscure.

The recent cloning and sequencing of this gene has finally provided a molecular perspective to the mutant phenotype (151). *EMB30* encodes a 163-kDa polypeptide with regions of sequence similarity to the yeast Sec7 protein and two putatively related proteins from humans and *C. elegans*. Sec7 functions in yeast as a cytosolic protein linked to the Golgi apparatus and secretory pathways. It therefore appears to function primarily in a cellular context rather than a developmental context. *EMB30* is not likely to perform precisely the same function in plants, based on sequence divergence and cell ultrastructure, but it probably is involved in some type of secretory pathway associated with the Golgi apparatus. The synthesis of complex polysaccharides destined for the cell wall is one example of a cellular function involving the Golgi apparatus that may be disrupted in mutant embryos. *EMB30* appears to be expressed at the same level throughout wild-type plants, not exclusively in developing embryos as might be predicted by the *D. melanogaster* paradigm of pattern-formation genes. Mutant cells also exhibit a variety of defects in size, shape, and division planes throughout development, a feature not addressed in previous studies. Many of these defects may result from disruption of the synthesis and secretion of cell wall components required for normal patterns of cell division and expansion (151). The irregular pattern of cell division observed in mutant embryos at early stages of development (96) may not reflect a unique function of this gene during early embryogenesis, but rather the most intriguing example of a wide range of defects in cell-division patterns. The production of fused cotyledons may result from a disruption of intracellular auxin transport or auxin binding at the cell surface. Thus, although the sequence of *EMB30* cannot eliminate a direct role of this gene in regulating early embryogenesis, it seems likely that this gene performs a more general cellular function throughout growth and development.

CONCLUSIONS

Three main factors have contributed to the renewed interest in plant embryo development in recent years: (*a*) the application of molecular techniques to studies of plant embryogenesis, (*b*) the establishment of model plant systems amenable to genetic analysis, and (*c*) the isolation and characterization of mutants with interesting defects in embryo development. Two complementary approaches to gene identification in *A. thaliana* (saturation mutagenesis and large-scale DNA sequencing) should near completion over the next 15 years. The entire *A. thaliana* genome will probably be sequenced within that period of time. As a result, a tremendous amount of information on genome structure and function will be generated in a relatively short period of time. The challenge for plant physiologists, biochemists, and embryologists will be to develop appropriate strategies to utilize large collections of developmental mutants and to interpret vast amounts of DNA sequence information within a cellular and developmental context. Meeting this challenge will require a large-scale multidisciplinary approach that combines advances in gene-knockout technology to establish the developmental significance of every cloned gene, enhanced imaging of gene-expression patterns to study gene function early in development, laser and genetic ablation of specific cell types to study cell interactions, further improvements in subcellular localization of gene products, and enhanced manipulation of immature embryos in culture to establish the role of environmental and developmental signals throughout plant embryogenesis. Results of these studies with model plant systems should then be viewed within an evolutionary context in order to understand not only how this developmental pathway functions at the molecular level, but also how it evolved through time. With this perspective, it should be possible to apply knowledge gained through basic research with model organisms to the solution of practical problems concerning seed development in a wide range of crop plants.

ACKNOWLEDGMENTS

I thank the members of my laboratory, past and present, for their significant contributions to this research effort, Brian Schwartz for helpful comments on the manuscript, and colleagues who provided unpublished information for this review. Research on embryo-defective mutants of *A. thaliana* in my laboratory at Oklahoma State University has been supported by grants from the National

Science Foundation (Developmental Biology, Special Projects, and EPSCoR Programs), the US Department of Agriculture (Competitive Grants Program), and the S.R. Noble Foundation (Plant Biology Division) in Ardmore, Oklahoma.

Any *Annual Review* chapter, as well as any article cited in an *Annual Review* chapter, may be purchased from the Annual Reviews Preprints and Reprints service.
1-800-347-8007; 415-259-5017; email: arpr@class.org

Literature Cited

1. Abbe EC, Stein OL. 1954. The growth of the shoot apex in maize: embryogeny. *Am. J. Bot.* 41:285–93
2. Asghar R, Fenton RD, DeMason DA, Close TJ. 1994. Nuclear and cytoplasmic localization of maize embryo and aleurone dehydrin. *Protoplasma* 177:87–94
3. Barton MK, Poethig RS. 1993. Formation of the shoot apical meristem in *Arabidopsis thaliana*: an analysis of development in the wild type and in the shoot meristemless mutant. *Development* 119:823–31
4. Bäumlein H, Boerjan W, Nagy I, Panitz R, Inzé D, Wobus U. 1991. Upstream sequences regulating legumin gene expression in heterologous transgenic plants. *Mol. Gen. Genet.* 225:121–28
5. Bäumlein H, Miséra S, Luersen H, Kölle K, Horstmann C, et al. 1994. The *FUS3* gene of *Arabidopsis thaliana* is a regulator of gene expression during late embryogenesis. *Plant J.* 6:379–87
6. Baus AD, Franzmann L, Meinke DW. 1986. Growth *in vitro* of arrested embryos from lethal mutants of *Arabidopsis thaliana*. *Theor. Appl. Genet.* 72:577–86
7. Beckett JB. 1978. B-A translocations in maize. I. Use in locating genes by chromosome arms. *J. Hered.* 69:27–36
8. Berleth T, Jürgens G. 1993. The role of the *monopteros* gene in organising the basal body region of the *Arabidopsis* embryo. *Development* 118:575–87
9. Bhattacharyya MK, Smith AM, Ellis THN, Hedley C, Martin C. 1990. The wrinkled-seed character of pea described by Mendel is caused by a transposon-like insertion in a gene encoding starch-branching enzyme. *Cell* 60:115–22
10. Bosnes M, Harris E, Aigeltinger L, Olsen OA. 1987. Morphology and ultrastructure of 11 barley shrunken endosperm mutants. *Theor. Appl. Genet.* 74:177–87
11. Brink RA, Cooper DC. 1947. The endosperm in seed development. *Bot. Rev.* 13:423–541
12. Brown JD, Beggs JD. 1992. Roles of PRP8 protein in the assembly of splicing complexes. *EMBO J.* 11:3721–29
13. Castle LA, Errampalli D, Atherton TL, Franzmann LH, Yoon ES, Meinke DW. 1993. Genetic and molecular characterization of embryonic mutants identified following seed transformation in *Arabidopsis. Mol. Gen. Genet.* 241:504–14
14. Castle LA, Meinke DW. 1993. Embryo-defective mutants as tools to study essential functions and regulatory processes in plant embryo development. *Semin. Dev. Biol.* 4:31–39
15. Castle LA, Meinke DW. 1994. A *FUSCA* gene of Arabidopsis encodes a novel protein essential for plant development. *Plant Cell* 6:25–41
16. Chasan R. 1992. Endosperm: food for thought. *Plant Cell* 4:235–36
17. Chasan R. 1993. Evolving developments. *Plant Cell* 5:363–69
18. Chasan R. 1993. Embryogenesis: new molecular insights. *Plant Cell* 5:597–99
19. Chasan R. 1994. A tale of two phenotypes. *Plant Cell* 6:571–73
20. Chasan R, Walbot V. 1993. Mechanisms of plant reproduction: questions and approaches. *Plant Cell* 5:1139–46
21. Chory J. 1993. Out of darkness: mutants reveal pathways controlling light-regulated development in plants. *Trends Genet.* 9:167–72
22. Clark JK, Sheridan WF. 1986. Developmental profiles of the maize embryo-lethal mutants *dek22* and *dek23. J. Hered.* 77:83–92
23. Clark JK, Sheridan WF. 1988. Characterization of the two maize embryo-lethal defective kernel mutants rgh*-1210 and fl*1253B: effects on embryo and gametophyte development. *Genetics* 120:279–90
24. Clark JK, Sheridan WF. 1991. Isolation and characterization of 51 *embryo-specific* mutations of maize. *Plant Cell* 3:935–51
25. Coe EH Jr, Neuffer MG, Hoisington DA. 1988. The genetics of corn. In *Corn and Corn Improvement*, ed. GF Sprague, JW

Dudley, pp. 81–258. Madison, WI: Am. Soc. Agron. 3rd ed.

26. Coen ES, Meyerowitz EM. 1991. The war of the whorls: genetic interactions controlling flower development. *Nature* 353:31–37

27. Conceicao AD, Krebbers E. 1994. A cotyledon regulatory region is responsible for the different spatial expression patterns of *Arabidopsis* 2S albumin genes. *Plant J.* 5:493–505

28. Cooke TJ, Racusen RH, Cohen JD. 1993. The role of auxin in plant embryogenesis. *Plant Cell* 5:1494–95

29. Decroocq-Ferrant V, Bianchi M, Van Went J, de Vries S, Kreis M. 1994. *Petunia hybrida* protein kinase genes homologous to *shaggy/zeste-white 3* from *Drosophila* expressed in anthers and ovules. *4th Int. Congr. Plant Mol. Biol., Amsterdam*, Abstr. No. 635

30. de Jong AJ, Cordewener J, Lo Schiavo F, Terzi M, Vandekerckhove J, et al. 1992. A carrot somatic embryo mutant is rescued by chitinase. *Plant Cell* 4:425–33

31. de Jong AJ, Heidstra R, Spaink HP, Hartog MV, Meijer EA, et al. 1993. *Rhizobium* lipooligosaccharides rescue a carrot somatic embryo mutant. *Plant Cell* 5:615–20

32. de Jong AJ, Schmidt EDL, de Vries SC. 1993. Early events in higher-plant embryogenesis. *Plant Mol. Biol.* 22:367–77

33. Deng X-W. 1994. Fresh view of light signal transduction in plants. *Cell* 76:423–26

34. Deng X-W, Matsui M, Wei N, Wagner D, Chu AM, et al. 1992. *COP1*, an Arabidopsis regulatory gene, encodes a protein with both a zinc-binding motif and a G-β homologous domain. *Cell* 71:791–801

35. Dennis L, Dean C, Flavell R, Goodman H, Koornneef M, et al. 1993. *The Multinational Coordinated Arabidopsis thaliana Genome Research Project, Progress Report: Year Three.* Washington, DC: Natl. Sci. Found. Publ. No. 93-173

36. Devic M, Delseny M, Gallois P. 1992. Promoter trapping in *Arabidopsis thaliana*: searching for embryo-specific genes. *CR. Soc. Biol.* 186:541–49

37. Dolan L, Janmaat K, Willemsen V, Linstead P, Poethig S, et al. 1993. Cellular organisation of the *Arabidopsis thaliana* root. *Development* 119:71–84

38. Dolfini SF, Sparvoli F. 1988. Cytological characterization of the embryo-lethal mutant *dek-1* of maize. *Protoplasma* 144:142–48

39. Dresselhaus T, Lörz H, Kranz E. 1994. Representative cDNA libraries from a few plant cells. *Plant J.* 5:605–10

39a. Dresselhaus T, Kranz E, Lörz H. 1994. Isolation of egg cell- and zygote-specific genes from maize (*Zea mays* L.). *4th Int.* *Congr. Plant Mol. Biol., Amsterdam*, Abstr. No. 611

40. Dumas C, Mogensen HL. 1993. Gametes and fertilization: maize as a model system for experimental embryogenesis in flowering plants. *Plant Cell* 5:1337–48

41. Dynlacht BD, Weinzierl ROJ, Admon A, Tjian R. 1993. The dTAF$_{II}$80 subunit of *Drosophila* TFIID contains β-transducin repeats. *Nature* 363:176–79

42. Errampalli D, Patton D, Castle L, Mickelson L, Hansen K, et al. 1991. Embryonic lethals and T-DNA insertional mutagenesis in *Arabidopsis*. *Plant Cell* 3:149–57

43. Faure J-E, Digonnet C, Dumas C. 1994. An *in vitro* system for adhesion and fusion of maize gametes. *Science* 263:1598–600

44. Fedoroff NV, Smith DL. 1993. A versatile system for detecting transposition in *Arabidopsis*. *Plant J.* 3:273–89

45. Feldmann KA. 1991. T-DNA insertion mutagenesis in *Arabidopsis*: mutational spectrum. *Plant J.* 1:71–82

46. Felker FC, Peterson DM, Nelson OE. 1987. Early grain development of the *seg2* maternal-effect shrunken endosperm mutant of barley. *Can. J. Bot.* 65:943–48

47. Fernandez DE, Turner FR, Crouch ML. 1991. *In situ* localization of storage protein mRNAs in developing meristems of *Brassica napus* embryos. *Development* 111:299–313

48. Fiedler U, Filistein R, Wobus U, Bäumlein H. 1993. A complex ensemble of cis-regulatory elements controls the expression of a *Vicia faba* non-storage seed protein gene. *Plant Mol. Biol.* 22:669–79

49. Finkelstein RR. 1993. Abscisic acid–insensitive mutations provide evidence for stage-specific signal pathways regulating expression of an *Arabidopsis* late embryogenesis-abundant (Lea) gene. *Mol. Gen. Genet.* 238:401–8

50. Franzmann L, Patton DA, Meinke DW. 1989. *In vitro* morphogenesis of arrested embryos from lethal mutants of *Arabidopsis thaliana*. *Theor. Appl. Genet.* 77:609–16

51. Franzmann LH, Yoon ES, Meinke DW. 1995. Saturating the genetic map of *Arabidopsis thaliana* with embryonic mutations. *Plant J.* 7: In press

52. Gasser CS, Robinson-Beers K. 1993. Pistil development. *Plant Cell* 5:1231–39

53. Giraudat J, Hauge BM, Valon C, Smalle J, Parcy F, Goodman HM. 1992. Isolation of the Arabidopsis *ABI3* gene by positional cloning. *Plant Cell* 4:1251–61

54. Goldberg RB. 1988. Plants: novel developmental processes. *Science* 240:1460–67

55. Goldberg RB, Barker SJ, Perez-Grau L. 1989. Regulation of gene expression during plant embryogenesis. *Cell* 56:149–60

55a. Goldberg RB, de Paiva G, Yadegari R.

1994. Plant embryogenesis: zygote to seed. *Science* 266:605–14

56. Greene BA, Hake S. 1993. The *Knotted-1* mutants of maize: investigating the circuitry of leaf development. *Semin. Dev. Biol.* 4:41–49

57. Guiltinan MJ, Marcotte WR, Quatrano RS. 1990. A leucine zipper protein that recognizes an abscisic acid response element. *Science* 250:267–70

58. Hake S. 1992. Unraveling the knots in plant development. *Trends Genet.* 8:109–14

59. Hattori T, Vasil V, Rosenkrans L, Hannah LC, McCarty DR, Vasil IK. 1992. The *Viviparous-1* gene and abscisic acid activate the *C1* regulatory gene for anthocyanin biosynthesis during seed maturation in maize. *Genes Dev.* 6: 609–18

60. Heath JD, Weldon R, Monnot C, Meinke DW. 1986. Analysis of storage proteins in normal and aborted seeds from embryo-lethal mutants of *Arabidopsis thaliana*. *Planta* 169:304–12

61. Heck GR, Fernandez DE. 1994. AGL-E, a MADS-box gene specifically expressed in embryos. *4th Int. Congr. Plant Mol. Biol., Amsterdam,* Abstr. No. 612

62. Höfte H, Desprez T, Amselem J, Chiapello H, Caboche M, et al. 1993. An inventory of 1152 expressed sequence tags obtained by partial sequencing of cDNAs from *Arabidopsis thaliana*. *Plant J.* 4:1051–61

63. Huang B-Q, Sheridan WF. 1994. Female gametophyte development in maize: microtubule organization and embryo sac polarity. *Plant Cell* 6:845–61

64. Hughes DW, Galau GA. 1991. Developmental and environmental induction of *Lea* and *LeaA* mRNAs and the postabscission program during embryo culture. *Plant Cell* 3:605–18

65. Irish VF. 1993. Cell fate determination in plant development. *Semin. Dev. Biol.* 4:73–81

66. Irish VF, Sussex IM. 1992. A fate map of the *Arabidopsis* embryonic shoot apical meristem. *Development* 115:745–53

67. Itoh Y, Kitamura Y, Arahira M, Fukazawa C. 1993. Cis-acting regulatory regions of the soybean seed storage 11S globulin gene and their interactions with seed embryo factors. *Plant Mol. Biol.* 21: 973–84

68. Jackson D, Veit B, Hake S. 1994. Expression of maize *KNOTTED1* related homeobox genes in the shoot apical meristem predicts patterns of morphogenesis in the vegetative shoot. *Development* 120:405–13

69. Jakobsen KS, Hughes DW, Galau GA. 1994. Simultaneous induction of postabscission and germination mRNAs in cultured dicotyledonous embryos. *Planta* 192:384–94

70. Johri BM, ed. 1984. *Embryology of Angiosperms*. Berlin: Springer-Verlag. 830 pp.

71. Josè-Estanyol M, Ruiz-Avila L, Puigdomènech P. 1992. A maize embryo-specific gene encodes a proline-rich and hydrophobic protein. *Plant Cell* 4:413–23

72. Jürgens G, Mayer U, Torres-Ruiz RA, Berleth T, Miséra S. 1991. Genetic analysis of pattern formation in the *Arabidopsis* embryo. In *Molecular and Cellular Basis of Pattern Formation*, ed. K Roberts, pp. 27–38. Cambridge, UK: Company of Biologists

73. Kamalay JC, Goldberg RB. 1980. Regulation of structural gene expression in tobacco. *Cell* 19:935–46

74. Kawagoe Y, Murai N. 1992. Four distinct nuclear proteins recognize *in vitro* the proximal promoter of the bean seed storage protein β-phaseolin gene conferring spatial and temporal control. *Plant J.* 2:927–36

75. Keith K, Kraml M, Dengler NG, McCourt P. 1994. *fusca3*: a heterochronic mutation affecting late embryo development in Arabidopsis. *Plant Cell* 6:589–600

76. Kiyosue T, Yamaguchi-Shinozaki K, Shinozaki K, Kamada H, Harada H. 1993. cDNA cloning of ECP40, an embryogenic-cell protein in carrot, and its expression during somatic and zygotic embryogenesis. *Plant Mol. Biol.* 21:1053–68

77. Koltunow AM. 1993. Apomixis: embryo sacs and embryos formed without meiosis or fertilization in ovules. *Plant Cell* 5: 1425–37

78. Koornneef M, Hanhart CJ, Hilhorst HWM, Karssen CM. 1989. *In vivo* inhibition of seed development and reserve protein accumulation in recombinants of abscisic acid biosynthesis and responsiveness mutants in *Arabidopsis thaliana*. *Plant Physiol.* 90:463–69

79. Lam E, Chua N-H. 1991. Tetramer of a 21-base pair synthetic element confers seed expression and transcriptional enhancement in response to water stress and abscisic acid. *J. Biol. Chem.* 266:17131–35

80. Léon-Kloosterziel KM, Keijzer CJ, Koornneef M. 1994. A seed shape mutant of Arabidopsis that is affected in integument development. *Plant Cell* 6:385–92

81. Lessard PA, Allen RD, Fujiwara T, Beachy RN. 1993. Upstream regulatory sequences from two β-conglycinin genes. *Plant Mol. Biol.* 22:873–85

82. Liang P, Pardee AB. 1992. Differential display of eukaryotic messenger RNA by means of the polymerase chain reaction. *Science* 257:967–71

83. Lindsey K, Topping JF. 1993. Embryogenesis: a question of pattern. *J. Exp. Bot.* 44:359–74

84. Lindsey K, Wei W, Clarke MC, McArdle HF, Rooke LM, Topping JF. 1993. Tagging

genomic sequences that direct transgene expression by activation of a promoter trap in plants. *Transgen. Res.* 2:33–47

85. Liu C-M, Xu Z-H, Chua N-H. 1993. Proembryo culture: *in vitro* development of early globular-stage zygotic embryos from *Brassica juncea*. *Plant J.* 3:291–300

86. Liu C-M, Xu Z-H, Chua N-H. 1993. Auxin polar transport is essential for the establishment of bilateral symmetry during early plant embryogenesis. *Plant Cell* 5:621–30

87. Lopes MA, Larkins BA. 1993. Endosperm origin, development, and function. *Plant Cell* 5:1383–99

88. Lo Schiavo F, Giuliano G, Sung ZR. 1988. Characterization of a temperature-sensitive cell mutant impaired in somatic embryogenesis. *Plant Sci.* 54:157–64

89. Lowe J, Nelson OE. 1946. Miniature seed: a study in the development of a defective caryopsis in maize. *Genetics* 31:525–33

90. Ma H, McMullen MD, Finer JJ. 1994. Identification of a homeobox-containing gene with enhanced expression during soybean (*Glycine max* L.) somatic embryo development. *Plant Mol. Biol.* 24:465–73

91. Maheshwari P. 1950. *An Introduction to the Embryology of Angiosperms.* New York: McGraw-Hill. 453 pp.

92. Malacinski GM, ed. 1988. *Developmental Genetics of Higher Organisms.* New York: Macmillan. 503 pp.

93. Mansfield SG, Briarty LG. 1991. Early embryogenesis in *Arabidopsis thaliana*. II. The developing embryo. *Can. J. Bot.* 69: 461–76

94. Mansfield SG, Briarty LG. 1992. Cotyledon cell development in *Arabidopsis thaliana* during reserve deposition. *Can. J. Bot.* 70:151–64

95. Marsden MPF, Meinke DW. 1985. Abnormal development of the suspensor in an embryo-lethal mutant of *Arabidopsis thaliana*. *Am. J. Bot.* 72:1801–12

96. Mayer U, Buttner G, Jürgens G. 1993. Apical-basal pattern formation in the *Arabidopsis* embryo: studies on the role of the *gnom* gene. *Development* 117:149–62

97. Mayer U, Ruiz RAT, Berleth T, Miséra S, Jürgens G. 1991. Mutations affecting body organization in the *Arabidopsis* embryo. *Nature* 353:402–7

98. McCarty DR. 1995. Genetic control and integration of maturation and germination pathways in seed development. *Annu. Rev. Plant Physiol. Plant Mol. Biol.* 46:71–93

99. McCarty DR, Hattori T, Carson CB, Vasil V, Vasil IK. 1991. The *viviparous-1* developmental gene of maize encodes a novel transcriptional activator. *Cell* 66:895–905

100. McNellis TW, von Arnim AG, Araki T, Komeda Y, Miséra S, Deng X-W. 1994. Genetic and molecular analysis of an allelic series of *cop1* mutants suggests functional

roles for the multiple protein domains. *Plant Cell* 6:487–500

101. Meinke DW. 1982. Embryo-lethal mutants of *Arabidopsis thaliana*: evidence for gametophytic expression of the mutant genes. *Theor. Appl. Genet.* 63:381–86

102. Meinke DW. 1985. Embryo-lethal mutants of *Arabidopsis thaliana*: analysis of mutants with a wide range of lethal phases. *Theor. Appl. Genet.* 69:543–52

103. Meinke DW. 1986. Embryo-lethal mutants and the study of plant embryo development. *Oxford Surv. Plant Mol. Cell. Biol.* 3:122–46

104. Meinke DW. 1991. Perspectives on genetic analysis of plant embryogenesis. *Plant Cell* 3:857–66

105. Meinke DW. 1991. Embryonic mutants of *Arabidopsis thaliana*. *Dev. Genet.* 12:382–92

106. Meinke DW. 1991. Genetic analysis of plant development. In *Plant Physiology: Growth and Development*, ed. FC Steward, RGS Bidwell, 10:437–90. New York: Academic

107. Meinke DW. 1992. A homoeotic mutant of *Arabidopsis thaliana* with leafy cotyledons. *Science* 258:1647–50

108. Meinke DW, ed. 1993. *Plant Developmental Genetics.* Special Issue. *Semin. Dev. Biol.* 4:1–89

109. Meinke DW. 1994. Seed development in *Arabidopsis*. See Ref. 113, pp. 253–95

110. Meinke DW, Franzmann LH, Nickle TC, Yeung EC. 1994. Leafy cotyledon mutants of Arabidopsis. *Plant Cell* 6:1049–64

111. Meinke DW, Sussex IM. 1979. Embryo-lethal mutants of *Arabidopsis thaliana*: a model system for genetic analysis of plant embryo development. *Dev. Biol.* 72:50–61

112. Meinke DW, Sussex IM. 1979. Isolation and characterization of six embryo-lethal mutants of *Arabidopsis thaliana*. *Dev. Biol.* 72:62–72

113. Meyerowitz EM, Somerville CR, eds. 1994. *Arabidopsis.* Cold Spring Harbor, NY: Cold Spring Harbor Lab. 1300 pp.

114. Miller ME, Chourey PS. 1992. The maize invertase-deficient *miniature-1* seed mutation is associated with aberrant pedicel and endosperm development. *Plant Cell* 4: 297–305

115. Miséra S, Müller AJ, Weiland-Heidecker U, Jürgens G. 1994. The *FUSCA* genes of *Arabidopsis*: negative regulators of light responses. *Mol. Gen. Genet.* 244:242–52

116. Müller AJ. 1963. Embryonentest zum Nachweis rezessiver Letalfaktoren bei *Arabidopsis thaliana*. *Biol. Zentralbl.* 82: 133–63

117. Müller AJ, Heidecker U. 1968. Lebensfähige und letale *fusca*-Mutanten bei *Arabidopsis thaliana*. *Arabidopsis Inf. Serv.* 5: 54–55

118. Nagato Y, Kitano H, Kamijima O, Kikuchi S, Satoh H. 1989. Developmental mutants showing abnormal organ differentiation in rice embryos. *Theor. Appl. Genet.* 78:11–15

119. Naito S, Hirai MY, Chino M, Komeda Y. 1994. Expression of a soybean (*Glycine max* [L.] Merr.) seed storage protein gene in transgenic *Arabidopsis thaliana* and its response to nutritional stress and to abscisic acid mutations. *Plant Physiol.* 104:497–503

120. Nambara E, Naito S, McCourt P. 1992. A mutant of *Arabidopsis* which is defective in seed development and storage protein accumulation is a new *abi3* allele. *Plant J.* 2:435–41

121. Neuffer MG, Sheridan WF. 1980. Defective kernel mutants of maize. I. Genetic and lethality studies. *Genetics* 95:929–44

122. Nuccio M, Hsieh TF, Lidiak R, Thomas T. 1993. Identification of stage-specific mRNAs in developing seeds of *Arabidopsis thaliana. 5th Int. Conf. Arabidopsis Res., Columbus, OH,* Abstr. No. 142

123. Ohad N, Pan S-M, Fischer R. 1994. Analysis of a T-DNA tagged *bio1* embryo arrested mutant in *Arabidopsis thaliana. 4th Int. Congr. Plant Mol. Biol., Amsterdam,* Abstr. No. 614

124. Okada K, Ueda J, Komaki MK, Bell CJ, Shimura Y. 1991. Requirement of the auxin polar transport system in early stages of *Arabidopsis* floral bud formation. *Plant Cell* 3:677–84

125. Ooms JJJ, Léon-Kloosterziel KM, Bartels D, Koornneef M, Karssen CM. 1993. Acquisition of desiccation tolerance and longevity in seeds of *Arabidopsis thaliana. Plant Physiol.* 102:1185–91

126. Or E, Boyer SK, Larkins BA. 1993. Opaque2 modifiers act posttranscriptionally and in a polar manner on gamma-zein gene expression in maize endosperm. *Plant Cell* 5:1599–609

127. Overvoorde PJ, Grimes HD. 1994. The role of calcium and calmodulin in carrot somatic embryogenesis. *Plant Cell Physiol.* 35:135–44

128. Paiva R, Kriz AL. 1994. Effect of abscisic acid on embryo-specific gene expression during normal and precocious germination in normal and *viviparous* maize (*Zea mays*) embryos. *Planta* 192:332–39

129. Pan Y-B, Peterson PA. 1989. Tagging of a maize gene involved in kernel development by an activated *Uq* transposable element. *Mol. Gen. Genet.* 219:324–27

130. Patton DA, Franzmann LH, Meinke DW. 1991. Mapping genes essential for embryo development in *Arabidopsis thaliana. Mol. Gen. Genet.* 227:337–47

131. Patton DA, Meinke DW. 1990. Ultrastructure of arrested embryos from lethal mutants of *Arabidopsis thaliana. Am. J. Bot.* 77:653–61

132. Pepper A, Delaney T, Washburn T, Poole D, Chory J. 1994. *DET1*, a negative regulator of light-mediated development and gene expression in *Arabidopsis*, encodes a novel nuclear-localized protein. *Cell* 78:109–16

133. Poethig S. 1989. Genetic mosaics and cell lineage analysis in plants. *Trends Genet.* 5:273–77

134. Raghavan V. 1976. *Experimental Embryogenesis in Vascular Plants.* New York: Academic. 603 pp.

135. Raghavan V. 1986. *Embryogenesis in Angiosperms.* Cambridge: Cambridge Univ. Press. 303 pp.

136. Randolph LF. 1936. Developmental morphology of the caryopsis in maize. *J. Agric. Res.* 53:881–916

137. Reiser L, Fischer RL. 1993. The ovule and the embryo sac. *Plant Cell* 5:1291–1301

138. Scanlon MJ, Stinard PS, James MG, Myers AM, Robertson DS. 1994. Genetic analysis of 63 mutations affecting maize kernel development isolated from *Mutator* stocks. *Genetics* 136:281–94

138a. Scheres B, Wolkenfelt H, Willemsen V, Terlouw M, Lawson E, et al. 1994. Embryonic origin of the *Arabidopsis* primary root and root meristem initials. *Development* 120:2475–87

139. Schiavone FM, Cooke TJ. 1987. Unusual patterns of somatic embryogenesis in the domesticated carrot: developmental effects of exogenous auxins and auxin transport inhibitors. *Cell Differ.* 21:53–62

140. Schmidt RJ, Ketudat M, Aukerman MJ, Hoschek G. 1992. Opaque-2 is a transcriptional activator that recognizes a specific target site in 22-kD zein genes. *Plant Cell* 4:689–700

141. Schnall JA, Hwang CH, Cooke TJ, Zimmerman JL. 1991. An evaluation of gene expression during somatic embryogenesis of two temperature-sensitive carrot variants unable to complete embryo development. *Physiol. Plant.* 82:498–504

142. Schneider T, Dinkins R, Robinson K, Shellhammer J, Meinke DW. 1989. An embryo-lethal mutant of *Arabidopsis thaliana* is a biotin auxotroph. *Dev. Biol.* 131:161–67

143. Schwartz BW, Yeung EC, Meinke DW. 1994. Disruption of morphogenesis and transformation of the suspensor in abnormal *suspensor* mutants of *Arabidopsis. Development* 120:3235–45

144. Shellhammer AJ. 1991. *Analysis of a biotin auxotroph of Arabidopsis thaliana.* PhD thesis. Oklahoma State Univ., Stillwater. 133 pp.

145. Shellhammer J, Meinke DW. 1990. Arrested embryos from the *bio1* auxotroph of

Arabidopsis contain reduced levels of biotin. *Plant Physiol.* 93:1162–67

146. Sheridan WF. 1988. Maize developmental genetics: genes of morphogenesis. *Annu. Rev. Genet.* 22:353–85

147. Sheridan WF, Clark JK. 1987. Maize embryogeny: a promising experimental system. *Trends Genet.* 3:3–6

148. Sheridan WF, Clark JK. 1993. Mutational analysis of morphogenesis of the maize embryo. *Plant J.* 3:347–58

149. Sheridan WF, Neuffer MG. 1980. Defective kernel mutants of maize. II. Morphological and embryo culture studies. *Genetics* 95:945–60

150. Sheridan WF, Neuffer MG. 1982. Maize developmental mutants. *J. Hered.* 73:318–29

151. Shevell DE, Leu W-M, Gillmor CS, Xia G, Feldmann KA, Chua N-H. 1994. *EMB30* is essential for normal cell division, cell expansion, and cell adhesion in *Arabidopsis* and encodes a protein that has similarity to Sec7. *Cell* 77:1051–62

152. Smith LG, Greene B, Veit B, Hake S. 1992. A dominant mutation in the maize homeobox gene, *Knotted-1*, causes its ectopic expression in leaf cells with altered fates. *Development* 116:21–30

153. Sollinger JD, Rivin C. 1993. *dks8*, a mutation specifically eliminating shoot formation during embryogenesis. *Maize Genet. Coop. Newsl.* 67:34–35

154. Steeves TA, Sussex IM. 1989. *Patterns in Plant Development*. Cambridge: Cambridge Univ. Press. 2nd ed.

155. Sterk P, Booij H, Schellekens GA, Van Kammen A, de Vries SC. 1991. Cell-specific expression of the carrot EP2 lipid transfer protein gene. *Plant Cell* 3:907–21

156. Sundaresan V, Springer P, Haward S, Volpe T, Dean C, et al. 1993. Transposon mutagenesis using gene trap and enhancer trap transposons. *5th Int. Conf. Arabidopsis Res., Columbus, OH*, Abstr. No. 292

157. Thomas TL. 1993. Gene expression during plant embryogenesis and germination: an overview. *Plant Cell* 5:1401–10

158. Topping JF, Agyeman F, Henricot B, Lindsey K. 1994. Identification of molecular markers of embryogenesis in *Arabidopsis thaliana* by promoter trapping. *Plant J.* 5:895–903

158a. Topping JF, Wei W, Lindsey K. 1991. Functional tagging of regulatory elements in the plant genome. *Development* 112:1009–19

158b. Torres-Ruiz R, Jürgens G. 1994. Mutations in the *FASS* gene uncouple pattern formation and morphogenesis in *Arabidopsis* development. *Development* 120:2967–78

159. Uwer U, Felix G, Altmann T, Willmitzer L. 1994. The Ac/Ds transposon system as a

tool to identify embryonic genes in *Arabidopsis thaliana*. *4th Int. Congr. Plant Mol. Biol., Amsterdam*, Abstr. No. 613

160. van Engelen FA, de Vries SC. 1992. Extracellular proteins in plant embryogenesis. *Trends Genet.* 8:66–70

161. Vellanoweth RL, Okita TW. 1993. Regulation of expression of wheat and rice seed storage protein genes. In *Control of Plant Gene Expression*, ed. DPS Verma, pp. 377–92. Boca Raton, FL: CRC

162. Vernon DM, Meinke DW. 1994. Embryogenic transformation of the suspensor in *twin*, a polyembryonic mutant of *Arabidopsis*. *Dev. Biol.* 165:566–73

163. Vilardell J, Martinez-Zapater JM, Goday A, Arenas C, Pages M. 1994. Regulation of the *rab17* gene promoter in transgenic *Arabidopsis* wild-type, ABA-deficient and ABA-insensitive mutants. *Plant Mol. Biol.* 24:561–69

164. Vivekananda J, Drew MC, Thomas TL. 1992. Hormonal and environmental regulation of the carrot lea-class gene Dc3. *Plant Physiol.* 100:576–81

165. Wang TL, Hedley CL. 1991. Seed development in peas: knowing your three 'r's' (or four, or five). *Seed Sci. Res.* 1:3–14

166. Wasson EA, West MAL, Matsudaira KL, Goldberg RB, Fischer RL, et al. 1994. Morphological development and cellular differentiation are uncoupled in embryo lethal mutants of *Arabidopsis*. *Plant Physiol.* 105s:52 (Abstr.)

167. Webb MC, Gunning BES. 1990. Embryo sac development in *Arabidopsis thaliana*. I. Megasporogenesis, including the microtubular cytoskeleton. *Sex. Plant Reprod.* 3:244–56

168. Webb MC, Gunning BES. 1991. The microtubular cytoskeleton during development of the zygote, proembryo, and free-nuclear endosperm in *Arabidopsis thaliana* (L.) Heynh. *Planta* 184:187–95

169. Weigl D. 1993. Patterning the *Arabidopsis* embryo. *Curr. Biol.* 3:443–45

170. West MAL, Harada JJ. 1993. Embryogenesis in higher plants: an overview. *Plant Cell* 5:1361–69

171. West MAL, Matsudaira KL, Danao JA, Goldberg RB, Fischer RL, et al. 1994. An embryo lethal mutant of *Arabidopsis* is impaired in the control of late developmental programs. *Plant Physiol.* 105s:43 (Abstr.)

172. Williams EG, Maheswaran G. 1986. Somatic embryogenesis: factors influencing coordinated behaviour of cells as an embryogenic group. *Ann. Bot.* 57:443–62

173. Wurtele ES, Wang H, Durgerian S, Nikolau BJ, Ulrich TH. 1993. Characterization of a gene that is expressed early in somatic embryogenesis of *Daucus carota*. *Plant Physiol.* 102:303–12

174. Yeung EC, Meinke DW. 1993. Embryogenesis in angiosperms: development of the suspensor. *Plant Cell* 5:1371–81

175. Zimmerman JL. 1993. Somatic embryogenesis: a model for early development in higher plants. *Plant Cell* 5:1411–23

Annu. Rev. Plant Physiol. Plant Mol. Biol. 1995. 46:395–418
Copyright © 1995 by Annual Reviews Inc. All rights reserved

PLANT GENOMES: A Current Molecular Description

Caroline Dean and Renate Schmidt

Department of Molecular Genetics, John Innes Centre, Biotechnology and Biological
Sciences Research Council, Norwich NR4 7UH, United Kingdom

KEY WORDS: repeated sequences, model species, comparative mapping, colinearity

CONTENTS

ABSTRACT

A major advance in plant genome analysis has been the development of
molecular marker maps. These, in combination with repeated sequence analysis, have given considerable insight into the organization and evolution of
many plant genomes. Physical mapping and sequencing of the genomes of the
model plant species *Arabidopsis thaliana* and rice are progressing rapidly. The
physical maps facilitate the isolation of genes by map-based cloning and will
enable the organization of whole chromosomes to be analyzed. Comparative
mapping studies between related plant species, using common sets of molecu-

lar markers, have revealed extensive collinearity over short chromosomal segments or whole chromosomes. Thus, in the future, collinearity as well as
homology may be used to clone genes from many plant species, maximally
exploiting the A. *thaliana* or rice physical maps.

INTRODUCTION

Plant genome size varies substantially even between closely related angio-
sperm species. Polyploidy is also a characteristic feature of plant genomes, and
even those species that remain diploid meiotically show a high degree of endopolyploidy in somatic tissues (49, 104). Why plant cells tolerate these size
differences and changes in ploidy levels and whether they confer a selective
advantage in some circumstances is unknown. Answers to these questions
may be elucidated once the organization and functioning of plant genomes is
better understood.

In this review, we briefly summarize work that has provided insights into
the structure of plant genomes. Readers are referred to other reviews in this
area for additional information (45, 85, 112, 142). We summarize what we
have learned so far from the analysis of model plant genomes and assess
current information about the conserved arrangement of genes (synteny) in
genomes of related plants. Our knowledge in this area is growing very quickly.
If the local order of genes in defined chromosomal regions is maintained
across a wide range of species, it will have far-reaching implications for plant
genome analysis as a whole. Analysis of the genomes of a few model species
could provide the basis for cloning genes from the majority of plant genomes
without the need to construct physical maps for every species.

ORGANIZATION OF PLANT GENOMES

Genome Size

Various methods have been used to estimate genome size. These include
microdensitometry of Feulgen-stained nuclei (12–14), DNA-reassociation kinetics (58, 79, 90), nuclear volume measurements (68), flow cytometry (6, 50),
and estimations from sampling genomic clone libraries (65). The diverse
methods give slightly different values for the same species (as illustrated in
Table 1), reflecting the different internal standards used in the individual
experiments and the reliability of the DNA content estimation of those internal
standards. Plant genomes vary in size between species by up to 1500-fold (12).
The fact that species with small genomes exist as fully functional flowering
plants has generated a great deal of speculation about the function of the extra
DNA in those species with large genomes. The DNA is nongenic (as discussed

in 45, 139), but whether it is nonfunctional is unknown. There has been similar speculation on the so-called junk DNA of mammalian genomes, but it is becoming clear that many regulatory sequences reside within this DNA (115).

Chromosome Structure and Genome Complexity

The earliest observations of genome organization at the chromosome level were made using cytogenetic techniques. These studies included plant species with large chromosomes, e.g. *Fritillaria* spp., *Tradescantia* spp., *Vicia* spp., and *Secale* spp. (34), or excellent genetics, e.g. maize (99). Several important concepts were first established by maize cytogenetics: association of linkage groups with chromosomes, chiasmata as points of genetic crossing-over, the breakage-fusion-bridge cycle, and the discovery of transposable elements (100, 123). Chromosome numbers, ploidy levels, positions of centromeres, the nucleolus-organizing region, and distribution of heterochromatic/euchromatic regions were established using light microscopy of chromosome spreads (34). These studies were extended by electron microscopy of sections through inter-phase nuclei (87) and through the use of synaptonemal complex spreads to examine chromosome pairing and recombination nodules (78). In combination with these studies, in situ hybridization techniques provided a powerful tool to map repeated DNA sequences onto the chromosomes (9, 72).

Significant advances in our understanding of plant genome complexity were made during the 1970s with the use of DNA-reassociation kinetic studies (reviewed in 45). These studies demonstrated that plant genomes are composed of repeated (rapidly reannealing) and low- or single-copy (slowly reannealing) DNA. Plants with large genomes were found to possess a much higher proportion of repeated DNA sequences than did those with smaller genomes (45, 48, 110). These studies also provided information on the arrangement of repeated sequences relative to low-copy ones. Genomes containing a high proportion of repeated sequences were shown to have only short stretches of single-copy sequence (< 2 kb). This interspersion pattern was shown to be quite different in plants with small genomes, where the low-copy sequences are present in much longer stretches [e.g. 120 kb in *A. thaliana* (119)]. Because the cloning was more straightforward, the repeated sequences of plant genomes were the first to be characterized. The sequences have been divided into two classes for descriptive convenience: tandemly and dispersed repeated sequences.

Organization of Repeated DNA Sequences

TANDEM REPEATS   One class of tandemly repeated sequences consists of rDNA—the genes encoding the ribosomal RNA. Tandem arrays of rDNA are found at one or a few sites in a given plant genome, and in most cases these clusters are associated with the nucleolus-organizing regions (46). The clusters contain several hundred to several thousand copies of the repeat unit, but not all copies are transcriptionally active (46). The repeat unit consists of a transcribed region of the 18S, 5.8S, and 25S cytoplasmic rDNAs, which is highly conserved, and an intergenic spacer, which is highly variable in sequence and in length (46). Repeat units as long as 18.5 kbp have been observed (159). The 5S rRNA genes are also found in tandemly repeated clusters uninterrupted by other sequences (44, 86), but they are separated physically from the rDNA clusters. As with the rDNA genes, the coding region is highly conserved, whereas the spacer sequences are variable in length and sequence.

A second major class of tandemly repeated sequences are satellite DNAs, so called because of the banding pattern of some repeats of this type in CsCl gradients. Satellite DNA sequences have been cloned from a number of species and the nucleotide sequences determined (9, 59, 98, 128, 132). They are composed of tandemly repeating units with a monomer size ranging from 150 to 350 bp long. There are often related families within the same species that differ in nucleotide sequence and monomer length. These related families are likely to have arisen from an early duplication, divergence, or insertion of sequences, followed by major amplification events (9, 74, 98). Satellite DNA sequences often colocalize with constitutive heterochromatin (56, 63, 76) and are found close to the telomeres and centromeres of plant chromosomes (9, 94, 158), although some are also found in interstitial regions of the chromosome (9, 51). Some satellite DNAs, most notably those localized around the centromere, are conserved across species (158), whereas others show no sequence conservation even between closely related species (9, 131).

Minisatellite and variable-number tandem repeats (VNTR) have also been found in plants (69). These are short tandem repeats that display a high degree of length variability because of changes in the copy number of the tandem repeats. In addition, microsatellite repeats have been characterized from a number of species. These are simple sequence repeats, e.g. mono-, di-, or trinucleotide repeats (10, 109). Both mini- and microsatellite repeats have proved extremely useful in mapping experiments (see below).

Telomeres are also composed of tandemly repeating units of the sequence $(T/A)nG1-8$. This composition was established by analysis of ciliate chromosomes (18), followed by isolation of the first plant telomeric sequence from *A. thaliana* (124). Telomeric repeat length has been analyzed extensively in a series of maize inbred lines and was found to vary from 1.8 to 40 kb. Half of the variation in telomere length could be accounted for by segregation of alleles at three loci (25). Closely linked to the telomeric sequences are other tandemly repeated sequences termed telomere-associated sequences. These sequences are also present in some interstitial regions of chromosomes (51).

DISPERSED REPEATED SEQUENCES   Active and inactive derivatives of retro-transposable elements form the most abundant class of dispersed repeats. Different families often constitute several percent of the genome. These elements are characterized by their mode of replication through an RNA intermediate. The retrotransposons fall into two categories depending on whether long terminal-repeats (LTRs) are produced during amplification. The non-LTR class may be related to the long interspersed-repeats (LINEs) found in mammalian genomes. An example of this class is the *del2* element in lily, which is present in approximately 240,000 copies (139). Members of the LTR class share a high degree of homology with viral retroelements and show very large variations in copy number and genome. An example of a low-copy-number element is *Ta1*, present in the *A. thaliana* genome at approximately 10 copies (150). In contrast, *WIRE-1* is present in the wheat genome at 10,000 copies (47), and *IFG* is present in pine at 10,000 copies (139).

The second major class of dispersed repeats are transposable elements that contain terminal inverted repeats and replicate through a DNA intermediate. Examples include *Ac, Spm,* and *Mu* in maize (41), *Tam* elements in *Antirrhinum majus* (31), *Tph1* in petunia (55), and *Tag1* in *A. thaliana* (147). These repeats generally have copy numbers ranging between 1 and 100. Two families of transposable elements have been described that appear to be present in significantly higher copy numbers. The *Tourist* element family has been identified in grasses such as maize, barley, rice, and sorghum (22). These elements are present in more than 10,000 copies, so on average every 30 kb of maize DNA contains a *Tourist* element (21). A second family of elements, *Stowaway,* has been found in monocots and dicots. These elements are characterized by a conserved terminal inverted repeat, small size, target-site specificity (TA), and potential to form stable DNA secondary structures (23).

Low-Copy Sequences

The low-copy sequences of genomes include the genic regions and have been analyzed predominantly by classical genetics following mutant phenotypes or natural variation. Phenotypic markers and then isozymes were the basis of the original maps. More recently, a number of DNA markers have been used; the first were RFLP markers (20). RFLP maps are available for several species (reviewed in 28). PCR-based markers have been developed to accelerate and simplify the mapping process. Randomly amplified polymorphic DNAs (RAPDs) (122, 155) and microsatellite markers (3, 10, 109) have been used extensively. In addition, minisatellites (69) and amplified fragment length polymorphisms (AFLPs) (161) have proved especially useful in analyzing multiple loci simultaneously.

A similarly large variety of mapping populations has been used to generate the linkage maps; the choice of population reflects the features of the plant species being analyzed. Examples include segregating F2 populations (e.g. lettuce, rice, *Brassica oleracea*), F3 or F4 families (*A. thaliana*), backcross families (potato), recombinant inbred lines (*A. thaliana,* maize), and dihaploids generated through microspore culture (barley, *Brassica napus*). RAPD maps have been developed for individual trees by analyzing megagametophyte tissue (148). The extensive range of aneuploid lines developed by Sears (133), including nullisomic/tetrasomic, monosomic, and substitution lines, has been used for mapping in hexaploid wheat (38).

Physical mapping of genes relative to repeated DNA sequences or other structural features of the chromosome has been achieved in some species using in situ hybridization analysis on metaphase or interphase nuclei (89, 137). However, as compared to its use in *Drosophila melanogaster* or human genome analysis, in situ hybridization analysis has been somewhat limited in plant genome analysis for localizing genes because of the difficulty of achieving successful single-copy hybridizations.

The development of pulsed-field gel electrophoresis (PFGE) to analyze large DNA molecules has considerably extended physical mapping in plant genomes. High-molecular-weight plant DNA is restricted with endonucleases that cut infrequently, because the nucleases have rare recognition sequences or are sensitive to methylation. The cleaved DNA is then separated using PFGE, and fragments ranging in size up to 1Mb or more can be analyzed (7, 52). Alternatively, genomic libraries have been constructed in vectors that can carry relatively large inserts, e.g. yeast artificial chromosome (YAC) clones that can carry up to 1000 kb (24), bacterial artificial chromosome (BAC) clones with inserts of approximately 200–300 kb (135), and bacteriophage P1 clones with inserts of up to 100 kb (143). Physical distribution of genes relative to each other or repeated sequences can then be analyzed by assessing the clones to which they hybridize. These libraries are also used to clone genes by map-based cloning. Genetically linked markers are hybridized to the libraries, followed, if necessary, by chromosome walking to generate larger contiguous regions (contigs) that cover the locus to be cloned.

Physical mapping has allowed considerable insight into the organization of plant genomes. For example, in *A. thaliana,* the genic regions appear to be distributed along most of the length of the chromosomes with dispersed repeats interrupting low-copy sequences on average every 120 kb (119). The majority of the 150-kb insert YAC clones analyzed, that were not chimeric with repeated sequences, also carried only low-copy sequences (7; R Schmidt, I Bancroft & C Dean, unpublished results). More details of the *A. thaliana* genome are given below. In contrast, the organization of the wheat genome, as analyzed by PFGE analysis and in situ hybridization, is very different. When the distribution of nonmethylated *NotI* and *MluI* restriction sites was determined in the wheat genome, these sites were more frequently associated with single-copy sequences than was expected. Also, the density of nonmethylated *NotI* and *MluI* sites was higher in distal, subtelomeric regions of the chromosomes. In addition, nonmethylated *NotI* and *MluI* were present in multicopy fragments of defined length. These analyses have led to a model for the organization of the wheat genome in which the region close to the centromere would be enriched for long stretches of repetitive DNA defined by regular spacing of unmethylated *MluI* sites. Other repeats would be distributed throughout the chromosome, but the highest density of single-copy sequences, characterized by a clustering of nonmethylated *NotI* and *MluI* islands would be localized preferentially to subtelomeric regions (107, 108). This localization coincides with the chromosomal region in which chiasmata are most frequently found and that is more decondensed at interphase (4).

Euchromatic DNA is thought to be wound into 10–15 nm "beads-on-a-string" fibers, which are then wound into loops. In HeLa cell and yeast chromosomes the organization of DNA into loops of 5–200 kb is maintained by the periodic attachment of the DNA to a proteinaceous chromosome scaffold. Scaffold-associated regions (SARs) from many eukaryotes, including pea and tobacco (64, 138), have been isolated and analyzed for common features. They are often found in AT-rich regions (53) and contain sequences related to the *D. melanogaster* topoisomerase II consensus sequence (53). Many regulatory proteins may bind to the base of the loops generated by the scaffold and exert their effects on all the genes within that loop (75). The organization and regulation of genes in this way would predict that low-copy sequences would be interrupted at some periodicity with repetitive sequences that form the scaffold attachment sites.

Comparison of Physical and Genetic Distances

The genetic distance between two markers is often quite different from the physical distance, reflecting the uneven distribution of recombination along a chromosome. This difference is exemplified by the nucleolus organizer region (NOR) in cereal chromosomes (61, 140). Cytogenetically, the NOR is located toward the distal end of the chromosome arm, whereas on the genetic map it is located very close to the centromere. This discrepancy is caused by the low level of recombination in the pericentromeric region of these chromosomes, causing clustering of markers genetically mapping to this region. Preferential recombination in distal chromosomal segments, and therefore expansion of the genetic map relative to the physical map, was shown for wheat chromosomes 7A, 7B, and 7D (153) and for the chromosome arms carrying the B genome (92). An extreme example of noncoincidence of cytogenetic and genetic maps was reported by Gustafson & Dille (62). They found that the RFLP markers mapping to rice chromosome 2 (101) were all present on one chromosome arm and not on the other.

In maize, physical vs genetic distance has been analyzed at a number of loci—*Bz1*, *Wx1*, *Adh1*, *A1*, and *Rl* (reviewed in 30). On average, there is 1456 kb/cM for the maize genome, in comparison to 1750 kb/cM for the *A1-Sh1* interval, 217 kb/cM within the *A1* locus, and 14 kb/cM within the *Bz1* locus. These studies have suggested that the coding regions themselves constitute recombination hotspots. Recent studies in yeast, however, have mapped recombination hotspots almost exclusively to promoter regions (157).

For the tomato genome as a whole, an average value of 550 kb/cM was found (52). However, PFGE analysis has demonstrated very large differences in the ratio of physical to genetic distances. Around the I_2 locus this value was 43 kb/cM (134), and around the *Tm2a* locus, 1 cM represented a minimum of 3.3 Mb (52). The suppression of recombination around *Tm2a* can perhaps be accounted for by the proximity of this locus to the centromere. In *A. thaliana*, a much lower range of kb/cM values has been observed. Using a population of 300 recombinant inbred lines, a 27 cM region was analyzed with 28 markers. The average ratio of physical to genetic distance was 200 kb/cM, varying from 70 to 500 kb/cM (M Stammers, R Schmidt & C Dean, unpublished results).

Duplication of Chromosome Segments Reflected in Low-Copy Sequence Duplication

The generation of the genetic linkage maps has proved useful not only for mapping loci, but it has also highlighted areas of the genome that are duplicated. Hybridization experiments using *B. oleracea* genome fragments or *A. thaliana* cDNA clones showed that approximately 12–15% of *A. thaliana* loci are duplicated (82, 102). There may also have been an ancestral chromosome (or segment) duplication in *A. thaliana* (82). Duplication of chromosomal segments has also occurred during the evolution of the genome in *Brassica* spp (82, 93). In maize, 28.6% of informative maize clones tested detected duplicated sequences arranged in a nonrandom order. Chromosomes 2 and 7 share 13 pairs of duplicate loci covering well over 50 cM. Chromosomes 3 and 8 share 10 duplicated loci and chromosomes 6 and 8 share 6, none of which are present on chromosome 3 (66). A smaller number of loci appear to be duplicated in the closely related species *Sorghum bicolor* (29, 154). Two models have been proposed to account for the origin of the duplicated sequences in the *Zea mays* and *S. bicolor* genomes. The first model involves polyploidization of the maize genome after the divergence of maize and sorghum and the second involves an ancestral duplication followed by sequence divergence in different genomic regions (154). Greater understanding of which regions become duplicated in the evolution of different species may elucidate the role of duplication and polyploidization in genome evolution.

Arrangement of Chromosomes in the Nucleus

Chromosomes are arranged in a nonrandom fashion in the nucleus. Telomeres in pea and *Vicia faba* nuclei are closely associated with the nuclear envelope or nucleolar periphery (120). In *V. faba* nuclei the telomeres are tightly clustered around one pole, whereas in pea the clustering is rather loose. These arrangements are thought to be caused by the way the chromosomes segregate at mitosis. Whether there are specific arrangements of chromosomes relative to each other in the nucleus is less clear. Despite early indications to the contrary, it now appears that homologous chromosomes are not closely associated in somatic tissues of plants (67, 121). There is, however, some defined arrangement of chromosomes within the nucleus. Evidence of this arrangement comes from the observation that two parental genomes in hybrids from wide crosses lie in distinct and separate domains throughout the cell cycle (88, 129). These results suggest a nonrandom chromosome arrangement in interphase nuclei, but more experiments with chromosome-specific probes are required to resolve interphase chromosome arrangement.

Individual nonhomologous chromosomes may associate with each other. Bennett (11) has noted that in cereals there are pairs of long and short chromosome arms, which tend to be similar in size. Close association of these similarly sized arms has been suggested to account for the nonrandom distribution of transposed *Ac* elements in maize, as measured by preferential transposition from the short arm of chromosome 9 to other chromosomal locations (40).

ANALYSIS OF MODEL SYSTEMS

There is considerable emphasis in eukaryotic genome research on the analysis of model organisms. Yeast will be the first eukaryotic genome to be sequenced fully, and information gained from this sequencing will be an important paradigm for other genome projects. The physical map of the 100 Mb genome of the nematode, *Caenorhabditis elegans,* is virtually complete and genomic sequencing is projected to be finished by the end of 1998. These projects are generating an enormous amount of new information on gene identification, density of genes, and organization and distribution of repetitive sequences (42, 77, 116, 144, 156). *A. thaliana* is considered the model species for dicots, and rice is the model species for monocots. In this section, we also summarize our understanding of the tomato genome because extensive mapping efforts in this species have allowed efficient map-based cloning.

Arabidopsis thaliana

Arabidopsis has one of the smallest genomes observed in higher plants with very low levels of repetitive DNA (Table 1). The rapidly reannealing fraction

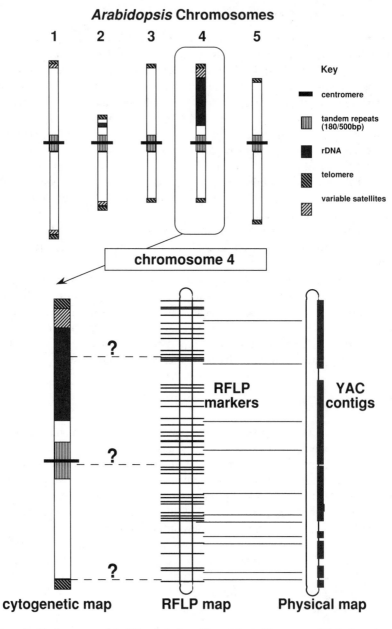

Figure 1   The karyogram of *Arabidopsis thaliana*. The cytological karyogram of *A. thaliana* shows the distribution of heterochromatin and the nucleolus organizer regions. Schematic drawings of the cytogenetic, the RFLP, and the physical maps are represented for chromosome 4. The RFLP map is well integrated with the physical map, although an integration of these maps with the cytogenetic map is pending.

Table 1 Haploid genome size for various plant species

Plant Species	Mb/1C
Arabidopsis thaliana (Arabidopsis)	70[a], 100[b], 145[c], 190[d]
Oryza sativa (rice)	430[c], 580[d]
Lycopersicon esculentum (tomato)	950[c], 965[d]
Zea mays (maize)	2500[c], 2300[d]
Hordeum vulgare (barley)	4900[c], 5300[d]
Triticum aestivum (wheat)	16000[c], 16700[d]
Fritillaria assyriaca (lily)	123000[d]

a. See Reference 90; b. See Reference 73;
c. See Reference 6; d. See Reference 12

accounts for 10% of the genome. Two related tandemly repeated sequences, the 180 bp and 500 bp repeats (98, 136), colocalize with the centromeric heterochromatin of all five *A. thaliana* chromosomes (94)—shown in the schematic illustration of the cytogenetic map in Figure 1 (adapted from Reference 130). Another tandem repeat (160 bp) appears to be adjacent to some of the chromocenters (8), but the satellite region to which this repeat corresponds has not yet been identified. The fourth component of the highly repeated DNA is the telomeric sequence. Each chromosome contains about 350 repeats of the 7-bp 5′-CCCTAAA-3′ sequence (124). A degenerate telomere sequence motif is found in a tandem array in the centromeric region of chromosome 1. Next to this sequence is a repetitive element that maps to five places in the genome (125).

The rDNA with a basic repeat unit of approximately 10 kb is present in several hundred copies and represents about 8% of the nuclear genome (119). In situ hybridization experiments localized the rDNA loci on chromosomes 2 and 4 (94), the latter carrying the larger block as illustrated in Figure 1. The 5S ribosomal RNA–encoding genes comprise 0.7% of the genome and are arranged in tandem arrays of a 497 bp unit, unlinked to the rDNA (26).

Pruitt & Meyerowitz (119) estimate that dispersed elements are found every 120 kb in single-copy nuclear DNA. A prominent class of dispersed repeats in other plant species are transposable elements. The elements so far identified in *A. thaliana,* the retrotransposon-like element *Ta1* and its related families as well as the transposable elements *Tat1* and *Tag1,* are only present in relatively low copy numbers (81, 117, 147, 150). More characterization of repeated sequence families is required in *A. thaliana,* in combination with in situ hybridization studies. The main limitation of this type of analysis in *A. thaliana* is the small size of the chromosomes. The position of labeled probes hybridized to denatured chromosomes can be determined only with limited resolution using conventional microscopy.

As of January 1994, there were 800 genetic loci identified by mutation in *A. thaliana,* about 200 of which had been mapped (35). The density of the molecular marker map is also increasing steadily. The two original RFLP maps contained approximately 100 markers each (27, 113). These markers have been supplemented by 100 RFLP markers generated from *Pst*I subclones (Y-G Liu, N Mitsukawa & RF Whittier, manuscript in preparation). PCR-based markers have also been developed: These are RAPDs (122), microsatellite markers (10), and CAPS (co-dominant cleaved amplified polymorphic sequences) markers (80). All together, over 500 molecular markers have been mapped. Two sets of recombinant inbred lines have also been developed (91, 122). Use of these recombinant inbred populations in future mapping experiments will allow direct comparison of mapping data from different laboratories.

Initially, an approach similar to that of the *C. elegans* genome project was adopted for physical mapping of the *A. thaliana* genome. Hauge et al (65) used the fingerprinting technique developed for *C. elegans* cosmids (32) to identify overlapping cosmids and thus generate a physical map. They analyzed 17,000 clones resulting in 750 cosmid contigs and estimated that 91–95% of the genome was covered. After fingerprinting cosmid clones corresponding to 10 genome equivalents, this cosmid-based effort was not continued. Several laboratories then began an international effort to construct an overlapping library in YAC clones. A number of YAC libraries are available for *A. thaliana.* The libraries vary in their average insert size and in the method used to cleave the insert plant DNA. Together, the libraries constructed from DNA of the Columbia ecotype cover more than 10 genome equivalents (33, 43, 60, 151). RFLP markers have been used to identify corresponding YAC clones and thereby place them onto the genetic map. In an early study, Hwang et al (73) used 125 RFLP markers for this type of analysis and established YAC contigs giving an approximate coverage of 30%. Furthermore, 25% of the markers were found to be physically linked and thus allowed the formation of YAC contigs encompassing several RFLP markers. Since then many more markers have been mapped. A particular focus on chromosome 4 has yielded almost complete YAC coverage in only nine contigs; the two largest cover 4.5 Mb each (R Schmidt, J West, K Love & C Dean, unpublished results).

The increasing YAC coverage facilitates map-based cloning approaches. Numerous positional cloning experiments are under way (127), and several have been completed (5, 57). These approaches have yielded local physical maps and provide important anchor points for the integration of the physical map with the classical genetic map. However, good contact points between the cytogenetic map and the physical and RFLP maps are still missing (see Figure 1).

Sequencing of the *A. thaliana* genome is under way, and a major expansion of activity is planned for the next 10 years. The first contribution to the large-scale genomic sequencing has been the completion of two 40-kb cosmid clones (65). A program to sequence 2 Mb of *A. thaliana* genomic DNA is under way and is being considered as a pilot program for a future large-scale project to be funded by the EC (16). In parallel to these genomic sequencing programs, expressed sequence tags (ESTs) are being generated (70, 114). ESTs are partial sequences of random cDNA clones isolated either from one cDNA library made from mRNA isolated from a mixture of tissues and conditions (114) or from different cDNA libraries made from a variety of tissues and cultured cells (70). To date, more than 8000 *A. thaliana* sequences have been submitted to the databases. A first analysis has shown that 40% of the clones show significant homology to known genes (114).

Rice

Rice has one of the smallest genomes among the monocots. The approximately 450-Mbp genome is distributed over 12 chromosomes. An extensive genome effort is under way, mainly in Japan, the goal of which is to isolate agronomically important genes in rice. A high-density RFLP map is being generated, extensive EST sequencing is currently under way, and YAC contigs are being generated to create a physical map of the rice genome (126). RFLP mapping has been carried out on an F2 population, consisting of 186 individuals. So far, 1533 markers have been placed onto the map, most of which are cDNA clones, but RAPDs and genomic clones are also being mapped. For the mapping of multigene families and isozyme genes, the divergence of 3′-untranslated regions in the different members in the gene family is being exploited. This has enabled the mapping of ribosomal protein genes and histone and peroxidase genes, all of which mapped to more than 15 loci (111).

The EST sequencing project has used eight cDNA libraries that have been prepared from various tissues as well as from cultured cells. Using different cDNA libraries for the EST sequencing approach provides some information on the expression pattern of the EST clones. The sequencing of 8800 randomly selected clones resulted in the generation of 4342 nonredundant sequences, 20% of which showed similarity to known proteins (141).

The physical mapping project in rice has used the same approach as that for *A. thaliana,* i.e. using RFLP markers as anchor points from which to build YAC contigs. YAC libraries, representing six genome equivalents, have been constructed from partially digested *EcoRI-* or *NotI*-cleaved DNA. The average insert size in these 7000 clones was found to be 350 kb. Using 186 markers, 578 YAC clones have been identified, giving an approximate genome coverage of 10–20%. The largest identified YAC contig covers 6.2 cM. Even at this early stage of the project the contigs cover two agronomically important loci, *Se-1,* which is responsible for photoperiod sensitivity, and *Xa-1,* which confers resistance to bacterial blight disease (149).

Tomato

The tomato genome is estimated to be around 950 Mbp (6). A minimum of 78% of the genome is composed of low-copy sequences (51). Four highly repeated families, two tandem and two dispersed, have been characterized (51). The dispersed repeats make up the bulk of the repetitive DNA (162). The two tandemly repeated sequences are the rDNA and a 162-bp satellite DNA of more than 75,000 copies (TGRI). TGRI sequences are located near the telomeres of most chromosomes. The arrays of repeats can be up to 1000-kb long and are separated from the telomeres by less than 150 kb of low-copy sequences. TGRI sequences are also found at interstitial sites and at two centromeres. The two dispersed high-copy repeated sequences of the tomato genome are termed TGRII and TGRIII. TGRII is present in approximately 4200 copies with an interspersion pattern of one every 130 kb. In situ hybridization has shown that this sequence hybridizes along the length of each chromosome, but not in the NOR. TGRI and TGRII do not cross-hybridize to the closely related potato genome, demonstrating that these repeated sequences are evolving at a rate higher than most genomic sequences. The TGRIII is present in approximately 2100 copies. In nine chromosomes it shows clustering around the centromeres, whereas in the other three it is present along the whole chromosome, except at the telomere. TGRIII hybridizes to all *Lycopersicon* species and some *Solanum* species (51).

A high-density RFLP map has been established with more than 1000 markers distributed over the 12 chromosomes (146). Overall, the number of markers on each chromosome correlates well with the pachytene length of the chromosomes (146). The distribution of markers within the chromosomes is not uniform. A higher marker density is found around the centromeres and in some cases near the telomeres. Results from physical mapping experiments and the RFLP mapping data support the idea that recombination is suppressed up to 10-fold in these areas of the genome, resulting in the observed clustering of markers (146). A 22,000 clone, 3-genome equivalent YAC library of the tomato genome has been prepared (96). This library, combined with the dense marker coverage, has opened up the tomato genome to map-based cloning strategies. Near-isogenic lines (97, 160) or bulked segregant analysis (106) have been used successfully to screen for large numbers of markers in a given interval of the genome. These additional markers are then used to identify YAC contigs in the region of interest. This approach avoids long-range chromosome walking experiments and has been termed chromosome landing. A gene encoding the race-specific resistance to *Pseudomonas syringae* pv. *tomato* was the first tomato gene to be isolated using this positional cloning strategy (95).

COMPARATIVE MAPPING

Genomes from closely related plant species show considerable DNA homology and often have the same number of chromosomes. It was therefore an interesting question to establish whether gene order is also conserved between related species. This problem was addressed readily as RFLP maps became available, some of which had been generated by exploiting the cross-homology of sequences from related species.

Plant species within the family Solanaceae show variation in haploid DNA content, although these species display the same basic chromosome number of 12. For example, pepper has a fourfold higher DNA content than does tomato (50). Reciprocal mapping of genomic and cDNA clones between tomato, potato, and pepper showed a high conservation of the gene-repertoire (19, 54, 145). A comparison of marker order between tomato and potato also showed a high degree of conservation. In 7 out of 12 chromosomes, the marker order in potato was unchanged compared to tomato for the markers tested. The remaining five chromosomes showed paracentric inversions with one breakpoint at or near the centromere (146). A comparison of marker order in tomato and pepper, however, showed only conservation of linkage segments rather than complete linkage groups. The pepper linkage groups 4, 5, and 10 corresponded to pieces of tomato chromosomes 1, 6, and 2, respectively. Prince et al (118) estimated that a minimum of 31% of the pepper molecular map is conserved with the tomato map; the sizes of the conserved blocks range from 2 to 67.6 cM. Interestingly, several regions of the tomato genome are not represented by any linkage groups of the pepper genome (118). Thus, substantial chromosome rearrangements, which are not restricted to the centromere region, must have taken place during the divergence of tomato and pepper. The tomato-pepper comparison also showed that 12% of the clones showed a change in locus number; however, this result was not confined to one species and, hence, could not account for the observed fourfold DNA content variation in these two species (145).

Similar studies are now being carried out in many other dicots (e.g. legumes). Comparative mapping of two north-temperate legumes, pea and lentil (152), and two tropical legume species, mungbean and cowpea (105), resulted in the identification of conserved linkage blocks. The genomes of *B. oleracea, B. rapa,* and *Brassica nigra* show largely homoeologous chromosomes that vary by a limited number of translocations (93). These genomes are being compared to *A. thaliana.* A recent study by Kowalski et al (82) has shown that extensive rearrangement of the chromosomes has occurred since the *Brassica* spp. and *A. thaliana* genomes diverged. There were 11 conserved regions extending from 3.7 to 49.6 cM, accounting for 24.6% of the *A. thaliana* genome and 29.9% of the *B. oleracea* genome. At least 17 translocation events and 9 inversions distinguished the two genomes, and there was evidence that the divergence in the genomes occurred prior to the duplication in the *B. oleracea* genome. Studies are in progress to investigate gene order at high resolution within 5–10 cM regions. So far, these studies have showed virtually complete collinearity on three regions of *A. thaliana* chromosomes 4 and 5 and on the corresponding genomic regions in *B. napus* (84; A Cavell, I Parkin, D Lydiate & M Trick, personal communication).

Conservation of gene-order has also been studied in monocots. Allohexaploid wheat contains three sets of seven homoeologous chromosome groups. RFLP mapping established a linkage map that is in fact a comparative map of three ancestral genomes. Overall, collinearity between chromosomes within a homoeologous group is conserved, although rearrangements are observed (36). Comparative mapping among the Triticeae (e.g. wheat, barley, and rye) showed extensive collinearity between the homoeologous chromosomes (37, 39). Although rye chromosome 1 showed complete collinearity when compared to the group 1 chromosomes of wheat, the comparison of other homoeologous chromosomes provided evidence for multiple translocations in the rye genome relative to the wheat genome (36).

Maize probes have been found to strongly cross-hybridize to sorghum, foxtail millet, and sugarcane; less strongly to pearl millet; and occasionally to barley (71). Maize and sorghum have 10 chromosomes, although the sorghum genome is 3–4 fold smaller than is the maize genome. Comparative mapping between these two species (15a, 17, 71, 154) was complicated because many loci are duplicated, often making the identification of orthologous sequences ambiguous. A higher level of duplication in the maize genome is likely to account for the 3–4 fold difference in genome size. Where map positions could be compared, many chromosomal segments clearly had conserved marker order, but multiple rearrangements must have occurred since the maize and sorghum genomes diverged. Interestingly, the maize repetitive sequences did not hybridize to sorghum DNA, suggesting that the repetitive sequences evolve faster than do the single-copy sequences (71). This difference may allow chromosome walks in the maize genome using sorghum libraries (15).

Extensive collinearity of conserved segments was also observed when the rice genome was compared to maize and wheat. These species have neither the same basic chromosome number nor a similar haploid DNA content. Furthermore, these species diverged more than 50 million years ago. Nevertheless, 85% of the rice clones tested cross-hybridized to maize sequences with a high proportion of the rice sequences duplicated in maize (72%). A maize linkage map was established using 250 rice loci (2). Overall, the comparative mapping of maize and rice detected 32 conserved linkage segments, ranging from 5 to 85 cM. These conserved segments account for 62% and 70% of the rice and maize genetic map, respectively (2). Even more conservation was found between the rice and the wheat genomes (1, 83). The single-copy probes from rice chromosomes 1–7 and 9 were effectively collinear with different wheat chromosomes (83). Rice chromosomes 10 and 12 showed some regions in common with wheat, but it is not clear how rice chromosomes 8 and 11 were related to wheat and maize.

The extensive collinearity of the wheat, rye, barley, rice, and maize genomes suggests that it may be possible to reconstruct a map of the ancestral cereal genome (Figure 2). This extensive comparative mapping will yield a

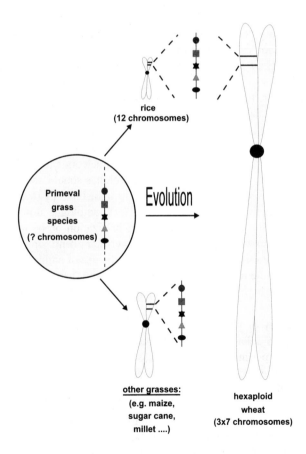

Figure 2   Genome evolution in grass species. The difference in genome sizes between grass species is represented graphically. Despite this variation of genome sizes, the order of markers in a given chromosome segment has remained largely unchanged during evolution.

large number of probes for any given region of any cereal genome. It will also allow information already available for the different cereals to be combined and exploited more effectively.

CONCLUSIONS

As more information accumulates on the structure of different plant genomes, the processes that occurred during their evolution become clearer. Genome sizes between closely related species vary enormously, but gene-repertoire and at least short-range collinearity are largely conserved. Most of the differences in genome sizes can be accounted for by repetitive sequences that show a much higher level of divergence than do the genic regions. The extent of the change and turnover of the repetitive sequences cannot be accounted for by simple amplification events creating variation in repeat numbers. Both tandem and dispersed repeats display amplification and deletion cycles that probably involve a variety of mechanisms including gene conversion and unequal crossing-over. However, these processes do not seem to have interfered with the basic genome organization.

Comparison of the cDNA (EST) sequences from *A. thaliana* and rice has demonstrated that corresponding genes show a much greater degree of homology at the predicted amino acid level than at the nucleotide level. The ever-increasing number of *A. thaliana* and rice EST clones should soon ensure that any gene from one species could be isolated from the other by first identifying, through a search of the EST database, the closest homologue at the amino acid level, and then using the EST clone as a hybridization probe to a genomic library. This process will be considerably easier than identifying homologues by cross-hybridization experiments or through the use of antibodies. The ability to easily cross the monocot-dicot boundary would then open up the possibility of isolating homologues of a given gene from a large range of plant species.

The collinearity of whole chromosomes or chromosome segments that is observed between related species has huge implications for plant genome analysis. Many experiments are in progress to assess the distances over which gene order is conserved between related species. One might predict that local gene order would be more conserved than would gene order over large chromosomal regions. This prediction needs to be tested in a wide range of dicots and monocots. Sets of common anchor markers need to be developed to aid this analysis. Extensive collinearity may allow map-based cloning in many more plant species. Genetic mapping would be done in the species with the large genome (and the interesting loci), and the cloning would be done in the related species, with the small genome, for which libraries and/or physical maps are available. Local gene order may be more conserved than is the nucleotide sequence of homologous genes in distantly related species. Therefore, the complete genomic sequence of the *A. thaliana* and rice genomes will accelerate gene cloning from many plant species based on relative genomic location as well as homology.

ACKNOWLEDGMENTS

We thank Mary Rayward-Smith for preparing Figure 1 and completing the references and Tracie Foote for Figure 2. We are extremely grateful to Dick Flavell, Jonathan Jones, Graham Moore, Mike Bevan, and Pat Heslop-Harrison for commenting on the manuscript. The authors acknowledge support from the EC Bridge program and BBSRC (Biotechnology and Biological Sciences Research Council) Plant Molecular Biology and Genome programs.

Literature Cited

1. Ahn S, Anderson JA, Sorrells ME, Tanksley SD. 1993. Homoeologous relationships of rice, wheat and maize chromosomes. *Mol. Gen. Genet.* 241:483–90
2. Ahn S, Tanksley SD. 1993. Comparative linkage maps of the rice and maize genomes. *Proc. Natl. Acad. Sci. USA* 90:7980–84
3. Akkaya MS, Bhagwat AA, Cregan PB. 1992. Length polymorphisms of simple sequence repeat DNA in soybean. *Genetics* 132:1131–39
4. Anamthawat-Jonsson K, Heslop-Harrison JS. 1990. Centromeres, telomeres and chromatin in the interphase nucleus of cereals. *Caryologia* 43:205–11
5. Arondel V, Lemieux B, Hwang I, Gibson S, Goodman HM, Somerville CR. 1992. Map-based cloning of a gene controlling omega-3 fatty acid desaturation in *Arabidopsis. Science* 258:1353–55
6. Arumuganathan K, Earle ED. 1991. Nuclear DNA content of some important plant species. *Plant Mol. Biol. Rep.* 9:208–18
7. Bancroft I, Westphal L, Schmidt R, Dean C. 1992. PFGE-resolved RFLP analysis and long range restriction mapping of the DNA of *Arabidopsis thaliana* using whole YAC clones as probes. *Nucleic Acids Res.* 20:6201–7
8. Bauwens S, Van Oostveldt P, Engler G, Van Montagu M. 1991. Distribution of the rDNA and the three classes of highly repetitive DNA in the chromatin of interphase nuclei of *Arabidopsis thaliana. Chromosoma* 101:41–48
9. Bedbrook JR, Jones J, O'Dell M, Thompson RD, Flavell RB. 1980. A molecular description of telomeric heterochromatin in *Secale* species. *Cell* 19:545–60
10. Bell CJ, Ecker JR. 1994. Assignment of 30 microsatellite loci to the linkage map of Arabidopsis. *Genomics* 19:137–44
11. Bennett MD. 1982. Nucleotypic basis of the spatial ordering of chromosomes in eukaryotes and the implication of the order for genomic evolution and phenotypic variation. In *Genome Evolution,* ed. GA Dover, RB Flavell, pp. 239–61. London: Academic
12. Bennett MD, Smith JB. 1976. Nuclear DNA amounts in angiosperms. *Philos. Trans. R. Soc. London Ser. B* 274:227–74
13. Bennett MD, Smith JB. 1991. Nuclear DNA amounts in angiosperms. *Philos. Trans. R. Soc. London Ser. B* 334:309–45
14. Bennett MD, Smith JB, Heslop-Harrison JS. 1982. Nuclear DNA amounts in angiosperms. *Philos. Trans. R. Soc. London Ser. B* 216:179–90
15. Bennetzen JL, Freeling M. 1993. Grasses as a single genetic system: genome composition, collinearity and compatibility. *Trends Genet.* 9:259–61
15a. Berhan AM, Hulbert SH, Butler LG, Bennetzen JL. 1993. Structure and evolution of the genomes of *Sorghum bicolor* and *Zea mays. Theor. Appl. Genet.* 86:598–604
16. Bevan M. 1994. Systematic sequencing of

the *Arabidopsis* genome—The "ESSA" Project. *4th Int. Congr. Plant Mol. Biol.*, Abstr. #1

17. Binelli G, Gianfranceschi L, Pe ME, Taramino G, Busso C, et al. 1992. Similarity of maize and *Sorghum* genomes as revealed by maize RFLP probes. *Theor. Appl. Genet.* 84:10–16

18. Blackburn EH, Budarf ML, Challoner PB, Cherry JM, Howard EA, et al. 1982. DNA termini in ciliate macronuclei. *Cold Spring Harbor Symp. Quant. Biol.* 47: 1195–207

19. Bonierbale MW, Plaisted RL, Tanksley SD. 1988. RFLP maps based on a common set of clones reveal modes of chromosomal evolution in potato and tomato. *Genetics* 120:1095–103

20. Botstein D, White RL, Skolnick MH, Davies RW. 1980. Construction of a genetic linkage map in man using restriction fragment length polymorphism. *Am. J. Hum. Genet.* 32:314–31

21. Bureau TE, Wessler SR. 1992. *Tourist:* a large family of small inverted repeat elements frequently associated with maize genes. *Plant Cell* 4:1283–94

22. Bureau TE, Wessler SR. 1994. Mobile inverted-repeat elements of the *Tourist* family are associated with the genes of many cereal grasses. *Proc. Natl. Acad. Sci. USA* 91:1411–15

23. Bureau TE, Wessler SR. 1994. *Stowaway:* a new family of inverted repeat elements associated with the genes of both monocotyledenous and dicotyledonous plants. *Plant Cell* 6:907–16

24. Burke DT, Carle GF, Olson MV. 1987. Cloning of large segments of exogenous DNA into yeast by means of artificial chromosome vectors. *Science* 236:806–12

25. Burr B, Burr FA, Matz EC, Romero-Severson J. 1992. Pinning down loose ends: mapping telomeres and factors affecting their length. *Plant Cell* 4:953–60

26. Campell BR, Song Y, Posch TE, Cullis CA, Town CD. 1992. Sequence and organization of 5S ribosomal RNA-encoding genes of *Arabidopsis thaliana. Gene* 112: 225–28

27. Chang C, Bowman JL, DeJohn AW, Lander ES, Meyerowitz EM. 1988. Restriction fragment length polymorphism linkage map for *Arabidopsis thaliana. Proc. Natl. Acad. Sci. USA* 85:6856–60

28. Chang C, Meyerowitz EM. 1991. Plant genome studies: restriction fragment length polymorphism and chromosome mapping information. *Curr. Opin. Genet. Dev.* 1: 112–18

29. Chittenden LM, Schertz KF, Lin Y-R, Wing RA, Paterson AH. 1994. A detailed RFLP map of *Sorghum bicolor* x *S. propinquum,* suitable for high-density mapping, suggests ancestral duplication of *Sorghum*

chromosomes or chromosomal segments. *Theor. Appl. Genet.* 87:925–33

30. Civardi L, Xia Y, Edwards KJ, Schnable PS, Nikolau BJ. 1994. The relationship between the genetic and physical distances in the cloned *a1-sh2* interval of the *Zea mays* L. genome. *Proc. Natl. Acad. Sci. USA* 91:8268–72

31. Coen ES, Carpenter R. 1986. Transposable elements in *Antirrhinum majus:* generators of genetic diversity. *Trends Genet.* 2:292–96

32. Coulson A, Sulston J, Brenner S, Karn J. 1986. Toward a physical map of the genome of the nematode *Caenorhabditis elegans. Proc. Natl. Acad. Sci. USA* 83: 7821–25

33. Creusot F, Billault A, Bouchez D, Caboche M, Camilleri C, et al. 1994. Construction of a large insert YAC library of *Arabidopsis thaliana. 4th Int. Congr. Plant Mol. Biol.*, Abstr. #88

34. Darlington CD, La Cour LF. 1969. *The Handling of Chromosomes.* London: Allen & Unwin. 5th ed.

35. Dennis E, Dean C, Flavell R, Goodman H, Koornneef M, et al. 1993. *The Multinational Coordinated* Arabidopsis thaliana *Genome Research Project. Progress Report Year 3.* Arlington, VA: Natl. Sci. Found. 71 pp.

36. Devos KM, Atkinson MD, Chinoy CN, Francis HA, Harcourt RL, et al. 1993. Chromosomal rearrangements in the rye genome relative to that of wheat. *Theor. Appl. Genet.* 85:673–80

37. Devos KM, Gale MD. 1993. Extended genetic maps of the homoeologous group 3 chromosomes of wheat, rye and barley. *Theor. Appl. Genet.* 85:649–52

38. Devos KM, Gale MD. 1993. The genetic maps of wheat and their potential in plant breeding. *Outl. Agric.* 22:93–99

39. Devos KM, Millan T, Gale MD. 1993. Comparative RFLP maps of homoeologous group-2 chromosomes of wheat, rye and barley. *Theor. Appl. Genet.* 85:784–92

40. Dooner HK, Belachew A, Burgess D, Harding S, Ralston M, Ralston E. 1994. Distribution of unlinked receptor sites for transposed *Ac* elements from the *bz-m2(Ac)* allele in maize. *Genetics* 136:261–79

41. Döring H-P, Starlinger P. 1986. Molecular genetics of transposable elements in plants. *Annu. Rev. Genet.* 20:175–200

42. Dujon B, Alexandraki D, Andre B, Ansorge W, Baladron V, et al. 1994. Complete DNA sequence of yeast chromosome XI. *Nature* 369:371–78

43. Ecker JR. 1990. PFGE and YAC analysis of the Arabidopsis genome. *Methods* 1: 186–94

44. Ellis THN, Lee D, Thomas CM, Simpson PR, Cleary WG, et al. 1988. 5S rRNA genes

in *Pisum:* sequence, long range and chromosomal organization. *Mol. Gen. Genet.* 214:333–42

45. Flavell R. 1980. The molecular characterization and organization of plant chromosomal DNA sequences. *Annu. Rev. Plant Phys.* 31:569–96

46. Flavell RB. 1986. The structure and control of expression of ribosomal RNA genes. *Oxford Surv. Plant Mol. Cell Biol.* 3:251–74

47. Flavell RB. 1986. Repetitive DNA and chromosome evolution in plants. *Philos. Trans. R. Soc. London Ser. B* 312:227–42

48. Flavell RB, Smith DB. 1976. Nucleotide sequence organisation in the wheat genome. *Heredity* 37:231–52

49. Galbraith DW, Harkins KR, Knapp S. 1991. Systemic endopolyploidy in *Arabidopsis thaliana. Plant Physiol.* 96: 985–89

50. Galbraith DW, Harkins KR, Maddox JR, Ayres NM, Sharma DP, Firoozabady E. 1983. Rapid flow cytometric analysis of the cell cycle in intact plant tissues. *Science* 220:1049–51

51. Ganal MW, Lapitan NLV, Tanksley SD. 1988. A molecular and cytogenetic survey of major repeated DNA sequences in tomato (*Lycopersicon esculentum*). *Mol. Gen. Genet.* 213:262–68

52. Ganal MW, Young ND, Tanksley SD. 1989. Pulsed field gel electrophoresis and physical mapping of large DNA fragments in the *Tm-2a* region of chromosome 9 in tomato. *Mol. Gen. Genet.* 215:395–400

53. Gasser SM, Laemmli UK. 1986. The organization of chromatin loops: characterization of a scaffold attachment site. *EMBO J.* 5:511–18

54. Gebhardt C, Ritter E, Barone A, Debener T, Walkemeier B, et al. 1991. RFLP maps of potato and their alignment with the homoeologous tomato genome. *Theor. Appl. Genet.* 83:49–57

55. Gerats AGM, Huits H, Vrijlandt E, Marana C, Souer E, Beld M. 1990. Molecular characterization of a nonautonomous transposable element (*dTph1*) of petunia. *Plant Cell* 2:1121–28

56. Gerlach WL, Peacock WJ. 1980. Chromosomal locations of highly repeated DNA sequences in wheat. *Heredity* 44: 269–76

57. Giraudat J, Hauge BM, Valon C, Smalle J, Parcy F, Goodman HM. 1992. Isolation of the Arabidopsis *ABI3* gene by positional cloning. *Plant Cell* 4:1251–61

58. Goldberg RB, Hoschek G, Kamalay JC, Timberlake WE. 1978. Sequence complexity of nuclear and polysomal RNA in leaves of the tobacco plant. *Cell* 14:123–31

59. Grellet F, Delcasso D, Panabieres F, Delseny M. 1986. Organization and evolution of a higher plant alphoid-like satellite DNA sequence. *J. Mol. Biol.* 187:495–507

60. Grill E, Somerville C. 1991. Construction and characterization of a yeast artificial chromosome library of Arabidopsis which is suitable for chromosome walking. *Mol. Gen. Genet.* 226:484–90

61. Gustafson JP, Butler E, McIntyre CL. 1990. Physical mapping of a low-copy DNA sequence in rye (*Secale cereale* L.). *Proc. Natl. Acad. Sci. USA* 87:1899–902

62. Gustafson JP, Dille JE. 1992. Chromosome location of *Oryza sativa* recombination linkage groups. *Proc. Natl. Acad. Sci. USA* 89:8646–50

63. Hagemann S, Scheer B, Schweizer D. 1993. Repetitive sequences in the genome of *Anemone blanda:* identification of tandem arrays and of dispersed repeats. *Chromosoma* 102:312–24

64. Hall G Jr, Allen GC, Loer DS, Thompson WF, Spiker S. 1991. Nuclear scaffolds and scaffold-attachment regions in higher plants. *Proc. Natl. Acad. Sci. USA* 88: 9320–24

65. Hauge BM, Hanley S, Giraudat J, Goodman HM. 1991. Mapping the Arabidopsis genome. In *Molecular Biology of Plant Development*, ed. GI Jenkins, W Schuch, pp. 45–56. Cambridge, UK: Company of Biologists

66. Helentjaris T, Weber D, Wright S. 1988. Identification of the genomic locations of duplicate nucleotide sequences in maize by analysis of restriction fragment length polymorphisms. *Genetics* 118:353–63

67. Heslop-Harrison JS, Bennett MD. 1990. Nuclear architecture in plants. *Trends Genet.* 6:401–5

68. Heslop-Harrison JS, Schwarzacher T. 1990. The ultrastructure of *Arabidopsis thaliana* chromosomes. *4th Int. Conf.* Arabidopsis *Res.* p. 3, (Abstr.)

69. Hillel J, Schaap T, Haberfeld A, Jeffreys AJ, Plotzky Y, et al. 1990. DNA fingerprints applied to gene introgression in breeding programs. *Genetics* 124:783–89

70. Höfte H, Desprez T, Amselem J, Chiapello H, Caboche M, et al. 1993. An inventory of 1152 expressed sequence tags obtained by partial sequencing of cDNAs from *Arabidopsis thaliana. Plant J.* 4: 1051–61

71. Hulbert SH, Richter TE, Axtell JD, Bennetzen JL. 1990. Genetic mapping and characterization of sorghum and related crops by means of maize DNA probes. *Proc. Natl. Acad. Sci. USA* 87:4251–55

72. Hutchinson J. 1983. In situ hybridisation mapping of plant chromosomes. In *Kew Chromosome Conference II*, ed. PE Brandham, MD Bennett, pp. 27–34. London: Allen & Unwin

73. Hwang I, Kohchi T, Hauge BM, Goodman H, Schmidt R, et al. 1991. Identification and map position of YAC clones compris-

ing one-third of the Arabidopsis genome. *Plant J.* 1:367–74

74. Ingham LD, Hanna WW, Baier JW, Hannah LC. 1993. Origin of the main class of repetitive DNA within selected *Pennisetum* species. *Mol. Gen. Genet.* 238:350–56

75. Jackson DA. 1991. Structure-function relationships in eukaryotic nuclei. *BioEssays* 13:1–10

76. Jamilena M, Rejon CR, Rejon MR. 1993. Repetitive DNA sequence families in *Crepis capillaris. Chromosoma* 102:272–78

77. Johnston M, Andrews S, Brinkman R, Cooper J, Ding H, et al. 1994. Complete nucleotide sequence of *Saccharomyces cerevisiae* chromosome VIII. *Science* 265:2077–82

78. Jones GH, Albini SM. 1988. Recombination nodules, chiasmata and crossing-over in the nucleolus organizing short arm of *Allium fistulosum. Heredity* 61:217–24

79. Kiper M, Bartels D, Herzfeld F, Richter G. 1979. The expression of a plant genome in hnRNA and mRNA. *Nucleic Acids Res.* 6:1961–78

80. Konieczny A, Ausubel FM. 1993. A procedure for mapping *Arabidopsis thaliana* mutations using co-dominant ecotype-specific PCR-based markers. *Plant J.* 4:403–10

81. Konieczny A, Voytas DF, Cummings MP, Ausubel FM. 1991. A superfamily of *Arabidopsis thaliana* retrotransposons. *Genetics* 127:801–9

82. Kowalski SP, Lan T-H, Feldmann KA, Paterson AH. 1994. Comparative mapping of *Arabidopsis thaliana* and *Brassica oleracea* chromosomes reveals islands of conserved organization. *Genetics* 138:1–12

83. Kurata N, Moore G, Nagamura Y, Foote T, Yano M, et al. 1994. Conservation of genome structure between rice and wheat. *Bio-Technology* 12:276–78

84. Lagercrantz U, Putterill J, Cavell A, Schmidt R, Parkin I, et al. 1994. The collinear genomes of *Brassica* and *Arabidopsis. ISHS Symp. Brassicas/9th Crucifer Genet. Workshop,* Abstr. #5

85. Lapitan NLV. 1992. Organization and evolution of higher plant nuclear genomes. *Genome* 5:171–81

86. Lapitan NLV, Ganal MW, Tanksley SD. 1991. Organization of the 5S ribosomal RNA genes in the genome of tomato. *Genome* 34:509–14

87. Leitch AR, Mosgöller W, Schwarzacher T, Bennett MD, Heslop-Harrison JS. 1990. Genomic in situ hybridization to sectioned nuclei shows chromosome domains in grass hybrids. *J. Cell Sci.* 95:335–41

88. Leitch AR, Schwarzacher T, Mosgöller W, Bennett MD, Heslop-Harrison JS. 1991. Parental genomes are seperated throughout the cell cycle in a plant hybrid. *Chromosoma* 101:206–13

89. Leitch IJ, Heslop-Harrison JS. 1993. Physical mapping of four sites of 5S rDNA sequences and one site of the α-amylase-2 gene in barley (*Hordeum vulgare*). *Genome* 36:517–23

90. Leutwiler LS, Hough-Evans BR, Meyerowitz EM. 1984. The DNA of *Arabidopsis thaliana. Mol. Gen. Genet.* 194:15–23

91. Lister C, Dean C. 1993. Recombinant inbred lines for mapping RFLP and phenotypic markers in *Arabidopsis thaliana. Plant J.* 4:745–50

92. Lukaszewski AJ, Curtis CA. 1993. Physical distribution of recombination in B-genome chromosomes of tetraploid wheat. *Theor. Appl. Genet.* 86:121–27

93. Lydiate D, Sharpe A, Lagercrantz U, Parkin I. 1993. Mapping the *Brassica* genome. *Outl. Agric.* 22:85–92

94. Maluszynska J, Heslop-Harrison JS. 1991. Localization of tandemly repeated DNA sequences in *Arabidopsis thaliana. Plant J.* 1:159–66

95. Martin GB, Brommonschenkel SH, Chunwongse J, Frary A, Ganal MW, et al. 1993. Map-based cloning of a protein kinase gene conferring disease resistance in tomato. *Science* 262:1432–36

96. Martin GB, Ganal MW, Tanksley SD. 1992. Construction of a yeast artificial chromosome library of tomato and identification of cloned segments linked to two disease resistance loci. *Mol. Gen. Genet.* 233:25–32

97. Martin GB, Williams JGK, Tanksley SD. 1991. Rapid identification of markers linked to a *Pseudomonas* resistance gene in tomato using random primers and near-isogenic lines. *Proc. Natl. Acad. Sci. USA* 88:2336–40

98. Martinez-Zapater JM, Estelle MA, Somerville CR. 1986. A highly repeated DNA sequence in *Arabidopsis thaliana. Mol. Gen. Genet.* 204:417-23

99. McClintock B. 1929. Chromosome morphology in *Zea mays. Science* 69:629

100. McClintock B. 1948. Mutable loci in maize. *Carnegie Inst. Wash. Yearb.* 47:155–69

101. McCouch SR, Kochert G, Yu ZH, Wang ZY, Khush GS, et al. 1988. Molecular mapping of rice chromosomes. *Theor. Appl. Genet.* 76:815–29

102. McGrath JM, Jancso MM, Pichersky E. 1993. Duplicate sequences with a similarity to expressed genes in the genome of *Arabidopsis thaliana. Theor. Appl. Genet.* 86:880–88

103. Deleted in proof

104. Melaragno JE, Mehrotra B, Coleman AW. 1993. Relationship between endopolyploidy and cell size in epidermal tissue of Arabidopsis. *Plant Cell* 5:1661–68

105. Menancio-Hautea D, Fatokun CA, Kumar L, Danesh D, Young ND. 1993. Comparative genome analysis of mungbean (*Vigna*

radiata L. Wilczek) and cowpea (*V. unguiculata* L. Walpers) using RFLP mapping data. *Theor. Appl. Genet.* 86:797–810

106. Michelmore RW, Paran I, Kessel RV. 1991. Identification of markers linked to disease resistance genes by bulked segregant analysis: a rapid method to detect markers in specific genomic regions using segregating populations. *Proc. Natl. Acad. Sci. USA* 88:9828–32

107. Moore G, Abbo S, Cheung W, Foote T, Gale M, et al. 1993. Key features of cereal genome organization as revealed by the use of cytosine methylation-sensitive restriction endonucleases. *Genomics* 15: 472–82

108. Moore G, Gale MD, Kurata N, Flavell RB. 1993. Molecular analysis of small grain cereal genomes: current status and prospects. *Bio-Technology* 11:584–89

109. Morgante M, Olivieri AM. 1993. PCR-amplified microsatellites as markers in plant genetics. *Plant J.* 3:175–82

110. Murray MG, Palmer JD, Cuellar RE, Thompson WF. 1978. DNA sequence organization in the pea genome. *Biochemistry* 18:5259–66

111. Nagamura Y, Yamamoto K, Harushima Y, Antonio BA, Sue N, et al. 1994. RFLP mapping of rice genome using cDNA clones. *4th Int. Congr. Plant Mol. Biol.*, Abstr. #27

112. Nagl W. 1976. Nuclear organization. *Annu. Rev. Plant Physiol.* 27:39–69

113. Nam H-G, Giraudat J, den Boer B, Moonan F, Loos WDB, et al. 1989. Restriction fragment length polymorphism linkage map of *Arabidopsis thaliana*. *Plant Cell* 1:699–705

114. Newman TC, Somerville C. 1993. Large scale sequencing of *Arabidopsis thaliana* var. Columbia cDNAs and generation of expressed sequence tags. *5th Int. Conf. Arabidopsis Res.* p. 39, (Abstr.)

115. Nowak R. 1994. Mining treasures from "junk DNA." *Science* 263:608–10

116. Oliver SG, van der Aart QJM, Agostoni-Carbone ML, Aigle M, Alberghina L, et al. 1992. The complete DNA sequence of yeast chromosome 111. *Nature* 357:38–46

117. Peleman J, Cottyn B, van Camp W, Van Montagu M, Inze D. 1991. Transient occurence of extrachromosomal DNA of an *Arabidopsis thaliana* transposon-like element, *Tat 1*. *Proc. Natl. Acad. Sci. USA* 88:3618–22

118. Prince JP, Pochard E, Tanksley SD. 1992. Construction of a molecular linkage map of pepper and a comparison of synteny with tomato. *Genome* 36:404–17

119. Pruitt RE, Meyerowitz EM. 1986. Characterisation of the genome of *Arabidopsis thaliana*. *J. Mol. Biol.* 187:169–83

120. Rawlins DJ, Highett MI, Shaw PJ. 1991. Localization of telomeres in plant interphase nuclei by in situ hybridization and 3D confocal microscopy. *Chromosoma* 100: 424–31

121. Rawlins DJ, Shaw PJ. 1988. Three dimensional organization of chromosomes of *Crepis capillaris* by optical tomography. *J. Cell Sci.* 91:401–14

122. Reiter RS, Williams JGK, Feldmann KA, Rafalski JA, Tingey SV, Scolnik PA. 1992. Global and local genome mapping in *Arabidopsis thaliana* by using recombinant inbred lines and random amplified polymorphic DNAs. *Proc. Natl. Acad. Sci. USA* 89:1477–81

123. Rhoades MM. 1984. The early years of maize genetics. *Annu. Rev. Genet.* 18:1–29

124. Richards EJ, Ausubel FM. 1988. Isolation of a higher eukaryotic telomere from *Arabidopsis thaliana*. *Cell* 53:127–36

125. Richards EJ, Goodman HM, Ausubel FM. 1991. The centromere region of *Arabidopsis thaliana* chromosome 1 contains telomere-similar sequences. *Nucleic Acids Res.* 19:3351–57

126. Sasaki T. 1994. An overview of rice genome research program: three year progress report. *4th Int. Congr. Plant Mol. Biol.*, Abstr. #2

127. Schmidt R, Dean C. 1992. Physical mapping of the *Arabidopsis thaliana* genome. In *Genome Analysis*. Vol. 4: *Strategies for Physical Mapping*, ed. KE Davies, SM Tilghman, pp. 71–98. New York: Cold Spring Harbor Lab.

128. Schmidt T, Heslop-Harrison JS. 1993. Variability and evolution of highly repeated DNA sequences in the genus *Beta*. *Genome* 36:1074–79

129. Schwarzacher T, Leitch AR, Bennett MD, Heslop-Harrison JS. 1989. In situ localization of parental genomes in a wide hybrid. *Ann. Bot.* 64:315–24

130. Schweizer D, Ambros P, Gründler P, Varga F. 1987. Attempts to relate cytological and molecular chromosome data of *Arabidopsis thaliana* to its genetic linkage map. *Arabidopsis Inf. Serv.* 25:27–34

131. Schweizer G, Borisjuk N, Borisjuk L, Stadler M, Stelzer T, et al. 1993. Molecular analysis of highly repeated genome fractions in *Solanum* and their use as markers for the characterization of species and cultivars. *Theor. Appl. Genet.* 85:801–8

132. Schweizer G, Ganal M, Ninnemann H, Hemleben V. 1988. Species-specific DNA sequences for identification of somatic hybrids between *Lycopersicon esculentum* and *Solanum acaule*. *Theor. Appl. Genet.* 75:679–84

133. Sears ER. 1954. The aneuploids of common wheat. *Mo. Agric. Exp. Stn. Res. Bull.* 572:1–59

134. Segal G, Sarfatti M, Schaffer MA, Ori N, Zamir D, Fluhr R. 1992. Correlation of

genetic and physical structure in the region surrounding the I$_2$ *Fusarium oxyspoprum* resistance locus in tomato. *Mol. Gen. Genet.* 231:179–85

135. Shizuya H, Birren B, Kim U-J, Mancino V, Slepak T, et al. 1992. Cloning and stable maintenance of 300-kilobase-pair fragments of human DNA in *Escherichia coli* using an F-factor based vector. *Proc. Natl. Acad. Sci. USA* 89:8794–97

136. Simoens CR, Gielen J, Van Montagu M, Inze D. 1988. Characterization of highly repetitive sequences of *Arabidopsis thaliana. Nucleic Acids Res.* 16:6753–66

137. Simpson PR, Newman M, Davies DR. 1988. Detection of legumin gene DNA sequences in pea by in situ hybridization. *Chromosoma* 96:454–58

138. Slatter RE, Dupree P, Gray JC. 1991. A scaffold-associated DNA region is located downstream of the pea plastocyanin gene. *Plant Cell* 3:1239–50

139. Smyth DR. 1991. Dispersed repeats in plant genomes. *Chromosoma* 100:355–59

140. Snape JW, Flavell RB, O'Dell M, Hughes WG, Payne PP. 1985. Intrachromosomal mapping of the nucleolus organiser region relative to three marker loci on chromosome 1B of wheat (*Triticum aestivum*). *Theor. Appl. Genet.* 69:263–70

141. Song J, Koga-Ban Y, Nagasaki H, Shinodzuka Y, Takiguchi T, et al. 1994. Analysis of rice cDNA. *4th Int. Congr. Plant Mol. Biol.*, Abstr. #28

142. Spiker S. 1985. Plant chromatin structure. *Annu. Rev. Plant Physiol.* 36:235–53

143. Sternberg N. 1990. Bacteriophage P1 cloning system for the isolation, amplification and recovery of DNA fragments as large as 100 kilobase pairs. *Proc. Natl. Acad. Sci. USA* 87:103–7

144. Sulston J, Du Z, Thomas K, Wilson R, Hillier L, et al. 1992. The *C. elegans* genome sequencing project: a beginning. *Nature* 356:37–41

145. Tanksley SD, Bernatzky R, Lapitan NL, Prince JP. 1988. Conservation of gene repertoire but not gene order in pepper and tomato. *Proc. Natl. Acad. Sci. USA* 85:6419–23

146. Tanksley SD, Ganal MW, Prince JP, de Vicente MC, Bonierbale MW, et al. 1992. High density molecular linkage maps of the tomato and potato genomes. *Genetics* 132:1141–60

147. Tsay Y-F, Frank MJ, Page T, Dean C, Crawford NM. 1993. Identification of a mobile endogenous transposon in *Arabidopsis thaliana. Science* 260:342–44

148. Tulsieram LK, Glaubitz JC, Kiss G, Carlson JE. 1992. Single tree genetic linkage mapping in conifers using haploid

DNA from megagametophytes. *Bio-Technology* 10:686–90

149. van Houten W, Umehara Y, Tanoue H, Momma T, Kurata N, et al. 1994. Selection of YAC clones for contig formation and gene isolation in rice. *4th Int. Congr. Plant Mol. Biol.*, Abstr. #29

150. Voytas DF, Ausubel FM. 1988. A copia-like transposable element family in *Arabidopsis thaliana. Nature* 336:242–44

151. Ward ER, Jen GC. 1990. Isolation of single-copy-sequence clones from a yeast artificial chromosome library of randomly-sheared *Arabidopsis thaliana* DNA. *Plant Mol. Biol.* 14:561–68

152. Weeden NF, Muehlbauer FJ, Ladizinsky G. 1992. Extensive conservation of linkage relationships between pea and lentil genetic maps. *Heredity* 83:123–29

153. Werner JE, Endo TR, Gill BS. 1992. Toward a cytogenetically based physical map of the wheat genome. *Proc. Natl. Acad. Sci. USA* 89:11307–11

154. Whitkus R, Doebley J, Lee M. 1992. Comparative genome mapping of *Sorghum* and maize. *Genetics* 132:1119–30

155. Williams JGK, Kubelik AR, Livak KJ, Rafalski JA, Tingey SV. 1990. DNA polymorphisms amplified by arbitrary primers are useful as genetic markers. *Nucleic Acids Res.* 18:6531–35

156. Wilson R, Ainscough R, Anderson K, Baynes C, Berks M, et al. 1994. 2.2Mb of contiguous nucleotide sequence from chromosome III of *C. elegans. Nature* 368:32–38

157. Wu T-C, Lichten M. 1994. Meiosis-induced double-strand break sites determined by yeast chromatin structure. *Science* 263:515–18

158. Xia X, Selvaraj G, Bertrand H. 1993. Structure and evolution of a highly repetitive DNA sequence from *Brassica napus. Plant Mol. Biol.* 21:213–24

159. Yakura K, Kato A, Tanfuji S. 1983. Structural organization of ribosomal DNA in four *Trillium* species and *Paris verticillata. Plant Cell Physiol.* 24:1231–40

160. Young ND, Zamir D, Ganal MW, Tanksley SD. 1988. Use of isogenic lines and simultaneous probing to identify DNA markers tightly linked to the *Tm-2a* gene in tomato. *Genetics* 120:579–85

161. Zabeau M, Kuiper M, Peleman J, van der Lee T, Reijans M, et al. 1994. Marker assisted breeding using AFLP^{TM}. *4th Int. Congr. Plant Mol. Biol.*, Abstr. #1844

162. Zamir D, Tanksley SD. 1988. Tomato genome is comprised largely of fast-evolving, low copy-number sequences. *Mol. Gen. Genet.* 213:254–61

Annu. Rev. Plant Physiol. Plant Mol. Biol. 1995. 46:419–44

HETEROLOGOUS EXPRESSION OF GENES IN BACTERIAL, FUNGAL, ANIMAL, AND PLANT CELLS

Wolf B. Frommer and Olaf Ninnemann

Institut für Genbiologische Forschung, D-14195 Berlin, Germany

KEY WORDS: expression cloning, functional expression, complementation, yeast, *Xenopus* oocyte

CONTENTS

419

ABSTRACT

Analysis of gene function is of central importance for the understanding of physiological processes. Expression of genes in heterologous organisms has allowed the isolation of many important genes (e.g. for nutrient uptake and transport) and has contributed a lot to the functional analysis of the gene products. For animal research, expression in *Xenopus* oocytes and cell cultures are predominant techniques, whereas in the plant domain, yeast has become the prevalent expression system. This review provides a survey of this quickly developing field and intends to assist researchers in determining appropriate experimental approaches for specific biological questions. Because heterologous expression technology is of special value for the analysis of proteins that are difficult to handle biochemically, the examples given concentrate on membrane proteins, i.e. transporter proteins. Also included is a detailed discussion of the functional expression methodology and its use in identifying and characterizing genes and proteins.

INTRODUCTION

In recent years plant biology has made rapid progress through the use of joint approaches from different disciplines such as physiology, biochemistry, genetics, and molecular biology. Significant progress has been made in the understanding of fundamental processes such as nutrition, metabolism, and transport. Whole plant studies are extremely powerful and have enabled a better understanding of the capacities of plants. However, these studies give only a crude picture of the underlying mechanisms because plants are so complex. Biochemical studies are often limited as well. Kinetic analysis often uncovers only the activity of the major stable components; the presence of multiple systems often makes it difficult to resolve complex kinetics. Thus, the analysis of single, isolated systems is desirable. In many cases, especially in transport physiology, methodology is limited. Genetics has always been instrumental for studying plant physiology, although the wealth of techniques available in bacterial or fungal genetics has not been routinely used because of the greater complexity of plants and because of technical limitations. The existence of plant mutants with distinct phenotypes does not always enable researchers to determine causes of the observed defects. Such determinations require new methods for isolating and characterizing individual proteins. Furthermore, the need for techniques to determine gene-product function has become more important because many genes with unknown functions are being identified through random and genome sequencing projects.

Heterologous expression of plant genes provides a new technique for determining gene-product function. This technique, which has been used successfully for the analysis of many mammalian genes, will be especially valuable for the analysis of plant functions for which no mutants are available and for which no screening scheme or phenotype is predictable. For example, until recently, most researchers thought that membrane permeable substances, such as water and ammonium, did not require transport proteins. Carriers for these substances were identified by heterologous expression cloning assays (115, 128). This review covers expression systems for both animal and plant genes. Emphasis is on functional expression and expression cloning techniques using yeast, *Xenopus* oocytes, and other animal cells. Because this approach has been most powerful in identifying genes that are otherwise difficult to define, such as integral membrane proteins, identification of transporters is covered in detail.

Ways to Isolate Genes of Interest

In the simplest case, PCR or heterologous screening techniques allow isolation of known genes from the organism of interest. The growing pool of information from random cDNA sequencing projects increases the probability of identifying genes by computer-based searches. Because it is not always possible to deduce gene function from the sequence, methods are still required to identify the role of the encoded proteins. Even highly related genes may have different functions as is the case with genes for plant nitrate-, amino acid-, and peptide transporters (59, 174, 182) or mammalian Na^+-coupled transport systems (134). If no related gene sequence is available, the classical biochemical approach for isolating the gene is to develop an in vitro assay for protein activity that enables purification. The purified protein can be used for immunoscreening of expression libraries or for N-terminal sequencing and subsequent isolation of the gene using oligonucleotides. The purification of some proteins can be problematic because of instability, low abundance, lack of appropriate assays, or difficulties in purification. Many of these problems occur during the purification of membrane proteins. The only biochemical detection system for transporters are transport assays, which demand the presence of two compartments. Functional reconstitution is therefore necessary, and this difficult technique has been successful in only a few cases. These difficulties may explain why little is known about transport proteins, especially in multicellular organisms.

Another tool for the isolation of transporters is the use of radiolabeled ligands or inhibitors that interact with the desired protein and hence facilitate protein purification. This technique allowed identification of the first transport protein from plants, the triose phosphate translocator (50). Methods that circumvent problems associated with biochemical identification procedures include, for example, isolation of membrane proteins by two-dimensional electrophoresis (56) from tissues that differ in their transport capacity or by immunological approaches with antisera directed against membrane vesicles (64).

Specific antibodies can be identified by their ability to inhibit transport processes and subsequently can be used to isolate the proteins from expression libraries (90).

Reverse genetics (i.e. the identification of genes starting from specific mutants) represents an alternative (163). Unfortunately, mutants or screening protocols are not available for all genes of interest. An efficient way to clone and simultaneously prove the function of a gene is functional expression in heterologous host cells. Complementation of mutants has been performed mainly in unicellular organisms. Essential prerequisites are the existence of suitable mutants and methods for selecting transformants. One advantage of expression cloning is the high probability of identifying full-length clones, thus allowing functional analysis.

How Similar Are Organisms?

Heterologous expression systems are based on the assumption that the basic principles of protein expression and function are similar in all organisms. The sequences of most eukaryotic proteins are well conserved (24). Eukaryotic organisms share many principles of cell compartmentation, intracellular transport, and regulation, such as vesicular trafficking along the secretory pathway (13, 17), cell-cycle control (131), signal transduction (8, 84), and chromatin structure (60). Nevertheless, important differences exist between fungal, plant, and animal cells in terms of the presence and composition of cell walls, and the presence of specialized organelles such as plastids and vacuoles. Regarding energization of secondary active transport processes at the plasma membrane, plants are more similar to yeast than to animal cells because plants and yeast use proton gradients, whereas animals use mainly sodium gradients. Multicellular organisms, however, have many properties for which no equivalent exists in unicellular organisms, such as intercellular communication across cell walls and through signals transferred in the vascular system in plants, or electrical and hormonal long-distance communication in animals.

EXPRESSION OF FOREIGN GENES IN BACTERIA

Classical molecular genetics has been developed for bacteria (especially *Escherichia coli*), and suitable expression vectors and hosts are available. The advantages of heterologous expression in *E. coli* include the availability of well-established molecular tools and defined mutants, as well as high growth rates and high yield of overproduced protein. As early as 1976 (177), an *E. coli* histidine auxotrophy was complemented by yeast DNA. Since then, several eukaryotic genes have been isolated and characterized by heterologous expression in bacteria (66, 76). In 1986, the first plant gene was identified by complementation of the *glnA E. coli* mutant (35). Since then, many other plant genes have been used to complement *E. coli* mutations (71, 137, 171). Disadvantages of the bacteria are the lack of organelles and cellular modification mechanisms responsible for the types of RNA and protein processing found in eukaryotes. Furthermore, eukaryotic polypeptides expressed in *E. coli* often denature and aggregate (112), and many membrane proteins, even from *E. coli* itself, are toxic when overexpressed in bacteria (160).

EXPRESSION OF FOREIGN GENES IN FUNGI

History of Yeast Complementation

In 1978, the development of yeast transformation provided a new way to isolate eukaryotic genes (77). Shortly thereafter, the development of shuttle vectors allowed complementation of the yeast *leu2* mutation with *E. coli* DNA (15). One year later, a *leu2/can1* mutant was utilized to isolate the yeast arginine permease gene (25). In 1981, the first heterologous complementation with eukaryotic DNA was performed with a genomic library from *Drosophila melanogaster* in the yeast *ade-8* mutant (75). The presence of intervening sequences in genomic DNA from higher eukaryotes turned out to be problematic (16, 118); therefore, cDNA libraries have become the predominant tool for complementation. The first gene of a higher eukaryote isolated from a cDNA library by complementation was the human homologue of a yeast gene controlling the cell cycle (106). Since then, functional expression has been used frequently to prove the function of genes or to isolate new genes (149). Functional expression in yeast also provides a source for preparative protein production for pharmaceuticals or for enzymatic synthesis of biochemicals.

How to Isolate a Gene by Complementation

Various approaches can be used to identify specific clones by transforming appropriate yeast mutants with plasmid-borne genomic or cDNA libraries (44, 151). The simplest strategy is the complementation of a recessive mutation, which includes the following steps: (*a*) Obtain or construct the appropriate strain that carries the mutation of interest and contains a selectable marker, normally an auxotrophy. (*b*) Establish a selection or screening system. (*c*) Clone the cDNA library into a shuttle vector carrying a gene that complements the auxotrophy (several libraries are already commercially available). (*d*) Transform the yeast mutant with the cDNA library. Several transformation protocols are available, but protocols that allow storage of competent cells are advantageous [transformation rates of approximately 10^4/g DNA (37)]. (*e*) Either select first for the presence of auxotrophic marker and then replica plate on selective media or directly perform a double selection. (*f*) Replate positive colonies, isolate plasmid DNA, and retransform to exclude artifacts resulting from reversion of the mutation in the yeast strain.

If appropriate mutants are not available, specific mutations can be introduced by gene disruption (152). Synthetic lethals can be used instead of mutants (118, 147, 149). Increased resistance to toxic compounds may provide an alternative selection procedure. The use of mutants is simple but requires a change in the selectable phenotype, such as differences in growth. Screening by functional assays can be an alternative if no selectable phenotype is available (33).

Numerous reports demonstrate the efficiency of functional expression of already identified genes in heterologous hosts, especially for analyzing mammalian genes involved in a broad range of cellular functions (Table 1). Heterologous expression in yeast has proved useful in the analysis of genes involved in human diseases (e.g. *p53*) (85), defects causing galactosemia (55), and enzyme testing with pharmacological agents (140). Comparable experiments using plant genes may provide a screen for potential herbicides based on their ability to interact specifically with a known protein. On the other hand, heterologous complementation has also been used successfully to isolate new genes, including those catalyzing specific metabolic and regulatory steps (Table 1).

Isolation of Enzyme-Encoding Plant Genes by Complementation

The first attempt in 1982 to complement yeast auxotrophic markers with a cosmid library from *Arabidopsis thaliana* was not successful (191). A decade later, two plant cDNAs were functionally expressed in yeast (3, 70) to confirm their assumed in vivo functions. One year later, Minet was able to show that several *A. thaliana* cDNAs could complement *Saccharomyces cerevisiae* auxotrophic marker mutations (123). Since then multiple reports have been published on cloning of genes that encode soluble proteins (Table 2).

Functional Expression of Membrane Proteins in Yeast

Since 1986, yeast cells have been used as functional expressions systems for membrane proteins of bacterial and animal origin (Table 1). For plant genes, yeast has become the preferred expression system (Table 2). Heterologous systems can be used if the respective function is lacking in the host as in the case of Na^+,K^+-ATPases in yeast (81) or active glucose transport in *Schizosaccharomyces pombe*. Glucose transporters were the first plant transport proteins to be functionally expressed in *S. pombe* (28, 156, 157). The proteins are targeted correctly to the plasma membrane. The *A. thaliana* H^+-ATPase AHA2 partially complemented the *S. cerevisiae* ATPase *pma1* mutation. The protein is functional, but a large proportion is trapped in the endoplasmic reticulum (ER). Removal of the C-terminal domain of AHA2 led to increased

Table 1   Expression of animal proteins in heterologous hosts

Functional expression in yeast	Reference
Vertebrate transcription factor JUN	176
Human cell cycle p34CDC2 and CDK2 kinases	183
Mammalian type IV phosphodiesterase	140
Chicken calmodulin	132
Human galactose 1 phosphate uridyltransferase	55
Rabbit α globin	15
Chicken actin ACT1	93
Mouse DNA topoisomerase—TOP2	2
Archaebacterial bacterio-opsin	105
Human multi-drug resistance protein MDR1	102
Mouse P-glycoprotein	154
Human estrogen receptor	120
Human mitochondrial anion channel	21
Mammalian Na^+/K^+ ATPase	81
Human Cytochrome P450	142
Rat M5 muscarinic acetylcholine receptor	83
Torpedo nicotinic acetylcholine receptor	61, 195
Nephrops 16-kDa ductin	80
Cloning by complementation in yeast	
Human purine biosynthetic proteins	123, 161
(ADE1, 2, 3, 8 homologues)	
Human dihydroorotate dehydrogenase	122
Human pyrroline 5 carboxylate reductase	38
Human glycogen branching enzyme	181
Rat cAMP phosphodiesterase	31
Human cell cycle control protein CDC2	106
Human DNA ligase I	11
Human CCAAT binding protein	12
Xenopus N-ras	172
Cloning by suppression of phenotype in yeast	
Human chaperonin-like protein	164
Functional expression of membrane proteins in oocytes	
E. coli glycerol transporter GLPF	116
Mammalian Na^+/glucose transporter SGLT1	72
Mammalian Na^+/myo-inositol transporter	104
Mammalian Na^+/nucleoside transporter	134
Mammalian glucose transporter GLUT1-7	167
Rabbit urea transporter UT2	196
Rabbit glutamate transporter	92

Table 1   (continued)

Rabbit oligopeptide transporter PEPT1	45
Mouse amino acid transporter	30, 97
Rat kidney $Na^+SO_4^2$ cotransporter	111
Rabbit sodium/phosphate cotransporter	190
Bovine cyclic nucleotide-gated Ca^{2+} channel	192
Rat kidney water channel	62
Rat gap junction proteins	40
Bovine Ca^{2+} sensing receptor	26
Mulitple neurotransmitter transporters	reviewed in 9
Multiple mammalian receptor proteins	reviewed in 166
Na^+/H^+ exchanger	187
Anion channels	reviewed in 89
Potassium channels	reviewed in 86
Expression of membrane proteins in insect cells	
Arabidopsis K^+ channels AKT1/KAT1	R Hedrich & H Sentenac, unpublished results
Functional expression of membrane proteins in COS cells	
Mammalian CD2/CD28	7
Plasmodium surface antigen	43
Mammalian interleukin receptors	67, 168, 194
Vacuolar adhesion molecule	133
Mouse receptor kinase	114

targeting to the plasma membrane and fully complemented *pma1*. The same yeast mutant was used to compare the biochemical properties of the three known major H^+-ATPase isoforms (135). Yeast expression was also used to demonstrate that the membrane-spanning subunit of the vacuolar pyrophosphatase is sufficient for proton translocation (96, 136). Organellar membrane proteins have also been functionally expressed in yeast. Although mitochondrial porins from potato expressed in yeast can be functionally reconstituted in lipid bilayers, only one of the isoforms is able to complement the respective yeast mutant (74; L Heins & UK Schmitz, unpublished data). In the case of plastids, yeast expression helped to resolve the debate over whether the major protein of the inner envelope of chloroplasts serves as the triose phosphate translocator (TPT) or as the import receptor for nuclear-encoded proteins. Expression, affinity purification from yeast, and subsequent reconstitution of the tagged protein in membrane vesicles demonstrated that TPT functions in triose phosphate transport in the same way that it functions in chloroplasts and that it is targeted to internal membranes (108).

Table 2   Expression of plant proteins in heterologous hosts

Functional expression in yeast	Reference
Wheat α-amylase	152
Tomato ethylene forming enzyme EFE	70
Rye SNF1 kinase homologue RKIN1	3
Tomato GTP-binding protein RAN	1
Arabidopsis GTP-binding protein RAB6	14
Arabidopsis protein phosphatase 1 (PP1A-AT)	46
Arabidopsis type I protein phosphatase homologue	129
Tobacco protein kinase homologue NPK	10
Alfalfa *cdc2* homologues	78
Maize mitochondrial T-URF 13	65
Maize TATA binding proteins (TBP)	184
Plant transcription factor TGA1	154
Maize transcriptional activator *Opaque-2*	117
Potato sucrose synthase	145
Functional expression of plant transport proteins in yeast	
Chlorella glucose transporter HUP1	156
Arabidopsis glucose transporter STP1	157
Spinach triose phosphate translocator	108
Potato mitochondria porins	74
Arabidopsis H$^+$-ATPases	135
Arabidopsis vacuolar pyrophosphatase subunit	96
Cloning of plant genes by complementation in yeast	
Arabidopsis chorismate mutase	41
Rape chloroplast 3-isopropylmalate dehydrogenase	42
Potato ATP-sulfurylase	99
Arabidopsis sterol synthesis cycloartenol synthase	33
Arabidopsis homologues of 8 auxotrophic mutants *URA1,2,4–10; ADE2, HIS3, LEU2, TRP1*	123
Arabidopsis homologue to Sec12 ER type II protein	36
Arabidopsis K$^+$-channel KAT1	4
Arabidopsis K$^+$-channel AKT1	165
Wheat K$^+$/H$^+$ symporter HKT1	159
Plant sucrose transporters SoSUT1, StSUT1	144, 145
Arabidopsis amino acid transporter AAP1–5, NAT2	49, 56, 82, 103
Arabidopsis peptide transporters AtPTR1–2, NTR1	58, 174
Arabidopsis ammonium transporter AMT1	128
Arabidopsis amino acid transporter AAT1	59

Table 2   (continued)

Cloning by suppression of phenotype in yeast	
Arabidopsis kinase	173
Arabidopsis kinase (amino acid transport)	M Kwart & WB Frommer, unpublished data
Potato kinase (phosphate transport)	G Leggewie, L Willmitzer & JW Riesmeier, unpublished data
Potato zinc finger protein (sucrose uptake)	Kühn & WB Frommer, unpublished data
Arabidopsis catalase (amino acid transport)	S Delrot, personal communication
Arabidopsis hydroxymethyl CoA synthase (amino acid transport)	S Delrot, personal communication
Functional expression of plant membrane proteins in oocytes	
Chlorella glucose transporter HUP1	6, 22
Arabidopsis hexose transporter STP1	22, 23
Arabidopsis vacuolar water channel γ-TIP	115
Arabidopsis maize and potato K$^+$-channels	27, 125, 158
Arabidopsis nitrate transporter CHL1	182
Spinach sucrose transporter SoSUT1	KJ Boorer & EM Wright, personal communication
Arabidopsis amino acid transporters AAP/NAT	KJ Boorer & EM Wright, personal communication
Arabidopsis plasma membrane water channels (PIP)	90
Expression cloning in mammalian cells	
Arabidopsis plasma membrane water channels (PIP)	90

Isolation of Transporter Genes

The major breakthrough in transport physiology was the isolation of carrier genes involved in the uptake and distribution of specific nutrients. Two different potassium channels and a high-affinity K$^+$/H$^+$ symporter were isolated by transformation of yeast potassium-uptake mutants (4, 159, 165). Similarly, a new family of amino acid permease (AAP) genes was isolated using a proline-uptake-deficient yeast strain (58, 103). The AAPs are high-affinity amino acid transporters with broad substrate specificity. Complementation of a yeast mutant deficient in histidine uptake and metabolism led to the identification of additional members of this family (49, 82, 103). A low-affinity amino acid transporter related to the mammalian counterpart and a protein (NTR1) related to the family of nitrate and peptide transporters were also identified (57, 59). To date, ten different amino acid transporters have been isolated by yeast expression cloning (WB Frommer, unpublished data). A plant peptide transporter, PTR2, which is related to NTR1, was able to complement a yeast mutant deficient in peptide uptake (102). Because mammalian multi-drug resistance proteins (MDRs; 174) can complement the *ste6* mutation in yeast, they were used to isolate plant homologues (L Covic & RR Lew, personal communication). Several genes affecting salt tolerance were identified in yeast. However, the use of these yeast sodium efflux mutants did not allow the identification of plant genes that encode efflux systems (63; A Rodriguez-Navarro, personal communication).

Extending the Scope of Complementation

Genetically well-characterized fungi such as *Neurospora crassa* and *Aspergillus nidulans* (138) provide a rich source of mutants. Sulfate transporters have been cloned from *N. crassa* by homologous complementation (95). These *N. crassa* genes were then used to isolate respective genes from *S. cerevisiae* and to construct a yeast deletion mutant. Selection of this mutant on low sulfate allowed the isolation of plant genes by complementation (169, 170; FW Smith, PM Ealing, M Hawkesford & DT Clarkson, personal communication). Heterologous complementation of a yeast ammonium-uptake mutant has led to the isolation of high-affinity ammonium transporter genes from plants (39, 128).

An extension of the method is to modify metabolic pathways in yeast for constructing strains suitable for complementation. Despite intense biochemical attempts, the major sugar carrier of plants were refractory to analysis. Yeast complementation seemed impossible because *S. cerevisiae* secretes an invertase that hydrolyzes the sucrose extracellularly and takes up the products by efficient hexose transporters. To circumvent this problem, an invertase-deficient mutant was constructed that functionally expressed a sucrose synthase that enabled the mutant to metabolize sucrose intracellularly. This strain was used in a manner analogous to classical mutant systems for complementation cloning of a sucrose transporter gene (144, 145). Analysis of yeast cells expressing the transporter indicated a proton symport mechanism and a suggested role in phloem loading for the cloned sucrose transporter. Direct evidence for the function of this protein comes from transgenic plants in which expression of the sucrose transporter was partially inhibited by antisense RNA (146).

Applications for Structure-Function Studies

After new genes are cloned by complementation, the biochemical properties of the proteins can be studied directly. The expression assay also allows analysis of structure-function relationships. A classical method in bacterial and fungal genetics is to select mutations from random mutagenized DNA rather than creating alterations by site-directed mutagenesis. Functional expression of heterologous proteins makes this technique applicable to many genes. For example, the use of a toxic glucose analogue as a selective agent allowed the identification of membrane-spanning domains of the glucose transporter that are relevant for substrate affinity and that may be involved in substrate recognition (193).

Studies on membrane transport processes in yeast and animal cells indicate that these processes are highly regulated (8, 68). Regulation acts at different levels, from targeting to the membrane to modification of protein activity and degradation of the gene products (100, 107). Heterologous complementation may allow isolation of regulatory plant genes in the relevant yeast mutants.

Heterologous Promoters in Yeast

Functional complementation with genomic libraries from animals has shown that heterologous promoters can function in yeast. Some plant and plant-virus promoters also function in yeast, and a high degree of conservation has been found for gene regulatory mechanisms (155). For example, the transcriptional regulator for maize zein proteins encoded by *Opaque-2* can substitute for the nitrogen starvation gene *GCN4* in yeast (117). Additional experiments are necessary to study these similarities in more detail.

Requirements for Heterologous Complementation of Mutations in Yeast

QUALITY OF THE MUTANT   For many yeast mutants used for complementation, the molecular basis of the mutation is not clear. The reversion frequency of point mutations can lead to large numbers of false positives in the selection. To circumvent these complications, gene disruption should be used to construct stable mutants in the gene of interest (152).

ROLE OF TARGETING AND STRUCTURE FOR FUNCTIONAL EXPRESSION   The principles of targeting seem to be conserved between organisms (17). The examples given above may indicate that many heterologously expressed proteins are targeted correctly in yeast. Although it seems probable that the plant proteins identified by complementation of yeast plasma-membrane transport mutations are targeted to the correct compartment, direct evidence for localization in the plant plasma membrane is required. Yeast targeting sequences can be used to improve heterologous expression (125). Only some of the plant proteins that complement the yeast functions are directly homologous to their yeast counterparts. Ammonium and peptide transporters as well as the amino acid transporter AAT1 from plants are related to their yeast counterparts. In contrast, no homologues of the plant AAP-amino acid permeases and the potassium transporters have been found in yeast. Thus sequence homology does not seem to be crucial for the targeting. Interestingly, if the heterologous proteins lack sequences required for assembly or targeting to the correct compartments in the cell, mistargeting can occur (135). For example, a plasma membrane gap junction proteolipid from arthropods functionally complements a yeast mutation in a vacuolar proteolipid gene because of incorrect targeting (80). Evidence has been presented that, in yeast, the tonoplast serves as a default compartment when targeting fails (32, 148). If true, then plant proteins localized to the plasma membrane of yeast are likely to be localized on the same membrane in plants.

In standard expression experiments, only a single cDNA is expressed. This does not allow the identification of genes encoding polypeptides that are subunits of multimeric proteins or that are dependent on specific cofactors. The correct folding of monomers and the subsequent assembly into oligomers often represents a prerequisite for protein transport from the ER (135). Misfolded or unassembled proteins tend to accumulate in the ER or are rapidly degraded. When factors necessary for the proper assembly of heterologous proteins are lacking or different in yeast, incorrect targeting can occur. The four subunits of the pentameric nicotinic acetylcholine receptor ($\alpha_2\beta\gamma\delta$) were coexpressed in yeast (87). The hydrophobic polypeptides entered the secretory pathway, where they were processed and glycosylated. However, in contrast to oocyte expression, no functional receptor was detected in yeast, possibly because of improper folding or assembly.

Because of our detailed knowledge of targeting processes and the large set of mutants in the protein secretion pathway, yeast complementation might provide a tool to isolate plant homologues. However, despite multiple trials, few cases of successful complementation of these processes have been reported (105a). Indications that the interaction domains in the multiprotein complexes are not sufficiently conserved between yeast and plants come from experiments in which a vacuolar targeting mutant could not be rescued by plant homologues. When the nonconserved C-terminal domain of the plant protein was replaced by the yeast counterpart, the chimeric protein was able to rescue the mutant (189). Alternative approaches have been developed to identify proteins that interact with a protein for which the gene has already been cloned. This two-hybrid system has been used to identify a number of genes such as DNA-binding proteins and kinases (47).

TOXICITY OF THE GENE PRODUCT FOR THE HOST CELL   Because the initial step for complementation of yeast mutations is the construction of a cDNA library in *E. coli*, the potential toxicity of proteins to *E. coli* can represent a problem. Toxic effects range from mildly deleterious to completely lethal (151). Expression of integral membrane protein genes can be toxic for *E. coli* (160). The toxicity may lead to elimination of certain genes during amplification of the cDNA library. The use of the shuttle vector pFL61 (123) provides an alternative because the vector prevents background expression from the cloned cDNAs in *E. coli*. For further subcloning, *E. coli* vectors with low plasmid copy number are recommended (O Ninnemann & WB Frommer, unpublished data).

YEAST AS A SOURCE FOR PURE PROTEIN   Affinity purification of functionally expressed tagged proteins is a simple method in principle (108, 175). Yet, one of the major disadvantages of yeast expression is the comparatively low yield. Frequently this results from low transcription of foreign genes (149).

SUPPRESSION OF THE MUTANT PHENOTYPE   Another complication can be genetic suppression, i.e. the phenomenon in which a gene functionally different from the mutation to be complemented can mimic functional complementation. Extracellular hydrolysis of substrates that cannot be taken up by the mutant strain, and subsequent uptake of the products by different carriers, can mimic functional complementation of a transport mutant. Chaperonin-like proteins from humans were found to complement a yeast histidine transport mutation (164). The chaperonin relieved the nitrogen repression of the yeast endogenous general amino acid permease, GAP, probably by stabilizing a GAP-activator protein. Complex selection schemes may also be sources for artifacts. Mutations in a glucose transporter gene can suppress potassium-uptake deficient mutations in yeast (5). Several other examples exist of cloned suppressor genes that complement different transport mutations (Table 2). Complementation of the putative branched-chain amino acid transport mutant, BAP1, is based on the presence of the toxic compound sulfometurone (39a). Attempts to complement the mutant with a plant cDNA library did not allow identification of amino acid transporters; instead, detoxifying enzymes such as catalase and hydroxymethyl CoA synthase were identified (S Delrot, unpublished data). Finally, we and others have observed artifactual complementation by complex plasmid rearrangements or chimeric cDNAs (124).

EXPRESSION OF FOREIGN GENES IN ANIMAL CELLS

Transient Expression in Xenopus laevis Oocytes

The X. laevis oocyte has become a major heterologous expression system because of its large size (34, 53, 69, 153, 160). Both soluble enzymes and integral membrane proteins have been expressed or cloned in oocytes. Some examples of transport proteins are shown in Table 2. Oocytes are ideal for studying transport processes because they are amenable to tracer-uptake studies and electrophysiological analysis. Similar to yeast, oocytes are frequently used to functionally characterize genes isolated by other means. The oocyte can also be used to identify unknown genes if suitable screening systems, such as uptake of radioactive tracers, are available. Alternatively, changes in ion flow can be determined by two-electrode voltage-clamp studies, provided the substrate is charged, cotransports a charged ion, or induces endogenous currents upon uptake. This technique is efficient, because transport can be measured with mixtures of substrates or substrates that are difficult to label. A major disadvantage of transport studies in yeast cells is the problem of maintaining the membrane potential during uptake experiments. In oocytes, tracer and clamp experiments can be combined, thus allowing the maintenance of membrane potential even if substrate uptake leads to depolarization of the membrane.

Oocyte expression has been used widely for animal gene research; however, in recent years it has been extended to the plant realm, demonstrating the general feasibility of this approach for cloning (162). The general approach of cloning by oocyte expression includes the following steps: (a) Capped mRNA is synthesized in vitro from a cDNA library. (b) RNA is injected into oocytes. (c) Tracer or voltage-clamp studies are used to detect the activity of interest. (d) RNA or cDNA library is fractionated stepwise, pools are injected separately, and this step is repeated until a single cDNA is identified.

Oocytes had been used successfully both for expression of cloned genes (69) and to isolate genes with novel functions (130). The Na^+/glucose symporter was isolated using ^{14}C tracer studies (72). Many channels and receptors were identified subsequently by electrophysiological approaches (54, 79, 179). G-protein-coupled receptors have been detected indirectly as a result of increasing intracellular calcium levels that activated calcium-dependent chloride currents (53, 113). Using a screening procedure with radiolabeled sulfate and phosphate, investigators were able to isolate genes for these carriers from kidney (111, 190). The same approach allowed the isolation of mammalian peptide transporters (45). The identification of other proteins, such as neurotransmitter transporters, has been reviewed elsewhere (9).

The cellular function of a mammalian membrane protein acting as a retrovirus receptor was demonstrated by an elegant electrophysiological study of oocytes expressing the protein. The addition and fractionation of a complex mixture of substrates provided evidence that the protein is an amino acid transporter (97). Further analysis allowed the study of glycosylation, structure-function relationship, and the facilitated diffusion mechanism for transport (30, 98). Despite the efficient use of this system for other organisms, no plant protein has been isolated by expression cloning in oocytes.

Characterization of Plant Transporters in Oocytes

In the case of plant transporters, oocytes have been used mainly as a tool to characterize electrophysiological properties. Such measurements are possible in yeast but do not yet represent a standard technique (18). Based on the finding that mammalian sodium-dependent glucose transporters could be characterized electrophysiologically in oocytes, the plant H^+/glucose transporter STP1 was expressed and studied in oocytes (6, 22, 23). The *A. thaliana KAT1* gene that was isolated by yeast complementation was shown to encode an inwardly rectifying potassium channel (73, 158, 165). The same holds true for a *KAT1* homologue predominantly expressed in guard cells that was identified by heterologous screening of a potato epidermal cDNA library (125). A detailed analysis in oocytes demonstrated that the potassium transporter HKT1 functions as a high-affinity K^+/H^+ symporter (159). Expression in oocytes showed that the *A. thaliana* Chl1 protein, which is responsible for chlorate resistance, is involved in low-affinity nitrate transport (182). The stoichiometry for proton cotransport of the *A. thaliana* sucrose transporter and AAPs has also been analyzed (KJ Boorer & EM Wright, unpublished data). The use of oocytes has allowed researchers to determine the function of integral membrane proteins of the tonoplast. Oocytes showed an increased water permeability when expressing the plant vacuolar γ-TIP, demonstrating a function in water transport (115). The same could be demonstrated for glucose transporters and for TIP-related plasma membrane proteins from plants and animals (48, 90, 197).

In yeast, substrate specificity can be analyzed easily by simple growth assays, by inhibition studies, or by radioactive-uptake measurements. Because radiolabeling of compounds can be difficult, substrate specificity is often determined by competition studies. The results are of limited value because substrates cannot be distinguished from inhibitors. Direct determination of uptake in oocytes by electrophysiological assays normally does not allow differentiation between inhibitors and nonsubstrates. This differentiation can be achieved either by competition assays or by studying pre-steady-state currents (109, 110). Such an analysis is currently used to determine substrate specificity and, in combination with site-specific mutagenesis, to elucidate structure-function relationships of sugar transporters from animals and plants (EM Wright, personal communication).

A limitation of oocytes is that measurements are performed with living cells in which regulatory and metabolic events can affect the function of the heterologous protein. Internal perfusion of the oocyte during transport studies circumvents this problem by decoupling the membrane from metabolism, leaving basically an in vitro system (20, 178). Analogous to the finding in yeast that heterologous expression of soluble proteins can mimic transporter expression, amino acid transport activities in oocytes can be induced by soluble proteins that may activate endogenous transporters (19, 180, 188). These proteins may represent regulatory subunits (88).

One of the advantages of oocyte expression is that proteins with multiple subunits can be identified by simultaneous expression of whole or fractionated cDNA libraries. After cloning and expression of one subunit, subsequent rounds of screening can be used to identify other subunits by looking for modulated activity (88). Coexpression in oocytes can also be used to study regulation, e.g. the activation of cloned K^+ channels by G-proteins (143). Such approaches will help researchers to characterize regulatory networks and to obtain a better understanding of signal transduction pathways.

The handling of oocytes is certainly more complicated than yeast cells. Efficient expression requires permanent access to high-quality oocytes. Furthermore, oocytes are not suitable for all purposes, e.g. they do not contain plastids or vacuoles. The finding that plant vacuolar proteins are functional at the oocyte plasma membrane indicates that the plasma membrane serves as the default compartment in these cells, thus extending the scope of the method (32, 148).

Expression in Insect Cell Cultures

Baculoviruses are valuable for producing preparative quantities of foreign proteins. Besides the high yield of protein (up to 150 mg/L for membrane proteins) (91), the proteins are posttranslationally modified. As in the other expression systems, several proteins may be expressed simultaneously (52). The utilization of baculovirus as a vehicle for protein expression for both soluble and integral membrane proteins has been demonstrated repeatedly, mainly for mammalian and viral genes (52). This method has allowed functional expression and electrophysiological characterization of a human K^+ channel (91). In addition, several soluble proteins of plant origin have been expressed: storage proteins, proteases (91), histidinol dehydrogenase (126), and mitochondrial proteins (101). The A. thaliana KAT1 potassium channel showed similar properties when expressed in insect cells and in oocytes (64, 73, 158; F Gaymard & H Sentenac, unpublished data; I Marten & R Hedrich, unpublished data). Functional expression of another A. thaliana K^+ channel, AKT1, in oocytes has been unsuccessful so far, but high quantities of pure protein have been isolated by expression in insect cells (H Sentenac, personal communication).

Expression in Mammalian Cells

Expression and cloning of genes in mammalian cell cultures has contributed to our understanding of signaling processes. Plasmids containing the SV40 origin of replication can be used for transient expression of heterologous genes in mammalian cell lines (7). Expression can be detected by immunoscreening, by using radiolabeled ligands, by phenotypic screening, or by complementation of mutants (133). A modification of the detection system allows the identification of intracellular proteins by immunostaining of permeabilized fixed cells (119). This expression system has led to the identification of many receptors and cell surface proteins (Table 1). COS and other cell lines have been used successfully (29, 94). Mammalian cell lines can also be used for expressing plant proteins. Expression of a cDNA library from A. thaliana roots in cell cultures and subsequent screening with antisera directed against the plasma membrane enabled the identification of plasma membrane water channels (90). In contrast to research in mammals, very few plant receptors have been identified at the molecular level (84). Expression in mammalian cells appears to represent an excellent tool to isolate new receptor genes and should help to overcome this lag. Furthermore, ectopic expression of foreign genes can be used to study defined physiological questions (130, 139).

EXPRESSION OF GENES IN PLANT CELLS

Heterologous expression in yeast and animal cells is well established. Surprisingly, plants or plant cells have not been used for this purpose, even though the technology is available. Well suited for this purpose are unicellular organisms such as green algae, for which transformation protocols have been developed. A number of mutants, e.g. those involved in nitrogen uptake and reduction, have been described that could be used for complementation (51, 141). Protoplasts or cell cultures from higher plants can be used as transient expression systems with transformation rates of up to 60% (G Neuhaus, personal communication). Even stable transformation rates for protoplasts are high enough for some species for complementation approaches (186). Several cell lines are available from different plants [e.g. tobacco BY-2 cells, which can be synchronized (127)]. Expression in plant cells would be particularly advantageous for resolving plant-specific problems, such as the targeting of vacuolar or plastidic proteins. Correct targeting of transiently expressed genes to chloroplasts was demonstrated by immunofluorescence (121). The feasibility of such approaches has been shown by expression of genes in higher plants with genetic backgrounds that lack a certain function, e.g. the demonstration that patatin encodes an esterase activity (150). Finally, ectopic expression of genes from heterologous species has proved to be an efficient tool for studying plant physiology (185).

CONCLUSIONS

Heterologous expression systems are powerful tools for isolating new genes and for characterizing proteins from all organisms. The major expression system for plant genes is yeast, which has allowed the isolation of more than 20 transporter genes. These genes represent only a small fraction of the transporters present in plants. Heterologous expression of plant genes in systems other than yeast will be important in the future. Combinations of the different expression systems—i.e. oocytes for electrophysiological characterization, yeast for the selection of mutant proteins with altered functions, and cell cultures for the production of protein for crystallization—will allow a better understanding of how plants and animals function.

ACKNOWLEDGMENTS

We are very much indebted to all those who have provided us with unpublished data. We want to thank especially Michèle Minet and Julian Schroeder for helpful discussions. In this context we would like to thank Jörg Riesmeier whose work was crucial for establishing yeast complementation in our group. We also would like to thank Doris Rentsch, Nicholas Provart, Frank Lauter, and Remi Lemoine for critical reading of the manuscript.

Any *Annual Review* chapter, as well as any article cited in an *Annual Review* chapter, may be purchased from the Annual Reviews Preprints and Reprints service. 1-800-347-8007; 415-259-5017; email: arpr@class.org

---

Literature Cited

1. Ach RA, Gruissem W. 1994. A small nuclear GTP-binding protein from tomato suppresses a *S. pombe* cell-cycle mutant. *Proc. Natl. Acad. Sci. USA* 91:5863–67
2. Adachi N, Miyaike M, Ikeda H, Kikuchi A. 1992. Characterization of cDNA encoding the mouse DNA topoisomerase II that can complement the budding yeast *top2* mutation. *Nucleic Acids Res.* 20: 5297–303
3. Alderson A, Sabelli PA, Dickinson JR, Cole D, Richardson M. 1991. Complementation of *snf1*, a mutation affecting global regulation of carbon metabolism in yeast, by a plant protein kinase cDNA. *Proc. Natl. Acad. Sci. USA* 88:8602–5
4. Anderson JA, Huprikar SS, Kochian LV, Lucas WJ, Gaber RF. 1992. Functional expression of a probable *A. thaliana* potassium channel in *S. cerevisiae. Proc. Natl. Acad. Sci. USA* 89:3736–40
5. Anderson JA, Nakamura RL, Gaber RF. 1994. Heterologous expression of K⁺-channels in *S. cerevisae*, strategies for molecular analysis of structure and function. In *Society of Exp. Biology: SEB SYMPOSIUM 48. Membrane Transport in Plants and Fungi.* In press
6. Aoshima H, Yamata M, Sauer N, Komor E, Schobert C. 1993. Heterologous expression of the H⁺ hexose cotransporter from *Chlorella* in *Xenopus* oocytes and its characterization with respect to sugar specificity, pH and membrane potential. *J. Plant Physiol.* 141:293–97
7. Aruffo A, Seed B. 1987. Molecular cloning of a CD28 cDNA by a high efficiency COS cell expression system. *Proc. Natl. Acad. Sci. USA* 84:8573–77
8. Assmann S. 1993. Signal transduction in guard cells. *Annu. Rev. Cell Biol.* 9:345–75
9. Attwell D, Bouvier M. 1992. Neurotransmitter transporters: cloners quick on the uptake. *Curr. Biol.* 2:541–43
10. Banno H, Hirano K, Nakamura T, Irie K, Nomoto S, et al. 1993. NPK1, a tobacco gene that encodes a protein with a domain homologous to yeast BCK1, STE11 and Byr2 protein kinases. *Mol. Cell. Biol.* 13: 4745–52
11. Barnes DE, Johnston LH, Kodama K, Tomkinson AE, Lasko DD, Lindahl T. 1990. Human DNA ligase I cDNA: cloning and functional expression in *S. cerevisiae. Proc. Natl. Acad. Sci. USA* 87:6679–83
12. Becker DM, Fikes JD, Guarente L. 1991. A cDNA encoding a human CCAAT-binding protein cloned by functional complementation in yeast. *Proc. Natl. Acad. Sci. USA* 8:1968–72
13. Bednarek SY, Raikhel NV. 1992. Intracellular trafficking of secretory proteins. *Plant Mol. Biol.* 20:133–50
14. Bednarek SY, Reynolds TL, Schroeder M, Grabowski R, Hengst L, et al. 1994. A small GTP-binding protein from *A. thaliana* functionally complements the yeast *YPT6* null mutant. *Plant Physiol.* 104:591–96
15. Beggs JD. 1978. Transformation of yeast by a replicating hybrid plasmid. *Nature* 275:105–8
16. Beggs JD, van den Berg J, van Ooyen A, Weissmann C. 1980. Abnormal expression of chromosomal rabbit β-globin gene in *S. cerevisiae. Nature* 283:835–40
17. Bennett MK, Scheller RH. 1993. The molecular machinery for secretion is conserved from yeast to neurons. *Proc. Natl. Acad. Sci. USA* 90:2559–63
18. Bertl A, Slayman CL, Gradmann D. 1993. Gating and conductance in an outward-rectifying K⁺ channel from the plasma mem-

brane of *S. cerevisiae. J. Membr. Biol.* 132: 183–99

19. Bertran J, Werner A, Moore ML, Stange G, Markovich D, et al. 1992. Expression cloning of a cDNA from rabbit kidney cortex that induces a single transport system for cystine and dibasic and neutral amino acids. *Proc. Natl. Acad. Sci. USA* 89:5601–5

20. Bezanilla F, Perozo E, Stefani E. 1994. Gating of shaker K^+ channels. *Biophys. J.* 66:996–1021

21. Blachly-Dyson E, Zambronicz EB, Yu WH, Adams V, McCabe ERB, et al. 1993. Cloning and functional expression in yeast of two human isoforms of the outer mitochondrial membrane channel, the voltage-dependent anion channel. *J. Biol. Chem.* 268(3):1835–41

22. Boorer KJ, Forde BG, Leigh RA, Miller AJ. 1992. Functional expression of a plant plasma membrane transporter in *Xenopus* oocytes. *FEBS Lett.* 302:166–68

23. Boorer KJ, Loo DDF, Wright EM. 1994. Steady-state and presteady-state kinetics of the H^+/hexose cotransporter (STP1) from *Arabidopsis thaliana* expressed in *Xenopus* oocytes. *J. Biol. Chem.* 269:20417–24

24. Botstein D, Fink GR. 1988. Yeast: an experimental organism for modern biology. *Science* 240:1439–43

25. Broach JR, Strathern JN, Hicks JB. 1979. Transformation in yeast: development of a hybrid cloning vector and isolation of the *CAN1* gene. *Gene* 8:121–33

26. Brown EM, Gamba G, Riccardi D, Lombardi M, Butters R, et al. 1993. Cloning and characterization of an extracellular Ca^{2+}-sensing receptor from bovine parathyroid. *Nature* 366:575–80

27. Cao YW, Anderova M, Crawford NM, Schroeder JI. 1992. Expression of an outward-rectifying potassium channel from maize mRNA and complementary RNA in *Xenopus* oocytes. *Plant Cell* 4:961–69

28. Caspari T, Stadler R, Sauer N, Tanner W. 1994. Structure/function relationship of the *Chlorella* glucose/H^+ symporter. *J. Biol. Chem.* 269:3498–502

29. Claudio T. 1992. Stable expression of heterologous multisubunit protein complexes established by calcium phosphate- or lipid-mediated cotransfection. *Methods Enzymol.* 207:391–408

30. Closs EI, Lyons CR, Kelly C, Cunningham JM. 1993. Characterization of the third member of the MCAT family of cationic amino acid transporters. *J. Biol. Chem.* 268:20796–800

31. Colicelli J, Birchmeier C, Michaeli T, O'Neill K, Riggs M, Wigler M. 1989. Isolation and characterization of a mammalian gene encoding a high affinity cAMP phosphodiesterase. *Proc. Natl. Acad. Sci. USA* 86:3599–603

32. Cooper A, Bussey H. 1992. Yeast Kex1p is a golgi-associated membrane protein. *J. Cell Biol.* 119:1459–68

33. Corey EJ, Matsuda SP, Bartel B. 1993. Isolation of an *A. thaliana* gene encoding cycloartenol synthase by functional expression in a yeast mutant lacking lanosterol synthase by the use of a chromatographic screen. *Proc. Natl. Acad. Sci. USA* 90: 11628–32

34. Dascal N. 1987. The use of *Xenopus* oocytes for the study of ion channels. *CRC Crit. Rev. Biochem.* 22:317–87

35. DasSarma S, Tischer E, Goodman HM. 1986. Plant glutamine synthetase complements a *glnA* mutation in *E. coli. Science* 232:1242–44

36. D'Enfert C, Gensse M, Gaillardin C. 1992. Fission yeast and a plant have functional homologues of the Sar1 and Sec12 proteins involved in ER to Golgi traffic in budding yeast. *EMBO J.* 11:4205–11

37. Dohmen RJ, Strasser AWM, Höner CB, Hollenberg CP. 1991. An efficient transformation procedure enabling long-term storage of competent cells of various yeast genera. *Yeast* 7:691–92

38. Dougherty KM, Brandriss MC, Valle D. 1992. Cloning human pyrroline-5-carboxylate reductase cDNA by complementation in *S. cerevisiae. J. Biol. Chem.* 267 (2):871–75

39. Dubois E, Grenson M. 1979. Methylamine/ ammonia uptake systems in *S. cerevisiae*: multiplicity and regulation. *Mol. Gen. Genet.* 175:67–76

39a. Dullin S, Germansen C, Kielland-Brandt MC. 1991. High affinity uptake system for branched chain amino acids in *Saccharomyces cerevisiae. Yeast* 7:933–41

40. Dunham B, Liu S, Taffet S, Trabka-Janik E, Delmar M, et al. 1992. Immunolocalization and expression of functional and nonfunctional cell-to-cell channels from wild type and mutant rat heart connexin 43 cDNA. *Circ. Res.* 70:1233–43

41. Eberhard J, Raesecke H, Schmid J, Amrhein N. 1993. Cloning and expression in yeast of a higher plant chorismate mutase. *FEBS Lett.* 334:233–36

42. Ellerström M, Josefsson L, Rask L, Ronne H. 1992. Cloning of a cDNA for rape chloroplast 3-isopropylmalate dehydrogenase by genetic complementation in yeast. *Plant Mol. Biol.* 18:557–66

43. Elliott JF, Albrecht GR, Gilladoga A, Handunnetti SM, Neequaye J, et al. 1990. Genes for *P. falsiparum* surface antigens cloned by expression in COS cells. *Proc. Natl. Acad. Sci. USA* 87:6363–67

44. Emr SD. 1990. Heterologous gene expression in yeast. *Methods Enzymol.* 185:231–34

45. Fei YJ, Kanai Y, Nussberger S, Ganapathy

V, Leibach FH, et al. 1994. Expression cloning of a mammalian proton-coupled oligopeptide transporter. *Nature* 368:563–66

46. Feirreira PC, Hemerly HS, Van Montagu M, Inze D. 1993. A protein phosphatase I from *A. thaliana* restores temperature sensitivity of a *S. pombe* cdc25ts/wee1 double mutant. *Plant J.* 4:81–87

47. Fields S, Sternglanz R. 1994. The two-hybrid system: an assay for protein-protein interactions. *Trends Genet.* 10:286–92

48. Fischbarg J, Kuang K, Vera JC, Arant S, Silverstein SC, et al. 1990. Glucose transporters serve as water channels. *Proc. Natl. Acad. Sci. USA* 87:3244–47

49. Fischer WN, Kwart M, Hummel S, Frommer WB. 1994. Organ-specific expression of three general amino acid permease genes from *Arabidopsis*. *J. Biol. Chem.* Submitted

50. Flügge UI, Fischer K, Gross A, Sebald W, Lottspeich F, Eckerskorn C. 1989. The triose phosphate-3-phosphoglycerate-phosphate translocator from spinach chloroplasts. *EMBO J.* 8:39–46

51. Franco AR, Cárdenas J, Fernández E. 1987. A mutant of *Chlamydomonas* altered in transport of ammonium and methylammonium. *Mol. Gen. Genet.* 206:414–18

52. Fraser MJ. 1992. The baculovirus-infected insect cell as a eukaryotic gene expression system. *Curr. Top. Microbiol. Immunol.* 158:131–72

53. Frech GC, Joho RH. 1992. Isolation of ion channel genes by expression cloning in *Xenopus* oocytes. *Methods Enzymol.* 207:592–604

54. Frech GC, VanDongen AMJ, Schuster G, Brown AM, Joho RH. 1989. A novel potassium channel with delayed rectifier properties isolated from rat brain by expression cloning. *Nature* 340:642–45

55. Fridovich-Keil JL, Jinks-Robertson S. 1993. A yeast expression system for human galactose-1-phosphate uridylyltransferase. *Proc. Natl. Acad. Sci. USA* 90:398–402

56. Frommer WB, Hummel S, Lemoine R, Delrot S. 1994. Developmental changes in the two-dimensional protein pattern of plasma membrane vesicles between sink and source leaves from sugar beet. *Plant Physiol. Biochem.* 32:205–9

57. Frommer WB, Hummel S, Rentsch D. 1994. Cloning of an *Arabidopsis* histidine transporting protein related to nitrate and peptide transporters. *FEBS Lett.* 347:185–89

58. Frommer WB, Hummel S, Riesmeier J. 1993. Expression cloning in yeast of a cDNA encoding a broad specificity amino acid permease from *A. thaliana*. *Proc. Natl. Acad. Sci. USA* 90:5944–48

59. Frommer WB, Hummel S, Unseld M, Ninnemann O. 1994. An amino acid transporter from *Arabidopsis* with low affinity and broad specificity related to mammalian cationic amino acid transporters. *Plant J.* Submitted

60. Frommer WB, Starlinger P. 1988. DNase I hypersensitive sites in the 5-region of the maize *Shrunken* gene in nuclei from different organs. *Mol. Gen. Genet.* 212:351–59

61. Fujita N, Nelson N, Fox TD, Claudio T, Lindstrom J, et al. 1986. Biosynthesis of the *Torpedo californica* acetylcholine receptor α subunit in yeast. *Science* 231:1284–86

62. Fushimi K, Uchida S, Hara Y, Hirata Y, Marumo F, Sasaki S. 1993. Cloning and expression of apical membrane water channel of rat kidney collecting tubule. *Nature* 361:549–52

63. Garciadeblas B, Rubio F, Quintero FJ, Bañuelos MA, Haro R, Rodríguez-Navarro A. 1993. Differential expression of two genes encoding isoforms of the ATPase involved in sodium efflux in *S. cerevisiae*. *Mol. Gen. Genet.* 236:363–68

64. Gaymard F, Thibaud JB, Touraine B, Simon-Plas F, Grouzis JP, et al. 1993. A test for screening monoclonal antibodies to membrane proteins based on their ability to inhibit protein reconstitution into vesicles. *Biochim. Biophys. Acta* 1150:73–78

65. Glab N, Petit PX, Slonimski PP. 1993. Mitochondrial disfunction in yeast expressing the cytoplasmic male sterility *T-urf13* gene from maize. *Mol. Gen. Genet.* 236:299–308

66. Gold L. 1990. Expression of heterologous proteins in *E. coli*. *Methods Enzymol.* 185:11–14

67. Goodwin RG, Friend D, Ziegler SF, Jerzy R, Falk BA, et al. 1990. Cloning of the human and murine interleukin-7 receptors: demonstration of a soluble form and homology to a new receptor family. *Cell* 60:941–51

68. Grenson M. 1992. Amino acid transporters in yeast. In *Molecular Aspects of Transport Proteins*, ed. JJLLM De Pont, pp. 219–45. Dordrecht: Elsevier Science

69. Gurdon JB, Wickens MP. 1983. The use of *Xenopus* oocytes for the expression of cloned genes. *Methods Enzymol.* 101:370–86

70. Hamilton AJ, Bouzayen M, Grierson D. 1991. Identification of a tomato gene for the ethylene-forming enzyme by expression in yeast. *Proc. Natl. Acad. Sci. USA* 88:7434–37

71. Hausmann L, Schell J, Töpfer R. 1993. Glycerol-3-phosphate dehydrogenase, a candidate for providing the backbone of glycerolipids. *Biol. Chem. Hoppe-Seyler* 374:525–26

72. Hediger MA, Coady MJ, Ikeda TS, Wright EM. 1987. Expression cloning and cDNA

sequencing of the Na$^+$/glucose co-transporter. *Nature* 330:379–81

73. Hedrich R, Moran O, Conti F, Busch H, Becker D, et al. 1994. Plant potassium channels differ from their animal counterparts in response to voltage and channel modulators. *Eur. J. Biophys.* In press

74. Heins L, Mentzel H, Schmid A, Benz R, Schmitz UK. 1994. Biochemical, molecular and functional characterization of porin isoforms from potato mitochondria. *J. Biol. Chem.* 269: In press

75. Henikoff S, Tatchell K, Hall BD, Nasmyth KA. 1981. Isolation of a gene from *Drosophila* by complementation in yeast. *Nature* 289:33–37

76. Henner DJ. 1990. Expression of heterologous proteins in *B. subtilis. Methods Enzymol.* 185:199–201

77. Hinnen A, Hicks JB, Fink GR. 1978. Transformation of yeast. *Proc. Natl. Acad. Sci. USA* 75:1929–33

78. Hirt H, Pay A, Bögre L, Meskiene I, Heberle-Bors E. 1993. *cdc2 MsB,* a cognate *cdc2* gene from alfalfa, complements the G1/S but not the G2/N transition of budding yeast *cdc28* mutants. *Plant J.* 4:61–69

79. Hollmann M, O'Shea-Greenfield A, Rogers SW, Heinemann S. 1989. Cloning by functional expression of a member of the glutamate receptor family. *Nature* 342: 643–48

80. Holzenburg A, Jones PC, Franklin T, Pali T, Heimburg T, et al. 1993. Evidence for a common structure for a class of membrane channels. *Eur. J. Biochem.* 213:21–30

81. Horowitz B, Eakle KA, Scheiner-Bobis G, Randolph GR, Chen CY, et al. 1990. Synthesis and assembly of functional mammalian Na$^+$,K$^+$-ATPase in yeast. *J. Biol. Chem.* 265:4189–92

82. Hsu LC, Chiou TJ, Chen L, Bush DR. 1993. Cloning a plant amino acid transporter by functional complementation of a yeast amino acid transport mutant. *Proc. Natl. Acad. Sci. USA* 90:7441–45

83. Huang HJ, Liao CF, Yang BC, Kuo TT. 1992. Functional expression of rat M5 muscarinic acetylcholine receptor in yeast. *Biochem. Biophys. Res. Commun.* 182:1180–86

84. Hughes DA. 1994. Histidine kinases hog the limelight. *Nature* 369:187–88

85. Ishioka C, Frebourg T, Yan YX, Vidal M, Friend SH, et al. 1993. Screening patients for heterozygous p53 mutations using a functional assay in yeast. *Nat. Genet.* 5(2): 124–29

86. Jan LY, Jan YN. 1989. Voltage sensitive ion channels. *Cell* 56:13–20

87. Jansen KU, Conroy WG, Claudio T, Fox TD, Fugita N, et al. 1989. Expression of the four subunits of the *T. californica* nicotinic

acetylcholine receptor in *S. cerevisiae. J. Biol. Chem.* 264:15022–27

88. Jaunin P, Jaisser F, Beggah AT, Takeyasu K, Mangeat P, et al. 1993. Role of the transmembrane and extracytoplasmic domain of β-subunits in subunit assembly, intracellular transport, and functional expression of Na, K-pumps. *J. Cell. Biol.* 123:1751–59

89. Jentsch TJ. 1994. Molecular physiology of anion channels. *Curr. Opin. Cell Biol.* 6: 600–6

90. Kammerloher W, Fischer U, Piechottka GP, Schäffner AR. 1994. Water channels in the plant plasma membrane cloned by immunoselection from a mammalian expression system. *Plant J.* 6:187–99

91. Kamp A, Korenbrot JI, Kitajewski J. 1992. Expression of ion channels in cultured cells using baculovirus. *Methods Enzymol.* 207: 423–31

92. Kanai Y, Hediger MA. 1992. Primary structure and functional characterization of a high-affinity glutamate transporter. *Nature* 360:467–71

93. Karlsson R, Aspenstrom P, Bystrom AS. 1991. A chicken β-actin can complement a disruption of the *S. cerevisiae* ACT1 gene. *Mol. Cell. Biol.* 11:213–17

94. Karschin A, Thorne BA, Thomas G, Lester HA. 1992. Vaccinia virus as vector to express ion channel genes. *Methods Enzymol.* 207:408–23

95. Ketter JS, Jarai G, Fu YH, Marzluf GA. 1991. Nucleotide sequence, mRNA stability, and DNA recognition elements of *cys*-14, the structural gene for sulfate permease II in *N. crassa. Biochemistry* 30:1780–87

96. Kim EJ, Zhen R, Rea PA. 1994. Heterologous expression of plant vacuolar pyrophosphatase in yeast demonstrates sufficiency of the substrate binding subunit for proton transport. *Proc. Natl. Acad. Sci. USA* 91:6128–32

97. Kim JW, Closs EI, Albritton LM, Cunningham JM. 1991. Transport of cationic amino acids by the mouse ecotropic retrovirus receptor. *Nature* 352:725–28

98. Kim JW, Cunningham JM. 1993. N-linked glycosylation of the receptor for murine ecotropic retroviruses altered in virus-infected cells. *J. Biol. Chem.* 268:16316–20

99. Klonus D, Höfgen R, Willmitzer L, Riesmeier J. 1994. Isolation and characterization of two cDNA clones encoding ATP-sulphurylases from potato by complementation of a yeast mutant. *Plant J.* 5: 105–12

100. Kölling R, Hollenberg CP. 1994. The ABC-transporter Ste6 accumulates in the plasma membrane in a ubiquitinated form in endocytosis mutants. *EMBO J.* 13:3261–71

101. Korth KL, Levings CS III. 1993. Baculovirus expression of the maize mitochondrial protein URF13 confers insecticidal

activity in cell cultures and larvae. *Proc. Natl. Acad. Sci. USA* 90:3388–92

102. Kuchler K, Thorner J. 1992. Functional expression of human mdr1 in the yeast *S. cerevisiae*. *Proc. Natl. Acad. Sci. USA* 89: 2302–6

103. Kwart M, Hirner B, Hummel S, Frommer WB. 1993. Differential expression of two related amino acid transporters with differing substrate specificty in *A. thaliana*. *Plant J.* 4:993–1002

104. Kwon HM, Yamauchi A, Uchida S, Preston AS, Garcia-Perez A, et al. 1992. Cloning of the cDNA for a Na^+/myo-inositol cotransporter, a hypertonicity stress protein. *J. Biol. Chem.* 267:6297–301

105. Lang-Hinrichs C, Queck I, Büldt G, Stahl U, Hildebrandt V. 1994. The archaebacterial membrane protein bacterio-opsin is expressed and N-terminally processed in the yeast *S. cerevisiae*. *Mol. Gen. Genet.* 244:183–88

105a. Lee HI, Gal S, Newman TC, Raikhel NV. 1993. The *Arabidopsis* ER retension receptor functions in yeast. *Proc. Natl. Acad. Sci. USA* 90:11433–37

106. Lee MG, Nurse P. 1987. Complementation used to clone a human homologue of the fission yeast cell cycle control gene *cdc2*. *Nature* 327:31–35

107. Ljungdahl PO, Gimeno CJ, Styles CA, Fink GR. 1992. *SHR3*: a novel component of the secretory pathway specifically required for localization of amino acid permeases in yeast. *Cell* 71:463–78

108. Loddenkötter B, Kammerer B, Fischer K, Flügge UI. 1993. Expression of the functional mature chloroplast triose phosphate translocator in yeast internal membranes and purification of the histidine-tagged protein by a single metal-affinity chromatography step. *Proc. Natl. Acad. Sci. USA* 90: 2155–59

109. Loo DDF, Hazama A, Supplisson S, Turk E, Wright EM. 1993. Relaxation kinetics of the Na^+/glucose cotransporter. *Proc. Natl. Acad. Sci. USA* 90:5767–71

110. Lostao MP, Hirayama BA, Loo DDF, Wright EM. 1994. Phenylglucosides and the Na^+/glucose cotransporter (SGLT1): analysis of interactions. *J. Membr. Biol.* In press

111. Markovich D, Forgo J, Stange G, Biber J, Murer H. 1993. Expression cloning of rat renal Na^+/SO_4^{2-} cotransport. *Proc. Natl. Acad. Sci. USA* 90:8073–77

112. Marston FAO. 1986. The purification of eucaryotic polypeptides synthesized in *E. coli*. *Biochem. J.* 240:1–12

113. Masu Y, Nakayama K, Tamaki H, Harada Y, Kuno M, Nakanishi S. 1987. cDNA cloning of bovine substance-K receptor though oocyte expression system. *Nature* 329: 836–38

114. Mathews LS, Vale WW. 1991. Expression cloning of an activin receptor, a predicted transmembrane serine kinase. *Cell* 65:973–82

115. Maurel C, Reizer J, Schroeder JI, Chrispeels MJ. 1993. The vacuolar membrane protein γ-TIP creates water specific channels in *Xenopus* oocytes. *EMBO J.* 12: 2241–47

116. Maurel C, Reizer J, Schroeder JI, Chrispeels MJ, Saier MH Jr. 1994. Functional characterization of the *E. coli* glycerol facilitator, GlpF, in *Xenopus* oocytes. *J. Biol. Chem.* 269:1–4

117. Mauri I, Maddaloni M, Lohma S, Motto M, Salamini F, et al. 1993. Functional expression of the transcriptional activator Opaque-2 of *Zea mays* in transformed yeast. *Mol. Gen. Genet.* 241:319–26

118. McKnight GL, McConaughy BL. 1983. Selection of functional cDNAs by complementation in yeast. *Proc. Natl. Acad. Sci. USA* 80:4412–16

119. Metzelaar MJ, Wijngaard PL, Perters PJ, Sixma JJ, Nieuwenhuist HK, Clevers HC. 1991. CD63 antigen, a novel lysosomal membrane glycoprotein cloned by a screening procedure for intracellular antigens in eucaryotic cells. *J. Biol. Chem.* 266:3239–45

120. Metzger D, White JH, Chambon P. 1988. The human oestrogen receptor functions in yeast. *Nature* 334:31–36

121. Mieszczak M, Klahre U, Levy JH, Goodall GJ, Filipowicz W. 1992. Multiple plant RNA binding proteins identified by PCR: expression of cDNAs encoding RNA binding proteins targeted to chloroplasts in *N. plumbaginifolia*. *Mol. Gen. Genet.* 234: 390–400

122. Minet M, Dufour ME, Lacroute F. 1992. Cloning and sequencing of a human cDNA coding for dihydroorotate dehydrogenase by complementation of the corresponding yeast mutant. *Gene* 121(2):393–96

123. Minet M, Dufour ME, Lacroute F. 1992. Complementation of *S. cerevisiae* auxotrophic mutants by *A. thaliana* cDNAs. *Plant J.* 2:417–22

124. Minet M, Lacroute F. 1990. Cloning and sequencing of a human cDNA coding for a multifunctional polypeptide of the purine pathway by complementation of the ade2-101 mutant in *S. cerevisiae*. *Curr. Genet.* 18:287–91

125. Müller-Röber BT, Ellenberg J, Provart N, Willmitzer L, Busch H, et al. 1994. Cloning and electrophysiological characterization of a voltage-dependent K^+ channel predominantly expressed in potato guard cells. *EMBO J.* Submitted

126. Nagai A, Suzuki K, Ward E, Moyer M, Hashimoto M, et al. 1992. Overexpression of plant histidinol dehydrogenase using a

baculovirus expression vector system. *Arch. Biochem. Biophys.* 295:235–39

127. Nagata T, Nemoto Y, Hasezawa S. 1992. Tobacco BY-2 cell line as the "HeLa" cell in the cell biology of higher plants. *Int. Rev. Cytol.* 132:1\-30

128. Ninnemann O, Jauniaux JC, Frommer WB. 1994. Identification of a high affinity NH_4^+ transporter from plants. *EMBO J.* 13:3464–71

129. Nitschke K, Fleig U, Schell J, Palme K. 1992. Complementation of the cs dis2-11 cell cycle mutant of *S. pombe* by a protein phosphatase from *A. thaliana. EMBO J.* 11:1327–33

130. Noma Y, Sideras P, Naito T, Bergstedt-Linquist S, Azuma C, et al. 1986. Cloning of cDNA encoding the murine IgG1 induction factor by a novel strategy using SP6 promoter. *Nature* 319:640–46

131. Nurse P. 1990. Universal control mechanism regulating onset of M-phase. *Nature* 344:503–7

132. Ohya Y, Anraku Y. 1989. Functional expression of chicken calmodulin in yeast. *Biochem. Biophys. Res. Commun.* 158 (2): 541–47

133. Osborn L, Hession C, Tizard R, Vassallo C, Luhowskyj S, et al. 1989. Direct expression cloning of vascular cell adhesion molecule 1, a cytokine-induced endothelial protein that binds to lymphocytes. *Cell* 59:1203–11

134. Pajor AM, Wright EM. 1992. Cloning and functional expression of a mammalian Na^+/nucleoside cotransporter. *J. Biol. Chem.* 267:3557–60

135. Palmgren MG, Christensen G. 1994. Functional comparisons between plant plasma membrane H^+-ATPase isoforms expressed in yeast. *J. Biol. Chem.* 269:3027–33

136. Pan YX, Gu HH, Dean GE. 1993. *S. cerevisiae* expression of exogenous vacuolar ATPase subunit B. *Biochim. Biophys. Acta* 1151:175–85

137. Pang QS, Hays JB, Rajagopal I, Schaefer TS. 1993. Selection of *Arabidopsis* cDNAs that partially correct phenotypes of *E. coli* DNA-damage-sensitive mutants. *Plant Mol. Biol.* 22:411–26

138. Perkins DD, Radford A, Newmeyer D, Björkmann M. 1982. Chromosomal loci of *N. crassa. Microbiol. Rev.* 46:426–570

139. Peterson EP, Martinez GM, Martinez-Zaguilan R, Perona R, Gillies RJ. 1994. NIH 3T3 cells transfected with a yeast H^+-ATPase have altered sensitivity to insulin, insulin growth factor-I and platelet-derived growth factor-AA. *J. Cell Physiol.* 159: 551–60

140. Pillai R, Kytle K, Reyes A, Colicelli J. 1993. Use of a yeast expression system for the isolation and analysis of drug-resistant mutants of a mammalian phosphodi-esterase. *Proc. Natl. Acad. Sci. USA* 90: 11970–74

141. Quesada A, Galván A, Schnell RA, Lefebvre PA, Fernández E. 1993. Five nitrate assimilation-related loci are clustered in *C. reinhardtii. Mol. Gen. Genet.* 240:387–94

142. Renaud JP, Cullin C, Pompon D, Beaune P, Mansuy D. 1990. Expression of human liver cytochrome P450 IIIA4 in yeast. *Eur. J. Biochem.* 194:889–96

143. Reuveny E, Slesinger PA, Inglese J, Morales JM, Iñiguez-Lluhli JA, et al. 1994. Activation of the cloned muscarinic potassium channel by G protein βγ subunits. *Nature* 370:143–46

144. Riesmeier JW, Hirner B, Frommer WB. 1993. Potato sucrose transporter expression in minor veins indicates a role in phloem loading. *Plant Cell* 5:1591–98

145. Riesmeier JW, Willmitzer L, Frommer WB. 1992. Isolation and characterization of a sucrose carrier cDNA from spinach by functional expression in yeast. *EMBO J.* 11:4705–13

146. Riesmeier JW, Willmitzer L, Frommer WB. 1994. Evidence for an essential role of the sucrose transporter in phloem loading and assimilate partitioning. *EMBO J.* 13:1–7

147. Rine J. 1991. Gene overexpression in studies of *S. cerevisiae. Methods Enzymol.* 194: 239–51

148. Roberts CJ, Notwehr SF, Stevens TH. 1992. Membrane protein sorting in the yeast secretory pathway. *J. Cell. Biol.* 119: 69–83

149. Romanos MA, Scorer CA, Clare JJ. 1992. Foreign gene expression in yeast: a review. *Yeast* 8:423–88

150. Rosahl S, Schell J, Willmitzer L. 1987. Expression of a tuber-specific storage protein in transgenic tobacco plants: demonstration of an esterase activity. *EMBO J.* 6:1155–59

151. Rose MD, Broach JR. 1991. Cloning genes by complementation in yeast. *Methods Enzymol.* 194:195–230

152. Rothstein RJ. 1983. One-step gene disruption in yeast. *Methods Enzymol.* 10:202–11

153. Rudy B, Iverson LE. 1992. Ion channels: expression of ion channels in *Xenopus* oocytes. *Methods Enzymol.* 207:225–390

154. Ruetz S, Gros P. 1994. Functional expression of P-glycoprotein in secretory vesicles. *J. Biol. Chem.* 269:12277–84

155. Rüth J, Schweyen RJ, Hirt H. 1994. The plant transcription factor TGA1 stimulates expression of the CaMV35S promoter in *S. cerevisiae. Plant Mol. Biol.* 25:323–28

156. Sauer N, Caspari T, Klebl F, Tanner W. 1990. Functional expression of the *Chlorella* hexose transporter in *S. pombe. Proc. Natl. Acad. Sci. USA* 87:7949–50

157. Sauer N, Friedländer K, Gräml-Wicke U.

1990. Primary structure, genomic organization and heterologous expression of a glucose transporter from *A. thaliana*. *EMBO J.* 9:3045–50

158. Schachtman DP, Gaber RF, Lucas WJ, Anderson JA, Schroeder JI. 1992. Expression of an inward-rectifying potassium channel by the *Arabidopsis* KAT1 cDNA. *Science* 258:1654–58

159. Schachtman DP, Schroeder JI. 1994. Cloning, transport mechanism and localization of a high affinity potassium uptake transporter from higher plants. *Nature* 370:655–58

160. Schertler GFX. 1992. Overproduction of membrane proteins. *Curr. Opin. Struct. Biol.* 2:534–44

161. Schild D, Brake AJ, Kiefer MC, Young D, Barr PJ. 1990. Cloning of three human multifunctional de novo purine biosynthetic genes by functional complementation of yeast mutations. *Proc. Natl. Acad. Sci. USA* 87:2916–20

162. Schroeder JI. 1994. Heterologous expression and functional analysis of higher plant transport proteins in *Xenopus* oocytes. *Methods Enzymol.* In press

163. Schulz B, Bennett MJ, Dilkes BP, Feldmann KA. 1995. T-DNA tagging in *Arabidopsis*. In *Plant Molecular Biology Manual II*, ed. SB Gelvin, RA Schilperoort, pp. K1–17. Dordrecht: Kluwer Academic. In press

164. Segel GB, Boal TR, Cardillo TS, Murant FG, Lichtman MA, Sherman F. 1992. Isolation of a gene encoding a chaperonin-like protein by complementation of yeast amino acid transport mutants with human cDNA. *Proc. Natl. Acad. Sci. USA* 89:6060–64

165. Sentenac H, Bonneaud N, Minet M, Lacroute F, Salmon JM, et al. 1992. Cloning and expression in yeast of a plant potassium ion transport system. *Science* 256:663–65

166. Sigel E. 1990. Use of *Xenopus* oocytes for the functional expression of plasma membrane proteins. *J. Membr. Biol.* 117:201–21

167. Silverman M. 1991. Structure and function of hexose transporters. *Annu. Rev. Biochem.* 60:757–94

168. Sims JE, March CJ, Cosman D. 1988. cDNA expression cloning of the IL-1 receptor, a member of the immunoglobulin superfamily. *Science* 241:585–89

169. Smith FW, Hawkesford MJ, Prosser IM, Clarkson DT. 1993. Approaches to cloning genes encoding for nutrient transporters in plants. In *Plant Nutrition—From Genetic Engineering to Field Practise*, ed. NJ Barrow, pp. 151–54. Dordrecht: Kluwer Academic

170. Smith FW, Hawkesford MJ, Prosser IM, Clarkson DT. 1994. Isolation of a cDNA from *S. cerevisae* that encodes a high affinity sulphate transporter at the plasma membrane. *Mol. Gen. Genet.* Submitted

171. Smith JK, Schloss JV, Mazur BJ. 1989. Functional expression of plant acetolactate synthase genes in *E. coli. Proc. Natl. Acad. Sci. USA* 86:4179–83

172. Spevak W, Keiper BD, Stratowa C, Castanon MJ. 1993. *S. cerevisiae cdc15* mutants arrested at a late stage in anaphase are rescued by *Xenopus* cDNAs encoding N-ras or a protein with β-transducin repeats. *Mol. Cell. Biol.* 13:4953–66

173. Stein JC, Nasrallah JB. 1993. A plant receptor-like gene, the S-locus receptor kinase of *Brassica oleraceae* L., encodes a functional serine/threonine kinase. *Plant Physiol.* 101:1103–6

174. Steiner HY, Song W, Zhang L, Naider F, Becker JM, Stacey G. 1994. An *Arabidopsis* peptide transporter is a member of a novel family of membrane transport proteins. *Plant Cell.* 6:1289–99

175. Stolz J, Stadler R, Operakova M, Sauer N. 1994. Functional reconstitution of the solubilized *A. thaliana* STP1 monosaccharide-H⁺ symporter in lipid vesicles and purification of the histidine tagged protein from transgenic *S. cerevisiae. Plant J.* 6:225–33

176. Struhl K. 1988. The JUN oncoprotein, a vertebrate transcription factor, activates transcription in yeast. *Nature* 332:649–50

177. Struhl K, Cameron JR, Davis RW. 1976. Functional genetic expression of eucaryotic DNA in *E. coli. Proc. Natl. Acad. Sci. USA* 73:1471–75

178. Taglialatella M, Toro L, Stefani E. 1992. Novel voltage clamp to record small, fast currents from ion channels expressed in *Xenopus* oocytes. *Biophys. J.* 61:78–82

179. Takumi T, Ohkubo H, Nakanishi S. 1988. Cloning of a membrane protein that induces a slow voltage-gated potassium current. *Science* 242:1042–45

180. Tate ST, Yan N, Udenfriend S. 1992. Expression cloning of a Na⁺-independent neutral amino acid transporter from rat kidney. *Proc. Natl. Acad. Sci. USA* 89:1–5

181. Thon VJ, Khalil M, Cannon JF. 1993. Isolation of human glycogen branching enzyme cDNAs by screening complementation in yeast. *J. Biol. Chem.* 268:7509–13

182. Tsay Y, Schroeder JI, Feldmann KA, Crawford NM. 1993. The herbicide sensitivity gene *CHL1* of *Arabidopsis* encodes a nitrate-inducible nitrate transporter. *Cell* 72:705–13

183. Tsuji N, Nomoto S, Yasuda H, Reed SI, Matsumoto K. 1991. Cloning of a human cDNA encoding a CDC2-related kinase by complementation of a budding yeast cdc28 mutation. *Proc. Natl. Acad. Sci. USA* 88:9006–10

184. Vogel JM, Roth B, Cigan M, Freeling M. 1993. Expression of the two maize TATA

binding protein genes and function of the encoded TBP proteins by complementation in yeast. *Plant Cell* 5:1627–38

185. von Schaewen A, Stitt M, Schmidt R, Sonnewald U, Willmitzer L. 1990. Expression of a yeast-derived invertase in the cell wall of tobacco and *Arabidopsis* plants. *EMBO J.* 9:3033–44

186. Walden R, Hayashi H, Schell J. 1991. T-DNA as a gene tag. *Plant J.* 1:281\-88

187. Wang Z, Orlowski J, Shull GE. 1993. Primary structure and functional expression of a novel gastrointestinal isoform of the rat Na^+/H^+ exchanger. *J. Biol. Chem.* 268: 11925–28

188. Wells RG, Hediger MA. 1992. Cloning of a rat kidney cDNA that stimulates dibasic and neutral amino acid transport and has sequence similarity to glucosidases. *Proc. Natl. Acad. Sci. USA* 89:5596–600

189. Welters P, Takegawa K, Emr SD, Chrispeels MJ. 1994. AtVPS34, a phosphatidylinositol 3-kinase of *A. thaliana* is an essential protein with homology to a calcium-dependent lipid binding (CaLB) domain. *Mol. Cell. Biol.* In press

190. Werner A, Moore ML, Mantei N, Biber J, Semenza G, Murer H. 1991. Cloning and expression of cDNA for a Na/Pi cotransport system of kidney cortex. *Proc. Natl. Acad. Sci. USA* 88:9608–12

191. Werner M, Crabeel M, Glansdorff N, Piérard A. 1982. Attempts at cloning *A. thaliana* genes by transformation of yeast auxotrophic mutants. *Arab. Inf. Serv.* 19:1–10

192. Weyand I, Godde M, Frings S, Weiner J, Müller F, et al. 1994. Cloning and functional expression of a cyclic-nucleotide-gated channel from mammalian sperm. *Nature* 368:859–63

193. Will A, Caspari T, Tanner W. 1994. K_m mutants of the *Chlorella* monosaccharide/H^+ cotransporter randomly generated by PCR. *Proc. Natl. Acad. Sci. USA* 91: 10163–67

194. Yamasaki K, Taga T, Hirata Y, Yawata H, Kawanishi Y, et al. 1988. Cloning and expression of the human interleukine-6 (BSF-2/IFNb2) receptor. *Science* 241:825–28

195. Yellen G, Migeon JC. 1990. Expression of *Torpedo* nicotinic acetylcholine receptor subunits in yeast is enhanced by use of yeast signal sequences. *Gene* 86:145–52

196. You G, Smith CP, Kanai Y, Lee W, Stelzner M, Hediger MA. 1993. Cloning and characterization of the vasopressin-regulated urea transporter. *Nature* 365:844–47

197. Zhang R, Logee KA, Verkman AS. 1990. Expression of mRNA coding for kidney and red cell water channels in *Xenopus* oocytes. *J. Biol. Chem.* 265:15375–78

Annu. Rev. Plant Physiol. Plant Mol. Biol. 1995. 46:445–74

LIGHT-REGULATED TRANSCRIPTION

William B. Terzaghi and Anthony R. Cashmore

Plant Science Institute, Department of Biology, Leidy Laboratory of Biology, Philadelphia, Pennsylvania 19104-6018

KEY WORDS: gene expression, photoreceptor, regulatory sequence, DNA-binding proteins, signal transduction

CONTENTS

ABSTRACT

This review focuses on light regulation of transcription, from light perception to changes in transcription, with an emphasis on regulatory elements in the promoters of light-regulated genes and proteins that bind to them. The abundance of over 100 mRNAs is regulated by light, and at least three photoreceptors influence transcription, but the importance of each photoreceptor may vary from gene to gene. Light-regulated promoters are composed of ubiquitous regulatory elements; the specific combination of elements appears to make a promoter light-regulated, and these combinations vary widely. Numerous proteins that bind to elements in light-regulated promoters have been identified and many have been cloned, but no cloned gene has been unequivocally assigned a function in light-regulated transcription. Substantial progress has been made in identifying steps in signal transduction from light perception to transcription, but these steps have yet to be assembled into complete pathways.

INTRODUCTION[1]

Plants use light as a source of information about their environments as well as a source of energy for photosynthesis. Plants sense both the quantity (fluence) and quality (wavelength) of light and respond in many ways, ranging from germination and de-etiolation of seedlings to timing of flowering. Many of these responses require changes in both nuclear and chloroplast gene expression. This review focuses exclusively on our current understanding of light-regulated transcription of nuclear genes, concentrating on *cis*-acting regulatory sequences and proteins that bind to them.

The general paradigm adopted for light-regulated transcription is that upon light perception, photoreceptors generate signals that are transduced via intermediates to activate transcription factors bound to their cognate sequences within regulatory regions of light-regulated genes. These factors then orchestrate the assembly of the transcription apparatus that commences transcription (154). Most research to date has identified end products (light-regulated transcripts) and then worked back toward the primary reactions in light perception.We take this approach in this review.

1

Abbreviations: B, blue light; R, red light; FR, far-red light; UV-A, ultraviolet A (320–400 nm); UV-B, ultraviolet B (280–320 nm); Pr, red-absorbing form of phytochrome; Pfr, far-red-absorbing form of phytochrome; LF, low fluence; VLF, very low fluence; HIR, high irradiance response; CRY, cryptochrome; CaMV 35S, Cauliflower Mosaic Virus 35S promoter; GUS, β-glucuronidase, encoded by bacterial vidA gene; CKII, casein kinase II. Nomenclature: Plant gene nomenclature is presently in flux, and several conventions have been proposed (e.g. 33a, 111a). We have used the names given by the original authors for all genes and mRNAs, except that for consistency we have placed all names of genes (except mutant alleles) and mRNAs in upper-case italics. Names of the gene products corresponding to the gene symbols used in this review are given in Reference 33a.

PHOTORECEPTORS

Three major classes of photoreceptors are known to influence transcription: phytochrome, blue/UV-A, and UV-B. In a sense, both protochlorophyllide and chlorophyll also function as photoreceptors, because light is required to convert protochlorophyllide to chlorophyllide, and the concentrations of chlorophyllide and chlorophyll also affect nuclear gene expression (152). Light-regulated genes may respond to one or more receptors, and these responses may be tempered by other environmental influences or developmental stages. See References 60, 80, 124, and 155 for reviews of the different photoreceptors.

Phytochrome

Phytochrome exists in two photochemically interconvertible forms (60, 124, 155). It is synthesized as Pr, the inactive form for most responses, whose absorption maximum is in the R (665 nm). Saturating R converts about 80% of Pr to the active form, Pfr, whose absorption maximum is in the FR (730 nm). Saturating FR results in an equilibrium of about 97% Pr and 3% Pfr, and both forms also absorb in the B. Phytochrome responses may be distinguished from those of other photoreceptors by their action spectrum and by FR reversibility (although not all Pfr responses are FR reversible). Several types of phytochrome apoproteins are encoded by a multigene family (60, 124, 155). Many researchers have demonstrated specific roles for individual phytochromes in various aspects of light-regulated plant development (31, 44, 60, 124). Three distinct classes of phytochrome responses can be identified based on fluence. VLF responses have a threshold of about 10^{-4} μmoles $\cdot$ m^{-2} R and are saturated by 10^{-1} μmoles $\cdot$ m^{-2} R; they may be triggered by FR flashes, which convert less than 0.1% of Pr to Pfr and thus are not FR reversible (13, 152). LF responses are saturated by about 1 mmole $\cdot$ m^{-2} R and are FR reversible. HIRs require continuous illumination of at least 1 μmole $\cdot$ m^{-2} $\cdot$ s^{-1} and may be more efficiently induced by FR.

Cryptochrome

The cryptochrome class of photoreceptors mediate responses to B/UV-A. Responses regulated by CRY are identified by their action spectrum, which typically shows a maximum in the B at about 450 nm, with shoulders in the UV-A at about 380 nm and in the B at about 480 nm (80). We have recently cloned the A. thaliana HY4 gene, and on the basis of genetic, photobiological, and sequence studies we concluded that this gene encodes the B-photoreceptor that mediates inhibition of hypocotyl elongation (1). An important component of our argument is the striking sequence homology between HY4 and microbial photolyases, a unique class of flavoproteins that are dependent for their activity on B/UV-A light (131).

There is evidence for more than one B/UV-A photoreceptor; and VLF, LF, and HIR responses to B have been identified (146). There also may be distinct UV-A receptors; for example, *hy4* mutants respond to UV-A (1, 146, 171). Nothing is presently known about such receptors; a possible candidate for a UV-A photoreceptor would be a HY4-related protein containing a pterin as a second chromophore, as is found in the UV-A-responsive class of pho-tolyases (131). We propose that HY4 and related proteins be called crypto-chromes. If certain responses to B or UV-A are demonstrated to be mediated by proteins unrelated to HY4, then we suggest that these also be called crypto-chromes and that some appropriate designation be used to distinguish these other families.

UV-B

Several responses, including the induction of flavonoid and anthocyanin bio-synthesis in many plant species (11, 148, 152) and, more specifically, the induction of flavonoid biosynthetic genes in cultured parsley cells and *A. thaliana,* are induced by UV-B (82, 116). UV-B excitation is needed for maximal response, but the amount of response can be tempered by CRY and Pfr (11, 82, 116, 152). The chemical nature of the UV-B receptor is unknown.

LIGHT-REGULATED TRANSCRIPTS

The abundance of at least 100 transcripts is known to change in response to light (11); many encode proteins involved in photosynthesis. Most of these transcripts increase in response to light, but there are also mRNAs, such as *PHYA, PCR, GBF3,* or the *NPR* genes of *Lemna gibba,* whose abundance decreases (19, 117, 137, 152, 155). Light requirements for transcript accumu-lation have not been examined for many of these genes, but for those studied, diverse responses have been observed in the photoreceptors involved, fluences required, and kinetics of accumulation of particular transcripts. For example, *CHS* mRNA in parsley cell cultures accumulates predominantly in response to UV-B, but this accumulation is tempered by CRY and Pfr (11, 116). By contrast, expression of most *RBCS* and *CAB* genes is regulated by Pfr, but this regulation is influenced by CRY (11, 41, 146, 152, 163), whereas the *GAP* genes of *A. thaliana* are regulated predominantly by CRY (47). Moreover, differing fluence requirements and kinetics of accumulation have been demon-strated for expression of individual members of the *CAB* and *RBCS* gene families in several plant species; some show VLF Pfr requirements, whereas others show a classic LF response (41, 163,167). The requirement for specific photoreceptors can also change during development. For example, *CHS* mRNA accumulation is controlled predominantly by Pfr in mustard seedlings and by CRY later in development (11, 58), whereas the expression of *RBCS* and *CAB* genes in pea is controlled predominantly by Pfr in etiolated seedlings and by CRY in green plants (152). Accumulation of light-regulated transcripts can also be influenced by growth regulators (10, 32, 132, 169), metabolic end products (144), cell type (9, 11, 152, 158), developmental state of the plant (22, 161) or of the plastids (150, 151), and by circadian rhythms (111, 122).

Most studies have demonstrated light-regulated changes in steady-state levels of mRNA as opposed to light-regulated transcription, and posttranscriptional regulation of mRNA abundance has also been demonstrated. One approach used to show light-induced transcription is in vitro nuclear run-on assays, in which nuclei isolated from plants given various light treatments are placed in conditions allowing the elongation of transcripts initiated prior to isolation, but in which de novo initiation is thought not to occur. Run-ons indicate the transcripts being made at the time of isolation, and the amounts of specific transcripts made in vitro can be quantitated by hybridization to suitable probes [although the use of run-ons to quantitate transcriptional activity has been criticized (152)]. Run-on assays have been used to demonstrate light-regulated transcription for a number of genes including *RBCS, CAB, CHS, PHYA, FEDA,* and *NPR*; discrepancies between run-on data and steady-state measurements have been interpreted as evidence for posttranscriptional regulation of transcript abundance (11, 15, 111, 145, 152, 155, 160, 161, 163, 167). Photoreceptors involved and fluence requirements have been determined for only a few of these genes.

Another approach used to infer light-regulated transcription is fusion of sequences from light-regulated genes to reporter genes, which are then introduced into plants by transient techniques or by *Agrobacterium tumefaciens–* mediated transformation. Changes in reporter activity after suitable light treatments are interpreted as showing light-regulated transcription. A vast literature has been amassed showing that sequences from a variety of genes are sufficient to confer light-responsiveness, including many genes involved in photosynthesis, such as *CAB* (9, 29, 79, 98, 111), *RBCS* (38, 50, 83, 134, 158), *FEDA* (27, 61, 160), *PETE* (99, 123, 159), *PETH* (99), *PSAF* (99), *RCA* (119), *GAP B* (85), and *GAPA* (34), as well as nitrate reductase (30). Gene fusions have also demonstrated that light negatively regulates *PHYA* (19) and *NPR* genes (118). Most studies have focused on promoter regions, but some studies show that nontranslated leader, coding, and 3′ regions can also influence transcript abundance (15, 27, 39, 48, 49, 55, 61, 123, 158) and, in the case of the maize *RBCSm3* gene, transcription (158). As with run-ons, discrepancies between light-regulated reporter gene activity and steady-state mRNA levels are thought to indicate posttranscriptional influences on transcript abundance, and the photoreceptors involved and fluence requirements have been determined for only a few of these genes.

REGULATORY DNA SEQUENCES

Regulatory *cis*-acting sequences have been identified by a combination of deletion and mutagenesis studies, comparing sequences for conserved elements, and by defining motifs specifically bound by proteins using footprinting, gel-retardation, methylation interference, and missing nucleoside assays. Although numerous *cis*-acting elements have been identified, relatively few have been functionally characterized, and our knowledge of sequences mediating light regulation is limited. Sequences required for Pfr responsiveness in any promoter have been carefully defined only recently (79), and identification of sequences required for Pfr vs CRY responsiveness is in its infancy (34, 47, 99, 128).

Numerous deletion studies have been performed on light-regulated promoters fused to reporter genes, or on fragments of light-regulated promoters fused to heterologous minimal promoters that do not respond to light. (Many minimal promoters used for such gain-of-function experiments are derived from the CaMV 35S promoter. The viral CCAAT and TATA elements and transcription start site are provided by −90 derivatives; −46 derivatives provide the viral TATA element and transcription start site. These minimal promoters typically are fused to reporters such as GUS or luciferase.) Positive and negative elements usually are identified by progressive deletions, but the general trend (with numerous exceptions; cf 156, 160) is that the magnitude of both reporter activity and light responsiveness is progressively lost to the extent that activity becomes difficult to measure, although deletions to −250 or less frequently retain light responsiveness. Thus, although the promoter is pared to a light-responsive core, much quantitative and perhaps qualitative regulatory information has been lost in the process. Much of this information may have been redundant: Sequence comparisons frequently show that elements found within the core region are reiterated multiple times within that region and further upstream. For example, three different fragments of the pea *RBCS3A* promoter were found to confer light responsiveness to a heterologous non-light-regulated promoter, and two GT-1-binding sites were found within each fragment (64, 83).

Deletion analyses typically pare light-regulatory regions down to 150–250 bp. To our knowledge the smallest sequence sufficient to confer light-responsiveness to a heterologous minimal promoter is the *CHS* unit I, a 52-bp fragment that makes a CaMV 35S −46:GUS fusion light-responsive (128, 166). These fragments are large enough to interact with many different DNA-binding proteins. Strategies adopted to identify regulatory elements within these fragments include sequence comparisons for conserved elements, gel-retardation assays, and footprinting. These approaches identify proteins present in the nuclear extract that bind to the element in vitro. They will identify the most abundant proteins, and further analysis is necessary to confirm the importance of this interaction. Therefore, elements identified by these approaches should be functionally characterized by mutation or by gain-of-function experiments in which the element is fused to a minimal non-light-regulated promoter.

Many *cis*-acting elements have been identified using the above-mentioned approaches, although relatively few have been studied in detail. The following generalizations can be made: (*a*) Similar elements are found in many different promoters. (*b*) Although promoters may be assembled from a relatively limited number of elements, the ways in which they are assembled vary widely, and even promoters of the same gene family can be quite different. (*c*) No element is found in all light-regulated promoters; i.e. no universal master switch has been identified. (*d*) No element is sufficient to confer light-responsiveness by itself to a non-light-regulated minimal promoter; i.e. two or more elements interact synergistically. (*e*) The same element can confer different expression characteristics in different contexts. (*f*) Monocots may use different elements for light regulation than do dicots. (*g*) Light regulation may involve turning genes off as much as turning them on, and some light-regulatory sequences may repress rather than activate transcription. For example, the products of *DET* and *COP* genes are thought to repress photomorphogenesis because mutations in these genes are recessive but result in light-grown morphology in mutant plants grown in the dark (31, 44).

Elements found repeatedly in light-regulated genes include the G-box, the GT-1 site, I-boxes and related motifs, TGACG, and A/T-rich elements. Many other elements have been described, but they rarely have been as well characterized (cf 5, 34, 100, 110).

GT-1 Sites

GT-1-binding sites, or boxes II and III of the pea *RBCS3A* gene, have the core sequence GGTTAA, but GT-1 will bind to many related sequences such that the consensus GT-1 site is GPu(T/A)AA(T/A) (63, 68, 133). Similar sequences are found in promoters from many light-regulated genes including *RBCS* from many species (40, 50, 158), *PHYA* from oats and rice (42, 84), *CAB* from *Nicotiana plumaginifolia* and *A. thaliana* (136, 149), spinach *RCA* (119) and *PETA* (91), and bean *CHS15* (100). GT-1 also binds to promoters of genes that are not light regulated, such as CaMV 35S and *N. plumbaginifolia* *ATP2-1* (68). A synthetic tetramer of the box II site conferred light regulation to a CaMV 35S −90:GUS fusion, although the receptor involved was not determined (88). Subsequent study showed that the box II tetramer was insufficient to confer light regulation to a CaMV 35S −46 or a minimal *RBCS* construct (38). This result suggested that sequences between −90 and −46 in the CaMV 35S promoter were binding a necessary activity, and it was later shown that both Box II and GATA sequences flanking Box III were needed for expression of a −166 deletion of the pea *RBCS3A* gene (133). GT-1 sites are usually found in tandem, and the spacing between the two sites is critical (64).

GT-1 sites thus seem important for light regulation. However, sequence comparisons find no obvious GT-1 sites in many light-regulated promoters, although the degeneracy of the GT-1 site makes it difficult to be certain [note that the L box is a perfect inverted Box II (cf 16, 65, 102, 136)]. Moreover, several studies have shown that deletion or mutation of GT-1 sites do not affect expression of a reporter gene, although in some cases this lack of effect is the result of redundant GT-1 sites (64). In other studies, redundancy has not been ruled out (50, 119).

GT-1 sites are also found in the rice PHYA promoter, where they act as constitutive activating elements (43), and in the bean *CHS15* gene, where they are involved in organ-specific expression and function as silencers in cultured cells (91). Moreover, deleting the GT-1 site in the *RCA* promoter increased root expression of the GUS reporter 100-fold, although no effect was observed on leaf expression (119). Similarly, inserting GT-1 sites into the full CaMV 35S promoter reduced expression in the dark (64). These results show that the same element can have quite different effects depending on its context.

I-Boxes

I-boxes, which we define as GATAA, are regulatory elements that are functionally important in many light-regulated promoters of both monocots and dicots (16, 23, 50, 62, 65, 79, 87, 98, 134). Indeed, the sequence complementary to ATGATAAGG was predicted to be a common light-regulatory element following a computer search for conserved elements in light-regulated genes (69). Related GATA motifs are found in many other promoters, some of which are light regulated but others of which are not, such as the CaMV 35S (11, 64, 87, 103, 133). Many *RBCS* genes have a single I-box found near a G-box 100–300 bp 5′ to the TATA box (11, 16), whereas in others and in many *CAB* promoters, two or three GATA (the bases that follow vary) elements arranged in tandem and separated by a few base pairs are found near the TATA box (11, 29, 62). Mutating the GATA elements in the petunia *CAB22R* promoter substantially decreased expression (62), whereas mutating the two I-boxes of the *A. thaliana RBCS1A* gene significantly reduced expression of two different reporters (50). Scanning mutagenesis showed that two GATA elements flanking the Box III GT-1 site of the pea *RBCS3A* gene, the 3AF3 and 3AF5 sites, were essential for high-level expression from the *RBCS3A* −166 truncated promoter (133). Deleting or mutating I-boxes in maize (134) or *L. gibba* (79, 155) *RBCS* and *CAB* genes affects light induction in transient assays in homologous systems. Similarly, a GATA element in the *L. gibba AB19* promoter, together with a redundant CCAAT element 5′ to the nominal CCAAT element, is required for phytochrome regulation of this gene in transient assays in *L. gibba* (79). The *as-2* element of the CaMV 35S promoter also includes a GATA motif essential for its activity. A tetramer of *as-2* elements fused to a CaMV 35S −90:GUS construct was expressed in leaves but was not responsive to light in transgenic tobacco (87).

Recent studies indicate that sequences flanking I-boxes rather than the core GATA motif determine binding specificity in vitro (16) and influence reporter activity from the spinach plastocyanin promoter in transgenic tobacco (100). If these results prove to be generally applicable, identification of I-boxes will become more difficult.

G-Boxes

G-boxes, elements with the core CACGTG, are found in the promoters of many genes that respond to a variety of different stimuli (57, 108). Many *CAB* and *RBCS* promoters have G-box sequences, as do many other light-regulated genes, including *CHS* (6, 14, 166) and *RCA* (119). Two *L. gibba* genes negatively regulated by Pfr also have G-boxes in the regulatory regions (118). Site-directed mutagenesis of the G-box in the *A. thaliana RBCS1A* promoter resulted in a severe drop in expression (50). Similarly, mutation of the G-box in the parsley *CHS* promoter resulted in a loss of expression and UV-inducibility (141), whereas deletion of the G box of the *RCA* promoter abolished leaf- and light-regulated expression of promoter:GUS fusions (119).

Numerous variations in sequences flanking the CACGTG core have been observed (57, 108). Functional differences for such variation were shown by the finding that tetramers of GCCACGTGGC directed root-specific expression in transgenic tobacco when fused to a 35S −90:GUS construct, whereas the G-box-like GTACGTGGCG directed developmentally regulated GUS accumulation in seeds (130). Other equally disparate examples will surely appear upon further study of G-box-like sequences.

TGACG

TGACG motifs are found in many promoters and are involved in transcriptional activation of several genes by auxin and/or salicylic acid (51, 57, 93, 94, 135). To date, no TGACG motif has been shown to be relevant to light-induced transcription, although leaf expression can be obtained by placing a tetramer of TGACG elements in front of a −90 CaMV 35S minimal promoter (88). However, some light-regulated promoters include elements with TGACG and TGACGT motifs (51, 103) that may be bound by two different classes of bZIP proteins (see below). TGACG motifs may therefore prove relevant to light regulation upon further study.

Box One (H-Box)

CHS promoters from several species have an element known as box I (one) or the H-box, which is essential for both light regulation (14, 128) and elicitor induction (95). It has the consensus CCTACC and is part of the unit 1, which is sufficient to confer light regulation to minimal promoters, and of the redundant unit 2 located further upstream (11). It has not been identified as essential in the context of other light-regulated promoters that are not involved in flavonoid synthesis (6).

AT-Rich Sequences

AT-rich sequences that are bound by nuclear proteins have been found in numerous light-regulated promoters, including the tomato *RBCS3A* (37, 156), pea *RBCS3A* (89), *N. plumbaginifolia CABE* (29, 136), maize *CAB-ml* (9), oat *PHYA3* (21), *L. gibba NPR2* (118), and pea *GS2* (153). The oat motif (PE1) acts as a general positive element in cells kept in the dark (20, 21). The 3AF1 site identified by footprinting in the pea *RBCS3A* promoter enhances expression in both roots and leaves when fused to CaMV 35S −90:GUS constructs but is inactive when fused to CaMV 35S −46 constructs, indicating that it must interact with other elements for activity (89). The 3AF1 site also includes a GATA motif, and mutagenesis of the GATA motif destroyed binding and transcriptional activation but also increased the GC content. Therefore, the question of whether the 3AF1 site is an AT-rich or a GATA motif remains unanswered. Deletion of a 33-bp AT-rich element from *GS2* promoter:GUS fusions resulted in a 10-fold decrease in GUS activity, but light induction was not investigated (153). Deletion of a region containing the AT-1 site in the tomato *RBCS3A* gene strongly inhibited reporter gene expression (156), whereas the AT-1 site in *N. plumbaginifolia CABE* is in a negative element (29). The AT-1-like element from maize *CAB-ml* is necessary for repressing expression in bundle sheath cells (9). AT-rich elements from other light-regulated genes have not been characterized functionally.

Other Sequences

Two other specific sequences are involved in Pfr-regulated gene expression. Site-directed mutagenesis demonstrated that a redundant CCAAT element in the *L. gibba AB19* promoter is essential for phytochrome responsiveness, in conjunction with a GATA element (79). Mutation of the CCAAT element resulted in an increase in dark expression to nearly light levels, rather than a loss of light expression. By contrast, mutation of a redundant CCAAT box in the spinach *PETE* promoter resulted in greatly reduced expression in transgenic tobacco leaves, although Pfr-responsiveness was not examined (100). An 11-bp element called RE-1 at −75 in the oat *PHYA3* promoter is involved in light-induced repression of *PHYA* transcription, and similar elements have been found in all *PHYA* genes studied (20, 155). Similarly, the CA-1 factor from *A. thaliana* appears to be required for dark suppression of *CAB140* expression and was not detected in *det1* mutants. Its recognition sequence has not been defined adequately to search for it in other promoters (149).

Another element of interest is the Gap-box, ATGAA(A/G)A, found in four copies in the *A. thaliana GAPB* promoter and three times in the *GAPA* promoter. This element is essential for light-regulated transcription, and a 109-bp fragment containing these four Gap-boxes could confer light regulation on a CaMV 35S −90 construct (34, 85).

GC-rich elements have been identified in many promoters, and a protein binding to a 21-bp GC-rich element from the *CABE* promoter was identified in *N. plumbaginifolia* nuclear extracts (29, 136). Although a specific role for GC-rich elements has not been demonstrated conclusively in dicots, elements identified as important for light regulation in monocots, but that are not recognized by dicots, are often GC rich (98, 134).

Sequences within nontranslated leaders have also been implicated as influencing transcription (15, 21, 27, 48), and a conserved sequence found in many light-regulated leaders has been proposed to act as a quantitative element (27).

Conclusion

How can promoters composed of a number of ubiquitous elements mediate light-regulated transcription? The evidence indicates that the specific combination of a number of general elements defines a light-regulated promoter (18). As described above, similar elements may have different effects depending on context. Moreover, although elements have been lumped together by classes, individual transcription factors may discriminate between related elements, as described below, and new elements will continue to be identified as more genes are studied. Consequently, plants have the potential to assemble a vast array of different promoters from a limited pool of elements.

DNA-BINDING PROTEINS

DNA-binding proteins have been identified, primarily using footprinting or gel-retardation assays, as factors that bind to regulatory elements within light-regulated promoters. Some have then been cloned from expression libraries using their cognate sequences as probes. Alternatively, homologs to transcription factors found in other systems have been cloned (74).

These studies identify proteins that bind to specific sequences in vitro. Studies with nuclear extracts will be biased toward the most abundant activities in the extract and will be influenced by considerations such as binding conditions used, interactions with other proteins, or modifications during the extraction protocol. Moreover, in vivo, factors will bind to their cognate sequence in chromatin. Neighboring proteins, or even more distant ones, are likely to influence binding specificities, yet these other proteins may not bind in vitro for reasons such as those cited above or because their binding sites are not present on the probe. These problems become more acute when studying cloned factors, where other plant proteins are not present. Consequently, although in vitro binding studies provide a powerful starting point, they do not allow conclusions to be drawn concerning whether a certain factor will bind to a given regulatory element within a specific promoter in vivo.

The surest approach to confirming the role of a factor is to identify mutants lacking the factor, correlate this lack with loss of gene expression, and then demonstrate that the cloned gene restores the missing activity. This approach has been used to study other plant transcription factors such as the B, C, P, and R gene products in maize (e.g. 66). However, no mutation affecting a DNA-binding protein involved in light-regulated transcription has been identified. One alternative is to generate mutants by genetic engineering. The best approach would be to generate specific mutants using targeted gene replacement by homologous recombination (126), but this is not yet routinely applicable in plants. Consequently, other approaches to generating mutants have been employed, including antisense, overexpression, and engineering dominant negative mutations that can interact with the binding site and/or endogenous protein yet fail to activate transcription.

Another problem is demonstrating that the factor is involved in transcription. The most definitive approach used in animals and yeast has been in vitro transcription, but simple and widely applicable procedures for preparing transcriptionally competent extracts have not been available in plants. Plant in vitro systems have been described (35, 127, 170), but they typically have high nonspecific background activity, show constitutive transcription of added constructs, and do not respond to exogenous stimuli. Consequently, in vitro transcription has not been widely used to study plant factors. Two recent studies describe contrasting systems that may improve this situation. In one study, *CHS* promoter fragments coupled to agarose beads were used to enrich transcription machinery in extracts from several different cell cultures (6). These fragments were transcribed from the appropriate start site by Pol II, and the system was used to demonstrate *cis*-acting sequences required for *CHS* transcription. In another approach, extracts prepared from evacuolated parsley protoplasts transcribed a *CHS* promoter fragment; transcription was sensitive to α-aminitin and was stimulated threefold by white or red light (59). However, neither system has progressed to the point where factors can be added and shown to activate transcription, and there have been no studies in which in vitro transcription has been used to show that a factor binding to light-regulatory sequences is in fact a transcriptional activator.

Most studies have relied on reporter assays to infer transcriptional activation, or they have simply relied on resemblance to known transcriptional activators. In reporter assays, a construct expressing the putative factor is introduced into cells along with a reporter construct that includes a binding site for the factor in the promoter of a reporter gene. Transcriptional activation is inferred if reporter activity increases above the reporter-only background when the factor is introduced, as shown for the *Opaque-2* gene product of maize (157). In a variation, the reporter gene may already be present. For example, GUS expression increased dramatically upon biolistic delivery of constructs expressing TAF-1 into plants transformed with a GUS construct driven by G-boxes fused to a CaMV 35S −90 minimal promoter (115). In an elegant variation on this approach, purified recombinant TGA1a has been microinjected into tobacco leaf cells transformed with a CaMV 35S −90:GUS fusion (which includes the TGA1a cognate element *as-1*); GUS was not detected until cells were injected with 10^4 molecules of recombinant TGA1a (113).

Sometimes these approaches fail, perhaps because of endogenous activity. One solution in this case is to make hybrid transcriptional activators in which various domains of the putative activator are fused to a DNA-binding domain that does not recognize plant sequences. Activation is inferred if significantly higher levels of reporter gene expression are observed when a fusion construct is cointroduced with a reporter in which the cognate sequence for this DNA-binding domain is an essential part of the promoter (138).

A problem with all these approaches is that they elevate the factor above its physiological concentration. Because many factors interact with one another, and some must dimerize to function, such overproduction will undoubtedly upset many equilibria. Consequently, although such studies demonstrate that a factor can perform a given function, whether it does so in vivo must be confirmed by other means. Bearing in mind these concerns, numerous proteins specifically binding to promoter fragments have been identified, and many have been cloned. We describe below the best characterized of these proteins.

GT Family

Proteins that bind to GT-1 sites have been identified in nuclear extracts from numerous species, and three proteins binding to GT elements have been cloned: one from tobacco (63, 121), one from rice (42), and one from *A. thaliana* (84). All contain trihelix (helix-loop-helix-loop-helix) structures predicted to bind DNA; the 43-kDa tobacco protein forms one, whereas both the *A. thaliana* and rice clones are predicted to form two. The tobacco clone is similar to the other two in the trihelix region, but not elsewhere. The 65-kDa *A. thaliana* protein was cloned using the rice gene (*GT-2*) as the probe and shows significant homology to rice GT-2 in several different domains. In vitro binding specificities determined for the tobacco clones were consistent with binding to the pea *RBCS3A* box II, but they did not bind to rice GT-2 sites. No differences in expression or in GT-1-binding activity were noted between light- and dark-grown material (63, 121). The rice clone has two DNA-binding domains, and the amino proximal domain preferentially bound GGTAAAT, whereas the carboxyl proximal domain preferred GGTAATT, and neither bound well to Box II (GGTTAA) (43). *GT-2* mRNA was not affected by treatment with R or FR, but declined in white light (42). The *A. thaliana* homolog appears to resemble rice *GT-2* in its binding-site preferences and shows constitutive expression (84). Functional assays have not been performed for any of the cloned GT factors. In conclusion, the data indicate that there may be multiple GT-binding factors, and it remains to be seen whether light-activation of transcription via GT-1 sites is mediated by one of the cloned factors or by others yet to be identified.

G-Box-Binding Factors (GBFs)

Proteins binding to G-boxes within many different promoters have been identified by competitive gel-retardation assays in extracts prepared from many different plant species, and in vitro and in vivo footprinting have demonstrated protein binding to G-box elements in *CHS, RBCS, and CAB* promoters (26, 57, 65, 101, 102, 108, 136, 142). A functional requirement for GBFs was demonstrated by in vitro transcription from a *CHS* promoter: Extracts depleted of GBFs using G-box probes were curtailed in *CHS* transcription (6). Many proteins that bind to sequences with the core CACGTG have been cloned from a variety of plant species (57, 77, 105, 108). Until recently, all cloned plant proteins binding to this sequence contained a bZIP motif (57, 105, 108). However, mammalian *myc* HLH-ZIP factors bind to some CACGTG motifs (54), and a tobacco HLH-bZIP factor binding to a CACGTG motif has been cloned (77).

No function in light regulation has been assigned unequivocally to any cloned GBF protein. Clones were isolated using probes derived from the promoter of interest; however, as described above, this provides no assurance that they bind to that promoter in vivo. The only G-box-binding protein that has been assigned a function is O2 from maize, which activates transcription of 22-kDa zein genes in maize (73, 139, 157). The proline-rich region of GBF1 can activate transcription in transient assays (138), and transient expression of TAF-1 activated transcription from a G-box tetramer fused to a CaMV 35S −90:GUS construct (115).

Expression patterns provide some clues to function. For example, genes such as TAF-1 (115) and GBF4 (107), which are expressed predominantly in roots, are unlikely to be positively involved in light regulation. However, only CPRF-1 and GBF3 transcripts have been shown to be light regulated, and only CPRF-1 mRNA increased in light (137, 166). Binding-site preferences provide another clue: Although they all bind to motifs with a CACGTG core, their affinities for different elements vary widely depending on the flanking sequences (7, 57, 73, 107, 138, 168). To further complicate the issue, a limited number of ACGT sequences are bound by members of both the GBF and TGA families (135). A final complication is that members of this family bind to DNA as dimers, and many promiscuously homo- and heterodimerize (7, 57, 107, 108, 137). The *A. thaliana* GBF4 does not homodimerize; instead it binds to DNA as a heterodimer with other members of the GBF family in a manner similar to the FOS oncoprotein (107). The net result is that there is probably a vast array of protein complexes that can bind to G-box elements. One can thus envisage hetero- or homodimers of GBF proteins that are involved specifically in light regulation; the difficulty and the challenge is figuring out which GBF is involved.

TGACG-Binding Proteins

As described above, although the genes shown to have necessary TGACG motifs are not light regulated, TGACG motifs are also found in light-regulated genes. Proteins binding to these motifs may therefore be relevant to light regulation. ASF1, an activity from tobacco that binds to the *as-1* element of the CaMV 35S promoter, has been identified by gel-retardation; and a tobacco factor, TGA1a, binding to this element has been cloned and shown to be a member of a small multigene family (75). TGA1a is expressed preferentially in roots, and the *as-1* element confers root-specific expression. Microinjection of recombinant TGA1a protein into leaf cells transformed with a CaMV 35S −90:GUS fusion resulted in GUS expression, consistent with the idea that TGA1a may be limiting in leaf cells (113). *A. thaliana* factors with high similarity to TGA1a have also been cloned (78, 109, 135). Detailed study of *A. thaliana* TGA1 showed that it only bound with high affinity to TGACG and did not need the T of the ACGT motif, enabling it to be distinguished from G-box-binding proteins. It also did not heterodimerize productively with GBF proteins (135). Other TGACG-binding proteins have been cloned from maize (56) and tobacco (74).

I-Box-Binding Proteins

Many factors binding to the I-box or related sequences have been identified in nuclear extracts from both monocots and dicots. Considerable variation has been observed in the types of sequences bound and in complex mobilities, as might be expected given the variation in I-box-like sequences described above. A factor from *L. gibba*, LRF-1, binds specifically to a sequence including a GATA motif and shows increased activity in the light (23). The tobacco protein ASF-2 has specificity for the *as-2* element of the CaMV 35S promoter and also binds to the tandem GATA elements of the petunia *CAB22R* promoter

(87). The *N. plumbaginifolia* protein GA-1 binds to the *A. thaliana RBCS1A* I-box, to tandem GATA repeats found in the *N. plumbaginifolia CABE* promoter, and to the *as-2* sequence (136). By contrast, the tobacco protein GAF-1 can bind only to the I-box (64), whereas the 3AF3 protein binds only to *as-2*-like sequences (133). Four separate I-box-binding activities have been identified in tomato, based on gel-retardation, UV cross-linking, sequence specificity, and biochemical properties (16). For most of these proteins, mutations in the flanking sequences abolished binding, whereas mutation of the core GAT had no effect. One of these activities, IBF-1a, showed a dramatic decrease in the dark. Curiously, of the I-boxes tested, IBF-1a bound exclusively to the *RBCS* element, yet binding was competitively inhibited by a fragment containing the G-box but no GATA motifs (16). Competition for binding in tobacco extracts has been observed between the *as-2* sequence and a wheat *CAB* promoter fragment with no clear similarity (67). It remains to be determined whether this ability of apparently unrelated sequences to compete for DNA-binding factors is caused by factors with dual specificities or requirements for common cofactors.

A tobacco protein with sequence similarity to a GATA-binding protein from fungi has been cloned, but DNA binding was not demonstrated (36). We have isolated a gene from *A. thaliana* that encodes a factor with sequence similarity in the zinc-finger domains to other GATA-binding proteins (RL Bertekap & AR Cashmore, unpublished data). Also, as described above, although 3AF1 generally is classified as an AT-rich-binding protein (cf 11, 64, 153), its cognate sequence includes a GATA motif, and GATA binding has not been excluded (89).

In summary, although numerous GATA-binding proteins have been described, with clear differences in specificities and suggestive changes in activity in response to light, the function of the proteins in light-regulated transcription remains to be determined.

AT-Rich-Binding Proteins

Studies of AT-rich sequences have lagged because these sequences are inherently more difficult to work with and because they are also bound by non-sequence-specific HMG proteins. Proteins that bind to AT-rich elements in light-regulated genes have been identified in nuclear extracts from many species (37, 136, 153) and have been cloned from tobacco (89, 153) and rice (114). The cloned metal-requiring factor 3AF-1 from tobacco binds to the pea *RBCS3A* promoter and is a constitutively expressed member of a small gene family, but it has not been functionally characterized (89). ATBP-1 is another cloned tobacco AT-rich-binding protein with little homology to 3AF-1 (153). It has an AT-hook (a DNA-binding domain found in HMGI) and an amino-terminal glutamine-rich potential activation domain. It is constitutively expressed but has not been functionally characterized. A tobacco nuclear activity that binds to the sequence used to clone ATBP1 resembles the pea nuclear AT-1 activity described by Datta & Cashmore (37) in that binding was inhibited by phosphorylation that appeared to be mediated by CKII (153). However, phosphorylation of the cloned factor was not demonstrated, and no CKII sites were identified within the (partial) sequence (153), illustrating the perils of correlating cloned factors to activities detected in extracts. A gene encoding a rice protein binding to the PE1 motif of oat *PHYA3* has been cloned (114). It is homologous to a soybean protein that binds to AT-rich tracts in soybean seed protein promoters (90); both contain the AT-hook and also have domains similar to histone H1 but have not been characterized further. DNA binding of these AT-binding proteins has not been characterized sufficiently to determine whether they are plant homologs of HMG proteins or sequence-specific factors with HMG-like DNA-binding domains.

Box One (H-Box) Factors

Proteins binding to the Box I (H-box) of *CHS* promoters have been identified by in vivo footprinting (11, 142) and gel-retardation analysis (172), and have been purified by affinity chromatography (172). In vitro transcription from immobilized templates was used to demonstrate that the H-box is required for *CHS* transcription. Depletion of H-box factors by competition in *trans* with H-box sequences reduced the activity (6). Affinity chromatography using H-box sequences yielded two different proteins with similar binding specificities but different molecular weights and other properties. Whether either is involved in light regulation remains to be determined (172).

Other Proteins

Evidence has also been presented for factors that act as repressors of transcription. As described above, the RE-1 element of oat *PHYA* genes represses PHYA transcription in the light. Footprinting analysis showed that the RE1 element is bound with equal activity in extracts from plants treated with R or FR, suggesting that regulation may be mediated by some means other than differential binding (20). Conversely, CA1, a factor that binds to the *A. thaliana CAB140* promoter, is proposed to be involved in down-regulation of this gene in the dark because this activity is lacking in *det1* mutants, which express *CAB140* aberrantly in the dark (149). There were no obvious differences in CA1 activity in light- and dark-grown wild-type plants.

Other interesting proteins have been found: (*a*) GAPF is a tobacco factor that binds to a repeated sequence found in the *GAP* genes of *A. thaliana* and that shows increased activity in extracts from light-grown plants (34, 85). (*b*) WF-1 is a wheat protein that binds to promoters for three Calvin cycle proteins and whose activity increases in light. The WF-1 binding-site was not determined, but conserved motifs from other light-regulated genes were not found in these genes (110). (c) ABF-1 is a pea factor that binds to an undefined site in the pea *AB80* promoter and whose activity also increases in the light (5).

Gruissem and colleagues have performed a comprehensive survey of protein-binding sites in tomato *RBCS* genes (26, 101, 102). Using in vitro footprinting, they uncovered numerous differences between tissues in which *RBCS* genes are known to be differentially expressed, but they did not detect differences in footprinting between tomato cotyledons kept in the dark and those given various light treatments (101). Although in vitro footprinting has limitations, this result suggests that the model for light regulation may need to be changed from the simple version in which regulatory proteins bind to their cognate sequences to activate transcription, to one in which either active forms displace inactive forms bound to that sequence or inactive factors are rendered active by other means.

Conclusion

Numerous proteins binding to regulatory elements in light-regulated promoters have been identified. Several proteins that bind to these elements have also been cloned, although no cloned protein has been identified positively as the factor identified in gel shifts, and the apparent degeneracy of recognition sequences makes this a daunting task. Because proteins interact with chromatin rather than naked DNA in vivo, neighboring or even distant proteins will undoubtedly influence the binding of a given factor, and the sum of the proteins complexed to a given promoter is most likely to orchestrate the assembly of the transcription apparatus (154). The challenge is to decipher the function of the cloned factors in the context of their native promoter.

FACTOR REGULATION

Most factors binding to light-regulated promoters appear to have constitutive binding activity, and most factors that have been cloned are also expressed constitutively. Some exceptions to this rule have been observed. Three factors have shown light-regulated changes in mRNA abundance: *CPRF-1* mRNA increases in the light (166), whereas rice *GT-2* (42) and *A. thaliana GBF3* mRNAs decrease (137). The mRNA abundance of two homeodomain-zip proteins of unknown function from *A. thaliana* also increases upon FR-illumination of dark-adapted plants (25). Five other nuclear factors have increased activities in nuclear extracts prepared from light-grown plants: LRF-1 from *L. gibba* (23), GAPF from tobacco (34, 85), IBF-1a from tomato (16), ABF-1 from pea (5), and WF-1 from wheat (110). Certain DNA-binding proteins may affect their own abundance, as shown for the mammalian *jun* proto-oncogene (4). We have found that both *GBF1* and *GBF3* transcripts (but not *GBF2*) are increased in transgenic *A. thaliana* expressing antisense *GBF1* mRNA. The *GBF1* promoter contains two G-box-like sequences, suggesting that *GBF1* may exert this effect while bound to its own promoter (J Chan & AR Cashmore, unpublished data). Similarly, *CPRF1* binds to a *cis*-acting sequence containing two ACGT elements in its own promoter, and deletion of this element resulted in increased expression in both light and dark (B Weisshaar & K Hahlbrock, personal communication).

Consequently, in many cases factors are likely to be regulated by some means other than changes in abundance. One possibility is to regulate DNA binding, for example by changing localization or affinities of the factors for their cognate sequences. Another is to alter the activation properties of factors that are already bound. Two general types of regulatory mechanisms have been described: phosphorylation/dephosphorylation and differential localization. Phosphorylation by CKII was inferred to inhibit binding of AT-1 and ATBP-1, because binding decreased in nuclear extracts treated with Mg^{2+} and ATP or GTP (37, 153). By contrast, dephosphorylation abolished DNA binding by SBF-1 and 3AF3 (70, 133). Dephosphorylation also reduced G-box binding by GBF, and this binding could be restored by phosphorylation with CKII (81). Inhibitors of protein kinases and phosphatases also affected the activity of cytosolic GBF in parsley (71). However, neither in vitro affinity nor binding specificity was affected by the phosphorylation states of CA-1 from *A. thaliana* (149) or 3AF5 (133). These studies were performed using crude extracts, and indirect effects mediated by other factors cannot be separated from direct effects of phosphorylation state. Effects of phosphorylation on GBF1 binding have been studied using purified components: Phosphorylation in vitro by nuclear CKII increased the G-box-binding activity of recombinant GBF-1, in keeping with the inhibitory effect of dephosphorylation observed in nuclear extracts (71). This result has been confirmed using recombinant GBF-1 and CKII reconstituted from recombinant plant catalytic and regulatory subunits (81a). Many animal transcription factors are also phosphorylated by CKII; the function of this phosphorylation remains to be determined (106). In summary, plants seem to regulate many factors by their phosphorylation state, but the actual role of phosphorylation needs to be determined for each protein.

Cytoplasmic retention by regulatory proteins has been demonstrated for transcription factors of the *rel* family, such as NF-KB (8). Cytoplasmic localization has been shown for GBF-related transcription factors in parsley, and the GBF factor was imported into the nucleus in response to light (71). Similarly, in *A. thaliana* and soybean, most of the G-box-binding activity and the material that cross-reacts with antisera to GBF1 is cytoplasmic (12). We have found that in the dark, GUS-GBF1 or GUS-GBF2 fusions are localized primarily in the cytoplasm of SB-P soybean cells, but GUS-GBF2 is transported into the nucleus in response to light, whereas GUS-GBF1 is not (WB Terzaghi & AR Cashmore, unpublished data).

Differential activity of accessory proteins that modify factor activity or localization but do not bind DNA directly is another potential mode of regulation. One candidate for such activity is GF14, a maize and *A. thaliana* protein that is homologous to mammalian brain 14-3-3 proteins. GF14 was cloned using a monoclonal antibody raised against partially purified GBF, but it does not bind DNA. It does bind Ca^{2+} and is phosphorylated in vitro by *A. thaliana* protein kinases (46, 96, 97). However, GF14 has never been shown unequivocally to be part of the GBF complex, and the alternative that it shares an epitope with the native GBF complex has not been ruled out. Another candidate is the DET1 protein, which is found in the nucleus but has not been demonstrated to bind DNA (120). Because *det1* mutants show aberrant expression of many light-regulated genes and morphogenic pathways in the dark, the likely function of DET1 is to repress transcription of light-regulated genes via some mechanism other than direct DNA binding (31, 44).

SIGNAL TRANSDUCTION

Studies using inhibitors and agonists indicate that G-proteins and Ca/calmodulin act as intermediates in phytochrome-regulated *CAB* gene expression (86, 129) and that Pfr regulates the activity of monomeric G-proteins in pea nuclear envelopes (33). Microinjection studies in cells of tomato PHYA-deficient mutants showed that either Pfr or GTPγS could activate expression of coinjected *CHS*:GUS or *CAB*:GUS fusions, whereas inhibitors of G-proteins prevented expression. Coinjection of Ca/calmodulin activated *CAB*:GUS but not *CHS*:GUS expression, whereas cGMP activated *CHS*:GUS expression, and both were needed to assemble functional chloroplasts (17, 112). These results support the general model that activation of GTP-binding proteins is an early step in Pfr signal transduction, whereas Ca/calmodulin and cGMP act later, on separate pathways.

Another possible step in signal transduction to Pfr-regulated transcription is protein phosphorylation, which has been implicated in other Pfr responses (44, 52, 53, 155). Phytochrome from a moss has a protein kinase domain at its C-terminus and appears to have a light-regulated autophosphorylating activity (2), whereas oat phytochrome antiserum coimmunoprecipitates a maize protein kinase specific for Pr (13). Moreover, there is sequence similarity between a C-terminal section of phytochromes and the histidine kinase domain of bacterial two-component system sensor proteins (140). Thus, there is evidence for protein kinase activity associated with phytochrome. However, evidence for Pfr-regulated phosphorylation effects on gene expression has been mixed. Protein kinase inhibitors did not affect *CAB* gene expression in photoautotrophic soybean cell cultures (129), but inhibitor studies indicated that protein phosphatase activity, most likely PP1, is required for light-induced gene expression in maize (143).

Both G-proteins and protein phosphorylation may also be involved in B-regulated gene expression (146). A B-activated G-protein has been identified in pea (162), and a 120-kDa B-activated protein kinase located in the plasma membrane has been characterized in several species (146). The localization and properties of this kinase indicate that it is either a B photoreceptor or is shortly downstream in the B signal transduction pathway in phototropism (146).

A clue to the likely mode of action of the HY4 B-photoreceptor comes from its similarity to photolyases, which mediate a redox reaction in response to the absorption of B/UV-A. We suggest that HY4 is also likely to mediate light-dependent electron transfer. B irradiation of hypocotyls has been shown to result in rapid plasma-membrane depolarization (147); our working model is that a light-dependent redox reaction initiated by HY4 in some manner results in plasma-membrane depolarization. The availability of HY4 protein and *A. thaliana hy4* mutants will facilitate testing of this model.

Insight into other steps in signal transduction comes from genetic studies. As described above, a series of mutants known as *cop* and *det* have been isolated that have light-grown morphology in the dark (31, 44). Many show aberrant accumulation of all light-regulated transcripts studied, although some do not significantly increase transcription of light-regulated genes in the dark, showing that photomorphogenesis can be separated genetically from light-regulated gene expression (24, 31, 44, 72, 165). Epistasis studies show that *cop1, cop8, cop9, cop10, fus6 (cop11), det1,* and *det2* act on both B and R responses, implying either that they work after B and R signal transduction pathways have merged or that they act on parallel pathways. Mutants affected in three different loci, known as *doc*, specifically accumulate *CAB* transcripts in the dark, showing that signal transduction to *CAB* and *RBCS* expression is distinct (92).

Because *cop, det,* and *doc* mutants are recessive, the normal function of their gene products may be to repress transcription of photomorphogenic genes in the dark (31, 44). The *COP1* (45), *DET1* (120), *COP9* (164), and *FUS6* (28) loci have been cloned. *COP1* encodes a novel protein with zinc-binding motifs, coiled-coil motifs, and elements similar to the WD-40 repeats found in Gβ-proteins, which are all important for COP1 function. COP1 is therefore suggested to be capable of both binding DNA and interacting with other proteins including the transcriptional apparatus (45, 104). *DET1* and *COP9* encode previously unknown nuclear proteins, and *FUS6* encodes a novel protein whose localization has not been determined. Although DET1 is found in the nucleus, it does not bind DNA, which suggests a role in repressing transcription of light-regulated genes by some other means (120). COP9 protein is also nuclear, where it forms part of a large complex that probably includes COP8 and FUS6 gene products (164).

Molecular-genetic analysis thus provides evidence for considerable complexity in signal transduction pathways. The experimental evidence showing that perturbation of G-proteins and modulation of calcium and cGMP are sufficient to trigger expression of light-regulated genes must be reconciled with the observed complexity of light responses and the genetic evidence for multiple loci involved in light signal transduction. Clearly, there are intermediate steps remaining to be elucidated, including the primary action of Pfr itself, and perhaps entire pathways to be discovered.

PERSPECTIVES

Unraveling the signal transduction pathways for the different photoreceptors will continue to be hotbed of activity, as will studies on the regulation of transcription factors and the characterization of proteins that interact with them. However, a great deal of filling-in remains to be done. The involved photoreceptors and fluence requirements remain to be determined for many transcripts, and the question of initiation vs posttranscriptional regulation requires an answer. Regulatory elements need to be defined more precisely in terms of photobiology and sequence. This will require site-directed mutagenesis and gain-of-function experiments, followed by determination of the quantity and quality of light required to induce the mutant constructs. To a great extent, more precise definition of regulatory elements will hinge on improving transient expression systems or *A. tumefaciens*–mediated transformation. Generating transgenic plants is a slow and labor-intensive process and has been difficult to do with many species. Moreover, high variation between transgenic plants requires evaluating many independent transformants for each construct, although flanking the transgene with scaffolding attachment regions appears to reduce this problem (e.g. 3). By contrast, obtaining appropriate amounts of light induction upon transient expression has been difficult, although there are numerous accounts (primarily in monocots) in which transient expression has been used to characterize light-regulated elements (e.g. 9, 14, 21, 79, 118, 128, 134, 155, 158). A consistent problem in transient assays is high levels of expression in the dark, perhaps resulting from basal expression, which may indicate that an important part of light regulation is switching genes off in the dark. However, present systems generally use stressed organisms such as protoplasts or cells with a (relatively) large metal object in their nucleus to attempt to infer something about normal gene regulation before the foreign DNA dissipates. One solution to this problem is to make autonomously replicating vectors such as viral vectors or mini-chromosomes that can be introduced by transient techniques but that would allow cells to recover fully before being analyzed. Of course, the most elegant solution will be a rapid and efficient system of targeted gene replacement, but development of this capability is probably a long way off for plants.

Perhaps the most daunting task is identifying the factors that interact with light-regulated promoters in vivo and their mode of action. Meeting this task will require better ways of identifying proteins bound to specific sequences, the isolation and characterization of more factors, and the development of better ways to analyze these factors (e.g. with better in vitro transcription systems). In particular, classical and reverse genetic approaches should be used to demonstrate the role in vivo of specific factors. By analogy with studies in other systems where many of these tools are already available, identifying the way in which transcription is activated in response to light will be a considerable challenge (154).

ACKNOWLEDGMENTS

We wish to thank colleagues who contributed reprints and preprints of relevant manuscripts. In particular, we thank Drs. A Batschauer, K Hahlbrock, R Oelmüller, E Tobin, and B Weisshaar for communicating data prior to publication. We also thank members of the Cashmore laboratory, especially Dr. L Klimczak and R Bertekap for critical reading of the manuscript.

Literature Cited

1. Ahmad M, Cashmore AR. 1993. *HY4* gene of *A. thaliana* encodes a protein with characteristics of a blue-light photoreceptor. *Nature* 366:162–66

2. Algarra P, Linder S, Thümmler F. 1993. Biochemical evidence that phytochrome of the moss *Ceratodon purpureus* is a light-regulated protein kinase. *FEBS Lett.* 315:69–73

3. Allen GC, Hall GE, Childs LC, Weissinger AK, Spiker S, Thompson WF. 1993. Scaffold attachment regions increase reporter gene expression in stably transformed plant cells. *Plant Cell* 5:603–13

4. Angel P, Hattori K, Smeal T, Karin M. 1988. The jun proto-oncogene is positively autoregulated by its product, Jun/AP-1. *Cell* 55:875–85

5. Argüello G, Garcia-Hernández E, Sánchez M, Gariglio P, Herrera-Estrella L, Simpson J. 1992. Characterization of DNA sequences that mediate nuclear protein binding to the regulatory region of the *Pisum sativum* (pea) chlorophyll a/b binding protein gene AB80: identification of a repeated heptamer motif. *Plant J.* 2:301–9

6. Arias JA, Dixon RA, Lamb CJ. 1993. Dissection of the functional architecture of a plant defense gene promoter using a homologous in vitro transcription initiation system. *Plant Cell* 5:485–96

7. Armstrong GA, Weisshaar B, Hahlbrock K. 1992. Homodimeric and heterodimeric leucine zipper proteins and nuclear factors from parsley recognize diverse promoter elements with ACGT cores. *Plant Cell* 4:525–37

8. Baeuerle PA, Baltimore D. 1988. Activation of DNA-binding activity in an apparently cytoplasmic precursor of the NF-KB transcription factor. *Cell* 53:211–17

9. Bansal KC, Bogorad L. 1993. Cell type-preferred expression of maize *cab-m1*: repression in bundle sheath cells and en-

hancement in mesophyll cells. *Proc. Natl. Acad. Sci. USA* 90:4057–61

10. Bartholomew DM, Bartley GE, Scolnik PA. 1991. Abscisic acid control of *rbcS* and *cab* transcription in tomato leaves. *Plant Physiol.* 96:291–96

11. Batschauer A, Gilmartin PM, Nagy F, Schäfer E. 1994. The molecular biology of photoregulated genes. See Ref. 80, pp. 559–99

12. Bertekap RL Jr, Terzaghi WB, Cashmore AR. 1994. Localization of GBF activity in *Arabidopsis* and cultured soybean cells. *Plant Physiol.* 105S:827

13. Biermann BJ, Pao LI, Feldman LJ. 1994. Pr-specific phytochrome phosphorylation in vitro by a protein kinase present in antiphytochrome maize immunoprecipitates. *Plant Physiol.* 105:243–51

14. Block A, Dangl JL, Hahlbrock K, Schulze-Lefert P. 1990. Functional borders, genetic fine structure, and distance requirements of *cis* elements mediating light responsiveness of the parsley chalcone synthase promoter. *Proc. Natl. Acad. Sci. USA* 87:5387–91

15. Bolle C, Sopory S, Lübberstedt T, Herrmann RG, Oelmüller R. 1994. Segments encoding 5′-untranslated leaders of genes for thylakoid proteins contain *cis*-elements essential for transcription. *Plant J.* 6:359–68

16. Borello U, Ceccarelli E, Giuliano G. 1993. Constitutive, light-responsive and circadian clock-responsive factors compete for the different I-box elements in plant light-regulated promoters. *Plant J.* 4:611–19

17. Bowler C, Neuhaus G, Yamagata H, Chua N-H. 1994. Cyclic GMP and calcium mediate phytochrome phototransduction. *Cell* 77:73–81

18. Britten RJ, Davidson EH. 1969. Gene regulation for higher cells: a theory. *Science* 165:349–57

19. Bruce WB, Christensen AH, Klein T, Fromm M, Quail PH. 1989. Photoregulation of a phytochrome gene promoter from oat transferred into rice by particle bombardment. *Proc. Natl. Acad. Sci. USA* 86:9692–96

20. Bruce WB, Deng X-W, Quail PH. 1991. A negatively acting DNA sequence element mediates phytochrome-directed repression of *phyA* gene transcription. *EMBO J.* 10:3015–24

21. Bruce WB, Quail PH. 1990. *cis*-Acting elements involved in photoregulation of an oat phytochrome promoter in rice. *Plant Cell* 2:1081–89

22. Brusslan JA, Tobin EM. 1992. Light-independent developmental regulation of cab gene expression in *Arabidopsis thaliana* seedlings. *Proc. Natl. Acad. Sci. USA* 89:7791–95

23. Buzby JS, Yamada T, Tobin EM. 1990. A light-regulated DNA-binding activity interacts with a conserved region of a *Lemna gibba rbcS* promoter. *Plant Cell* 2:805–14

24. Cabrera y Poch HL, Peto CA, Chory J. 1993. A mutation in the *Arabidopsis DET3* gene uncouples photoregulated leaf development from gene expression and chloroplast biogenesis. *Plant J.* 4:671–82

25. Carabelli M, Sessa G, Baima S, Morelli G, Ruberti I. 1993. The *Arabidopsis Athb-2* and -4 genes are strongly induced by far-red-rich light. *Plant J.* 4:469–79

26. Carrasco P, Manzara T, Gruissem W. 1993. Developmental and organ-specific changes in DNA-protein interactions in the tomato *rbcS3B* and *rbcS3C* promoter regions. *Plant Mol. Biol.* 21:1–15

27. Caspar T, Quail PH. 1993. Promoter and leader regions involved in the expression of the *Arabidopsis* ferredoxin A gene. *Plant J.* 3:161–74

28. Castle LA, Meinke DW. 1994. A *FUSCA* gene of *Arabidopsis* encodes a novel protein essential for plant development. *Plant Cell* 6:25–41

29. Castresana C, Garcia-Luque I, Alonso E, Malik VS, Cashmore AR. 1988. Both positive and negative regulatory elements mediate expression of a photoregulated *CAB* gene from *Nicotiana plumbaginifolia*. *EMBO J.* 7:1929–36

30. Cheng CL, Acedo GN, Cristinsin M, Conkling MA. 1992. Sucrose mimics the light induction of *Arabidopsis* nitrate reductase gene transcription. *Proc. Natl. Acad. Sci. USA* 89:1861–64

31. Chory J. 1993. Out of darkness: Mutants reveal pathways controlling light-regulated development in plants. *Trends Genet.* 9:167–72

32. Chory J, Reinecke D, Sim S, Washburn T, Brenner M. 1994. A role for cytokinins in de-etiolation in *Arabidopsis*. *Plant Physiol.* 104:339–47

33. Clark GB, Memon AR, Tong C-G, Thompson GA Jr. 1993. Phytochrome regulates GTP-binding protein activity in the envelope of pea nuclei. *Plant J.* 4:399–402

33a. Commission on Plant Gene Nomenclature. 1994. Nomenclature of sequenced plant genes. *Plant Mol. Biol. Rep.* 12:S1–S109 (CPGN Suppl.)

34. Conley TR, Park SC, Kwon HB, Peng HP, Shih MC. 1994. Characterization of *cis*-acting elements in light regulation of the nuclear gene encoding the A subunit of chloroplast isozymes of glyceraldehyde 3-phosphate dehydrogenase from *Arabidopsis thaliana*. *Mol. Cell. Biol.* 14:2523–33

35. Cooke R, Penon P. 1990. In vitro transcription from cauliflower mosaic virus promot-

ers by a cell-free extract from tobacco cells. *Plant Mol. Biol.* 14:391–405

36. Daniel-Vedele F, Caboche M. 1993. A tobacco cDNA clone encoding a GATA-1 zinc finger protein homologous to regulators of nitrogen metabolism in fungi. *Mol. Gen. Genet.* 240:365–73

37. Datta N, Cashmore AR. 1989. Binding of a pea nuclear protein to promoters of certain photoregulated genes is modulated by phosphorylation. *Plant Cell* 1:1069–77

38. Davis MC, Yong M-H, Giulmartin PM, Goyvaerts E, Kuhlemeier C, et al. 1990. Minimal sequence requirements for the regulated expression of *rbcS-3A* from *Pisum sativum* in transgenic tobacco plants. *Photochem. Photobiol.* 52:43–50

39. Dean C, Favreau M, Bond-Nutter D, Bedbrook J, Dunsmuir P. 1989. Sequences downstream of translation start regulate quantitative expression of two petunia *rbcS* genes. *Plant Cell* 1:201–8

40. Dean C, Pichersky E, Dunsmuir P. 1989. Structure, evolution and regulation of *rbcS* genes in higher plants. *Annu. Rev. Plant Physiol.* 40:415–39

41. Dedonder A, Rethy R, Frederiq H, Van Montagu M, Krebbers E. 1993. *Arabidopsis rbcS* genes are differentially regulated by light. *Plant Physiol.* 101:801–8

42. Dehesh K, Bruce WB, Quail PH. 1990. A *trans*-acting factor that binds to a GT-motif in a phytochrome gene promoter. *Science* 250:1397–99

43. Dehesh K, Hung H, Tepperman JM, Quail PH. 1992. GT2: a transcription factor with twin autonomous DNA-binding domains of closely related but different target sequence specificity. *EMBO J.* 11:4131–44

44. Deng X-W. 1994. Fresh view of light signal transduction in plants. *Cell* 76:423–26

45. Deng X-W, Matsui M, Wei N, Wagner D, Chu AM, et al. 1992. *COP1*, an *Arabidopsis* regulatory gene, encodes a protein with both a zinc-binding motif and a Gβ homologous domain. *Cell* 71:791–801

46. de Vetten NC, Guihua L, Ferl RJ. 1992. A maize protein associated with the G-box binding complex has homology to brain regulatory proteins. *Plant Cell* 4:1295–307

47. Dewdney J, Conley TR, Shih MC, Goodman HM. 1993. Effects of blue and red light on expression of nuclear genes encoding chloroplast glyceraldehyde-3-phosphate dehydrogenase of *Arabidopsis thaliana*. *Plant Physiol.* 103:1115–21

48. Dickey LF, Gallo-Meagher M, Thompson WF. 1992. Light regulatory sequences are located within the 5′ portion of the *Fed-1* message sequence. *EMBO J.* 11:2311–17

49. Dickey LF, Nguyen T-T, Allen GC, Thompson WF. 1994. Light modulation of ferredoxin mRNA abundance requires an open reading frame. *Plant Cell* 6:1171–76

50. Donald RGK, Cashmore AR. 1990. Mutation in either G box or I box sequences profoundly affects expression from the *Arabidopsis thaliana rbcs-1A* promoter. *EMBO J.* 9:1717–26

51. Ellis JG, Tokuhisa JG, Llewellyn DJ, Bouchez D, Singh K, et al. 1993. Does the *ocs*-element occur as a functional component of the promoters of plant genes? *Plant J.* 4: 433–43

52. Fallon KM, Shacklock PS, Trewavas AJ. 1993. Detection in vivo of very rapid red light-induced calcium-sensitive protein phosphorylation in etiolated wheat (*Triticum aestivum*) leaf protoplasts. *Plant Physiol.* 101:1039–45

53. Fallon KM, Trewavas AJ. 1994. Phosphorylation of a renatured protein from etiolated wheat leaf protoplasts is modulated by blue and red light. *Plant Physiol.* 105: 253–58

54. Fisher F, Crouch DH, Jayaraman PS, Clark W, Gillespie DA, Goding CR. 1993. Transcription activation by *Myc* and *Max*: Flanking sequences target activation to a subset of CACGTG motifs in vivo. *EMBO J.* 12: 5075–82

55. Flieger K, Wicke A, Herrmann RG, Oelmüller R. 1994. Promoter and leader sequences of the spinach *psaD* and *psaF* genes direct an opposite light response in tobacco cotyledons. *PsaD* sequences downstream of the ATG codon are required for a positive light response. *Plant J.* 6: 359–68

56. Foley RC, Grossman C, Ellis JG, Llewellyn DJ, Dennis ES, et al. 1993. Isolation of a maize bZIP protein subfamily: candidates for the *ocs*-element transcription factor. *Plant J.* 3:669–79

57. Foster R, Izawa T, Chua N-H. 1994. Plant bZIP proteins gather at ACGT elements. *FASEB J.* 8:192–200

58. Frohnmeyer H, Ehmann B, Kretsch T, Rocholl M, Harter K, et al. 1992. Differential usage of photoreceptors for chalcone synthase gene expression during plant development. *Plant J.* 2:899–906

59. Frohnmeyer H, Hahlbrock K, Schäfer E. 1994. A light-responsive in vitro transcription system from evacuolated parsley protoplasts. *Plant J.* 5:437–49

60. Furuya M. 1993. Phytochromes: their molecular species, gene families, and functions. *Annu. Rev. Plant Physiol. Plant Mol. Biol.* 44:617–45

61. Gallo-Meagher M, Sowinski DA, Elliott RC, Thompson WF. 1992. Both internal and external regulatory elements control expression of the pea *Fed-1* gene in transgenic tobacco seedlings. *Plant Cell* 4:389–95

62. Gidoni D, Brosio P, Bond-Nutter D, Bedbrook J, Dunsmuir P. 1989. Novel *cis*-act-

ing elements in petunia *Cab* gene promoters. *Mol. Gen. Genet.* 215:337–44

63. Gilmartin PM, Memelink J, Hiratsuka K, Kay SA, Chua N-H. 1992. Characterization of a gene encoding a DNA binding protein with specificity for a light-responsive element. *Plant Cell* 4:839–49

64. Gilmartin PM, Sarokin L, Memelink J, Chua N-H. 1990. Molecular light switches for plant genes. *Plant Cell* 2:369–78

65. Giuliano G, Pichersky E, Malik VS, Timko MP, Scolnik PA, Cashmore AR. 1988. An evolutionarily conserved protein binding sequence upstream of a plant light-regulated gene. *Proc. Natl. Acad. Sci. USA* 85:7089–93

66. Goff SA, Cone KC, Chandler VL. 1992. Functional analysis of the transcriptional activator encoded by the maize *B* gene: evidence for a direct functional interaction between two classes of regulatory proteins. *Genes Dev.* 6:864–75

67. Gotor C, Romero LC, Inouye K, Lam E. 1993. Analysis of three tissue-specific elements from the wheat *Cab-1* enhancer. *Plant J.* 3:509–18

68. Green PJ, Yong M-H, Cuozzo M, Kano-Murakami Y, Silverstein P, Chua N-H. 1988. Binding site requirements for pea nuclear protein factor GT-1 correlate with sequences required for light-dependent transcriptional activation of the *rbcS-3A* gene. *EMBO J.* 7:4035–44

69. Grob U, Stuber K. 1987. Discrimination of phytochrome dependent light-inducible from non-light-inducible plant genes. Prediction of a common light-responsive element (LRE) in phytochrome dependent light-inducible genes. *Nucleic Acids Res.* 15:9957–72

70. Harrison MJ, Lawton MA, Lamb CJ, Dixon RA. 1991. Characterization of a nuclear protein that binds to three elements within the silencer region of a bean chalcone synthase promoter. *Proc. Natl. Acad. Sci. USA* 88:2515–19

71. Harter K, Kircher S, Frohnmeyer H, Krenz M, Nagy I, Schäfer E. 1994. Light-regulated modification and nuclear translocation of cytosolic G-box binding factors in parsley. *Plant Cell* 6:545–59

72. Hou Y, von Arnim AG, Deng X-W. 1993. A new class of *Arabidopsis* constitutive photomorphogenic genes involved in regulating cotyledon development. *Plant Cell* 5:329–39

73. Izawa T, Foster R, Chua N-H. 1993. Plant bZIP protein DNA binding specificity. *J. Mol. Biol.* 230:1131–44

74. Katagiri F, Chua N-H. 1992. Plant transcription factors: present knowledge and future challenges. *Trends Genet.* 8:22–27

75. Katagiri F, Lam E, Chua N-H. 1989. Two tobacco DNA-binding proteins with ho-

mology to the nuclear factor CREB. *Nature* 340:727–30

76. Kaufman LS, Thompson WF, Briggs W. 1984. Different red light requirements for phytochrome-induced accumulation of *cab* RNA and *rbcS* RNA. *Science* 226:1447–49

77. Kawaoka A, Kawamoto T, Sekine M, Yoshida K, Takano M, Shinmyo A. 1994. A *cis*-acting element and a *trans*-acting factor involved in the wound-induced expression of a horseradish peroxidase gene. *Plant J.* 6:87–97

78. Kawata T, Imada T, Shiraishi H, Okada K, Shimura Y, Iwabuchi M. 1992. A cDNA clone encoding HPB-1b homologue in *Arabidopsis thaliana*. *Nucleic Acids Res.* 20:1141

79. Kehoe DM, Degenhardt J, Winicov I, Tobin EM. 1994. Two 10-bp regions are critical for phytochrome regulation of a *Lemna gibba Lhcb* gene promoter. *Plant Cell* 6:1123–94

80. Kendrick RE, Kronenberg GHM, eds. 1994. *Photomorphogenesis in Plants*. Dordrecht: Kluwer Academic. 828 pp. 2nd ed.

81. Klimczak LJ, Schindler U, Cashmore AR. 1992. DNA binding activity of the *Arabidopsis* G-box binding factor GBF1 is stimulated by phosphorylation by casein kinase II from broccoli. *Plant Cell* 4:87–98

81a. Klimczak LJ, Collinge MA, Farini D, Giuliano G, Walker JC, Cashmore AR. 1994. Reconstitution of *Arabidopsis* casein kinase II from recombinant subunits and phosphorylation of transcription factor GBF1. *Plant Cell.* In press

82. Kubasek WL, Shirley BW, McKillop A, Goodman HM, Briggs W, Ausubel FM. 1992. Regulation of flavonoid biosynthetic genes in germinating *Arabidopsis* seedlings. *Plant Cell* 4:1229–36

83. Kuhlemeier C, Cuozzo M, Green P, Goyvaerts E, Ward K, Chua N-H. 1988. Localization and conditional redundancy of regulatory elements in *rbcS-3A*, a pea gene encoding the small subunit of ribulose-bisphosphate carboxylase. *Proc. Natl. Acad. Sci. USA* 85:4662–66

84. Kuhn RM, Caspar T, Dehesh K, Quail PH. 1993. DNA binding factor GT-2 from *Arabidopsis*. *Plant Mol. Biol.* 23:337–48

85. Kwon H-B, Park S-C, Peng H-P, Goodman HM, Dewdney J, Shih M-C. 1994. Identification of a light-responsive region of the nuclear gene encoding the B subunit of chloroplast glyceraldehyde 3-phosphate dehydrogenase from *Arabidopsis thaliana*. *Plant Physiol.* 105:357–67

86. Lam E, Benedyk M, Chua N-H. 1989. Characterization of phytochrome-regulated gene expression in a photoautotrophic cell suspension: possible role for calmodulin. *Mol. Cell Biol.* 9:4819–23

87. Lam E, Chua N-H. 1989. ASF-2: a factor

that binds to the cauliflower mosaic virus 35S promoter and a conserved GATA motif in *Cab* promoters. *Plant Cell* 1:1147–56

88. Lam E, Chua N-H. 1990. GT-1 binding site confers light responsive expression in transgenic tobacco. *Science* 248:471–74

89. Lam E, Kano-Murakami Y, Gilmartin P, Niner B, Chua N-H. 1990. A metal-dependent DNA-binding protein interacts with a constitutive element of a light-responsive promoter. *Plant Cell* 2:857–66

90. Laux T, Seurinck J, Goldberg RB. 1991. A soybean embryo cDNA encodes a DNA binding protein with histone and HMG-protein-like domains. *Nucleic Acids Res.* 19:4768

91. Lawton MA, Dean SM, Dron M, Kooter JM, Kragh KM, et al. 1991. Silencer region of a chalcone synthase promoter contains multiple binding sites for a factor, SBF-1, closely related to GT-1. *Plant Mol. Biol.* 16:235–49

92. Li H, Altschmied L, Chory J. 1994. *Arabidopsis* mutants define downstream branches in the phototransduction pathway. *Genes Dev.* 8:339–49

93. Liu X-J, Lam E. 1994. Two binding sites for the plant transcription factor ASF-1 are auxin-responsive elements. *J. Biol. Chem.* 269:668–75

94. Liu Z-B, Ulmasov T, Shi X, Hagen G, Guilfoyle TJ. 1994. Soybean *GH3* promoter contains multiple auxin-inducible elements. *Plant Cell* 6:645–57

95. Loake GJ, Faktor O, Lamb CJ, Dixon RA. 1992. Combination of H-box [CCTACC (N)7CT] and G-box (CACGTG) *cis* elements is necessary for feed-forward stimulation of a chalcone synthase promoter by the phenylpropanoid-pathway intermediate p-coumaric acid. *Proc. Natl. Acad. Sci. USA* 89:9230–34

96. Lu G, DeLisle AJ, De Vetten NC, Ferl RJ. 1992. Brain proteins in plants: An *Arabidopsis* homolog to neurotransmitter pathway activators is part of a DNA binding complex. *Proc. Natl. Acad. Sci. USA* 89: 11490–94

97. Lu G, Sehnke PC, Ferl RJ. 1994. Phosphorylation and calcium binding properties of an *Arabidopsis* GF14 brain protein homolog. *Plant Cell* 6:501–10

98. Luan S, Bogorad L. 1992. A rice *cab* gene promoter contains separate *cis*-acting elements that regulate expression in dicot and monocot plants. *Plant Cell* 4:971–81

99. Lübberstedt T, Bolle CEH, Sopory S, Flieger K, Herrmann RG, Oelmüller R. 1994. Promoters from genes for plastid proteins possess regions with different sensitivities toward red and blue light. *Plant Physiol.* 104:997–1006

100. Lübberstedt T, Oelmüller R, Wanner G,

Herrmann RG. 1994. Interacting *cis* elements in the plastocyanin promoter from spinach ensure regulated high-level expression. *Mol. Gen. Genet.* 242:602–13

101. Manzara T, Carrasco P, Gruissem W. 1991. Developmental and organ-specific changes in promoter DNA-protein interactions in the tomato *rbcS* gene family. *Plant Cell* 3:1305–16

102. Manzara T, Carrasco P, Gruissem W. 1993. Developmental and organ-specific changes in DNA-protein interactions in the tomato *rbcS1*, *rbcS2* and *rbcS3A* promoter regions. *Plant Mol. Biol.* 21:69–88

103. McGrath JM, Terzaghi WB, Sridhar P, Cashmore AR, Pichersky E. 1992. Sequence of the fourth and fifth photosystem II type I chlorophyll a/b-binding protein genes of *Arabidopsis thaliana* and evidence for the presence of a full complement of the extended *CAB* gene family. *Plant Mol. Biol.* 19:725–33

104. McNellis TW, von Arnim AG, Araki T, Komeda Y, Miséra S, Deng X-W. 1994. Genetic and molecular analysis of an allelic series of *cop1* mutants suggests functional roles for the multiple protein domains. *Plant Cell* 6:487–500

105. Meier I, Gruissem W. 1994. Novel conserved sequence motifs in plant G-box binding proteins and implications for interactive domains. *Nucleic Acids Res.* 22: 470–78

106. Meisner H, Czech MP. 1991. Phosphorylation of transcriptional factors and cell-cycle-dependent proteins by casein kinase II. *Curr. Opin. Cell Biol.* 3:474–83

107. Menkens AE, Cashmore AR. 1994. Isolation and characterization of a fourth *Arabidopsis thaliana* G-box-binding factor, which has similarities to Fos oncoprotein. *Proc. Natl. Acad. Sci. USA* 91:2522–26

108. Menkens AE, Schindler U, Cashmore AR. 1994. The G-box (CACGTG): a ubiquitous regulatory DNA element bound by the GBF family of bZIP proteins. *Trends Biochem.* In press

109. Miao Z-H, Liu X, Lam E. 1994. TGA3 is a distinct member of the TGA family of bZIP transcription factors in *Arabidopsis thaliana*. *Plant Mol. Biol.* 25:1–11

110. Miles AJ, Potts SC, Willingham NM, Raines CA, Lloyd JC. 1993. A light- and developmentally-regulated DNA-binding interaction is common to the upstream sequences of the wheat Calvin cycle bisphosphatase genes. *Plant Mol. Biol.* 22:507–16

111. Millar AJ, Kay SA. 1991. Circadian control of *cab* gene transcription and mRNA accumulation in *Arabidopsis*. *Plant Cell* 3:541–50

111a. National Science Foundation. 1993. *The Multinational Coordinated* Arabidopsis thaliana *Genome Research Project. Pro-*

gress Report: Year Three. (NSF93-173). Washington, DC: US Gov. Print. Office

112. Neuhaus G, Bowler C, Kern R, Chua N-H. 1993. Calcium/calmodulin-dependent and -independent phytochrome signal transduction pathways. *Cell* 73:937–52

113. Neuhaus G, Neuhaus-Url G, Katagiri F, Seipel K, Chua N-H. 1994. Tissue-specific expression of *as-1* in transgenic tobacco. *Plant Cell* 6:827–34

114. Nieto-Sotelo J, Ichida A, Quail PH. 1994. PF1: an A-T hook-containing DNA binding protein from rice that interacts with a functionally defined d(AT)-rich element in the oat phytochrome *A3* gene promoter. *Plant Cell* 6:287–301

115. Oeda K, Salinas J, Chua N-H. 1991. A tobacco bZip transcription activator (TAF-1) binds to a G-box-like motif conserved in plant genes. *EMBO J.* 10:1793–802

116. Ohl S, Hahlbrock K, Schafer E. 1989. A stable blue-light-derived signal modulates ultraviolet-light-induced activation of the chalcone-synthase gene in cultured parsley cells. *Planta* 177:228–36

117. Okubara PA, Tobin EM. 1991. Isolation and characterization of three genes negatively regulated by phytochrome action in *Lemna gibba. Plant Physiol.* 96:1237–45

118. Okubara PA, Williams SA, Doxsee RA, Tobin EM. 1993. Analysis of genes negatively regulated by phytochrome action in *Lemna gibba* and identification of a promoter region required for phytochrome responsiveness. *Plant Physiol.* 101:915–24

119. Orozco BM, Ogren WL. 1993. Localization of light-inducible and tissue-specific regions of the spinach ribulose bisphosphate carboxylase/oxygenase (rubisco) activase promoter in transgenic tobacco plants. *Plant Mol. Biol.* 23:1129–38

120. Pepper A, Delaney T, Washburn T, Poole D, Chory J. 1994. *DET1*, a negative regulator of light-mediated development and gene expression in *Arabidopsis*, encodes a novel nuclear-localized protein. *Cell* 78: 109–16

121. Perisic O, Lam E. 1992. A tobacco DNA binding protein that interacts with a light-responsive box II element. *Plant Cell* 4: 831–38

122. Piechulla B. 1993. 'Circadian clock' directs the expression of plant genes. *Plant Mol. Biol.* 22:533–42

123. Pwee K-H, Gray JC. 1993. The pea plastocyanin promoter directs cell-specific but not full light-regulated expression in transgenic tobacco plants. *Plant J.* 3:437–49

124. Quail PH. 1994. Phytochrome genes and their expression. See Ref. 80, pp. 71–104

125. Reed JW, Nagatani A, Elich TD, Fagan M, Chory J. 1994. Phytochrome A and phytochrome B have overlapping but distinct functions in *Arabidopsis* development. *Plant Physiol.* 104:1139–49

126. Robbins J. 1993. Gene targeting: the precise manipulation of the mammalian genome. *Circ. Res.* 73:3–9

127. Roberts MW, Okita TW. 1991. Accurate in vitro transcription of plant promoters with nuclear extracts prepared from cultured plant cells. *Plant Mol. Biol.* 16:771–86

128. Rocholl M, Talke-Messerer C, Kaiser T, Batschauer A. 1994. Unit 1 of the mustard chalcone synthase promoter is sufficient to mediate light responses from different photoreceptors. *Plant Sci.* 97:189–98

129. Romero LC, Lam E. 1993. Guanine nucleotide binding protein involvement in early steps of phytochrome-regulated gene expression. *Proc. Natl. Acad. Sci. USA* 90: 1465–69

130. Salinas J, Oeda K, Chua N-H. 1992. Two G-box-related sequences confer different expression patterns in transgenic tobacco. *Plant Cell* 4:1485–93

131. Sancar A. 1994. Structure and function of DNA photolyase. *Biochemistry* 33:2–9

132. Sano H, Youssefian S. 1994. Light and nutritional regulation of transcripts encoding a wheat protein kinase homolog is mediated by cytokinins. *Proc. Natl. Acad. Sci. USA* 91:2582–86

133. Sarokin LP, Chua N-H. 1992. Binding sites for two novel phosphoproteins, 3AF5 and 3AF3, are required for *rbcS-3A* expression. *Plant Cell* 4:473–83

134. Schäffner AR, Sheen J. 1991. Maize *rbcS* promoter activity depends on sequence elements not found in dicot *rbcS* promoters. *Plant Cell* 3:997–1012

135. Schindler U, Beckmann H, Cashmore AR. 1992. TGA1 and G-box binding factors: two distinct classes of *Arabidopsis* leucine zipper proteins compete for binding to the G-box-like element TGACGTGG. *Plant Cell* 4:1309–19

136. Schindler U, Cashmore AR. 1990. Photoregulated gene expression may involve ubiquitous DNA binding proteins. *EMBO J.* 9:3415–27

137. Schindler U, Menkens AE, Beckmann H, Ecker JR, Cashmore AR. 1992. Heterodimerization between light-regulated and ubiquitously expressed *Arabidopsis* GBF bZip proteins. *EMBO J.* 11:1261–73

138. Schindler U, Terzaghi WB, Beckmann H, Kadesch T, Cashmore AR. 1992. DNA binding site preferences and transcriptional activation properties of the *Arabidopsis* transcription factor GBF1. *EMBO J.* 11: 1275–89

139. Schmidt RJ, Ketudat M, Aukerman MJ, Hoschek G. 1992. Opaque-2 is a transcriptional activator that recognizes a specific site in 22-kD zein genes. *Plant Cell* 4:689–700

140. Schneider-Poetsch HAW. 1992. Signal transduction by phytochrome: phytochromes have a module related to the transmitter modules of bacterial sensor proteins. *Photochem. Photobiol.* 56:839–46

141. Schulze-Lefert P, Becker-Andre M, Schulz W, Hahlbrock K, Dangl JL. 1989. Functional architecture of the light-responsive chalcone synthase reporter from parsley. *Plant Cell* 1:707–14

142. Schulze-Lefert P, Dangl JL, Becker-Andre M, Hahlbrock K, Dangl JL. 1989. Inducible in vivo DNA footprints define sequences necessary for UV light activation of the parsley chalcone synthase gene. *EMBO J.* 8:651–56

143. Sheen J. 1993. Protein phosphatase activity is required for light-inducible gene expression in maize. *EMBO J.* 12:3497–505

144. Sheen J. 1994. Feedback control of gene expression. *Photosynth. Res.* 39:427–38

145. Shirley BW, Meagher RB. 1990. A potential role for RNA turnover in the light regulation of plant gene expression: ribulose-1,5-bisphosphate carboxylase small subunit in soybean. *Nucleic Acids Res.* 18: 3377–85

146. Short TW, Briggs WR. 1994. The transduction of blue light signals in higher plants. *Annu. Rev. Plant Physiol. Plant Mol. Biol.* 45:143–71

147. Spalding E, Cosgrove D. 1989. Large plasma-membrane depolarization precedes rapid blue-light-induced growth inhibition in cucumber. *Planta* 178:407–10

148. Stapleton AE. 1992. Ultraviolet radiation and plants: burning questions. *Plant Cell* 4:1353–58

149. Sun L, Doxsee RA, Harel E, Tobin EM. 1993. CA-1, a novel phosphoprotein, interacts with the promoter of the *cab*140 gene in *Arabidopsis* and is undetectable in *det1* mutant seedlings. *Plant Cell* 5:109–21

150. Susek RE, Ausubel FM, Chory J. 1994. Signal transduction mutants of *Arabidopsis* uncouple nuclear *cab* and *rbcS* gene expression from chloroplast development. *Cell* 74:787–99

151. Taylor WC. 1989. Regulatory interactions between nuclear and plastid genomes. *Annu. Rev. Plant Physiol. Plant Mol. Biol.* 40:211–33

152. Thompson WF, White MJ. 1991. Physiological and molecular studies of light-regulated nuclear genes in higher plants. *Annu. Rev. Plant Physiol. Plant Mol. Biol.* 42:423–66

153. Tjaden G, Coruzzi GM. 1994. A novel AT-rich DNA binding protein that combines an HMG I-like DNA binding domain with a putative transcription domain. *Plant Cell* 6:107–18

154. Tjian R, Maniatis T. 1994. Transcriptional activation: a complex puzzle with few easy pieces. *Cell* 77:5–8

155. Tobin EM, Kehoe DM. 1994. Phytochrome regulated gene expression. *Semin. Cell Biol.* In press

156. Ueda T, Pichersky E, Malik VS, Cashmore AR. 1989. The level of expression of the tomato *rbcS-3A* gene is modulated by a far-upstream promoter element in a developmentally regulated manner. *Plant Cell* 1:217–27

157. Ueda T, Waverczak W, Ward W, Sher N, Ketudat M, et al. 1992. Mutations of 22- and 27-kD zein promoters affect transactivation by the Opaque-2 protein. *Plant Cell* 4:701–9

158. Viret JF, Mabrouk Y, Bogorad L. 1994. Transcriptional photoregulation of cell-type preferred expression of maize *rbcS-m3:3'* and 5' sequences are involved. *Proc. Natl. Acad. Sci. USA* 91:8577–81

159. Vorst O, Kock P, Lever A, Weterings B, Weisbeek P, Smeekens S. 1993. The promoter of the *Arabidopsis thaliana* plastocyanin gene contains a far upstream enhancer-like element involved in chloroplast-dependent expression. *Plant J.* 4: 933–45

160. Vorst O, van Dam F, Weisbeek P, Smeekens S. 1993. Light-regulated expression of the *Arabidopsis thaliana* ferredoxin A gene involves both transcriptional and post-transcriptional processes. *Plant J.* 3:793–803

161. Wanner LA, Gruissem W. 1991. Expression dynamics of the tomato *rbcS* gene family during development. *Plant Cell* 3: 1289-1303

162. Warpeha KMF, Hamm HE, Rasenick MM, Kaufman LS. 1991. A blue-light-activated GTP-binding protein in the plasma membranes of etiolated peas. *Proc. Natl. Acad. Sci. USA* 88:8925–29

163. Wehmeyer B, Cashmore AR, Schafer E. 1990. Photocontrol of the expression of genes encoding chlorophyll a/b binding proteins and small subunit of ribulose-1,5-bisphosphate carboxylase in etiolated seedlings of *Lycopersicon esculentum* (L.) and *Nicotiana tabacum* (L.). *Plant Physiol.* 93: 990–97

164. Wei N, Chamovitz DA, Deng X-W. 1994. *Arabidopsis COP9* is a component of a novel signaling complex mediating light control of development. *Cell* 78:117–24

165. Wei N, Kwok SF, Arnim AG, Lee A, McNellis TW, et al. 1994. *Arabidopsis COP8, COP10,* and *COP11* genes are involved in repression of photomorphogenic development in darkness. *Plant Cell* 6: 626–43

166. Weisshaar B, Armstrong GA, Block A, de Costa e Silva O, Hahlbrock K. 1991. Light-inducible and constitutively expressed DNA-binding proteins recognizing a plant

promoter element with functional relevance in light responsiveness. *EMBO J.* 10:1777–86

167. White MJ, Fristensky BW, Falconet D, Childs LC, Watson JC, et al. 1992. Expression of the chlorophyll-a/b-protein multigene family in pea (*Pisum sativum* L.). *Planta* 188:190–98

168. Williams ME, Foster R, Chua N-H. 1992. Sequences flanking the hexameric G-box core CACGTG affect the specificity of protein binding. *Plant Cell* 4:485–96

169. Williams SA, Weatherwax SC, Bray EA, Tobin EM. 1994. NPR genes, which are negatively regulated by phytochrome action in *Lemna gibba* L. G-3, can also be positively regulated by abscisic acid. *Plant Physiol.* 105:949–54

170. Yamazaki K, Katagiri F, Imaseki H, Chua N-H. 1990. TGA1a, a tobacco DNA-binding protein, increases the rate of preinitiation complex formation in a plant in vitro transcription system. *Proc. Natl. Acad. Sci. USA* 87:7035–39

171. Young JC, Liscum EL, Hangarter RP. 1992. Spectral dependence of light stimulated hypocotyl elongation in photomorphogenic mutants of *Arabidopsis*: evidence for a UV-A photosensor. *Planta* 188:106–14

172. Yu LM, Lamb CJ, Dixon RA. 1993. Purification and biochemical characterization of proteins which bind to the H-box *cis*-element implicated in transcriptional activation of plant defense genes. *Plant J.* 3:805–16

Annu. Rev. Plant Physiol. Plant Mol. Biol. 1995. 46:475–96

STARCH SYNTHESIS IN MAIZE ENDOSPERMS

Oliver Nelson and David Pan

Laboratory of Genetics, University of Wisconsin, Madison, Wisconsin 53705

KEY WORDS: amylopectin, amylose, maize starch mutants, phytoglycogen

CONTENTS

ABSTRACT

Research over the past three decades has greatly increased our understanding of starch biosynthesis in storage tissues, but our knowledge is still incomplete. Advances and areas of uncertainty are discussed with a focus on the maize mutants that have been useful in elucidating the process. This emphasis on mutant genes promises additional insights when we discover the role of several mutants known to affect starch synthesis and when we isolate mutations in genes encoding enzymes that have important roles in the process. The goal is a complete understanding of starch synthesis. This understanding will also facilitate the design of transgenic plants that might produce unique starches with promise as raw materials for industrial processes.

INTRODUCTION

Starch biosynthesis in storage organs and tissues has become the subject of investigations in numerous laboratories, following a period during which only a few laboratories displayed substantial interest in the topic. This historical lack of interest is surprising because not only do the cereals, with their starchy endosperms, together with the starchy roots and tubers, feed the human population of the world and supply a major portion of the food for many domestic animals, but starch is also an important industrial commodity. The resurgence in interest may be attributed to a combination of factors. For example, the academic community is realizing that intriguing questions about starch biosynthesis remain unsolved. The wet millers, who produce starch commercially, are interested in learning more about the genetic basis for starch properties. The maize mutants, *waxy* (*wx*) and *amylose extender* (*ae*), have for years been grown on extensive acreages for their unique starches.

Genetic engineering might allow production of additional unique starches by transgenic plants. Genetic engineering has already allowed the enhancement of total starch production in transgenic potato plants (101) through the transfer from *Escherichia coli* of a mutant allele of the gene encoding the enzyme adenosine diphosphate glucose (ADPGlc) pyrophosphorylase (E.C. 2.7.7.27). A basic knowledge of how this enzyme plays a key role in the regulation of glycogen synthesis in bacteria and starch synthesis in some plants is required to design such an approach, and this knowledge has been supplied largely by research in the Preiss laboratory (see 85 for a comprehensive review). A more complete understanding of the entire process of starch synthesis is a prerequisite to any attempts to manipulate the process to our benefit. Hannah et al (38) emphasized this point in their discussion of possible biotechnological modification of polysaccharides.

This review of starch synthesis relies heavily on evidence based on maize mutants. The biochemical lesions responsible for altered starches and/or reduced quantities of starch were all identified first in maize (12, 17, 66, 73, 79, 113). Although research over the past 30 years has identified the most important steps in starch synthesis in storage organs, our information is incomplete. A more complete picture will require identifying biochemical lesions in maize mutants such as *dull* (*du*), *sugary enhancer* (*se*) (32), and *sugary2* (*su2*), as well as finding mutants in important genes such as those encoding the two soluble starch synthases, SSSI and SSSII, so that we can establish their respective roles in starch synthesis. This deficiency in our knowledge extends also to the two branching enzymes, BEI and BEIIa, for which no null or hypomorphic mutants have been reported. We must obtain definitive evidence as to whether enzymes such as starch phosphorylase (E.C. 2.4.1.1) and the disproportionating enzyme (D-enzyme) (81) play any role in starch synthesis. In the following discussion, we use the mutant symbol to designate a homozygous mutant plant or endosperm, e.g. *wx* to specify *wx/wx* plants, *wx/wx/wx* endosperms, or the starch produced by endosperms of that genetic constitution. When we refer to heterozygotes, we use explicit genetic designations.

This review also focuses on the recent advances in our understanding of starch synthesis in potato tubers (45, 70, 115, 116, 118, 119). This research offers insight into starch synthesis in a somatic tissue, in contrast to the endosperms of cereals.

Other uncertainties concerning starch synthesis remain to be clarified: What is the genesis of the phosphate groups covalently linked to some of the glucose molecules of starch, predominantly to C6? Why is potato starch more heavily phosphorylated than the cereal starches (106)? The starch synthases require an oligosacccharide primer to which they add glucose units in α-1,4 linkages, and the best candidate for this function has been found in potato tubers (52, 109, 110). The question of which metabolite(s) enter the amyloplasts as precursor(s) to the substrate(s) that will be utilized for starch synthesis has received different answers in experiments with amyloplasts isolated from several organisms. Seeds that are homozygous mutants for the most severe blocks in starch synthesis still produce 15–20% as much starch as do nonmutant controls, even when the mutant produces no protein product (36). This result raises the important question of whether the small amount of residual enzyme activity observed (in most cases encoded by another locus) is sufficient to account for the starch synthesis observed, or is the mutant revealing a secondary route of synthesis?

COMPONENTS OF STARCH

Amyloses

The starch produced in nonmutant endosperms or other storage organs consists of two principal components, amylose and amylopectin. Although amylose is usually described as a linear α-1,4 glucan, some of the molecules are sparsely branched. Takeda et al (107) found that about half the molecules in an amylose preparation, which was verified as pure amylose by gel filtration, were branched, with an average of 5.3 branches per molecule. The number average degree of polymerization, or d.p.n., indicates the average number of glucose units per polymer in a starch sample. In these lightly branched amyloses, d.p.n. varied between 930 and 990. The variability depends on the steeping condition to which the corn kernels were subjected prior to isolation of the starch, and the maximum degree of polymerization was between 1930 and 2220. The average chain length varied between 295 and 335 glucose molecules. The iodine affinity was 20.1g/100g with an absorption maximum (λ_{max}) of 644 nm, following reaction with iodine. The iodine affinity or iodine-binding capacity measures the weight of iodine bound by a stated weight of starch. Under standard conditions, amylose binds approximately 20% of its own weight of iodine, whereas amylopectin binds none, so the ratio of these two components in any sample can be ascertained (10). The amount of monoacyl lipids within the starch granules is positively correlated with the amount of amylose that is present (67, 99), and failure to remove the lipids before adding the KI/I_2 agent results in an estimate of amylose that may be 3.5–7.4% lower than the true value (68).

The Hizukuri laboratory (108) has examined the heterogeneity of structure in corn amylose. Incubation of the amylose in 10% aqueous 1-butanol separated the sample into soluble and insoluble fractions. These fractions had linear and branched molecules in the proportions of 84:16 and 60:40, respectively. The branched molecules in the soluble fraction had a d.p.n. of 1620, about 20 branches per molecule, with short (d.p.n. ~ 18), long (d.p.n. > 230), and very long (d.p.n. not stated) chains. The large branched molecules in the soluble fraction may be immature amylopectin molecules. The branched molecules in the insoluble fraction had a d.p.n. of 2200 and a mean of 6 chains per molecule. These chains were short (d.p.n. ~ 18) or very long (d.p.n. > 370).

Amylopectins

The amylopectins from the starch preparation discussed above (107) had an iodine affinity of 1.05–1.25g/100g with a λ_{max} between 554 and 556 nm. The d.p.n. varied from 4800 to 10,200 with an average chain length of 21 for all samples. The low estimate of 4800 is for starch that had been steeped in cold water and may reflect some degradation by amylases. The percentages of amylopectin chains that fell into different average weights of polymerization (d.p.w.) as measured by gel filtration for commercially prepared starch were F1 (d.p.w. not stated) 10%, F2 (d.p.w. 47) 20%, and F3 (d.p.w. 18) 70%, so that there is a preponderance of short chains. All preparations had similar distributions of chain lengths, which, together with the relatively high iodine affinities for amylopectins, suggests that there are long B-chains together with well-separated side chains. Manners (61) has discussed the advances in our understanding of amylopectins as well as the techniques of elucidating these structures.

Both the amylose and amylopectin components can be degraded completely by the combined action of β-amylase and pullulanase, indicating that the branch points result from α-1,6 linkages. With β-amylase alone, the amylose samples are about 82% hydrolyzed, and the amylopectin samples are 59% hydrolyzed.

GENETIC CONTROL OF STARCH SYNTHESIS

Mutants Affecting the Supply of Substrate

In considering the loci known to affect starch synthesis, it is useful to discuss separately those loci that have a role in the synthesis of substrate(s) and those loci that actually participate in the synthesis of the polysaccharides.

The primary source of carbon skeletons for starch synthesis is sucrose translocated to the endosperm. In maize, most of the sucrose is hydrolyzed as it enters the endosperm, where sucrose is then resynthesized (89). Chourey et al (19) have detected the presence of sucrose phosphate synthase in developing maize endosperms. This enzyme is required for sucrose synthesis because although sucrose synthase can synthesize sucrose in vitro, it does not appear to do so in the maize endosperm (18). It has also been reported (66) that the maize mutant *miniature1* (*mn1*) seed, which has a grossly defective seed phenotype, lacks both soluble and wall-bound invertase activity (E.C. 3.2.1.26) in the basal portion of the developing seed. The lack of invertase activity in the seed is correlated with the degeneration of maternal cells in the pedicel, which was noted in the original report of the mutant (58). The *mn1* endosperms contain only 40% as much soluble sugars as do *Mn1* endosperms, and the percentage that is sucrose is much higher in *mn1* than in *Mn1* at short intervals following a ^{14}C pulse. It is clear that some sucrose can enter maize endosperms without being hydrolyzed, but it is proposed that the monosaccharide gradient between pedicel apoplasm and endosperm cells is a driving force in assimilate movement into the endosperm (90).

Sucrose in the developing endosperm is converted to uridine diphosphoglucose and fructose by the major isoform of sucrose synthase (E.C. 2.4.1.13), which is encoded by nonmutant alleles of the *shrunken-1* (*sh1*) locus (17). Homozygous *sh1* seeds produce only 55–60% as much starch as do nonmutant seeds in comparable genetic background; mutant seeds display a distinctive shrunken phenotype and frequently have a cavity in the endosperm. The *sh1* locus is one of the few loci in higher plants at which interallelic complementation—in which certain mutant alleles of independent origin are capable of interacting in heterozygous endosperms to produce a nonmutant phenotype—has been reported. An increase to only 10% of nonmutant activity from the 7% encoded by a second locus encoding the minor isoform of sucrose synthase (16) is sufficient to change the phenotype from a markedly defective one to one that is completely normal (18).

The question of what metabolite(s) enter the amyloplasts in starch-synthesizing sink organs has been difficult to settle definitively. Intact amyloplasts are difficult to isolate because the heavy starch granules tend to rupture the membranes; however, there has been some success in making amyloplast preparations from both maize (15, 25) and wheat (29). ^{14}C-labeled triose phosphates were reported to be taken up by isolated maize amyloplasts, with the label subsequently found in an alcohol-precipitable product presumed to be starch (25). This finding was discounted by the same laboratory (110a) after discovering that the product was not starch. The general consensus (75, 85, 110a) is that hexose phosphates are the compounds that are taken up by amyloplasts. Only Glc-1-P was effective for amyloplasts from wheat endosperms (3), whereas Glc-6-P was taken up by pea embryo amyloplasts (43). Many maize inbreds of commercial importance lack an endosperm cytosolic phosphoglucomutase with no discernible consequences on yield or viability (80), suggesting that maize amyloplasts may import both Glc-1-P and Glc-6-P.

Heldt et al (42) proposed that amyloplasts contain a Pi translocator with altered specificity such that not only Pi and triose phosphates but also Glc-6-P can be transported. Sullivan et al (104) cloned and sequenced a cDNA for a nonmutant allele, *Bt1-R802*, of the *bt1* locus in maize. The inferred amino acid sequence has at its amino terminus the characteristics of a chloroplast transit peptide. The mature sequence has its greatest similarity to a yeast ATP/ADP carrier protein and also has considerable similarity to a family of mitochondrial inner membrane translocator proteins. Li et al (55) showed that nonmutant BT1 protein synthesized in vitro can be imported by pea chloroplasts and become part of the inner membrane. The BT1 protein has been reported to be the most abundant protein in *Bt1* endosperm amyloplast membranes, but it is missing from *bt1* amyloplasts, from amyloplasts of a *Bt1* endosperm suspension culture, and from membrane polypeptides from maize chloroplasts or root mitochondria (15). Seeds homozygous for mutant alleles of this locus are markedly defective in endosperm development and have only about 14% as much starch as do nonmutant seeds (71). The function of this major amyloplast membrane protein is still unresolved.

Mutations in both the *shrunken-2* (*sh2*) and *brittle-2* (*bt2*) genes result in the loss of ADPGlc pyrophosphorylase (E.C. 2.7.7.b) activity from developing maize endosperms (113). Using a more efficient assay, Dickinson & Preiss (23) showed that both mutants had low activity for this enzyme (~ 5–7% of nonmutant activity). Mutant seeds produce only about 20% as much starch as do nonmutant seeds, and they are extremely defective. The developing endosperms of *sh2* and *bt2* are much higher in sucrose plus reducing sugars (59% of dry weight for both *bt2* and *sh2* at 21 d after pollination) than is *sugary1* (*su1*), which has about 16% (71). Further, *bt2* and *sh2* mutants retain a considerably higher proportion of their sugars after harvest at three weeks postpollination than does *su1* (38). The *su1* mutant is the basis for ordinary sweet corn. The *sh2* mutant has been used as the basis for a second type of sweet corn, the so-called super sweet corn.

Early research (39, 40) on the genetic control of ADPGlc pyrophosphory-lase activity led to the conclusion that both *bt2* and *sh2* are structural genes for the enzyme, which is now known to be a heterotetramer (98). This contrasts with the enzyme of enteric bacteria, which is homotetrameric. The *sh2* locus encodes the larger subunit of ADPGlc pyrophosphorylase in maize, and the sequence homology of the two genes, although not exceptionally high, indicates that they may have evolved from an ancestral gene (76, 98). Extensive research by Preiss and collaborators (85) has shown that ADPGlc pyrophos-phorylase plays a key role in the regulation of starch synthesis in the maize endosperm through its stimulation by 3-phosphoglycerate and its inhibition by orthophosphate. This regulatory role for ADPGlc pyrophosphorylase is appar-ently not universal among the cereals because the barley endosperm enzyme is insensitive to 3-phosphoglyerate and inorganic phosphate (49). The enzyme in potato tubers is regulated in the same way as the maize endosperm enzyme (70). In addition, it has been possible to enhance starch synthesis in transgenic potatoes by using a mutant ADPGlc pyrophosphorylase gene from *E. coli* that is less sensitive to its regulatory factors (fructose-1,6-bisphosphate activation and AMP inhibition) than is the nonmutant enzyme (101). The possibility of further modifications of starch synthesis in transgenic potato plants to produce novel carbohydrates or to alter amylose-to-amylopectin ratios has been dis-cussed (91).

A challenge to the concept that ADPGlc as a substrate for starch biosynthe-sis is produced by ADPGlc pyrophosphorylase (84) has prompted consider-able discussion. Researchers in Akazawa's laboratory have found that amy-loplasts in cell cultures of *Acer pseudoplatanus* can import ADPGlc via an ADP-ATP translocator, and the label present in the ADPGlc is found sub-sequently in the starch. This observation has led to the suggestion that ADPGlc is synthesized in the cytosol from sucrose by sucrose synthase, utiliz-ing ADP as an acceptor rather than UDP, which is the preferred substrate. Okita (75) has summarized the evidence that is incompatible with this hy-pothesis. Among the salient points are the great decrease in starch synthesis of mutants in *bt2* and *sh2*, genes that encode the two subunits of ADPGlc pyro-phosphorylase (113). No starch is synthesized in the leaves of *Arabidopsis thaliana* mutants that have lost activity in either of the subunit genes, *ADG1* or *ADG2*, nor in the tubers of potato plants expressing antisense constructs of an ADPGlc pyrophosphorylase gene (70). In addition, ADPGlc pyrophosphory-lase activity is apparently confined to amyloplasts in storage tissues (26, 28). There is little convincing evidence that this alternate path of ADPGlc synthesis is a major source of ADPGlc. However, the examination of nine *sh2* and four *bt2* mutants of independent origin showed that all these mutant genotypes had low residual endosperm ADPGlc pyrophosphorylase activity and synthesized starch in equivalent quantities to that synthesized by seeds homozygous for the reference alleles, *bt2-R* and *sh2-R*. The double mutant seeds, *bt2-R and sh2-R,* have activity as high as either of the single mutants (40). This observation suggests that the residual activity is not encoded by the mutant alleles, and this has been confirmed in a recent report (36). Shannon and colleagues (57) have shown that in vitro ADPGlc is transported into maize amyloplasts and incorporated into starch. They suggest that the cytosolic synthesis of ADPGlc via sucrose synthase may provide a source of ADPGlc in early endosperm development (13 d after pollination) before much ADPGlc pyrophosphorylase is present. Developing rice endosperms can produce ADPGlc from sucrose and ADP at a rate that is about half that at which ADPGlc is produced by the pyrophosphorylase (82). Shannon's group also reported (83) that developing *bt1* endosperms have 10 times as much ADPGlc present as do nonmutant endosperms but that *bt1* amyloplasts do not have unusually high quantities present, which suggests that in the absence of the BT1 protein(s) from the amyloplast membrane, ADPGlc can be synthesized in the cytosol.

Enzymes Catalyzing Steps in Starch Synthesis

AMYLOSE SYNTHESIS   Sprague et al (100) showed that *waxy* (*wx*) maize lacked the amylose component of starch. Nelson & Rines (73) reported that developing *wx* endosperms lacked a granule-bound starch synthase (GBSS) that transferred Glc from UDPGlc to the nonreducing ends of starch molecules. They concluded that there must be two pathways for starch synthesis, one leading to amylose and the other to amylopectin. An enzyme with the specificity of the enzyme missing from *wx* endosperms was reported to be present in starch granules from developing bean cotyledons and maize endosperms (54). ADPGlc was the preferred substrate for the enzyme (87) in starch granules from maize endosperms, because incorporation into starch from ADPGlc is about three times as rapid as when UDPGlc is the substrate (1).

Some researchers have been skeptical about whether the *wx* locus is indeed the structural locus for this enzyme (85) and whether the enzyme is responsible for amylose synthesis (2), despite sound genetic and biochemical reasons for these conclusions. The evidence for *Wx* encoding a GBSS includes the observation that the activity of this GBSS (E.C. 2.4.1.b) increases linearly with the number of nonmutant (*Wx*) alleles present in the endosperms from which the starch granule preparation is made. This response extends from one copy in a diploid series (*wx/wx/Wx*) through six copies (*Wx/Wx/Wx/Wx/Wx/Wx*) in tetraploid endosperms (111, 112). The reference allele for the *wx* locus, *wx-C*, results in starch granules lacking the *Wx* protein (27). Wessler et al (123) have reported that two intermediate *wx* alleles, *wx-S5* and *wx-S9*, which have resulted from the excision of a transposable element, *Ds*, from the mutable *wx* allele, *wx-ml*, condition the production of starch with about 5% and 9% amylose, respectively. These alleles contain extra nucleotides in the *Wx* gene, remnants of the eight base-pair duplication of *wx* DNA created at the site of the *Ds* insertion. This report (123) shows the possibility of isolating *wx* alleles capable of producing starches with a range of amylose contents by screening the intermediate phenotype revertants from transposable element–mutants. There are also *waxy* mutants in barley, rice, and sorghum (30). A GBSS is implicated as the gene product in all these cereals.

After identifying a GBSS in potato tubers (118), the same group (45) isolated a potato mutant that produces starch consisting solely of amylopectin. The mutant was isolated following X-irradiation of the excised leaves of a monoploid (1X) *Solanum tuberosum* clone, which were then induced to produce adventitious shoots. After tentative identification of an amylose-free minituber, the other minitubers from that shoot were used to propagate the line by producing adventitious shoots. Spontaneous chromosome doubling restored 2X and 4X forms. This route was followed because all the waxy (amylose-free) variants known in other species are the result of recessive mutations. In this instance, the mutation was sought in somatic cells of a tetraploid species. The isolation of this mutant in the potato is a considerable achievement. Subsequent research has demonstrated that the *amylose-free* mutant (*amf*) can be repaired by the gene that encodes the GBSS (115) and that the DNA sequence of the *amf* allele differs from the nonmutant *Amf* sequence by the deletion of a single base pair (116).

Given this evidence linking amylose synthesis to the GBSS, it was a surprise when Smith (97) reported that in the pea embryo, a protein homologous to the WX protein was not the GBSS. The GBSS activity was ascribed to a 77-kDa protein that had only weak homology to the WX protein, which was abundant in the pea starch preparations but seemingly without activity. Preiss and colleagues (95) investigated the GBSS of pea by solubilizing proteins bound to pea embryo starch followed by chromatography on a Mono-Q column. Preiss et al identified two activity peaks: The major peak, GBSSI, which had 80% of the starch synthase activity, contained only the 59-kDa WX protein. The second activity peak, GBSSII, also contained the 59-kDa WX protein together with other proteins but not the 77-kDa polypeptide. The authors suggested that the partition of some WX protein into the second peak might result from that protein having more linked α-glucan primer. This hypothesis is based on the observation that starch-synthesizing activity in the second peak without added primer was higher relative to the activity with primer present than was the primerless activity in similar tests of activities in the first peak. This report does clarify that GBSS activity is in the WX protein.

SYNTHESIS OF AMYLOPECTIN   The laboratory that first detected the existence of the GBSSs was also the first to report the existence of soluble starch synthase (SSS; E.C. 2.4.1.b) in maize (34) and to show that in contrast to the endosperm GBSS, which can utilize all four nucleoside diphosphate glucoses as substrates (1), SSS requires ADPGlc. Ozbun et al (78) found that there are two SSSs in maize endosperms, SSSI and SSSII. SSSI is capable of unprimed synthesis if the reaction medium contains citrate and bovine serum albumin, whereas SSSII requires a primer. Current activity in the purification of the isoforms (56, 69) should aid our understanding of their relative roles in amylopectin synthesis, although the greatest help would be mutations that abolish the activity of each isoform. No mutation affecting the activity of either enzyme has been identified in a higher plant, which is surprising when one considers the numerous independent mutations that have disrupted other loci affecting starch synthesis. Three possible reasons for this observation can be suggested. First, such mutations may be lethal. Second, loss of either enzyme does not reduce the amount of starch so that a mutation is not detectable. Third, each enzyme is encoded by duplicate loci and mutations in each locus are required before starch synthesis is affected. The first two possibilities are not likely, but the third is plausible.

Mutations have been identified in the unicellular green alga *Chlamydomonas reinhardtii* that inactivate the more active of the two SSSs present in the large, single chloroplast of nonmutant algae (33). These mutants, *st3,* lack the SSS that is partly homologous to maize endosperm SSSII. *st3* synthesize only 20–40% of the starch present in nonmutant cells, the mutant starch has fewer intermediate-size glucans that constitute clusters making up the bulk of the amylopectin fraction, and there are more very short chains (d.p. 2–7) in this fraction than in nonmutant amylopectin.

Delrue et al (22) have isolated *C. reinhardtii* mutants that are missing the GBSS. These mutants, *st2,* lack not only amylose but also a minor amylopectin fraction that is present in nonmutant starch. This amylopectin fraction (type II, ~ 2–10% of the starch in nonmutant cells) has very long unbranched chains and intermediate branches with chain lengths of 8–40. Delrue et al conclude that the GBSS is also responsible for the synthesis of this fraction, implying that some linear molecules are not protected from branching enzymes. This conclusion is compatible with reports that the amylose from maize has some branched molecules (107, 108). The process may go farther in *C. reinhardtii* so that some molecules are isolated in the amylopectin fraction. Most laboratories investigating the amylopectins synthesized by *wx* maize mutants do not report finding polysaccharides different from the amylopectin component of *Wx* starch. For this reason, Fuwa et al (35) examined the amylopectins produced by the *bt1, du, su1,* and *su2* mutants in double mutants with *wx* so that an anomalous amylose could not complicate assessment of changes in mutant amylopectins.

Branching enzymes   Because the basic reaction in the synthesis of starch macromolecules is the formation of α-1,4 glucosidic linkages, leading to the synthesis of a linear polymer, the activity of branching enzymes (E.C. 2.4.1.18) in introducing α-1,6 branch points into the straight chain polymers or unbranched portions of partially branched polymers are important in amylopectin synthesis. Lavintman (51) reported the presence of two branching enzymes in developing *su1* endosperms. One was the Q-enzyme that can introduce branch points into amylose to form a branched compound similar to amylopectin. The other was an enzyme that starting with amylose, could make phytoglycogen, the highly branched water-soluble polysaccharide that is the principal polysaccharide in *su1* endosperms (20). Manners et al (62) also found that in addition to a Q-enzyme in *su1* endosperms, there was an activity that could further branch amylopectin. Boyer & Preiss (12, 13) reported that developing maize endosperms had three branching enzymes, BEI, BEIIa, and BEIIb, which they concluded were under separate genetic control based on their finding that the *ae* mutants of maize lack the IIb isoform but have undiminished I and IIa activity. The reduced amount of starch produced by *ae* seeds has a higher amylose content ($\sim 40\%$) and anomalous amylopectin with fewer branches than are present in nonmutant starch (44, 105). The anomalous amylopectin of *ae* seeds has been reported by several laboratories beginning with Mercier's (65). Purification of the isozymes has shown that their action spectrum differs in vitro (37). The BEI isozyme has 10-fold greater activity in branching amylose than either BEIIa or IIb but less than one sixth of their activity when the substrate is amylopectin. The authors suggest that the lightly branched polysaccharides produced by BEI are the substrates for the action of BEIIa and b. The two BEII isozymes are similar in immunological properties. Both are 80-kDa proteins and produce the same tryptic and chymotryptic digests, yet are separable on a DEAE-Sepharose column.

The reference allele, *ae1-Ref* (117), and, with one exception, all *ae1* alleles isolated subsequently are recessive. Stinard et al (102) have reported the isolation of a dominant mutant allele, *Ae1-5180*. This allele arose in a genotype containing the transposable element *Mutator* (*Mu*) and is associated with a complex rearrangement. Cloning of the *ae1* locus has made it possible to show that nonmutant *Ae1-+* alleles and the *ae1-Ref* allele produce a 2.7-kb transcript that is not found in endosperms that have one, two, or three copies of the *Ae1-5180* allele. The mechanism of suppression of nonmutant expression in heterozygous endosperms, *Ae1-+/Ae1-+/Ae1-5180* or *Ae1-5180/Ae1-5180/Ae1-+*, is not yet understood, but it opens a route for the utilization of elite hybrids to produce *ae1*-type starch without an extended breeding program. The investigations of this unusual mutant support the conclusion (41) that the *ae1* locus is the structural locus for BEIIb. No mutation has been identified that specifically affects the activity of BEI or IIa.

In contrast to the branching activity found in maize kernels, the activity present in potato tubers involves only a single 79-kDa enzyme (119). An antibody against the native potato enzyme recognizes maize BEI and BEIIb, but the antibody fully inhibits the activity of only BEI. Antibodies against the denatured potato BE react with denatured forms of all the maize BEs but are unable to inhibit the activity of any of the maize enzymes.

There are two other loci, *dull-1* (*du1*) and (*su2*) at which mutations result in the synthesis of starch with higher percentages of amylose (about 35–40%, as measured colorimetrically following reaction of the starch with an KI/I$_2$ solution) (24). The endosperms of both mutants synthesize less starch than do nonmutant endosperms, and the amylopectin component is reduced more than is the amylose component (71). It is hypothesized that the biochemical lesions have their primary effect on the synthesis of amylopectin, and the effect on amylose synthesis is secondary. Boyer & Preiss (13) reported that in *du1* endosperms, BEIIa and SSSII activity are both much reduced. It is difficult to assess this tentative identification of the primary lesion in the *du1* mutants in view of reports (120, 121) that *du1* amylopectin has the highest degree of branching of any sample tested. The double mutant *du1 su2* produces less starch than either mutant alone but with a somewhat higher amylose content (47%). This datum (50) suggests that the nonmutant alleles of the two loci do not catalyze sequential steps in a pathway leading to the production of amylopectin.

The *su1* mutants also result in the production of starch with a higher percentage of amylose (20) in most genetic backgrounds. This mutant is the genotypic basis of the sweet corn varieties and hybrids. It is also the only locus at which the principal polysaccharide storage product of mutants is not starch but the highly branched, high molecular weight, water-soluble polysaccharide, phytoglycogen. Because the amylopectin component is reduced the most, the percentage of starch that is amylose is enhanced; however, the content of amylose is reduced in these mutants (20). The action of branching enzymes that introduce the α-1,6 branch points are important in the formation of amylopectin and phytoglycogen. Boyer et al (14) found that maize BEI had the properties of the enzyme capable of converting amylose or amylopectin to phytoglycogen (51, 62) and that a homogeneous BEI preparation could attack *su1* starch granules with the release of a phytoglycogen-like glucan. No change was noted when nonmutant starch granules were challenged with the enzyme. This report assigns greatly different properties to BEI than does that of Guan & Preiss (37). Because both *su1* and nonmutant endosperms have all three BEs present, what prevents BEI from producing the same effect in nonmutant endosperms as it does in *su1* endosperms in vivo? The answer may lie in the report (79) that developing *su1* endosperms have much reduced debranching activity measured as the ability to hydrolyze pullulan. There are also *sugary* mutants in rice (6) and in sorghum (92). These variants accumulate large amounts of a water-soluble polysaccharide and presumably represent mutations in genes that are homologous to the *su1* gene of maize.

Erlander (31) suggested that phytoglycogen is an intermediate on the pathway to starch, and the phytoglycogen is subsequently debranched with the debranched chains forming the amylose component of starch. This is not the route by which amylose is synthesized (73), and no enzyme present in developing maize endosperms is capable of debranching phytoglycogen (63). Boyer et al (14) also noted that the A:B chain ratio of phytoglycogen, which is 1:1, could not produce the 2:1 chain ratio of amylopectin by debranching. Pan & Nelson (79) suggested that the primary effect of the *su1* mutation was a deficiency of a debranching enzyme, emphasizing that the role of the enzyme in nonmutant endosperms was a dynamic one with both branching and debranching occurring simultaneously. The attenuation of debranching activity by the *su1* mutation allows the branching activity to proceed to the production of phytoglycogen. Additional evidence for this role of the nonmutant alleles at the *su1* locus comes from the observation (20) that the *su1* alleles are epistatic to mutant alleles at almost all other loci affecting starch synthesis (i.e. the double mutant endosperms produce large quantities of phytoglycogen). One exception is the *ae su1* double mutant. The *ae* mutant lacks BEIIb (12, 13) and result in the production of amylopectin with longer and less-branched chains (46, 47, 105). The other exception is the *sh2* mutant, and the double mutant *sh2 su1* also lacks the copious production of phytoglycogen (20, 44). This observation raises an interesting question because the nonmutant alleles at the *sh2* locus, in concert with those at the *bt2* locus, catalyze the synthesis of ADPGlc, the substrate for the starch synthases. Is there a pathway, other than through ADPGlc, that leads to the synthesis of unbranched starch? Or is it simply that one or more of the branching enzymes is inhibited by the high concentration of reducing sugars resulting from the presence of the *sh2* allele?

The *su1* locus has now been cloned, and a partial cDNA has been sequenced (M James, personal communication). The deduced amino acid sequence has homologies to enzymes that hydrolyze starch and, in particular, to two domains that are conserved in such hydrolases. In BEI, the catalytic regions of amylolytic enzymes are conserved (8), and a debranching enzyme might also contain these regions.

Almost all the enzymes that catalyze the formation of α-1,4 linkages between glucose molecules require an oligosaccharide primer. When no added primer is required, it seems that a previously synthesized primer is already attached to the enzyme. The most likely candidates for supplying primer molecules in storage tissue have been found in potato tubers. Lavintman & Cardini (52) reported that a particulate fraction isolated from potato tuber juice could transfer glucose from UDPGlc to a glucoprotein, which then could act as an acceptor for Glc from either ADPGlc or Glc-1-P using the same fraction. Subsequent investigations (53, 109) suggested that a two-step reaction is involved. In the first step, an acceptor protein accepts Glc from micromolar concentrations of UDPGlc, but this reaction plateaus after 15 min with only one Glc molecule attached to the protein. The second step transfers Glc from micromolar concentrations of ADPGlc, UDPGlc, or Glc-1-P to the glucoprotein acceptor, resulting in the formation of α-1,4-glucans. The two-step nature of the reaction was confirmed by showing that 1,5-D-gluconolactone, a potent inhibitor of the starch synthases and phosphorylase, does not affect the glucosylation of the acceptor (110). The acceptor protein catalyzes addition of the single Glc residue to itself to form the primer utilized in the second step (4). The acceptor protein is designated as a UDPGlc-protein transglucosylase (UPTG; E.C. 2.4.1.112). Sivak et al (93, 94) have also reported that a starch phosphorylase (phosphorylase II) is capable of synthesizing long α-1,4-glucans in the absence of a primer. Glucoamylase treatment of the enzyme to eliminate a possible endogenous primer has no effect on its activity. As yet, such activities have not been reported in other starch-synthesizing organs.

Investigations Of Mutant Starches

Several laboratories have investigated the structures of starches produced by the mutants that affect synthesis (7, 11, 46, 47, 65, 105, 120, 121, 124). It is difficult to compare the reports rigorously because the methods of starch isolation and separation of amylose and amylopectin fractions, the genetic background in which the mutants are present, and the chromatographic separations of native or debranched starches may differ from laboratory to laboratory. Further, Asaoka et al (5) have reported that the environmental conditions under which rice is grown can modify starches significantly. It would be useful if the data from these laboratories included the amounts of each fraction per endosperm in addition to the fraction's percentage of the total. Nevertheless, there are general points of agreement. The *wx* mutants, which synthesize starch consisting wholly of amylopectin that is the same as that produced by nonmutant endosperms, are epistatic to all other mutants in double mutant combinations. Only amylopectin is synthesized; however, the *ae, du1*, and *su1* mutants all modify the amylopectin synthesized in double mutants with *wx*. The *ae wx* double mutants produce starches, which have apparent amylose contents of 15–25%, measured by iodine affinity or by blue value. These double mutants do not produce a true amylose fraction. The high iodine affinity is the result of an anomalous amylopectin with longer chain lengths than nonmutant amylopectin (11). Starch produced by nonmutant endosperms has a small amount of an intermediate fraction (IM) that is neither amylose nor amylopectin but appears between these fractions. This fraction may be an immature amylopectin whose synthesis was interrupted by the endosperm reaching physiological maturity. In the *ae, dul,* and *sul* mutants, this IM fraction is a larger proportion of the starch (121).

In contrast to the lower amount of branching in *ae* amylopectin and IM, the amylopectins of *dul* and *sul* are highly branched (120). The *dul* amylopectin and IM has the highest degree of branching of any sample assayed (121), and in the double mutant *ae dul,* the elongation of the amylopectin chains that is characteristic of *ae* amylopectin is not present.

GENES ACTIVE ONLY IN ENDOSPERM STARCH SYNTHESIS

Some genes affecting starch synthesis in developing maize endosperms may have no role in developing embryos or in chloroplasts. The first case of this specialization was the report (122) that *Wx* activity is not required for embryo amylose synthesis. It was later shown (74) that the starch granules isolated from the embryos of immature *wx* seeds had GBSS activity as high as that of granules isolated from the embryos of *Wx* seeds at the same stage of development. Akatsuka & Nelson (1) demonstrated that the embryo enzyme could be distinguished on biochemical grounds from the endosperm enzyme and was presumably encoded by a different gene.

The evidence that nonmutant alleles at the *bt2* and *sh2* loci encode subunits of endosperm ADPGlc pyrophosphorylase has been discussed. Preiss et al (86) showed that the embryos of nonmutant maize seeds contain an ADPGlc pyrophosphorylase that is distinguishable from the enzyme found in the endosperms. They also showed that the activity in embryos of *sh2* seeds is equivalent to the activity in the embryos of nonmutant seeds, despite the great disparity in the ADPGlc pyrophosphorylase activities in the endosperms of the two genotypes.

The specialization of sucrose synthase isozymes is different and more complicated. Chourey & Nelson (17) reported that *sh1* endosperms have greatly attenuated sucrose synthase activity and that the embryos of mutant and nonmutant seeds have equivalent sucrose synthase activity. The embryo activity is contributed by a second sucrose synthase gene that is also responsible for a minor share of the activity present in the endosperm, and the activity is unaffected by *sh1* mutations. Further, this activity is encoded by nonmutant alleles at a locus now designated as *sucrose synthase (sus),* which has a high degree of homology with the *sh1* locus (64). This isozyme is present in many plant tissues, including the meristematic tips of roots and shoots (88). The expression of the *sh1* locus likewise is not confined to the endosperm; the enzyme it encodes is present in maize seedlings (16) and in the phloem of maize plants.

RESIDUAL ENZYME ACTIVITY IN MUTANT ENDOSPERMS

When the function of a locus implicated in starch synthesis is identified through a mutant, that activity is usually not completely lost in mutant endosperms (71). This observation is true even when the mutation involved can be shown to be a null mutation (27, 36). The basis for this residual activity, where known, is varied. Regardless, the question posed previously remains unanswered: Is the residual activity sufficient to account for the amount of starch synthesized or is another pathway operative that partially circumvents the block?

Preparations of starch granules from wx endosperms have low GBSS activity, less than 10% of the activity in Wx starch granules (74). This is true even of starch granules from a wx mutant that was subsequently shown to have a complete deletion of the wx locus. This activity (72) depends on a second GBSS not encoded by the Wx locus; this enzyme has a lower apparent K_m (7×10^{-5} M) vs WX (at 3×10^{-3}). As a consequence of this disparity in K_m, the glucosyl transferase activity of wx starch granules is equal to that of Wx starch at the lowest substrate concentrations tested (0.1 and 1.0 µM). The existence of the second GBSS was confirmed by MacDonald & Preiss (59); they solubilized GBSS activity from both Wx and wx endoperms and compared them with the two soluble starch synthases (SSSI and SSSII). They found two GBSSs (NGBI and NGBII) in Wx starch granules. The NGBI isozyme differed from both SSSI and SSSII, although the authors were unsure whether NGBII differed from SSSII. If NGBII were the same as SSSII, it might explain the observation (54) that maize endosperm starch granules in vitro incorporated glucose from labeled UDPGlc equally into amylose and amylopectin.

The wx starch granules also had two GBSSs. The lesser of these two activities was indistinguishable from SSSI and could have resulted from contamination by SSSI. The other activity was clearly distinguishable from NGBI and the two SSSs present in both genotypes and presumably accounts for the activity detected earlier in wx starch granules (74). The absence of the minor GBSS from Wx indicates that it is expressed only when NGBI is not. The role of this minor GBSS in starch synthesis, if any, is unknown. Because the null wx mutants do not synthesize any amylose, this enzyme does not compensate for the loss of the major GBSS (NGBI).

The $bt2$ and $sh2$ mutants all have 5–7% of nonmutant activity present in the endosperm, as does the double mutant, $bt2$-R $sh2$-R (40). One explanation for these observations is that the complete absence of ADPGlc pyrophosphorylase is lethal, so only partially functioning mutants are viable. This explanation seems unlikely because this amount of activity is found even in $sh2$ mutants that do not produce a protein product (36.) Therefore, the activity is encoded by another locus (or loci) in the genome, which may be a pyrophosphorylase that normally functions in the synthesis of another nucleoside diphosphate glucose but that can provide weak ADPGlc pyrophosphorylase activity. Is the observed ADPGlc pyrophosphorylase activity sufficient to account for the 20–25% of nonmutant starch synthesis that is typical of *bt2* or *sh2* endosperms? Is there another (minor) pathway for the synthesis of ADPGlc, as has been suggested, or is there a modicum of starch synthesis proceeding by a pathway that does not utilize a nucleoside diphosphate sugar?

IS THERE A ROLE FOR PHOSPHORYLASE IN STARCH SYNTHESIS?

Until the discovery of the glucosyl transferases that transfer glucose from nucleoside diphosphate glucoses to the nonreducing ends of growing starch or glycogen molecules, it was assumed that the enzyme responsible for lengthening the chains was starch phosphorylase (E.C. 2.4.1.1). Because the action of this enzyme ($Glc_n + Glc\text{-}1\text{-}P \leftrightarrow Glc_{n+1} + Pi$) is readily reversible and the amount of Pi in homogenates of starch-synthesizing storage tissues would be inimical to starch synthesis, it was necessary to postulate that much of the Pi was effectively sequestered away from the sites of starch synthesis. Since the discovery of these glucosyl transferases (34, 54), many investigators have tacitly assumed that they are responsible for all starch synthesis. The GBSS and the SSSs, which catalyze essentially irreversible reactions, clearly are better suited to fulfill the synthetic role. The mutations (*bt2* and *sh2*) that so drastically lower the ADPGlc pyrophosphorylase activity attest to the major role of the ADPGlc to starch glucosyl transferases. Yet there is no evidence to demonstrate conclusively that an α-glucan phosphorylase does not make a contribution. Phosphorylase activity in the developing endosperm increases and then decreases over time roughly in parallel with enzymes that are known to participate in starch synthesis (114). Ozbun et al (77) also confirmed the large increase in phosphorylase activity during endosperm development and concluded that there was no evidence for or against its involvement in starch synthesis. Badenhuizen (9) was convinced of a role in starch synthesis for phosphorylase. The most forceful evidence was that normal starch could be synthesized in sterile potato tubers even when most SS activity was lost during growth at 30°C, because phosphorylase is much more thermostable (60). Slabnik & Frydman (96) reported that slices of potato tubers contained a phosphorylase that they characterized as being nonclassical because it did not require a primer and was inactivated by heating at 55°C for five minutes. In conjunction with a branching enzyme (Q) and with Glc-1-P as a substrate, the phosphorylase formed an amylopectin-like polysaccharide. When the Q-enzyme activity was inhibited by ATP and Mg, amylose was synthesized. We have already noted evidence (93, 94, 109, 110) that preparations from potato tubers have both a phosphorylase that can add Glc to the glucoprotein acceptor and a phosphorylase that can initiate primer synthesis and then lengthen that primer.

FUTURE PROSPECTS

The coordinated biochemical and genetic analysis of mutants conditioning specific phenotypes has been an invaluable biological tool, although it has been utilized less in plant biochemistry and physiology than has been possible or desirable. The potential exists for future investigations to clarify aspects that are still unclear and possibly to uncover unsuspected pathways and reactions. Null mutations in the soluble starch synthases and branching enzymes I and IIa would be the most expedient route to clarify their roles. A more intensive search of the numerous mutants induced by mutagen treatments and transposon insertions may identify such mutants. The transposon mutants would have the advantage of providing a route to cloning the gene. We also look forward to the future possibility of targeted gene replacement that would allow the generation of mutants at a locus where none now exist. An understanding of starch synthesis would allow alterations of the process by plant breeding and genetic engineering to produce unique and valuable starches.

Literature Cited

1. Akatsuka T, Nelson OE. 1966. Starch granule-bound adenosine diphosphate glucose-starch glucosyl tranferases of maize seeds. *J. Biol. Chem.* 241:2280–86
2. Akazawa T, Murata T. 1965. Adsorption of ADPG-starch transglucosylase by amylose. *Biochem. Biophys. Res. Commun.* 19:21–26
3. ap Rees T, Entwistle G. 1989. Entry into the amyloplasts of carbon for starch synthesis. In *Physiology, Biochemistry, And Genetics of Nongreen Plastids,* ed. CD Boyer, RC Hardison, JC Shannon, pp. 49–61. Rockville, MD: Am. Soc. Plant Physiol. 292 pp.
4. Ardita F, Tandecarz JS. 1992. Potato tuber UDP-glucose; protein transglucosylase catalyzes its own glucosylation. *Plant Physiol.* 99:1342–47
5. Asaoka M, Okuno K, Fuwa H. 1985. Genetic and environmental control of starch properties in rice seeds. In *New Approaches*

to *Research on Cereal Carbohydrates,* ed. D Hill, L Munck, pp. 29–38. Amsterdam: Elsevier
6. Asaoka M, Okuno K, Sugimoto Y, Yano M, Omura T, Fuwa H. 1985. Structure and properties of endosperm starch and water soluble polysaccharides from sugary mutant of rice (*Oryza sativa* L.). *Starch/Staerke* 37:364–66
7. Baba T, Arai Y, Amano E, Itoh T. 1981. Role of the recessive amylose-extender allele in starch biosynthesis of maize. *Starch/Staerke* 33:79–83
8. Baba T, Kimura K, Mizuno K, Etoh H, Ishida Y, et al. 1991. Sequence conservation of the catalytic regions of amylolytic enzymes in maize branching enzyme-I. *Biochem. Biophys. Res. Commun.* 181:87–94
9. Badenhuizen NP. 1973. Fundamental problems in the biosynthesis of starch granules. *Ann. NY Acad. Sci.* 210:11–15

10. Banks W, Greenwood CT. 1975. The reaction of starch and its components with iodine. In *Starch and its Components*, pp. 67–112. New York: Halsted. 342 pp.

11. Boyer CD, Garwood DL, Shannon JC. 1976. The interaction of the *amylose extender* and *waxy* mutants of maize (*Zea mays* L.): the fine structure of *amylose extender waxy* starch. *Starch/Staerke* 28:405–10

12. Boyer CD, Preiss J. 1978. Multiple forms of starch branching enzyme of maize: evidence for independent genetic control. *Biochem. Biophys. Res. Commun.* 80:169–75

13. Boyer CD, Preiss J. 1981. Evidence for independent genetic control of the multiple forms of maize endosperm branching enzymes and starch synthases. *Plant Physiol.* 67:1141–45

14. Boyer CD, Simpson EKG, Damewood PA. 1982. The possible relationship of starch and phytoglycogen in sweet corn. II. The role of branching enzyme I. *Starch/Staerke* 34:81–85

15. Cao H, Boyer CD, Shannon JC. 1993. *Bt1* gene encodes the most abundant and organelle-specific protein(s) in maize amyloplast membranes. *Plant Physiol.* 102 (Suppl.):51 (Abstr.)

16. Chourey PS. 1981. Genetic control of sucrose synthetase in maize endosperm. *Mol. Gen. Genet.* 184:372–76

17. Chourey PS, Nelson OE. 1976. The enzymatic deficiency conditioned by the *shrunken-1* mutations in maize. *Biochem. Genet.* 14:1041–55

18. Chourey PS, Nelson OE. 1979. Interallelic complementation at the *sh* locus in maize—at the enzyme level. *Genetics* 91:317–25

19. Chourey PS, Tallercio EW, Im KH. 1993. Sucrose phosphate synthase (SPS) in developing kernels of maize. *Plant Physiol.* 102 (Suppl.):6 (Abstr.)

20. Creech RG. 1965. Genetic control of carbohydrate synthesis in maize endosperm. *Genetics* 52:1175–86

21. Deleted in proof

22. Delrue B, Fontaine T, Routier F, Decq A, Wieruszeski J-M, et al. 1992. Waxy *Chlamydomonas reinhardtii*: monocellular algal mutants defective in amylose biosynthesis and granule-bound starch synthase activity accumulate a structurally modified amylopectin. *J. Bacteriol.* 174:3612–20

23. Dickinson DB, Preiss J. 1969. Presence of ADP-glucose pyrophosphorylase activity in *shrunken2* and *brittle2* mutants of maize. *Plant Physiol.* 49:1058–62

24. Dunn GM, Kramer HH, Whistler RL. 1953. Gene dosage effects on corn endosperm carbohydrates. *Agron. J.* 45:101–4

25. Echeverria E, Boyer CD, Liu K-C, Shannon JC. 1985. Isolation of amyloplasts from developing maize endosperm. *Plant Physiol.* 77:513–19

26. Echeverria E, Boyer CD, Thomas PA, Liu K-C, Shannon JC. 1988. Enzyme activities associated with maize kernel amyloplasts. *Plant Physiol.* 86:786–92

27. Echt CS, Schwartz D. 1981. Evidence for the inclusion of controlling elements within the structural gene at the waxy locus in maize. *Genetics* 99:275–84

28. Entwistle G, ap Rees T. 1988. Enzymic capacities of amyloplasts from wheat (*Triticum aestivum*) endosperm. *Biochem. J.* 255:391–96

29. Entwistle G, Tyson RH, ap Rees T. 1988. Isolation of amyloplasts from wheat endosperm. *Phytochemistry* 27:993–96

30. Eriksson G. 1969. The waxy character. *Hereditas* 63:180–204

31. Erlander S. 1958. Proposed mechanism for the synthesis of starch by glycogen. *Enzymologia* 19:273–83

32. Ferguson JE, Rhodes AM, Dickinson DB. 1978. The genetics of *sugary enhancer* (*se*), an independent modifier of sweet corn (*su*). *J. Hered.* 69:377–80

33. Fontaine T, D'Hulst C, Maddelein M-L, Routier F, Pepin TM, et al. 1993. Toward an understanding of the biogenesis of the starch granule. Evidence that *Chlamydomonas* soluble starch synthase II controls the synthesis of intermediate size glucans of amylopectin. *J. Biol. Chem.* 268:16223–30

34. Frydman RB, Cardini CE. 1964. Soluble enzymes related to starch synthesis. *Biochem. Biophys. Res. Commun.* 171:407–11

35. Fuwa H, Glover DV, Miyaura K, Inouchi N, Konishi Y, Sugimoto Y. 1987. Chain length distribution of amylopectins in double- and triple-mutants containing the waxy gene in the inbred Oh43 background. *Starch/Staerke* 39:295–98

36. Giroux MJ, Hannah LC. 1994. ADP-glucose pyrophosphorylase genes in *shrunken2* and *brittle2* mutants of maize. *Mol. Gen. Genet.* 243:400–8

37. Guan HP, Preiss J. 1993. Differentiation of the properties of the branching isozymes from maize. *Plant Physiol.* 102:1269–73

38. Hannah LC, Giroux MJ, Boyer CD. 1993. Biotechnological modification of carbohydrates for sweet corn and maize improvement. *Sci. Horticult.* 55:177–97

39. Hannah LC, Nelson OE. 1975. Characterization of adenosine diphosphate glucose pyrophosphorylases from developing maize seeds. *Plant Physiol.* 55:297–302

40. Hannah LC, Nelson OE. 1976. Characterization of ADP-glucose pyrophosphorylase from the *shrunken-2* and *brittle-2* mutants of maize. *Biochem. Genet.* 14:547–60

41. Hedman KD, Boyer CD. 1982. Gene dosage at the *amylose-extender* locus of maize: effects on the levels of starch branching enzymes. *Biochem. Genet.* 20:483–92

42. Heldt HW, Flügge U-I, Borchert S. 1991. Diversity of specificity and function of phosphate translocators in various plastids. *Plant Physiol.* 95:341–43

43. Hill LM, Smith AM. 1991. Evidence that glucose-6-phosphate is imported as the substrate for starch synthesis by the plastids of developing pea embryos. *Planta* 185:91-96

44. Holder DG, Glover DV, Shannon JC. 1974. Interaction of *shrunken-2* with five other carbohydrate genes in corn endosperm. *Crop Sci.* 14:643–46

45. Hovenkamp-Hermelink JHM, Jacobsen E, Ponstein AS, Visser RGF, Vos-Scheperkeuter GH, et al. 1987. Isolation of an amylose-free starch mutant of the potato (*Solanum tuberosum* L.) *Theor. Appl. Genet.* 75:217–21

46. Ikawa Y, Glover DV, Sugimoto Y, Fuwa H. 1981. Some structural characteristics of starches of maize having a specific genetic background. *Starch/Staerke* 33:9–13

47. Inouchi N, Glover DV, Fuwa H. 1987. Chain length distribution of amylopectins of several single mutants and the normal counterpart, and *sugary-1* phytoglycogen in maize (*Zea mays* L.). *Starch/Staerke* 39:259–66

48. Deleted in proof

49. Kleczkowski LA, Villand P, Luthi P, Olsen O-A, Preiss J. 1993. Insensitivity of barley endosperm ADP-glucose pyrophosphorylase to 3-phosphoglycerate and orthophosphate regulation. *Plant Physiol.* 101:179–86

50. Kramer HH, Pfahler PL, Whistle RL. 1958. Gene interactions in maize affecting endosperm properties. *Agron. J.* 50:207–10

51. Lavintman N. 1966. The formation of branched glucans in sweet corn. *Arch. Biochem. Biophys.* 116:1–8

52. Lavintman N, Cardini CE. 1973. Particulate UDP-glucose protein transglucosylase from potato tuber. *FEBS Lett.* 29:43–46

53. Lavintman N, Tandecarz J, Carceller M, Mendiara S, Cardini CE. 1974. Role of uridine diphosphate glucose in the biosynthesis of starch. Mechanism of formation and enlargement of a glucoproteic acceptor. *Eur. J. Biochem.* 50:145–55

54. Leloir LF, DeFekete MAR, Cardini CE. 1961. Starch and oligosaccharide synthesis from uridine diphosphate glucose. *J. Biol. Chem.* 236:636–41

55. Li HM, Sullivan TD, Keegstra K. 1992. Information for targeting to the chloroplastic inner envelope membrane is contained in the mature region of the maize Bt1-encoded protein. *J. Biol. Chem.* 267:18999–19004

56. Libal-Weksler Y, Sivak MN, Preiss J. 1993. Partial purification and characterization of two isoforms of soluble starch synthase from W64A maize kernels. *Plant Physiol.* 102 (Suppl.):51 (Abstr.)

57. Liu K-C, Boyer CD, Shannon JC. 1991. Evidence for an adenylate translocator in maize amyloplast membranes. In *Molecular Approaches to Compartmentation and Metabolic Regulation,* ed. AHC Huang, L Taiz, pp. 236–37. Rockville, MD: Am. Soc. Plant Physiol.

58. Lowe J, Nelson OE. 1946. *Miniature seed*-a study in the development of a defective caryopsis in maize. *Genetics* 31:525–33

59. MacDonald FD, Preiss J. 1985. Partial purification and characterization of granule-bound starch synthases from normal and waxy maize. *Plant Physiol.* 78:849–52

60. Mangat BS, Badenhuizen NP. 1971. Studies on the origin of amylose and amylopectin in starch granules. III. The effect of temperature on enzyme activities and amylose content. *Can. J. Bot.* 49:1787–92

61. Manners DJ. 1989. Recent developments in our understanding of amylopectin structure. *Carbohydr. Polymers* 89:87-112

62. Manners DJ, Rowe JJM, Rowe KL. 1969. Studies on carbohydrate-metabolizing enzymes. Part XIX. Sweet-corn branching enzymes. *Carbohydr. Res.* 8:72–81

63. Manners DJ, Rowe KL. 1969. Studies on carbohydrate-metabolising enzymes. Part XX. Sweet-corn debranching enzymes. *Carbohydr. Res.* 9:107–21

64. McCarty DR, Shaw JR, Hannah LC. 1986. The cloning, genetic mapping, and expression of the constitutive sucrose synthase locus of maize. *Proc. Natl. Acad. Sci. USA* 83:9099–103

65. Mercier C. 1973. The fine structure of corn starches of various amylose percentages: waxy, normal and amylomaize. *Starch/Staerke* 25:78–83

66. Miller ME, Chourey PS. 1992. The maize invertase-deficient *miniature-1* seed mutation is associated with aberrant pedicel and endosperm development. *Plant Cell* 4:297–305

67. Morrison WR. 1981. Starch lipids: a reappraisal. *Starch/Staerke* 33:408–10

68. Morrison WR, Laignelet B. 1983. An improved colorimetric procedure for determining apparent and total amylose in cereal and other starches. *J. Cereal Sci.* 1:9–20

69. Mu C, Ko YT, Wasserman BP. 1993. Resolution of soluble starch synthase isoforms in dent maize by gel filtration chromatography and isoelectric focusing. *Plant Physiol.* 102 (Suppl.):51 (Abstr.)

70. Muller-Rober B, Sonnewald U, Willmitzer L. 1992. Inhibition of the ADP-glucose pyrophosphorylase in transgenic potatoes leads to sugar-storing tubers and influences tuber formation and expression of tuber storage protein genes. *EMBO J.* 11:1229–38

71. Nelson OE. 1988. Genetic control of polysaccharide and storage protein synthesis in the endosperms of barley, maize, and sorghum. In *Advances in Cereal Science and Technology,* ed. Y Pomeranz, 3:41–71. St. Paul, MN: Am. Assoc. Cereal Chem.

72. Nelson OE, Chourey PS, Chang MT. 1978. Nucleoside diphosphate sugar-starch glucosyl transferase activity of *wx* starch granules. *Plant Physiol.* 62:383–86

73. Nelson OE, Rines HW. 1962. The enzymatic deficiency in the *waxy* mutant of maize. *Biochem. Biophys. Res. Commun.* 9:297–300

74. Nelson OE, Tsai CY. 1964. Glucose transfer from adenosine diphosphate glucose to starch in preparations of *waxy* seeds. *Science* 145:1194–95

75. Okita TW. 1992. Is there an alternative pathway for starch synthesis? *Plant Physiol.* 100:560–64

76. Okita TW, Nakata PA, Anderson JM, Sowokinos J, Morell, Preiss J. 1990. The subunit structure of potato tuber ADPglucose pyrophosphorylase. *Plant Physiol.* 93: 785–90

77. Ozbun JL, Hawker JS, Greenberg E, Lammel C, Preiss J. 1973. Starch synthetase, phosphorylase, ADPglucose pyrophosphorylase, and UDPglucose pyrophosphorylase in developing maize kernels. *Plant Physiol.* 51:1–5

78. Ozbun JL, Hawker JS, Preiss J. 1971. Adenosine diphosphoglucose-starch glucosyl transferases from developing kernels of *waxy* maize. *Plant Physiol.* 78:765–69

79. Pan D, Nelson OE. 1984. A debranching enzyme deficiency in endosperms of the *sugary1* mutants of maize. *Plant Physiol.* 74:324–28

80. Pan D, Strelow LI, Nelson OE. 1990. Many maize inbreds lack an endosperm cytosolic phosphoglucomutase. *Plant Physiol.* 93: 1650–53

81. Peat S, Whelan WJ, Rees WR. 1953. D-enzyme: a disproportionating enzyme in potato juice. *Nature* 172:158–60

82. Perez CM, Perdon AA, Resurrecion AP, Villareal RM, Juliano BO. 1975. Enzymes of carbohydrate metabolism in the developing rice grain. *Plant Physiol.* 56: 579–83

83. Pien F-M, Boyer CD, Shannon JC. 1993. Compartmentation of carbohydrate intermediates in the endosperm of starch-deficient maize genotypes. *Plant Physiol.* 102: 52 (Abstr.)

84. Pozueta-Romero J, Ardita F, Akazawa T. 1991. ADP-glucose transport by the chloroplast adenylate translocator is linked to starch biosynthesis. *Plant Physiol.* 97: 1565–72

85. Preiss J. 1991. Biology and molecular biology of starch synthesis and its regulation.
Oxford Surv. Plant Mol. Cell Biol. 7:59–114

86. Preiss J, Lammel C, Sabraw A. 1971. A unique adenosine diphosphoglucose pyrophosphorylase associated with maize embryo tissue. *Plant Physiol.* 47:104–8

87. Recondo E, Leloir LF. 1961. Adenosine diphosphate glucose and starch synthesis. *Biochem. Biophys. Res. Commun.* 6:85–88

88. Rowland LJ, Chen Y-C, Chourey PS. 1989. Anaerobic treatment alters the cell-specific expression of *Adh-1, Sh,* and *Sus* genes in roots of maize seedlings. *Mol. Gen. Genet.* 218:33–40

89. Shannon JC. 1968. Carbon-14 distribution in carbohydrates of immature *Zea mays* kernels following $^{14}CO_2$ treatment of intact plants. *Plant Physiol.* 43:1215–20

90. Shannon JC, Knievel DP, Chourey PS, Liu S-Y, Liu KC. 1993. Carbohydrate metabolism in the pedicel and endosperm of miniature maize kernels. *Plant Physiol.* 102 (Suppl.):42 (Abstr.)

91. Shewmaker C, Stalker D. 1992. Modifying starch biosynthesis with transgenes in potatoes. *Plant Physiol.* 100:1083–86

92. Singh R. 1973. *Effect of high lysine* (hl) *and sugary* (su) *mutant genes on improved nutritional quality of sorghum grain.* PhD thesis. Purdue Univ., Lafayette, Ind. 95 pp.

93. Sivak MN, Tandecarz JS, Cardini CE. 1981. Studies on potato tuber phosphorylase-catalyzed reaction in the absence of an exogenous accceptor I. Characterization and properties of the enzyme. *Arch. Biochem. Biophys.* 212:525–36

94. Sivak MN, Tandecarz JS, Cardini CE. 1981. Studies on potato tuber phosphorylase-catalyzed reaction in the absence of an exogenous acceptor II. Characterization of the reaction product. *Arch. Biochem. Biophys.* 212:537–45

95. Sivak MN, Wagner M, Preiss J. 1993. Biochemical evidence for the role of the waxy protein from pea (*Pisum sativum* L.) as a granule-bound starch synthase. *Plant Physiol.* 103:1355–59

96. Slabnik E, Frydman RB. 1970. A phosphorylase involved in starch biosynthesis. *Biochem. Biophys. Res. Commun.* 38:709–14

97. Smith AM. 1990. Evidence that the "waxy" protein of pea (*Pisum sativum* L.) is not the major starch granule-bound starch synthase. *Planta* 182:599–604

98. Smith-White BJ, Preiss J. 1992. Comparison of proteins of ADP-glucose pyrophosphorylase from diverse sources. *J. Mol. Evol.* 34:449–64

99. South JB, Morrison WR, Nelson OE. 1991. A relationship between the amylose and lipid contents of starches from various mutants for amylose content in maize. *J. Cereal Sci.* 14:267–78

100. Sprague GF, Brimhall B, Hixon RM. 1943.

Some effects of the *waxy* gene in corn on properties of the endosperm starch. *J. Am. Soc. Agron.* 35:817–22

101. Stark DM, Timmerman KP, Barry GF, Preiss J, Kishore GM. 1992. Regulation of the amount of starch in plant tissues by ADP glucose pyrophosphorylase. *Science* 258:287–92

102. Stinard PS, Robertson DS, Schnable PS. 1993. Genetic isolation, cloning, and analysis of a *Mutator*-induced, dominant antimorph of the maize *amylose extender1* locus. *Plant Cell* 5:1555–66

103. Deleted in proof

104. Sullivan TD, Strelow LI, Illingworth CA, Phillips RL, Nelson OE. 1991. Analysis of maize *Brittle-1* alleles and a defective Suppressor-mutator-induced mutable allele. *Plant Cell* 3:1337–48

105. Takeda C, Takeda Y, Hizukuri S. 1993. Structure of the amylopectin fraction of amylomaize. *Carbohydr. Res.* 246:273–81

106. Takeda Y, Hizukuri S. 1982. Location of phosphate groups in potato amylopectin. *Carbohydr. Res.* 102:321-27

107. Takeda Y, Shitaozono T, Hizukuri S. 1988. Molecular structure of corn starch. *Starch/Staerke* 40:51–54

108. Takeda Y, Shitaozono T, Hizukuri S. 1990. Structures of sub-fractions of corn amylose. *Carbohydr. Res.* 199:207–14

109. Tandecarz JS, Cardini CE. 1978. A two-step enzymatic formation of a glucoprotein in potato tuber. *Biochim. Biophys. Acta* 543:423–29

110. Tandecarz JS, Cardini CE. 1979. Effect of 1,5-gluconolactone on the formation of a glucoprotein in potato tuber. *Plant Sci. Lett.* 15:151–55

110a. Tobias RB, Boyer CD, Shannon JC. 1992. Alterations in carbohydrate intermediates in the endosperm of starch-deficient maize (*Zea mays* L.) genotypes. *Plant Physiol.* 99:146–52

111. Tsai CY. 1965. Correlation of enzymatic activity with *Wx* dosage. *Maize Genet. Coop. Newslett.* 39:153–56

112. Tsai CY. 1974. The function of the *Waxy* locus in starch synthesis in maize endosperm. *Biochem. Genet* 11:83–96

113. Tsai CY, Nelson OE. 1966. Starch-deficient maize mutant lacking adenosine diphosphate glucose pyrophosphorylase activity. *Science* 151:341–43

114. Tsai CY, Salamini F, Nelson OE. 1970. Enzymes of carbohydrate metabolism in the developing endosperm of maize. *Plant Physiol.* 46:299–306

115. Van Der Leij FR, Visser RGF, Oosterhaven K, Van Der Kop DAM, Jacobsen E, Feenstra WJ. 1991. Complementation of the amylose-free starch mutant of potato (*Solanum tuberosum*) by the gene encoding granule-bound starch synthase. *Theor. Appl. Genet.* 82:289–95

116. Van Der Leij FR, Visser RGF, Ponstein AS, Jacobsen E, Feenstra WJ. 1991. Sequence of the structural gene for granule-bound starch synthase of potato (*Solanum tuberosum* L.) and evidence for a single point deletion in the *amf* allele. *Mol. Gen. Genet.* 228:240–48

117. Vineyard ML, Bear RP. 1952. Amylose content. *Maize Genet. Coop. News Lett.* 26:5

118. Vos-Scheperkeuter GH, De Boer W, Visser RGF, Feenstra WJ, Witholt B. 1986. Identification of granule-bound starch synthase in potato tubers. *Plant Physiol.* 82:411–16

119. Vos-Scherperkeuter GH, De Wit JG, Ponstein AS, Feenstra WJ, Witholt B. 1989. Immunological comparison of the starch branching enzymes from potato tubers and maize kernels. *Plant Physiol.* 90:75–84

120. Wang Y-J, White P, Pollak L, Jane J. 1993. Characterization of starch structures of 17 maize endosperm mutant genotypes with Oh43 inbred line background. *Cereal Chem.* 70:171–79

121. Wang Y-J, White P, Pollak L, Jane J. 1993. Amylopectin and intermediate materials in starches from mutant genotypes of the Oh43 inbred line. *Cereal Chem.* 70:521–25

122. Weatherwax P. 1922. A rare carbohydrate in waxy maize. *Genetics* 7:568–72

123. Wessler S, Baran G, Varagona MJ, Dellaporta SL. 1986. Excision of *Ds* produces *waxy* proteins with a range of enzymatic activities. *EMBO J.* 5:2427–32

124. Yeh J-Y, Garwood DL, Shannon JC. 1981. Characterization of starch from maize endosperm mutants. *Starch/Staerke* 33:222–30

Annu. Rev. Plant Physiol. Plant Mol. Biol. 1995. 46:497–520

POLYSACCHARIDE-MODIFYING ENZYMES IN THE PLANT CELL WALL

Stephen C. Fry

ICMB, Division of Biological Sciences, University of Edinburgh, Edinburgh EH9 3JH, United Kingdom

KEY WORDS: xyloglucan, transglycosylation, hydrolases, plant cell expansion, metabolism in muro, cell wall enzymes

CONTENTS

ABSTRACT

Enzymes found in the cell walls of higher plants are surveyed briefly, especially those in the primary cell walls of growing tissue. Attention is focused on

0066-4294/95/0601-0497$05.00

497

hydrolases and transglycosylases that attack the structural carbohydrates of the wall. The reactions catalyzed by these enzymes in vitro are described. Exo-O-glycosylhydrolases are divided into two categories based on specificity for the aglycone group of the substrate: Those with low specificity are in the majority over those with high specificity. Endo-O-glycosylhydrolase-catalyzed reactions are discussed, especially the ability of cellulase to hydrolyze cellulose, xyloglucan, and mixed-linkage glucan. Transglycosylation is defined, and the distinction is drawn between exo- and endo-transglycosylation. Enzymes that catalyze only exo-transglycosylation have not been found in plant cell walls, although some exo-O-glycosylhydrolases can catalyze limited exo-transglycosylation under suitable conditions. In contrast, xyloglucan endo-transglycosylase (XET) catalyzes endo-transglycosylation in the absence of hydrolysis. The evidence that these enzymes act in the walls of living cells is evaluated critically, and the natural and biotechnological regulation of wall enzyme action is discussed. Many enzymes occur in cell walls, and many biological roles have been proposed for them, but we still have far to go in investigating the reactions catalyzed by wall enzymes in vivo and in testing the proposed roles of these enzymes.

BACKGROUND

The concept is now well established that the primary cell wall is a metabolically active compartment of the cell (53). Part of the evidence supporting this concept is the occurrence in cell walls of enzymes whose substrate specificities indicate that they could modify the structural components of the wall. However, the mere existence of wall enzymes does not prove that they act in vivo. They could be at a site that precludes access to the postulated substrate, or inhibitors of the enzyme could be present. Thus, a complementary part of the evidence for metabolically active walls is the demonstration that wall components actually undergo chemical modifications in vivo.

Understanding the biochemistry of primary cell walls is important because of the crucial roles these walls play in determining (a) the rate and direction of cell expansion, (b) the tenacity of cell-cell bonding, (c) the susceptibility of the tissue to external enzymic attack, (d) the water- and ion-binding capacity of the cell wall, and (e) the production of oligosaccharins (wall fragments with signaling roles). The existence of wall enzymes also offers exciting biotechnological opportunities—whether through the rational design of novel agrochemicals that target wall enzymes or by genetic manipulation of plants to modify the expression of wall enzymes.

Plant cell walls are composed mainly of polysaccharides (9, 75). Structural proteins present in the cell wall and the genes coding for them have also been described (56, 99, 100). This review discusses wall enzymes, which will be defined as enzymes located in the cell wall. On plasmolysis of the cell, a wall enzyme thus defined would remain with the wall or dissolve in the plasmolyzing solution rather than stay with the protoplast. This review concentrates on those enzymes that potentially act on the structural carbohydrates of primary cell walls, and it critically explores the evidence that these enzymes act in the walls of living cells. Finally, it outlines our understanding of the regulation and biological roles of wall enzymes. The emphasis is on rapidly growing vegetative tissue rather than fruits or seeds (see 32 for a discussion of wall enzymes of ripening fruit) and on wall enzymes that appear likely to modify the carbohydrates of the plant's own walls rather than those of pathogens.

PROTEINS IN THE PLANT CELL WALL

Several types of evidence support the presence of particular enzymes in the wall. The greatest spatial precision is provided by microscopy. Enzymes can be made visible by labeled antibodies (7, 50, 57), which should preferably recognize the polypeptide backbone of the enzyme rather than any carbohydrate side-chains because the latter are more likely to be common to several unrelated wall enzymes. An alternative is to use artificial substrates that, when acted on by the enzyme of interest, yield an immobile product that can be detected by its color, fluorescence, or electron-opacity. For example, X-GlcA (5-bromo-4-chloro-3-indolyl β-D-glucuronide) is used to detect β-D-glucuronidase (the *gus* gene product) (99); similar methods can be used for endogenous enzymes.

Other types of evidence (35–37) for wall enzymes include (*a*) co-purification of the enzyme with the wall during cell fractionation, (*b*) showing that the enzyme is lost into the medium during protoplast isolation, (*c*) showing that cells can enzymically modify substrates that cannot permeate the plasma membrane, and (*d*) leaching of enzymes from the surfaces of intact cells under conditions that fail to release cytosolic marker enzymes. Each of these methods has limitations, and, whenever possible, several approaches should be used in parallel. For example, a problem with the first method is that a vacuolar enzyme may bind to the cell wall, e.g. ionically (79), during homogenization. Artifactually wall-bound β-glucosidase (79) and invertase (70) may be more readily eluted with salt solutions than are the corresponding naturally wall-bound enzymes. However, the greater resistance to salt elution of naturally wall-bound enzymes than artifactually wall-bound enymes is a feature that has to be established individually for each enzyme of interest.

REACTIONS CATALYZED BY WALL ENZYMES IN VITRO

Broad Classes of Enzymes

Four broad classes of reactions are catalyzed by wall enzyme activities (14, 35): (a) hydrolysis (of glycosidic, ester, and peptide bonds), (b) transglycosylation, (c) transacylation (possibly involved in the synthesis of cutin from acyl-CoAs), and (d) redox reactions [with the electron acceptors O_2 (5, 22, 69), H_2O_2 (115, 119), and NAD^+ (44), and electron donors such as phenols (22, 119), ascorbate (69, 104), polyamines (5), and malate (44)].

Wall enzymes thus fall into three of the Enzyme Commission's six classes (112): Most are hydrolases (EC 3.–.–.–); fewer are oxidoreductases (EC 1.–.–.–); and only recently has a transferase (EC 2.–.–.–) been demonstrated. There is no obvious reason why lyases (EC 4.–.–.–) and isomerases (EC 5.–.–.–) should not exist and act in the wall, but they have not yet been found. Ligases (EC 6.–.–.–) would be unlikely to act because the apoplast lacks the ATP required to drive bond synthesis. Because there is no strong evidence for in muro transacylation or redox reactions involving the structural carbohydrates of the plant cell wall, this review focuses on hydrolysis and transglycosylation.

Hydrolases

Of the known hydrolases, most act on glycosidic bonds. Of these, the majority are O-glycosylhydrolases (EC 3.2.1.–), but a few may be N-glycosylhydrolases (EC 3.2.2.–) (64). Other wall hydrolases act on phosphate esters [phosphatases (EC 3.1.3.–) (31)], carboxy-esters [pectin methylesterases (EC 3.1.1.11) (107, 113, 117) and cutinase (E.C. 3.1.1.–) (50)], and peptide bonds [proteinases (EC 3.4.–.–) (109)].

Exo-O-Glycosylhydrolases

EXO-O-GLYCOSYLHYDROLASES WITH LOW AGLYCONE SPECIFICITY   Exo-O-glycosylhydrolases are often simply called glycosidases, but this name is ambiguous because the Enzyme Commission also classifies endo-O-glycosylhydrolases as glycosidases (112). Exo-O-glycosylhydrolases attack poly- and oligosaccharides progressively from the nonreducing termini, releasing monosaccharides or, in one case, disaccharides. In general, exo-O-glycosylhydrolases do not hydrolyze the glycosidic bond of a sugar residue that is itself substituted, e.g. with one of the noncarbohydrate groups linked to wall polysaccharides such as O-acetyl or O-feruloyl esters or O-methyl ethers. However, many exo-O-glycosylhydrolases will attack almost any compound that possesses the appropriate sugar residue as an unmodified nonreducing terminus, regardless of the chemical nature of the aglycone. For example, a β-D-glucosidase (EC 3.2.1.21) from the cell walls of cultured *Glycine max* cells will readily hydrolyze *p*-nitrophenyl β-D-glucopyranoside (but not methyl β-D-glucopyranoside), at least four disaccharides [D-Glc*p*-β-(1→4)-D-Glc, D-Glc*p*-β-(1→3)-D-Glc, D-Glc*p*-β-(1→2)-D-Glc, and D-Glc*p*-β-(1→6)-D-Glc], or the nonreducing end of the polysaccharides β-(1→3)-D-glucan and mixed-linkage β-(1→3),(1→4)-D-glucan (MLG) (15). Therefore, many exo-*O*-glycosylhydrolases can be assayed conveniently using chromogenic or fluorogenic substrates such as 4-nitrophenyl or 4-methylumbelliferyl β-D-glucopyranoside, which when hydrolyzed yield 4-nitrophenol (yellow) or 4-methylumbelliferone (fluorescent), respectively.

Many exo-*O*-glycosylhydrolases have low aglycone specificity, and the wall polymers that can act as substrates for them are also numerous (Table 1). It is therefore difficult to be precise about the natural substrates and roles of these enzymes. One possible role is in hydrolysis of sugars that pass through the wall as translocated carbon sources such as sucrose [hydrolyzed to glucose and fructose by β-D-fructofuranosidase (invertase)] and raffinose (hydrolyzed to fructose, galactose, and glucose by α-D-galactopyranosidase and invertase). Another probable role is in phenolic metabolism: Cell wall β-D-glucosidase hydrolyzes coniferin (4-β-D-glucopyranosyloxy-3-methoxy-*trans*-cinnamyl alcohol, which is secreted into the walls of xylem cells) to liberate coniferyl alcohol, an immediate precursor of lignin (110). Another enzyme of secondary metabolism, linamarase, a β-D-glucosidase of clover leaf cell walls, can hydrolyze cyanogenic glucosides (e.g. linamarin) present in the vacuoles of clover leaves (57) to release 2-hydroxyisobutyronitrile, which decomposes nonenzymically to HCN. The wall enzyme cannot attack vacuolar linamarin until tissue is crushed, e.g. in the mouth of a herbivore, whereupon HCN is produced, which may be a feeding deterrent.

Other exo-*O*-glycosylhydrolases may trim particular nonreducing terminal sugar residues off cell wall polysaccharides, oligosaccharides, and glycoproteins. Table 1 lists nonreducing termini of wall polymers that in theory are targets of each enzyme. In reality, however, few of these polymers have been tested as substrates. Nonreducing termini of the backbones of polysaccharides (see note *a* in Table 1) will be rare because polysaccharide chains are long. For this reason and because the backbone is often substituted with other sugar residues, digestion of backbones by exo-*O*-glycosylhydrolases seems unlikely to be extensive for most polysaccharides.

Nonreducing termini of the side chains of branched polysaccharides are far more abundant in the cell wall and seem likely to be major targets of exo-*O*-glycosylhydrolases in vivo. Enzymic removal of side chains could affect the physical properties of a polysaccharide (4); nevertheless, it is hard to imagine any dramatic effect of limited exo-hydrolysis on wall loosening in contrast to limited endo-hydrolysis. Generally, the nearer a sugar residue is to the backbone of the polysaccharide, the harder it is for an enzyme to hydrolyze it—pre-

Table 1   Cell wall exo-O-glycosylhydrolases and the wall polymers that are their potential natural substrates

Enzyme	Wall polymers that are possible substrates
Enzymes with high specificity for the aglycone of the substrate	
α-D-Xylosidase	xyloglucan
α-L-Fucosidase	xyloglucan, N-linked glycoproteins (?)
α-D-Galacturonidase	homogalacturonan (pectic acid)
Enzymes with low specificity for the aglycone of the substrate	
α-D-Mannopyranosidase	N-linked glycoproteins (e.g. peroxidase)
β-D-Mannopyranosidase	mannans[a]
α-L-Arabinofuranosidase	RG-I, xylans, extensins[b]
β-L-Arabinofuranosidase	extensins[c]
β-D-Xylopyranosidase	xylans[a], xylans[d]
α-D-Glucopyranosidase	none known
β-D-Glucopyranosidase	cellulose[a], xyloglucan[a], callose[a], MLG[a,e]
α-D-Galactopyranosidase	mannans, extensin
β-D-Galactopyranosidase	RG-I[e], xyloglucan
α-L-Rhamnosidase	RG-II[e]
β-D-Fructofuranosidase[f]	none known
α-D-Glucuronidase	xylans
β-D-Glucuronidase	none known
N-Acetyl-β-D-glucosaminidase	N-linked glycoproteins
N-Acetyl-β-D-galactosaminidase	N-linked glycoproteins
β-D-Cellobiosylhydrolase	cellulose[a]

[a]Indicates that the potential substrate is the nonreducing terminal sugar residue of the backbone of a polysaccharide.

[b]Nonreducing terminus of hydroxyproline tetra-arabinoside.

[c]Nonreducing terminus of hydroxyproline mono- to tri-arabinosides.

[d]Some arabinoxylans have β-D-Xyl$p$ residues attached to some of the Ara$f$ side-chains.

[e]Abbreviations: RG = rhamnogalacturonan; MLG = mixed linkage $(1\rightarrow3),(1\rightarrow4)$-β-D-glucan.

[f]Invertase.

sumably for steric reasons. A β-D-galactosidase of *Tropaeolum majus* cotyledon cell walls is unusual in that it can remove most of the β-D-galactose residues of xyloglucan (23); in contrast, an α-L-fucosidase and an α-D-xylosidase have little or no effect on xyloglucan (see below). Many other enzymes have not been rigorously tested in this way.

Sometimes, the significant end product of exo-hydrolysis may be the monosaccharide rather than the modified polysaccharide. β-D-Galactosidase is often suggested to be involved in fruit ripening (17, 113), although it is not clear what significant effect the trimming of a few galactose residues from (presumably pectic) polysaccharides could have; perhaps the relevant product is free galactose, an agent that does affect fruit ripening (92).

Cellulose cellobiohydrolase attacks the nonreducing terminus of cellulose, releasing the disaccharide, β-D-Glc*p*-(1→4)-D-Glc (cellobiose), but there are few compelling reports of its occurrence in plants (65). It has not been determined whether plant cellobiohydrolases can also attack appropriate nonreducing termini of MLG or whether it is specific for cellulose.

EXO-*O*-GLYCOSYLHYDROLASES WITH HIGH AGLYCONE SPECIFICITY   Some primary walls contain an α-D-xylosidase that acts on oligosaccharides such as XXXG and XXFG (Figure 1) to remove the single xylose residue farthest from the reducing terminus (26, 61, 89). This enzyme does not hydrolyze 4-nitrophenyl α-D-xylopyranoside or the disaccharide α-D-Xyl*p*-(1→6)-D-Glc (61). The α-D-xylosidase does not release measurable amounts of xylose from polymeric xyloglucan, but it may release the single xylose residue farthest from the reducing terminus of the polysaccharide. In a typical xyloglucan (M_r = 10^5–10^6), this residue would be less than 0.1% of the mass and might be overlooked.

A similar story applies to a wall-located α-L-fucosidase (7, 28): It removes the α-L-fucose residue from XXFG (Figure 1) but will not attack 4-nitrophenyl α-L-fucopyranoside. It apparently fails to release fucose from polymeric xyloglucan, but it may remove the single fucose group farthest from the reducing terminus of a long xyloglucan chain, releasing an undetectable amount of fucose.

Cell walls contain an α-D-galacturonidase [exo-polygalacturonase (EC 3.2.1.67)], which liberates galacturonic acid from the nonreducing terminus of homogalacturonan and oligogalacturonides larger than the pentasaccharide

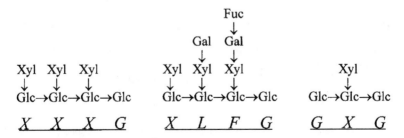

Figure 1   Representative structures of xyloglucan-derived oligosaccharides (XGOs), illustrating the main features of the naming system used (see 40). Each oligosaccharide is based on a backbone of β-(1→4)-linked D-glucose residues (Glc). Each Glc unit of the backbone is given a one-letter code (e.g. G, X, L, or F), based on which other sugars (if any) are attached as side chains. To name an oligosaccharide, the relevant code letters are listed sequentially from the nonreducing end. Thus, the three oligosaccharides shown here are named XXXG, XLFG, and GXG. (XXXG and XLFG have previously been referred to as XG7 and XG10, respectively.)

(41, 60). The enzyme gradually shortens these substances until the pentasaccharide is formed, which then accumulates.

The three enzymes discussed above may convert oligosaccharins to biologically inactive products. α-D-Xylosidase would cut the growth-promoting oligosaccharide XXXG (74) to GXXG, which is expected not to be an acceptor substrate for xyloglucan endo-transglycosylase (XET) (67) and thus (see below) not a growth promoter. The ability of α-L-fucosidase to convert XXFG to XXLG likewise destroys its growth-inhibiting activity, which depends on the α-L-fucose residue (8, 73, 111). Similarly, although oligogalacturonides at or about the size of the dodecasaccharide are highly effective elicitors of phytoalexin synthesis, they lack this effect after α-D-galacturonidase has trimmed them to the pentasaccharide size (55).

α-D-Xylosidase could play an important role in cell expansion if it can act on a high-molecular-weight, wall-bound xyloglucan molecule, removing the single xylose residue furthest from the reducing terminus. This action would be expected to prevent the polysaccharide molecule from acting as an acceptor substrate of XET. Thus, the α-D-xylosidase could have an impact out of proportion to the magnitude of the reaction catalyzed. The inactivated xyloglucan chain could later be reactivated as an XET acceptor substrate by the action of β-D-glucosidase. The sequential inactivation and reactivation of a xyloglucan chain by the alternating action of α-D-xylosidase and β-D-glucosidase would stall when an F (Fuc→Gal→Xyl→Glc; Figure 1) or L (Gal→Xyl→Glc) unit was reached. To pass this block, α-L-fucosidase and β-D-galactosidase action would be required. Furthermore, if the galactose residue were O-acetylated, as frequently appears to be the case, the action of an esterase might also be required. There may thus be a complex battery of exo-O-glycosylhydrolases and esterases in the cell wall, constantly modulating the ability of xyloglucan chains to act as the acceptor substrate for XET.

Endo-O-Glycosylhydrolases

Endo-O-glycosylhydrolases (endo-glycanases) attack polysaccharides at any position except at or very near the termini. They can thus have an immediate and large impact on molecular weight, potentially halving it at a stroke. Such enzymes have been studied intensively in ripening fruit (32), but they are also found in other tissues.

Cellulase [endo-(1→4)-β-D-glucanase (EC 3.2.1.4)] has been found in many plant cell walls, including those of growing tissues (18, 48, 65, 98), ripening fruits (1, 58), and abscising and dehiscing organs (10, 20). Cellulases hydrolyze D-Glc-β-(1→4)-D-Glc bonds, which occur in cellulose, xyloglucan, and MLG. The action of plant cellulases is difficult to demonstrate on native cellulose because this substrate is semicrystalline and only very slowly attacked, and because the substrate and products are both water insoluble. In contrast, xyloglucan and MLG are water soluble and thus readily accessible to cellulase, and their degradation is easily measured by a loss in viscosity (45). Viscometric assays are also frequently made with carboxymethylcellulose (CMC), a cheap soluble derivative of cellulose that is an excellent substrate for cellulases (3, 48).

Other endo-O-glycosylhydrolases found in plant cell walls include xylanases (66), pectinases (endo-polygalacturonases) (82, 118), mannanases (46), MLGases (47, 101), and an endo-xyloglucanase that differs from cellulase in not attacking CMC (24). Also, much work has dealt with endo-$(1\rightarrow3)$-β-D-glucanases and chitinases (71), which seem likely to act mainly on the polysaccharides of invading microbes.

Transglycosylases in the Plant Cell Wall

Recently, interest has focused on wall enzymes that catalyze transglycosylation. O-Transglycosylation is a reaction in which a sugar residue is transferred from its aglycone (A) to an alcohol (HO—A′), so that A′ then becomes the new aglycone:

$$\text{sugar—A} + \text{HO—A}' \leftrightarrow \text{sugar—A}' + \text{HO—A}. \qquad 1.$$

The molecule that is cut (sugar—A) is called the donor substrate, and HO—A′ is the acceptor substrate. If the cleaved sugar—A bond in the donor substrate is a nonreducing terminus, the reaction is described as exo-transglycosylation; otherwise, it is endo-transglycosylation. Many O-glycosylhydrolases catalyze both hydrolysis and transglycosylation: The latter is favored by the presence of high concentrations of an appropriate acceptor substrate (21). Glycosylhydrolases (EC 3.2.–.–) and transglycosylases (EC 2.4.–.–) are classified separately: Only those enzymes exhibiting a high transglycosylation:hydrolysis ratio at low acceptor substrate concentrations are classified as transglycosylases (112).

EXO-TRANSGLYCOSYLATION   Some exo-transglycosylation is catalyzed by an exo-O-glucosylhydrolase found in the walls of cultured *G. max* cells (15). With a β-$(1\rightarrow3)$-D-glucan (laminarin) as donor and D-[^{3}H]glucose as acceptor, this enzyme catalyzes the transfer of single glucose residues, presumably from the nonreducing terminus of the donor, to the [^{3}H]glucose to form ^{3}H-disaccharides:

$$Glc \rightarrow Glc \rightarrow Glc \rightarrow Glc \rightarrow Glc \rightarrow Glc \rightarrow Glc \rightarrow Glc \rightarrow Glc \rightarrow Glc \rightarrow \ldots + [^3\textbf{H}]\textbf{Glc}$$
$$\Downarrow$$
$$Glc \rightarrow [^3\textbf{H}]\textbf{Glc} + Glc \rightarrow Glc \rightarrow Glc \rightarrow Glc \rightarrow Glc \rightarrow Glc \rightarrow Glc \rightarrow Glc \rightarrow Glc \rightarrow \ldots$$

In this reaction, the disaccharides formed ($Glc\rightarrow$[^{3}H]Glc) had β-linkages and were $(1\rightarrow2)$-, $(1\rightarrow3)$-, $(1\rightarrow4)$-, and $(1\rightarrow6)$-linked in the ratio of approximately

1:1:1:10. The same enzyme also catalyzed the hydrolysis of laminarin. Cline & Albersheim found that with 0.6% (w/v) laminarin and 1.5% (w/v) [³H]glucose, the transglycosylation:hydrolysis ratio of the enzyme was roughly 0.3 (see Figure 12 of Reference 15). The enzyme is thus classified as a hydrolase. A similar enzyme was shown to transfer glucose residues from *p*-nitrophenyl β-D-glucoside (donor) to gentiobiose or cellobiose (acceptors) (80).

The biological role of exo-β-transglucosylation is unclear. Nari et al (80) proposed that it elongates hemicellulose chains in muro; however, this seems improbable because, during transglycosylation, for every glycosidic bond made another must be cut, so that net bond synthesis is impossible. Where polysaccharide chain elongation occurs (e.g. in the Golgi membranes), a constant supply of low-molecular-weight donor substrates (NDP-sugars) has to be provided by the cell. No low-molecular-weight β-glucosides are known to be secreted by growing cells; thus, net elongation of hemicelluloses in the apoplast seems unlikely.

ENDO-TRANSGLYCOSYLATION   Endo-transglycosylation was proposed as an attractive theoretical explanation for the molecular rearrangements that must take place within the wall during cell expansion (2). The first experimental evidence that, with the benefit of hindsight, can be interpreted as resulting from endo-transglycosylation was the observation (29) that xyloglucan oligosaccharides (XGOs) such as XXXG promoted the depolymerization (assayed viscometrically) of xyloglucan by an enzyme preparation from pea seedlings. The promotion of loss of viscosity was initially interpreted as resulting from allosteric activation of cellulase by the XGOs; however, this would be unusual because XGOs are the end products of cellulase action on xyloglucan.

Independent work (11, 74, 102, 103) showed that cultured plant cells perform a transglycosylation resulting in the attachment of endogenous xyloglucan to exogenous ³H-XGOs, producing a ³H-labeled polymer. The reaction does not go via free ³H-monosaccharides (which would be reincorporated readily into new ³H-polysaccharides). For example, when [*xylosyl*-³H]XXFG (i.e. XXFG carrying ³H in the xylose residues) was supplied to cultures, all the radioactivity could be isolated from the ³H-polymer as [³H]xylosyl-glucose after Driselase-hydrolysis (11); if the cells had first hydrolyzed the [³H]XXFG to free [³H]xylose and then taken this up, it would have been converted, via [³H]xylulose 5-phosphate (13), to a wide range of metabolites. The bond formed between xyloglucan and XXFG appeared to be glycosidic because it was stable to alkali but not to cellulase (11, 103). When [*reducing terminal glucose*-³H]XXFG was supplied, the label became the new reducing terminus of the polymer, as shown by the fact that the [³H]glucose remained NaBH₄-reducible to [³H]glucitol (103). Therefore, the XXFG must have been the acceptor substrate and the polymeric xyloglucan the donor. Because the moiety attached to the nonreducing terminus of the XXFG moiety was of high molecular weight, the group transferred could not have been a single sugar residue; thus, the reaction was endo-transglycosylation. The reaction is shown schematically in Figure 2.

The new bond formed during xyloglucan endo-transglycosylation is probably chemically identical [Glc-β-(1→4)-Glc] to that broken in the donor substrate. Supporting this conclusion is the fact that the acceptor substrate (e.g. [³]XXFG) can be released from the reaction product by digestion with *Trichoderma viride* cellulase (103), an enzyme known to cleave Glc-β-(1→4)-Glc bonds. NMR studies also indicate that no other bonds are generated during XET-catalyzed transglycosylation of xyloglucan (86).

When two samples of xyloglucan differing in molecular weight (the smaller molecules being ³H-labeled) were mixed in vitro in the presence of an enzyme from *Phaseolus vulgaris* leaves, some of the ³H was transferred from the low- to the high-molecular-weight class (74). A fall in the molecular weight of [³H]xyloglucan can be the result of hydrolysis, but a rise cannot be, again indicating enzymic transglycosylation.

Conversely, a single population of xyloglucan, initially of a fairly narrow molecular weight range [mean ~ 420 kDa], was converted in vitro by a *Vigna angularis* stem enzyme to two populations (high and low molecular weight) (85). Transglycosylation conserves bonds, so it cannot cause a change in mean molecular weight. In this experimental situation, the largest molecules appeared to be selectively removed from the system (by precipitation), thus pulling the total population of xyloglucan molecules away from a normal distribution of chain lengths. Transglycosylation would proceed exergonically (i.e. $\Delta G^{\circ\prime}$ would be negative) if it broke a (1→4) bond and made a (1→6) bond, as is catalyzed (with α-Glc linkages) by starch-branching enzyme (EC 2.4.1.18). This is because the $\Delta G^{\circ\prime}$ of hydrolysis of a (1→6) bond is smaller

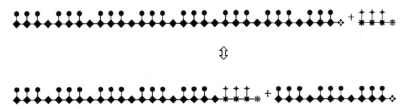

Figure 2   Transglycosylation reaction whereby a segment of high-molecular-weight xyloglucan becomes attached to a radioactive oligosaccharide, XXXG (see Figure 1). ◆ = glucose residue of polysaccharide; ❖ = reducing terminal glucose moiety of polysaccharide; ● = xylose residue of polysaccharide; ✳ = glucose residue of oligosaccharide; ✲ = reducing terminal glucose moiety of oligosaccharide, ✚ = xylose residue of oligosaccharide. Both of the reducing termini (❖, ✲) remain as reducing termini throughout the reaction. Galactose, fucose, and other residues that are not essential for the reaction are not shown.

than that of a ($1\rightarrow4$) bond. However, in xyloglucan transglycosylation, where a ($1\rightarrow4$) bond is replaced by a new but identical ($1\rightarrow4$) bond, the reaction is neither exergonic nor endergonic ($\Delta G^{\circ\prime} \approx 0$). Thus, transglycosylation can only change the mean molecular weight of xyloglucan if some process (e.g. precipitation) removes a specific size class from the equilibrium.

The acceptor substrate does not require its reducing terminus; thus, [*glucitol*-^{3}H]XXXGol (39) is an acceptor. The use of these radioactively or fluorescently labeled acceptor substrates has provided the basis of a simple in vitro assay for XET activity. Nonradioactive, high-molecular-weight xyloglucan is mixed with a labeled XGO and incubated in the presence of the putative enzyme. The reaction is then assessed by monitoring the formation of labeled products with a molecular weight greater than that of the starting XGO. Monitoring can be by dialysis, ethanol-precipitation, gel-permeation chromatography, or, most simply, by filter-paper binding: The labeled high-molecular-weight product adheres firmly to chromatography paper by hydrogen-bonding, whereas remaining unreacted XGO is rapidly washed off the paper by running tap water (39). A fluorescent pyridylamino-XXXGol (83) can be used in a similar way.

This type of assay has allowed exploration of the substrate specificity of the reaction. Xyloglucans of the primary cell wall contain Glc, Xyl, Gal, Fuc, and other residues (75). However, Fuc is not required in the donor substrate (39, 86), and the acceptor requires only Glc and Xyl (67), although the presence of Gal increases the affinity of the enzyme (i.e. decreases the K_m) (39). The requirement for α-Xyl is absolute: Plant extracts completely failed to catalyze transglycosylation (using XGOs as acceptor substrates) with polysaccharides other than xyloglucan such as β-xylans and several β-glucans (39). The minimum acceptor substrate that exhibits a strong activity is the pentasaccharide XXG; neither of the tetrasaccharides, GXG and XGG, was effective (67).

The radiochemical assay showed the transglycosylation to be catalyzed by an enzyme activity [xyloglucan endo-transglycosylase (XET)] that can be extracted readily from plant tissues. XET may be universal in land plants (embryophytes): It has been detected in bryophytes (sporophytes and gametophytes), graminaceous and liliaceous monocots, and numerous dicots (39). Members of the Gramineae often are particularly rich in XET activity, which may be surprising because these plants generally contain relatively little xyloglucan. XET activity is highest in growing stems, but it is also found in roots (93), leaves (I Potter & SC Fry, unpublished data), some fruits [e.g. kiwifruit (94) and persimmon (16)], and in cell cultures (49, 91, 103).

Farkaš et al (30) showed that during XGO-induced promotion of xyloglucan depolymerization by a *T. majus* seed xyloglucanase, the XGOs were consumed in a transglycosylation reaction; however, these authors maintained that the pea enzyme influenced by XGOs (29) was a hydrolase.

XET from pea stems can be extracted by grinding in water; salts or detergents do not appreciably enhance yield of enzyme activity, which suggests that XET is not ionically bonded to the wall or trapped in the protoplast (39). XET can be extracted from *V. angularis* stems by vacuum infiltration/centrifugation (86), so it is soluble in the apoplast. In carrot cell cultures, most of the activity is found in the culture filtrate, again indicating that XET is a soluble apoplastic enzyme (49). However, a proportion of the XET activity in kiwifruit (94) and spinach cell cultures (91) seems to be ionically bonded to the wall and can be eluted only by use of a high pH or ionic strength.

An XET was purified 87-fold from *V. angularis* stem apoplastic fluid by $(NH_4)_2SO_4$-precipitation, and by concanavalin-A-affinity, gel-permeation, and ion-exchange chromatography (86). The XET is a glycoprotein of M_r 33,000. It is highly active in transglycosylation and lacks hydrolase activity, so it is a true transglycosylase (EC 2.4.1.–) rather than a hydrolase that catalyzes some transglycosylation at suitably high substrate concentrations.

Another enzyme with XET activity was purified, and the gene sequenced, from *T. majus* seeds (19, 27). The cotyledons of this plant store xyloglucan as a carbon reserve that is mobilized after germination. The enzyme from these seeds differs from the XET of growing stems in that it will catalyze either transglycosylation or hydrolysis of xyloglucan, depending on substrate concentration (27). The enzyme is referred to here as xyloglucan endo-transglycosylase/hydrolase (XETH).

The *V. angularis* stem XET gene, and similar genes from *Arabidopsis thaliana*, *G. max*, *Lycopersicon esculentum,* and *Triticum aestivum*, have also been sequenced (88). All five genes encode a set of four cysteine residues near the C-terminal. They exhibit moderate homology with *T. majus* seed XETH (19), and with three genes of unknown function—*meri*-5 of *A. thaliana* (78), *brul* of *G. max* (120), and *TCH4* of *A. thaliana* (J Braam, personal communication). The *A. thaliana* XET gene is much more similar to the *T. aestivum* XET gene than to *A. thaliana meri*-5. It will be interesting to see whether the *meri*-5 protein has XET (or other transglycosylase) activity. XET, XETH, *meri*-5, *brul*, and five MLGases from *Bacillus subtilis* all share a highly conserved sequence [Asp-Glu-Ile-Asp-(Ile/Phe)-Glu-Phe-Leu-Gly], which is probably at or near the catalytic site (88).

Are wall polysaccharides other than xyloglucan subject to endo-transglycosylation in vivo? Oligo-$[^{14}C]$galacturonides did not participate in exo- or endo-transglycosylation reactions in the apoplast of cultured rose cells or in extracts of fruit tissues (41), but other polysaccharides may have their own transglycosylases.

EVIDENCE THAT WALL ENZYMES ACT IN VIVO

The mere presence of an enzyme in plant cell walls does not prove that it catalyzes any in vivo reaction there. Even the partial wall autolysis frequently observed (87, 108) when purified wall samples are incubated in vitro without exogenous enzymes fails to prove that similar reactions occur in vivo.

Microscopical Evidence

Perhaps the most direct evidence for chemical changes occurring in the walls of living cells is obtained by microscopy. Light and electron microscopy show that during the differentiation of xylem vessel elements, end walls are lysed to form perforation plates. In the process, all structural wall material vanishes from quite a large area, and one must conclude that, among other things, the cellulose has been lysed (98). This observation suggests the action of both enzymes (cellulase and cellobiohydrolase) thought to be required for cellulolysis. Extensive wall lysis can also be seen in the thick walls of certain seeds, e.g. tamarind (95), during and after germination: It is assumed that the degradation products are used by the seedling as a carbon source. Seed reserve polysaccharides include xyloglucan [e.g. in *T. majus* and tamarind (24, 95)], mannans [e.g. in lettuce, fenugreek, guar, and dates (46, 63, 72)], and arabinogalactans [e.g. in lupin and *G. max* (6)]. Observed thinning of these seed cell walls is thus evidence for the action of enzymes (e.g. endo-glucanases, endo-mannanases, galactanases) that attack these polysaccharides.

The microscopical approach is less definitive in situations where a wall thins and at the same time expands in area, e.g. in growing cereal coleoptiles, because the thinning could be the result of a rearrangement of existing wall material rather than the result of lysis. In such situations, measurements of wall mass per organ (or per cell if cell division has ceased) are required for evidence of wall degradation. Data can be gathered, e.g. during hormone-stimulated growth, by weighing the yield of wall material isolated per coleoptile sampled at intervals, or by interferometry (12). The latter approach showed that in pea stem segments treated with auxin in the absence of glucose, the outer epidermal walls suffered a net loss of dry mass, indicating lysis, presumably catalyzed by wall enzymes.

Metabolic Evidence

Other evidence for wall reactions in vivo comes from analyses of wall polysaccharides. Auxin- and H^+-treated tissues often exhibit a rapid fall in the mean molecular weight of wall-bound xyloglucan (68, 84, 105) and generate some soluble apoplastic xyloglucan (106). The critical requirement here is to demonstrate that existing wall-bound xyloglucan molecules change in molecular weight or in strength of wall binding. Alternatively, the cell may respond to auxin by synthesizing and/or secreting new xyloglucan molecules that differ from those already located in the wall. Pulse-chase experiments are required to distinguish between these alternatives, of the type performed by Franz for a glucan (33) and by Labavitch & Ray for xyloglucan (62).

Changes other than those in molecular weight have also been claimed, e.g. in the degree of esterification of pectic polysaccharides in the walls of *G. max* hypocotyls (116). Again, a critical requirement is to show that existing wall-bound pectins lose ester groups, rather than that older cells synthesize and/or secrete more low-methoxy pectins than do younger cells.

A different line of evidence for wall reactions in vivo is the observation that exogenous substrates unable to cross the plasma membrane undergo reactions in vivo. Such substrates would permeate the wall matrix and be accessible to wall enzymes. The data (see above) on transglycosylation of XGOs fed to cultured plant cells fall into this category (103), as does the observation that exogenous O-[^{14}C]feruloyl-[^{3}H]oligosaccharides become oxidatively bonded to structural wall polymers (38). The remaining doubt left by this type of evidence is whether the enzymes would be able to act on endogenous wall-bound substrates—a wall-bound enzyme and a wall-bound substrate may be spatially separated, which would preclude any reaction. A major goal remaining in work on XET is to discover whether high-molecular-weight, wall-bound xyloglucan can act as both donor and acceptor substrate in vivo. Such a demonstration is difficult, even with pulse-chase radiolabeling experiments, because there is no inevitable difference between the substrates and the products.

In conclusion, much work remains to be done on wall metabolism to determine whether known wall enzymes act on wall-bound substrates in vivo and possibly to lead to the discovery of novel wall enzymes. It is essential to study the substrates and their behavior as well as the enzymes and their genes.

CONTROL OF WALL ENZYME ACTION

Evidence from Measurements of Enzyme Activity

Regulation of wall enzyme action is a biological imperative; uncontrolled lysis is the alternative. One aspect of regulation is the modulation of the amount of active enzyme present in the wall. Little is known of the relative importance of the potential control points—synthesis of catalytically active enzyme (by pre- or posttranslational control), its secretion into the cell wall, and its subsequent degradation. Pitfalls in measuring changes in wall enzyme activity include the possibility that the enzyme is present but not extractable, or that it is degraded or otherwise inactivated during extraction.

Cellulase levels are increased dramatically in pea stems by application of auxin, especially the high doses that cause stem swelling. Hydrolysis of xyloglucan catalyzed by the induced cellulase may be the molecular basis of the observed lateral expansion of the cells (48). Modest increases in cellulase may also be induced by the very low doses of auxin that cause elongation in pea stem segments (18). However, there appears to be no powerful experimental evidence that auxin-induced turnover of xyloglucan is caused by elevated cellulase levels (53).

Extractable XET activity is regulated by many factors, supporting the idea that it plays an important physiological role. Should the amount of an enzyme (such as XET), suspected to be involved in cell elongation, be expressed as activity per mg fresh weight, per mg dry weight, per mg protein, per mg cellulose, per mm^2 of cell wall, per mm of organ length, per cell, or per organ? Because wall elongation is thought to be limited by the extensibility of a thin (inner) layer of the cell wall, activity per mm^2 of side wall is particularly relevant, though difficult to measure. Fortunately, in the many experimental subjects that grow uniaxially and without concurrent cell division (e.g. coleoptiles and internodes at the stages usually used for hormone bioassays), sidewall area is proportional to fresh weight. Thus, XET activity per unit fresh weight of tissue is a useful parameter to express data.

XET activity per unit fresh weight varies along the length of a stem (39, 85) and a root (93), often peaking at or near the zone of maximal cell expansion, thus providing circumstantial evidence for a role in cell expansion. Of various hormone treatments that affect cell elongation, auxin often fails to affect XET activity [e.g. in excised pea stem segments (39)]. However, in intact stems of dwarf pea plants (90) and in lettuce hypocotyls (91), gibberellic acid (GA$_3$) promotes elongation and concomitantly increases extractable XET activity. In cucumber hypocotyls, the effect of GA$_3$ on XET was slight but that of indole-3-acetic acid (IAA) was more pronounced (91). It thus appears that growth-regulating hormone treatments often affect XET levels, but there is no simple relationship between extractable XET activity and elongation rate. XET production in carrot cell cultures was also affected by hormones: In one cell line, withdrawal of 2,4-D caused dramatic change from isodiametric to uniaxial cell expansion and concomitantly promoted XET output; in another line, withdrawal evoked embryogenesis (i.e. suppressed cell expansion) and suppressed XET output (49). Thus, both lines exhibited a positive correlation between cell elongation and XET secretion.

In an attempt to gain direct evidence that XET enhances wall extension, XET was applied to heat-inactivated hypocotyl segments of cucumber, and their ability to creep was tested mechanically (77). The XET had no effect, whereas applied expansins [nonenzymic proteins that break cellulosic hydrogen bonds (76)] promoted elongation. Negative results are difficult to interpret. The lack of effect of applied XET could have been either because XET alone was not sufficient to promote wall creep in these denatured segments or because the exogenous XET failed to gain access to the relevant xyloglucan molecules within the wall matrix. XET may act synergistically with other wall proteins such as expansins to bring about cell-wall loosening.

In nongrowing kiwifruit pericarp tissue, XET is strongly enhanced by a dose of ethylene that initiates fruit ripening. The rise in XET activity precedes a dramatic wall swelling, which suggests a causative role in fruit softening (94).

Other factors that can regulate the action of a wall enzyme are the presence of an inhibitor or the control of enzyme secretion. Tobacco callus cell walls contain a protein ($M_r = 17,000$) that, at least in vitro, inhibits the acid invertase present in the same walls (114). Hoson (53) has also suggested the presence of an inhibitor of xyloglucan degradation (but not of cellulase). Protoplast-to-apoplast secretion of invertase is regulated by light in radish seedlings (42), and the secretion of peroxidase may be inhibited by GA_3 (34).

Evidence from Use of Molecular Probes

Other studies have reported the timing and localization of accumulation of mRNAs encoding wall enzymes and of the immunocytochemically detected enzyme protein itself. Such studies also have pitfalls that preclude any simplistic interpretation of when, where, and at what rate the enzyme acts: The mRNA may not be translated or the translation product may not be secreted, and the antigenic enzyme protein may not have been modified to its active form. An antibody raised against *T. majus* seed XETH recognized traces of an antigen in growing tissue (19); it remains to be seen whether this antigen represents a small amount of the seed enzyme present in the growing tissue or a small amount of cross-reactivity between the seed XETH and the vegetative tissue XET.

The transcription of XET-related genes can suggest when and where XET-like proteins are expressed. For instance, *meri-5* is transcribed in the *A. thaliana* shoot meristematic dome (but not in the surrounding leaf primordia) and in vascular branch-points in both shoot and root (78). This observation is difficult to relate to the observed distribution of XET activity in growing plants, where the activity peaks in the rapidly elongating zones. *bru1* transcription is induced in young *G. max* stems in response to growth-promoting doses of brassinosteroids (120). It remains to be seen whether the proteins encoded by *meri-5* and *bru1* possess XET activity.

MANIPULATION OF WALL ENZYME ACTION

Chemical Manipulation

To test critically the biological roles of wall enzymes, it would be valuable to have the means of changing their activity and observing the consequences. Administration of specific inhibitors of enzyme action is one approach. Unfortunately, few such inhibitors are known. The exo-glycosylhydrolases can often be inhibited by appropriate aldonolactones: Galactonolactone inhibits β-D-galactosidase, gluconolactone inhibits β-D-glucosidase (15, 25), and arabinonolactone inhibits α-L-arabinofuranosidase (59). Aldonolactones do not interfere with the action of auxin in coleoptiles, which suggests that the action of the targeted exo-glycosylhydrolases on the cell wall is not essential for auxin-induced wall loosening (25). Nojirimycin and related compounds are relatively specific inhibitors of exo-glycosylhydrolases, and in oat coleoptiles, nojirimycin inhibited both auxin-induced exo-hydrolysis of MLG and auxin-induced elongation (81). Curiously, nojirimycin also inhibited auxin-induced elongation in pea stem segments (81), which do not contain MLG. Nojirimycin and gluconolactone do not inhibit pea XET (39; SC Fry, unpublished data).

Antibodies to enzymes often inhibit the enzyme activity; thus, they can be used to test biological function. Antibodies against exo- or endo-MLGases (or, better, both together) inhibited auxin-induced growth in maize coleoptiles (54), which again suggests a role for MLG hydrolysis in cell elongation.

Although no specific inhibitors of XET are known, appropriate exogenous XGOs should act competitively to decrease the frequency with which the enzyme selects a high-molecular-weight endogenous xyloglucan chain as acceptor substrate. It is thus of interest that exogenous XGOs promote the elongation of pea stem segments (74). This observation is compatible with the idea that polysaccharide-to-polysaccharide endo-transglycosylation is normally required in order to reanneal cleaved xyloglucan chains and thereby restore the strength of the cell wall after incremental expansion. When the apoplast is flooded with low-molecular-weight XGOs, the cleaved chains are reannealed to the oligomers, which, not being anchored in the wall, cannot contribute to wall strength. The consequence is unrestrained cell expansion, at least while turgor remains high.

Epoxyalkyl oligoglucosides have been reported to be active site-directed inhibitors of the plant endo-glucanases that attack MLG (51, 52). Determination of the effect of these inhibitors on physiological processes in plants (Gramineae) that possess MLG, as well as the development of corresponding inhibitors of XET and other important wall enzymes, are exciting future prospects for this field.

Genetic Manipulation

Another approach to testing biological function is genetic transformation, including antisense insertions and sense-gene silencing (43, 97, 107). The application of this technology to understanding the physiological significance of wall enzymes will be of great interest. The power of the technology can be illustrated by recent work on the role of pectinase (endo-polygalacturonase) in fruit ripening. Many ripening fruits, including tomatoes, accumulate pectinase (32), which was thought to play a major role in fruit softening by weakening the middle lamella. However, in transgenic tomatoes possessing negligible pectinase activity, the fruit still softened during ripening. The only major physiological consequence of knocking out pectinase was a prolongation of the fruit's shelf-life before the final stages of senescence rendered the fruit unsellable (96). Similar technology is currently being applied to other wall enzymes, including XET.

CONCLUSION

Many wall enzymes are known and many biological roles have been proposed for them, but we have far to go in determining the nature and rate of the in vivo reactions catalyzed by these enzymes and in testing their proposed roles. The methods exist.

ACKNOWLEDGMENT

Work in the author's laboratory has been supported as a Project of Technological Priority by the Commission of the European Communities.

Literature Cited

1. Abeles FB, Biles CL. 1991. Cellulase activity in developing apple fruits. *Sci. Hortic.* 47:77–87
2. Albersheim P. 1974. Structure and growth of walls of cells in culture. In *Tissue Culture and Plant Science*, ed. HE Street, pp. 379–404. London: Academic
3. Almin KE, Eriksson KE, Janson C. 1967. Enzymic degradation of polymers. II. Viscometric determination of cellulase activity in absolute terms. *Biochim. Biophys. Acta* 139:248–53
4. Andrewartha KA, Phillips DR, Stone BA. 1979. Solution properties of wheat flour arabinoxylans and enzymically modified arabinoxylans. *Carbohydr. Res.* 77:191–204
5. Angelini R, Manes F, Federico R. 1990. Spatial and functional correlation between diamine-oxidase and peroxidase activities and their dependence upon de-etiolation and wounding in chick-pea stems. *Planta* 182:89–96
6. Aspinall GO, Begbie R, Hamilton A, Whyte JNC. 1967. Polysaccharides of soybeans. III. Extraction and fractionation of polysaccharides from cotyledon meal. *J. Chem. Soc. C* 1967:1065–70

7. Augur C, Benhamou N, Darvill AG, Albersheim P. 1993. Purification, characterization, and cell wall localization of an α-fucosidase that inactivates a xyloglucan oligosaccharin. *Plant J.* 3:415–26

8. Augur C, Yu L, Sakai K, Ogawa T, Sinaÿ P, et al. 1992. Further studies of the ability of xyloglucan oligosaccharides to inhibit auxin-stimulated growth. *Plant Physiol.* 99:180–85

9. Bacic A, Harris PJ, Stone BA. 1988. Structure and function of plant cell walls. In *The Biochemistry of Plants. A Comprehensive Treatise*, ed. J Preiss, 14:297–371. San Diego: Academic

10. Baird LM, Reid PD. 1992. Cellulase activity and localization during induced abscission of *Coleus blumei*. *J. Plant Growth Regul.* 11:129–34

11. Baydoun EA-H, Fry SC. 1989. In vivo degradation and extracellular polymerbinding of xyloglucan nonasaccharide, a natural anti-auxin. *J. Plant Physiol.* 134:453–59

12. Bret-Harte MS, Baskin TI, Green PB. 1991. Auxin stimulates both deposition and breakdown of material in the pea outer epidermal cell wall as measured interferometrically. *Planta* 185:462–71

13. Carpita NC, Brown RA, Weller KM. 1982. Uptake and metabolic fate of glucose, arabinose and xylose by *Zea mays* coleoptiles in relation to cell wall synthesis. *Plant Physiol.* 69:1173–80

14. Cassab GI, Varner JE. 1988. Cell wall proteins. *Annu. Rev. Plant Physiol. Plant Mol. Biol.* 39:321–53

15. Cline K, Albersheim P. 1981. Host-pathogen interactions. XVII. Purification and characterization of a β-glucosyl hydrolase/transferase present in the walls of soybean cells. *Plant Physiol.* 68:207–20

16. Cutillas-Iturralde A, Zarra I, Fry SC, Lorences EP. 1994. Implication of persimmon fruit hemicellulose metabolism in the softening process. Importance of xyloglucan endotransglycosylase. *Physiol. Plant.* 91:169–76

17. Cutillas-Iturralde A, Zarra I, Lorences EP. 1993. Metabolism of cell wall polysaccharides from persimmon fruit. Pectin solubilization during fruit ripening occurs in the apparent absence of polygalacturonase activity. *Physiol. Plant.* 89:369–75

18. Davies E, Özbay O. 1980. A potential role for cellulase in hormone-controlled elongation. *Z. Pflanzenphysiol.* 99:461–69

19. del Campillo E, Lewis LN. 1992. Occurrence of 9.5 cellulase and other hydrolases in flower reproductive organs undergoing major cell wall disruption. *Plant Physiol.* 99:1015–20

20. de Silva J, Jarman CD, Arrowsmith DA, Stronach MS, Chengappa S, et al. 1993. Molecular characterisation of a xyloglucan-specific endo 1,4-β-D-glucanase (xyloglucan endo-transglycosylase) from nasturtium seeds. *Plant J.* 3:701–11

21. Dey PM. 1979. Transglycosylation activity of sweet almond α-galactosidase; synthesis of saccharides. *Phytochemistry* 18:35–38

22. Driouich A, Laine A-C, Vian B, Faye L. 1992. Characterization and localization of laccase forms in stem and cell cultures of sycamore. *Plant J.* 2:13–24

23. Edwards M, Bowman YJL, Dea ICM, Reid JSG. 1988. A β-D-galactosidase from nasturtium (*Tropaeolum majus* L.) cotyledons. *J. Biol. Chem.* 263:4333–37

24. Edwards M, Dea ICM, Bulpin PV, Reid JSG. 1986. Purification and properties of a novel xyloglucan-specific endo-(1→4)-β-D-glucanase from germinated nasturtium seeds (*Tropaeolum majus* L.). *J. Biol. Chem.* 261:9489–94

25. Evans ML. 1974. Evidence against the involvement of galactosidase or glucosidase in auxin- or acid-stimulated growth. *Plant Physiol.* 54:213–15

26. Fanutti C, Gidley MJ, Reid JSG. 1991. A xyloglucan-oligosaccharide-specific α-D-xylosidase or exo-oligoxyloglucan-α-xylohydrolase from germinated nasturtium (*Tropaeolum majus* L.) seeds. Purification, properties and its interaction with a xyloglucan-specific endo-(1→4)-β-D-glucanase and other hydrolases during storage-xyloglucan mobilisation. *Planta* 184:137–47

27. Fanutti C, Gidley MJ, Reid JSG. 1993. Action of a pure xyloglucan endo-transglycosylase (formerly called xyloglucan-specific endo-(1→4)-β-D-glucanase) from the cotyledons of germinated nasturtium seeds. *Plant J.* 3:691–700

28. Farkaš V, Hanna R, Maclachlan G. 1991. Xyloglucan oligosaccharide α-L-fucosidase activity from growing pea stems and germinating nasturtium seeds. *Phytochemistry* 30:3203–7

29. Farkaš V, Maclachlan G. 1988. Stimulation of pea 1,4-β-glucanase activity by oligosaccharides derived from xyloglucan. *Carbohydr. Res.* 184:213–20

30. Farkaš V, Sulová Z, Stratilová E, Hanna R, Maclachlan G. 1992. Cleavage of xyloglucan by nasturtium seed xyloglucanase and transglycosylation to xyloglucan subunit oligosaccharides. *Arch. Biochem. Biophys.* 298:365–70

31. Ferte N, Moustacas A-M, Nari J, Teissere M, Borel M, et al. 1993. Characterization and kinetic properties of a soya-bean cell-wall phosphatase. *Eur. J. Biochem.* 211:297–304

32. Fischer RL, Bennett AB. 1991. Role of cell wall hydrolases in fruit ripening. *Annu.*

Rev. Plant Physiol. Plant Mol. Biol. 42: 675–703

33. Franz G. 1972. Polysaccharidmetabolismus in den Zellwänden wachsender Keimlinge von *Phaseolus aureus*. *Planta* 102: 334–47

34. Fry SC. 1980. Gibberellin-controlled pectinic acid and protein secretion in growing cells. *Phytochemistry* 19:735–40

35. Fry SC. 1985. Primary cell wall metabolism. *Oxford Surv. Plant Mol. Cell Biol.* 2:1–42

36. Fry SC. 1988. *The Growing Plant Cell Wall: Chemical and Metabolic Analysis.* Harlow, Essex: Longman. 333 pp.

37. Fry SC. 1991. Cell wall-bound proteins. In *Methods in Plant Biochemistry,* ed. LJ Rogers, 5:30–31. London: Academic

38. Fry SC, Miller JG. 1987. H_2O_2-dependent cross-linking of feruloyl-pectins *in vivo*. *Food Hydrocolloids* 1:395–97

39. Fry SC, Smith RC, Renwick KF, Martin DJ, Hodge SK, Matthews KJ. 1992. Xyloglucan endotransglycosylase, a new wall-loosening enzyme activity from plants. *Biochem. J.* 282:821–28

40. Fry SC, York WS, Albersheim P, Darvill A, Hayashi T, et al. 1993. An unambiguous nomenclature for xyloglucan-derived oligosaccharides. *Physiol. Plant.* 89:1–3

41. García-Romera I, Fry SC. 1994. Absence of transglycosylation with oligogalacturonides in plant cells. *Phytochemistry* 35: 67–72

42. Ghorbel A, Mouatassim B, Faye L. 1984. Studies on β-fructosidase from radish seedlings. V. Immunochemical evidence for an enzyme photo-regulated transfer from cytoplasm to cell wall. *Plant Sci. Lett.* 33:35–41

43. Gray J, Picton S, Shabeer J, Schuch W, Grierson D. 1992. Molecular biology of fruit ripening and its manipulation with antisense genes. *Plant Mol. Biol.* 19:69–87

44. Gross GG. 1977. Cell wall-bound malate dehydrogenase from horseradish. *Phytochemistry* 16:319–21

45. Guignard R, Pilet P-E. 1978. Action d'une cellulase extraite du *Pisum sativum* sur des xyloglucanes. *C. R. Acad. Sci. D* 286:855–57

46. Halmer P. 1989. *De novo* synthesis of mannanase by the endosperm of *Lactuca sativa*. *Phytochemistry* 28:371–78

47. Hatfield RD, Nevins DJ. 1986. Purification and properties of an endo-glucanase isolated from the cell walls of *Zea mays* seedlings. *Carbohydr. Res.* 148:265–78

48. Hayashi T, Maclachlan G. 1984. Pea xyloglucan and cellulose. III. Metabolism during lateral expansion of pea epicotyl cells. *Plant Physiol.* 76:739–42

49. Hetherington PR, Fry SC. 1993. Xyloglucan endotransglycosylase activity in carrot cell suspensions during cell elongation and somatic embryogenesis. *Plant Physiol.* 103:987–92

50. Hiscock SJ, Dewey FM, Coleman JOD, Dickinson HG. 1994. Identification and localization of an active cutinase in the pollen of *Brassica napus* L. *Planta* 193: 377–84

51. Høj PB, Rodriguez EB, Iser JR, Stick RV, Stone BA. 1991. Active site-directed inhibition by optically pure epoxyalkyl cellobiosides reveals differences in active site geometry of two 1,3-1,4-β-D-glucan 4-glucanohydrolases. The importance of epoxide stereochemistry for enzyme inactivation. *J. Biol. Chem.* 266:11628–31

52. Høj PB, Rodriguez EB, Stick RV, Stone BA. 1989. Differences in active site structure in a family of β-glucan endohydrolases deduced from the kinetics of inactivation by epoxyalkyl β-oligoglucosides. *J. Biol. Chem.* 264:4939–47

53. Hoson T. 1993. Regulation of polysaccharide breakdown during auxin-induced cell wall loosening. *J. Plant Res.* 106:369–81

54. Inouhe M, Nevins DJ. 1991. Inhibition of auxin-induced cell elongation of maize coleoptiles by antibodies specific for cell wall glycanases. *Plant Physiol.* 96:426–31

55. Jin DF, West CA. 1984. Characteristics of galacturonic acid oligomers as elicitors of casbene synthetase activity in castor bean seedlings. *Plant Physiol.* 74:989–92

56. José M, Puigdomènech P. 1993. Structure and expression of genes coding for structural proteins of the plant cell wall. *New Phytol.* 125:259–82

57. Kakes P. 1985. Linamarase and other β-glucosidases are present in the cell walls of *Trifolium repens* L. leaves. *Planta* 166: 156–60

58. Kanellis AK, Kalaitzis P. 1992. Cellulase occurs in multiple active forms in ripe avocado fruit mesocarp. *Plant Physiol.* 98: 472–79

59. Konno H, Yamasaki Y, Katoh K. 1987. Purification of an α-L-arabinofuranosidase from carrot cell cultures and its involvement in arabinose-rich polymer degradation. *Physiol. Plant.* 69:405–12

60. Konno H, Yamasaki Y, Katoh K. 1989. Extracellular exo-polygalacturonase secreted from carrot cell cultures. Its purification and involvement in pectic polymer degradation. *Physiol. Plant.* 76:514–20

61. Koyama T, Hayashi T, Kato Y, Matsuda K. 1983. Degradation of xyloglucan by wall-bound enzymes from soybean tissue. II. Degradation of the fragment heptasaccharide from xyloglucan and the characteristic action pattern of the α-D-xylosidase in the enzyme system. *Plant Cell Physiol.* 24: 155–62

62. Labavitch JM, Ray PM. 1974. Turnover of

cell wall polysaccharides in elongating pea stem segments. *Plant Physiol.* 53:669–73

63. Leung DWM, Bewley JD. 1983. A role for α-galactosidase in the degradation of the endosperm cell walls of lettuce seeds cv Grand Rapids. *Planta* 157:274–77

64. Lhernould S, Karamanos Y, Bougerie S, Strecker G, Julien R, Morvan H. 1992. Peptide-*N*-(*N*-acetylglucosaminyl)-asparagine amidase activity could explain the occurrence of extracellular xylomannosides in a plant cell suspension. *Glyco-conjugate J.* 9:191–97

65. Liénart Y, Barnoud F. 1985. β-D-Glucanase activities in pure cell-wall-enriched fractions from *Valerianella olitoria* cells. *Planta* 165:68–75

66. Liénart Y, Comtat J, Barnoud F. 1985. Purification of cell-wall β-D-xylanases from *Acacia* cultured cells. *Plant Sci.* 41:91–96

67. Lorences EP, Fry SC. 1993. Xyloglucan oligosaccharides with at least two α-D-xylose residues act as acceptor substrates for xyloglucan endotransglycosylase and promote the depolymerisation of xyloglucan. *Physiol. Plant.* 88:105–12

68. Lorences EP, Zarra I. 1987. Auxin-induced growth in hypocotyl segments of *Pinus pinaster* Aiton. Changes in molecular weight distribution of hemicellulosic polysaccharides. *J. Exp. Bot.* 38:960–67

69. Marchesini A, Kroneck PMH. 1979. Ascorbate oxidase from *Cucurbita pepo* medullosa. New method of purification and reinvestigation of properties. *Eur. J. Biochem.* 101:65–76

70. Masuda H, Sugawara S. 1978. Adsorption of cytoplasmic and wall-bound saccharase of sugar beet seedlings to cell wall. *Arch. Biochem. Biophys.* 42:1479–83

71. Mauch F, Meehl JB, Staehelin LA. 1992. Ethylene-induced chitinase and β-1,3-glucanase accumulate specifically in the lower epidermis and along vascular strands of bean leaves. *Planta* 186:367–75

72. McCleary BV. 1983. Enzymic interactions in the hydrolysis of galactomannan in germinating guar: the role of exo-β-mannanase. *Phytochemistry* 22:649–58

73. McDougall GJ, Fry SC. 1989. Structure-activity relationships for xyloglucan oligosaccharides with anti-auxin activity. *Plant Physiol.* 89:883–87

74. McDougall GJ, Fry SC. 1990. Xyloglucan oligosaccharides promote growth and activate cellulase: evidence for a role of cellulase in cell expansion. *Plant Physiol.* 93: 1042–48

75. McNeil M, Darvill AG, Fry SC, Albersheim P. 1984. Structure and function of the primary cell walls of plants. *Annu. Rev. Biochem.* 53:625–63

76. McQueen-Mason SJ, Cosgrove DJ. 1994. Disruption of hydrogen bonding between wall polymers by proteins that induce plant wall extension. *Proc. Natl. Acad. Sci. USA* 91:6574–78

77. McQueen-Mason SJ, Fry SC, Durachko DM, Cosgrove DJ. 1993. The relationship between xyloglucan endotransglycosylase and *in-vitro* cell wall extension in cucumber hypocotyls. *Planta* 190:327–31

78. Medford JI, Elmer JS, Klee HJ. 1991. Molecular cloning and characterization of genes expressed in shoot apical meristems. *Plant Cell* 3:359–70

79. Nagahashi G, Lassiter GD, Patterson DL. 1992. Unique properties of cell wall-associated β-glucosidases. *Plant Sci.* 81:163–68

80. Nari J, Noat G, Ricard J, Franchini E, Moustacas A-M. 1983. Catalytic properties and tentative function of a cell wall β-glucosyltransferase from soybean cells cultured in vitro. *Plant Sci. Lett.* 28:313–20

81. Nevins DJ. 1975. The effect of nojirimycin on plant growth and its implications concerning a role for exo-β-glucanases in auxin-induced cell expansion. *Plant Cell Physiol.* 16:347–56

82. Niogret M-F, Dubald M, Mandaron P, Mache R. 1991. Characterization of pollen polygalacturonase encoded by several cDNA clones in maize. *Plant Mol. Biol.* 17:1155–64

83. Nishitani K. 1992. A novel method for detection of endo-xyloglucan transferase. *Plant Cell Physiol.* 33:1159–64

84. Nishitani K, Masuda Y. 1982. Acid pH-induced structural changes in cell wall xyloglucans in *Vigna angularis* epicotyl segments. *Plant Sci. Lett.* 28:87–94

85. Nishitani K, Tominaga R. 1991. In vitro molecular weight increase in xyloglucans by apoplastic enzyme preparation from epicotyls of *Vigna angularis*. *Physiol. Plant.* 82:490–97

86. Nishitani K, Tominaga R. 1992. Endo-xyloglucan transferase, a novel class of glycosyltransferase that catalyzes transfer of a segment of xyloglucan molecule to another xyloglucan molecule. *J. Biol. Chem.* 267: 21058–64

87. Nock LP, Smith CJ. 1987. Identification of polysaccharide hydrolases involved in autolytic degradation of *Zea* cell walls. *Plant Physiol.* 84:1044–50

88. Okazawa K, Sato Y, Nakagawa T, Asada K, Kato I, et al. 1993. Molecular cloning and DNA sequencing of endo-xyloglucan transferase, a novel class of glycosyltransferase that mediates molecular grafting between matrix polysaccharides in plant cell walls. *J. Biol. Chem.* 268:25364–68

89. O'Neill RA, Albersheim P, Darvill AG. 1989. Purification and characterization of a xyloglucan oligosaccharide-specific xy-

losidase from pea seedlings. *J. Biol. Chem.* 264:20430–37

90. Potter I, Fry SC. 1993. Xyloglucan endotransglycosylase activity in pea internodes: effects of applied gibberellic acid. *Plant Physiol.* 103:235–41

91. Potter I, Fry SC. 1994. Changes in xyloglucan endotransglycosylase (XET) activity during hormone-induced growth in lettuce and cucumber hypocotyls and spinach cell suspension cultures. *J. Exp. Bot.* 45:287–94

92. Priem B, Gross KC. 1992. Mannosyl- and xylosyl-containing glycans promote tomato (*Lycopersicon esculentum* Mill.) fruit ripening. *Plant Physiol.* 98:399–401

93. Pritchard J, Hetherington PR, Fry SC, Tomos AD. 1993. Xyloglucan endotransglycosylase activity, microfibril orientation and the profiles of cell wall properties along growing regions of maize roots. *J. Exp. Bot.* 44:1281–89

94. Redgwell RJ, Fry SC. 1993. Xyloglucan endotransglycosylase activity increases during kiwifruit (*Actinidia deliciosa*) ripening: implications for fruit softening. *Plant Physiol.* 103:1399–406

95. Reis D, Vian B, Derzens D, Roland JC. 1987. Sequential patterns of intramural digestion of galactoxyloglucan in tamarind seedlings. *Planta* 170:60–73

96. Schuch W, Kanczler J, Robertson D, Hobson G, Tucker G, et al. 1991. Fruit quality characteristics of transgenic tomato fruit with altered polygalacturonase activity. *Hortic. Sci.* 26:1517–20

97. Sheehy RE, Kramer M, Hiatt WR. 1988. Reduction of polygalacturonase activity in tomato fruit by antisense RNA. *Proc. Natl. Acad. Sci. USA* 85:8805–9

98. Sheldrake AR. 1970. Cellulase and cell differentiation in *Acer pseudoplatanus*. *Planta* 95:167–78

99. Shirsat AH, Wilford N, Evans IM, Gatehouse LN, Croy RRD. 1991. Expression of a *Brassica napus* extensin gene in the vascular system of transgenic tobacco and rape plants. *Plant Mol. Biol.* 17:710–19

100. Showalter AM. 1993. Structure and function of plant cell wall proteins. *Plant Cell* 5:9–23

101. Slakeski N, Fincher GB. 1992. Developmental regulation of $(1{\rightarrow}3,1{\rightarrow}4)$-β-glucanase gene expression in barley. Tissue-specific expression of individual isoenzymes. *Plant Physiol.* 99:1146–50

102. Smith RC, Fry SC. 1989. Extracellular transglycosylation involving xyloglucan oligosaccharides in vivo. In *5th Cell Wall Meeting Book of Abstracts and Programme*, ed. SC Fry, CT Brett, JSG Reid, p. 139. Edinburgh: Scottish Cell Wall Group

103. Smith RC, Fry SC. 1991. Endotransglyco-

sylation of xyloglucans in plant cell suspension cultures. *Biochem. J.* 279:529–35

104. Takahama U, Oniki T. 1992. Regulation of peroxidase-dependent oxidation of phenolics in the apoplast of spinach leaves by ascorbate. *Plant Cell Physiol.* 33:379–87

105. Talbott LD, Ray PM. 1992. Changes in size of previously deposited and newly synthesized pea cell wall matrix polysaccharides. Effects of auxin and turgor. *Plant Physiol.* 98:369–79

106. Terry ME, Jones RL, Bonner BA. 1981. Soluble cell wall polysaccharides released from pea stems by centrifugation. I. Effect of auxin. *Plant Physiol.* 68:531–37

107. Tieman DM, Harriman RW, Ramamohan G, Handa AK. 1992. An antisense pectin methylesterase gene alters pectin chemistry and soluble solids in tomato fruit. *Plant Cell* 4:667–79

108. Valero P, Labrador E. 1993. Inhibition of cell wall autolysis and auxin-induced elongation of *Cicer arientinum* epicotyls by β-galactosidase antibodies. *Physiol. Plant.* 89:199–203

109. van der Wilden W, Segers JHL, Chrispeels MJ. 1983. Cell walls of *Phaseolus vulgaris* leaves contain the Azocoll-digesting proteinase. *Plant Physiol.* 73:576–78

110. Wallace G, Fry SC. 1994. Phenolic components of the plant cell wall. *Int. Rev. Cytol.* 151:229–67

111. Warneck H, Seitz HU. 1993. Inhibition of gibberellic acid-induced elongation-growth of pea epicotyls by xyloglucan oligosaccharides. *J. Exp. Bot.* 44:1105–9

112. Webb EC. 1992. *Enzyme Nomenclature 1992: Recommendations (1992) of the Nomenclature Committee of the International Union of Biochemistry and Molecular Biology*. San Diego: Academic. 862 pp.

113. Wegrzyn TF, Macrae EA. 1992. Pectinesterase, polygalacturonase and β-galactosidase during softening of ethylene-treated kiwifruit. *Hortic. Sci.* 27:900–2

114. Weil M, Krausgrill S, Schuster A, Rausch T. 1994. A 17-kDa *Nicotiana tabacum* cell-wall peptide acts as an in-vitro inhibitor of the cell-wall isoform of acid invertase. *Planta* 193:438–45

115. Welinder KG. 1992. Plant peroxidases: structure-function relationships. In *Progress and Prospects in Biochemistry and Physiology*, ed. T Gaspar, C Penel, H Greppin, pp. 1–24. Geneva: Univ. Geneva Press

116. Yamaoka T, Chiba N. 1983. Changes in the coagulating ability of pectin during the growth of soybean hypocotyls. *Plant Cell Physiol.* 24:1281–90

117. Yamaoka T, Tsukada K, Takahashi H, Yamauchi N. 1983. Purification of a cell-wall bound pectin-gelatinizing factor and examination of its identity with pectin

methylesterase. *Bot. Mag. (Tokyo)* 96:139–44

118. Zheng L, Heupel RC, Dellapenna D. 1992. The β subunit of tomato fruit polygalacturonase isoenzyme. 1: Isolation, characterization, and identification of unique structural features. *Plant Cell* 4:1147–56

119. Zheng X, van Huystee RB. 1992. Peroxidase-regulated elongation of segments from peanut hypocotyls. *Plant Sci.* 81:47–56

120. Zurek DM, Clouse SD. 1994. Molecular cloning and characterization of a brassinosteroid-regulated gene from elongating soybean epicotyls. *Plant Physiol.* 104:161–70

Annu. Rev. Plant Physiol. Plant Mol. Biol. 1995. 46:521–47

BIOCHEMISTRY AND MOLECULAR BIOLOGY OF THE ISOPRENOID BIOSYNTHETIC PATHWAY IN PLANTS

Joseph Chappell

Plant Physiology/Biochemistry/Molecular Biology Program, Agronomy Department, University of Kentucky, Lexington, Kentucky 40546-0091

KEY WORDS: prenyltransferases, terpene synthases and cyclases, gene comparisons, metabolic channels

CONTENTS

ABSTRACT

This review first summarizes the diverse nature of isoprenoids found in plants, emphasizing the wide range of physiological functions these compounds

0066-4294/95/0601-0521$05.00

serve. The biosynthetic origins of isoprenoids have occupied chemists and biochemists for decades, and the second section of this review recaps some of the conceptual models used to rationalize key biosynthetic reactions in the isoprenoid pathway. The third section describes briefly some of the recently developed experimental systems that have helped researchers uncover much of the biochemistry and molecular biology of isoprenoids. The fourth section compares the deduced amino acid sequences of enzymes with similar catalytic functions and attempts to correlate these sequences with the known enzymology. The fifth and final section focuses on our limited understanding of how isoprenoid biosynthesis is regulated in plants.

INTRODUCTION

Isoprenoids constitute a large family of compounds with over 20,000 members identified to date. The isoprenoid structures are diverse and range from relatively simple linear hydrocarbon chains to some of the most complex ring structures known. Some isoprenoids are compounds likely to be familiar to everyone (e.g. rubber and menthol), others are somewhat more obscure (e.g. the antimicrobial sesquiterpene phytoalexins ipomeamarone and capsidiol), and others are well known to plant physiologists as vital for plant growth and development (e.g. gibberellic acid and sterols). Terpenes and terpenoids, older terms for these compounds, are perhaps more descriptive terms because they conjure up memories of aromatic fragrances like the turpentine oils from which the first isoprenoids were isolated and hence named. However, the common denominator for this diverse array of compounds is their biosynthetic origin, starting from isoprene, the universal five-carbon building block, and the "biogenic isoprene rule," the key theorem put forth by Ruzicka in 1953 (71) to rationalize the biosynthetic origins of all isoprenoids. The polymerization of two diphosphorylated isoprene building blocks generates geranyl diphosphate (GPP), a linear C10 intermediate that can be converted to cyclic or linear end-products representing the monoterpenes, or used in another round of polymerization. The addition of a third isoprene unit to GPP generates farnesyl diphosphate (FPP), which can also be converted to cyclic or linear products representing the sesquiterpene class. Continuing the polymerization and chemical differentiation cycle can thus lead to the production of other classes of isoprenoids named according to the number of isoprene building blocks that have gone into their biosynthesis. The term isoprenoid is therefore reflective of the biosynthetic origins of this family of compounds and is inclusive of all compounds derived from one or more isoprene units.

The topic of isoprenoids encompasses a rich and broad subject area, derived from a fascination researchers have had with the diversity of isoprenoids found in nature and from over five decades of intense investigations into isoprenoid biosynthesis. There are many excellent reviews of particular classes of isoprenoids (9, 10, 12, 20, 83) and of various aspects of their biosynthesis (3, 33, 49). However, only Gray (40) attempted to integrate the chemistry and biochemistry of all isoprenoids produced in plants as a means of identifying the many mechanisms possibly regulating isoprenoid biosynthesis in plants. Gray noted in 1987 that although activities had been measured for many of the enzymes constituting the central isoprenoid biosynthetic pathway, activities for many of the enzymes making up the branch pathways had not. He also provided a critical assessment of how various branch pathways of the isoprenoid biosynthetic pathway may be localized to intracellular compartments, and how branch pathways may be integrated through metabolite transport mechanisms and mechanisms of feedback regulation. Finally, Gray listed criteria necessary for determining rate-limiting steps in the biosynthesis of any isoprenoid and implied that although modulations in the levels of various biosynthetic enzymes had been documented, such documentation did not constitute proof of a rate-limiting step. This review revisits many of these same issues and topics.

ISOPRENOIDS ARE FAMILIAR METABOLITES

Plant isoprenoids comprise a structurally diverse group of compounds that can be divided into classes of primary and secondary metabolites (Figure 1). Isoprenoids that are primary metabolites include sterols, carotenoids, growth regulators, and the polyprenol substituents of dolichols, quinones, and proteins. These compounds are essential for membrane integrity, photoprotection, orchestration of developmental programs, and anchoring essential biochemical functions to specific membrane systems, respectively. Isoprenoids classified as secondary metabolites include monoterpenes, sesquiterpenes, and diterpenes. Compounds within this category are considered secondary because they are not essential for viability; however, they mediate important interactions between plants and their environment. For example, specific terpenoids have been correlated with plant-plant (76), plant-insect (35), and plant-pathogen (77) interactions (see 33 for additional references).

ENZYMOLOGY OF THE ISOPRENOID PATHWAY

Early Enzymatic Steps in the Pathway

The isoprenoid biosynthetic pathway is sometimes referred to as the mevalonate pathway. From a technical viewpoint, this makes some sense. Mevalonate is a six-carbon intermediate in the pathway, arising from the sequential condensation of three acetyl-CoA units to generate 3-hydroxy-3-methylglutaryl

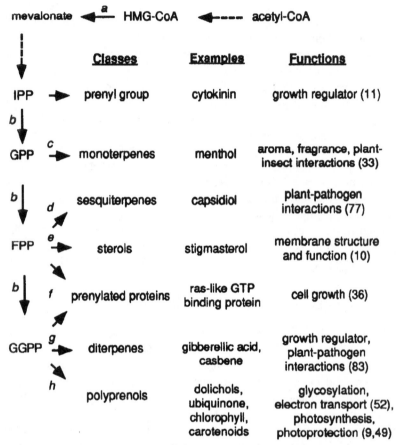

Figure 1 A depiction of the isoprenoid biosynthetic pathway with respect to the types of end products and their physiological significance. Broken arrows indicate multiple steps or reactions. The rate-limiting step for sterol metabolism in mammals resides at the early step in the pathway catalyzed by 3-hydroxy-3-methylglutaryl Coenzyme A (HMG-CoA) reductase (*a*) (37). In plants, the control of isoprenoid metabolism remains controversial (6, 7, 14, 15, 58). Other enzyme reactions discussed in this review include prenyltransferases (*b*), monoterpene synthases (*c*), sesquiterpene synthases (*d*), squalene synthase (*e*), prenyl:protein transferases (*f*), diterpene synthases (*g*), and phytoene synthase (*h*).

Coenzyme A (HMG-CoA). HMG-CoA is converted to mevalonate in an irreversible reaction catalyzed by HMG-CoA reductase (HMGR). Owing to the irreversible nature of this reaction, early workers surmised correctly that this step was a likely regulatory point of sterol biosynthesis in mammalian systems and eventually correlated the absolute rate of cholesterol biosynthesis with the level of this enzyme activity (37). Whether this enzyme plays a similar rate-limiting role in controlling plant isoprenoid biosynthesis remains unresolved

(6, 7, 14, 15, 58; see below). However, as pointed out by Gray (40), an obvious omission in the literature is any systematic measurement of the enzyme activities leading up to HMG-CoA and any calculations for how limiting these activities might be for the central biosynthetic pathway. For example, using an in vitro assay for the conversion of acetyl-CoA to HMG-CoA, Gondet et al (38) recently concluded that the enzymes responsible for this conversion were not up-regulated in tobacco cells selected for enhanced sterol biosynthesis. Such coupled assays may or may not be an accurate measurement of a biosynthetic segment consisting of multiple enzymes.

The six-carbon mevalonate is sequentially phosphorylated and decarboxylated to generate isopentenyl diphosphate (IPP) by the enzymes mevalonate kinase, mevalonate 5-phosphate kinase, and mevalonate 5-diphosphate decarboxylase. These enzyme activities have been measured directly or indirectly in plant extracts and the in vitro activities appear sufficient to account for the in vivo accumulation rates of isoprenoids (40). However, based on in vitro assays, the mevalonate kinase reaction may be sensitive to feedback modulation by mevalonate phosphate and other downstream allylic diphosphate intermediates (40), and the level of mevalonate diphosphate decarboxylase is induced several-fold in sweet potato root tissue challenged with a fungal pathogen or treated with toxic chemicals (59). An *Arabidopsis thaliana* mevalonate kinase cDNA clone was isolated recently by genetic complementation in yeast (68) and provides a means for investigating mechanisms controlling this enzyme activity (see below). IPP along with DMAPP (dimethylallyl diphosphate), an interconvertible isomer of IPP, represent the activated monomer building blocks for all other isoprenoids. IPP isomerase, the enzyme catalyzing the isomerization reaction, has been characterized and partially purified from a number of plant sources (25, 61), and a yeast gene has been isolated (4). However, the ready incorporation of IPP into isoprenoids by cell-free extracts (40) and the calculated level of isomerase activity in germinating pea seedlings (41) have been taken as evidence that the isomerase activity is not limiting for isoprenoid biosynthesis. This assumption needs to be reinvestigated in light of the modulation of this enzyme activity in yeast (4).

Intermediate Steps: Common Features of Prenyltransferases and Terpene Cyclases

Condensation of DMAPP with one IPP in a head-to-tail fashion generates geranyl diphosphate (GPP); addition of a second IPP unit generates farnesyl diphosphate (FPP); a third IPP generates geranylgeranyl diphosphate (GGPP); and so on. These polymerization reactions are catalyzed by prenyltransferases, which direct the attack of a carbocation (i.e. an electron-deficient carbon atom resulting from the loss of the diphosphate moiety) of one substrate molecule to an electron-rich carbon atom of a double bond on the second IPP molecule

(Figure 2). The electrophilic nature of these reactions is unusual relative to the more general nucleophilic condensations occurring in other biosynthetic pathways, but this characteristic appears to be a common reaction mechanism among isoprenoid biosynthetic enzymes and especially among those catalyzing the cyclization of various isoprenoid intermediates (33). The enzymes responsible for the cyclization of GPP, FPP, and GGPP are referred to as monoterpene, sesquiterpene, and diterpene synthases or cyclases, and represent reactions committing carbon from the central isoprenoid pathway to end products in the monoterpene, sesquiterpene, and diterpene classes, respectively. Two important biochemical distinctions between the prenyltransferase and the cyclase reactions should be kept in mind. The prenyltransferases catalyze carbon-carbon bond formation between two substrate molecules, whereas the cyclases catalyze an intramolecular cyclization. The prenyltransferases also catalyze reactions with very little variance in the stereochemistry or length of the ensuing polymer. Different prenyltransferases vary only in the

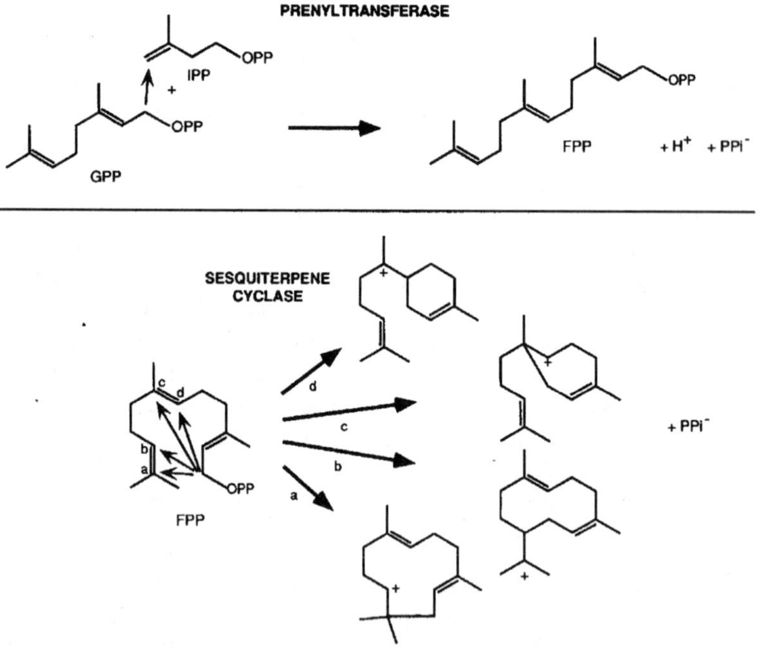

Figure 2   The reactions catalyzed by prenyltransferases (e.g. FPP synthase) and terpene cyclases (e.g. sesquiterpene cyclases) utilize a similar reaction mechanism. In both reactions, one carbon atom is rendered electron deficient because of the loss of the diphosphate substituent, a strong electron-withdrawing group. The resulting carbocation then attacks another carbon atom that is electron-rich by virtue of its association with a double bond. (Adapted from Reference 12.)

length of the allylic substrates they can accept in initiating these reactions. The cyclases are also very substrate specific. However, different sesquiterpene cyclases, for example, can utilize the same substrate to produce dramatically different reaction products (Figure 2).

Earlier confusion has been resolved as to whether one prenyltransferase was responsible for the synthesis of both GPP and FPP, or whether there were two specific transferases. Based on biochemical and molecular evidence, only one prenyltransferase seems necessary. Ogura et al (61) and Green & West (42) reported that FPP synthases partially purified from pumpkin and castor bean endosperm, respectively, were capable of using either DMAPP or GPP as its allylic diphosphate acceptor. More recently, the Karst laboratory has reported the cloning of a single *Arabidopsis thaliana* gene capable of complementing a GPP/FPP synthase mutant in yeast (24). The deduced amino acid sequence of the plant enzyme is similar to the deduced protein sequence for both the rat (17) and yeast (5) enzymes.

Squalene synthase (63a) and phytoene synthase (26) represent two other commitment steps diverting carbon from the central isoprenoid pathway to sterol or carotenoid biosynthesis, respectively (Figure 3). Overall, the reactions catalyzed by these two enzymes can be rationalized as an electrophilic attack of a double bond similar to those discussed above for the prenyltransferase and terpene cyclases. However, the squalene synthase and phytoene synthase reactions are complicated by proceeding in two distinct steps and catalyzing a head-to-head condensation. Prenyltransferases, in contrast, catalyze head-to-tail condensations. In the schematic representation of the squalene synthase reaction shown in Figure 3, the strong electron-withdrawing capacity of one diphosphate substituent sufficiently ionizes the C1' atom of one allylic substrate, which in turn attacks the electron-rich double bond between C2 and C3 of the second allylic diphosphate substrate to form an unusual cyclopropyl intermediate. The reaction intermediates formed in this first step of the reaction are either presqualene or prephytoene and arise concomitant with the loss of one diphosphate group and one proton. Presqualene or prephytoene are converted to the corresponding C30 and C40 reaction products in a second, rate-limiting step of the reaction. The second reaction step entails a complex rearrangement requiring the breakage of the cyclopropyl ring and formation of a new C1-C1' bond. Although the second reaction step is not well understood, the squalene synthase step requires NADPH, but the phytoene synthase does not, and the final reaction product phytoene retains a double bond between the C1 and C1' atoms, which is not found in squalene.

Although the terpene synthases, prenyltransferases, and squalene and phytoene synthases discussed above appear to rely on the diphosphate substituent as an initial ionizing agent in the proposed reaction mechanisms, a second type

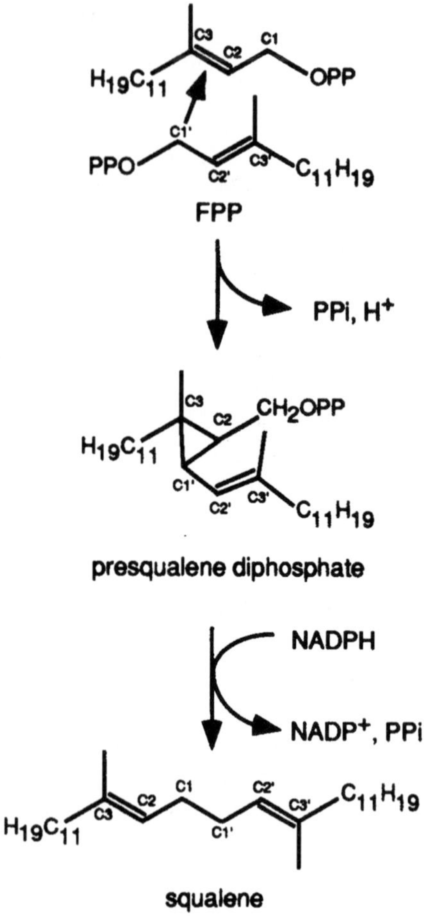

Figure 3 A proposed reaction mechanism for squalene synthase. Like the terpene cyclases and prenyltransferases discussed in Figure 2, this mechanism relies on the electrophilic attack of a double bond by a carbocation intermediate. Loss of the diphosphate substituent ionizes C1′ of one FPP molecule, and the resultant carbocation attacks the electron-rich double bond between C2 and C3 of the second FPP molecule. Unlike those mechanisms discussed in Figure 2, this electrophilic attack generates a cyclopropyl intermediate that is converted to squalene in a second partial reaction. A similar mechanism for phytoene synthase has also been proposed. (Adapted from Reference 63a.)

of terpene synthase (cyclase) relies on a proton-addition mechanism to generate the first carbocation intermediate needed to initiate the reaction (Figure 4). Examples of cyclase enzymes utilizing this mechanism include diterpene synthases (43, 50, 77a, 83) and oxidosqualene cyclase (cycloartenol and lanosterol synthases) (1, 57), the enzyme converting the linear oxidosqualene intermediate into cycloartenol, a tetracyclic C30 intermediate in sterol biosynthesis. Although the electrophilic attacks by carbocation intermediates appear to be a common feature of all the cyclase and prenyltransferase reactions discussed above, the utilization of a proton-addition mechanism vs reliance on the ionizing nature of the diphosphate moiety to generate a carbocation intermediate distinguishes these two reaction mechanisms. An implication of this distinction is that the cyclase enzymes representing the two reaction mechanisms may not be similar to one another, although cyclases and prenyltransferases using a similar reaction mechanism may have common features reflecting their common enzymology.

Late Steps: Secondary Modifications and Transformations

Rarely are the reaction products of the prenyltransferase and terpene cyclases end-products in themselves. One exception is abienol cyclase, a diterpene synthase, which simultaneously cyclizes GGPP and introduces a hydroxyl function (43). More typically, the polyprenol diphosphates either are committed to branch pathways by the initial action of the cyclases and undergo various degrees of modification, or are covalently attached to other cellular constituents by very different types of prenyltransferase reactions. The transferases responsible for attachment of the polyprenols to quinones and chlorophylls in plants have been partially characterized and the importance of their localization to mitochondria and plastids discussed (40, 52). The recent discovery of prenyltransferase(s) catalyzing the thioether linkage of isoprenol groups to cysteine residues at the carboxy terminus of proteins in mammalian and yeast cells has stimulated searches for similar reactions in plant cells. The significance of protein prenylation in plants has yet to be established, but in

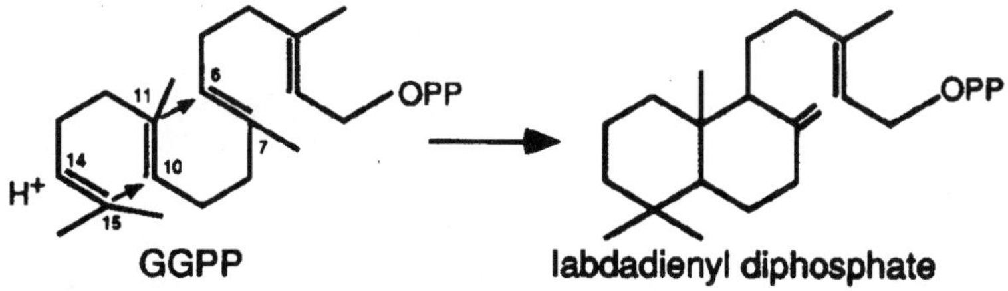

GGPP   labdadienyl diphosphate

Figure 4   A proposed reaction mechanism for the cyclization of geranylgeranyl diphosphate (GGPP) to the bicyclic diterpene labdadienyl diphosphate. This mechanism, unlike those in Figures 3 and 4, relies on a proton addition at C14 of the GGPP molecule to initiate a complex reaction cycle. The initial addition of a proton at C14 creates a carbocation at C15; this carbocation then attacks electron-rich C10 and results in a new carbocation at C11. The C11 carbocation attacks electron-rich C6 followed by a proton elimination at C7 to terminate the reaction. A similar proton-addition mechanism has been proposed for other diterpene and oxidosqualene cyclases. (Adapted from Reference 83.)

mammalian cells it is thought to be a regulatory mechanism controlling several cellular functions such as cell cycle activity (36). The initial evidence for protein prenylation was that the transition of mammalian cells through the cell cycle could be blocked by mevinolin, a competitive inhibitor of HMGR activity. Growth of the cells was not fully restored by the exogenous application of sterols, but it was when both mevalonate and sterols were supplied (37). This result led to speculations that sterol and another nonsterol isoprenoid were required for cell cycle activity. Subsequent studies have shown that ras kinase association with the inner surface of the cell membrane requires farnesylation and that geranylgeranylation of other proteins serves a similar function (36). Randall et al (67) recently demonstrated the prenylation of proteins in tobacco cell cultures, but the function of these prenylated proteins was not clarified. Protein:prenyltransferase activities specific for FPP or GGPP, and characterization of the carboxy terminal sequence requirements for the two prenyltransferases, were also reported at that time. A pea protein:farnesyl transferase cDNA clone has been reported (84).

Most secondary modifications occur after the synthesis of the cyclic product(s) and can include hydroxylation, methylation, isomerization, demethylation, hydration, and reduction (33). Impressive progress in characterizing some of the enzymes responsible for the modification of monoterpenes produced in sage and diterpenes produced in conifers has been reported. Synthesis of abietic acid in conifers requires the action to two stereo-specific, membrane-associated, P-450 hydroxylases and a soluble dehydrogenase (31). Funk et al (30, 32) have also described a cytochrome P-450–dependent hydroxylase in sage that they believe is important in camphor catabolism. Karp et al (48) characterized a P-450 hydroxylase specific for monoterpene modification in sage and mint species, and Dehal & Croteau (23) described two distinct, soluble dehydrogenases responsible for the oxidation of the hydroxyl functions of borneol and sabinol.

SYSTEMS FOR STUDYING ISOPRENOID BIOCHEMISTRY

Detailed biochemical studies of isoprenoid metabolism in plants have been limited for several reasons. Until recently, suitable radiolabeled substrates were not commercially available, and investigators had to prepare the substrates themselves (22). Although synthesis of these precursors is not particularly difficult, their synthesis was something out of the ordinary for many plant physiologists. Isoprenoids also tend to accumulate over long developmental periods, which means the enzymes responsible for their biosynthesis are either present in low abundance or have low turnover rates (rate of substrate to product conversion per enzyme molecule per unit time). Sterols are an example of slowly accumulating isoprenoids that can accumulate to impressive levels by the time a tissue reaches maturity (0.1–0.4% of leaf dry weight). Nonetheless, the sterol biosynthetic machinery is not abundant, and many of the enzymes catalyzing the latter steps in this pathway have not been measured directly. Other classes of isoprenoids are synthesized in very low amounts in specific cell types or tissues. Such rare or low-abundance isoprenoids include growth regulators synthesized in meristematic cells, and the monoterpenes, sesquiterpenes, and diterpenes synthesized in glands, ducts, and trichomes (33, 34). The difficulties associated with biochemical characterization of the enzymes responsible for the synthesis of such low-abundance compounds include the detection limits of the enzyme assays and difficulties encountered in trying to purify the corresponding low-abundance biosynthetic enzymes.

Recent developments such as a procedure for the large-scale isolation of trichomes (34), the use of flowers and fruits for studies of sterol and carotenoid biosynthesis (9, 58), and the use of tuber tissues and plant cell cultures for molecular studies of sterol accumulation and elicitation of sesquiterpene phytoalexins (14, 15, 59, 60, 79, 80, 86) have proved useful for carrying out biochemical and molecular investigations of isoprenoid metabolism. For example, when fungal elicitors are added to rapidly growing tobacco-cell suspension cultures, the cultures cease sterol production and, instead, synthesize and secrete antimicrobial sesquiterpenes (81). The decline in sterol biosynthesis is correlated with a suppression of squalene synthase enzyme activity, and the induction of sesquiterpene biosynthesis with an induction of a sesquiterpene cyclase enzyme activity (79, 81). Because these two enzymes are positioned at a putative branch point in the isoprenoid pathway, the induction of one enzyme and the suppression of the other were interpreted as an important mechanism controlling carbon flow and, hence, end-product formation. Zook & Kuć (86) recently documented a similar finding in pathogen- or elicitor-challenged potato tuber disks. The rapid changeover in the isoprenoid metabolism of the potato tubers and cell cultures, within 12 h of elicitor addition, and the substantial changes in the amounts and synthesis rates of sterols and sesquiterpenes have facilitated efforts to purify enzymes (44, 81a), to clone induced vs suppressed cDNAs using differential screening techniques (29), and to investigate specific mechanisms controlling the level of the biosynthetic machinery (15, 15a, 15b, 29, 82, 82a, 85).

GENE COMPARISONS

Monoterpene, Sesquiterpene, and Diterpene Cyclases

The development of systems suitable for detailed biochemical studies also provided a means for cloning the biosynthetic genes. Recently, sequences of a monoterpene cyclase (18), a sesquiterpene cyclase (29), and a diterpene cyclase (55) have been reported. These three genes were cloned using conventional cloning strategies, such as preparing cDNA libraries with mRNA isolated from tissue or cells enriched for a cyclase mRNA, then screening the library with antibodies prepared to the purified cyclase protein (51) or with oligonucleotide probes designed from the amino acid sequence of the purified cyclase protein (18).

All three cyclases catalyze reactions typified by the electrophilic attack of a carbocation generated by the ionizing effect of the diphosphate substituent, except they utilize GPP, FPP, or GGPP as their substrates, and their reaction products are different. In Figure 5, the deduced amino acid sequences from these genes are aligned schematically relative to one another as well as to a fungal (*Penicillium roqueforti*) sesquiterpene cyclase gene coding for aristolochene synthase (64). The reactions catalyzed by the fungal and tobacco enzymes are essentially identical, except with respect to the orientation of one methyl substituent. One might predict that the fungal and plant sesquiterpene cyclases should share sequence similarity based on their enzymatic similarity, but these two enzymes are not very similar to one another at all. Instead, a much more significant level of similarity between the plant monoterpene, sesquiterpene, and diterpene cyclase proteins is observed. Overall, the similarity and identity (calculated by the GCG software program) between the tobacco and mint enzymes are 53% and 33%, and between the tobacco and castor bean enzymes are 64% and 42%, respectively. This surprising degree of similarity extends to the conservation of a domain and amino acid residues of putative importance in the reaction mechanisms. An aspartate-rich domain suggested to coordinate a metal cofactor (Mg^{+2} or Mn^{+2}) at the substrate-binding site, effectively neutralizing the negatively charged diphosphate group of the otherwise hydrophobic substrates, has been found in a number of isoprenoid biosynthetic enzymes, including prenyltransferases and cyclases (29, 47, 74). The position and sequence of the aspartate-rich motif is conserved among all three plant cyclases but is somewhat less conserved in the fungal enzyme. Chemical modifying agents have also been used to demonstrate that histidine and cysteine residues are important for enzyme catalysis (65, 66). The positions of one cysteine and three histidine residues are also conserved within the three plant enzymes. Although the genomic organization of exons and introns has been determined only for the castor bean (casbene synthase) and tobacco (epi-aristolochene synthase) genes, the gene organizations are strikingly similar with regard to intron positions and exon sizes. The only exception seems to be at the amino terminus where both the mint and castor bean enzymes are obviously different from the tobacco enzyme. This difference may be explained by the fact that the castor bean and mint enzymes are targeted to plastids, whereas the tobacco enzyme is cytoplasmic, and their

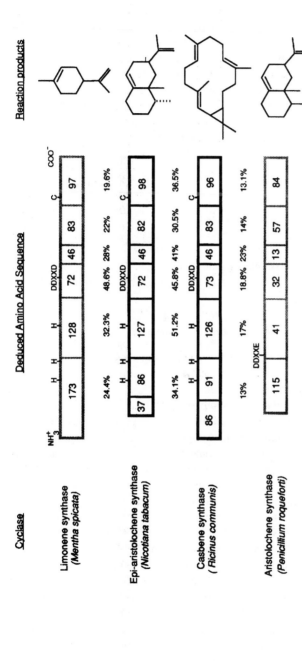

Figure 5 A schematic representation of an amino acid sequence alignment between a mint monoterpene cyclase (18), a castor bean diterpene cyclase (55), a tobacco sesquiterpene cyclase (29), a fungal sesquiterpene cyclase (64). Sequence alignments used deduced amino acid sequences corresponding to exons or analogous regions within proteins and were performed using the Clustal algorithm in the DNASTAR software package. Solid vertical bars correspond to intron positions within the tobacco and castor bean genes. The hatched bars in the mint and fungal genes delimit the corresponding protein domains used to calculate similarity scores. Numbers within the boxes indicate the number of amino acids encoded by an exon (tobacco and castor bean only) or corresponding region of the mint and fungal proteins. Percentages refer to identity scores between the indicated domains; and H, C, and DDXXD refer to conserved histidine, cysteine, and aspartate-rich residues within the plant cyclase proteins. (Adapted from Reference 55.)

additional amino terminal sequences show some similarity to plastid-targeting motifs (18, 55).

The significance of the similarity between the plant cyclases is not clear. It may signify evolution of plant cyclases from a single ancestral gene or may be a reflection of convergent evolution. The importance of the similarity in intron placement within the tobacco and castor bean genes and the conservation of exon size is also unclear. One possibility is that it represents a mechanism to conserve functional domains that are responsible for discrete partial steps of the overall cyclase reaction. Nonetheless, the similarities between the plant sequences may also be of practical value for cloning additional cyclase genes. PCR amplification of other cyclase genes involved in other terpene biosynthetic pathways should be attempted using PCR primers designed to regions of the cyclase proteins that are absolutely conserved between the three plant genes. Any sequence information of additional cyclase genomic clones will also provide insight into the conservation of gene structure as well as conserved protein motifs within the enzyme proteins. If intron-exon junctions delimit functional domains of the cyclase proteins, then such a gene organization should be widely conserved. Ultimately, more critical tests of any conserved protein sequence should include domain-motif swapping between cyclase genes and detailed examination of the resulting hybrid protein expressed in bacteria (8, 12a, 13).

Squalene Synthase and Phytoene Synthase

Detailed amino acid comparisons between other plant isoprenoid biosynthetic genes coding for enzymes of similar catalytic function, as done above for the cyclases, is currently not possible. However, comparisons of other animal, bacterial, and fungal isoprenoid biosynthetic genes have been important for the cloning of analogous plant genes, and subsequent sequence comparisons have resulted in functional inferences. In comparing the deduced amino acid sequences of the yeast and mammalian squalene synthases proteins, Robinson et al (69) identified several conserved domains convenient for the design of PCR primers and suggested using these primers for PCR cloning strategies. We also recognized this possibility in comparisons of the yeast (45) and rat (56) squalene synthase genes. The feasibility of this approach was tested recently using tobacco cell culture mRNA to isolate and sequence a partial squalene synthetase cDNA clone (J Chappell, unpublished data).

The comparison of the deduced amino acid sequence of this clone to other squalene synthase sequences and a tomato phytoene synthase sequence indicates the conservation of several sequence motifs and amino acid residues (Figure 6). The amino acid residues were previously implicated in the enzyme reaction mechanism by tagging experiments using chemical modifying reagents (69). The comparison in Figure 6 is limited to emphasize the similarity

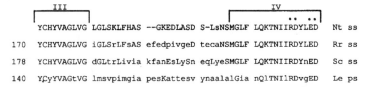

```
        III                               IV
    ┌─────────────┐              ┌──────────── •• ••┐
    YCHYVAGLVG LGLSKLFHAS --GKEDLASD S-LsNSMGLF LQKTNIIRDYLED     Nt ss
170 YCHYVAGLVG iGLSrLFsAS efedpivgeD tecaNSMGLF LQKTNIIRDYLED     Rr ss
178 YCHYVAGLVG dGLtrLivia kfanEsLySn eqLyeSMGLF LQKTNIIRDYnED     Sc ss
140 YℓyYVAGtVG lmsvpimgia pesKattesv ynaalalGia nQlTNIlRDvgED     Le ps
```

Figure 6 Partial amino acid sequence alignment of a phytoene synthase from tomato (Leps) (8a) and squalene synthase proteins from tobacco (Ntss), rat (Rrss) (56), and yeast (Scss) (45). The tobacco amino acid sequence was obtained from a reverse transcription–polymerase chain reaction cDNA clone (J Chappell, unpublished data), isolated using degenerate oligonucleotide primers designed to domains III and V (69). Similarity between phytoene synthase and squalene synthase is maximal between domains III and IV, and the marked arginine, aspartate, and glutamate residues are considered highly conserved (69).

found in domains III and IV in the nomenclature proposed by Robinson et al (69). Interestingly, domain IV contains a DXXED motif, similar to the aspartate-rich motif found in the cyclases discussed above, and one of four highly conserved arginine residue domains found in squalene synthase and phytoene synthase sequences. Robinson et al also observed a second aspartate-rich motif in domain II, but this domain and motif are less well conserved between the squalene synthase proteins and phytoene synthase, and our current information for the plant squalene synthase does not extend into this region.

Oxidosqualene Synthases

Corey et al (19) recently used a modified yeast expression system to isolate a full-length *A. thaliana* oxidosqualene cyclase (cycloartenol synthase) cDNA. The system took advantage of yeast mutants lacking the lanosterol synthase gene (*erg7*), which codes for an oxidosqualene cyclase similar to the plant cycloartenol synthase. Both the plant and yeast enzymes convert an epoxide form of squalene through a complex series of reactions mediated by carbocation cyclizations and other carbon and hydrogen rearrangements to generate either cycloartenol or lanosterol, the polycyclic precursors for sterol biosynthesis in the respective organisms (Figure 7). Pools of *erg7* mutant yeast transformed with an *A. thaliana* cDNA library were lysed, and lysates were assayed in vitro for cycloartenol synthase activity using a simple TLC detection system. Single transformants capable of cycloartenol synthesis were subsequently isolated and the correponding cDNA sequenced. The amino acid sequence predicted for the *A. thaliana* cycloartenol synthase protein was very similar to that predicted for the yeast lanosterol synthase protein and much less so for the bacterial hopene-squalene cyclase. Several domains of extended homology were identified between the plant and fungal enzyme, which might indicate functional significance (19).

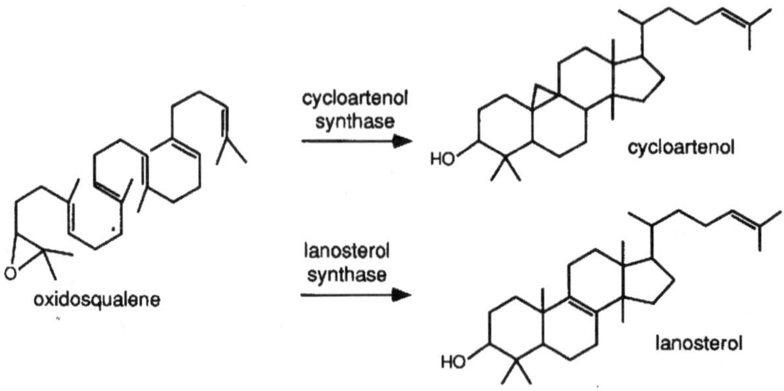

Figure 7   Reactions catalyzed by oxidosqualene cyclases from plants (cycloartenol synthase) and from yeast and animals (lanosterol synthase). The proposed cyclizations are initiated by proton addition to the epoxide function, followed by a wave of electrophilic attack cyclizations, and are terminated by elimination of a proton. Note the similarity between these cyclizations and the proton-addition cyclizations of diterpenes in Figure 4.

Abe & Prestwich (2) used a radiolabeled, suicide-inhibitor of oxido-squalene cyclase to isolate and sequence a peptide fragment of the rat protein thought to represent part of the enzyme's active site. The radioactivity was associated equally with two adjacent Asp residues within a DDXXEX domain. This domain is conserved in the *A. thaliana* cycloartenol synthase protein and in a modified form (DCXXE) in yeast (2, 19). This motif shares some similarity to the DDXXD/E motif found in the prenyltransferases and monoterpene, sesquiterpene, and diterpene cyclases discussed above. And as mentioned earlier, this motif has been implicated in coordinating a Mg^{+2} diphosphate bridge to support the ionization of the respective allylic diphosphate substrates. The essentiality of this motif has been confirmed at least for prenyl-transferases, wherein mutations of the first aspartate residue in the motif eliminate all enzyme activity (47, 74). Although such aspartate-rich domains may be essential for enzyme catalysis, the conservation of this motif in cy-clases that do not require diphosphorylated substrates indicates that this do-main may play a role in catalysis other than to coordinate a metal cation with the diphosphate substituent. An alternative role for these acidic domains may be in filling what Johnson (46) described as the point-charge model for stabi-lizing tertiary and allylic carbocations in the oxidosqualene cyclase reactions.

Poralla et al (63) identified an additional but more complex motif when they aligned two bacterial hopene-squalene cyclases with the fungal and plant oxidosqualene cyclases. The consensus sequence of this motif is [K/R][G/A] X2-3[F/Y/W][L/I/V]X3QX2-5GXW and is found four to five times within the aligned sequences with various degrees of conservation. The consensus sequence is referred to as the QW motif because of the absolute conservation of Gln (Q) at position 10 and Trp (W) at position 16. The QW motif was not found within the plant monoterpene, sesquiterpene, or diterpene cyclase sequences mentioned above, nor did the authors find this motif conserved in multiple copies in any other proteins registered in the Swissprot library. Although no experimental data has yet been presented, one of these QW motifs apparently is labeled to some extent by the suicide-inhibitor discussed above. Poralla et al suggest that the QW motif, like the aspartate-rich domain, may play some role in stabilizing the carbocation-electrophilic interactions described by the Johnson point-charge model, by virtue of the electron-rich aromatic amino acids found within this sequence. Further testing of mutant forms of both the QW and aspartate-rich domains along with the use of other substrate analogs should help to determine the roles these domains play in catalysis.

REGULATION OF ISOPRENOID BIOSYNTHESIS

Transcriptional Control of the Biosynthetic Machinery

Regulation of the isoprenoid pathway can be divided conveniently into coarse- and fine-control mechanisms (40). Coarse control refers to mechanisms controlling the absolute level of the biosynthetic machinery and to a lesser extent the catabolic machinery. Mechanisms of coarse control imply the existence of regulatory genes that orchestrate the coordinated expression (transcription) of structural genes coding for the isoprenoid biosynthetic enzymes, structural genes controlling the catabolic rate of an isoprenoid, and structural genes that control secretion or intracellular targeting of an isoprenoid. Fine control refers to posttranslational mechanisms that are not on-off switches for isoprenoid biosynthesis but rather are fine-tuning mechanisms that ensure that the synthesis rate of an isoprenoid is consistent with immediate demands and resources of a cell. Fine-control mechanisms include modulation of enzyme activities by protein modifications (e.g. protein phosphorylation), feedback regulation via a reaction product or pathway end-product, and other kinetic controls that affect the catalytic efficiency of a biosynthetic enzyme(s). These divisions of coarse and fine control are artificial, and one must recognize that controls at various levels are likely to work in concert.

Goldstein & Brown's studies on the regulation of HMGR in mammalian cells illustrate these points well (reviewed in 37). When compactin or mevinolin, two potent inhibitors of HMGR enzyme activity, are added to cultured mammalian cells, mevalonate biosynthesis is blocked, and within a few hours the level of HMGR protein increases 200-fold. The increase in HMGR protein results from an eightfold increase in the transcription rate of the reductase gene, a fivefold increase in the translational efficiency of the HMGR mRNA, and a fivefold decrease in the turnover rate of the HMGR protein. The increase in transcription rate apparently is mediated by a decrease in the level of cytosolic oxysterols interacting with *trans*-acting factors that bind to the sterol-regulatory elements (SREs) within the HMGR gene promoter. The increased translational efficiency of the HMGR mRNA results, in part, from alternatively spliced HMGR mRNA species that occur in mevalonate-starved cells. Amino acid sequences within the N-terminal, membrane-spanning domain of the HMGR protein somehow regulate the degradation rate of the enzyme, although this process is less well understood. Posttranslational modification of the mammalian HMGR protein is a well-documented mechanism controlling this key enzyme activity. The HMGR activity is inactivated by a reversible phosphorylation that does not seem related to any feedback regulatory mechanism but rather is related to an AMP or ADP-stimulated kinase.

Much of the recent information concerning the coarse control of isoprenoid metabolism in plants has come from developmental studies of fruits (9, 49, 58) and flowers (9, 49) and from studies in which isoprenoid metabolism has been induced by pathogens or elicitors (15, 29, 85). Much of the recent progress has consisted of the cloning of genes for putative rate-limiting enzymes and in the use of these molecular probes to measure changes in the expression patterns of the respective genes. Examples include the changes in HMGR (15, 58, 85), sesquiterpene cyclase (29), diterpene cyclase (51), phytoene synthase (9), and phytoene desaturase (62) mRNAs in developmental or stress-induced tissues. Little is known about how these changes in mRNAs come about, except in the case of the sesquiterpene cyclase, where changes in the in vivo transcription rate of the cyclase gene were demonstrated using the thiouridine-labeling method (82). Additional dissection of the promoter elements controlling the expression patterns of these genes and the isolation of the *trans*-activating proteins interacting with these promoter elements should provide important insight into several regulatory mechanisms.

In all cases, regulatory genes (i.e. genes that orchestrate the transcription rate of structural genes encoding biosynthetic enzymes) must control isoprenoid biosynthesis, but little information about such genes is currently available. Croteau & Gershenzon (21) critically reviewed the genetic and biochemical basis of monoterpene composition in mint species and were able to correlate genetically defined loci with specific structural genes or putative regulatory genes. Isolation of these regulatory genes will likely remain problematic, because the mint family lacks many of the powerful tools, such as transposon tags and visual screens, that have been useful in the isolation of other regulatory genes [e.g. those genes regulating anthocyanin biosynthesis (27)]. Although several genetic mutants of maize, tomato, and *A. thaliana* have been identified that affect fruit ripening and carotenoid accumulation (9), none of the possible regulatory genes have been isolated to date.

The Benveniste laboratory has used an alternative approach to screen for mutants in isoprenoid biosynthesis. These investigators selected UV-muta-genized, haploid protoplasts for growth in the presence of inhibitors of sterol biosynthesis (54, 73). Both of the inhibitors affect relatively late steps in the pathway: a triazole compound, which inhibits the demethylation of obtusifo-liol, and an N-alkyl morpholine, which inhibits the isomerization of cycloeu-calenol to obtusifoliol (Figure 8). The sterol composition of the selected to-bacco mutants was altered dramatically and exhibited significant increases in the proportion of cyclopropylsterols (i.e. intermediates in the sterol biosyn-

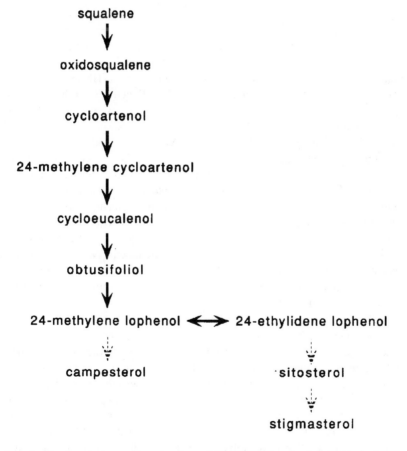

Figure 8 Late steps in the sterol branch pathway. LAB 170250F is a triazole herbicide that inhibits the demethylation of obtusifoliol to 24-methylase lophenol, whereas the N-alkyl morpholine herbicides inhibit the isomerase step converting cycloeucalenol to obtusifoliol.

thetic pathway). Interestingly, the extra sterol accumulating in the *LAB1-4* mutant accumulates predominately in sterol esters associated with novel cytoplasmic lipid droplets (39). Additional studies of the *LAB1-4* mutant demonstrated that the phenotype of the mutation does not require continuous selection with the demethylase inhibitor; but the altered sterol composition is attenuated in regenerated plants and dampened over generations (53). Genetic studies also indicated that the alteration in sterol composition segregated as a single semi-dominant mutation (53). Biochemical analysis suggested that the enzyme activities up to the step involving HMG-CoA synthase may not be enhanced, although the HMGR activity is increased approximately threefold (38). Whether the *LAB1-4* mutation directly or indirectly affects the reductase or other downstream biosynthetic enzyme activities remains to be determined.

Metabolic Channels

The observation of gene families has been one of the more intriguing findings arising from the molecular biology of the isoprenoid biosynthetic genes in plants. For example, *A. thaliana* contains two or more HMGR genes (28), *Hevea brasiliensis* contains at least three HMGR genes (16), and *Solanum tuberosum* has three or more genes (78, 85). Recent experiments have suggested unique roles for the multiple HMGR genes. Narita & Gruissem (58) found that the increase in fruit size of developing tomatoes was sensitive to mevinolin, a potent competitive inhibitor of HMGR enzyme activity, but carotenoid accumulation was not sensitive to the inhibitor. Supplementing the mevinolin-treated fruits with exogenous mevalonate restored normal fruit development. The investigators noted that fruit enlargement is correlated with cell division and enlargement, developmental phases requiring a significant level of sterol biosynthesis. As to why fruit enlargement but not carotenoid accumulation was sensitive to the mevinolin treatment, Narita & Gruissem speculated that carotenoid biosynthesis must be regulated independently of sterol biosynthesis and that HMGR activity must be limiting only for sterol biosynthesis under these conditions.

Choi et al (15) have taken this concept one step further. These investigators have used potato tuber slices to measure a metabolic changeover similar to that described above for tobacco cell cultures. In potato, steroid accumulation is induced upon wounding or slicing the tuber tissue but is arrested if wounded tissue is treated with a fungal elicitor. Instead, sesquiterpene phytoalexins accumulate in the elicitor-treated tuber slices (80). Using gene-specific probes for three potato HMGR genes (HMG1, -2, and -3), Choi et al correlated wound-inducible steroid accumulation with the induction of HMG1 gene expression, and the elicitor-inducible accumulation of sesquiterpene phytoalexins with HMG2 and HMG3 gene expression. HMG1 gene expression was not induced in the elicitor-treated tuber slices.

The results of these two studies are not consistent with the traditional view that isoprenoid metabolism occurs in a homogeneous environment with intermediates mixing freely and accessible to successive or competing enzymes. The results are more consistent with the idea that there are metabolic channels or arrays of isozymes dedicated to the production of specific classes of isoprenoids, with each array regulated independently from one another (Figure 9). Regulation of end-product formation would thus depend on the regulation of each metabolic unit by the accumulation of select isozymes, the insertion of those isozymes into the correct metabolic unit, and various posttranslational modifications modulating isozyme activity levels.

Consistent with the metabolic-channel concept are the recent results of Chappell, Wolf, Proulx, Cuellar, and Saunders (unpublished data). To investi-

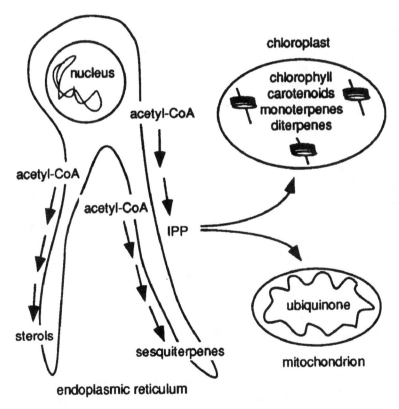

Figure 9 A hypothetical model suggesting that classes of isoprenoids are synthesized by discrete metabolic channels. Entire metabolic channels for sterol and sesquiterpene biosynthesis are localized to the endoplasmic reticulum. Synthesis of other isoprenoids, such as carotenoids and long-chain prenols of ubiquinone, require the coordinated activity of cytoplasmic and organellar biosynthetic channels.

gate whether HMGR activity could be limiting in plants, a constitutively expressing, truncated form of a hamster HMGR gene was engineered into tobacco plants. The impact of the resulting enzyme activity on the accumulation of sterols and elicitor-inducible sesquiterpenes (cytosolic-synthesized isoprenoids), and carotenoids and phytols (chloroplastic-synthesized isoprenoids) was determined. Surprisingly, only the level of sterols was increased significantly in the transgenic plants. However, the principal sterol accumulating was cycloartenol, an intermediate in the sterol biosynthetic pathway (Figure 8). This result suggests that HMGR activity is not limiting for the synthesis of campesterol, stigmasterol, and sitosterol, the major end-product sterols found in plants. Also, because the carotenoid and phytol levels were unaltered in the transgenic plants, their synthesis is likely to occur independent of sterol biosynthesis. As depicted in Figure 9, this may entail the coordination between a cytosolic and a plastidic pathway (49), or the plastids could be entirely self-sufficient for their own isoprenoid biosynthetic needs. The recent report by Rohmer et al (70) lends some support to the latter possibility.

Some predictions regarding such metabolic channels or metabolons (75) are obvious. Domains of the endoplasmic reticulum or some other cytoskeletal structure must be dedicated to the assembly of unique metabolons. Enzymes destined for metabolons must also have specific targeting information within their amino acid sequence or their secondary or tertiary structures. Genes coding for enzymes making up a specific metabolon must have similar transcriptional networks to coordinate expression of the metabolic unit. The recently cloned genes discussed above provide ample opportunities for testing some of these predictions. For example, if metabolic units are regulated coordinately at the transcriptional level, then one would expect to find conserved regulatory sequences or *cis*-sequence motifs within the promoters of elicitor-inducible HMGR and sesquiterpene cyclase genes. Also, if particular HMGR isozymes are limiting for each metabolic channel, then engineering high-level expression of each HMGR isoform should alter only one class of isoprenoids. These types of experiments are currently being pursued in a number of laboratories.

FUTURE PROSPECTS

I have tried to emphasize some of the practical and fundamental values of isoprenoids, to convey some appreciation for the wealth of chemical and biochemical information concerning this biosynthetic pathway, and to highlight some of the recent advances and speculations concerning our molecular understanding of isoprenoid metabolism. Advances in the molecular biology of isoprenoid metabolism promise to complement and extend both the applications and scientific value of this important pathway in plants. Such advances will, however, require a more complete understanding of the structure-function relationships of the isoprenoid biosynthetic enzymes, and a much better appreciation for metabolic channels and the regulation of isoprenoid biosynthesis. Ultimately, I fully expect to see the design, construction, and expression of chimeric isoprenoid biosynthetic genes in plants, resulting in the accumulation of novel isoprenoids in transgenic plants.

ACKNOWLEDGMENTS

I am grateful to the Cane, Croteau, Hohn, and West laboratories for sharing information prior to publication, and to Rick Bostock, Rod Croteau, and Charles West for many stimulating and challenging discussions concerning the chemistry and biochemistry of isoprenoids. I also thank Rick Bostock and Jeff Newman for taking the time to read an earlier version of this review. Work in my laboratory has been consistently supported by the National Science Foundation. This is journal article 95-06-001 from the Kentucky Agricultural Experiment Station.

Any *Annual Review* chapter, as well as any article cited in an *Annual Review* chapter, may be purchased from the Annual Reviews Preprints and Reprints service.
1-800-347-8007; 415-259-5017; email: arpr@class.org

Literature Cited

1. Abe I, Ebizuka Y, Seo S, Sankawa U. 1989. Purification of squalene-2,3-epoxide cyclases from cell suspension cultures of *Rabdosia japonica* Hara. *FEBS Lett.* 249 100–4

2. Abe I, Prestwich GD. 1994. Active site mapping of affinity-labeled rat oxidosqualene cyclase. *J. Biol. Chem.* 269:802–4

3. Alonso WR, Croteau R. 1993. Prenyltransferases and cyclases. In *Methods in Plant Biochemistry*, ed. PG Lea, 9:239–60. London: Academic

4. Anderson MS, Muehlbacher M, Street IP, Proffit J, Poulter CD. 1989. Isopentenyl diphosphate:dimethylallyl diphosphate isomerase: an improved purification of the enzyme and isolation of the gene from *Saccharomyces cerevisae. J. Biol. Chem.* 264: 19169–75

5. Anderson MS, Yarger JG, Burck CL, Poulter CD. 1989. Farnesyl diphosphate synthatase. Molecular cloning, sequence, and expression of an essential gene from *Saccharomyces cerevisiae. J. Biol. Chem.* 264: 19176–84

6. Bach TJ. 1986. Hydroxymethylglutaryl-CoA reductase, a key enzyme in phytosterol synthesis? *Lipids* 21:82–88

7. Bach TJ, Boronat A, Caelles C, Ferrer A, Weber T, Wettstein A. 1991. Aspects related to mevalonate biosynthesis in plants. *Lipids* 26:637–48

8. Back K, Yin S, Chappell J. 1994. Expression of a plant sesquiterpene cyclase gene in *Escherichia coli. Arch. Biochem. Biophys.* 315:527–32

8a. Bartley GE, Viitanen PV, Bacot KO, Scolnik PA. 1992. A tomato gene expressed during fruit ripening encodes an enzyme of the carotenoid biosynthesis pathway. *J. Biol. Chem.* 267:5036–39

9. Bartley GE, Scolnik PA. 1994. Molecular biology of carotenoid biosynthesis in plants. *Annu. Rev. Plant Physiol. Plant Mol. Biol.* 45:287–301

10. Benveniste P. 1986. Sterol biosynthesis. *Annu. Rev. Plant Physiol.* 37:275–308

11. Binns AN. 1994. Cytokinin accumulation and action: biochemical, genetic and molecular approaches. *Annu. Rev. Plant Physiol. Plant Mol. Biol.* 45:173–96

12. Cane DE . 1981. Biosynthesis of sesquiterpenes. In *Biosynthesis of Isoprenoid Compounds*, ed. JW Porter, SL Spurgeon, pp. 283–374. New York: Wiley

12a. Cane DE, Wu Z, Oliver JS, Hohn TM.

1993. Overproduction of soluble trichodi-ene synthase from *Fusarium sporotri-chioides* in *Escherichia coli*. *Arch. Bio-chem. Biophys.* 300:416–22

13. Cane DE, Wu Z, Proctor RH, Hohn TM. 1993. Overexpression in *Escherichia coli* of soluble aristolochene synthase from *Penicillium roqueforti*. *Arch. Biochem. Biophys.* 304:415–19

14. Chappell J, VonLanken C, Vögeli U. 1991. Elicitor-inducible 3-hydroxy-3-methylglu-taryl coenzyme A reductase activity is re-quired for sesquiterpene accumulation in tobacco cell suspension cultures. *Plant Physiol.* 97:693–98

15. Choi D, Ward BL, Bostock RM. 1992. Dif-ferential induction and suppression of po-tato 3-hydroxy-3-methylglutaryl coen-zyme A reductase genes in response to *Phytophthora infestans* and to its elicitor arachidonic acid. *Plant Cell* 4:1333–44

15a. Choi D, Bostock RM. 1994. Involvement of de novo protein synthesis, protein ki-nase, extracellular Ca^{2+}, and lipoxygenase in arachidonic acid induction of 3-hydroxy-3-methylglutaryl coenzyme A reductase genes and isoprenoid accumulation in po-tato (*Solanum tuberosum* L.). *Plant Physiol.* 104:1237–44

15b. Choi D, Bostock RM, Avdiushko S, Hilde-brand DF. 1994. Lipid-derived signals that discriminate wound- and pathogen-respon-sive isoprenoid pathways in plants: methyl jasmonate and the fungal elicitor arachi-donic acid induce different 3-hydroxy-3-methylglutaryl-coenzyme A reductase genes and antimicrobial isoprenoids in *So-lanum tuberosum* L. *Proc. Natl. Acad. Sci. USA* 91:2329–33

16. Chye ML, Tan CT, Chua N-M. 1992. Three genes encode 3-hydroxy-3-methylglutaryl-coenzyme A reductase in *Hevea brasilien-sis*: hmg1 and hmg3 are differentially ex-pressed. *Plant Mol. Biol.* 19:473–84

17. Clarke CF, Tanaka RD, Svenson K, Wam-sley M, Fogelman AM, Edwards PA. 1987. Molecular cloning and sequencing of a cholesterol-repressible enzyme related to prenyltransferase in the isoprene biosyn-thetic pathway. *Mol. Cell Biol.* 7:3138–46

18. Colby SM, Alonso WR, Katahira EJ, McGarvey DJ, Croteau R. 1993. 4S-Li-monene synthase from the oil glands of spearmint (*Mentha spicata*): cDNA isola-tion, characterization, and bacterial expres-sion of the catalytically active monoterpene cyclase. *J. Biol. Chem.* 268:23016–24

19. Corey EJ, Matsuda SPT, Bartel B. 1993. Isolation of an *Arabidopsis thaliana* gene encoding cycloartenol synthase by func-tional expression in a yeast mutant lacking lanosterol synthase by the use of a chroma-tographic screen. *Proc. Natl. Acad. Sci. USA* 90:11628–32

20. Croteau R. 1987. Biosynthesis and catabo-lism of monoterpenoids. *Chem. Rev.* 87: 929–54

21. Croteau R, Gershenzon J. 1994. Genetic control of monoterpene biosynthesis in mints (*Mentha*: Lamiaceae). In *Recent Ad-vances in Phytochemistry*, ed. H Stafford, pp. 193–229. New York: Plenum

22. Davisson VJ, Woodside AB, Poulter CD. 1986. Synthesis of allylic and homoallylic isoprenoid pyrophosphate. *Methods Enzy-mol.* 110:130–44

23. Dehal SS, Croteau R. 1987. Metabolism of monoterpenes: specificity of the dehydro-genases responsible for the biosynthesis of camphor, 3-thujone and 3-isothujone. *Arch. Biochem. Biophys.* 258:287–91

24. Delourme D, Lacroute F, Karst F. 1993. EMBL accession number x75789

25. Dogbo O, Camara B. 1987. Purification of isopentenyl pyrophosphate isomerase and geranylgeranyl pyrophosphate synthase from *Capsicum* chromoplasts by affinity chromatography. *Biochem. Biophys. Acta* 920:140–48

26. Dogbo O, Laferrière A, D'Harlingue A, Camara B. 1988. Carotenoid biosynthesis: isolation and characterization of a bifunc-tional enzyme catalyzing the synthesis of phytoene. *Proc. Natl. Acad. Sci. USA* 85: 7054–58

27. Dooner HK, Robbins TP, Jorgensen RA. 1991. Genetic and developmental control of anthocyanin biosynthesis. *Annu. Rev. Genet.* 25:173–99

28. Enjuto M, Balcells L, Campos N, Caelles C, Arró M, Boronat A. 1994. *Arabidopsis thaliana* contains two differentially ex-pressed HMG-CoA reductase genes which encode microsomal forms of the enzyme. *Proc. Natl. Acad. Sci. USA* 91:927–31

29. Facchini PJ, Chappell J. 1992. Gene family for an elicitor-induced sesquiterpene cy-clase in tobacco. *Proc. Natl. Acad. Sci. USA* 89:11088–92

30. Funk C, Croteau R. 1993. Induction and characterization of a cytochrome P-450-de-pendent camphor hydroxylase in tissue cul-tures of common sage (*Salvia officinalis*). *Plant Physiol.* 101:1231–37

31. Funk C, Croteau R. 1994. Diterpenoid resin acid biosynthesis in conifers: charac-terization of two cytochrome P450-de-pendent monooxygenases and an aldehyde dehydrogenase involved in abietic acid bio-synthesis. *Arch. Biochem. Biophys.* 308: 258–66

32. Funk C, Koepp AE, Croteau RB. 1992. Catabolism of camphor in tissue cultures and leaf disks of common sage (*Salvia officinalis*). *Arch. Biochem. Biophys.* 294: 306–13

33. Gershenzon J, Croteau R. 1993. Terpenoid biosynthesis: the basic pathway and forma-

tion of monoterpenes, sesquiterpenes, and diterpenes. In *Lipid Metabolism in Plants,* ed. TS Moore, pp. 340–88. Boca Raton, FL: CRC

34. Gershenzon J, McCaskill D, Rajaonarivony JIM, Mihaliak C, Karp F, Croteau R. 1992. Isolation of secretory cells from plant glandular trichomes and their use in biosynthetic studies of monoterpenes and other gland products. *Anal. Biochem.* 200: 130–38

35. Gibson RW, Pickett JA. 1983. Wild potato repels aphids by release of aphid alarm pheromone. *Nature* 302: 608–9

36. Glomset JA, Gelb MH, Farnsworth CC. 1990. Prenyl proteins in eukaryotic cells: a new type of membrane anchor. *Trends Biochem. Sci.* 15:139–42

37. Goldstein JL, Brown MS. 1990. Regulation of the mevalonate pathway. *Nature* 343: 425–30

39. Gondet L, Weber T, Maillot-Vernier P, Benveniste P, Bach TJ. 1992. Regulatory role of microsomal 3-hydroxy-3-methylglutaryl-coenzyme A reductase in a tobacco mutant that overproduces sterols. *Biochem. Biophys. Res. Commun.* 186:888–93

38. Gondet L, Bronner R, Benveniste P. 1994. Regulation of sterol content in membranes by subcellular compartmentation of sterylesters accumulating in a sterol-overproducing tobacco mutant. *Plant Physiol.* 105: 509–18

40. Gray JC. 1987. Control of isoprenoid biosynthesis in higher plants. *Adv. Bot. Res.* 14:25–91

41. Green TR, Baisted DJ. 1972. Development of the activities of enzymes of the isoprenoid pathway during early stages of pea-seed germination. *Biochem. J.* 130: 983–95

42. Green TR, West CA. 1974. Purification and characterization of two forms of geranyl transferase from *Ricinus communis*. *Biochemistry* 13:4720–29

43. Guo Z, Severson RF, Wagner GJ. 1994. Biosynthesis of the diterpene *cis*-abienol in cell-free extracts of tobacco trichomes. *Arch. Biochem. Biophys.* 308:103–8

44. Hanley KM, Chappell J. 1992. Solubilization, partial purification and immunodetection of squalene synthetase from tobacco cell suspension cultures. *Plant Physiol.* 98: 215–20

45. Jennings SM, Tsay YH, Fisch TM, Robinson GW. 1991. Molecular cloning and characterization of the yeast gene for squalene synthetase. *Proc. Natl. Acad. Sci. USA* 88: 6038–42

46. Johnson WS. 1991. Fifty years of research. *Tetrahedron* 47(41):R11–R50

47. Joly A, Edwards PA. 1993. Effect of site-directed mutagenesis of conserved aspartate and arginine residues upon farnesyl

diphosphate synthase activity. *J. Biol. Chem.* 268: 26983–89

48. Karp F, Harris JL, Croteau R. 1987. Metabolism of monoterpenes: demonstration of the hydroxylation of (+)-sabinene to (+)-cis-sabinol by an enzyme preparation from sage (*Salvia officinalis*) leaves. *Arch. Biochem. Biophys.* 256:179–93

49. Kleinig H. 1989. The role of plastids in isoprenoid biosynthesis. *Annu. Rev. Plant Physiol. Plant Mol. Biol.* 40:39–59

50. LaFever RE, Vogel BS, Croteau R. 1994. Diterpenoid resin acid biosynthesis in conifers: enzymatic cyclization of geranylgeranyl pyrophosphate to abietadiene, the precursor of abietic acid. *Arch. Biochem. Biophys.* 313:139–49

51. Lois AF, West CA. 1990. Regulation of expression of the casbene synthetase gene during elicitation of castor bean seedlings with pectic fragments. *Arch. Biochem. Biophys.* 276:270–77

52. Lütke-Brinkhaus F, Liedvogel B, Kleinig H. 1984. On the biosynthesis of ubiquinones in plant mitochondria. *Eur. J. Biochem.* 141:537–41

53. Maillot-Vernier P, Gondet L, Schaller H, Benveniste P, Belliard G. 1991. Genetic study and further biochemical characterization of a mutant that overproduces sterols. *Mol. Gen. Genet.* 231:33–40

54. Maillot-Vernier P, Schaller H, Benveniste P, Belliard G. 1989. Biochemical characterization of a sterol mutant plant regenerated from a tobacco callus resistant to a triazole cytochrome-P-450-obtusifoliol-14-demethylase inhibitor. *Biochem. Biophys. Res. Commun.* 165: 125–30

55. Mau CJD, West CA. 1994. Cloning of casbene synthase cDNA: evidence for conserved structural features among terpenoid cyclases in plants. *Proc. Natl. Acad. Sci. USA* 91:8497–501

56. McKenzie TL, Jiang G, Straubhaar JR, Conrad DG, Shechter I. 1992. Molecular cloning, expression and characterization of the cDNA for the rat hepatic squalene synthase. *J. Biol. Chem.* 267:21368–74

57. Moore WR, Schatzman GL. 1992. Purification of 2,3-oxidosqualene cyclase from rat liver. *J. Biol. Chem.* 267:22003–6

58. Narita JO, Gruissem W. 1989. Tomato hydroxymethylglutaryl-CoA reductase is required early in fruit development but not during ripening. *Plant Cell* 1:181–90

59. Ôba K, Tatematsu H, Yamashita K, Uritani I. 1976. Induction of furano-type production and formation of the enzyme system from mevalonate to isopentenyl pyrophosphate in sweet potato root tissue injured by *Ceratocystis fimbriata* and by toxic chemicals. *Plant Physiol.* 58:51–56

60. Ôba K, Uritani I. 1979. Biosynthesis of

furano-terpenes by sweet potato cell culture. *Plant Cell Physiol.* 20:819–26

61. Ogura K, Nishino T, Seto S. 1968. The purification of prenyltransferase and isopentenyl pyrophosphate isomerase of pumpkin fruit and their some properties. *J. Biochem.* 64:197–203

62. Pecker I, Chamovitz D, Linden H, Sandmann G, Hirschberg J. 1992. A single polypeptide catalyzing the conversion of phytoene to ξ-carotene is transcriptionally regulated during tomato fruit ripening. *Proc. Natl. Acad. Sci. USA* 89:4962–66

63. Poralla K, Hewelt A, Prestwich GD, Abe I, Reipen I, Sprenger G. 1994. A specific amino acid repeat in squalene and oxidosqualene cyclases. *Trends Biochem. Sci.* 19:157–58

63a. Poulter CD, Rilling HC. 1981. Conversion of farnesyl pyrophosphate to squalene. In *Biosynthesis of Isoprenoid Compounds*, ed. JW Porter, SL Spurgeon, pp. 413–41. New York: Wiley

64. Proctor RH, Hohn TM. 1993. Aristolochene synthase. Isolation, characterization, and bacterial expression of a sesquiterpenoid biosynthetic gene (*Ari1*) from *Penicillium roqueforti. J. Biol. Chem.* 268:4543–48

65. Rajaonarivony JIM, Gershenzon J, Croteau R. 1992. Characterization and mechansim of (4S)-limonene synthase, a monoterpene cyclase from the glandular trichomes of peppermint (*Mentha x piperita*). *Arch. Biochem. Biophys.* 296:49–57

66. Rajaonarivony JIM, Gershenzon J, Miyazaki J, Croteau R. 1992. Evidence for an essential histidine residue in 4S-limonene synthase and other terpene cyclases. *Arch. Biochem. Biophys.* 299:77–82

67. Randall SK, Marshall MS, Crowell DN. 1993. Protein isoprenylaion in suspension-cultured tobacco cells. *Plant Cell* 5:433–42

68. Riou C, Tourte Y, Lacroute F, Karst F. 1994. EMBL accession number x77793

69. Robinson GW, Tsay YH, Kienzle BK, Smith-Monroy CA, Bishop RW. 1993. Conservation between human and fungal squalene synthetases: similarities in structure, function, and regulation. *Mol. Cell. Biol.* 13:2706–17

70. Rohmer M, Knani M, Simonin P, Sutter B, Sahm H. 1993. Isoprenoid biosynthesis in bacteria: a novel pathway for the early steps leading to isopentenyl diphosphate. *Biochem. J.* 295:517–24

71. Ruzicka L. 1953. The isoprene rule and the biogenesis of terpenic compounds. *Experientia* 10:357–67

72. Deleted in proof

73. Schaller H, Maillot-Vernier P, Benveniste P, Belliard G. 1991. Sterol composition of tobacco calli selected for resistance to fenpropimorph. *Phytochemistry* 30:2547–54

74. Song L, Poulter CD. 1994. Yeast farnesyl-diphosphate synthase: site-directed mutagenesis of residues in highly conserved prenyltransferase domains I and II. *Proc. Natl. Acad. Sci. USA* 91:3044–48

75. Srere PA. 1987. Complexes of sequential metabolic enzymes. *Annu. Rev. Biochem.* 56:89–124

76. Stevens KL. 1984. Biological activity and chemistry of sesquiterpene lactones. In *Isopentenoids in Plants*, ed. WD Nes, G Fuller, L-S Tsai, pp. 65–80. New York: Marcel Dekker

77. Stoessl A, Stothers JB, Ward EWB. 1976. Sesquiterpenoid stress compounds of the *Solanaceae. Phytochemistry* 15:855–73

77a. Sun T-P, Kamiya Y. 1994. The Arabidopsis *GA1* locus encodes the cyclase ent-kaurene synthetase A of gibberellin biosynthesis. *Plant Cell* 6:1506–18

78. Takeuchi A, Ôba K, Uritani I. 1977. Changes in acetyl CoA synthetase activity of sweet potato in response to infection by *Ceratocystis fimbriata* and injury. *Agric. Biol. Chem.* 41:1141–45

79. Threlfall DR, Whitehead IM. 1988. Co-ordinated inhibition of squalene synthetase and induction of enzymes of sesquiterpenoid phytoalexin biosynthesis in cultures of *Nicotiana tabacum. Phytochemistry* 27:2567–80

80. Tjamos EC, Kuć JA. 1982. Inhibition of steroid glycoalkaloid accumulation by arachidonic and eicosapentaenoic acids in potato. *Science* 217:542–44

81. Vögeli U, Chappell J. 1988. Induction of sesquiterpene cyclase and suppression of squalene synthetase activities in plant cell cultures treated with fungal elicitor. *Plant Physiol.* 88:1291–96

81a. Vögeli U, Freeman JW, Chappell J. 1990. Purification and characterization of an inducible sesquiterpene cyclase from elicitor-treated tobacco cell suspension cultures. *Plant Physiol.* 93:182–87

82. Vögeli U, Chappell J. 1990. Regulation of a sesquiterpene cyclase in cellulase-treated tobacco cell suspension cultures. *Plant Physiol.* 94:1860–66

82a. Vögeli U, Vögeli-Lange R, Chappell J. 1992. Inhibition of phytoalexin biosynthesis in elicitor-treated tobacco cell suspension cultures by calcium/calmodulin antagonists. *Plant Physiol.* 100:1369–76

83. West CA. 1981. Biosynthesis of diterpenes. In *Biosynthesis of Isoprenoid Compounds*, ed. JW Porter, SL Spurgeon, pp. 375–441. New York: Wiley

84. Yang Z, Cramer CL, Watson JC. 1993. Protein farnesyltransferase in plants. Molecular cloning and expression of a homolog of the beta subunit from the garden pea. *Plant Physiol.* 101:667–74

85. Yang Z, Park H, Lacy GH, Cramer CL.

1991. Differential activation of potato 3-dydroxy-3-methylglutaryl coenzyme A reductase genes by wounding and pathogen challenge. *Plant Cell* 3:397–405

86. Zook MN, Kuć JA. 1991. Induction of sesquiterpene cyclase and suppression of squalene synthetase activity in elicitor treated or fungal infected potato tube tissue. *Physiol. Mol. Plant Pathol.* 39:377–90

Annu. Rev. Plant Physiol. Plant Mol. Biol. 1995. 46:549–75

MOLECULAR BIOLOGY OF RHODOPHYTE AND CHROMOPHYTE PLASTIDS[1]

M. Reith

Marine Biology Section, NRC Institute for Marine Biosciences, Halifax, Nova Scotia B3H 3Z1, Canada

KEY WORDS: algae, evolution, plastid genes, plastid genomes, plant metabolism

CONTENTS

---

[1] The Canadian government has the right to retain a nonexclusive royalty-free license in and to any copyright covering this paper.

ABSTRACT

Recent information on the plastid genes and genomes of rhodophyte (red algae) and chromophyte (yellow and brown algae) plastids are summarized. The plastid genomes of these algae contain many more genes than those of metaphyte (land plant) plastids, and the encoded proteins are involved in a much wider range of metabolic activities. Rhodophyte and chromophyte plastid genomes contain very few introns and maintain many ancestral gene operons, which suggests that they are more primitive than those of metaphytes. Hypotheses of plastid evolution (monophyletic or polyphyletic primary origins) are discussed and assessed. The data available at present support the idea that all plastids arose from a single endosymbiotic event that involved a cyanobacterium and a eukaryote host (monophyletic origin). Information from rhodophyte and chromophyte plastid genomes also provides insights into the evolution of other organisms. Finally, the implications for plastid metabolism that arise from the presence of several genes on rhodophyte and chromophyte plastid genomes are discussed.

INTRODUCTION

Nearly ten years ago, the first complete nucleotide sequences of plastid genomes, those of tobacco (95) and the liverwort *Marchantia polymorpha* (72), were published. These sequences provided a detailed picture of the photosynthetic apparatus and plastid gene expression. Subsequent isolation and N-terminal sequencing of photosystem proteins has allowed the identification of several open reading frames (ORFs) conserved on all these genomes and has added to our knowledge of photosystem components and structure. These studies, and the subsequent sequencing of the rice plastid genome (38), indicated that the plastid genome was dedicated to encoding proteins involved in two processes: photosynthesis and gene expression (74). The genes encoding proteins involved in other aspects of plastid metabolism, as well as some of the photosynthetic and gene expression proteins, had been transferred to the nuclear genome. On the basis of the evolutionary distance between liverworts and Angiosperms, some authors (13, 74) suggested that the similar complement of genes in these two plastid genomes indicated that nearly all of the genes in the cyanobacterial endosymbiont that was the progenitor of plastids were lost or transferred to the host nucleus at an early stage in plastid evolution.

At about the same time as the publication of these first plastid genome sequences, the first observations that algal, particularly red algal (rhodophyte) and brown or yellow algal (chromophyte), plastid genomes differed from those of land plants (metaphytes) were being published (37, 82). These observations, which indicated that *rbcS*, the gene encoding the small subunit of ribulose-1,5-

bisphosphate carboxylase, was a plastid-encoded gene in these organisms, did not alter significantly the view that plastid genomes primarily encoded proteins for photosynthesis and gene expression. In the past five years, extensive investigations of the plastid genomes of rhodophytes and chromophytes have produced complete sequences for two rhodophyte plastid genomes (85; D Bryant, V Stirewalt, C Michalowski, W Löffelhardt & H Bohnert, unpublished data) and significant amounts of information on two chromophyte plastid genomes (17, 24). These studies present a startlingly different view of plastid genomes from that given by metaphytes and provide some interesting insights into plastid evolution and metabolism.

This review focuses on those molecular biological characteristics that distinguish rhodophyte and chromophyte plastids from those of green algae (chlorophytes) and metaphytes. For the purposes of this review, the term rhodophyte refers to an alga with a light-harvesting system that contains chlorophyll *a* and phycobilisomes, and chromophyte refers to an alga that contains both chlorophyll *a* and chlorophyll *c*. These simplifications may associate phyla with distinct evolutionary origins and, therefore, do not imply that each group is monophyletic. Rhodophytes include the red algae (Rhodophyceae); the pre-rhodophytes (Cyanidiophyceae), a group that mainly contains species isolated from the mixture once called *Cyanidium caldarium* and related organisms (92); and the glaucocystophytes (Glaucocystophyceae) such as the cyanelle-containing alga *Cyanophora paradoxa*. Chromophytes, whose plastids are usually bounded by four, rather than two, membranes, are an assemblage of eight phyla, some with notable distinctions. The Cryptophyta (cryptomonads) and Eustigmatophyta both contain phycobiliproteins in addition to chlorophyll *c*. The cryptomonad plastid also contains a nucleomorph, a vestigial nucleus apparently derived from the nucleus of a eukaryotic endosymbiont, located between the second and third bounding membranes. The Dinophyta (dinoflagellates) have plastids with three, and in some groups only two, bounding membranes and a light-harvesting complex based on the xanthophyll peridinin rather than fucoxanthin.

Hypotheses of Plastid Evolution

For a number of years, it has been generally accepted that plastids arose through one or more endosymbiotic events in which a photosynthetic prokaryote entered a eukaryotic host cell (33). More recently, the case for endosymbioses in which photosynthetic eukaryotes were the progenitors of some plastid types has become very convincing. Thus, one needs to distinguish between these two types of endosymbioses, which are usually referred to as primary (prokaryote-eukaryote) or secondary (eukaryote-eukaryote) endosymbioses. The current versions of the two main hypotheses of plastid evolution are distinguished mainly by the number of primary symbioses invoked: only one in monophyletic schemes for the origins of plastids as compared to two or more in polyphyletic schemes. Monophyletic hypotheses propose that a single endosymbiotic cyanobacterium was the ancestor of all plastids (12, 18, 49). The number of secondary endosymbioses postulated differs in various monophyletic schemes. Key arguments for monophyletic schemes of plastid origin are the relatively simple biochemical differences between chlorophylls *a, b,* and *c* and the assumed difficulty of establishing an organelle from an endosymbiont (12).

Polyphyletic arguments are based on the idea that plastid pigment composition reflects the pigment composition of the endosymbiont (29, 80, 104). Polyphyletic hypotheses thus propose that at least two primary endosymbioses occurred: one with a cyanobacterial symbiont leading to rhodophytes and one involving a prochlorophyte (a prokaryote that contains chlorophyll *a* and chlorophyll *b*) resulting in chlorophytes. Some polyphyletic hypotheses suggest a third type of endosymbiont, the brownish photoheterotrophic eubacterium, *Heliobacterium chlorum,* as a progenitor of chromophyte plastids (61). Other variations have invoked two or more cyanobacteria as the progenitors of plastids (55). The number of secondary endosymbioses proposed varies among polyphyletic hypotheses. Several recent reviews have covered the controversies of plastid evolution in more detail (10, 17, 18, 30–32, 48, 49, 55, 81, 110).

The obvious difficulty in the field of plastid evolution is how one might go about testing these two hypotheses, because the events under consideration occurred on the order of a billion years ago (35, 46). Molecular phylogenetic analyses, which attempt to extract evolutionary information from sequence data, have become the most commonly used tools. However, the number and choice of species used, the correctness of the sequence alignment, and the analytical method can all influence phylogenetic topology. In addition, such analyses generate phylogenetic trees for an individual gene or protein and do not necessarily reflect the evolution of an organelle or organism. Only if the analyses of multiple genes or proteins result in similar trees in most cases can one begin to assume a similar evolutionary pattern for the organelle or organism in question.

An alternative method for assessing hypotheses of plastid origins is to apply the principles of parsimony to individual molecular characters. The key to this approach is the determination of shared and derived characters that would suggest a common ancestor for all individuals having these traits (98). In the case of plastid evolution, the identification of characters that are shared among all plastids but differ from those of cyanobacteria or other photosynthetic prokaryotes would support a monophyletic origin of plastids from a single photosynthetic eubacterium. On the other hand, a polyphyletic scheme in which plastids arose from several different photosynthetic eubacteria would require that each plastid type and its ancestral prokaryote share characters that differ in other such pairs.

PLASTID GENES AND GENOMES

Genome Organization

Compared to the amount of information available for metaphyte plastid genomes, the data on rhodophyte and chromophyte plastid genomes are quite sparse. Physical maps are available for only five rhodophyte and ten chromophyte plastid genomes (Table 1). As for most other plastid genomes, rhodophyte and chromophyte plastid genomes are circular molecules of 100–200 kbp that are usually divided by repeated rRNA-encoding regions into large and small single-copy regions. Rhodophyte plastid genomes (except for that of *C. paradoxa*) tend to be larger (175–190 kbp) than those of chromophytes and metaphytes. Chromophyte genomes are usually in the range of 115–135 kbp, with the exception of that of *Heterosigma akashiwo* (154 kbp). For comparison, the characteristics of some chlorophyte and metaphyte plastid genomes are indicated in Table 1.

At least some of the variation of genome size within either the rhodophytes or chromophytes appears to be the result of changes in the repeated rRNA-encoding regions of these genomes. The larger genome of *H. akashiwo* appears to result in part from the expansion of the repeats to include additional genes. The smaller genome sizes of *Griffithsia pacifica, Chondrus crispus,* and *Antithamnion* sp. among red algae can be accounted for by the loss of one of the repeats. Interestingly, these species are placed near the crown of the Rhodophycean tree in a recent phylogenetic analysis (79), suggesting that the loss of one of the rRNA operons may have occurred just once in the rhodophyte lineage, prior to the divergence of these species from a common ancestor. Because the *Palmaria palmata* plastid has two rRNA copies (RK Singh, unpublished data) and the Palmariales are one of the earlier diverging groups in the Floridiophycean branch, the loss of the rRNA operon must have occurred after the divergence of the Palmariales. The expansion or deletion of the repeat regions in metaphyte plastid genomes is well documented (75). These events have apparently occurred several times in the evolution of plastids.

Another notable characteristic of rhodophyte plastid genomes is the orientation of the repeat regions. In both metaphytes and all chromophytes investigated, the rRNA operons are organized as inverted repeats with the 5S rRNA gene closest to the small single-copy region of the genome. In the rhodophytes, three different organizations have been found: inverted repeats with the 5S gene near the small single-copy region in *G. sulphuraria* (60), inverted repeats with the 16S gene near the small single-copy region in *C. paradoxa*

Table 1 Plastid genome organization and gene content

Species	Size (kbp)	rRNA repeat number and orientation[a]	Repeat size (kbp)	Number of genes mapped	References
RHODOPHYTES					
Cyanophora paradoxa	134	2i	12	150[c]	54
Galdieria sulphuraria[b]	?	2i	5	60	110
Porphyra purpurea	191	2d	4.8	255[c]	83, 85
Porphyra yezoensis	185	2i	6.6	25	96, 97
Griffithsia pacifica	178	1	-	24	97
Chondrus crispus	173	1	-	15	8[d]
Antithamnion sp.[b]	180	1	-	55	110[e]
CHROMOPHYTES					
Cryptomonas Φ	118	2i	6	70	17
Pyrenomonas salina	127	2i	5	20	58
Coscinodiscus granii	118	2i	10	35	48
Cyclotella menghiniana	128	2i	17	25	7
Odontella sinensis	118	2i	8.7	96	24
Ochromonas danica	121	2i	7.5	19	97
Heterosigma akashiwo	154	2i	22	27	97
Vaucheria bursata	124	2i	7.5	34	48
Dictyota dichotoma	123	2i	4.7	30	48
Pylaiella littoralis	133	2i	5	23	56
CHLOROPHYTES					
Chlamydomonas reinhardtii	196	2i	20	75	6
Chlamydomonas moewusii	292	2i	41	74	6
Euglena gracilis	143	3.5d	5.9	108[c]	34
METAPHYTES					
Marchantia polymorpha	121	2i	10	129[c]	72
Nicotiana tabacum	156	2i	25	138[c]	95
Oryza sativa	135	2i	21	133[c]	38

[a]Orientation of the rRNA repeats is either inverted (i) or direct (d)

[b]The plastid genomes of these species have not been completely mapped

[c]These are completely sequenced genomes

[d]See also C Boyen & C LeBlanc, unpublished data

[e]See also K Zetsche, unpublished data

(54) and *Porphyra yezoensis* (96), and nontandem, direct repeats in *Porphyra purpurea* (84). Some of this diversity of the repeat organizations may represent reorganizations that occurred following the separation of the three rhodophyte lineages. The contradiction in rRNA operon organization between the two *Porphyra* species suggests that some of the *P. yezoensis* mapping data have been interpreted incorrectly. Some evidence for this misinterpretation is suggested by the mapping of the 23S rRNA gene adjacent to *psbC/D* (96): Sequence data from both *P. purpurea* (84, 85) and *G. sulphuraria* (60) demonstrate that the 16S rRNA gene is closer to *psbC/D* than is the 23S gene. Inversion of this rRNA operon in the *P. yezoensis* map would result in a direct repeat organization. Further analysis of the *P. yezoensis* chloroplast genome as well as those of other red algae should help to clarify the evolution of these different repeat organizations.

Coding Capacity

Recently, substantial progress has been made on the sequencing of several rhodophyte and chromophyte plastid genomes. Complete sequences are now available for the plastid genomes of two rhodophytes, *P. purpurea* (85) and *C. paradoxa* (D Bryant, V Stirewalt, C Michalowski, W Löffelhardt & H Bohnert, unpublished data), and significant data have also been accumulating for the *G. sulphuraria* and *Antithamnion* sp. plastids (110). Among chromophytes, the plastid genomes of *Odontella sinensis* (24) and *Cryptomonas* Φ (17) are the best characterized, with 50–70% of the genes mapped.

The most striking characteristic of rhodophyte and chromophyte plastid genomes is that a substantially larger number of genes are encoded on these genomes than are encoded on metaphyte plastid genomes. The *P. purpurea* plastid genome is less than 40 kbp larger than that of tobacco, yet it encodes nearly twice as many genes (Table 1). The *Cryptomonas* Φ and *O. sinensis* plastid genomes are approximately 35 kbp smaller than that of tobacco, yet they contain more genes. These comparisons are based on total numbers of genes; the large inverted repeats in metaphytes often carry duplicate copies of ten or more genes; thus, the number of unique genes in metaphyte plastid genomes is less than 120. Rhodophyte and chromophyte plastid genomes, except that of *H. akashiwo,* tend to have small repeat regions with, at most, only a few genes in addition to the rRNA operon. Thus, the differences in coding capacity between rhodophytes or chromophytes and metaphytes is even more pronounced if one eliminates the number of duplicated genes in these comparisons.

Rhodophyte and chromophyte plastid genomes differ greatly from those of chlorophytes in their absence or near absence of introns. The *M. polymorpha* plastid genome has 21 introns that occupy approximately 10% (12,756 bp) of the genome, whereas *C. paradoxa* (D Bryant, V Stirewalt, C Michalowski, W

Löffelhardt & H Bohnert, unpublished data) has a single intron [in *trnL*(UAA), as in cyanobacteria (50)], and introns are completely absent in *P. purpurea* (85). The only other intron known in rhodophyte or chromophyte plastid genomes is in the *cpeB* gene of *Rhodella violacea* (3). Because this intron is not present in four other rhodophyte *cpeB* genes, it appears to have entered the *R. violacea cpeB* after the separation of *R. violacea* from most or all other rhodophytes.

Introns are also abundant in chlorophyte algae (6), and *Euglena gracilis* has more than 150 introns in its 143-kbp plastid genome (34). Interestingly, the metaphyte introns are mostly group II introns, whereas those of chlorophytes are mostly group I introns. The *E. gracilis* introns are both group II and group III introns. Thus, it appears that the rhodophyte and chromophyte plastids, like cyanobacteria, have avoided the introduction of introns into their genomes, whereas the plastid genomes of the chlorophyte/metaphyte lineage have expanded through multiple insertions of introns.

Gene Content

Metaphyte plastid genomes contain primarily genes for either gene expression or photosynthesis. A few other genes are present on the plastid genomes of most metaphytes, including the nonphotosynthetic plant *Epifagus virginiana*, that do not fall into either of these categories. Notable examples include *clpP*, which encodes a subunit of the clp protease, and *accD*, which encodes a subunit of acetyl CoA carboxylase. However, *accD* is absent from the rice plastid genome. In addition to *clpP* and *accD*, the *M. polymorpha* plastid genome encodes two genes for sulphur transport (*mbpX* and *mbpY*) and three genes for chlorophyll biosynthesis (*chlB, chlL,* and *chlN*). In comparison to metaphyte plastids, rhodophyte and chromophyte plastids generally have more genes encoding photosynthesis and gene expression proteins and more genes for biosynthesis or other functions (Table 2).

To date, the richness of rhodophyte plastid gene content is best demonstrated in *P. purpurea*. The *P. purpurea* plastid genome contains more than twice as many genes as in metaphyte plastid genomes, including approximately 150 genes not found in metaphyte plastid genomes. The functions of about 20% of the plastid genes from *P. purpurea* have not yet been determined. The genes of known function include 50 genes for photosynthesis proteins and 102 for gene expression, in comparison to metaphytes, which have approximately 30 and 60 genes, respectively. Although 10 of the 50 *P. purpurea* photosynthesis genes encode phycobilisome proteins, and thus are absent from the plastid genomes of groups other than rhodophytes (except cryptomonads, which contain only *cpeB*, encoding the β subunit of phycoerythrin), nearly all known proteins of the thylakoid membrane complexes, as well as the mobile electron carriers cytochrome c_{553} and ferredoxin, are encoded in the *P. purpurea* plastid genome. Only one ATPase protein gene (*atpC*), one photosystem II protein gene (*psbM*), one cytochrome b6/f complex protein gene (*petC*), and two subunits of the oxygen-evolving complex (*psbO* and the gene encoding a 9–12 kDa protein) are absent.

In metaphytes, genes for ribosomal proteins, rRNAs, tRNAs, RNA polymerase subunits, and a single translation factor (*infA*) make up the gene expression class of plastid genes. In *P. purpurea*, there are significantly more ribosomal proteins (46 vs 21), a few more tRNA genes, and four translation factor genes. However, more intriguing is the presence of two tRNA synthetases (*syfB*, *syh*), a DNA replication factor (*dnaB*), two genes involved in tRNA and rRNA maturation (*rnpB*, *rne*), and five genes encoding apparent transcriptional regulatory proteins (*trsA*, *trsB*, *trsC*, *trsD*, *trsE*). These genes are members of the bacterial *envZ*, *ompR*, *crp*, *ompR*, and *lysR* gene families, respectively. *trsB* has been identified on the plastid genomes of *G. sulphuraria* (111), *Antithamnion* sp. (111), *Cryptomonas* Φ (17), *R. violacea* (Genbank Acc. No. U01831), and *C. paradoxa* (D Bryant, V Stirewalt, C Michalowski, W Löffelhardt & H Bohnert, unpublished data), whereas the *C. paradoxa* (D Bryant, V Stirewalt, C Michalowski, W Löffelhardt & H Bohnert, unpublished data) and *O. sinensis* (24) plastid genomes contain *trsE*.

The diverse functions of the genes of the *P. purpurea* plastid genome are most apparent when one looks at the genes that encode products other than those for photosynthesis or gene expression. One finds genes that encode proteins involved in the biosynthesis of amino acids (*argB*, *carA*, *gltB*, *ilvB*, *ilvH*, *trpA*, *trpG*), fatty acids (*accA*, *accB*, *accD*, *acpA*, *fabH*), chlorophyll (*chlB*, *chlI*, *chlL*, *chlN*), carotenoids (*preA*), phycobilins (*pbsA*), and thiamine (*thiG*). In addition, there are genes encoding proteins involved in nitrogen assimilation (*glnB*, *hesB*), redox reactions [thioredoxin (*trxA*) and a subunit of ferredoxin-thioredoxin reductase (*ftrB*)], protein transport (*secA*, *secY*), and glycolysis (*pgmA*), as well as protease (*clpC*, *ftsH*), chaperonin (*dnaK*, *groEL*), and pyruvate dehydrogenase subunits (*odpA*, *odpB*). Many of these genes have been detected in other rhodophyte and chromophyte genomes. In addition, several other genes not present on the *P. purpurea* plastid genome have been detected on that of *C. paradoxa*. These include genes for a ribosomal protein (*rpl12*), a chaperonin (*groES*), a histidine transport protein (*hisP*), and a cell-division protein (*ftsW*), as well as genes for enzymes involved in the biosynthesis of NAD (*nadA*), heme (*hemA*), histidine (*hisH*), and carotenoids (*crtE*) (54; D Bryant, V Stirewalt, C Michalowski, W Löffelhardt & H Bohnert, unpublished data). In addition, a gene encoding a histone-like protein (*hlpA*) has been found on the *Cryptomonas* Φ plastid genome (103). Obviously, fewer rhodophyte and chromophyte plastid genes have been translocated to the nucleus than have metaphyte plastid genes.

Table 2   Genes of rhodophyte and chromophyte plastid genomes not found in metaphyte plastid genomes

Gene	Organisms[a]	Gene	Organisms[a]
Photosynthesis		rps10	P, C, R, A, G, O
atpD	P, C, A, G, O	rps13	P, C
atpG	P, C, A, G, O	rps17	P, O
petF	P, C, A, G, R	rps20	P, C
petJ	P	syfB	P
petK	P, C	syh	P
phycobiliprotein genes	rhodophytes	trsA	P
psaD	P, R, O	trsB	P, C, A, G, R[b]
psaE	P, C, O	trsC	P
psaF	P, C, O	trsD	P
psaL	P, R, O	trsE	P, C, O
rbcS	rhodophytes, chromophytes	tsf	P, G
		Biosynthesis	
Gene Expression		accA	P, A
dnaB	P, O	accB	P
hlpA	R	acpA	P, C, O, R[e]
infB	P, G	argB	P
infC	P	carA	P
rne	P	crtE	C
rnpB	P, C	fabH	P
rpl1	P, C	ftsW	C
rpl3	P, C	glnB	P
rpl4	P	gltB	P, A[d]
rpl6	P, C	hemA	C
rpl9	P	hesB	P
rpl11	P, C	hisH	C
rpl12	C	hisP	C
rpl13	P, O	ilvB	P
rpl18	P, C, O	ilvH	P
rpl19	P, C	nadA	C
rpl24	P	odpA	P
rpl27	P, O[c]	odpB	P
rpl29	P, O	pbsA	P
rpl31	P, O	pgmA	P
rpl34	P, C	preA	P, C
rpl35	P, C, O	thiG	P
rpsl	P	trpA	P, A
rps5	P, C	trpG	P, C
rps6	P		

Table 2   (*continued*)

		Gene	Organisms[a]
Miscellaneous		*groEL*	P, C, G[d]
cfxQ	P, O	*groES*	C
clpC	P[f]	*secA*	P, A, O[h]
dnaK	P, C, R, G, O[g]	secY	P, C, R, O, A[i]
ftsH	P, O	trxA	P[j]

[a]Symbols used (and references): P, *Porphyra purpurea* (83, 85); C, *Cyanophora paradoxa* (54; D Bryant, V Stirewalt, C Michalowski, W Löffelhardt & H Bohnert, unpublished data); A, *Antithamnion* sp. (110; K Valentin, unpublished data); G, *Galdieria sulphuraria* (110); R, *Cryptomonas* Φ (17); O, *Odontella sinensis* (24)

[b]Also found in *Porphyridium aerugineum* (110) and Rhodella violacea (Genbank Acc. No. U01831)

[c]Also found in *Pleurochrysis carterae* (26)

[d]Also found in *Cyanidium* RK-1 (110)

[e]Also found in *Cylindrotheca* sp. (39)

[f]Also found in *Cricosphaera carterae* (110) and *Heterosigma akashiwo* (Genbank Acc. No. Z25810)

[g]Also found in *C. carterae* (110), *Cyanidium* RK-1 (110), and Pavlova lutherii (91)

[h]Also found in *C. carterae* (110), *P. lutherii* (89), *Ochromonas danica* (110), and *Heterosigma akashiwo* (K Valentin, unpublished data)

[i]Also found in *P. lutherii* (90)

[j]Also found in *Cyanidium* RK-1 (110), *Griffithsia pacifica* (86), and *Porphyra yezoensis* (86)

Nearly as interesting as the diversity of the *P. purpurea* plastid genes is the list of those genes not found on the *P. purpurea* plastid genome that are present on metaphyte plastid genomes. These genes include *psbM, rps15, infA, clpP, trnP*(GGG), *mbpX, mbpY,* and several conserved ORFs (*ycf1, ycf2, ycf14, ycf15*). Also absent are all 11 genes that encode NADH-plastoquinone oxidoreductase subunits. These 11 genes also appear to be absent from plastid genomes of chromophytes (24), *E. gracilis* (34), and two *Chlamydomonas* species (6). Presumably, most of these genes were transferred to the nucleus in *P. purpurea,* although this is likely not the case for *ycf14,* which is located in the *trnK*(UUU) intron, encodes an RNA maturase, and probably entered an ancestral metaphyte plastid genome along with the intron. Likewise, *trnP*(GGG) has not been transferred to the nucleus, but rather its function has been replaced by expanded wobble base-pairing in another proline tRNA. Thus, in comparing the approximately 250 genes of the *P. purpurea* plastid genome with the 120 of metaphyte plastid genomes, three groups of genes are delineated: about 100 common genes, about 150 genes that are present only on the *P. purpurea* plastid genome, and 20 that are only found in metaphyte plastid genomes.

PLASTID PROTEIN IMPORT

Rhodophytes

Rhodophyte and chromophyte nuclear genes that encode plastid proteins and RNAs have also provided some interesting insights into plastid evolution. Only four rhodophyte nuclear genes for plastid proteins have been characterized, and each protein has an amino-terminal transit peptide resembling that used by metaphytes for directing proteins synthesized in the cytoplasm to the plastid (Figure 1A). As in metaphytes (44), these transit peptides are rich in hydoxylated amino acids and are predicted to have a random-coil configuration. The four known transit peptides show some conservation of amino acids, particularly at the N-terminal end of the transit peptide. Except in the *Aglaothamnion* sp. γ-phycoerythrin transit peptide, there is a conserved region of eight amino acids near the middle of the transit peptides where five amino acids are identical and two are similar. Although intriguing, many more transit peptide sequences need to be examined before these similarities can be considered meaningful.

A.

```
                        **::* :                       ::   :: :          : :
Aglaothamnion γPE       MASPAFAVNMFTPVKL-------------------------------------------
Cyanophora FNR          MAFVASVPV----------------FANASGLKT-EAKVCQKPALKNSFFRGE
Gracilaria GAPA         MAFVAPVSSVFSTSSKSAVC--SGRSSFAQFSGLKKVNNTARLQTAEQGSAFGGV
Chondrus GAPA           MAFVAPVATVRATT-KSSVCQVQGRSTFAQFSGMKKVNQSSRLQPAQSGSAFGGY
Pea GAPA                MASATFSVAKPAIK--------ANGKGFSEFSGLRNSSRHLPFSRKSSDDFHSLV
```

```
                        *       :   :* :
Aglaothamnion γPE       ---SGSFTASMPVDSKPAASATGVRM
Cyanophora FNR          EVTSRSFFASQAVSAKPATTFEVDTTIRA
Gracilaria GAPA         SDANDAFFNAVNTMGAPARTSNAPS
Chondrus GAPA           SDANDAFY--TRVSGIVAAT-FGPT
Pea GAPA                TFQTNA----VGSSGGHKKSLVVEA
```

B.

```
Odontella AtpC          MKFFCVAGLLASAAAFQAQPAAFTTYSPAVGGATSNVFSESSSPAHRNRRATIVM
Phaeodactylum FCP       MKFAVFAFLLASAAAFAPAWQSARTSVATNM
Chroomonas CpeA         MFAKTLASLAVIGSAAAYVPMMSMDMGRREVVQAGAAAAAVTPFLSGAPAGA
Symbiodinium PCP        MVRGARKAIAVGVAVAVACGLQKHLNFVFGPRHAAPVAAAAASMMMAPAAFA
```

Figure 1 Transit peptides and signal peptides of plastid-localized proteins from rhodophytes and chromophytes. *A.* Transit peptides of rhodophyte plastid proteins. Residues conserved in all four transit peptides are indicated by an asterisk over the alignment; those conserved in three of the four are indicated by a colon [*Aglaothamnion* sp. γPE (1), *C. paradoxa* FNR (42), *Gracilaria chilensis* GAPA (111), *C. crispus* GAPA (52), and pea GAPA (9)]. *B.* Prepeptides of chromophyte plastid proteins. Signal peptide regions are underlined, and presumptive lumen-targeting sequences are double underlined [*O. sinensis* AtpC (78), *Phaeodactylum tricornutum* FCP (4, 5), *Chroomonas* sp. CpeA (43), and *Symbiodinium* sp. PCP (71)].

The rhodophyte GapA transit peptides can also be aligned with metaphyte GapA transit peptides (52, 111) (Figure 1A). In addition, all known rhodophyte and metaphyte GapA transit peptides have an intron that occurs at nearly the same position (112). These introns are positioned no more than six nucleotides apart, similar to the difference seen for intron 2 in pea *GapA* and *GapB*, genes that appear to have arisen within the metaphyte lineage by duplication (9). These observations suggest that the *GapA* intron 1 was present in a common eukaryotic ancestor (because spliceosomal introns are apparently absent in cyanobacteria) of both rhodophytes and metaphytes.

In addition, Apt et al (1) have demonstrated functional interchangeability between rhodophyte and metaphyte transit peptides. The sequence encoding the *Aglaothamnion* sp. γ-phycoerythrin transit peptide was inserted in place of the usual transit peptide–encoding sequence of a pea *rbcS* gene. The protein encoded by the chimeric gene was synthesized in vitro and found to be capable of transport into isolated pea chloroplasts and assembly into the holoenzyme. This experiment indicates a similar import mechanism in both rhodophytes and metaphytes.

Chromophytes

Unlike rhodophytes, chromophyte nuclear genes for plastid-localized proteins do not have a transit peptide, but instead have a signal peptide (Figure 1B). Again, only a handful of genes of this type have been analyzed, but all have both a 17–18 amino acid signal peptide followed by a more variable region of 14–38 amino acids that is present before amino terminus of the mature peptide. Because the outer membranes of chromophyte plastids are usually contiguous with the endoplasmic reticulum (ER), it appears that the signal peptide directs transport of the protein into the ER lumen, where the signal peptide is removed. Presumably, the remaining portion of the prepeptide then directs the protein to the interior of the plastid (through the inner plastid membranes). The *Phaeodactylum tricornutum* fucoxanthin-chlorophyll *a*-binding protein (FCP) signal peptide is capable of directing the preprotein to the ER lumen of canine pancreatic microsomes (4).

Two chromophyte nucleus-encoded plastid proteins, *Chroomonas* sp. phycoerythrin α subunit (CpeA) (44) and *Symbiodinium* sp. peridinin-chlorophyll *a*-binding protein (PCP) (73), have what appears to be a second signal sequence immediately before the mature protein (doubly underlined in Figure 1B). Cryptomonad CpeA has been shown to be located in the thylakoid lumen (57) and cyanobacterial and metaphyte proteins are directed to the thylakoid lumen by a signal peptide–like sequence (43). Thus, the prepeptides for these two proteins appear to have three parts: a signal peptide for ER-targeting, a middle region that presumably directs the preprotein to the plastid, and a second signal peptide for targeting to the thylakoid lumen.

THE EVOLUTION OF PLASTIDS

Molecular Phylogenetic Analyses

Although molecular phylogenetic analyses of rhodophyte or chromophyte plastid–encoded proteins and RNAs have often suffered from inadequate numbers of sequences, the expanding data on rhodophyte and chromophyte plastids have produced several phylogenetic trees with a sufficient number of taxa to be useful in assessing the origins of plastids. In general, most trees based on plastid proteins have two main branches: rhodophytes/chromophytes and chlorophytes/metaphytes. Both branches are more closely associated with each other than cyanobacteria and are more closely associated with cyanobacteria than any other eubacteria, including prochlorophytes (e.g. 55, 68, 110).

The exceptions to this observation are the trees generated from *rbcL* and *rbcS* genes. In all trees published for these genes (19, 25, 62, 102), regardless of the alignments or analysis methods used, rhodophytes and chromophytes, except *C. paradoxa*, associate with α and β proteobacteria rather than cyanobacteria. *C. paradoxa* and all chlorophytes and metaphytes are clearly associated with cyanobacteria. This unusual result is inconsistent with either monophyletic or polyphyletic schemes of plastid origins. The best explanation of these results that is still consistent with other data requires a lateral transfer of *rbcL* and *rbcS* genes from an α or β proteobacterium into an ancestral plastid of the rhodophyte/chromophyte lineage (62), after the divergence of *C. paradoxa* from this lineage. The probable origin of mitochondria from an α proteobacteria (32) raises the possibility that the rhodophyte/chromophyte *rbcL* and *rbcS* genes might have been transferred from the symbiotic progenitor of mitochondria (62).

Plastid Gene Clusters

One aspect of plastid genome organization that can be assessed for shared and derived characters is the clustering of genes or operons. A number of plastid gene operons, such as *atpBE, psbEFLJ, psbDC,* and *psaAB,* are conserved in all plastids investigated, with the exception of chlorophytes (6). These operons are present in cyanobacteria and thus reflect the ancestral organization. More interesting are operons that are conserved in plastids but that differ from the organization of their genes in cyanobacteria. Such operons would indicate a common ancestor among all plastid types. Two gene clusters of this type have been identified (83), and several more are possible candidates.

The first of these gene clusters occurs as three different operons in cyanobacteria (and *Escherichia coli*): *rpoB/C1/C2, rps2/tsf,* and *atpI/H/G/F/D/A.* In *P. purpurea,* these genes are assembled into a single large cluster, although they may be organized as two transcriptional units. Most of this cluster has also been found in other red algae and in *C. paradoxa,* although in *C. para-*

doxa the *tsf* and *atpI* genes apparently have moved to the nucleus (54). In at least some chromophytes the *rpoB/C1/C2* and *atpI/H/G/F/D/A* operons are separate (7, 24). Metaphytes, however, still maintain the basic arrangement seen in *P. purpurea,* although the *tsf, atpG,* and *atpD* genes have been transferred to the nucleus (37, 72, 95). In *E. gracilis* (34) and chlorophytes (6), the same gene transfers have occurred, but the remaining gene cluster has been either split into two pieces (*E. gracilis*) or completely disassembled and spread around the chloroplast genome (chlorophytes).

A second operon common to all known plastids is that containing *psbB/ycf8/psbN/psbH*. Although the *ycf8* gene has not been detected in cyanobacteria, *psbB* is known to be separated from *psbN* and *psbH* (63). In *P. purpurea,* this operon is preceded by ORF291 (83, 85), and a similar organization occurs in *C. paradoxa* (54), except that *psbH* has been inverted and placed at the beginning of this gene cluster. Otherwise, in all other plastids investigated, ORF291 has been removed from the plastid genome, but the operon with remaining genes has been conserved.

Several other gene clusters might also indicate plastid-specific gene arrangements, but either missing data from cyanobacteria and some plastids or gene transfers to the nucleus have complicated the analysis of these operons. One of these clusters is a large ribosomal protein operon that contains homologues of the *E. coli* S10, *spc,* alpha, and *str* operons. This operon is present in *P. purpurea* and chromophytes, but in *C. paradoxa,* chlorophytes, and metaphytes it is organized differently, and several genes have been transferred to the nucleus. The *str* operon has been characterized in two cyanobacteria (11, 67), and the other ribosomal protein operons do not appear to be immediately adjacent, but further analysis is required. The members of two other gene pairs, *rpl33/rps18* and *ycf9/trnG*(GCC), are adjacent and encoded on the same DNA strand in all known plastids (except *E. gracilis*), but there is no information on the cyanobacterial organization of these genes. Further analyses of cyanobacterial as well as chromophyte and chlorophyte plastid genomes are required to clarify the evolution of these operons.

Relatedness of Light-Harvesting Complexes

One of the major questions of plastid evolution is how the light-harvesting complexes (LHCs) of the different algal groups evolved. The recent isolation (108), from the rhodophyte *Porphyridium cruentum,* of a photosystem I LHC that contains chlorophyll *a,* xanthophylls, and carotenoids begins to clarify this question. Antibodies to the chlorophyll *a/b*-containing photosystem I complex of barley and the *P. tricornutum* fucoxanthin-chlorophyll *a/c* complex recognize the rhodophyte photosystem I LHC proteins (108). Interestingly, these antibodies do not detect LHC-type proteins in either a cyanobacterium or a prochlorophyte. These observations suggest a eukaryotic origin for LHC proteins and support the hypothesis that the ancestral plastid contained both phycobilisomes and LHCs (10).

Secondary Endosymbioses in Plastid Evolution

Before assessing the evidence for plastid origin hypotheses, the role of secondary endosymbioses in plastid evolution should be considered. The strongest evidence for secondary endosymbioses has come from phylogenetic studies that used cryptomonad 18S rRNA gene sequences. Douglas et al (20) first demonstrated the presence of two 18S rRNA genes in *Cryptomonas* Φ, and phylogenetic analysis showed that one of these clustered with the 18S rRNA genes of rhodophytes, whereas the other was located in a branch containing chlorophytes and *Acanthamoeba castellanii*. These authors proposed that the rRNA gene similar to that of rhodophytes was present in the nucleomorph of the plastid and that the nucleomorph was derived from the nucleus of a rhodophyte endosymbiont. Recent in situ hybridization studies (65) have demonstrated that this 18S rRNA is associated specifically with the nucleomorph. Similar phylogenetic results were obtained independently in an investigation of 18S rRNA genes from another cryptomonad, *Pyrenomonas salina* (22).

A similar in situ hybridization study has been conducted on the rRNA genes of a chlorarachniophyte (66), which is an amoeboid alga with plastids that are bounded by four membranes and that contain chlorophyll *b* and a nucleomorph. As for cryptomonads, two rRNA genes were detected, one of which was localized in the nucleomorph. The cryptomonad and chlorarachniophyte data suggest strongly that secondary endosymbioses produced the plastids of these groups and, by extension, all other plastids with more than two bounding membranes. This would include other chromophyte plastids and those of euglenoids, because the plastids of this latter group are bounded by three membranes.

Although the 18S gene phylogenies indicate that a rhodophyte was the progenitor of cryptomonad plastids, similar studies cannot be done to establish the source of other chromophyte plastids, because of the absence of the nucleomorph. Phylogenetic trees based on plastid genes and proteins are the best means to approach this question, and in general, these trees also indicate that rhodophyte plastids are closely related to chromophyte plastids (e.g. 55, 68, 110). Particularly strong evidence for this association comes from the *rbcL/rbcS* trees that indicate an α or β proteobacterial origin for these genes in both rhodophytes and chromophytes, including cryptomonads (19, 25, 62, 102). Only a single lateral gene transfer event is required to explain this result if it occurred in an ancestral rhodophyte and subsequent rhodophytes were the progenitors of chromophyte plastids.

Another observation from the phylogenetic studies of cryptomonad 18S rRNA genes is that the nuclear genes do not cluster with the nuclear genes of other chromophytes. Similarly, more extensive 18S rRNA gene phylogenies (14) clearly separate euglenoids and dinoflagellates from all other photosynthetic eukaryotes (and each other) and place chlorarachniophytes, cryptomonads, and chromophytes (chrysophytes, phaeophytes, xanthophytes, eustigmatophytes, and diatoms) on separate branches of the phylogenetic tree. Another taxon traditionally grouped with the chromophytes, the Haptophyta, also appears to be separated from the remaining chromophytes. The critical concern that arises is the number of secondary endosymbioses that have occurred in the evolution of plastids. The above phylogenetic analyses indicate as many as five, and possibly six, such events.

Consistent with the 18S rRNA gene phylogenies and a eukaryotic origin of LHCs is the observation that the LHCs of cryptomonads and dinoflagellates differ from those of other chromophytes, which indicates independent origins. The LHC of the main chromophyte group uses a fucoxanthin-chlorophyll c LHC, whereas that of the cryptomonads contains phycobiliproteins in addition to chlorophyll c, and dinoflagellates use a peridinin-chlorophyll c LHC. The dinoflagellate peridinin-chlorophyll a protein is completely unrelated to the fucoxanthin-chlorophyll a protein of chromophytes (71). Additional characterization of nuclear genes from the various chromophyte and other algal groups will be required to sort out the relationships between them and to better establish the number of secondary endosymbiotic events responsible for chromophyte plastids.

A Monophyletic Origin of Plastids

If all chromophytes, euglenoids, and chlorarachniophytes arose through secondary endosymbioses involving rhodophytes or chlorophytes, the debate between monophyletic vs polyphyletic primary origins of plastids is reduced to distinguishing between monophyletic and diphyletic (chlorophyte vs rhodophyte) options. The observation that most phylogenetic trees associate all plastids in one clade that is associated with cyanobacteria strongly supports the monophyly of plastids. The detection of related LHC proteins in the three different types of light-harvesting systems also strongly supports a monophyletic origin of plastids. The common occurrence of several derived plastid operons, the functional interchangeability of rhodophyte and metaphyte transit peptides, and the conserved GapA transit peptide and its first intron indicate that rhodophyte and metaphyte plastids are both derived from a single primary endosymbiotic event. Taken together, the evidence for a monophyletic origin of plastids is becoming quite impressive.

In addition, there is substantial evidence against a polyphyletic origin. Phylogenetic trees based on both rRNA genes (99, 100) and protein sequences (73) indicate that prochlorophytes are polyphyletic within the cyanobacteria and show no specific affiliation with chlorophyte or metaphyte plastids. Similarly, phylogenetic analyses do not support the suggestion that chromophyte plastids evolved from *Heliobacterium chlorum* (107). A characteristic that has been interpreted as indicating a polyphyletic origin of plastids is the seven-amino-acid gap at the carboxyl terminus of PsbA that is found in chlorophytes, metaphytes, *E. gracilis,* and the prochlorophyte *Prochlorothrix hollandica,* but not in rhodophytes, chromophytes, and another prochlorophyte, *Prochloron didemni* (53, 69). However, phylogenetic trees of PsbA fail to show a clear association of prochlorophytes with chlorophytes and metaphytes (69, 70). Although not yet definitive, the current evidence points toward a monophyletic origin of plastids, a conclusion reached recently by several authors (18, 49, 77, 81), but not without its opponents (55, 110).

Of course, it would be difficult to distinguish between a monophyletic origin and a polyphyletic origin involving two or more closely related cyano-bacteria. Although this possibility cannot be ruled out, it seems unnecessarily complex and there is no evidence to support it. Another alternative proposal, that *C. paradoxa* and other glaucocystophytes arose through an independent symbiosis (55), is based primarily on the observation that *C. paradoxa* has cyanobacterial *rbcL* and *rbcS* genes, but otherwise has many of the charac-teristics of rhodophytes, which have the α or β proteobacterial *rbcL* and *rbcS*. Because the most likely explanation for the origin of the rhodophyte and chromophyte *rbcL* and *rbcS* involves a lateral transfer at some early point in the rhodophyte lineage, this apparent contradiction can be resolved within a monophyletic origin of plastids by assuming that *C. paradoxa* diverged from the rhodophyte lineage prior to this gene transfer event.

INSIGHTS INTO THE EVOLUTION OF OTHER ORGANISMS

Although information from rhodophyte and chromophyte plastids is most directly relevant to the evolution of plastids, two observations provide insights into the evolution of other organisms. The first of these observations is associ-ated with the discovery of a 35-kbp circular molecular in *Plasmodium falci-parum,* the human malaria parasite (105). This highly A+T-rich molecule has been shown to have rRNA-encoding inverted repeats, *rpoBC* genes (encoding subunits of RNA polymerase), and several ribosomal protein genes. Phyloge-netic analyses of these genes suggest a distant relationship to plastids (28). Nuclear rRNA gene phylogenies suggest that *P. falciparum* and other mem-bers of the Apicomplexa evolved from or shared a common ancestor with dinoflagellates and ciliates (27). An ORF has been identified on the 35-kbp molecule that is more than 50% identical to an ORF detected in several red algal plastid genomes (106), an observation that further strengthens the idea that the 35-kbp molecule is a remnant of a plastid genome (76). These data imply that a red alga was the progenitor of dinoflagellate plastids. It will be interesting to see whether the 35-kbp molecule is also present in other members of the Apicomplexa.

A second evolutionary insight is derived from the similarity of rhodophyte plastid heme oxygenase to those of birds and mammals. In cyanobacteria and rhodophytes, heme oxygenase is a biosynthetic enzyme that opens the heme ring to produce linear tetrapyrroles that are further modified into phycobilins (reviewed in 2). This enzyme also presumably functions in chlorophytes and metaphytes to produce the phytochrome chromophore. In birds and mammals, however, heme oxygenase is a catabolic enzyme that degrades heme generated during hemoglobin breakdown. The plant and animal enzymes also have different energy sources: The plant heme oxygenase receives electrons through ferredoxin, whereas the mammalian enzyme uses the microsomal cytochrome P-450 as an electron donor. In addition, the mammalian enzyme is located in the microsomal membranes, whereas the plant heme oxygenase is a soluble, plastid enzyme. A heme oxygenase gene has been identified on the plastid genome of *P. purpurea,* and the encoded protein is approximately 35% identical to animal heme oygenases (Figure 2). A particularly highly conserved region of 26 amino acids occurs around the presumed active site histidine.

This remarkable similarity of the vertebrate and rhodophyte plastid heme oxygenases indicates that these genes have descended from a common ancestral gene. However, neither heme oxygenase nor its product, biliverdin IXα, have been observed in protozoa, fungi, or invertebrate animals (29a; S Beale, personal communication), and thus, the evolutionary origins of the vertebrate heme oxygenase are unclear. Possibly, heme oxygenase is, or was, present in lower animals, but it has not yet been detected. If so, such a possibility might support the idea that animals arose from a photosynthetic ancestor (14). Alternatively, the vertebrate heme oxygenase may have arisen through a relatively recent lateral transfer of a cyanobacterial, algal, or plant heme oxygenase. Either of these possibilities has interesting implications for the evolution of vertebrates.

IMPLICATIONS FOR PLASTID METABOLISM

Rhodophyte and chromophyte plastid genes can also contribute to our understanding of plastid physiology. With more than 150 genes not found in metaphyte plastid genomes, including more than 30 ORFs unknown in any other organism, the *P. purpurea* plastid genome, for example, can be used for the identification of genes that encode plastid proteins. The ORFs can be matched with N-terminal sequences of isolated proteins to provide complete amino acid sequence information and to identify a function for the ORF. At least one ORF from the *P. purpurea* plastid genome has already been identified in this way

```
                                            :*:: *:* *:: *  ***    *    * :*
        HUMAN     MERPQPDSMP.........QDLSEALKEATKEVHTQAENAEFMRNFQKG
          RAT     MERPQLDSMS.........QDLSEALKEATKEVHIRAENSEFMRNFQKG
      CHICKEN     METSQPHNAESMS.........QDLSELLKEATKEVHEQAENTPFMKNFQKG
       HUMAN2     MSAEVETSEGVDESEKKNSGALEKENQMRMADLSELLKEGTKEAHDRAENTQFVKDFLKG
         RAT2     MSSEVETSEGVDESENNSTAP.EKENHTKMADLSELLKEGTKEAHDRAENTQFVKDFLKG
     PORPHYRA     MVSSLANELREGTTKSHSMAENVSFVKSFLGG

                       :::        **    * *:***   :**   *   :* **  **:*   :*  *      *
        HUMAN     QVTRDGFKLVMASLYHIYVALEEEIERNKESPVFAPVYFPEELHRKAALEQDLAFWYGPR
          RAT     QVSREGFKLVMASLYHIYTALEEEIERNKQNPVYAPLYFPEELHRRAALEQDMAFWYGPH
      CHICKEN     QVSLHEFKLVTASLYFIYSALEEEIERNKDNPVYAPVYFPMELHRKAALEKDLEYFYGSN
       HUMAN2     NIKKELFKLATTALYFTYSALEEEMDRNKDHPAFAPLYFPMELHRKEALTKDMEYFFGEN
         RAT2     NIKKELFKLATTALYFTYSALEEEMDRNKDHPAFAPLYFPTELHRKEALIKDMEYFFGEN
     PORPHYRA     VVDKKSYRKLIANLYFVYSAIEEEILLNKDHPAIKPIYF.TELNRKTSLAKDLNYYYGPD

                        *        *  * **   *    *       ************ ***********  *:  *   *  *
        HUMAN     WQEVIPYTPAMQRYVKRLHEVGRTEPELLVAHAYTRYLGDLSGGQVLKKIAQKALDLPSS
          RAT     WQEAIPYTPATQHYVKRLHEVGGTHPELLVAHAYTRYLGDLSGGQVLKKIAQKAMALPSS
      CHICKEN     WRAEIPCPEATQKYVERLHVVGKKHPELLVAHAYTRYLGDLSGGQVLKKIAQKALQLPST
       HUMAN2     WEEQVQCPKAAQKYVERIHYIGQNEPELLVAHAYTRYMGDLSGGQVLKKVAQRALKLPST
         RAT2     WEEQVKCSEAAQKYVDRIHYVGQNEPELLVAHAYTRYMGDLSGGQVLKKVAQRALKLPST
     PORPHYRA     WLNIIEPSSATQVYVNRIHNIGNKQPELLVAHAYTRYLGDLSGGQVLKKIARGAMNL.SD
                                                                          ↑

                        ::* *   *             *: **  ::: :          :  :*   :*    *   *   *
        HUMAN     GEGLAFFTFPNIASATKFKQLYRSRMNSLEMTPAVRQRVIEEAKTAFLLNIQLFEELQEL
          RAT     GEGLAFFTFPSIDNPTKFKQLYRARMNTLEMTPEVKHRVTEEAKTAFLLNIELFEELQAL
      CHICKEN     GEGLAFFTFDGVSNATKTKQLYRSRMNALEMDHATKKRVLEEAKKAFLLNIQVFEALQKL
       HUMAN2     GEGTQFYLFENVDNAQQFKQLYRARMNALDLNMKTKERIVE.ANKAFEYNMQIFNELDQA
         RAT2     GEGTQFYLFEHVDNAQQFKQFYRARMNALDLSMKTKERIVEEANKAFEYNMQIFSELDQA
     PORPHYRA     ERGTKFYDFDEIEDDKIFKNNYRSALDTIPLSDEQVQNVVAEANISFTLNMKMFEELNSS

                                                                              :        :
        HUMAN     LTHDTKDQSPSRAPGLRQRASNKVQDSAPVETPRGKPPLNT...RS.Q....APLLRWVL
          RAT     LTEEHKDQSPSQTEFLRQRPASLVQDTTSAETPRGKSQIST...SSSQ....TPLLRWVL
      CHICKEN     VSKSQENGHAVQPKAELTRTSVNKSHENSPAAGKESERTST...MQADMLTTSPLVRWLL
       HUMAN2     GSTLARETLEDGFPVHDGKGDMRKCPFYAAEQDKGLEGSLS..LPTSYAVLRKPSLQFIL
         RAT2     GSMLTKETLEDGLPVHDGKGDVRKCPFYAAQPDKGTLGGSNCPFRTAMAVLRKPSLQLIL
     PORPHYRA     IVKIITMIIVSTVRKFTLKSILATAD

                                    :
        HUMAN     TLSFLVATVAVGLYAM
          RAT     TLSFLLATVAVGIYAM
      CHICKEN     ALGFIATTVAVGLFAM
       HUMAN2     AAGVALAAGLLAWYYM
         RAT2     AASVALVAGLLAWYYM
     PORPHYRA
```

Figure 2 Alignment of heme oxygenase sequences [human (109), rat (94), chicken (23), human2 (64), rat2 (87), and *Porphyra* (85)]. Amino acids conserved in all heme oxygenases are indicated by an asterisk over the alignment. Residues conserved among all the animal sequences, but not *P. purpurea,* are indicated by a colon. The arrow indicates the active site histidine.

(85). ORF39 encodes a protein similar to a 4.1-kDa photosystem II core component that has been isolated from cyanobacteria (40), spinach, and wheat (41) and for which N-terminal sequence data are available. Rhodophyte and chromophyte plastid genomes can also provide hybridization probes for the isolation of the homologous genes from cyanobacteria and metaphytes.

In conjunction with accumulating evidence for a monophyletic origin of plastids, rhodophyte and chromophyte plastid genes can also provide insights into the enzymes and biochemical processes that may occur in all plastids. For example, the identification in rhodophyte and chromophyte plastid genomes of genes that encode proteins that are homologous to components of the bacterial protein transport complex (*secA* and *secY;* Table 2) suggested that all plastids might utilize a homologous complex for the transport of proteins through thylakoid membranes. The detection of a SecA homolog in pea chloroplasts (109a) and the demonstration of its involvement in protein transport into thylakoids have confirmed this hypothesis. Other possibilities arising from the presence of specific genes in the plastid genomes of rhodophytes and chromophytes are described below.

Acetyl CoA Carboxylase

Acetyl CoA carboxylase (ACC), the first enzyme in the fatty acid biosynthesis pathway, occurs in two forms: a prokaryotic form consisting of four separate subunits and a eukaryotic form consisting of a single, large polypeptide chain (36). Plastids are thought to be the major site of fatty acid synthesis in plants, and until recently, the plastid enzyme was assumed to be of the eukaryotic form because that was the only form of ACC that had ever been purified from plants (36). However, the identification of the genes for the four subunits of the *E. coli* ACC (51) demonstrated that one of the subunits, the β subunit of the carboxytransferase portion of ACC (encoded by *accD*), is homologous to a conserved ORF (*ycf11*) found on dicot plastid genomes. Sasaki and coworkers (47, 88) have shown that antibodies to the *ycf11* protein detect a plastid-localized protein in pea and inhibit plastid-associated ACC activity. These antibodies precipitate a biotin-containing complex consisting of several proteins. These observations suggest that the pea plastid ACC is of the prokaryotic type. Interestingly, AccD is absent from monocot plastids (as is its gene from the corresponding plastid genomes), apparently replaced by a nuclear-encoded eukaryotic ACC. The identification of genes for three of the four ACC subunits in the *P. purpurea* plastid genome (85) provides further support for the prokaryotic nature of plastid ACC and may be helpful in the characterization of the dicot enzyme.

Nitrogen Assimilation and Amino Acid Biosynthesis

Several genes have been identified on rhodophyte and chromophyte plastid genomes that are involved in nitrogen assimilation and amino acid biosynthesis. One gene found on the *P. purpurea* plastid genome is *glnB*, which encodes a protein known as P_{II} in eubacteria. P_{II} detects the nitrogen status of the cell and through a series of other protein interactions regulates both the activity and expression of key enzymes in the nitrogen assimilation pathway (59). The presence of this gene in the plastid genome of *P. purpurea* suggests that a similar nitrogen regulatory protein may be present in metaphytes, even though none of the nitrogen assimilation proteins are encoded on the metaphyte plastid genome.

ilvH encodes the small subunit of acetolactate synthase (ALS), one of the initial enzymes in the leucine/isoleucine/valine biosynthetic pathway and that is sensitive to herbicides of the sulfonylurea, imidazoline, and triazolopyrimidine classes (45). In *E. coli,* the small subunit of ALS is implicated in the sensitivity of the enzyme to feedback inhibition by valine (21). However, in metaphytes, the small subunit of ALS has yet to be described. The presence of *ilvH* in *P. purpurea* raises the possibility that the metaphyte ALS might also have a second subunit involved in the regulation of this enzyme.

Transcriptional Regulators

As mentioned above, five genes encoding proteins involved in transcriptional regulation have been identified in the *P. purpurea* plastid genome (85), and at least two of these have been found on other rhodophyte or chromophyte plastid genomes (Table 2). In metaphyte plastids, transcriptional regulation of gene expression has been suggested for some time, but until recently, there has been no definitive evidence to support this contention. Mullet and coworkers (15, 16, 93) have defined a blue-light-responsive promoter for *psbD-psbC* that is present on several monocot and dicot plastid genomes. Whether or not a homologue of one of the rhodophyte DNA-binding proteins binds to this promoter and affects transcription remains to be tested, but it is certainly an interesting possibility.

CONCLUDING REMARKS

The plastids of rhodophytes and chromophytes are only beginning to be characterized at a molecular level, yet they have already contributed important information concerning the evolution and physiology of plastids. The plastid genomes of rhodophytes and chromophytes provide an interesting contrast to those of metaphytes, having retained many more primitive characteristics. The information from rhodophyte and chromophyte plastids also indicates that plastid evolution has been an ongoing process, certainly with a substantial amount of gene transfer, intron introduction, and gene rearrangement, especially in the chlorophyte/metaphyte lineage after its separation from rhodophytes. Rhodophyte and chromophyte plastid genomes have also provided evidence supporting a monophyletic rather than polyphyletic origin of plastids. Additional information from rhodophytes, chromophytes, and chlorophytes is still required before this conclusion can be completely accepted and before some of the many interesting questions that remain about plastid evolution can be addressed. Just as the huge gene content of rhodophyte plastids was a surprise when only metaphyte plastid genomes were well characterized, there will undoubtedly be many more surprises as the plastid genomes of chlorophyte algae and the various chromophyte groups become well known.

ACKNOWLEDGMENTS

I would like to thank Drs. Catherine Boyen, Don Bryant, Catherine LeBlanc, Klaus Kowallik, Rama K. Singh, Klaus Valentin, and Klaus Zetsche for communication of results prior to publication; Drs. Mark Ragan and Sam Beale for helpful discussions; and Dr. Ron M. MacKay for critical comments on the manuscript. This is NRCC publication number 38052.

> Any *Annual Review* chapter, as well as any article cited in an *Annual Review* chapter, may be purchased from the Annual Reviews Preprints and Reprints service.
> 1-800-347-8007; 415-259-5017; email: arpr@class.org

Literature Cited

1. Apt KE, Hoffman NE, Grossman AR. 1993. The γ subunit of R-phycoerythrin and its possible mode of transport into the plastid of red algae. *J. Biol. Chem.* 268:16206–15

2. Beale SI. 1993. Biosynthesis of phycobilins. *Chem. Rev.* 93:785–802

3. Bernard C, Thomas JC, Mazel D, Mousseau A, Castets AM, et al. 1992. Characterization of the genes encoding phycoerythrin in the red alga *Rhodella violacea*: evidence for a splitting of the *rpeB* gene by an intron. *Proc. Natl. Acad. Sci. USA* 89:9564–68

4. Bhaya D, Grossman A. 1991. Targeting proteins to diatom plastids involves transport through an endoplasmic reticulum. *Mol. Gen. Genet.* 229:400–4

5. Bhaya D, Grossman A. 1993. Characterization of gene clusters encoding the fucoxanthin chlorophyll proteins of the diatom *Phaeodactylum tricornutum*. *Nucleic Acids Res.* 21:4458–66

6. Boudreau E, Otis C, Turmel M. 1994. Conserved gene clusters in the highly rearranged chloroplast genomes of *Chlamydomonas moewusii* and *Chlamydomonas reinhardtii*. *Plant Mol. Biol.* 24:585–602

7. Bourne CM, Palmer JD, Stoermer EF. 1992. Organization of the chloroplast genome of the freshwater centric diatom *Cyclotella meneghiniana*. *J. Phycol.* 28:347–55

8. Boyen C, Somerville CC, Le Gall Y, Kloareg B, Loiseaux-de Goër S. 1991. Physical mapping of the plastid genome from the rhodophyte *Chondrus crispus*. *J. Phycol.* 27:s11

9. Brinkmann H, Cerff R, Salomon M, Soll J. 1989. Cloning and sequence analysis of cDNAs encoding the cytosolic precursors of subunits GapA and GapB of chloroplast glyceraldehyde-3-phosophate dehydrogenase from pea and spinach. *Plant Mol. Biol.* 13:81–94

10. Bryant DA. 1992. Puzzles of chloroplast ancestry. *Curr. Biol.* 2:240–42

11. Buttarelli FR, Calogero RA, Tiboni O, Gualerzi CO, Pon CL. 1989. Characterization of the *str* operon genes from *Spirulina platensis* and their evolutionary relationship to those of other prokaryotes. *Mol. Gen. Genet.* 217:97–104

12. Cavalier-Smith T. 1982. The origins of plastids. *Biol. J. Linn. Soc.* 17:289–306

13. Cavalier-Smith T. 1987. The simultaneous symbiotic origin of mitochondria, chloroplasts, and microbodies. *Ann. NY Acad. Sci.* 503:55–71

14. Cavalier-Smith T. 1993. Kingdom Protozoa and its 18 phyla. *Microbiol. Rev.* 57:953–94

15. Christopher DA, Kim M, Mullet JE. 1992. A novel light-regulated promoter is conserved in cereal and dicot chloroplasts. *Plant Cell* 4:785–98

16. Christopher DA, Mullet JE. 1994. Separate photosensory pathways co-regulate blue light/ultraviolet-A-activated *psbD-psbC* transcription and light-induced D2 and CP43 degradation in barley (*Hordeum vulgare*) chloroplasts. *Plant Physiol.* 104:1119–29

17. Douglas SE. 1992. Eukaryote-eukaryote endosymbioses: insights from studies of a cryptomonad alga. *BioSystems* 28:57–68

18. Douglas SE. 1994. Chloroplast origins and evolution. In *The Cyanobacteria*, ed. DA Bryant, pp. 91–118. Dordrecht: Kluwer

19. Douglas SE, Durnford DG, Morden CW. 1990. Nucleotide sequence of the gene for the large subunit of ribulose-1,5-bisphosphate carboxylase/oxygenase from *Cryptomonas* Φ: evidence supporting the polyphyletic origin of plastids. *J. Phycol.* 26:500–8

20. Douglas SE, Murphy CA, Spencer DA, Gray MW. 1991. Cryptomonad algae are evolutionary chimaeras of two phylogenetically distinct unicellular eukaryotes. *Nature* 350:148–51

21. Eoyang L, Silverman PM. 1986. Role of small subunit (IlvN polypeptide) of acetohydroxyacid synthase I from *Escherichia coli* K-12 in sensitivity of the enzyme to valine inhibition. *J. Bacteriol.* 166: 901–4

22. Eschbach S, Wolters J, Sitte P. 1991. Primary and secondary structure of the nuclear small subunit ribosomal RNA of the cryptomonad *Pyrenomonas salina* as inferred from the gene sequence: evolutionary implications. *J. Mol. Evol.* 32:247–52

23. Evans CO, Healey JF, Greene Y, Bonkovsky HL. 1991. Cloning, sequencing and expression of cDNA for chick liver haem oxygenase. Comparison of avian and mammalian cDNAs and deduced proteins. *Biochem. J.* 273: 659–66

24. Freier U, Stöbe B, Kowallik KV. 1994. *The diatom* Odontella sinensis: *a missing link in chloroplast genome evolution.* Presented at 4th Int. Congr. Plant Mol. Biol., Amsterdam. Abstr. 116

25. Fujiwara S, Iwahashi H, Someya J, Nishikawa S, Minaka N. 1993. Structure and cotranscription of the plastid-encoded *rbcL* and *rbcS* genes of *Pleurochrysis carterae* (Prymnesiophyta). *J. Phycol.* 29: 347–55

26. Fujiwara S, Kawachi M, Inouye I, Someya J. 1994. The gene for ribosomal protein L27 is located on the plastid rather than the nuclear genome of the chlorophyll c-containing alga *Pleurochrysis carterae. Plant Mol. Biol.* 24:253–57

27. Gajadhar AA, Marquardt WC, Hall R, Gunderson J, Ariztia-Carmona EV, Sogin ML. 1991. Ribosomal RNA sequences of *Sarcocystis muris, Theileria annulata* and *Crypthecodinium cohnii* reveal evolutionary relationships among apicomplexans, dinoflagellates and ciliates. *Mol. Biochem. Parasitol.* 45:147–54

28. Gardner MJ, Feagin JE, Moore DJ, Rangachari K, Williamson DH, Wilson RJM. 1993. Sequence and organization of large subunit rRNA genes from the extrachromosomal 35 kb circular DNA of the malaria parasite *Plasmodium falciparum. Nucleic Acids Res.* 21:1067–71

29. Gibbs SP. 1981. The chloroplast of some algal groups may have evolved from endosymbiotic eukaryotic algae. *Ann. NY Acad. Sci.* 361:193–208

29a. Granick S, Beale SI. 1978. Hemes, chlorophylls, and related compounds: biosynthesis and metabolic regulation. *Adv. Enzymol.* 46:33–203

30. Gray MW. 1989. The evolutionary origins of organelles. *Trends Genet.* 5:294–99

31. Gray MW. 1991. Origin and evolution of plastid genomes and genes. In *Cell Culture and Somatic Cell Genetics of Plants,* ed. L Bogorad, IK Vasil, pp. 303–30. San Diego: Academic

32. Gray MW. 1992. The endosymbiont hypothesis revisited. *Int. Rev. Cytol.* 141: 233–357

33. Gray MW, Doolittle WF. 1982. Has the endosymbiont hypothesis been proven? *Microbiol. Rev.* 46:1–42

34. Hallick RB, Hong L, Drager RG, Favreau MR, Monfort A, et al. 1993. Complete sequence of *Euglena gracilis* chloroplast DNA. *Nucleic Acids Res.* 21:3537–44

35. Han TM, Runnegar B. 1992. Megascopic eukaryotic algae from the 2.1-billion-year-old negaunee iron-formation, Michigan. *Science* 257:232–35

36. Harwood JL. 1988. Fatty acid metabolism. *Annu. Rev. Plant Physiol. Plant Mol. Biol.* 39:101–38

37. Heinhorst S, Shively JM. 1983. Encoding of both subunits of ribulose-1,5-bisphosphate carboxylase by organelle genome of *Cyanophora paradoxa. Nature* 304:373–74

38. Hiratsuka J, Shimada H, Whittier R, Ishibashi T, Sakamoto M, et al. 1989. The complete sequence of the rice (*Oryza sativa*) chloroplast genome: Intermolecular recombination between distinct tRNA genes accounts for a major plastid DNA inversion during the evolution of the cereals. *Mol. Gen. Genet.* 217:185–94

39. Hwang S-R, Tabita FR. 1991. Acyl carrier protein-derived sequence encoded by the chloroplast genome in the marine diatom *Cylindrotheca* sp. strain N1. *J. Biol. Chem.* 266:13492–94

40. Ikeuchi M, Koike H, Inoue Y. 1989. N-terminal sequencing of low-molecular-mass components in cyanobacterial photosystem II core complex. Two components correspond to unidentified open reading frames of plant chloroplast DNA. *FEBS Lett.* 253: 178–82

41. Ikeuchi M, Takio K, Inoue Y. 1989. N-terminal sequencing of photosystem II low-molecular-mass proteins. 5 and 4.1 kDa components of the O_2-evolving core complex from higher plants. *FEBS Lett.* 242: 263–69

42. Jakowitsch J, Bayer MG, Maier TL, Luttke A, Gebhart UB, et al. 1993. Sequence analysis of pre-ferredoxin-NADP(+)-reductase cDNA from *Cyanophora paradoxa* specifying a precursor for a nucleus-encoded cyanelle polypeptide. *Plant Mol. Biol.* 21:1023–33

43. Jenkins J, Hiller RG, Speirs J, Godovac-Zimmerman J. 1990. A genomic clone encoding a cryptophyte phycoerythrin α-subunit. Evidence for three α-subunits and

an N-terminal membrane transit sequence. *FEBS Lett.* 273: 191–94

44. Keegstra K, Olsen LJ, Theg SM. 1989. Chloroplastic precursors and their transport across the envelope membranes. *Annu. Rev. Plant Physiol. Plant Mol. Biol.* 40: 471–501

45. Kishore GM, Shah DM. 1988. Amino acid biosynthesis inhibitors as herbicides. *Annu. Rev. Biochem.* 57:627–63

46. Knoll AH. 1992. The early evolution of eukaryotes: a geological perspective. *Science* 256:622–27

47. Konishi T, Sasaki Y. 1994. Compartmentalization of two forms of acetyl-CoA carboxylase in plants and the origin of their tolerance toward herbicides. *Proc. Natl. Acad. Sci. USA* 91:3598–601

48. Kowallik KV. 1992. Origin and evolution of plastids from chlorophyll-*a*+*c*-containing algae: suggested ancestral relationships to red and green algal plastids. In *Origins of Plastids*, ed. RA Lewin, pp. 223–63. New York: Chapman & Hall

49. Kowallik KV. 1994. From endosymbionts to chloroplasts: evidence for a single prokaryotic/eukaryotic endocytobiosis. *Endocytobiosis Cell Res.* 10:137–49

50. Kuhsel MG, Strickland R, Palmer JD. 1990. An ancient group I intron is shared by eubacteria and chloroplasts. *Science* 250: 1570–73

51. Li SJ, Cronan JE. 1992. The genes encoding the two carboxyltransferase subunits of *Escherichia coli* acetyl-CoA carboxylase. *J. Biol. Chem.* 267:16841–47

52. Liaud M-F, Valentin C, Brandt U, Bouget F-Y, Kloareg B, Cerff R. 1993. The GAPDH gene system of the red alga *Chondrus crispus*: promoter structures, intron/exon organization, genomic complexity and differential expression of genes. *Plant Mol. Biol.* 23:981–94

53. Lockhart PJ, Penny D, Hendy MD, Larkum ADW. 1993. Is *Prochlorothrix hollandica* the best choice as a prokaryotic model for higher plant Chl a/b photosynthesis? *Photosynth. Res.* 37:61–68

54. Löffelhardt W, Bohnert HJ. 1994. Structure and function of the cyanelle genome. *Int. Rev. Cytol.* 151:29–65

55. Loiseaux-de Goër S. 1995. Plastid lineages. *Prog. Phycol. Res.* In press

56. Loiseaux-de Goër SL, Markowicz Y, Dalmon J, Audren H. 1988. Physical maps of the two circular plastid DNA molecules of the brown alga *Pylaiella littoralis* (L.) Kjellm. *Curr. Genet.* 14:155–62

57. Ludwig M, Gibbs SP. 1989. Localization of phycoerythrin at the lumenal surface of the thylakoid membrane in *Rhodomonas lens*. *J. Cell Biol.* 108:875–84

58. Maerz M, Wolters J, Hofmann CJ, Sitte P, Maier UG. 1992. Plastid DNA from *Py-*

renomonas salina (Cryptophyceae): physical map, genes, and evolutionary implications. *Curr. Genet.* 21:73–81

59. Magasanik B. 1988. Reversible phosphorylation of an enhancer binding protein regulates the transcription of bacterial nitrogen utilization genes. *Trends Biochem. Sci.* 13:475–79

60. Maid U, Zetsche K. 1992. A 16 kb small single-copy region separates the plastid DNA inverted repeat of the unicellular red alga *Cyanidium caldarium*: physical mapping of the IR-flanking regions and nucleotide sequences of the *psbD-psbC*, *rps16*, 5S rRNA and *rpl21* genes. *Plant Mol. Biol.* 19:1001–10

61. Margulis L, Obar R. 1985. *Heliobacterium* and the origin of chrysoplasts. *BioSystems* 17:317–25

62. Martin W, Somerville CC, Loiseaux-de Goër S. 1992. Molecular phylogenies of plastid origins and algal evolution. *J. Mol. Evol.* 35:385–404

63. Mayes SR, Barber J. 1991. Primary structure of the *psbN-psbH-petC-petA* gene cluster of the cyanobacterium *Synechocystis* PCC 6803. *Plant Mol. Biol.* 17:289–93

64. McCoubrey WK, Ewing JF, Maines MD. 1992. Human heme oxygenase-2: characterization and expression of a full-length cDNA and evidence suggesting that the two HO-2 transcripts may differ by choice of polyadenylation signal. *Arch. Biochem. Biophys.* 295:13–20

65. McFadden GI, Gilson PR, Douglas SE. 1994. The photosynthetic endosymbiont in cryptomonad cells produces both chloroplast and cytoplasmic-type ribosomes. *J. Cell Sci.* 107:649–57

66. McFadden GI, Gilson PR, Hofmann CJ, Adcock GJ, Maier UG. 1994. Evidence that an amoeba acquired a chloroplast by retaining part of an engulfed eukaryotic alga. *Proc. Natl. Acad. Sci. USA* 91:3690–94

67. Meng BY, Shinozaki K, Sugiura M. 1989. Genes for the ribosomal proteins S12 and S7 and elongation factors EF-G and EF-Tu of the cyanobacterium, *Anacystis nidulans*: structural homology between 16Sr RNA and S7 mRNA. *Mol. Gen. Genet.* 216:25–30

68. Morden CW, Delwiche CF, Kuhsel M, Palmer JD. 1992. Gene phylogenies and the endosymbiotic origin of plastids. *BioSystems* 28:75–90

69. Morden CW, Golden SS. 1989. *psbA* genes indicate common ancestry of prochlorophytes and chloroplasts. *Nature* 337:382–85

70. Morden CW, Golden SS. 1989. *psbA* genes indicate common ancestry of prochlorophytes and chloroplasts (corrigendum). *Nature* 339:400

71. Norris BJ, Miller DJ. 1994. Nucleotide se-

quence of a cDNA clone encoding the precursor of the peridinin-chlorophyll *a*-binding protein from the dinoflagellate *Symbiodinium* sp. *Plant Mol. Biol.* 24:673–77

72. Ohyama K, Fukuzawa H, Kohchi T, Shirai H, Sano T, et al. 1986. Chloroplast gene organization deduced from complete sequence of liverwort *Marchantia polymorpha. Nature* 322:572–74

73. Palenik BP, Haselkorn R. 1992. Multiple evolutionary origins of prochlorophytes, the chlorophyll *b*-containing prokaryotes. *Nature* 355:265–67

74. Palmer JD. 1985. Comparative organization of chloroplast genomes. *Annu. Rev. Genet.* 19:325–54

75. Palmer JD. 1991. Plastid chromosomes: structure and evolution. In *Cell Culture and Somatic Cell Genetics of Plants,* ed. L Bogorad, IK Vasil, pp. 5–53. San Diego: Academic

76. Palmer JD. 1992. Green ancestry of malarial parasites? *Curr. Biol.* 2:318–20

77. Palmer JD. 1993. A genetic rainbow of plastids. *Nature* 364:762–64

78. Pancic PG, Strotmann H. 1993. Structure of the nuclear encoded γ subunit of CF_0CF_1 of the diatom *Odontella sinensis* including its presequence. *FEBS Lett.* 320:61–66

79. Ragan MA, Bird CJ, Rice EL, Gutell RR, Murphy CA, Singh RK. 1994. A molecular phylogeny of the marine red algae (Rhodophyta) based on the nuclear small-subunit rRNA gene. *Proc. Natl. Acad. Sci. USA* 91:7276–80

80. Raven P. 1970. A multiple origin of plastids and mitochondria. *Science* 169:641–46

81. Reith M. 1995. The evolution of chloroplasts and the photosynthetic apparatus. In *Oxygenic Photosynthesis: The Light Reactions,* ed. CF Yocum, D Ort. Dordrecht: Kluwer. In press

82. Reith M, Cattolico RA. 1986. Inverted repeat of *Olisthodiscus luteus* chloroplast DNA contains genes for both subunits of ribulose-1,5-bisphosphate carboxylase and the 32,000 dalton Q_B protein: phylogenetic implications. *Proc. Natl. Acad. Sci. USA* 83:8599–603

83. Reith M, Munholland J. 1993. A high-resolution gene map of the chloroplast genome of the red alga *Porphyra purpurea. Plant Cell* 5:465–75

84. Reith M, Munholland J. 1993. The ribosomal RNA repeats are non-identical and directly oriented in the chloroplast genome of the red alga *Porphyra purpurea. Curr. Genet.* 24:443–50

85. Reith M, Munholland J. 1994. *The complete nucleotide sequence of the chloroplast genome of* Porphyra purpurea. Presented at 4th Int. Congr. Plant Mol. Biol., Amsterdam. Abstr. 118

86. Reynolds AE, Chesnick JM, Woolford J,

Cattolico RA. 1994. Chloroplast encoded thioredoxin genes in the red algae *Porphyra yezoensis* and *Griffithsia pacifica:* evolutionary implications. *Plant Mol. Biol.* 25:13–21

87. Rotenberg MO, Maines MD. 1990. Isolation, characterization, and expression in *Escherichia coli* of a cDNA encoding rat heme oxygenase-2. *J. Biol. Chem.* 265:7501–6

88. Sasaki Y, Hakamada K, Suama Y, Nagano Y, Furusawa I, Matsuno R. 1993. Chloroplast-encoded protein as a subunit of acetyl-CoA carboxylase in pea plant. *J. Biol. Chem.* 268:25118-23

89. Scaramuzzi CD, Hiller RG, Stokes HW. 1992. Identification of a chloroplast-encoded *secA* gene homologue in a chromophytic alga: possible role in chloroplast protein translocation. *Curr. Genet.* 22:421–27

90. Scaramuzzi CD, Stokes HW, Hiller RG. 1992. Characterisation of a chloroplast-encoded *secY* homologue and *atpH* from a chromophytic alga. Evidence for a novel chloroplast genome organisation. *FEBS Lett.* 304:119–23

91. Scaramuzzi CD, Stokes HW, Hiller RG. 1992. Heat shock Hsp70 protein is chloroplast-encoded in the chromophytic alga *Pavlova lutherii. Plant Mol. Biol.* 18:467–76

92. Seckbach J. 1992. The cyanidiophyceae and the "anomolous symbiosis" of *Cyanidium caldarium.* In *Algae and Symbiosis,* ed. W Reisser, pp. 339–426. New York: Biopress

93. Sexton TB, Christopher DA, Mullet JE. 1990. Light-induced switch in barley *psbD-psbC* promoter utilization: a novel mechanism regulating chloroplast gene expression. *EMBO J.* 9:4485–94

94. Shibahara S, Muller R, Taguchi H, Yoshida T. 1985. Cloning and expression of cDNA for rat heme oxygenase. *Proc. Natl. Acad. Sci. USA* 82:7865–69

95. Shinozaki K, Ohme M, Tanaka M, Wakasugi T, Hayashida N, et al. 1986. The complete nucleotide sequence of the tobacco chloroplast genome: its gene organization and expression. *EMBO J.* 5:2043–49

96. Shivji MS. 1991. Organization of the chloroplast genome in the red alga *Porphyra yezoensis. Curr. Genet.* 19:49–54

97. Shivji MS, Li N, Cattolico RA. 1992. Structure and organization of rhodophyte and chromophyte plastid genomes: implications for the ancestry of plastids. *Mol. Gen. Genet.* 232:65–73

98. Stewart C-B. 1993. The powers and pitfalls of parsimony. *Nature* 361:603–7

99. Turner S, Burger-Wiersma T, Giovannoni SJ, Mur LR, Pace NR. 1989. The relationship of a prochlorophyte *Prochlorothrix*

hollandica to green chloroplasts. *Nature* 337:380–82

100. Urbach E, Robertson DL, Chisholm SW. 1992. Multiple evolutionary origins of prochlorophytes within the cyanobacterial radiation. *Nature* 355:267–69

101. Deleted in proof

102. Valentin K, Zetsche K. 1990. Rubisco genes indicate a close phylogenetic relation between the plastids of Chromophyta and Rhodophyta. *Plant Mol. Biol.* 15:575–84

103. Wang S, Liu X-Q. 1991. The plastid genome of *Cryptomonas* Φ encodes an hsp70-like protein, a histone-like protein, and an acyl carrier protein. *Proc. Natl. Acad. Sci. USA* 88:10783–87

104. Whatley JM, Whatley FR. 1981. Chloroplast evolution. *New Phytol.* 87:233–47

105. Williamson DH, Gardner MJ, Preiser P, Moore DJ, Rangachari K, Wilson RJM. 1994. The evolutionary origin of the 35 kb circular DNA of *Plasmodium falciparum*: new evidence supports a possible rhodophyte ancestry. *Mol. Gen. Genet.* 243:249–52

106. Wilson RJM, Gardner MJ, Feagin JE, Williamson DH. 1991. Have malaria parasites three genomes? *Parasitol. Today* 7:136–38

107. Witt D, Stackebrandt E. 1988. Disproving the hypothesis of a common ancestry for the *Ochromonas danica* chrysoplast and *Heliobacterium chlorum*. *Arch. Microbiol.* 150:244–48

108. Wolfe GR, Cunningham FX, Durnford D, Green BR, Gantt E. 1994. Evidence for a common origin of chloroplasts with light-harvesting complexes of different pigmentation. *Nature* 367:566–68

109. Yoshida T, Biro P, Cohen T, Muller RM, Shibahara S. 1988. Human heme oxygenase cDNA and induction of its mRNA by hemin. *Eur. J. Biochem.* 171:457–61

109a. Yuan J, Henry R, McCaffery M, Cline K. 1994. SecA homolog in protein transport within chloroplasts: evidence for endosymbiont-derived sorting. *Science* 266:796–98

110. Zetsche K, Valentin K. 1994. Structure, coding capacity and gene sequences of the plastid genome from red algae. *Endocytobiosis Cell Res.* 10:107–27

111. Zhou YH, Ragan MA. 1993. cDNA cloning and characterization of the nuclear gene encoding chloroplast glyceraldehyde-3-phosphate dehydrogenase from the marine red alga *Gracilaria verrucosa*. *Curr. Genet.* 23:483–89

112. Zhou YH, Ragan MA. 1994. Cloning and characterization of the nuclear gene encoding plastid glyceraldehyde-3-phosphate dehydrogenase from the marine red alga *Gracilaria verrucosa*. *Curr. Genet.* 26:79–86

AUTHOR INDEX

SUBJECT INDEX

CUMULATIVE INDEXES

CONTRIBUTING AUTHORS, VOLUMES 37–46

CHAPTER TITLES, VOLUMES 37–46

GENETICS & MOLECULAR BIOLOGY

Structure and Function of Nucleic Acids

Role/Regulation/Organization of Nuclear Genes

ACCLIMATION AND ADAPTATION

Economic Botany

Physiological Ecology

Plant Genetics/Evolution

ANNUAL REVIEWS

a nonprofit scientific publisher
4139 El Camino Way
P.O. Box 10139
Palo Alto, CA 94303-0139 • USA

Annual Reviews publications may be ordered directly from our office; through stockists, booksellers and subscription agents, worldwide; and through participating professional societies. **Prices are subject to change without notice. We do not ship on approval.**

- **Individuals:** Prepayment required on new accounts. in US dollars, checks drawn on a US bank.

- **Institutional Buyers:** Include purchase order. Calif. Corp. #161041 • ARI Fed. I.D. #94-1156476

- **Students / Recent Graduates:** $10.00 discount from retail price, per volume. *Requirements:* **1.** be a degree candidate at, or a graduate within the past three years from, an accredited institution; **2.** present proof of status (photocopy of your student I.D. or proof of date of graduation); **3.** Order direct from Annual Reviews; **4.** prepay. This discount **does not** apply to standing orders, *Index on Diskette*, Special Publications, ARPR, or institutional buyers.

- **Professional Society Members:** Many Societies offer *Annual Reviews* to members at reduced rates. Check with your society or contact our office for a list of participating societies.

- **California orders** add applicable sales tax. • **Canadian orders** add 7% GST. Registration #R 121 449-029.

- **Postage paid** by Annual Reviews (4th class bookrate/surface mail). UPS ground service is available at $2.00 extra per book within the contiguous 48 states only. UPS air service or US airmail is available to any location at actual cost. UPS requires a street address. P.O. Box, APO, FPO, not acceptable.

- **Standing Orders:** Set up a standing order and the new volume in series is sent automatically each year upon publication. Each year you can save 10% by prepayment of prerelease invoices sent 90 days prior to the publication date. Cancellation may be made at any time.

- **Prepublication Orders:** Advance orders may be placed for any volume and will be charged to your account upon receipt. Volumes not yet published will be shipped during month of publication indicated.

N O T E	For copies of individual articles from any *Annual Review*, or copies of any article cited in an *Annual Review*, call **Annual Reviews Preprints and Reprints (ARPR)** toll free 1-800-347-8007 (fax toll free 1-800-347-8008) from the USA or Canada. From elsewhere call 1-415-259-5017.

ANNUAL REVIEWS SERIES *Volumes not listed are no longer in print*	**Prices, postpaid, per volume.** **USA/other countries**	Regular Order Please send Volume(s):	Standing Order Begin with Volume:
☐ *Annual Review of* **ANTHROPOLOGY**			
Vols. 1-20 (1972-91) ....$41 / $46			
Vols. 21-22 (1992-93) ....$44 / $49			
Vol. 23-24 (1994 and Oct. 1995) ....$47 / $52	Vol(s). _____	Vol. _____	
☐ *Annual Review of* **ASTRONOMY AND ASTROPHYSICS**			
Vols. 1, 5-14, 16-29 (1963, 67-76, 78-91) ....$53 / $58			
Vols. 30-31 (1992-93) ....$57 / $62			
Vol. 32-33 (1994 and Sept. 1995) ....$60 / $65	Vol(s). _____	Vol. _____	
☐ *Annual Review of* **BIOCHEMISTRY**			
Vols. 31-34, 36-60 (1962-65,67-91).......................$41 / $47			
Vols. 61-62 (1992-93) ....$46 / $52			
Vol. 63-64 (1994 and July 1995) ....$49 / $55	Vol(s). _____	Vol. _____	
☐ *Annual Review of* **BIOPHYSICS AND BIOMOLECULAR STRUCTURE**			
Vols. 1-20 (1972-91) ....$55 / $60			
Vols. 21-22 (1992-93) ....$59 / $64			
Vol. 23-24 (1994 and June 1995) ....$62 / $67	Vol(s). _____	Vol. _____	

❏ *Annual Review of* **CELL AND DEVELOPMENTAL BIOLOGY** (new title beginning with volume 11)

Vols.	1-7	(1985-91) ....\$41 / \$46		
Vols.	8-9	(1992-93) ....\$46 / \$51		
Vol.	10-11	(1994 and Nov. 1995) ....\$49 / \$54	Vol(s). _____	Vol. _____

❏ *Annual Review of* **COMPUTER SCIENCE** (Series suspended)

Vols.	1-2	(1986-87) ....\$41 / \$46	
Vols.	3-4	(1988-89/90) ....\$47 / \$52	Vol(s). _____

Special package price for

Vols.	1-4	(if ordered together) ....\$100 / \$115	❏ Send all four volumes.

❏ *Annual Review of* **EARTH AND PLANETARY SCIENCES**

Vols.	1-6, 8-19	(1973-78, 80-91) ....\$55 / \$60		
Vols.	20-21	(1992-93) ....\$59 / \$64		
Vol.	22-23	(1994 and May 1995) ....\$62 / \$67	Vol(s). _____	Vol. _____

❏ *Annual Review of* **ECOLOGY AND SYSTEMATICS**

Vols.	2-12, 14-17, 19-22..(1971-81, 83-86, 88-91) ..\$40 / \$45			
Vols.	23-24	(1992-93) ....\$44 / \$49		
Vol.	25-26	(1994 and Nov. 1995) ....\$47 / \$52	Vol(s). _____	Vol. _____

❏ *Annual Review of* **ENERGY AND THE ENVIRONMENT**

Vols.	1-16	(1976-91) ....\$64 / \$69		
Vols.	17-18	(1992-93) ....\$68 / \$73		
Vol.	19-20	(1994 and Oct. 1995) ....\$71 / \$76	Vol(s). _____	Vol. _____

❏ *Annual Review of* **ENTOMOLOGY**

Vols.	10-16, 18, 20-36 (1965-71, 73, 75-91) ....\$40 / \$45			
Vols.	37-38	(1992-93) ....\$44 / \$49		
Vol.	39-40	(1994 and Jan. 1995) ....\$47 / \$52	Vol(s). _____	Vol. _____

❏ *Annual Review of* **FLUID MECHANICS**

Vols.	2-4, 7	(1970-72, 75)		
	9-11, 16-23	(1977-79, 84-91) ....\$40 / \$45		
Vols.	24-25	(1992-93) ....\$44 / \$49		
Vol.	26-27	(1994 and Jan. 1995).... ....\$47 / \$52	Vol(s). _____	Vol. _____

❏ *Annual Review of* **GENETICS**

Vols.	1-12, 14-25	(1967-78, 80-91) ....\$40 / \$45		
Vols.	26-27	(1992-93) ....\$44 / \$49		
Vol.	28-29	(1994 and Dec. 1995) ....\$47 / \$52	Vol(s). _____	Vol. _____

❏ *Annual Review of* **IMMUNOLOGY**

Vols.	1-9	(1983-91) ....\$41 / \$46		
Vols.	10-11	(1992-93) ....\$45 / \$50		
Vol.	12-13	(1994 and April 1995) ....\$48 / \$53	Vol(s). _____	Vol. _____

❏ *Annual Review of* **MATERIALS SCIENCE**

Vols.	1, 3-19	(1971, 73-89) ....\$68 / \$73		
Vols.	20-23	(1990-93) ....\$72 / \$77		
Vol.	24-25	(1994 and Aug. 1995) ....\$75 / \$80	Vol(s). _____	Vol. _____

❏ *Annual Review of* **MEDICINE: Selected Topics in the Clinical Sciences**

Vols.	9, 11-15, 17-42	(1958, 60-64, 66-42) ....\$40 / \$45		
Vols.	43-44	(1992-93) ....\$44 / \$49		
Vol.	45-46	(1994 and April 1995) ....\$47 / \$52	Vol(s). _____	Vol. _____